CW01249793

Applications of Synthetic Resin Latices
Volume 1

Applications of Synthetic Resin Latices

Volume 1

Fundamental Chemistry of Latices and Applications in Adhesives

by

H. WARSON
Chemical Consultant, UK

and

C.A. FINCH
Pentafin Associates, UK

JOHN WILEY & SONS, LTD
Chichester • New York • Weinheim • Brisbane • Singapore • Toronto

Copyright © 2001 John Wiley & Sons Ltd,
Baffins Lane, Chichester,
West Sussex PO19 1UD, England

National 01243 779777
International (+44) 1243 779777
e-mail (for orders and customer service enquiries): cs-books@wiley.co.uk
Visit our Home Page on http://www.wiley.co.uk
or http://www.wiley.com

All Rights Reserved. No part of this publication may be reproduced, stored in a retrieval system, or transmitted, in any form or by any means, electronic, mechanical, photocopying, recording, scanning or otherwise, except under the terms of the Copyright Designs and Patents Act 1988 or under the terms of a licence issued by the Copyright Licensing Agency, 90 Tottenham Court Road, London W1P 9HE, UK, without the permission in writing of the Publisher

Other Wiley Editorial Offices

John Wiley & Sons, Inc., 605 Third Avenue,
New York, NY 10158-0012, USA

WILEY-VCH Verlag GmbH, Pappelallee 3,
D-69469 Weinheim, Germany

John Wiley & Sons Australia, Ltd, 33 Park Road, Milton,
Queensland 4064, Australia

John Wiley & Sons (Asia) Pte Ltd, Clementi Loop #02-01,
Jin Xing Distripark, Singapore 129809

John Wiley & Sons (Canada) Ltd, 22 Worcester Road,
Rexdale, Ontario M9W 1L1, Canada

Library of Congress Cataloging-in-Publication Data

Warson, Henry.
   Applications of synthetic resin latices / by H. Warson and C.A. Finch.
     p. cm.
   Includes bibliographical references and indexes.
   Contents: v. 1. Fundamental chemistry of latices and applications of adhesives—v. 2. Latices in surface castings : emulsion paints—v. 3. Latices in surface coatings : diverse applications.
   ISBN 0-471-95268-0 (v. 1 : acid-free paper)—ISBN 0-471-95461-6 (v. 2 : acid-free paper)—ISBN 0-471-95462-4 (v. 3 : acid-free paper)
   1. Gums and resins, Synthetic. 2. Polymers. 3. Emulsions. I. Finch, C.A. II. Title.

TP1120.W34 2001
668'.374—dc21
                                                                                          00-044922

British Library Cataloguing in Publication Data

A catalogue record for this book is available from the British Library

ISBN 0 471 95268 0

Typeset in 10/12pt Times by Laser Words Pvt Ltd, Madras, India
Printed and bound in Great Britain by Biddles Ltd, Guildford and King's Lynn
This book is printed on acid-free paper responsibly manufactured from sustainable forestry, in which at least two trees are planted for each one used for paper production

Contents

Preface	xvii
Applications—A Note on References	xix
Patent Validity	xxi
Introduction	xxiii
1 Fundamentals of Polymer Chemistry	1
1 The Concept of a Polymer	1
1.1 Historical introduction	1
1.2 Definitions	2
2 Addition Polymerisation	4
2.1 Free radical polymerisation	5
2.1.1 Retardation and inhibition	8
2.1.2 Free radical initiation	9
2.1.3 Redox polymerisation	11
2.2 Copolymerisation	13
2.2.1 The Q, e scheme	15
3 Chain Branching; Block and Graft Copolymers	16
3.1 Chain branching	16
3.2 Graft copolymers	18
4 Polymer Structure and Properties	19
4.1 Polymer structure	19
4.2 Molecular weight effects	24
4.3 Transition points	26
5 Technology of Polymerisation	27
5.1 Bulk polymerisation	28
5.2 Solution polymerisation	28
5.3 Suspension polymerisation	29
6 The Principal Monomers and Their Polymer	30
6.1 Hydrocarbons	30
6.2 Vinyl esters	31

	6.3	Chlorinated monomers	31
	6.4	Acrylics	32
		6.4.1 Acrylic and methacrylic acids	32
		6.4.2 Individual acrylic and methacrylic esters	33
		6.4.3 Acrylics based on the amide group	38
		6.4.4 Cationic acrylic monomers	38
		6.4.5 Acrylonitrile	38
	6.5	Polymerisable acids and anhydrides	39
	6.6	Self-emulsifying monomers	39
	6.7	Esters for copolymerisation	40
	6.8	Monomers with several double bonds	41
	6.9	Allyl derivatives	43
	6.10	Vinyl ethers	43
	6.11	Miscellaneous monomers containing nitrogen	44
	6.12	Toxicity and handling	45
7	Physical Properties of Monomers	45
8	Appendix	46
References	47

2 Emulsions and Colloids   49

1	Introduction to Colloid Chemistry		49
	1.1	Historical	49
	1.2	Emulsions	49
		1.2.1 Properties of surfaces	50
		1.2.2 Micelles, surfactants	51
		1.2.3 Emulsion particles, emulsifier efficiency	54
		1.2.4 Emulsion viscosity	55
2	Surfactants and Stabilisers		58
3	A Survey of Technical Surfactants		59
	3.1	Anionic types—fatty acid soaps and allied salts	59
	3.2	Sulfates and sulfonates	60
		3.2.1 Sulfated fatty alcohols	61
		3.2.2 Sulfates of ethoxylated alcohols	61
		3.2.3 Sulfated natural oils and esters	61
		3.2.4 Sulfonates—general	62
		3.2.5 Sulfonated aromatic and condensed ring compounds	62
		3.2.6 Aliphatic chain sulfonates	63
		3.2.7 The sulfosuccinates	64
	3.3	Hydrotropes	66
	3.4	Phosphates	66
	3.5	Fluorochemical emulsifiers	67
	3.6	Non-ionic types	68
		3.6.1 Biodegradable emulsifiers	71
	3.7	Cationic types	71

3.8	Amphoteric emulsifiers	73
3.9	Polymerisable surfactants	74
3.10	Dispersing agents	74
3.11	Solid dispersants	75

4 Colloids and High Molecular Weight Stabilisers — 76
  4.1 Starch and dextrin — 76
  4.2 Cellulose ethers — 77
    4.2.1 Properties of technical cellulose derivatives — 78
    4.2.2 Methyl cellulose — 78
    4.2.3 Hydroxyethyl cellulose — 79
    4.2.4 Hydroxypropyl cellulose — 82
    4.2.5 Carboxymethyl cellulose — 83
    4.2.6 Other cellulose derivatives — 83
  4.3 Other semi-synthetic colloids — 84
  4.4 Polyvinyl alcohol — 84
    4.4.1 Properties of polyvinyl alcohol as emulsifier — 85
    4.4.2 Chemically modified polyvinyl alcohols — 86
  4.5 Polyelectrolytes — 86
  4.6 Other synthetic water-soluble polymers — 87
  4.7 Associative thickeners — 88
References — 88
Supplementary references for Section 3.9 — 89

3 Emulsion Polymerisation — 90

1 General Principles — 90
  1.1 Chemistry and fundamentals — 90
    1.1.1 Stabilisers — 91
    1.1.2 Free radical initiators — 93
    1.1.3 Water-soluble monomers — 95
2 Technology, Laboratory and Large Scale — 96
  2.1 Monomers — 96
    2.1.1 Liquid monomers — 96
    2.1.2 Volatile monomers — 97
  2.2 Pilot plants and large-scale reactors — 98
    2.2.1 On-line control of polymerisation — 102
  2.3 Special techniques; continuous processes — 102
    2.3.1 The cascade process — 104
    2.3.2 The loop reactor — 106
    2.3.3 Special techniques; procedural variations — 108
  2.4 Varied polymerisation techniques — 108
    2.4.1 Seed latices — 108
    2.4.2 Monodisperse latices — 108
    2.4.3 Core–shell latices — 111
    2.4.4 Emulsifier-free latices — 112

2.4.5 Mini-emulsions	113
2.4.6 Graft copolymerisation	114
3 Vinyl Acetate and Copolymer Latices	115
3.1 Historical development of latices	115
3.2 Polyvinyl acetate and related copolymers	116
3.2.1 Vinyl acetate, polyvinyl alcohol stabilised, delayed addition of monomer [174]	118
3.2.2 Vinyl acetate, redox method with heel	118
3.2.3 Vinyl acetate and copolymers, mixed surfactant and colloid stabilisers; delayed addition of monomer	119
3.2.4 Vinyl acetate polymers and copolymers, modified starch emulsifiers	121
3.2.5 Vinyl acetate; ethylene oxide–propylene oxide block copolymer emulsifier	122
3.2.6 Vinyl acetate, positively charged polymers	124
3.3 Copolymers of vinyl acetate	124
3.3.1 Vinyl acetate–ethyl acrylate copolymer, pre-emulsion feed, sulfosuccinate emulsifier	125
3.3.2 Copolymers of vinyl acetate and ethylene	127
3.3.3 Veova® copolymer	127
3.4 Alkali-soluble polymers	128
3.5 Equivalence of 'external' and 'internal' plasticisation	131
4 Acrylic Latices	133
4.1 Typical acrylic formulations	134
4.1.1 Acrylic copolymer; delayed addition procedure	135
4.1.2 Acrylic ester latex; two-stage redox process	135
4.1.3 Acrylic ester–styrene copolymer	135
4.1.4 Alkali-soluble polymer; acrylic polymer and seed latex	136
4.1.5 Core–shell polymerisation	138
4.2 Developments in polymerisation of acrylics	138
5 Styrene Latices	140
5.1 Processes—general	140
5.1.1 Soap-type surfactants	141
5.1.2 Vinyl acetate–maleic acid stabiliser	142
5.1.3 Cationic emulsifiers	143
6 Diene Polymers and Copolymers	143
6.1 Introduction	143
6.2 Dienes: non-rubber applications	144
6.3 Carboxylated latex formulations	144
7 Vinyl Halides	146
7.1 Vinyl chloride polymers and copolymers	146
7.1.1 Early processes	147
7.1.2 Graft copolymer	147
7.2 Vinylidene chloride polymers and copolymers	148

	7.3	Monomers containing bromine	148
	7.4	Vinyl compounds containing fluorine	148
8	Emulsion Polymerisation—Theoretical Considerations		150
	8.1	Types of mechanism	151
	8.2	Theories of Harkins and Smith and Ewart	153
		8.2.1 Modifications by Gardon	157
	8.3	Theory of Medvedev	157
	8.4	Theoretical developments—Fitch and Ugelstadt	158
	8.5	Copolymerisation—theoretical	161
	8.6	Specificity of surfactants and stabilisers	162
	8.7	Latex particle stabilization	164
	8.8	Further quantitative investigations	166
		8.8.1 Styrene	166
	8.9	Emulsion polymerisation, acrylic and methacrylic esters	167
		8.9.1 Emulsion polymerisation of methyl methacrylate	167
		8.9.2 Emulsion polymerisation of butyl acrylate	169
		8.9.3 Copolymers of butyl acrylate	169
	8.10	Emulsion polymerisation of vinyl acetate	170
		8.10.1 Copolymerisation—mainly of vinyl acetate	173
		8.10.2 Vinyl acetate and acrylic copolymers	175
	8.11	Styrene copolymers (except with dienes)	175
	8.12	Butadiene copolymers	176
	8.13	Copolymerisation including a water-soluble monomer	177
9	Emulsion Polymerisation in Supercritical Carbon Dioxide		178
10	Inverse Emulsions		179
11	Interpenetrating Networks (IPN)		180
12	Conclusions		181
References			182

4 Latex Properties Relative to Applications   **192**

1	Introduction		192
2	Various Properties and Descriptions		194
	2.1	Emulsion particles and resultant properties	194
		2.1.1 Particle size distribution	194
	2.2	Particle characterization	194
	2.3	Core–shell copolymers—surface and buried groups	196
	2.4	Adsorption of surfactants on latex particles	197
	2.5	Emulsion viscosity	197
		2.5.1 Definitions and theory	197
		2.5.2 The measurement of viscosity	199
		2.5.3 Viscosity, dilatancy and rheopexy phenomena	200
		2.5.4 Viscosity phenomena with latices	202
3	Freeze–Thaw and Tropical Stability		203
	3.1	Freeze–thaw stability—general principles	203

3.2 Freeze–thaw stability—technical methods of control	205
3.3 Freeze–thaw stability—tests	207
3.4 Tropical stability	208
4 Emulsion Films	209
4.1 Film formation	209
4.1.1 Methods of study—film formation	213
4.2 Theoretical developments	215
4.3 Minimum film-forming temperature	220
4.4 Specific studies—film formation	221
4.4.1 Vinyl acetate and copolymer latices	221
4.4.2 Acrylic latices	224
4.4.3 Styrene–butadiene latices	227
4.5 Summary	227
5 Addition of Plasticisers and Coalescing Solvents	228
5.1 Introduction	228
5.2 The mechanism of platicisation in emulsion	230
5.3 Plasticisation of principal polymers	231
5.3.1 Plasticisation of polystyrene	231
5.3.2 Plasticisation of polyvinyl acetate	231
5.3.3 Plasticisation of polyvinyl chloride	237
5.4 Transient plasticisers (coalescing solvents)	238
6 Pigmentation	242
6.1 Introduction	242
6.1.1 The dried composition	244
6.2 Pigmentary power	245
6.3 Some common pigments and extenders	246
6.3.1 Titanium dioxide	246
6.3.2 Other white pigments	248
6.3.3 Extenders	248
6.4 Pigmentation of latices	249
6.4.1 Treatment of china clay	251
6.5 General properties of pigmented systems	252
6.5.1 Critical pigment volume concentration	252
6.6 Pigment encapsulation	254
7 Organic Extenders	255
References	261

5 Crosslinking and Curing — **268**

1 Introduction	268
1.1 Problems of crosslinking in latices	270
1.2 A survey of crosslinking	272
2 Estimation of Crosslinking	273
2.1 Cure studies	273
2.1.1 Measurement of the degree of swelling	274

2.1.2 Viscosity characteristics 274
2.1.3 Gel permeation chromatography 275
2.2 Cure tests 275
3 Multiple Double Bonds 275
3.1 Polymers including divinylbenzene and allied monomers 275
3.2 Polymers derived from acrylic and allied esters—multiple double bonds 277
3.3 Allyl polymers—multiple double bonds 278
3.4 Diene polymers 280
3.5 Other vinyl-type polymers with residual unsaturation 282
4 The Carboxyl Function in Polymers 284
4.1 Introduction 284
4.2 Combinations with hydroxyl and epoxide groups 286
4.3 Metal oxides 288
5 The Hydroxyl Function 288
5.1 Reactions with aldehydes 289
5.1.1 Dioxalane derivatives 290
6 The Epoxide Function 290
6.1 Internal epoxide groups in unsaturated monomer 292
6.1.1 Two-pack latex system in single vessel 295
6.1.2 Allyl glycidyl ethers 296
6.1.3 Miscellaneous formulations including glycidyl esters 297
6.2 External epoxide compounds 298
7 Di-Isocyanates and Polyurethanes In Vinyl-Type Cures 300
8 Polymers Containing (Meth)Acrylamide 301
8.1 Chemistry of the reactions 301
8.2 Simple derivatives of acrylamide 304
8.2.1 Formulations mainly based on acrylic esters and/or styrene 305
8.2.2 Formulations with vinyl acetate as the principal monomer 306
8.2.3 Formulations including vinyl chloride 307
8.2.4 Formulations including isoprogyl-, butyl- and isobutyl, isobutoxy acrylamide 308
8.3 Diacetoneacrylamide (DAAM) 308
8.4 Methylene bis acrylamide (MBAM) 310
8.5 The aminimides 310
8.6 Acrylamidoglycollates 314
8.7 Other amide derivatives with unsaturation 315
8.8 Cure with aminoplasts 316
9 Polymers Containing Halogens 321
10 Radiation Curing 325
10.1 Ultraviolet curing 326

	10.1.1 Photoinitiators	327
	10.1.2 Photoinitiators for radical cure	328
	10.1.3 Monomers and polymers in UV cured systems	328
	10.1.4 Practical examples of UV cure	329
10.2	Cure by electron beam (EB)	331
11	Incorporation of Siloxanes	332
12	Cure of Eva and Saturated Polymers	333
13	Other Crosslinking Systems	334
	13.1 Acid and alkali induced crosslinking	334
	13.2 Crosslinking via β-ketonic acid esters	334
	13.3 Crosslinking with reactive nitrogen compounds	338
	13.3.1 Carbodi-imides	340
	13.4 Intra- and inter-molecular reactions	341
	13.5 Unspecified	342
References		343

6 Technical Polymer Latices — 351

1	Introduction	351
2	Polyvinyl Acetate and Related Polymers	353
	2.1 Polyvinyl acetate	353
	2.2 Copolymers of dinyl acetate	355
	2.2.1 Types of copolymer	355
	2.2.2 Equivalence of 'internal' and 'external' plasticisation	361
	2.2.3 Alkali-soluble copolymers	363
	2.3 Typical commercial products	364
3	Acrylic Polymers and Copolymers	365
	3.1 Properties of acrylic polymers	365
	3.2 Properties of acrylic resin films	367
4	Vinyl Halide Polymers	371
	4.1 Polyvinyl chloride	371
	4.1.1 Compounding of polyvinyl chloride and copolymer latices	372
	4.2 Polymers and copolymers of vinylidene chloride	375
	4.3 Fluorocarbon latices	380
5	Styrene Polymers and Copolymers	381
	5.1 Polystyrene	381
	5.2 Styrene–butadiene copolymers	383
6	Polymers Based on Aliphatic Hydrocarbons	385
7	Vinylpyrrolidone Copolymers	388
8	The Testing of Latices	389
	8.1 Total solids content	389
	8.2 Monomer content	390

		8.2.1 Chemical methods—vinyl acetate	390
		8.2.2 Other monomers	391
	8.3	Specific gravity	392
	8.4	pH stability	392
	8.5	Freeze–thaw stability	392
	8.6	Emulsion viscosity	393
	8.7	Particle size	393
	8.8	Soap titration; surface tension measurements	396
	8.9	Settling and sedimentation	398
	8.10	Mechanical stability	398
	8.11	Minimum film temperature (MFT)	399
	8.12	Relative molecular weight (viscosity)	399
		8.12.1 Viscosity measurements	400
	8.13	General film measurements	400
	8.14	Determination of charge	401
References			401

7 Unsaturated Polyester and Non-vinyl Emulsions — 404

1	Introduction		404
2	Alkyds and Polyesters		405
	2.1	Polyesters	405
		2.1.1 Polyester emulsions with vinyl monomers	407
		2.1.2 Water-in-oil emulsions	408
	2.2	Alkyds and allied products	410
		2.2.1 Theoretical and miscellaneous preparations	410
		2.2.2 Pigmented alkyd emulsions	413
		2.2.3 Alkyd emulsion blends	413
		2.2.4 Styrenated alkyd emulsions	415
		2.2.5 Ecological and environmental studies	417
3	Aminoplasts		417
4	Polyamide Emulsions		419
5	Epoxide Resins and Blends		421
	5.1	Initial condensation in emulsion	423
	5.2	Amine salt emulsification	423
	5.3	Non-ionic emulsifiers	424
	5.4	Miscellaneous emulsifiers	425
	5.5	Cycloaliphatic epoxides	426
	5.6	Water-in-oil emulsions	427
	5.7	Epoxides containing halogens	427
	5.8	Epoxide–polyamide emulsions	427
		5.8.1 Epoxide–polyamide blends in coatings	428
	5.9	Epoxide emulsions—miscellaneous blends and reactions	429
		5.9.1 Reaction with vinyl-type monomers	429
		5.9.2 Miscellaneous	430

	5.10 Applications	432
6	Phenoplasts	434
7	Silicones	437
	7.1 Elementary chemistry	437
	7.2 Some methods of emulsification	439
	7.2.1 General principles	439
	7.2.2 Cationic emulsifiers	439
	7.2.3 Anionic emulsifiers—mainly sulfates and sulfonates	440
	7.2.4 Non-ionic surfactants and phosphate surfactants	441
	7.2.5 Hydrophilic silicones as emulsifiers	441
	7.2.6 Problems of hydrogen release	442
	7.2.7 Theoretical developments	443
	7.3 Textile and paper applications of silicone (organopolysiloxane) latices	444
	7.3.1 General properties required	444
	7.3.2 Glass fibre treatment	446
	7.3.3 Paper treatment	446
	7.4 Silicone latices in surface coating and building	447
	7.5 Miscellaneous applications	448
8	Polyurethanes	449
	8.1 Chemistry of isocyanate adducts	449
	8.2 Polyurethane latex	450
	8.2.1 Emulsions via blocked copolymers	451
	8.2.2 Emulsification by salt formation—cationic	452
	8.2.3 Emulsification by salt formation—anionic	454
	8.2.4 Direct (chain extension) emulsification	456
	8.2.5 Vinyl–urethane block copolymers	457
	8.2.6 Vinyl polymerisation in a polyurethane latex	457
	8.3 Applications of polyurethane latices	458
	8.3.1 General properties, including adhesives	458
	8.3.2 Textiles and allied applications	459
	8.3.3 Paper applications	460
	8.3.4 Coatings	461
	8.3.5 Leather treatment	462
	8.3.6 Sizes for glass fibres and miscellaneous	462
References		462

8 Polymer Latices in the Formulation of Adhesives   470

1 Introduction   470
    1.1 Terminology of adhesives   471
2 Survey of Theories of Adhesion   473
    2.1 Basic theory   473
    2.2 The function of adhesives: latex adhesives   477

3 Latex-Based Adhesives: Some Practical Aspects	478
3.1 Introduction	478
3.2 Functions of adhesives	479
3.3 Practical requirements of adhesives	481
4 Specific Adhesive Types	482
4.1 Polyvinyl acetate homopolymer-based adhesives	484
4.1.1 Adhesive latices for compounding	486
4.1.2 Plasticisers	488
4.1.3 Tackifiers, adhesion promoters and other additives	495
4.1.4 Fillers	500
4.1.5 Some typical formulations	500
4.2 Polyvinyl acetate copolymer-based adhesives	505
4.2.1 Polyvinyl acetate copolymer latices with other esters	505
4.2.2 Ethylene–vinyl acetate copolymer latex-based adhesives	506
4.2.3 Crosslinking latex-based adhesives	512
4.3 Polyvinyl propionate latices in adhesives	517
4.4 Acrylic copolymer latices	518
4.4.1 Some typical formulations	523
4.4.2 Crosslinking systems	525
4.5 Vinylidene chloride and other halogenated polymer latices in adhesives	526
4.5.1 Some typical formulations	530
4.6 Butadiene copolymer latices	533
4.6.1 Some typical formulations	536
4.7 Other latex types	537
5 Latex-Based Adhesives with Specific Functions	538
5.1 Pressure-sensitive adhesives	538
5.1.1 Pressure-sensitive adhesion and terminology	538
5.1.2 Types of latices formulated into pressure-sensitive adhesives	540
5.2 Removable pressure-sensitive adhesives	567
5.2.1 Acrylic and other copolymer latices	567
5.3 Contact adhesives	568
5.3.1 Some commercial formulations	569
5.4 Heat-seal adhesives	572
5.4.1 General	572
5.4.2 Ethylene–vinyl acetate latex copolymers	572
5.4.3 Acrylate latex copolymers	573
5.5 Delayed tack adhesives	573
5.5.1 Some commercial formulations	573
5.6 Remoistenable adhesives	574
5.6.1 General	574
5.6.2 Some commercial formulations	574

	5.7 Quick-tack and contact-grab adhesives	575
	5.7.1 Some commercial formulations	575
	5.8 Redispersible components for pressure-sensitive adhesives	576
6	Adhesives for Specific Applications	577
	6.1 Wood adhesives	577
	6.1.1 Polymer latices for wood adhesives	582
	6.1.2 Aminoplasts and phenoplasts in wood adhesives	583
	6.1.3 Glycidyl and epoxy additives	584
	6.1.4 Addition of metal complexes	584
	6.1.5 Addition of aziridines, oxiranes and related compounds	586
	6.1.6 Addition of isocyanates	587
	6.1.7 Diene copolymers with crosslinking	589
	6.1.8 Some commercial formulations	590
	6.2 Latex-based adhesives for paper bonding	593
	6.2.1 General	593
	6.2.2 Some commercial formulations	594
	6.3 Adhesives for packaging	595
	6.4 Miscellaneous latex-based adhesives	599
	6.4.1 Adhesives for organic substrates	600
	6.4.2 Adhesives for inorganic substrates	603
	6.4.3 Other applications	606
7	Rubber-to-Fabric Adhesives	606
	7.1 Introduction	607
	7.2 Vinylpyridine copolymer adhesives	609
	7.3 Dip additives and composition	610
	7.4 Other rubber adhesive systems	614
	7.4.1 Blocked isocyanates	614
	7.4.2 Epoxides and glycidyl groups	615
	7.4.3 Dip additives	616
	7.5 Modifications of resorcinol–formaldehyde resins	618
	7.6 Latex modifications	619
	7.6.1 Latex dip additions	621
8	Test Methods for Adhesives	624
9	Glossary of Terms Relating to Adhesives	624
10	Appendix: Some Technical Products for Water-Based Adhesives	633
References		664

Index   **687**

Preface

A word, allow us, sweet ladies and gentlemen
We pray you hear, while we alone appear,
We are the Prologue

In 1972 a volume by one of the current authors appeared under the title *Applications of Synthetic Resin Emulsions*. This has long been out of print. The many advances that have been made, both in the development of polymer emulsions (described more correctly as latices, by analogy with natural rubber latices) and their application in various industries, merit a new trilogy under the title *Applications of Synthetic Resin Latices*. In the words of the slightly modified quotation from an English version of the Prologue to Pagliacci, which was quoted in the Prologue to the earlier book, the current volumes are still the only comprehensive ones on the subject.

With the present trends in ensuring environmental safety, water-based latices are having an ever-increasing role in a diverse range of applications. Many of these overcome the limitations of water-based polymers since in film forming a minimum temperature is necessary. In the case of some polymers, there are limitations due to insufficient film hardness, with a tendency of films to 'creep' on a vertical surface.

Developments in the formation of latices have included increasing molecular weights of polymers and the use of 'core–shell' copolymers, which often have the effect of reducing the minimum film-forming temperature, therefore enabling harder films to be formed and, above all, the use of crosslinking reactions, often described as 'curing'. As this has now become a major part of the technology of the preparation and application of these latices, it is the subject of a separate chapter. The minimum use of water-soluble emulsion stabilisers, or in some cases the use of stabilisers that can become water insoluble, is another feature of developments in latex technology.

The general structure of these volumes is similar to that of the earlier book. Each chapter is complete in itself, but a limited amount of repetition occurs between the earlier chapters, which describe the chemical basis of the processes and also additives required in applications, e.g. pigments as well as some typical commercial products, and the specific chapters on the various application industries.

Like the previous text, these volumes are intended for three classes of readers: chemists and other scientists who are performing the necessary research and development, technical service personnel who examine latices for potential applications and also, not least, the technologists in the application industries who may wish to use these water-based systems.

The standardization of units and their presentation has caused some problems, particularly with the adoption of the various SI units. The hybrid units sometimes found in patent applications have been avoided as far as possible, e.g. pressures in pounds per square centimetre and area dimensions given in square inches but with the thickness in millimetres. Temperatures are normally quoted in degrees Celsius (°C) but exceptionally in degrees Fahrenheit (°F), especially when results were originally quoted in rounded figures such as 250 °F and 300 °F. In most cases, however, units quoted in papers and patents are quoted as in the originals.

Sufficient information on technology is included for the processes involved to be understood, but no pretence is made that the chapters concerned provide comprehensive expertise on a wide range of industries. It is to be hoped that the texts will also point the way to future improvements and that these volumes, produced at the start of the new century, will be useful for many years to come.

Henry Warson
26 Blythe Court, 4 Grange Road,
Solihull B91 1BL

C.A. Finch
18 West End,
Weston Turville,
Aylesbury HP22 5TT

Applications—A note on References

Austral. P = Australia Patent
BP = British Patent
Belg. P = Belgium Patent
Can. P = Canadian Patent
Cz. P = Czech Patent
East G. P = East German Patent
Eur. P = European Patent
FP = French Patent
GP = German Patent
Ind. P = Indian Patent
Neth. P = Netherland Patent
Pol. P = Polish Patent
Rom. P = Romania Patent
S. Afr. P = South African Patent
Sp. P = Spanish Patent
USP = United States Patent
USSRP = Soviet Union Patent

There is a complication in references to Japanese specifications (JP). From about 1977 all references quoted are to the Japan Kokai, patents which are published before examination. Earlier references may be to the Kokai series or to full patents. These numbers do not correspond in the Japanese system, and if copies are required of earlier patents, it is necessary to check for the required series. The appropriate *Index Volume to Chemical Abstracts* is useful for this purpose.

Patent Validity

Patents are often published in countries other than that in which the original application is applied; hence the country of the patentee may be different from that of the named country. European and World order patents are valid in a number of countries, as indicated in the specifications. In many countries, e.g. Great Britain, France and Germany, patents are now published without examination unless the applicant requires this. It is up to anyone objecting to do so formally, but objection can only be made by a person or persons directly affected by the publication.

Publication of extracts from, or abstracts of, patents in these volumes must not be construed as an invitation to operate them, where the use is protected by laws of the country involved.

Introduction

1 SCOPE

With the growth of the synthetic resin (polymer) industry in the past five decades, a considerable amount of technology has been compiled. Whilst there are many general and specialised volumes dealing both with the chemistry of polymers and their technological applications, we believe that the only single volume which endeavoured to co-ordinate the literature on applications of polymers in emulsion form was that published by one of the current authors in 1972. After nearly 30 years, with the vast increase in available information, it has been found necessary to publish a new compilation in three volumes.

Whilst the terms *resin* and *polymer* do not have exactly the same meaning, they are often used as synonyms where no confusion is likely to occur. Most of the materials with which these volumes are concerned are more properly described as polymers–a general term applied to compounds of high molecular weight featuring substantially repeating units without loss of simple units such as water or ammonia by condensation. This definition applies to vinyl polymers generally, and often, also, to polyepoxides and to polyurethanes. The term resin implies a condensation product of comparatively low molecular weight. As an example of this, apart from uncured aminoplasts and phenoplasts, the term resin is also used to describe products such as the condensates of rosin (of natural origin, usually from trees) with maleic anhydride. A further term, *macromolecule*, is more generally applied to both resins and polymers, but its use tends to be restricted to academic and 'pure' scientific publications. In this text, the terms *polymer emulsion* and *polymer latex* have been used interchangeably throughout, usually depending on the source of the information discussed. The term *latex* is often restricted to rubbers, either natural or synthetic. In the United States, the term *latex* is almost universal, and, indeed, may be considered as more correct. Many products developed in Germany are referred to as *dispersions ('dispersionen')*, so this term is retained in some cases. A *dispersion* may also imply a stabilised solid of a fine particle size often a pigment) in a liquid. However, the term *dispersion polymerization* is applied both to suspension polymerization in aqueous media, and *dispersion polymer* to stabilized macromolecular emulsions in non-aqueous media.

There is no rigid distinction between the properties of a rubber and a resin or polymer. Nevertheless, since rubber latices form a technology of their own, they are not considered here, unless they are proposed as functional alternatives to polymers with more 'plastic' properties. For this reason, the applications of only a limited number of polymers of butadiene, styrene and isoprene are discussed.

It must be admitted that this nomenclature is rather confusing, but, since it will be encountered elsewhere, distinction is made is made between the various terms only where it occurs in technical practice. Otherwise, as long as no ambiguity is likely to be caused, the terms *resin* and *polymer* are used virtually as synonyms.

In nearly every case, the feature of a polymer emulsion is that it is applied as an auxiliary factor rather than as a primary material e.g. it is a medium for a paint, or a major (functional) component of an adhesive. As such, latices have tended, in some ways, to be the 'Cinderella' of the world of polymer science, especially as progress in their applications has been evolutionary, rather than taking place in 'quantum style' leaps.

Information on the preparation, and properties of latices is found in numerous patents, theoretical papers and the literature of the principal manufacturers of monomers and other raw materials. Information on applications of latices is often found in patents and the technical literature of polymer manufacturers. The preparation of a group of latices and their application is often described in the same patent specification. Selected information from these sources is presented in each of these volumes. Some technical journals contain useful data, and much information of considerable technical value is scatter throughout the trade literature, but is seldom abstracted adequately and is often issued undated.

This first volume surveys the fundamentals of polymer chemistry, the principles and practice of polymer latices, including pigmented systems and some current technical latices. The properties of alkali-soluble latices are considered. Volumes 2 and 3 describe in some detail the numerous applications of latices.

Crosslinking and curing of polymers has become important in the last few decades, and is the subject of Chapter 5. Some non-vinyl latices, such as those from alkyds and polyurethanes are also the subject of a separate Chapter, since these have assumed some industrial importance.

2 HISTORICAL AND GENERAL SURVEY

Natural rubber latex has been known in the West since 1736, when it was first introduced to Europe by de la Condamine. Rubber latex is a milky white liquid obtained from *Hevea braziliensis*. This was coagulated, and the product sold as *gutta percha*. After seeds of this tree had been smuggled out of Brazil and propagated in the UK, large scale cultivation commenced after 1900 in Malaysia, Natural rubber is an example of a natural polymer. It is now known to be a polymer of isoprene, $CH_2:C(CH_3).CH:CH_2$, and may be written in a simple manner as $(C_5H_8)_n$. Its biosynthesis is not from isoprene, but by a complex route through isopentenyl pyrophosphate $CH_3.C(:CH_2)_2.O.P(:O)(OH).O.P(:O)(OH)_2$ [1]. It is necessary to consider the term latex in greater depth. A latex is a special case of an emulsion, which is a special case of a suspension of an insoluble liquid or 'oil' in water prepared so that it is stable for long periods, of the order of months or, even, years, rather than minutes. To ensure that an emulsion remains stable, and that the two phases do not separate, it is necessary to add suitable emulsifying agents, alone or in combination (see Chapter 2) These emulsifiers function by their surface activity at the interface of the two phases. They are essentially balanced molecules with a hydrophilic or water-loving part, and a *hydrophobic* or water repellent portion, such as a long-chain hydrocarbon. A typical soap for this purpose consists of a sodium ion, Na^+, together with the carboxyl group COO^-, whilst the hydrophobic 'tail' is C_nH_{2n+1}, typically where $n = 15-20$, preferably 18. Another type of emulsifier or stabilizer is a water–soluble colloid, including high molecular weight water–soluble compounds such as natural gums, e.g. gum arabic (gum acacia), a naturally occurring carbohydrate, natural proteins such as casein (which is solubilised by alkali) or synthetic or semi-synthetic polymers such as polyvinyl alcohol and hydroxyethyl cellulose respectively. Natural latex is stabilised by phospholipids, also known as phosphatides, and also by proteins. Phospholipids are components of all animal and vegetable cells: they are triglycerides containing two long-chain fatty acid groups such as those derived from oleic or stearic acid, and a phosphoric acid residue attached to a base.

Why is an essentially solid material, such as natural latex considered as an emulsion, which is a fine dispersion of one liquid in another in which it is insoluble? This is connected to the manufacture of polymer emulsions by the process of emulsion polymerisation (see Chapter 3). The simplest process consists of forming a stable emulsion of one liquid (a monomer) in water, and subjecting this monomer to a polymerisation process, maintaining the stability of the emulsion throughout until all of the monomer is converted to polymer. In a modification described later, the monomers may be added gradually to the polymerising emulsion.

The final product has all the characteristics of an emulsion, but, on drying, produces a film, or possibly a powder, of a polymer, depending on drying conditions (see Chapter 1, Section 4.3). It differs from a stable dispersion of a solid, such as a pigment, in that a good dispersion of the latter consists of the 'ultimate' particles of an agglomerate, which is broken down on forming the dispersion. Polymers in emulsions, being in general non-crystalline, may be subject to quite extensive further change whilst maintaining the emulsion (latex) state, e.g. they may be directly plasticised.

This illustrates the connection between the various synthetic latices (polymer emulsions) and natural rubber latex. Physically, they represent the same type of product. Chemically, all these latices are prepared by substantially the same route, although there are major technological variation and developments (see Chapter 3). Polymerisation occurs substantially by addition polymerisation of a vinyl monomer, $CH_2{:}CHR$, the simplest of which is ethylene, $CH_2{:}CH_2$. Polymers formed by emulsion polymerisation have, in general, plastic or elastic properties. A polymer is described as having elastic properties when a strip, extended to double its length, regains its original shape and length, after the applied force which has caused it to stretch has been removed. Polymers with this property are arbitrarily defined as rubbers. In most synthetic rubbers, as formed by emulsion polymerisation, all, or a large part, of the polymer molecule consists of diene units, e.g. the polymer is formed from a monomer containing two double bonds, but it substantially retains one such double bond in each unit after polymerisation. This tends, for steric reasons, to give the polymer molecule the helical chain shape characteristic of the rubber molecule.

The properties of the main classes of synthetic polymers, as described in Chapters 2 and 3, vary widely, from rigid glasses to extremely soft and sticky products, varying to some extent on the molecular weight of the polymer. Many polymers in the intermediate range also some show some degree of elasticity.

The terms copolymer and copolymerisation imply that several monomers have been polymerised together, and are usually used when it is wished to emphasise that several monomers have been included, but is often restricted to the use of two monomers. Terpolymer is occasionally used when three monomers have been copolymerised together. Polymer is used as a general term, to describe the final product, and does not necessarily imply that only one monomer is present.

3 A GENERAL SURVEY OF POLYMER LATEX APPLICATIONS

Many polymers are used either directly, as in injection moulding, or in calendering processes such as the manufacture of polyvinyl chloride sheeting. Alternatively, they may be applied in liquid form from either solution or from emulsion. In a special case, prepolymers can be polymerised, often by crosslinking (see Chapter 6) from a liquid or fusible state to a solid infusible form.

Solvents have a number of disadvantages. They are costly, and only in rare cases will a solvent recovery plant justify its original outlay. Solvents are either flammable in which case special fire precautions must be taken (e.g. in industrial establishments where large scale spraying operations take place, as in coating automobile bodies), or toxic when non-flammable chlorinated hydrocarbons are substituted, again making special precautions necessary. The use of polymers based on aqueous media therefore has obvious advantages. Water, the cheapest possible solvent, presents no special fire or toxicity hazard, although it has the disadvantage of a high latent heat of evaporation (538 cal/g^{-1}) and a comparatively high bp (100°), compared with that of ethanol (78°) or acetone (56°), both of which have much lower latent heats of evaporation. The evaporation of water is also comparatively slow, because of the relative humidity of the atmosphere.

The evaporation of a thin film of many latices–almost any thickness from ~50 μ–1 mm–produces a continuous film of polymer (but see 'Conditions for film formation' in Chapter 4). As an example, an ethyl acrylate copolymer latex will deposit a continuous elastic and rather tacky film. On the other hand, latices based on vinyl acetate or butyl methacrylate form slightly brittle, continuous films if the water is removed by evaporation at ~25°. These latices, however, if allowed to evaporate near 0°, produce weak, somewhat 'crazed' films of little coherence. These properties may be considerably modified with differing molecular weight of the polymers.

Various methods of modifying polymers to produce compositions with specific properties are employed. The addition of a non-volatile liquid, known as a plasticiser, has the effect of softening or flexibilising the polymers. This process of plasticisation suggests that there is a solubility of the plasticiser in the polymer, known as the compatibility of the polymer with the plasticiser.

Polymer latices may be pigmented. The pigments are wetted out or dispersed in water with suitable additives, and the resulting mixtures used as paints and coatings. Typically, a methyl methacrylate polymer latex may be used as a surface coating on polyvinyl chloride sheet to avoid a slight after-tack.

However, nearly every useful application of latex polymers is based on either a coating application (including adhesives—see Chapter 8), or as a general binding and filling agent, typically in paper coating or in non-woven fabric manufacture (see Volume 3). Many latices may be precipitated by treatment with acid or alkali, salts of polyvalent metals, or by tannic acid. For some purposes, especially in paper manufacture, controlled precipitation of the latex takes place. Similar controlled precipitation is employed with some textile processes in which positively-charged latices are exhausted on to a fabric. In certain conditions, especially with coatings based on vinyl chloride copolymer latices, the dried films are heat treated so that the polymer particles fuse together satisfactorily, imparting the characteristic tough properties of the polymer to the film.

In some latices, a polymerisable acid may be copolymerised with a monomer e.g. crotonic acid, CH_3. CH: CH. COOH, with vinyl acetate or methacrylic acid, CH_2 : $CH(CH_3)$.COOH, with acrylic esters. The resulting latices can form aqueous solutions when made alkaline with a strong alkali, such as sodium hydroxide, or, more usually, with a volatile base such as ammonia or an amine. The alkali causes the carboxylic groups to ionise. There is usually a considerable increase in viscosity with a latex of this type on solubilisation, since it is sensitive to changes in the molecular weight of the polymer and, sometimes, to small variations in pH. These solutions are used in strippable coatings, and are sometimes added to paint formulations to improve brushability or final gloss. In some cases, where there is no risk of alkaline attack, or application is only for a short time, e.g. in polishes, the ammonium, or amine salts of such copolymers may be used as a major component of the formulation (see Chapter 16). Because of their high viscosity, these products can seldom be obtained in the alkaline state at more than ~25% concentration. Acidification results in complete precipitation of the product – the initial emulsion state cannot be restored.

The use of crosslinking reactions is of increasing importance in polymer latex technology (see Chapter 5) where this is treated more fully.

3.1 The principal applications of polymer latices

Since adhesive behaviour governs almost all other applications, the first chapter on applications (Chapter 8) is devoted to this subject. Adhesive applications are discussed in relation to the specific substrates involved including wood and other cellular materials, paper and packaging, and a variety of miscellaneous substrates, including adhesives suitable for bonding different polymers, in relation to the methods of assembly and the performance required.

Polymer latices are widely employed in surface coatings (paints), both decorative ('architectural') and industrial. Surface coatings comprise the major industrial outlet: their use is described in detail in Volume 2.

Volume 3 describes the principles and practice of the use of polymer latex-based systems in many other industries, including building and construction products, in cement-based products and additives, sealants, flooring compositions, decorative finishes and redispersible powders, paper impregnation and coating, including beater addition, wallpaper coatings, and many speciality coatings for pressure sensitive papers, release papers, electroconductive, photographic and security papers. Discussion of the use of speciality acrylic and non-vinyl latices in leather finishing is followed by a chapter on textile applications including warp sizing, fabric coating impregnation and lamination. Finishing for specific objectives such as flame- and water-proofing, use in upholstery applications (including tufted carpets and automobile interiors), polishes, leather finishing and many other specialised applications is also discussed. Other applications mentioned include the use of latices in vibration damping systems, and in electrographic toner manufacture and other imaging applications. Use of polymer latices in polishes is described in Chapter 16, whilst the final chapter considers use of speciality polymer latices as a binder (including pyrotechnics, inks and graphic arts materials, horticultural and ground stabilising applications), in foods, glass fibre treatments, photographic and copying systems. Barrier and antistatic coatings, gravure and offset coatings and many other miscellaneous applications are also discussed.

Surface coatings and adhesives probably provide the largest industrial outlets for synthetic latices, in terms of tonnage, but these remarkable materials have many other uses as indicated in this summary.

REFERENCES

1. Blackley, D.C., *Polymer Latices: Science and Technology.* 2nd Edition, Vol. 2, pp. 118–131, 1997: Chapman & Hall, London

1
Fundamentals of Polymer Chemistry

H. Warson

1 THE CONCEPT OF A POLYMER

1.1 Historical introduction

The differences between the properties of crystalline organic materials of low molecular weight and the more indefinable class of materials referred to by Graham in 1861 as 'colloids' has long engaged the attention of chemists. This class includes natural substances such as gum acacia, which in solution are unable to pass through a semi-permeable membrane. Rubber is also included among this class of material.

The idea that the distinguishing feature of colloids was that they had a much higher molecular weight than crystalline substances came fairly slowly. Until the work of Raoult, who developed the cryoscopic method of estimating molecular weight, and Van't Hoff, who enunciated the solution laws, it was difficult to estimate even approximately the polymeric state of materials. It also seems that in the nineteenth century there was little idea that a colloid could consist, not of a product of fixed molecular weight, but of molecules of a broad band of molecular weights with essentially the same repeat units in each.

Vague ideas of partial valence unfortunately derived from inorganic chemistry and a preoccupation with the idea of ring formation persisted until after 1920. In addition chemists did not realise that a process such as ozonisation virtually destroyed a polymer as such, and the molecular weight of the ozonide, for example of rubber, had no bearing on the original molecular weight.

The theory that polymers are built up of chain formulae was vigorously advocated by Staudinger from 1920 onwards [1]. He extended this in 1929 to the idea of a three-dimensional network copolymer to account for the insolubility and infusibility of many synthetic polymers, for by that time technology had by far outstripped theory. Continuing the historical outline, mention must be made of Carothers, who from 1929 began a classical series of experiments which indicated that polymers of definite structure could be obtained by the use of classical organic chemical reactions, the properties of the polymer being controlled by the starting compounds [2]. Whilst this was based on research in condensation compounds (see Section 1.2) the principles hold good for addition polymers.

The last four decades have seen major advances in the characterisation of polymers. Apart from increased sophistication in methods of measuring molecular weight, such as the cryoscopic and vapour pressure methods, almost the whole range of the spectrum has been called into service to elucidate polymer structure. Ultraviolet and visible spectroscopy, infrared spectroscopy, Raman and emission spectroscopy, photon correlation spectroscopy, nuclear magnetic resonance and electron spin resonance all play a part in our understanding of the structure of polymers; X-ray diffraction and small-angle X-ray scattering have been used with solid polymers. Thermal behaviour in its various aspects, including differential thermal analysis and high-temperature pyrolysis followed by gas–liquid chromatography, has also been of considerable value. Other separation methods include size exclusion and hydrodynamic chromatography. Electron microscopy is of special interest with particles formed in emulsion polymerisation. Thermal and gravimetric analysis give useful information in many cases. There are a number of standard works that can be consulted [3–6].

1.2 Definitions

A polymer in its simplest form can be regarded as comprising molecules of closely related composition of molecular weight at least 2000, although in many cases typical properties do not become obvious until the mean molecular weight is about 5000. There is virtually no upper end to the molecular weight range of polymers since giant three-dimensional networks may produce crosslinked polymers of a molecular weight of many millions.

Polymers (macromolecules) are built up from basic units, sometimes referred to as 'mers'. These units can be extremely simple, as in addition polymerisation, where a simple molecule adds on to itself or other simple molecules, by methods that will be indicated subsequently. Thus ethylene $CH_2:CH_2$ can be converted into polyethylene, of which the repeating unit is —CH_2CH_2—, often written as —$(CH_2CH_2)_n$—, where n is the number of repeating units, the nature of the end groups being discussed later.

The major alternative type of polymer is formed by condensation polymerisation in which a simple molecule is eliminated when two other molecules condense. In most cases the simple molecule is water, but alternatives include ammonia, an alcohol and a variety of simple substances. The formation of a condensation polymer can best be illustrated by the condensation of hexamethylenediamine with adipic acid to form the polyamide best known as nylon®:

$$H_2N(CH_2)_6\overset{H}{\underset{|}{N}}H + HOOC(CH_2)_4CO.OH + H\overset{H}{\underset{|}{N}}(CH_2)_6NH_2 \qquad (1)$$
$$= H_2N(CH_2)_6NH.OC(CH_2)_4CONH(CH_2)_6NH_2$$
$$+ H_2O \qquad\qquad + H_2O$$

This formula has been written in order to show the elimination of water. The product of condensation can continue to react through its end groups of hexamethylenediamine and adipic acid and thus a high molecular weight polymer is prepared.

Monomers such as adipic acid and hexamethylenediamine are described as bifunctional because they have two reactive groups. As such they can only form linear polymers. Similarly, the simple vinyl monomers such as ethylene $CH_2{:}CH_2$ and vinyl acetate $CH_2{:}CHOOCCH_3$ are considered to be bifunctional. If the functionality of a monomer is greater than two, a branched structure may be formed. Thus the condensation of glycerol $HOCH_2CH(OH)CH_2OH$ with adipic acid $HOOC(CH_2)_4COOH$ will give a branched structure. It is represented diagrammatically below:

$$\begin{array}{l}
HOOC(CH_2)_4COOCH_2CHCH_2OOC(CH_2)_4COOCHCH_2O{-} \\
\qquad\qquad\qquad\quad | \qquad\qquad\qquad\qquad\qquad\qquad\quad CH_2O{-}\\
\qquad\qquad\qquad\quad O\\
\qquad\qquad\qquad\quad |\\
\qquad\qquad\qquad\quad CO\\
\qquad\qquad\qquad\quad |\\
\qquad\qquad\qquad\quad (CH_2)_4\\
\qquad\qquad\qquad\quad |\\
\qquad\qquad\qquad\quad CO\\
\qquad\qquad\qquad\quad |\\
\qquad\qquad\qquad\quad O\\
\qquad\qquad\qquad\quad |\qquad\qquad\qquad\qquad\qquad\quad |\\
\qquad\qquad\qquad\quad CH_2 \qquad\qquad\qquad\qquad\quad O\\
\qquad\qquad\qquad\quad |\qquad\qquad\qquad\qquad\qquad\quad |\\
\qquad\qquad H\ COOC(CH_2)_4COOCH_2CHCH_2O{-}\\
\qquad\qquad\qquad\quad |\\
\qquad\qquad\qquad\quad CH_2\\
\qquad\qquad\qquad\quad |\\
\qquad\qquad\qquad\quad O\\
\qquad\qquad\qquad\quad |\\
\qquad\qquad\qquad\quad CO(CH_2)_4COO{-}
\end{array}$$

The condensation is actually three dimensional, and ultimately a three-dimensional structure is formed as the various branches link up.

Although this formula has been idealised, there is a statistical probability of the various hydroxyl and carboxyl groups combining. This results in a network being built up, and whilst it has to be illustrated on the plane of the paper, it will not necessarily be planar. As functionality increases, the probability of such networks becoming interlinked increases, as does the probability with increase in molecular weight. Thus a gigantic macromolecule will be formed which is insoluble and infusible before decomposition. It is only comparatively recently that structural details of these crosslinked or 'reticulated' polymers have been elucidated with some certainty. Further details of crosslinking are given in Chapter 5.

Addition polymers are normally formed from unsaturated carbon-to-carbon linkages. This is not necessarily the case since other unsaturated linkages including only one carbon bond may be polymerised.

Addition polymerisation of a different type takes place through the opening of a ring, especially the epoxide ring in ethylene oxide $\mathrm{CH_2{-}CH_2}$ with an O bridge. This opens as —CH_2CH_2O—; ethylene oxide thus acts as a bifunctional monomer forming a polymer as $H(CH_2CH_2O)_n\ CH_2CH_2OH$, in this case a terminal water molecule being added. A feature of this type of addition is that it is much easier to control the degree of addition, especially at relatively low levels, than in the vinyl polymerisation described above.

Addition polymerisations from which polymer emulsions may be available occur with the silicones and diisocyanates. These controlled addition polymerisations are sometimes referred to as giving 'stepwise' addition polymers. This term may also refer to condensation resins. Further details are given in Chapter 7.

2 ADDITION POLYMERISATION

Addition polymerisation, the main type with which this volume is concerned, is essentially a chain reaction, and may be defined as one in which only a small initial amount of initial energy is required to start an extensive chain reaction converting monomers, which may be of different formulae, into polymers. A well-known example of a chain reaction is the initiation of the reaction between hydrogen and chlorine molecules. A chain reaction consists of three stages, initiation, propagation and termination, and may be represented simply by the progression:

$$\text{Activation} \quad +M \quad +M \quad +nM$$
$$M{-}M^*{-}M_2^*{-}M_3^*{-}M_{n+3}{-}\text{etc.}$$

The termination reaction depends on several factors, which will be discussed later.

The mechanism of polymerisation can be divided broadly into two main classes, free radical polymerisation and ionic polymerisation, although there are some others.[†] Ionic polymerisation was probably the earliest type to be noted, and is divided into cationic and anionic polymerisations. Cationic polymerisation depends on the use of catalysts which are good electron acceptors. Typical examples are the Friedel–Crafts catalysts such as aluminium chloride $AlCl_3$ and boron trifluoride BF_3.

Monomers that polymerise in the presence of these catalysts have substituents of the electron releasing type. They include styrene $C_6H_5CH{:}CH_2$ and the vinyl ethers $CH_2{:}CHOC_nH_{2n+1}$ [7].

Anionic initiators include reagents capable of providing negative ions, and are effective with monomers containing electronegative substituents such

---

[†] Some modern sources prefer to refer to addition polymerisation and stepwise polymerisation.

Addition polymerisation

as acrylonitrile $CH_2{:}CHCN$ and methyl methacrylate $CH_2{:}C(CH_3)COOCH_3$. Styrene may also be polymerised by an anionic method. Typical catalysts include sodium in liquid ammonia, alkali metal alkyls, Grignard reagents and triphenylmethyl sodium $(C_6H_5)_3C\text{-Na}$.

Amongst other modern methods of polymerisation are the Ziegler–Natta catalysts [8] and group transfer polymerisation catalysts [9]. Ionic polymerisation is not of interest in normal aqueous polymerisation since in general the carbonium ions by which cationic species are propagated and the corresponding carbanions in anionic polymerisations are only stable in media of low dielectric constant, and are immediately hydrolysed by water.

2.1 Free radical polymerisation

A free radical may be defined as an intermediate compound containing an odd number of electrons, but which do not carry an electric charge and are not free ions. The first stable free radical, triphenylmethyl $(C_6H_4)_3C\cdot$, was isolated by Gomberg in 1900, and in gaseous reactions the existence of radicals such as methyl $CH_3\cdot$ was postulated at an early date.

The decomposition of oxidizing agents of the peroxide type, as well as compounds such as azodiisobutyronitrile

$$(CH_3)_2\underset{NC}{C}.N{:}N\underset{CN}{C}(CH_3)_2$$

which decomposes into two radicals, $(CH_3)_2\underset{CN}{C}\cdot$ and nitrogen N_2, is well-known. Thus a free radical mechanism is the basis of addition polymerisation where these types of initiator are employed. For a transient free radical the convention will be used of including a single dot after or over the active element with the odd electron.

A polymerisation reaction may be simply expressed as follows. Let R be a radical from any source. $CH_2{:}CHX$ represents a simple vinyl monomer where X is a substituent, which may be H as in ethylene $CH_2{:}CH_2$, Cl as in vinyl chloride $CH_2{:}CHCl$, $OOC.CH_3$ as in vinyl acetate $CH_2{:}CHOOCCH_3$ or many other groups, which will be indicated in lists of monomers.

The first stage of the chain reaction, the initiation process, consists of the attack of the free radical on one of the doubly bonded carbon atoms of the monomer. One electron of the double bond pairs with the odd electron of the free radical to form a bond between the latter and one carbon atom. The remaining electron of the double bond shifts to the other carbon atom which now becomes a free radical. This can be expressed simply in equation form:

$$R + CH_2{:}CHX \longrightarrow R.CH_2\underset{X}{\overset{H}{C}}\cdot \qquad (2)$$

The new free radical can, however, in its turn add on extra monomer units, and a chain reaction occurs, representing the propagation stage:

$$\text{R.CH}_2\overset{\text{H}}{\underset{\text{X}}{\text{C}\cdot}} + n\,(\text{CH}_2\text{CHX}) \longrightarrow \text{R:}(\text{CH}_2\text{CHX})_n\,\text{CH}_2\overset{\text{H}}{\underset{\text{X}}{\text{C}\cdot}} \qquad (3)$$

The final stage is termination, which may take place by one of several processes. One of these is combination of two growing chains reacting together:

$$\text{R}(\text{CH}_2\text{CHX})_n\text{CH}_2\dot{\text{C}}\text{HX} + \text{R}(\text{CH}_2\text{CHX})_m\text{CH}_2\dot{\text{C}}\text{HX}$$
$$= \text{R}(\text{CH}_2\text{CH})_n\text{CH}_2\text{CHXCH}_2\text{CHX}(\text{CH}_2\text{CHX})_m\text{R} \qquad (4\text{a})$$

An alternative is disproportionation through transfer of a hydrogen atom:

$$\text{R}(\text{CH}_2\text{CHX})_n\text{CH}_2\dot{\text{C}}\text{HX} + \text{R}(\text{CH}_2\text{CHX})_m$$
$$= \text{R}(\text{CH}_2\text{CHX})_n\text{CH}_2\text{CH}_2\text{X} + \text{R}(\text{CH}_2\text{CHX})_m\text{CH:CHX} \qquad (4\text{b})$$

A further possibility is chain transfer. This is not a complete termination reaction, but it ends the propagation of a growing chain and enables a new one to commence. Chain transfer may take place via a monomer, and may be regarded as a transfer of a proton or of a hydrogen atom:

$$\text{Z-CH}_2\overset{\text{X}}{\underset{\text{H}}{\text{C}\cdot}} + \text{CH}_2\text{CHX} = \text{Z-CH:CHX} + \text{CH}_3\overset{\text{X}}{\underset{\text{H}}{\text{C}\cdot}} \qquad (5)$$

where Z is a polymeric chain.

Chain transfer takes place very often via a fortuitous impurity or via a chain transfer agent which is deliberately added. Alkyl mercaptans with alkyl chains C_8 or above are frequently added for this purpose in polymerisation formulations. A typical reagent is *t*-dodecyl mercaptan, which reacts as in the following equation:

$$\text{R}(\text{CH}_2\text{CHX})_n\text{CH}_2\overset{\text{H}}{\underset{\text{X}}{\text{C}\cdot}} + t\text{-C}_{12}\text{H}_{25}\text{SH} \qquad (6\text{a})$$
$$= \text{R}(\text{CH}_2\text{CHX})_n\text{CH}_2\text{CH}_2\text{X} + \text{C}_{12}\text{H}_{25}\text{S}\cdot$$

Chlorinated hydrocarbons are also commonly used as chain transfer agents, and with carbon tetrachloride it is a chlorine atom rather than a hydrogen atom that takes part in the transfer:

$$R(CH_2CHX)_n\,CH_2\underset{X}{\overset{H}{C}}\cdot + CCl_4 = R(CH_2CHX)_n\,CHXCl + Cl_3C\cdot \qquad (6b)$$

Most common solvents are sufficiently active to take part in chain transfer termination, the aliphatic straight-chain hydrocarbons and benzene being amongst the least active. The effect of solvents is apparent in the following equation, where SolH denotes a solvent:

$$R(CH_2CHX)_n\,CH_2\underset{X}{\overset{H}{C}}\cdot + SolH = RCH_2CHX)_n\,CH_2X + Sol\cdot \qquad (6c)$$

In all the cases mentioned, the radicals on the right-hand side of the equations must be sufficiently active to start a new chain; otherwise they act as a retarder or inhibitor (see the next section)

Derivatives of allyl alcohol $CH_2{:}CHCH_2OH$, although polymerisable by virtue of the ethylenic bond, have marked chain transfer properties and produce polymers of low molecular weight relatively slowly (see also Section 2.1.2). Stable intermediate products do not form during a polymerisation by a free radical chain reaction, and the time of formation of each polymer molecule is of the order of 10^{-3} s.

Kinetic equations have been deduced for the various processes of polymerisation. These have been explained simply in a number of treatises [10–13]. The classic book by Flory [10] derives these equations in greater detail.

A useful idea which may be introduced at this stage is that of the order of addition of monomers to a growing chain during a polymerisation. It has been assumed in the elementary discussion that if a growing radical $M\text{-}CH_2C\cdot$ is considered, the next unit of monomer will add on to produce

$$M-CH_2-\underset{X}{\overset{H}{C}}-\underset{H}{\overset{H}{C}}-\underset{X}{\overset{H}{C}}\cdot$$

It is theoretically possible, however, for the next unit of monomer to add on, producing

$$M-CH_2-\underset{X}{\overset{H}{C}}-\underset{X}{\overset{H}{C}}-\underset{H}{\overset{H}{C}}\cdot$$

8   Fundamentals of polymer chemistry

The latter type of addition in which similar groups add in adjacent fashion is known as 'head-to-head' addition in contrast to the first type above, known as 'head-to-tail' addition. The head-to-tail addition is much more usual in polymerisations, although in all cases head-to-head polymerisation occurs at least to some extent.

There are various ways of estimating head-to-head polymerisation, both physical and chemical. Nuclear magnetic resonance data should be mentioned amongst the former. The elucidation of polyvinyl acetate $(CH_2CH(OOCCH_3)_n)$ may be taken as representative of a chemical investigation. A head-to-tail polymer when hydrolysed to polyvinyl alcohol would typically produce units of $(CH_2CHOHCH_2CHOH)$. A head-to-head unit is $(CH_2CHOHCHOHCH_2)$. In the latter case there are two hydroxyl groups on adjacent carbon atoms, and the polymer is therefore broken down by periodic acid HIO_4, which attacks this type of unit. It is possible to estimate the amount of head-to-head addition from molecular weight reduction or by estimation of the products of oxidation.

2.1.1 Retardation and inhibition

If the addition of a chain transfer agent to a polymerising system works efficiently, it will both slow the polymerisation rate and reduce the molecular weight. This is because the free radical formed in the equivalent of equation (6a) may be much less active than the original radical in starting new chains, and when these are formed, they are terminated after a relatively short growth.

In some cases, however, polymerisation is completely inhibited since the inhibitor reacts with radicals as soon as they are formed. The most well known is p-benzoquinone.

$$O=C \begin{matrix} C=C \\ \\ C=C \end{matrix} C=O$$

This produces radicals that are resonance stabilised and are removed from a system by mutual combination or disproportionation. Only a small amount of inhibitor is required to stop polymerisation of a system. A calculation shows that for a concentration of azodiisobutyronitrile of 1×10^{-3} mole^{-1} in benzene at 60 °C, a concentration of 8.6×10^{-5} mole L^{-1} h^{-1} of inhibitor is required [10]. p-Hydroquinone $C_6H_4(OH)_2$, probably the most widely used inhibitor, only functions effectively in the presence of oxygen which converts it to a quinone–hydroquinone complex giving stable radicals. One of the most effective inhibitors is the stable free radical 2 : 2-diphenyl-1-picryl hydrazyl:

$$\begin{matrix} C_6H_5 \\ \diagdown \\ \end{matrix} N-N- \underset{NO_2}{\overset{NO_2}{\bigcirc}} -NO_2$$

Addition polymerisation

This compound reacts with free radicals in an almost quantitative manner to give inactive products, and is used occasionally to estimate the formation of free radicals.

Aromatic compounds such as nitrobenzene $C_6H_5NO_2$ and the dinitrobenzenes (*o*-, *m*-, *p*-)$C_6H_4(NO_2)_2$ are retarders for most monomers, e.g. styrene, but tend to inhibit vinyl acetate polymerisation, since the monomer produces very active radicals which are not resonance stabilised. Derivatives of allyl alcohol such as allyl acetate are a special case. Whilst radicals are formed from this monomer, the propagation reaction (equation 3) competes with that shown in the following equation:

$$M_x + CH_2{:}CHCH_2OOCCH_3 = M_xH + H_2C.CH{:}CHOOCCH_3 \qquad (7)$$

In this case the allylic radical is formed by removal of an alpha hydrogen from the monomer, producing an extremely stable radical which disappears through bimolecular combination. Reaction (7) is referred to as a degradative chain transfer [11–14].

2.1.2 Free radical initiation

Initiators of the type required for vinyl polymerisations are formed from compounds with relatively weak valency links which are relatively easily broken thermally. Irradiation of various wavelengths is sometimes employed to generate the radicals from an initiator, although more usually irradiation will generate radicals from a monomer as in the following equation:

$$CH_2CHX \xrightarrow{\eta v} CH_2CHX^* \qquad (8)$$

The activated molecule then functions as a starting radical. Since, however, irradiation is not normally a method of initiation in emulsion polymerisation, it will only be given a brief mention. The decomposition of azodiisobutyronitrile has already been mentioned (see Section 2.1), and it may be noted that the formation of radicals from this initiator is accelerated by irradiation.

Another well-known initiator is dibenzoyl peroxide, which decomposes in two stages:

$$C_6H_5CO.OO.OCC_6H_5 \longrightarrow 2\ C_6H_5COO\cdot \qquad (9a)$$

$$C_6H_5COO. \longrightarrow C_6H_5\cdot + CO_2 \qquad (9b)$$

Studies have shown that under normal conditions the decomposition proceeds through to the second stage, and it is the phenyl radical $C_6H_5.$ that adds on to the monomer. Dibenzoyl peroxide decomposes at a rate suitable for most direct polymerisations in bulk, solution and aqueous media, whether in emulsion or bead form, since most of these reactions are performed at 60–100 °C. Dibenzoyl peroxide has a half-life of 5 h at 77 °C.

A number of other diacyl peroxides have been examined. These include o-, m- and p-bromobenzoyl peroxides, in which the bromine atoms are useful as markers to show the fate of the radicals. Dilauroyl hydroperoxide $C_{11}H_{23}CO.OO.OCC_{11}H_{23}$ has been used technically.

Hydroperoxides as represented by t-butyl hydroperoxide $(CH_3)_3C.O.O.H$ and cumene hydroperoxide $C_6H_5C(CH_3).O.O.H$ represent an allied class with technical interest. The primary dissociation

$$R.CX.O.O.H. \longrightarrow R.CXO\cdot + OH\cdot$$

is by secondary decompositions, which may include various secondary reactions of the peroxide induced by the radical in a second-order reaction and by considerable chain transfer. These hydroperoxides are of interest in redox initiators (see Section 2.1.3).

Dialkyl peroxides of the type di-t-butyl peroxide $(CH_3)_3C.O.O.C(CH_3)_3$ are also of considerable interest, and tend to be subject to less side reactions except for their own further decomposition, as shown in the second equation below:

$$(CH_3)_3COOC(CH_3)_3 \longrightarrow 2(CH_3)_3CO\cdot \tag{10a}$$

$$(CH_3)_3CO. \longrightarrow (CH_3)_2CO + CH_3\cdot \tag{10b}$$

These peroxides are useful for polymerisations that take place at 100–120 °C, whilst di-t-butyl peroxide, which is volatile, has been used to produce radicals for gas phase polymerisations.

A number of peresters are in commercial production, e.g. t-butyl perbenzoate $(CH_3)_3C.O.O.OC.C_6H_5$, which acts as a source of t-butoxy radicals at a lower temperature than di-t-butyl hydroperoxide, and also as a source of benzoyloxy radicals at high temperatures. The final decomposition, apart from some secondary reactions, is probably mainly

$$(CH_3)_3C.O.O.OCC_6H_5 \longrightarrow (CH_3)_3CO + CO_2 + C_6H_5\cdot \tag{11}$$

For a more detailed description of the decomposition of peroxides a monograph by one of the current authors should be consulted [15]. Whilst some hydroperoxides have limited aqueous solubility, the water-soluble initiators are a major type utilised for polymerisations in aqueous media. In addition, some peroxides of relatively high boiling point such as $tert$-butyl hydroperoxide are sometimes added towards the end of emulsion polymerisations (see Chapter 2) to ensure a more complete polymerisation. These peroxides are also sometimes included in redox polymerisation (see Section 2.1.3), especially to ensure rapid polymerisation of the remaining unpolymerised monomers.

Hydrogen peroxide H_2O_2 is the simplest compound in this class and is available technically as a 30–40 % solution. (This should not be confused with the 20–30 volume solution available in pharmacies.) Initiation is not caused by the simple decomposition $H_2O_2 = 2\ OH\cdot$, but the presence of a trace of ferrous ion, of the order of a few parts per million of water present, seems to be essential, and radicals are generated according to the Haber–Weiss mechanism:

$$H_2O_2 + Fe^{2+} \longrightarrow HO^- \cdot + Fe^{3+} + HO\cdot$$

The hydroxyl radical formed commences a polymerisation chain in the usual manner and is in competition with a second reaction that consumes the radical:

$$Fe^{2+} + HO\cdot = Fe^{3+} + OH^-\cdot$$

When polymerisations are performed it seems of no consequence whether the soluble iron compound is in the ferrous or ferric form. There is little doubt that an equilibrium exists between the two states of oxidation, probably due to a complex being formed with the monomer present.

The other major class of water-soluble initiators consists of the persulfate salts, which for simplicity may be regarded as salts of persulfuric acid $H_2S_2O_8$. Potassium persulfate $K_2S_2O_8$ is the least soluble salt of the series, between 2 and 4 % according to temperature, but the restricted solubility facilitates its manufacture at a lower cost than sodium persulfate $Na_2S_2O_8$ or ammonium persulfate $(NH_4)_2S_2O_8$. The decomposition of persulfate may be regarded as thermal dissociation of sulfate ion radicals:

$$S_2O_8^{2-} \longrightarrow 2\ SO_4^-.$$

A secondary reaction may, however, produce hydroxyl radicals by reaction with water, and these hydroxyls may be the true initiators:

$$SO_4^-\cdot + H_2O \longrightarrow HSO_4^- + HO\cdot$$

Research using ^{35}S-modified persulfate has shown that the use of a persulfate initiator may give additional or even sole stabilization to a polymer prepared in emulsion. This may be explained by the polymer having ionised end groups from a persulfate initiator, e.g. $ZOSO_3Na$, where Z indicates a polymer residue.

A general account of initiation methods for vinyl acetate is applicable to most monomers [16].

2.1.3 Redox polymerisation

The formation of free radicals, which has already been described, proceeds essentially by a unimolecular reaction, except in the case where ferrous ions are included. However, radicals can be formed readily by a bimolecular reaction, with the added advantage that they can be formed *in situ* at ambient or even subambient temperatures. These systems normally depend on the simultaneous reaction of an oxidizing and a reducing agent, and often require in addition a transition element that can exist in several valency states. The Haber–Weiss mechanism for initiation is the simplest case of a redox system.

Redox systems have assumed considerable importance in water-based systems, since most components in systems normally employed are water soluble. This type of polymerisation was developed simultaneously during the Second World War in Great Britain, the United States and Germany, with special reference to the manufacture of synthetic rubbers. For vinyl polymerisations, as distinct from those where dienes are the sole or a major component, hydrogen peroxide or a persulfate is the oxidizing moiety, with a sulfur salt as the reductant. These include sodium metabisulfite $Na_2S_2O_5$, sodium hydrosulfate (also known as hyposulfite or dithionite) $Na_2S_2O_4$, sodium thiosulfate $Na_2S_2O_3$ and sodium formaldehyde sulfoxylate $Na(HSO_3.CH_2O)$. The last named is one of the most effective and has been reported to initiate polymerisations, in conjunction with a persulfate, at temperatures as low as 0 °C. In almost all of these redox polymerisations, a complete absence of oxygen seems essential, possibly because of the destruction by oxygen of the intermediate radicals that form.

However, in redox polymerisations operated under reflux conditions, or in otherwise unsealed reactors, it is often unnecessary in large-scale operations to continue the nitrogen blanket after polymerisation has begun, probably because monomer vapour acts as a sealant against further oxygen inhibition.

There have been relatively few detailed studies of the mechanisms of redox initiation of polymerisation. A recent survey of redox systems is available [17]. The review already quoted [15] gives a number of redox initiators, especially suitable for vinyl acetate, most of which are also suitable for other monomers.

Since almost all such reactions take place in water, a reaction involving ions may be used as an illustration. Hydrogen peroxide is used as the oxidizing moiety, together with a bisulfite ion:

$$H_2O_2 + S_2O_5^{2-} = HO\cdot + (HS_2O_6)\cdot$$

The HS_2O_6 represented here is not the dithionate ion, but an ion radical whose formula might be

$$\begin{array}{cc} O^- & O^- \\ | & | \\ -S-O-S.OH \\ \| & \| \\ O & O \end{array}$$

Alternatively, a hydroxyl radical may be formed together with an acid dithionate ion. Some evidence exists for a fragment of the reducing agent rather than the oxidizing agent acting as the starting radical for the polymerisation chain. This seems to be true of many phosphorus-containing reducing agents; e.g. hypophosphorous acid with a diazonium salt activated by a copper salt when used as an initiating system for acrylonitrile shows evidence of a direct phosphorus bond with the polymer chain and also shows that the phosphorus is present as one atom per chain of polymer [18]. Many of the formulations for polymerisation quoted in the various application chapters are based on redox initiation.

2.2 Copolymerisation

There is no reason why the process should be confined to one species of monomer. In general, a growing polymer chain may add on most other monomers according to a general set of rules which, with some exceptions, will be enunciated later.

If we have two monomers denoted by M_i and M_n and $M_i \cdot$ and $M_n \cdot$ denote chain radicals having M_i and M_{ii} as terminal groups, irrespective of chain length four reactions are required to describe the growth of polymer:

$$M_i \cdot + M_i \xrightarrow{K_{11}} M_i \cdot$$
$$M_i \cdot + M_n \xrightarrow{K_{12}} M_{ii} \cdot$$
$$M_{ii} \cdot + M_{ii} \xrightarrow{K_{22}} M_{ii} \cdot$$
$$M_{ii} \cdot + M_i \xrightarrow{K_{21}} M_i \cdot$$

where K has the usual meaning of a reaction rate constant. These reactions reach a 'steady state' of copolymerisation in which the concentration of radicals is constant; i.e. the copolymerisation is constant and the rates of formation of radicals and destruction of radicals by chain termination are constant. Under these conditions the rates of formation of each of the two radicals remain constant and without considering any elaborate mathematical derivations we may define the monomer reactivity rations r_1 and r_2 by the expressions

$$r_1 = \frac{K_{11}}{K_{12}} \quad \text{and} \quad r_2 = \frac{K_{22}}{K_{21}}$$

These ratios represent the tendency of a radical formed from one monomer to combine with itself rather than with another monomer. It can be made intelligible by a practical example. Thus, for styrene $C_6H_5CH:CH_2$ (r_1) and butadiene $CH_2:CHCH:CH_2$ (r_2), $r_1 = 0.78$ and $r_2 = 1.39$. These figures tend to indicate that if we start with an equimolar mixture, styrene radicals tend to copolymerise with butadiene rather than themselves, but butadiene has a slight preference for its radicals to polymerise with each other. This shows that if we copolymerise an equimolar mixture of styrene and butadiene, a point occurs at which only styrene would remain in the unpolymerised state. However, for styrene and methyl methacrylate, $r_1 = 0.52$ and $r_2 = 0.46$ respectively. These two monomers therefore copolymerise together in almost any ratio. As the properties imparted to a copolymer by equal weight ratios of these two monomers are broadly similar, it is often possible to replace one by the other on cost alone, although the inclusion of styrene may cause yellowing of copolymer films exposed to sunlight.

Nevertheless, if an attempt is made to copolymerise vinyl acetate with styrene, only the latter will polymerise, and in practice styrene is an inhibitor for vinyl acetate. The reactivity ratios, r_1 and r_2 for styrene and vinyl acetate respectively have been given as 55 and 0.01. However, vinyl benzoate $CH_2{:}CHOOCC_6H_5$ has a slight tendency to copolymerise with styrene, probably because of a resonance effect. If we consider the case of vinyl acetate and *trans*-dichlorethylene (TDE) *trans*-$CH_2Cl{:}CH_2Cl$, r_1(vinyl acetate) $= 0.85$ and $r_2 = 0$. The latter implies that TDE does not polymerise by itself, but only in the presence of vinyl acetate. Vinyl acetate, on the other hand, has a greater tendency to copolymerise with TDE than with itself, and therefore if the ratios are adjusted correctly all of the TDE can be copolymerised.

Let us consider the copolymerisation of vinyl acetate and maleic anhydride:

$$\begin{array}{c} CH-C \overset{O}{\underset{}{\diagup}} \\ \diagdown O \\ CH-C \underset{O}{\diagup} \end{array} \quad r_1 = 0.055, \; r_2 = 0$$

Sometimes a very low r_2 is quoted for maleic anhydride, e.g. 0.003. Vinyl acetate thus has a strong preference to add on to maleic anhydride in a growing radical rather than on to another vinyl acetate molecule, whilst maleic anhydride, which has practically no tendency to add on to itself, readily adds to a vinyl acetate unit of a growing chain. (Note that homopolymers of maleic anhydride have been made by drastic methods.) This is a mathematical explanation of the fact that vinyl acetate and maleic anhydride tend to alternate in a copolymer whatever the starting ratios. Excess maleic anhydride, if present, does not homopolymerise. Surplus vinyl acetate, if present, forms homopolymer, a term used to distinguish the polymer formed from a single monomer in contradistinction to a copolymer. Styrene also forms an alternating copolymer with maleic anhydride.

Only in one or two exceptional cases has both r_1 and r_2 been reported to be above 1. Otherwise it is a general principle that at least one of the two ratios is less than 1. It will be readily seen that in a mixture of two monomers the composition of the copolymer gradually changes unless an 'azeotropic' mixture is used, i.e. one balanced in accord with r_1 and r_2, provided that r_1 and r_2 are each <1.

Polymers of fixed composition are sometimes made by starting with a small quantity of monomers, e.g. 2–5 % in the desired ratios, and adding a feedstock which will maintain the original ratio of reactants. This is especially noted, as will be shown later, in emulsion polymerisation. If it is desired to include the more sluggishly polymerising monomer, and an excess is used, this must be removed at the end, by distillation or extraction.

Addition polymerisation

However, as a general principle it should not be assumed that, because two or more monomers copolymerise completely, the resultant copolymer is reasonably homogeneous. Often, because of compatibility variations amongst the constantly varying species of polymers formed, the properties of the final copolymer are liable to vary very markedly from those of a truly homogeneous copolymer.

The term 'copolymer' is sometimes confined to a polymer formed from two monomers only. In a more general sense, it can be used to cover polymers formed from a larger number of monomers, for which the principles enunciated in this section apply. The term 'terpolymer' is sometimes used when three monomers have been copolymerised.

When copolymerisation takes place in a heterogeneous medium, as in emulsion polymerisation (see Chapter 2), whilst the conditions for copolymerisation still hold, the reaction is complicated by the environment of each species present. Taking into account factors such as whether the initiator is water or monomer soluble (most peroxidic organic initiators are soluble in both), the high aqueous solubility of monomers such as acrylic acid $CH_2:CHCOOH$ and, if partition between water-soluble and water-insoluble monomers is significant, the *apparent* reactivities may differ markedly from those in a homogeneous medium. Thus, in an attempted emulsion polymerisation, butyl methacrylate $CH_2:C(CH_3)COOC_4H_9$ and sodium methacrylate $NaOOCC(CH_3):CH_2$ polymerise substantially independently. On the other hand, methyl methacrylate and sodium methacrylate will copolymerise together since methyl methacrylate has appreciable water solubility [19, 20].

More unusually vinyl acetate and vinyl stearate $CH_2:CHOOCC_{17}H_{35}$ will only copolymerise in emulsion if a very large surface is present due to very small emulsion particles (of order <0.1 m) or a class of emulsifier known as a 'solubiliser' is present, which has the effect of solubilising vinyl stearate to a limited extent in water, increasing the compatibility with vinyl acetate which is about 2.3 % water soluble.

Problems relating to copolymerisation in emulsion will be found in Chapter 3 and Sections 2.2.1 and 8.5. For more advanced texts, see the Appendix, Section 8.

2.2.1 The Q, e scheme

Several efforts have been made to place the relative reactivities of monomers on a chemical–mathematical basis. The chief of these has been due to Alfrey and Price [21]. Comparison of a series of monomers with a standard monomer is most readily made by using the reciprocal of r with respect to that monomer; i.e. the higher the value of $1/r$ the poorer the copolymerisation characteristics. Thus, taking styrene as an arbitrary 1.0, methyl methacrylate 2.2 and acrylonitrile 20, vinyl acetate is very high on this scale. However, the relative scale of reactivities is not interchangeable using different radicals as references [22].

It has been observed that the product $r_1 r_2$ tends to be smallest when one of the two monomers concerned has strongly electropositive (electron-releasing) substituents and there are electronegative (electron-attracting) substituents on the other. Thus alternation tends to occur when the polarities of the monomers are opposite.

Alfrey and Price therefore proposed the following equation:

$$K_{12} = P_1 Q_2 \exp(-e_1 e_2)$$

where P_1 and Q_2 are constants relating to the general activity of the monomers M_1 and M_2 respectively and e_1 and e_2 are proportional to the residual electrostatic charges in the respective reaction groups. It is assumed that each monomer and its corresponding radical has the same reactivity. Hence, from the reactions in Section 2.2,

$$r_1 r_2 = \exp[-(e_1 - e_2)^2]$$

The product of the reactivity ratios is thus independent of Q. The following equation is also useful:

$$Q_2 = \frac{Q_1}{r_1} \exp[-e, (e_1 - e_2)]$$

A series of Q and e values has been assigned to a series of monomers by Price [23]. Typical e values are -0.8 for butadiene, -0.8 for styrene, -0.3 for vinyl acetate, $+0.2$ for vinyl chloride, $+0.4$ for methyl methacrylate and $+1.1$ for acrylonitrile.

Whilst the Q, e scheme is semi-empirical, it has proved highly useful in coordinating otherwise disjointed data.

3 CHAIN BRANCHING; BLOCK AND GRAFT COPOLYMERS

3.1 Chain branching

Occasionally chain transfer (Section 2.1) results in a hydrogen atom being removed from a growing polymer chain. Thus in a chain that might be represented as $(CH_2CHX)_n$, the addition of further units of $CH_2{:}CHX$ might produce an intermediate as $(CH_2CHX)_n.CH_2\dot{C}.X.C.CH_2CHX$. A short side chain is thus formed by hydrogen transfer. For simplicity, this has been shown on the penultimate unit, but this need not be so; nor is there any reason why there should only be one hydrogen abstraction per growing chain. From the radicals formed branched chains may grow.

Chain branching occurs most readily from a tertiary carbon atom, i.e. a carbon atom to which only one hydrogen atom is attached, the other groupings depending on a carbon to carbon attachment, e.g. an alkyl or an aryl group. The mechanism is based on abstraction of a hydrogen atom, although of course abstraction can also occur with a halogen. atom. With polyvinyl acetate, investigations have shown that limited chain transfer can occur through the methyl grouping of the acetoxy group ·OOCCH$_3$. The result of this type of branching is a drastic reduction of molecular weight of the polymer during hydrolysis, since the entire branch is hydrolysed at the acetoxy group at which branching has occurred, producing an extra fragment for each branch of the original molecule. It has also been shown that in a unit of a polyvinyl acetate polymer the ratio of the positions marked (1), (2) and (3) is 1 : 3 : 1.

$$-CH_2-CH.OOCCH_3$$

(1)   (2)   (3)

It is now known that there is significant chain transfer on the vinyl H atoms of vinyl acetate [24].

Another method of forming branched chains involves the retention of a vinyl group on the terminal unit of a polymer molecule, either by disproportionation or by chain transfer to monomer. The polymer molecule with residual unsaturation could then become the unit of a further growing chain. Thus a polymer molecule of formula CH$_2$CHX—(Mp), where Mp represents a polymer chain, may become incorporated into another chain to give a structure (Mq).CH$_2$.CX—(Mp)(Mr), where Mq and Mr represent polymer chains of various lengths, that may be of the same configuration or based on different monomers, depending on conditions.

Ethylene, CH$_2$:CH$_2$, which is normally a gas (b.p. 760 mm Hg: -104 °C, critical temperature 9.5 °C) is prone to chain branching when polymerised by the free radical polymerisation process at high temperatures and pressures, most branches having short chains. In this case intramolecular formation of short chains occurs by chain transfer, and is usually known as 'back biting'.

$$\begin{array}{c} \quad\quad CH_2 \\ CH_2 \quad CH_2 \\ -\overset{*}{C}H_2 \quad CH_2 \end{array} \longrightarrow \begin{array}{c} CH_3 \\ (CH_2)_3 \\ -\overset{*}{C}H \end{array}$$

$$\begin{array}{c} CH_3 \\ (CH_2)_3 \\ -\overset{*}{C}H \end{array} + n\, C_2H_4 \longrightarrow \begin{array}{c} CH_3 \\ (CH_2)_3 \\ \overset{*}{C}H-(CH_2CH_2)_nE \end{array}$$

where E represents an end group. The carbon with the asterisk is the same throughout to illustrate the reaction.

Excessive chain branching can lead to crosslinking and insolubility (Chapter 5). It is possible for chain branching to occur from completed or 'dead' molecules by hydrogen abstraction, and although this impinges on grafting, it is treated as a chain branching phenomenon if it occurs during a polymerisation.

3.2 Graft copolymers

The idea of a graft copolymer is a natural extension of the concept of chain branching and involves the introduction of active centres in a previously prepared chain from which a new chain can grow. In most cases this is an added monomer, although two-polymer molecules can combine directly to form a graft. The graft base need not be an ethylene addition polymer. Various natural products, including proteins and water-soluble gums, have been used as a basis for graft copolymers by formation of active centres.

A block copolymer differs from a graft in only that the active centre is always at the end of the molecule. In the simplest case, an unsaturated chain end arising from a chain transfer can act as the basis for the addition of a block of units of a second monomer, whilst successive monomers or the original may make an additional block. Another possibility is the simultaneous polymerisation of a monomer which is soluble only in water with one which is water insoluble, provided that the latter is in the form of a fine particle size emulsion. Whether the initiator is water soluble or monomer soluble, an extensive transfer through the surface is likely, with the continuation of the chain in the alternate medium.

There are a number of ways of achieving active centres, many of which depend on an anionic or cationic mechanism, especially the former. However, since in water-based graft polymerisation only free radical polymerisations and possibly a few direct chemical reactions involving an elimination are of interest, the discussion here will be confined to these topics.

Graft centres are formed in much the same manner as points of branching, with the difference that the graft base is preformed. It may be possible to peroxidise a polymer directly with oxygen, to provide hydroperoxide O.OH groups directly attached to carbon. This is facilitated, particularly, where numerous tertiary carbon occur as, for example, in polypropylene $\left(-CH_2.\underset{\underset{CH_3}{|}}{CH}-\right)$. In other cases the direct use of a peroxide type of initiator encourages the formation of free radicals on existing polymer chains. Particularly useful in this respect is $tert$-butyl hydroperoxide, $tert$-C_4H_9.O.OH, because of the strong tendency of the radical formed from it to abstract hydrogen atoms. Dibenzoyl peroxide $(C_6H_5CO.O)_2$ is also frequently used as a graft initiator. In aqueous systems initiators such as $tert$-butyl hydroperoxide may be used in conjunction with a salt of a sulfur-reducing acid to lower the temperature at which radicals are generated.

Graft methods make it possible to add to polymers such as butadiene–styrene chains of a monomer that is not normally polymerisable, such as vinyl acetate. The polymerisation medium in which a graft can take place is in general not restricted; the process may take place fairly readily in emulsion. There is a vast amount of literature available on the formation and properties of graft copolymers [25].

There are very often special considerations in respect of graft copolymerisations that take place in emulsion form, with particular reference to water-soluble stabilisers of the polyvinyl alcohol type [26]. In some cases halogen atoms may be removed by a radical. This occurs particularly with polymers and copolymers based on vinyl chloride CH_2:CHCl, vinylidene chloride $CH_2.CCl_2$ and chloroprene $CH_2.CClCH$:CH_2. Ultraviolet light and other forms of irradiation are particularly useful in this respect.

Properties of graft copolymers are sometimes unique, and not necessarily an intermediate or balance between those of polymers derived from the respective monomers. This is particularly noticeable with solubility properties and transition points. A brief reference may be made here to the more direct chemical types of graft formation that do not involve free radicals. These depend on the direct reaction of an active group on the polymer. The simplest group is hydroxyl ·OH, which under suitable conditions may react with carboxyl ·COOH, carboxyanhydride ·C:OOO:C· or carbochloride ·COCl to form esters or polyesters depending on the nature of the side chain. Equally, hydroxyl groups may react with oxirane $CH_3.CHX$ groups. This applies especially
$$\underset{O}{\diagdown\diagup}$$
with ethylene oxide $CH_2\!-\!CH_2$ to form oxyethylene side chains, giving graft
$$\underset{O}{\diagdown\diagup}$$
copolymers of the type

$$\begin{array}{c} CHOH\!-\!CH_2\!-\!CH\!-\!CH_2\!- \\ \diagdown \\ OC_2H_4(OC_2H_4)_n\,OH \end{array}$$

This will be of special interest in dealing with emulsions.

4 POLYMER STRUCTURE AND PROPERTIES

4.1 Polymer structure

The physical properties of a polymer are determined by the configuration of the constituent atoms, and to some extent by the molecular weight. The configuration is partly dependent on the main chain, and partly on the various side groups. Most of the polymers which we are considering are based on long chains of carbon atoms. In representing formulae we are limited by the plane of the paper, but a three-dimensional structure must be considered. The C—C internuclear distance is 1.54 Å, and where free rotation occurs the C—C—C bond is fixed at 109° (the tetrahedral angle).

By tradition, we represent the polyethylene chain in the full extended fashion:

$$\diagup\!\!\!\!CH_2\diagup\!\!\!\!CH_2\diagup\!\!\!\!CH_2$$
$$\diagdown\!\!\!\!CH_2\diagup\diagdown\!\!\!\!CH_2\diagup$$

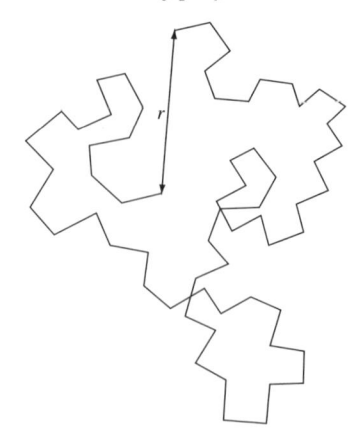

Figure 1.1 Diagrammatic molecular coil. (Reproduced from Moore [27].)

In practice the polymer is an irregular coil, as shown in Figure 1.1. The dimension most frequently used to describe an 'average' configuration is the 'root mean square', symbolised as r, which can be symbolised mathematically as

$$\frac{[\sum n_1 r_1^2]^{0.5}}{[\sum n_1]}$$

where there are n individual polymer molecules, and the distance apart of the chain ends is r_1, r_2, etc. This concept of root mean square is necessary in dealing with certain solution properties, and also certain properties of elasticity.

No real polymer molecule can have completely free and unrestricted rotation, although an unbranched polythene $(C_2H_4)_n$ approaches closest to this ideal. (The theoretical polymethylene $(CH_2)_m$ has been prepared by the polymerisation of diazomethane CH_2N_2, with elimination of nitrogen.) The properties of polyethylene over a wide range of molecular weights are, at ambient temperatures, those of a flexible, relatively inelastic molecule, which softens fairly readily. Chain branching hinders free rotation and raises the softening point of the polymer. Even a small number of crosslinks may, however, cause a major hindrance to the free rotation of the internal carbon bonds of the chain, resulting in a sharp increase in stiffness of the resulting product.

Many side chain groups cause steric hindrance and restrictions in the free rotation about the double bonds. A typical example is polystyrene, where the planar zigzag formulation is probably modified by rotations of 180° a round alternate double bonds to produce a structure of minimum energy, such as

```
              C6H5                C6H5
               \                   \
                CH—CH2              CH—CH2
    --CH2           CH—CH2              CH—CH2
                     |                   |
                    C6H5                C6H5
```

Because of the steric hindrance, polystyrene is a much harder polymer than polyethylene.

Other molecular forces that effect the physical state of the polymers are the various dipole forces and the London or dispersion forces. If different parts of a group carry opposite charges, e.g. the carbonyl :C=O and hydroxyl —O—H$^+$, strong interchain attraction occurs between groups on different chains by attraction of opposite charges. This attraction is strongly temperature dependent. A special, case of dipole forces is that of hydrogen bonding, by which hydrogen atoms attached to electronegative atoms such as oxygen or nitrogen exert a strong attraction towards electronegative atoms on other chains. The principal groups of polymers in which hydrogen bonding occurs are the hydroxyl and the amino .NHX or amide .CONH$_2$ groups and are illustrated by the following:

$$-O\overset{H}{\underset{H}{\diagdown\diagup}}O- \qquad -O-H\cdots O-CH-\underset{H-O}{\diagdown} \qquad \overset{\diagdown}{\underset{\diagup}{N}}-H\cdots O-C$$

The net effect of dipole forces, especially hydrogen bonding, is to stiffen and strengthen the polymer molecules, and in extreme cases to cause crystalline polymers to be formed (see below). Examples of polymers with strong hydrogen bonding are polyvinyl alcohol (—CH$_2$CHOH—)$_n$, polyacrylamide (CH$_2$CH–)$_n$ and all polymers including carboxylic acid
|
CONH$_2$

groups, e.g. copolymers including units of acrylic acid CH$_2$:CHCOOH and crotonic acid CH$_2$:CH.CH$_2$COOH.

The London forces between molecules come from time-varying dipole moments arising out of the continuously varying configurations of nuclei and electrons which must, of course, average out to zero. These forces, which are independent of temperature, vary inversely as the seventh power of the distance between the chains, as do dipole forces, and only operate at distances below 5 Å.

Forces between chains lead to a cohesive energy, approximately equal to the energies of vaporisation or sublimation. A high cohesive energy is associated with a high melting point and may be associated with crystallinity. A low cohesive energy results in a polymer having a low softening point and easy deformation by stresses applied externally.

Whilst inorganic materials often crystallise and solid organic polymers generally possess crystallinity, X-ray diffraction patterns have shown that in some polymers there are non-amorphous and crystalline regions, or crystallites. Whilst crystallinity is a characteristic of natural products such as proteins and synthetic condensation products such as the polyamide fibres, crystallinity sometimes occurs in addition polymers. Even if we discount types prepared by special methods, such as use of the Ziegler–Natta catalysts [8], which will not be discussed further here since they are not formed by classical free radical reactions, a number of polymers prepared directly or indirectly by free radical methods give rise to crystallinity.

One of these already mentioned is polyvinyl alcohol, formed by hydrolysis of polyvinyl acetate. It must, however, be almost completely hydrolysed, of the order of 99.5 %, to be effectively crystalline, under which conditions it can be oriented and drawn into fibres. If hydrolysis is partial, the resulting disorder prevents crystallinity. This is the case with the so-called 'polyvinyl alcohol' of saponification value about 120, which is used for emulsion polymerisation. This polymer consists, by molar proportions, of about 88 % of vinyl alcohol and 12 % of vinyl acetate units.

Polymers of vinylidene chloride CH_2CCl_2 are strongly crystalline. Polymers of vinyl chloride CH_2CHCl and acrylonitrile CH_2CHCN are partially crystalline, but crystallinity can be induced by stretching the polymer to a fibre structure to induce orientation. Polyethylene, when substantially free from branching, is crystalline and wax-like because of the simple molecular structure. It does not, of course, have the other properties associated with crystallinity caused by hydrogen bonding, such as high cohesive strength.

Another type of crystallinity found in polymers is side chain crystallinity, e.g. in polyvinyl stearate $(-CH_2CH-)_n$ or polyoctadecyl acrylate
$$\qquad\qquad\qquad\qquad\qquad\qquad |$$
$$\qquad\qquad\qquad\qquad\qquad OOC.C_{17}H_{35}$$

$(-CH_2CH-)_n$ This type of crystallinity has relatively little application,
$\quad\ \ |$
$\ COOC_{16}H_{37}$
since the products tend to simulate the crystalline properties of a wax. However, this property may be useful in connection with synthetic resin-based polishes, the subject of a later chapter.

In considering the effect of side chains on polymer properties, it is convenient to take a series of esters based on acrylic acid and compare the derived polymers. These are most readily compared by the second-order transition points (T_g). Technical publications show some variation in these figures, probably because of variations in molecular weight. However, polymers prepared under approximately the same conditions have much the same degree of polymerisation (DP), and emulsion polymers are preferred as standards in this connection.

Figure 1.2 shows the variations in T_g of a series of homologous polymers based on acrylic acid $CH_2{:}CHCOOH$ and methacrylic acid $CH_2{:}C(CH_3)COOH$. The striking difference in T_g of the polymers based on the methyl esters should be noted, being almost 100 °C. This is due to the steric effect of the angular methyl .CH_3 group on the carbon atom to which the carboxyl group is attached. Polymethyl methacrylate is an extremely hard solid, used *inter alia* for 'unbreakable' glass.

The effect of the angular methyl group slowly diminishes as the alcohol side chains become longer; these latter keep the chains apart and reduce the polar

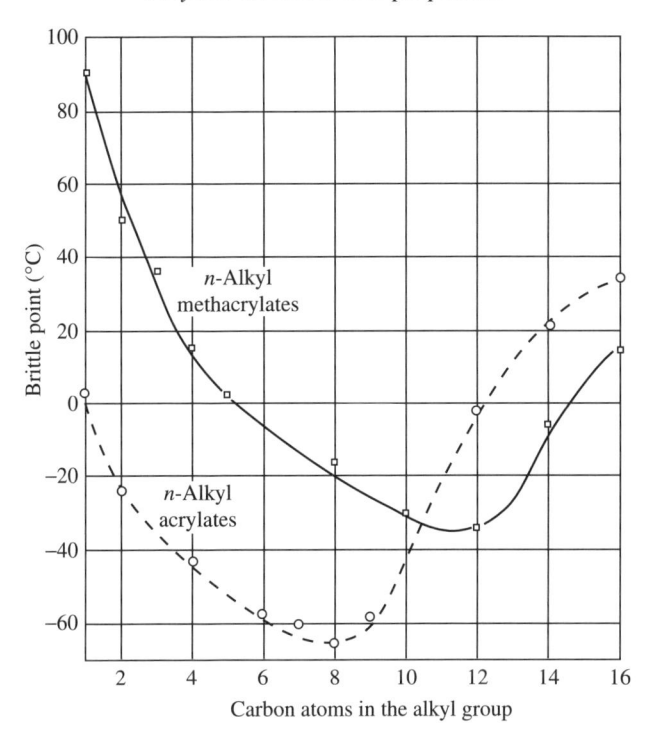

Figure 1.2 Brittle points of polymeric n-alkyl acrylates and methacrylates. (Reproduced with permission from Riddle [28].)

forces. In consequence the T_g diminishes in the case of alkyl ester polymers of acrylic acid until the alkyl chain reaches about 10 carbon atoms. It then increases again with side chain crystallinity. The methacrylate ester polymers, however, continue to drop in T_g, usually until a C_{13} alkyl group is reached, since the steric effect of the angular methyl group on the main chain also prevents side chain crystallinity at first.

Similar conditions prevail in the homologous series of vinyl esters of straight chain fatty acids based on the hypothetical vinyl alcohol $CH_2.CHOH$. From polyvinyl formate $(-CH_2CH-)_n$ with side group OOCH through polyvinyl acetate $(-CH_2CH-)_n$ with side group $OOCCH_3$ to vinyl laurate $(-CHCH-)_n$ with side group $OOC_{11}H_{23}$ there is a steady fall in T_g, the polymers varying from fairly brittle films derived from a latex at ambient temperature to viscous sticky oils as the length of the alcohol chain increases. Note, however, that the polymerisation and even copolymerisation of monomers with long side chains, above about C_{12}, becomes increasingly sluggish.

The above examples, in both the acrylic and the vinyl ester series, have considered the effect of straight chains inserted as side chains in polymer molecules. The effect of branched chains, however, is different. As chain branching increases, the effect of the overall size of the side chain diminishes. An example of this will be illustrated in Chapter 9 when considering specific examples of monomers that might be the basis of emulsions for paints. Thus polyisobutyl methacrylate has a higher T_g than polybutyl methacrylate. Polymers based on *tert*-butyl acrylate or *tert*-butyl methacrylate have a higher softening point than the corresponding *n*-butyl esters.

Another interesting example of the effect of branched chains is that of the various synthetic branched chain acids in which the carbon atom in the α position to the carbon of the carboxyl is quaternary, corresponding to a general formula $HOOC.C(R_1)(R_2)(R_3)$., where R_1 is CH_3, R_2 is CH_3 or C_2H_5 and R_3 is a longer chain alkyl group, which may be represented as $C_{4-6}H_{9-13}$. These form vinyl esters which correspond in total side chain length to vinyl caprate $CH_2:CHOOCC_9H_{19}$ but do not impart the same flexibility in copolymers [29].

It is often more practical to measure the effect of monomers of this type by copolymerising them with a harder monomer such as vinyl acetate and measuring the relative effects. Thus the vinyl esters of these branches chain acids, although they are based on C_{10} acids on average, are similar to a C_4 straight chain fatty acid as far as lowering of the T_g is concerned. It is also interesting to note that polymers and copolymers of these acids afford much greater resistance to hydrolysis than polymers of vinyl esters of *n*-alkyl acids. In copolymers these highly branched groups have a shielding effect on neighbouring ester groups, reducing their ease of hydrolysis by alkali [30, 31]. In this connection the angular methyl group in methacrylate ester polymers has the effect of making hydrolysis of these products extremely difficult.

4.2 Molecular weight effects

The molecular weight scatter formed as a result of any polymerisation is typical of a Gaussian type. Thus a fractionation of polystyrene is shown in Figure 1.3, in which the distribution and cumulative weight totals are shown as a percentage.

Before discussing the general effect of molecular weight on polymer characteristics, some further definitions are desirable. The number average molecular weight M_n is the simple arithmetical average of each molecule as a summation, divided by the number of molecules, the 'popular' idea of an average. Another measurement of average is the 'weight' average, and is an expression of the fact that the higher molecular weight fractions of a polymer play a greater role in determining the properties than do the fractions of lower molecular weight. Its definition is based on multiplying the number of identical molecules of molecular weight M_n by the overall weight of molecules of that weight and

Polymer structure and properties

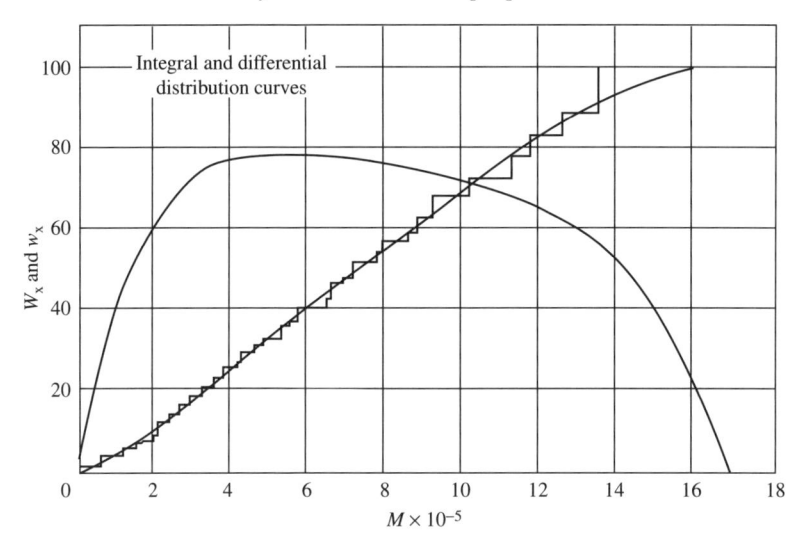

Figure 1.3 Molecular weight distribution for thermally polymerised polystyrene as established by fractionation. (From the results of Merz and Raetz [32].)

dividing by the sum total of the weights. Mathematically, this is given by

$$M_w = \frac{\sum w_1 M_1}{\sum w_1}$$

where w_1 represents the overall weight of molecules of molecular weight M_1. The weight average molecular weight M_w is invariably greater than the number average as its real effect is to square the weight figure. For certain purposes, the z average is used in which M_1 in the equation above is squared, giving even higher prominence to the higher molecular weight fractions.

In practice all the viscosity characteristics of a polymer solution depend on M_w rather than M_n. Thus nine unit fragments of a monomer of molecular weight 100 individually pulled off a polymer of molecular weight 1 000 000 reduces its M_n to 100 000. The M_w is just over 999 000. This corresponds to a negligible viscosity change.

A number of methods of measuring molecular weight are used and are summarised here:

(a) *Osmometry*. This is a vapour pressure method, useful for polymers of molecular weight up to about 25 000; membrane osmometry is used for molecular weights from 20 000 to 1 000 000. These are number average methods.
(b) *Viscometry*. This is a relative method, but the simplest, and its application is widespread in industry. Viscometry is approximately a weight average method.

(c) *Light scattering.* This is a weight average method.
(d) *Gel permeation chromatography.* This is a direct fractionation method using molecular weight. It is relatively rapid and has proved to be one of the most valuable modern methods.
(e) *Chemical methods.* These usually depend on measuring distinctive end groups. They are number average methods.

In some cases selective precipitation can be used to fractionate a polymer according to molecular weight. This is essentially a relative method based on known standards. This method also differentiates between varying species in a copolymer.

The properties of polymers are governed to some extent by molecular weight as well as molecular structure. Properties also depend partly on the distribution of molecular weights, and in copolymers on the distribution of molecular species. The differences in solubility in solvents in exploited in fractionation where blended solvents are used, only one being a good solvent for the polymer. The added poor solvent will tend to precipitate the higher molecular weight fractions first. Thus polyvinyl acetate may be fractionated by the gradual addition of hexane C_6H_{14} to dilute solutions of the polymer in benzene.

In some cases molecular weight variations have an extreme effect on polymer properties. This is particularly significant in the polyvinyl ethers $(-CH_2CH\underset{|}{}OC_nH_{2n+1})$ in which a polymer can vary from an oil at a molecular weight of about 5000 to a rubbery material if the molecular weight is above 100 000. The polyvinyl ethers, however, are not prepared as homopolymers by a free radical mechanism. The differences are usually illustrated by the change in the second-order transition point (see the next section). The softening points, which correspond approximately to melting ranges, and which are estimated by standard methods [33], are also affected by molecular weight.

The overall effect of solvents on polymers is too complex to be considered here. However, the reader is referred to the treatise by Flory [10] or the simpler treatment as shown in references [3] to [6], [11] and [12].

4.3 Transition points

Although when dealing with a crystalline substance there is a sharp melting point, sometimes denoted T_m, when dealing with a polymer containing molecules with a range of molecular weights it is not possible to describe the changes in state on heating in a similar manner. Amorphous materials, unless crosslinked or decomposing at a relatively low temperature, will soften gradually, and although a softening point or range may be quoted, this depends on an arbitrarily chosen test, usually on the time taken for a steel ball to penetrate a known thickness of the polymer.

Technology of polymerisation

However, an amorphous polymer has a number of physical changes of condition, the most important being the second-order transition point, usually referred to as T_g, already mentioned previously. Physically this transition point is connected with the mobility of the polymer chains. Below T_g, the chains may be regarded as substantially immobile, except for movements around an equilibrium position. Above this temperature appreciable movement of segments occurs in the polymer chains. Below the T_g, the polymer is a hard, brittle solid; above this temperature increased flexibility and possibly rubber-like characteristics are observed.

The second-order transition point may be measured in various ways; e.g. the rate of change of polymer density varies with temperature, as does the rate of change of other properties such as specific heat. Most useful is differential thermal analysis (DTA), which indicates the differential in the heating capacity of a substance. Modern DTA instruments are extremely sensitive.

The significance of T_g in emulsion polymers is indicated in Chapter 4 (Section 4.4). It may be noted that an alternative temperature, known as the minimum film formation temperature (MFT) is frequently a substitute for T_g This is the lowest temperature at which a drying emulsion containing polymer particles will form a continuous film. Because of the conditions of film formation, this temperature is usually 3–5 °C higher than the T_g. DTA results have shown that many polymers have transition points other than T_g. These are associated with the thermal motion of the molecules.

In many cases where a polymer has practical utility, it may be desirable that T_g should be reduced to achieve reasonable flexibility for the polymer. This is accomplished by plasticisation which reduces T_g to a level below ambient, or below the MFT in the case of a latex. As an example of plasticisation, about 40 parts of di-2-ethylhexyl phthalate are required to transform 60 parts of polyvinyl chloride from a hard, horny material to a flexible sheet. 'Internal plasticisation' is a term used for the formation of a copolymer, the auxiliary monomer of which gives increased flexibility to the polymer formed from the principal monomer.

5 TECHNOLOGY OF POLYMERISATION

Monomers may be polymerised by free radical initiation by one of five methods: polymerisation in bulk, in solution, dispersed as large particles in water or occasionally in another non-solvent (suspension polymerisation), or dispersed as fine particles, less than 1.5 μm, usually less than 1 μm in diameter. The last-named process is usually known as emulsion polymerisation. As the applications of polymers in emulsion is the basis of this series of volumes, emulsion polymerisation is the subject of Chapter 3. A variant of suspension polymerisation may be described as solution precipitation. It is often applied to copolymers, e.g. a copolymer of methyl methacrylate and methacrylic acid. In concentrated solution, the acid solubilises the methyl methacrylate. On polymerisation a fine water-insoluble powder is produced, which, depending on the monomer ratios, is usually alkali soluble.

In the past two decades a variation of emulsion polymerisation has been introduced, the polymers being known as 'dispersymers'. To form dyspersymers, a liquid monomer forms an emulsion-like product in an organic liquid, usually a liquid hydrocarbon, in which the polymer is insoluble. The final emulsion closely resembles an aqueous polymer emulsion in physical appearance [34].

5.1 Bulk polymerisation

Whilst in most cases laboratory experiments may be performed on undiluted monomers, or on controlled dilutions with solvents do not affect the polymerisation seriously, this process produces difficulties in large-scale production, which may be of the order of 5 tonnes in a single batch. Problems are caused by an increase in viscosity of the mass during polymerisation, and in particular the removal of the heat of polymerisation, which for most monomers is of the order of 20 kg cal gm^{-1} $mole^{-1}$. Special equipment with a high surface–volume ratio is desirable, as with the polymerisation of methyl methacrylate which is polymerised in thin sheets with a very low initiator ratio. Bulk polymerisation of vinyl acetate was described in reports of German factory production after the Second World War [35]. In this case the hot polymer is sufficiently fluid to be discharged directly from a cylindrical reactor. In an alternative continuous process the monomer is passed down a polymerisation tower [36]. Processes have been developed for the bulk polymerisation of vinyl chloride which is insoluble in its own monomer [37].

5.2 Solution polymerisation

Polymerisation with a solvent diluent can be readily accomplished as the major problems of bulk polymerisation are overcome with increasing dilution. Some practical problems persist, however. Commercial solvents are seldom pure and the impurities may have an inhibiting or retarding effect on polymerisation; this is especially so with monomers such as vinyl acetate which are not resonance stabilised. In addition, many solvents have a chain transfer effect (Section 2.1). Towards the end of a polymerisation the degree of dilution of the monomer is extremely high; the efficiency of initiator therefore falls, and it is lost by chain transfer with the solvent or by mutual destruction of the radicals. Thus several repeat initiations are necessary towards the end of a practical polymerisation in solvent to achieve the 99+ % polymerisation generally desired.

Practical experience has shown that molecular weights in solution polymerisation are also susceptible to a number of other factors, such as the type and nature of stirring, and the type and nature of the reactor, including its shape and the surface–volume ratio. There is likely to be a 'wall effect', which may terminate growing radicals. In addition, stirring conditions affect the rate

Technology of polymerisation

of attainment of equilibrium, whilst the amount of reflux, where present, also affects the nature of the final polymer. Since it is not usually desirable and may be difficult to distill unpolymerised monomer, even if a satisfactory azeotrope with the solvent exists, direct solvent polymerisation has limited practical application and is of principal interest where the solutions are used directly, either as solvent-based coatings or as adhesives [37]. Solvent polymerisation can normally be used only to prepare polymers of relatively low molecular weight.

5.3 Suspension polymerisation

Suspension polymerisation may be described as a water-cooled bulk polymerisation, although initiators that are water soluble may create some variations. The fundamental theory is simple and depends on the addition to the water of a dispersing agent. This may be a natural water-soluble colloid such as gum acacia, gum tragacanth, a semi-synthetic such as many cellulose derivatives (see Chapter 2) or a fully synthetic polymer. These include polyvinyl alcohol, or alternatively a water-soluble salt derived from a styrene/maleic anhydride 1 : 1 copolymer

$$[-\underset{\underset{C_6H_5}{|}}{CH}-CH_2-\underset{\underset{OC}{|}}{CH}-\underset{\underset{CO}{|}}{CH}-]$$
$$\underset{O}{\diagdown\diagup}$$

These dispersing agents may be mixed; occasionally a small quantity of surfactant of the order of 0.01 % is added. The normal concentration of dispersing agents is about 0.1 %, based on the water present. Monomer or monomers are added so that overall concentration is 25–40 %, although occasionally specific formulations claim 50 %.

The function of the dispersing agent is that of forming an 'envelope' around the beads as formed by stirring and preventing their coagulation and fusion during polymerisation. An intermediate or 'sticky' state occurs in almost all polymerisations in which a solution of polymer in monomer of high viscosity is formed, and the beads would fuse together very readily, except for the energy supplied by the stirring in keeping them apart and the stabilising action of the dispersing agent. Certain monomers, e.g. vinyl chloride, which are not solvents for their own monomers are an exception to this rule, but the same principle applies to the dispersant acting as a particle stabiliser. This type of polymerisation is sometimes referred to as 'bead' polymerisation.

There is a very clear distinction between suspension polymerisation and emulsion polymerisation. Whilst emulsion polymerisation produces particles usually $\leqslant 1$ μm in diameter, occasionally up to 2.5 μm, suspension particles are at least ten times larger in diameter, often of the order of 1 mm, although they are not necessarily spherical in shape. The kinetics of polymerisation of the two types are often quite different. To ensure that beads or 'pearls' (another term used) are formed, the second-order transition point T_g (see Chapter 4, Section 4.4.2) must be below the ambient temperature; otherwise the beads will flow together as soon as stirring is stopped. This tendency can be reduced somewhat, e.g. by coating a bead dispersion of low molecular weight polyvinyl acetate with cetyl alcohol, which is present during the polymerisation. Bead polymerisation is only practicable as a general rule, where the T_g is above about 25 °C and preferably above 35 °C. A summary of some suspension polymerisation processes for vinyl chloride is available [38].

It is possible to perform suspension polymerisation using solid dispersants. Thus styrene may be polymerised in suspension with organic peroxidic initiators with a tricalcium ortho-phosphate dispersant and sodium dodecylbenzene sulfonate [39].

6 THE PRINCIPAL MONOMERS AND THEIR POLYMER

6.1 Hydrocarbons

The simplest hydrocarbon capable of free radical addition polymerisation is ethylene C_2H_4, which as a gas is treated under pressure. Higher aliphatic hydrocarbons such as propylene $CH_3CH:CH_2$, 1-butene $CH_3CH_2CH:CH_2$ and a number of longer chain aliphatic ethenes cannot in general be polymerised by themselves by free radical, as distinct from ionic methods, because of their allylic character. However, they are capable of copolymerisation, and some specifications have claimed their copolymerisation with vinyl acetate in emulsion. Only hydrocarbons with their unsaturation in the 1-position can be copolymerised satisfactorily in this manner.

Styrene $C_6H_5CH:CH_2$ is the simplest aromatic hydrocarbon monomer. Others are vinyl-toluene and o-, m- and p-methylstyrene $CH_3C_6H_4CH:CH_2$. α-Methylstyrene $C_6H_4C(CH_3):CH_2$ is also a technical product, but its polymerisation has problems because it has a low ceiling temperature; i.e. the propagation and depropagation rates during formation tend to become equal and hence no polymer is formed unless a low-temperature initiator system is used.

The divinyl-benzenes, written

$$CH_2:CH-\underset{}{\bigcirc}-CH:CH_2$$

a notation used when it is desired to leave the positions of the substituents undecided, are a by-product of styrene manufacture and are used for crosslinking.

The dienes are described in Section 6.8. There have been a few other specialised monomers based on condensed rings, but as they are generally solids their use in emulsion systems in very limited, if at all.

6.2 Vinyl esters

Vinyl esters are derived from the hypothetical vinyl alcohol, $CH_2{:}CHOH$, an isomer of acetaldehyde CH_3CHO, which is normally formed when an attempt is made to prepare the monomer. The esters, however, whether derived from acetylene or ethylene (see later), are of major importance in many latex applications. The principal ester of commerce is vinyl acetate $CH_2{:}CHOOCCH_3$, a liquid which is fairly readily hydrolysed, of b.p. 73 °C. Vinyl acetate has the advantage of being one of the cheapest monomers to manufacture.

Vinyl propionate $CH_3CH_2COOCH{:}CH_2$ is fairly well established as a monomer, probably by direct acetylene preparation. Other esters are encountered less frequently, and in most cases are probably prepared by vinylolysis rather than directly, using vinyl acetate as an intermediate [40–42]. Vinylolysis is not the same as *trans*-esterification and involves a mercury salt such as *p*-toluene sulfonate as an intermediate. Thus vinyl caprate is prepared by reacting vinyl acetate with capric acid in the presence of a mercuric salt, using an excess of vinyl acetate; the reaction is reversible:

$$CH_3.CO.O.CH{:}CH_2 + C_9H_{19}COOH \rightleftharpoons C_9H_{19}CO.O.CH{:}CH_2 + CH_3COOH$$

The capric acid can be conveniently removed with sodium carbonate after removal of excess acetic acid with sodium bicarbonate, which does not react with the higher fatty acids. The vinyl esters of mixed C_8, C_{10} and C_{12} fatty acids have been used technically in forming copolymers with vinyl acetate [43].

Vinyl butyrate $CH_3CH_2CH_2OOCCH{:}CH_2$ is referred to in the literature, but is probably not in commercial production. Vinyl laurate $CH_3(CH_2)_8COOCH{:}CH_2$ has been in technical production in Germany. Of other esters of fatty acids, only vinyl stearate $C_{17}H_{35}COOCH{:}CH_2$, a solid, has been manufactured on a technical scale. The most interesting vinyl esters have been derivatives of pivalic acid $(CH_3)_3C.COOH$, the simplest branched chain fatty acid in which the carbon atom adjacent to the carboxyl group is quaternary. In these vinyl esters, one methyl group may be replaced by ethyl, and a second by a longer alkyl chain, and thus the general formula for the esters is $CH_3C_{7-8}H_{14-16}C(CH_3)(C_{1-2}H_{2-4})OOCCH{:}CH_2$ [44].

Vinyl chloroacetate $CH_3ClCOOCH{:}CH_2$ is occasionally quoted. Because of the relatively labile atom, copolymers including this monomer take part in a number of crosslinking reactions.

Vinyl benzoate $C_6H_5COOCH{:}CH_2$ is the only aromatic vinyl ester that finds some application if relatively hard polymers with some alkali resistance are required [43].

6.3 Chlorinated monomers

Vinyl chloride $CH_2{:}CHCl$, a gas, is the cheapest monomer of the series, and has widespread commercial use. It may be polymerised in bulk with specialised apparatus, but it is also polymerised both in suspension and in emulsion. It is frequently copolymerised, especially with vinyl acetate.

Vinylidene chloride $CH_2:CCl_2$, a low boiling liquid, is also a relatively low cost monomer. It forms a polymer with a marked tendency to crystallise because of its relatively symmetrical structure. In most cases it is copolymerised, especially with vinyl chloride or methyl acrylate.

The corresponding symmetrical compound in the *trans* form, *trans*-dichloroethylene CHCl:CHCl, has been used in limited quantities as a comonomer with vinyl acetate [45]. Trichlorethylene $CHCl:CCl_2$, although not normally considered as a monomer, may take part in some copolymerisations, especially with vinyl acetate.

Chloroprene $CH_2:CCl.CH:CH_2$ a diene (see Section 5.8), is mainly used in the formulation of elastomers, but occasionally a polymer containing chloroprene is used as an alternative to a hydrocarbon diene.

Vinyl bromide $CH_2:CHBr$ is technically available, and finds some application in specialist polymers with fire-resistant properties. The boiling point is 15.8 °C. A number of highly chlorinated or brominated alcohol esters of acrylic and methacrylic acids have been described, and are probably in limited production, often for captive use for the production of fire-resistant polymers.

Vinyl fluoride CH_2CHF, vinylidene fluoride $CH_2:CF_2$, tetrafluoroethylene $F_2C:CF_2$ and hexafluoropropylene $CF_3CF:CF_2$ have found industrial applications in recent years. The monomers are gaseous. For perfluoroalkyl acrylates see the next section.

6.4 Acrylics

6.4.1 Acrylic and methacrylic acids

The most numerous class of monomers are the acrylics, viz. esters of acrylic acid $CH_2:CHCOOH$ and methacrylic acid $CH_2:C(CH_3)COOH$. Both are crystalline solids at low ambient temperatures, becoming liquid at slightly higher temperatures (see Figure 1.2). These acids polymerise and copolymerise extremely readily, being frequently employed in copolymers to obtain alkali-soluble polymers. Whilst both acids are water soluble, methacrylic acid, as might be expected because of its angular methyl group, is more soluble in ester monomers, and to some extent in styrene, and as such is more useful in copolymerisation, especially if water based.

Whilst esters of acrylic acid give soft and flexible polymers, except for those with long alkyl chains, methyl methacrylate polymerises to an extremely hard polymers. The polymers in this series become softer with increasing alkyl chain lengths up to C_{12}. The highest alkyl chain acrylics in both series tend to give side chain crystallisation.

6.4.2 Individual acrylic and methacrylic esters

A range of esters of acrylic acid are available commercially from methyl acrylate through ethyl acrylate to n-heptyl acrylate and 2-ethylhexyl acrylate $CH_2{:}COOCH_2CH(C_2H_5)C_4H_9$. They vary from the fairly volatile, pungent liquids of the lowest member of the series to characteristic, but not necessarily unpleasant, odour of the higher members of the series. The highest members require distillation under reduced pressure to avoid simultaneous decomposition and polymerisation.

The methacrylic ester series closely parallels the acrylics, but boiling points tend to be somewhat higher, especially with the short chain esters (Table 1.1). Methyl methacrylate $CH_2{:}C(CH_3)COOCH_3$ is by far the most freely available and least costly of the monomers of the series.

As an alternative to the simple alkyl esters, several alkoxyethyl acrylates are available commercially, e.g. ethoxyethyl methacrylate $CH_3{:}C(CH_3)COOC_2H_4OC_2H_5$ and the corresponding acrylate. The ether oxygen which interrupts the chain tends to promote rather more flexibility than a simple carbon atom.

Some technical perfluorinated alkyl acrylates have been described. They include N-ethylperfluorooctanesulfonamido)ethyl acrylate $C_nF_{2n+1}SO_2N$ (C_2H_5)–CH_2O–$C(O)$–$CH{=}CH_2$ (n approximately 7.5, fluorine content 51.7 %), the corresponding methacrylate and the corresponding butyl derivatives. The ethyl derivatives are waxy solids, the ethyl acrylate and the corresponding methacrylate derivative having a melting range of 27–42 °C. The butyl acrylic derivative is a liquid, freezing at -10 °C.

Various glycol diacrylates and dimethacrylates are available. Ethylene glycol dimethacrylate $CH_2{:}C(CH_3)COOC_2H_4OOC(CH_3){:}CH_2$ is extremely reactive, and is sometimes marketed as a solution in methyl methacrylate. It polymerises extremely readily and acts as a powerful crosslinking agent. The dimethacrylates of triethylene glycol and higher glycols, some of which are also readily available, are less reactive, retain better flexibility and are more controllable in their polymerisation characteristics.

Glycidyl methacrylate $CH_2CHCH_2OOCC(CH_3){:}CH_2$ with epoxide $\underset{O}{\diagdown\diagup}$ has two reactive groups, the epoxide group being distinct in nature from the vinyl double bond (see also Chapters 5 and 7). The epoxide group is only slowly reactive in water, and even in emulsion polymerisation does not hydrolyse excessively. However, the presence of the group makes the methacrylate moiety much more prone to ready polymerisation.

The half esters of both ethylene glycol and propylene glycol are now monomers of commerce, propylene glycol monoacrylate $CH_3CHOHCH_2OOCCH{:}CH_3$ being typical (the primary alcohol unit is the

34   *Fundamentals of polymer chemistry*

active one in the formula). These monomers provide a source of the mildly reactive and hydrophilic groups on polymer chains. A problem with these monomers is that traces of a glycol dimethacrylate may be present as impurities at a low level.

6.4.3 Acrylics based on the amide group

Acrylamide $CH_2{:}CHCONH_2$ and methacrylamide $CH_2.C(CH_3)CONH_2$ are articles of commerce, especially the former. Polymers of the former are water soluble, but the solubility of the latter depends on conditions of preparation, e.g. molecular weight of the polymers. Both are very frequently used in copolymerisation. Polyacrylamide is often used as a flocculating agent. Certain derivatives, viz. methylolacrylamide $CH_2{:}CHCONHCH_2OH$, methoxymethylacrylamide $CH_2{:}CHCONHCH_2OCH_3$ and isobutoxyacrylamide $CH_2{:}CHCONHCH_2OC_4H_9$-iso, are of interest in crosslinking. The last named has the advantage of being monomer soluble but water insoluble, making it more amenable to handling in emulsion polymerisation. Diacetoneacrylamide N-(1,1,-dimethyl-3-oxobutyl) acrylamide [also known as 1-dimethyl-3-oxobutyl)acrylamide] $(CH_2{:}CHNHC(O)(CH_3)_2CH_2COCH_3$ has the advantage of both water and monomer solubility [46].

6.4.4 Cationic acrylic monomers

If a compound such as dimethylaminoethyl alcohol $(CH_3)_2.NC_2H_4OH$ is esterified via the hydroxyl groups with acrylic or methacrylic acids instead of neutralising the amino group, a cationic monomer, e.g. $(CH_3)_2NC_2H_4OOCCH{:}CH_2$., is formed. At acid pH levels this monomer is cationic, with the amino group forming salts that polymerise and copolymerise in the normal way via the acrylic double bond. Another typical monomer is t-butylaminoethyl methacrylate t-$C_4H_9NHC_2H_4OOCC(CH_3){:}CH_2$. At neutral or higher pH levels, ionisation of this weak base is suppressed, and it acts as a nonionic monomer. However, hydrolysis tends to be rapid in aqueous media at high pH, forming the alkanolamine salts of the acids. Cationic monomers from the corresponding aminopropyl compound are also known.

6.4.5 Acrylonitrile

Acrylonitrile $CH_2{:}CHCN$, and the less frequently used methacrylonitrile $CH_2{:}C(CH_3)CN$, give extremely hard polymers and are employed as comonomers to give solvent resistance. Acrylonitrile monomer, like vinyl chloride, is not a solvent for its own polymer and is about 7 % soluble in water, although its polymer is insoluble, making it of interest in theoretical studies. These monomers are unusually toxic because of the nitrile group.

6.5 Polymerisable acids and anhydrides

Besides acrylic and methacrylic acid, crotonic acid (strictly the *cis* acid) $CH_3CH{:}CHCOOH$, a white powder, often takes part in copolymerisations, especially with vinyl acetate, but it only self-polymerises at low pH and with great difficulty. Itaconic acid (methylenesuccinic acid) $CH_2{:}C(CH_3COOH)COOH$, a water-soluble solid, also readily takes part in copolymerisation, although it will only homopolymerise at about pH2.

Maleic acid *cis*-$HOOCCH{:}CHCOOH$, the simplest dibasic acid, is rarely copolymerised on its own, but frequently as the anhydride, maleic anhydride:

$$\begin{matrix} CHCO \\ CHCO \end{matrix}\!\!>\!\!O$$

which is much more reactive. However, it cannot be directly polymerised in water, although its rate of hydrolysis is slow. It readily forms copolymers, e.g. with styrene, ethylene or vinyl acetate, most readily as alternating (equimolar) copolymers, irrespective of the initial molar ratios. These copolymers are water soluble in their alkaline form after hydrolysis and frequently occur as stabilisers in emulsion polymerisation.

Fumaric acid *trans*-$HOOCCH{:}CHCOOH$, the isomer of maleic acid and thermodynamically the most stable form, is occasionally used as a comonomer, although there is some doubt as to its reactivity, and it may do little more than provide end groups, thus acting as a chain transfer agent.

Aconitic acid, an unsaturated carboxylic acid of formula $HOOCCH_3C(COOH){:}CHCOOH$, obtained by removing the elements of water from citric acid, is occasionally quoted as a monomer in patents and theoretical studies. Citraconic acid (methylmaleic acid) $CH_3C(COOH){:}CHCOOH$, its isomer mesaconic acid(methylfumaric acid) and citraconic anhydride are also occasionally used for copolymerisation. The acids in this paragraph are not articles of commerce as far as has been ascertained.

Various alkyl and alkoxy diesters of itaconic acid have been introduced, but as far as is known, production has not been sustained, although they are extremely good internal plasticisers for polyvinyl acetate. Their relatively high cost migitated against their use.

6.6 Self-emulsifying monomers

A number of monomers have the property of stabilising emulsions without the assistance of emulsifiers (see Chapter 2) or with a minimal quantity. Their polymers are generally water soluble and often so are their copolymers, depending on monomer ratios. These monomers contain strongly hydrophilic groups, the sulfonate .SO_3Na being the most usual. They usually copolymerise readily with most of the standard monomers used in emulsion polymerisation.

Amongst the earliest was sodium vinyl sulfonate $CH_2{:}CHSO_3Na$, which was in use in Germany in the 1940s. Other monomers of this class include sodium sulfoethyl methacrylate $CH_2{:}CHC({:}O)NH(CH_3)_2CH_2SO_3Na$.

Of unusual interest is 2-acrylamido-2-methylpropanesulfonic acid (AMPS monomer®) $CH_2{:}CHC({:}O)NHC(CH_3)_2CH_2SO_3H$, normally used as the sodium salt. This monomer also copolymerises readily. A monograph describes these compounds in greater detail [47].

The salts of the polymerisable acids have appreciable self-emulsifying powers when used as comonomers, especially when they are about 10 % or more by weight. The alkylolamine unsaturated esters (Section 6.4.3), when used in the form of their alkali or amine salts, come into this category [48].

6.7 Esters for copolymerisation

Esters of maleic and fumaric acids are often used in copolymerisation, both the diesters and more unusually the monoesters being reacted. The fumarate diesters, which are rather non-volatile liquids, have a feeble tendency to form homopolymers on prolonged heating with initiators, but little, if any, evidence exists to suggest that maleic esters can homopolymerise. Copolymerisation characteristics of fumarate esters are more favourable than those of maleate esters, and they are mainly copolymerised with vinyl acetate to impart internal plasticisation. It has been suggested that maleate and fumarate esters isomerise to identical products during a polymerisation reaction, but this has not been proved. Although in theory the units entering a polymer should become identical with the disappearance of the double bond, there are many steric factors associated with the polymer molecules as a whole.

The principal esters are those of n-butyl alcohol, 2-ethylhexyl alcohol, a technical mixture of C9–11 alcohols and 'nonyl alcohol', which is 1,3,3-trimethylhexanol.

The half esters of maleic acid and their salts are occasionally quoted in patents and other technical literature, and seem, probably because of their polar–non-polar balance to polymerise fairly readily. The methyl half ester cis-$CH3OOCCH{:}CHCOOH$ is a solid; some of the higher alkali half esters are liquids. Half esters of other polymerisable acids such as fumaric and itaconic acids have been reported, but are more difficult to prepare. The half esters tend to disproportionate fairly readily, especially in the presence of water, to the free acid and the diester:

$$2CH_3OOCCH{:}CHCOOH = HOOCCH{:}COOH + CH_3OOCCH{:}CHCOOCH_3$$

A number of successful copolymerisations in emulsion of half esters of long-chain alcohol and sterically hindered alcohols have been disclosed [49].

6.8 Monomers with several double bonds

Unsaturated hydrocarbons containing two double bonds constitute a special class of monomer. The principal representatives of this class are butadiene $CH_2:CH.CH:CH_2$, isoprene $CH_2:C(CH_3).CH:CH_2$ and chloroprene $CH_2CCl.CH:CH_2$.

When a monomer contains more than one double bond which can polymerise approximately equally freely, crosslinking can occur readily, and small quantities of this type of monomer are added to other polymerising systems to obtain controlled crosslinking. Examples are *p*-divinyl benzene (see Section 6.1), and ethylene glycol dimethacrylate $CH_2:C(CH_3)COOCH_2CH_2OOCC(CH_3):CH_2$.

In these cases the radicals formed are resonance stabilised, so that two chains can form simultaneously, and when a biradical is added to a growing chain, two points occur from which the chain can continue, resulting in rapid branching and crosslinking.

The dienes are a special class in distinction to monomers such as the divinyl benzenes and the diesters such as a glycol acrylate. If a monomer such as butadiene is polymerised, the monoradical formed is highly stabilised by resonance. The two resonance forms can be represented as

$$RH-CH_2=CH-CH_2 \quad \text{and} \quad R-CH_2-CH-CH=CH_2$$

where R represents a residual monovalent group.

As a result, two methods of addition are possible, one being known as 1 : 2 addition, the other as 1 : 4 addition, and may be represented by the following:

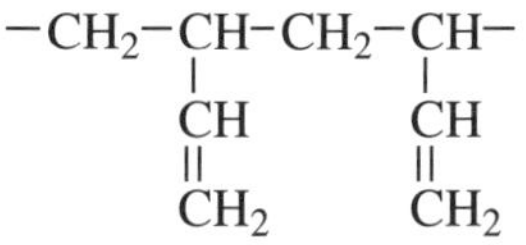

$$\begin{array}{c} -CH_2-CH-CH_2-CH- \\ | | \\ CH CH \\ || || \\ CH_2 CH_2 \end{array}$$

Bi-unit of a 1 : 2 addition

$$-CH_2-CH=CH-CH_2-CH_2-CH=CH-CH_2-$$

Bi-unit of 1 : 4 addition

During a free radical polymerisation in emulsion, about 20 % of a 1 : 2 polymer addition and 80 % of 1.4 addition takes place. Copolymerisation with other monomers such as styrene tends to increase 1 : 2 units at the expense of 1 : 4 units.

A further possibility of variation occurs because the 1 : 2 unit possesses an asymmetric carbon atom, while due to the double bond, 1 : 4 addition may occur in the *cis* or *trans* positions, giving the following isomers:

$$\begin{array}{cc} \underset{H}{\overset{-H_2C}{>}}C=C\underset{H}{\overset{CH_2-}{<}} & \underset{H}{\overset{-H_2C}{>}}C=C\underset{CH_2-}{\overset{H}{<}} \\ cis & trans \end{array}$$

It has been found possible to deduce various structures by infrared absorption bonds, *trans* formation having been shown to decrease with temperature.

During a polymerisation including butadiene, there is a greater than usual tendency for side reactions to occur. These involve the residual double bonds in completed molecules or growing chains. This often causes gel formation, as measured by the insoluble fractions in acetone, or another standard solvent. Gel formation and other crosslinking reactions occur with increasing frequency as the degree of polymerisation increases. In consequence, when solid products of controlled properties are required, polymerisations and copolymerisations involving butadiene are not taken to completion. The reaction is inhibited before polymerisation is complete and surplus monomer is removed by distillation. Possibilities for isomerism in the polymerisation of chloroprene and isoprene are even more complex than with butadiene.

The application of diene polymers and copolymers is largely associated with synthetic rubber, but these copolymers have other applications; e.g. copolymers with a styrene content of 40 % and above have been used for coatings and for carpet backing. In these copolymers the residual double bonds render them prone to degradative oxidation.

The structure of butadiene copolymers is interesting and accounts for their physical properties. A polymer molecule may be considered to be a randomly coiled chain—an irregular spiral—in the unstretched state. Elastomers in the fully stretched state, particularly natural rubber, i.e. polyisoprene $(C_5H_8)_n$, tend to crystallise, this crystallisation being lost when the stress causing the extension is removed. Ideally a limited number of crosslinks is desirable for elastic recovery to occur. Because of their less regular structure, copolymers of butadiene do not tend to crystallise.

Modern work has shown that where polymerisation takes place by methods that produce a highly stereoregular or stereospecific products, the tendency is for crystallisation to occur on stretching. In most copolymers that we will consider in these volumes, the high quantity of comonomer causes the normal plastic type of property to predominate, rather than the rubber-like extensibility. Thus the bulky phenyl C_6H_5 groups in the styrene copolymers effectively prevent crystallisation, and the copolymers in film form tend to approximate more closely in properties to other vinyl-type polymers.

The double bonds in polymers involving dienes facilitate crosslinking, which in rubber technology is known as vulcanisation. The utilization of the double bonds for crosslinking has increased in recent years.

6.9 Allyl derivatives

Allyl alcohol $CH_3{:}CHCH_2OH$ and its simple derivatives, such as allyl acetate $CH_2{:}CH_2OOCCCH_3$, have little practical application in vinyl polymerisation, because of their powerful tendency to degradation chain transfer (p. xx). Similar considerations apply to methallyl alcohol $CH_2{:}C(CH_3)CH_2OH$ and its derivatives. A practical difficulty also arises with allyl alcohol and its more volatile derivatives because of their extreme lachrimatory character.

Certain other allyl derivatives, however, are of greater utility. Diallyl o-phthalate $o\text{-}CH_2{:}CHCH_3OOCC_6H_4OOCCH_2CH{:}CH_2$ contains two vinyl groups, and as such the tendency to crosslink is in competition with that of chain transfer. Whilst this diester is not normally used in emulsion polymerisation, it is frequently included in the thermosetting polyesters, especially in conjunction with a monomer such as styrene, which will reduce the tendency to premature crosslinking. These derivatives find particular application in reinforced polyesters, viz. those reinforced with glass fibres.

Allyl derivatives containing epoxide groups seem to copolymerise somewhat more readily, probably because the nucleophilic epoxide group reduces the tendency to resonance. These derivatives are of interest as they are potentially crosslinking monomers. They include allyl glycidyl ether

$$CH_2{:}CHCH_2OCH_2CH\underset{O}{\overset{}{-}}CH_2$$

and allyl dimethyl glycidate

$$\underset{CH_3}{\overset{CH_3}{\diagdown}}C\underset{O}{\overset{}{-}}CHCOOCH_2CH{:}CH_2$$

which is formed by Darzen's reaction.

This little-known reaction would repay further study, at any rate as far as polymer production is concerned. It is fundamentally the reaction of a chlorinated ester, such as allyl chloroacetate with acetone in the presence of a stoichiometric quantity of alkali near 0 °C, sodium hydride being particularly effective [49].

Monomers such as allyl methacrylate $CH_2{:}CHCH_2OOCC(CH_3){:}CH_2$ are occasionally quoted, having mild crosslinking properties. A useful volume describing allyl compounds is available [14].

6.10 Vinyl ethers

Whilst the vinyl ethers have long been known as monomers, they have been unimportant in aqueous polymerisation. By themselves they only form copolymers, not homopolymers under free radical conditions, and ionic catalysts are used when a homopolymer is required. Although the vinyl ethers copolymerise readily with many vinyl monomers under free radical conditions, difficulty arises during polymerisation in the presence of water since they hydrolyse readily to acetaldehyde and alcohols below a pH of about 5.5. This makes emulsion polymerisation with a monomer such as vinyl acetate difficult, except under careful control of pH.

Except for the tendency to hydrolysis, the physical properties of the vinyl ether monomers closely resemble that of the corresponding saturated compounds. Available monomers, including vinyl methyl ether CH_2CHOCH_3, vinyl ethyl ether $CH_2:CHOC_2H_5$, both n- and isobutyl vinyl ethers $CH_2:CHOC_4H_9$ and a long-chain alkyl ether, vinyl cetyl ether $CH_3:CHOC_{16}H_{33}$, have been available.

6.11 Miscellaneous monomers containing nitrogen

N-Vinylpyrrolidone is a completely water-miscible cyclic monomer which can be regarded as a cyclic imide. It readily forms polymers and copolymers, the water soluble types being used as protective colloids. The monomer is

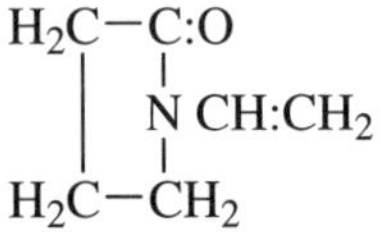

$$\begin{array}{c} H_2C-C{:}O \\ | \quad\quad | \\ | \quad\; N\;CH{:}CH_2 \\ | \quad\quad | \\ H_2C-CH_2 \end{array}$$

2-Vinylpyridine, 4-vinylpyridine and to a lesser extent 2-methyl-5-vinylpyridine have been prepared commercially, and polymerise to give products that are the basis of polymeric cationic electrolytes. They are most frequently in copolymers with butadiene and styrene in tyre cord adhesives (see Chapter 8). The physical properties of the polymers tend to resemble those of styrene. The formulae of the monomers are shown below:

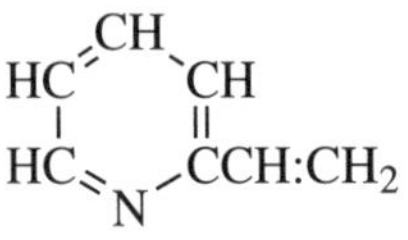

$$\begin{array}{ccc}
\;\;\;\;CH & \;\;\;\;CCH{:}CH_2 & \;\;\;\;CH \\
HC^{\diagup}\;\;^{\diagdown}CH & HC^{\diagup}\;\;^{\diagdown}CH & HC^{\diagup}\;\;^{\diagdown}CH \\
|\;\;\;\;\;\;\;\;|| & |\;\;\;\;\;\;\;\;|| & |\;\;\;\;\;\;\;\;|| \\
HC_{\diagdown N\diagup}CCH{:}CH_2 & HC_{\diagdown N\diagup}CH & H_2C{:}CHC_{\diagdown N\diagup}CCH_3 \\
\text{2-Vinylpyridine} & \text{4-Vinylpyridine} & \text{2-Methyl-5-vinylpyridine}
\end{array}$$

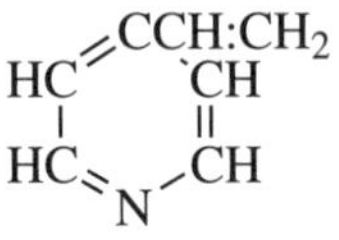

1-Vinyl imidazole and the allied 1-vinyl 2-methylimidazole are basic monomers produced on a small scale, and their major function is improvement of adhesion:

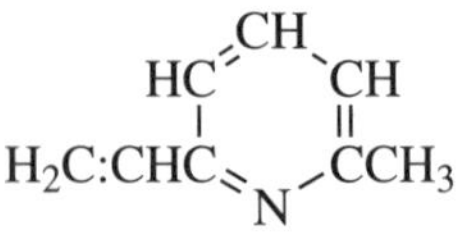

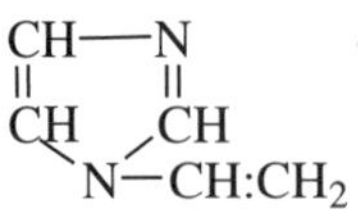

$$\begin{array}{cc}
CH\!-\!N & CH\!-\!N \\
||\;\;\;\;\;\; || & ||\;\;\;\;\;\; || \\
CH\;\;\;\;CH & CH\;\;\;\;C\!-\!CH_3 \\
\;\;^{\diagdown}N\!-\!CH{:}CH_2 & \;\;^{\diagdown}N\!-\!CH{:}CH_2 \\
\text{1-Vinylimidazole} & \text{1-Vinyl-2-methylimidazole}
\end{array}$$

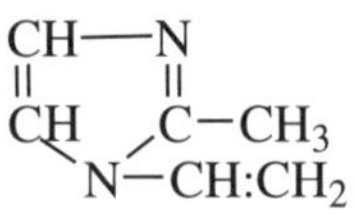

The principal monomers and their polymer 41

Vinyl caprolactam is occasionally used as a reactive thinner:

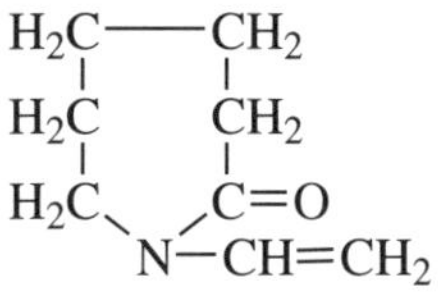

$$\begin{array}{c} H_2C-CH_2 \\ | \quad\quad | \\ H_2C \quad CH_2 \\ | \quad\quad | \\ H_2C \quad C=O \\ \diagdown N-CH=CH_2 \end{array}$$

It has a melting point of 34 °C, and may be distilled under reduced pressure. It also has the property of improving adhesion.

Divinylethylene urea and divinylpropylene urea, with melting points of 66 and 65 °C respectively, find utilization as reactive thinners:

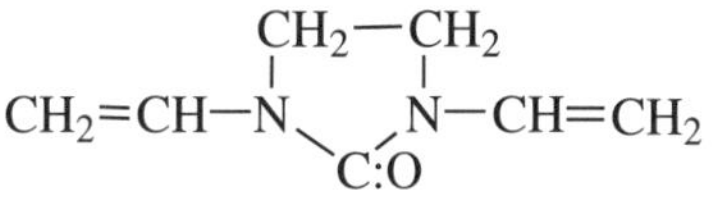

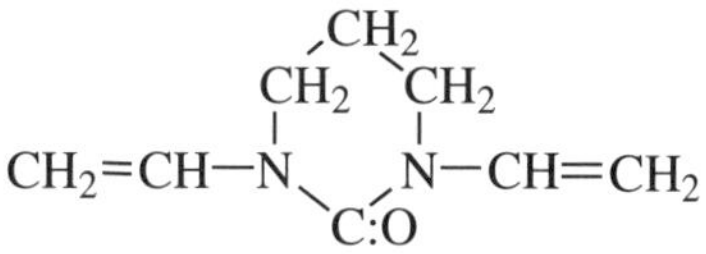

$$\begin{array}{cc} \begin{array}{c} CH_2-CH_2 \\ | \quad\quad | \\ CH_2=CH-N \quad N-CH=CH_2 \\ \diagdown C{:}O \diagup \end{array} & \begin{array}{c} \quad\quad CH_2 \\ CH_2 \quad CH_2 \\ | \quad\quad | \\ CH_2=CH-N \quad N-CH=CH_2 \\ \diagdown C{:}O \diagup \end{array} \\ \text{Divinylethylene urea} & \text{Divinylpropylene urea} \end{array}$$

6.12 Toxicity and handling

As a general rule, all the quoted monomers should be handled with at least the precautions associated with the corresponding saturated compounds. In some cases, e.g. the acrylic esters, the toxicity, in particular the vapour, is more toxic than the corresponding saturated esters. The lower acrylic esters, but not the methacrylic esters, have an extremely unpleasant odour, but the level of intolerance is well below the maximum safety level recommended. Some precautions are advised in handling acrylamide.

Acrylonitrile and methacrylonitrile have the characteristic toxicity of cyanides. On the laboratory scale they should be handled in well-ventilated fume cupboards and prevented from coming into direct contact with the skin. Special precautions, including the wearing of oilskins and fresh air breathing apparatus, are required for large-scale manufacturing processes.

Allyl alcohol and some of its derivatives are lachrimatory.

The above comments are of a general character only. In all cases manufacturers' literature and official literature should be consulted, safety information being obligatory in many countries.

The following are synthetic monomers based on the vinyl esters of mixed branched chain acids, known as Versatic® acids, the feature being that the carbon atom is in the alpha position of quaternary:

Veova 9 is the vinyl ester of acids averaging 9 carbon atoms; b.p. 185–200 °C, s.g. 0.89. The T_g of the homopolymer is 60 °C. Veova 10 is the vinyl ester of acids averaging 10 carbon atoms and is less branched than Veova 9; b.p. 133–136 °C, s.g. 0.875–0.885. Some perfluorinated acrylic derivatives are described in Section 6.4.

Table 1.1 Physical properties of the principal monomers

Monomer	Formula	b.p. (°C)	m.p. (°C)	s.g. Pressure (mm) ($d_{20/20}$)
Ethylene	$CH_2{:}CH_2$	−104		0.5139 ($d_{20/4}$)
Propylene	$CH_2{:}CHCH_3$	−31		0.5951 ($d_{20 4}$)
1-Butene	$CH_2{:}CHCH_2CH_3$	−6.3		0.6734 ($d_{20/4}$)
1-Hexene	$CH_2{:}CHC_3H_6CH_3$	+63.5		0.7194($d_{15.5/15.5}$)
1-Octene	$CH_2{:}CHC_5H_{10}CH_3$	121.3		0.905 ($d_{25/25}$)
Styrene	$C_6H_5CH{:}CH_2$	145.2		0.8930
Vinyl toluene	$CH_3C_6H_4CH{:}CH_2$	167.7		0.9062 ($d_{25/25}$)
o-Methyl styrene	$C_6H_4C(CH_3){:}CH_2$	163.4		
Divinyl benzene (55 % technical product)	$C_6H_4(CH{:}CH_2)_2$	195		
m-Diisopropenylbenzene	$m\text{-}C_6H_4[C(CH_3){:}CH_2]_2$	231		0.925
p-diisopropenylbenzene (sublimes at 64.5 °C)	$p\text{-}C_6H_4[C(CH_3){:}CH_2]_2$			0.965
Butadiene 1:3	$CH_2{:}CHCH{:}CH_2$	−4.7		0.6205[a]
Isoprene	$CH_2{:}C(CH_3)CH{:}CH_2$	34.07		0.686($d_{15.6/15.6}$)
Vinyl chloride	$CH_2{:}CHCl$	−14		0.912
Vinylidene chloride	$CH_2{:}CCl_2$	31.7		1.1219($d_{20/4}$)
trans-Dichloroethylene	$CHCl{:}CHCl$	49		1.265($d_{15/4}$)
Chloroprene	$CH_2{:}CClCH{:}CH_2$	59.4		0.9583
Vinyl fluoride	$CH_2{:}CHF$	−57		
Vinylidene fluoride	$CH_2{:}CF_2$	−84		
Tetrafluoroethylene	$CF_2{:}CF_2$	−76		
Vinyl formate	$CH_2{:}CHOOCH_3$	46.6		0.9651
Vinyl acetate	$CH_2{:}CHOOCCH_3$	72.7		0.9338

[a] under own vapour at 21°C

Vinyl propionate	CH$_2$:CHOOCC$_2$H$_5$	94.9		0.9173
Vinyl butyrate	CH$_2$:CHOOCC$_3$H$_7$	116.7		0.9022
Vinyl caprate	CH$_2$:CHOOCC$_9$H$_{19}$	148/50		
Vinyl laurate	CH$_2$:CHOOCC$_{11}$H$_{23}$	142/10		
Vinyl stearate	CH$_2$:CHOOCC$_{17}$H$_{35}$	187–188/4.3	35-36	
Vinyl chloroacetate	CH$_2$:CHOOCCH$_2$Cl		44-46/20	1.1888
Vinyl benzoate	CH$_2$:CHOOCC$_6$H$_5$	203		1.0703
Acrylic acid	CH$_2$:CHCOOH	141.3	12.3	1.0472
Methacrylic acid	CH$_2$:C(CH$_3$)COOH	161	15	1.015
Methyl acrylate	CH$_2$:CHCOOCH$_3$	80		0.950
Ethyl acrylate	CH$_2$:CHCOOC$_2$H$_5$	99.6		0.9230
n-Butyl acrylate	CH$_2$:CHCOOC$_4$H$_9$	148.8		0.9015
n-Heptyl acrylate	CH$_2$CHCOOC$_7$H$_{15}$	106/25		0.8794($d_{25/4}$)
2-Ethylhexyl acrylate	CH$_2$:CHCOOC$_8$H$_{17}$	128/50		0.8869
Lauryl acrylate	CH$_2$:CHCOOC$_{11}$H$_{23}$			
Methyl methacrylate	CH$_2$:C(CH$_3$)COOCH$_3$	100.5		0.939
Ethyl methacrylate	CH$_2$:C(CH$_3$)COOC$_2$H$_5$	118.4		0.909
n-Butyl methacrylate	CH$_2$:C(CH$_3$)COOC$_4$H$_9$	166		0.893
Lauryl methacrylate	CH$_2$:C(CH$_3$)COOC$_{12}$H$_{23}$		−8	0.868($d_{25/15.6}$)
Ethoxyethyl acrylate	CH$_2$:CHCOOC$_2$H$_4$OC$_2$H$_5$	174.1		0.9834
Ethoxyethyl methacrylate	CH$_2$:C(CH$_3$)COOC$_2$H$_4$OC$_2$H$_5$	91–93/35		0.971($d_{15.5/15.5}$)
Ethylene glycol dimethacrylate (96 % technical)	CH$_2$:C(CH$_3$)COOC$_2$H$_4$OOCC (CH$_3$):CH$_2$	96–98/4		1.06($d_{15.5/15.5}$)
Glycidyl acrylate	CH$_2$:CHCOOCH$_2$CHCH$_2$\\/\\O	57/2		1.1074

(continued overleaf)

Table 1.1 (continued)

Monomer	Formula	b.p. (°C)	m.p. (°C)	s.g. Pressure (mm) ($d_{20/20}$)
Glycidyl methacrylate	$CH_2{:}C(CH_3)COOCH_2CHCH_2$ \ O	75/10		1.073(25)
Ethylene glycol monoacrylate	$CH_2{:}CHCOOC_2H_4OH$	76/8		1.11
Ethylene glycol monomethacrylate	$CH_2{:}C(CH_3)COOC_2H_4OH$	84/5		1.07
Propylene glycol monoacrylate	$CH_2{:}CHCOOC_3H_6OH$	85/9		1.05
Propylene glycol monomethacrylate	$CH_2{:}C(CH_3)COOC_3H_6OH$	92/8		1.03
Acrylamide	$CH_2{:}CHCONH_2$	125/25	85	1.222(30°)
Methacrylamide	$CH_2{:}C(CH_3)CONH_2$		110	
Acrylonitrile	$CH_2{:}CHCN$	77.3		0.8060
Methacrylonitrile	$CH_2C(CH_3)CN$	90.3		0.8001($d_{20/4}$)
Methylolacrylamide (available as 60 % solution)	$CH_2CHCONHCH_2OH$			
Methylenediacrylamide	$(CH_2CHCONH)_2CH_2$	97.5/40		0.933($d_{25/4}$)
Diethylaminoethyl methacrylate	$(C_2H_5)_2NC_2H_4OOCC(CH_3){:}CH_2$	103/12		0.914($d_{20/4}$)
t-Butyl aminoethyl methacrylate	$t{-}C_4H_9NHC_2H_4OOCCHCH_3){:}CH_2$	97.5/40		0.933
trans-Crotonic acid	$CH_3CH{:}CHCOOH$	72		0.963($d_{80/4}$)
Itaconic acid	$CH_2{:}C(CH_2COOH)COOH$	167		1.6
Maleic acid	cis-(:CHCOOH)$_2$	200	130	1.609($d_{20/4}$)
Fumaric acid	trans-(:CHCOOH)$_2$	290	286	1.635($d_{20/4}$)
Aconitic acid	$HOOCCH_2C(COOH){:}CHCOOH$		191	
Maleic anhydride	$({:}CHCO)_2O$	202	52.5	1.48
Di-n-butyl maleate	$({:}CHCOOC_4H_9)_2$	280.6		0.9964
Di-2-ethylhexyl maleate	$({:}CHCOOC_8H_{17})_2$	209/10		0.9436($d_{15.5/15.5}$)
Dinonyl maleate	$({:}CHCOOC_9H_{19})_2$			0.9030($d_{15.5/15.5}$)
Di-n-butyl fumarate	$({:}CHCOOC_4H_9)_2$	138/8		0.9869($d_{20/4}$)

Name	Formula	bp	density
Di-2-ethylhexyl fumarate	(:CHCOOC$_8$H$_{17}$)$_2$		
Methyl acid maleate	HOOCCH:CHCOOCH$_3$		
Butyl acid maleate	HOOCCH:CHCOOC$_4$H$_9$		
Dimethyl itaconate	CH$_3$OOCCH$_2$C(:CH$_2$)COOCH$_3$	91.5/10	1.27($d_{2/4}$)
Dibutyl itaconate	C$_4$H$_9$OOCCH$_2$C(:CH$_2$)COOC$_4$H$_9$	140/10	0.9833($d_{2/2}$)
Allyl alcohol	CH$_2$:CHCH$_2$OH	96	0.8540($d_{20/4}$)
Allyl chloride	CH$_2$:CHCH$_2$Cl	45	0.9397($d_{20/4}$)
Allyl acetate	CH$_2$:CHCH$_2$OOCCH$_3$	103.5	0.928
Diallyl phthalate	(CH$_2$:CHCH$_2$OOC)$_2$C$_6$H$_4$	290(150/1)	
Allyl glycidyl ether	CH$_2$:CHCH$_2$OCH$_2$CHCH$_2$O	50–52/15	0.967($d_{20/4}$)
Allyl dimethyl glycidate	CH$_2$:CHCH$_2$OCCHC(CH$_3$)$_2$	89/8	
	‖		
	O		
Vinyl methyl ether	CH$_2$:CHOCH$_3$	6	0.7500
Vinyl ethyl ether	CH$_2$:CHOC$_2$H$_5$	35.5	0.7541
Vinyl n-butyl ether	CH$_2$:CHOC$_4$H$_9$	94	0.7803
Vinyl isobutyl ether	CH$_2$:CHOC$_4$H$_9$	83	0.7706
Vinyl cetyl ether	CH$_2$:CHOC$_{16}$H$_{33}$		
N-Vinyl pyrrolidone	CH$_2$C(O)N(CH:CH$_2$)CH$_2$CH$_2$	148/100	13.5 · 1.04
2-Vinyl pyridine	CHCHCHCHC(CH:CH$_2$)N	110/150	0.9746
4-Vinyl pyridine	CHCHC(CH:CH$_2$)CHCHN	121/150	0.988
2-Methyl 5-vinylpyridine	CHC(CH:CH$_2$)CHCHC(CH$_3$)N	75/15	

7 PHYSICAL PROPERTIES OF MONOMERS

Table 1.1 is not intended to be exhaustive, but gives the b.p and m.p. values where they are above about $-5\ °C$ and density (s.g.) of the principal monomers. The order in which they are given is that of previous sections.

8 APPENDIX

This section is devoted to a list of references to which the reader may refer if more information is required on the subjects listed in this chapter, with special, but not exclusive, reference to monomers and their general and polymerisation properties. It is not intended to be exhaustive, and as far as possible is based on monographs and surveys, to avoid the necessity of obtaining copies of numerous original works.

There have been few, if any, publications in English or other Western languages in the quarter century specifically devoted to monomers. A few earlier works, still useful, are quoted here. For a general account of monomers, the following is suggested:

C.E. Schildknecht, *Vinyl and Related Monomers*, Wiley, New York, 1952

Although more than 40 years old, this volume is still very valuable.

R.H. Boundy and R.F. Boyer (eds.), *Styrene, Its Polymers, Copolymers and Derivatives*, Reinhold, New York, 1952

E.H. Riddle, *Monomeric Acrylic Esters*, Reinhold, New York, 1954

S.A. Miller (ed.), *Acetylene, Its Properties, Manufacture and Uses*, Vol. 1, Ernest Benn, London, 1965

S.A. Miller (ed.), *Ethylene and Its Industrial Derivatives*, Chapters 6 and 11, Ernest Benn, London, 1969

J.V. Koleske and L.H. Wartman (eds.), *Polyvinyl Chloride, Its Preparation and Properties*, Gordon and Breach, New York, and Macdonald Technical and Scientific, London, 1969

Many of the major producers issue bulletins on properties and polymerisation of the various monomers. The following multivolume Encyclopaedias have many articles of interest:

Encyclopedia of Polymer Science, 2nd edn, eds. H.F. Mark, G. Gaylord and N.M. Bikales, Wiley, New York

Comprehensive Polymer Science, eds. C. Booth and C. Price, Pergamon Press, Oxford, 1989

Handbook of Polymer Science and Technology, eds. G. Allen and J.C. Bevington, Marcel Dekker, New York

P.A. Lovell, M.S. El-Aasser, eds, *Emulsion Polymerisation, and Emulsion Polymers*, John Wiley, 1997.

REFERENCES

1. H. Staudinger *et al.*, *Ber.*, **53** 1073 (1929); *Angew. Chem.*, **42**, 37–40 (1929)
2. W.H. Carrothers, in *Collected Papers on High Polymeric Substances*, H. Mark and G. Stafford Whitby (eds.), Interscience Publishers, 1940
3. C. Booth and C. Price (eds.), *Comprehensive Polymer Science*, Vol. 1, Pergamon Press, 1989
4. H.R. Allcock and F.W. Lampe, *Contemporary Polymer Chemistry*, Vol. 1, 2nd edn, Prentice-Hall, 1990
5. M.P. Stevens, *Polymer Chemistry*, 2nd edn., Oxford University Press, 1990
6. G. Odian, *Principles of Polymer Chemistry*, 3rd edn., Wiley, 1991
7. P.E.M. Allen, in, *The Chemistry of Cationic Polymerisation*, Plesch (ed.), Pergamon Press, 1963, Ch. 3; As ref 3; Vol. 3, Part 1, G. Sauvet and P. Sigwalt, pp. 579–637; H.st.D. Nuyken, Pask, pp. 639–710
8. G. Natta and F. Danusso, *Stereoregular Polymers and Stereospecific Polymerisation*, Pergamon Press, Oxford, 1967; Y.V. Kissin in *Handbook of Polymer Science and Technology*, Chereminosoff (ed.), Vol. 8, Marcel Dekker, 1989, pp. 9–14
9. M.A. Doherty, P. Gores and A.H.E. Mueller, *Polym. Prepr. (Am. Chem. Soc.)*, **29**(2), 72–3 (1988)
10. P.J. Flory, *Principles of Polymer Chemistry*, Cornell University Press, 1953, Ch. 4, p. 106 *et seq.*
11. G.C. Eastmond, A. Ledwith, S. Russo and P. Sigwalt, *Comprehensive Polymer Science*, Sec. 1, Pergamon Press, 1989
12. F.W. Billmeyer, *Textbook of Polymer Science*, 3rd edn, Wiley, 1985
13. C.H. Bamford, *Encyclopedia of Polymer Science and Engineering*, 2nd edn, Vol. 13, pp. 729–35 1988 (refers specifically to retardation and inhibition)
14. C.E. Schildknecht, *Allyl Compounds and Their Polymers*, Ch. 1, pp. 195 *et seq.*
15. H. Warson, *Per-Compounds and Per-Salts in Polymer Processes*, Solihull Chemical Services, 1980
16. M. El-Aaser and J.W. Vanderhoff (eds.), *Emulsion Polymerisation of Vinyl Acetate*, Applied Science Publishers, 1981
17. G.S. Misra, and U.D.-N. Bajai, *Progress in Polymer Science*, Vol. 8, Pergamon Press, 1982–3
18. H. Warson, *Makromol. Chemie*, **105**, 228–45 (1967)
19. H. Warson and R.J. Parsons, *J. Polym. Sci.*, **34**(127), 251–269 (1959)
20. H. Warson, *Peintures, Pigments, Vernis*, **43**(7), 438–446 (1967)
21. T. Alfrey and D. Price, *J. Polym. Sci.*, **2**, 101–106 (1947)
22. P.J. Flory, *Principles of Polymer Chemistry*, Cornell University Press, 1953, Table 20, p. 188
23. D. Price, *J. Polym. Sci.*, **3**, 772, (1949)
24. H.N. Friedlander, H.E. Harris and H.E. Pritchard, *J. Polym. Sci.*, **4A**(1) 649–64 (1966)
25. J.M.G. Cowie, *Comprehensive Polymer Science*, Vol. 3, Pergamon Press, 1989, pp. 33–42
26. H. Warson, *Chem. Ind.*, **983**, 220–2 (21 March 1983)
27. W.R. Moore, *An Introduction to Polymer Chemistry*, University of London Press, 1967, p. 278
28. E.H. Riddle, *Monomeric Acrylic Esters*, Reinhold, 1964
29. H. Warson, in *Properties and Applications of Polyvinyl Alcohol*, Society of Chemical Industry, 1968, pp. 46–76
30. R.K. Tubbs, H.K. Inskip and P.M. Subramanian, in *Properties and Applications of Polyvinyl Alcohol*, Society of Chemical Industry, 1968, pp. 88–103
31. K. Noro, in *Polyvinyl Alcohol, Properties and Applications*, C.A. Finch (ed.), Wiley, 1973, Ch. 7, pp. 147–166
32. E.H. Merz and R.W. Raetz, *J. Polym. Sci.*, **5**, 587 (1950)
33. American Society for Testing Materials (ASTM), Vicat Softening Point D-1525-65T; Ring and Ball Apparatus E28-67
34. K.E.J. Barrett (ed.), *Dispersion Polymerisation in Organic Media*, Wiley, 1975

35. S.J. Baum and R.D. Dunlop, FIAT No. 1102, US Department of Commerce, 1947, pp. 13–17 and 42–3
36. S.J. Baum and R.D. Dunlop, FIAT No. 1102, US Department of Commerce, 1947, p. 49
37. L.I. Nass, C.A. Heiberger and M. Langsam (eds.), *Encyclopedia of Polymer Science*, 2nd edn, Vol. 1, Wiley, 1985, pp. 127–38
38. H. Warson, *Polym., Paint, Col. J.*, **178**, 625–7 and 865–7 (1988)
39. Y. Kobayashi and T. Yoshikawa (Hitachi Chemical), JP 92 339,805–6, 1993; *Chem. Abstr.*, **119**, 118076, 140017 (1993)
40. W.J. Toussaint, and L.G. McDowell (Carbide and Carbon), USP 2,299,862, 1942
41. R.L. Adelman, *J. Org. Chem.*, **14**, 1057 (1949)
42. J.E.O. Mayne, H. Warson, and R.J. Parsons (Vinyl Products), BP 827,718, 1960; equivalent to USP 2,989,544, 1961
43. J.E.O. Mayne and H. Warson (Vinyl Products), BP 877,103, 1961
44. Shell Chemicals, Technical Manuals, VM 1.1, VM1.2, 1991
45. F. Brown and C.D. Mitchell (Dunlop), BP 701,258, 1951
46. H. Warson, *Derivatives of Acrylamide*, 1990, pp. 46–51
47. H. Warson, *Polymerisable Surfactants and Their Applications (Self-Emulsification)*, Solihull Chemical Sevices, 1989
48. H. Warson, *The Polymerisable Half Esters; Their Polymers and Applications*, Solihull Chemical Services, 1978
49. H. Warson *et al.* (Vinyl Products), BP 995,726, 1965

2
Emulsions and Colloids

H. Warson

1 INTRODUCTION TO COLLOID CHEMISTRY

1.1 Historical

The idea of a colloidal form of matter dates from Thomas Graham in 1861. In its original form a distinction was made between crystalloids and colloids, the former being the classical crystalline materials whilst the latter included various natural gums and resins and less well defined materials. A result has been an overlap between the chemistry of resins, and hence polymers, and the various forms of physical chemistry with which colloid chemistry deals. As a broad generalisation, colloids may be considered to embrace all forms of matter in which the surface is very large compared to the mass and in which surface properties play a large part. Thus the properties of emulsions generally and paints, both emulsion and oil, are covered by colloid chemistry. Fogs and cigarette smoke also come under the same heading as they are very fine dispersions of solids in air and thus have a large surface area.

Many properties of synthetic polymers are best described in terms of colloid chemistry. These include the physical phenomena at the various transition points (Chapter 1, Section 4.3). Solutions of polymers are often referred to as colloidal solutions, although the term is rather vague and has been used to include dispersions and emulsions of such fine particle size that they are below the wavelength of visible light (0.76–0.4 µm). A distinction should be made between particles that are at their ultimate molecular size and larger particles that are aggregates of molecules. With the very largest polymer molecules, of molecular weight above 10^6, colloidal particles may consist of single macromolecules.

1.2 Emulsions

An emulsion is a system containing two liquid phases, one of which, the dispersed phase, is dispersed in a globular state in the other, the continuous phase. Although the two liquids are normally mutually insoluble, an emulsion may be formed where there is partial mutual solubility, the equilibrium solution becoming the continuous phase.

An inversion of phases is possible under some conditions, the disperse phase becoming the continuous one and vice versa. It is not essential for the liquid in the major proportion by volume to be the continuous phase. If small spheres of identical diameter are packed as closely as possible into a given space so that each sphere is just touching 12 others, they will occupy 74 % of the total volume, this being the theoretical maximum concentration of an emulsion consisting of uniform spheres, although in practice the latter is unlikely to be the case. Particles are capable of deformation, which increases the maximum quantity of the disperse phase.

Furthermore, particles are seldom of the same size. In a close-packed system of even spheres the gaps can be depicted as containing smaller spheres, which partially fill the spaces, just touching the sides of the larger spheres. Their diameter $r' = r(\sqrt{2} - 1)$, where r is the diameter of the main spheres and r' of the smaller spheres. In an emulsion with particles of varying sizes, the concentration of the disperse phase may be very high. Disperse phases of over 80 % have been prepared.

The main classes of emulsion are oil in water (O/W) and water in oil (W/O). Milk is an example of the former and butter of the latter. However, it is possible to prepare an emulsion of one organic liquid in another in which it is immiscible; this occurs under some conditions if polar liquids, e.g. ethyl alcohol, are dispersed in mineral oils.

In order that emulsions remain stable and not separate into two layers, a suitable stabiliser or emulsifier must be present. These form a protective film on the disperse phase. Details are given in the following sections.

1.2.1 Properties of surfaces

Below the surface of a liquid, molecular attractions balance, but on the surface an imbalance of forces occurs. Since there are virtually none in the gaseous phase, the molecules at the surface experience an inward pull which would cause the drops of liquid to assume the smallest possible surface area, i.e. a spherical shape. The conditions at the surface of a liquid are thus different from those in the bulk. The condition of tension at the surface of the liquid is known as surface tension. Thus when spheres of identical or miscible types touch they will coalesce, further reducing the overall surface. These cohesional forces causing coalescence are of interest in considering problems of adhesion.

The actual nature of the surface layer has been a matter of some conjecture, but it is probably several molecules deep, varying with the liquid and with conditions such as temperature. Surface tension is a linear force, the unit of surface tension being the dyne cm^{-1} (in SI units, millinewtons per metre or mN m^{-1}; 1mN m^{-1} = 1 dyn cm^{-1}). Problems connected with interfacial tensions are essentially identical with those of surface tension. Whilst surface tension should be theoretically measured in a vacuum, in practice the presence of air has a negligible effect. There is, however, a strong interfacial tension between immiscible liquids. Dissolved substances markedly affect the surface tension of a liquid. Inorganic salts tend to increase the surface tension of water, whilst organic compounds which include hydrophilic groups such as hydroxyl, —OH, tend to reduce it.

If a solute reduces the value of the surface tension, i.e. the potential energy at the interface, it can be shown on thermodynamic grounds that it will concentrate at the interface. This phenomenon is known as adsorption. In this connection the Gibbs adsorption isotherm may be mentioned. It is expressed as

$$G = -\frac{c}{RT}\frac{ds}{dc}$$

where s represents surface tension, G the interfacial excess of the adsorbed component, c the concentration and R the gas constant. From the above equation it can be seen that G is positive when the added substance lowers the surface tension.

The Freundlich adsorption isotherm and empirical equation is also of interest in considering adsorption at an interface:

$$\frac{x}{m} = aC^{1/n}$$

where a and n are constants, n being >1 for a given solute and adsorbent, m is the mass of adsorbing material (in g), x the amount of solute adsorbed by this mass and C the equilibrium concentration of the solute. The equation implies that relative adsorption is greater in dilute solution.

Elementary ideas of surface tension and surface adsorption should suffice to obtain a background to the technology and science of emulsion polymerisation. There are a number of reference books [1, 2].

Two other ideas must be mentioned before a more detailed discussion is given on emulsions. One is the study of unimolecular films of substances containing a hydrophilic group on the surface of water, with which the names of A. Pockels, I. Langmuir, N.K. Adam, D.A. Wilson and J.W. McBain are associated. Under pressure the molecules become oriented with the hydrophilic head in the water layer and a hydrophobic tail. Suitable molecules include stearic acid $C_{17}H_{35}COOH$ (the COOH being the hydrophilic portion), esters of fatty acids of general formula R^1COOH and phenols, $R_2C_6H_4OH$. The surface area per single molecule approximates to 20.5 Å in many cases. Phenols with tertiary alkyl groups have a molar area of about 24 Å, as do many other benzene derivatives such as amines.

1.2.2 Micelles, surfactants

The second of the key ideas is due to McBain, who explained certain irregularities in physical properties such as the electrical conductivity of solutions of soaps, e.g. sodium stearate, by postulating that they exist in aqueous solution as micelles or aggregates which can be considered to exist with the hydrophobic parts of the molecule of soap turned inwards to each other, while the polar heads are turned outwards. A micelle can be considered to consist of 50–100 ions, which could be regarded as a single, large polyvalent ion. It is surrounded by a cloud of gegen ions of opposite charge.

There is, however, a threshold level with all micelle-forming materials below which they are in ordinary solution. This is known as the critical micelle concentration (CMC). As the overall concentration increases and micelles form, they are capable of solubilising liquids that are normally insoluble, by a type of insertion mechanism. Figure 2.1 shows the solubilisation of ethyl benzene [3].

To form micelles effectively a molecule must have a hydrophilic and a hydrophobic portion and is said to be amphipathic. The most well known are the soaps. For simplicity, long straight chain aliphatic hydrocarbon chains on a molecule will be considered. In the soap series there is virtually no tendency to form micelles where n is 8 or less in the series $C_n H_{2n+1} COO^- Na^+$. At $n = 9$ there is a weak tendency to form micelles, but it is strong over the range $n = 12$ to $n = 18$, weakening to $n = 22$, beyond which the hydrophobic tail governs the properties of the 'soap', which becomes virtually insoluble.

The balanced compounds, which are capable of forming micelles, have a powerful effect in reducing both the interfacial tension of the solvent (water) with air and the interfacial tension between water and otherwise immiscible liquids. This is hardly surprising in view of their physical properties, and such compounds are generally known as emulsifiers, surfactants or surface-active materials.

When a water or a non-polar material such as a hydrocarbon is shaken with water containing a surfactant, the surfactant tends to become absorbed on the non-polar liquid, with its hydrophobic portion on the surface and the hydrophilic 'tail' in the water. This is especially true when its concentration is above the CMC. This results in a number of energy changes. The reduction in surface tension, or to be more accurate interfacial tension, facilitates the break of the particles into very small drops, whilst the charge given by the adsorbed charge helps the drops to repel one another rather than coalesce. Thus we have a stable system of a hydrophobic liquid dispersed in water, known as an emulsion.

This a drastic simplification of a complex phenomenon. Electrical phenomena such as the 'zeta' (ζ) potential, the charge at the surface, and the Helmholtz electrical double layer all affect stability of the emulsion. The surfactant need not necessarily be ionic and various non-ionic stabilisers will be described later. Equally, very high molecular weight water-soluble products, often referred to as 'colloids', such as the natural gums, gum acacia and gum tragacanth, assist in giving the emulsion stability, probably by helping to provide a rigid envelope around the particles. Synthetic high molecular weight products such as polyvinyl alcohol are not merely stabilising 'colloids' but are, in many cases, direct emulsifiers. The polyelectrolytes such as sodium

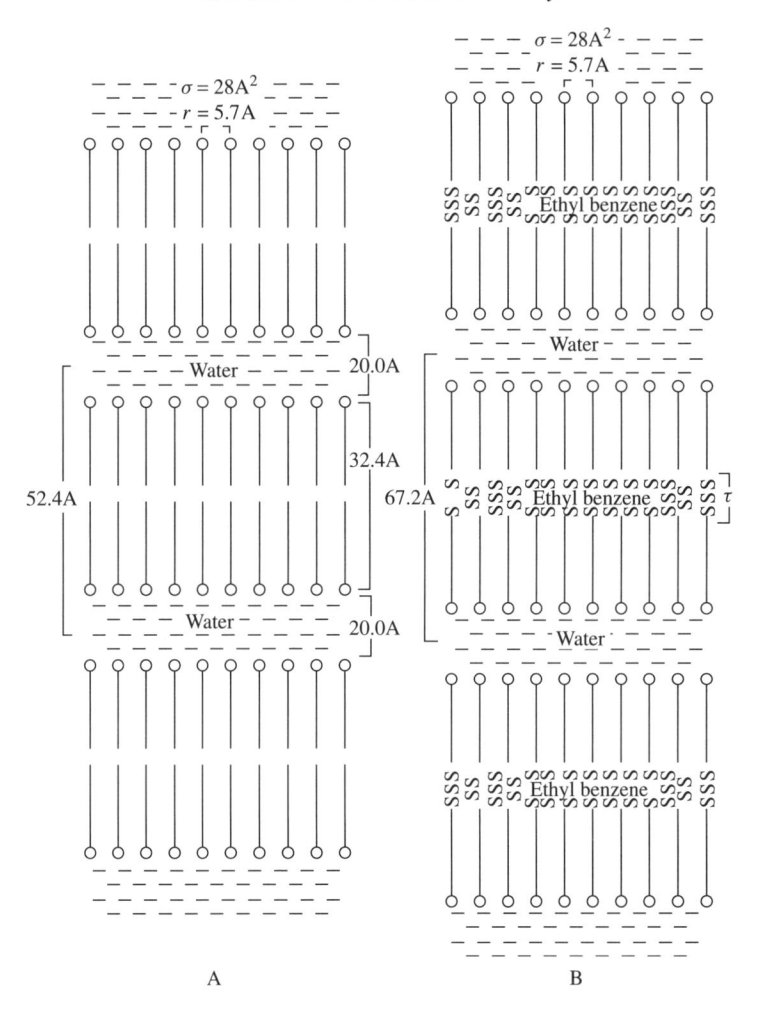

Figure 2.1 Idealised diagrammatic cross-section of soap micelles in 15 % aqueous potassium laurate. (A) Without additive oil. (B) Saturated with ethylbenzene. The actual structures are undoubtedly far less ordered. The only spacings actually measured are the long and short spacings.

polyacrylate, $n\text{-Na}^+(\text{OOCHCH}_2)_n{}^-$, are also emulsifiers, and sometimes have powerful effects in reducing surface tension.

In contrast, some water-soluble products, such as sodium hexametaphosphate $\text{Na}_{14}(\text{P}_{12}\text{O}_{37})$ and some organic products that are water soluble and are strongly adsorbed on to many surfaces, particularly on to pigments, have no surfactant properties and are known as dispersants. These are of interest when a pigment is compounded with an emulsion.

1.2.3 Emulsion particles, emulsifier efficiency

The particle sizes of emulsions vary and may be as high as 5 microns (1 micron, abbreviated μm = 10^{-4} cm), although at this level they are not very stable and tend to settle and coalesce if the disperse phase is more dense than water, or to cream to the surface if the density is above one. Whilst emulsion droplets seldom even approximate to the same size and follow a Gaussian distribution, it is customary to quote an average diameter, based on the droplets, or 'particles' as they are referred to after emulsion polymerisation. These are spherical, and a photomicrograph will readily show their diameters. The surface of the droplets will be proportional to the square of the indicated diameters and the volume, or mass, to the cube of the diameters. Thus, if an emulsion of standard droplet diameter 1 μm is compared with a very finely dispersed type of droplet diameter 0.1 μm, a fairly common variation, the surface areas will vary by a factor of 100 and the volumes by a factor of 1000. An increase of surfactant will produce a smaller particle size emulsion, as more stabiliser will be available to cover the increased surface available.

Mathematically, 1 kg of dispersed phase, with a mean particle radius of 1 μm and s.g. 1.0, would have a surface area of 3000 m², while for a particle radius of 0.1 μm this rises to 300 000 m², indicating the enormous potential of the surface forces involved. If it is assumed that the surface is covered in the former case by a surfactant of cross-sectional area 20 Å, then approximately 7.5 gm of surfactant will be required if the molecular weight of the latter is 300, representing about 3–4 % by weight, and would not be far from the amount required in most cases. With the fine particle size emulsion, the amount would be multiplied by 100, giving a level far above that normally used, which would be about 5–10 % in most cases. Thus, it may be assumed that for very fine particle size emulsions it is unnecessary for the surface to be completely covered by surfactant.

It is not necessary for the hydrophobic portion of an emulsifier to have a negative charge at one end as in the carboxylate ions. It may have a positive charge, i.e. it may be cation active, the term sometimes used. This must not be confused with the type of polymerisation known as cationic, as distinct from the free radical. The salts of many amines may be used as emulsifying agents, such as octadecylamine hydrochloride $C_{18}H_{37}NH_3^+Cl^-$. In this case it is the cation that is absorbed on the dispersed particles. The so-called non-ionic surfactants, often based on a balance of ethylene oxide units and a hydrophobic chain, as in $C_{18}H_{35}(OC_2H_4)_{15}OH$, usually carry a small negative charge. Some water-soluble polymers, e.g. polyvinyl alcohol, also carry a negative charge.

A system has evolved of estimating the effect of a surfactant by the hydrophobic/hydrophilic balance (hydrophobic = water repelling, hydrophilic = water attracting). Values have been ascribed to the various groups making up the surfactant, such as stearate $C_{17}H_{35}COO$—, and the units making up non-ionic surfactants, these being known as HLB values. They are useful in selecting emulsifiers, as will be shown later, especially for non-ionic types. Whilst there are a number of rules by which to select emulsifiers, the choice remains to a considerable extent empirical.

A development of the idea of a simple emulsifier was made by Schulman and Cockbain [4]. Based on studies of molecular associations by the injection of certain surfactant-type substances under monolayers of cholesterol, they postulated that a strong polar interaction exists between substances with a hydrophilic group dissolved in a hydrophobic liquid such as liquid paraffin and the hydrophilic group of a surfactant. There were a number of conditions, including steric ones. Among the oil-soluble components cholesterol (a natural alcohol with a fused ring formula, empirically $C_{27}H_{45}OH$) is the most efficient; cetyl alcohol $C_{16}H_{33}OH$ is also very efficient; oleyl alcohol $C_{17}H_{35}OH$ is not efficient because of the steric reasons illustrated in Figure 2.2.

Among the water-soluble components, sodium cetyl sulfate was efficient, the sulfonates rather less so, followed by carboxylates. The strong interactions enabled emulsions to be formed very readily by hand-shaking alone. The complex formed may be regarded as highly stable, and in equimolecular proportions. It can be regarded as a condensed liquid film, with a resulting extremely low interfacial tension at the interface. The general principle of an interfacial condensed film is of interest in some aspects of emulsion polymerisation (Chapter 3).

Under specific conditions, especially in the presence of a cosurfactant, sometimes referred to as an amphiphile, microemulsions may be formed. They are transparent or translucent and thermodynamically stable, and may be O/W or W/O. They may be defined as having particle sizes in the range 75–1500 Å diameter. In exceptional cases a polymer latex may come into this category. For further information reference [5] should be consulted.

1.2.4 Emulsion viscosity

As a formal definition, the coefficient of viscosity is the shearing stress across an area when there is a unit velocity gradient normal to the area. It may be measured in poises or centipoises, but in the SI system it is measured in pascal-seconds, Pa s (to convert centipoise to Pa s multiply by 10^{-3}). The viscosity of an emulsion depends substantially on that of the external phase. It is increased in a complex manner by the dispersed phase, depending on the volume fraction of the system as a whole, the deformability of the particles under stress, which itself may be more dependent on the stabiliser sheath round the particles than on the nature of the disperse phase, and on any 'bridging' of particles by the stabilisers present. Dilute emulsions which are mechanically stable tend to be somewhat Newtonian in characteristics. As concentration increases, the Newtonian characteristics disappear and the strain tends to increase disproportionately to the stress; in other words, viscosity decreases with agitation. The normal phenomena of non-Newtonian liquids are followed. There is often a yield value, i.e. a definite stress which must be

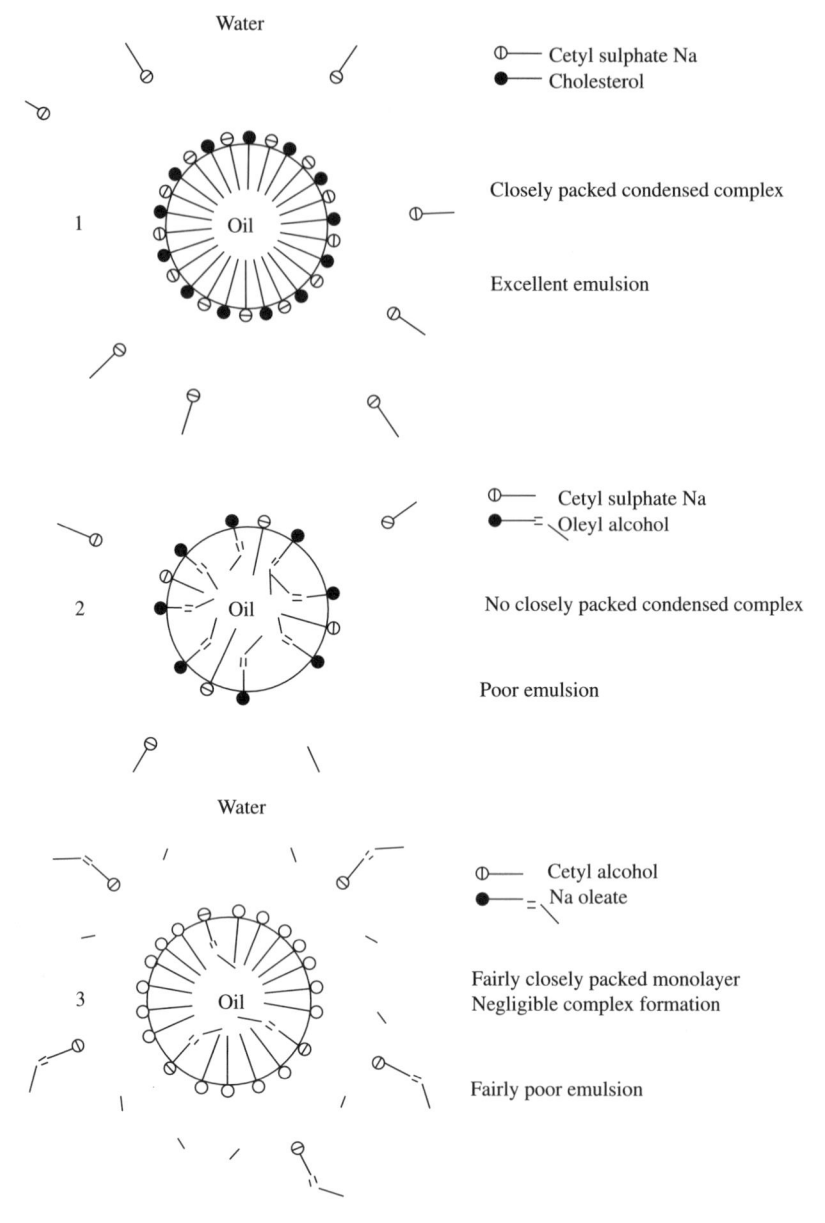

Figure 2.2 Complex formation at the oil-water interface in emulsions, according to schulman and Cockbain [4]. (1) Sodium cetyl sulfate and cholesterol form a closely packed complex, giving a good emulsion. (2) Sodium sulfate and oleyl alcohol form a poorly packed complex because of the double bond in the alcohol, giving a poor emulsion. (3) Cetyl alcohol and sodium oleate form a fairly close packed complex monolayer, giving a fair emulsion.

Introduction to Colloid Chemistry 57

imparted before definite movement is caused by shear forces. In the case of pseudoplastic flow, no yield value is exhibited, but the viscosity depends on the shear rate.

Under some conditions, an emulsion of reversed phase, i.e. water-in-oil, can be prepared, especially if the organic phase is in excess. The particles in this case are much more irregular in shape and emulsion inversion, i.e. the change from O/W and W/O and vice versa, will depend on the nature of the hydrophobic phase, the surfactants and the temperature. It is often possible to prepare an emulsion of a material that is solid at ambient temperature, e.g. a natural wax such as beeswax, by melting it, the m.p. of these materials usually being circa 80 °C, adding emulsifier to either or both phases, followed by adding the water phase to the molten wax with vigorous stirring until the emulsion, which is at first of a W/O type, suddenly inverts to O/W, after which it is rapidly cooled with continued stirring. Although the disperse phase is a solid, the product is regarded as an emulsion, since its properties are identical with that of an emulsion formed from a liquid disperse phase.

Thixotropy is a time-dependent variation of viscosity, the viscosity falling with increased time at constant shear. The phenomenon is reversed on removing the shear, but there is always a hysteresis, a delay in the return to the original condition (Figure 2.3).

In dilatant flow, the viscosity increases with the rate of shear. It is a fairly rare phenomenon, but occasionally occurs in polymer emulsions and can be

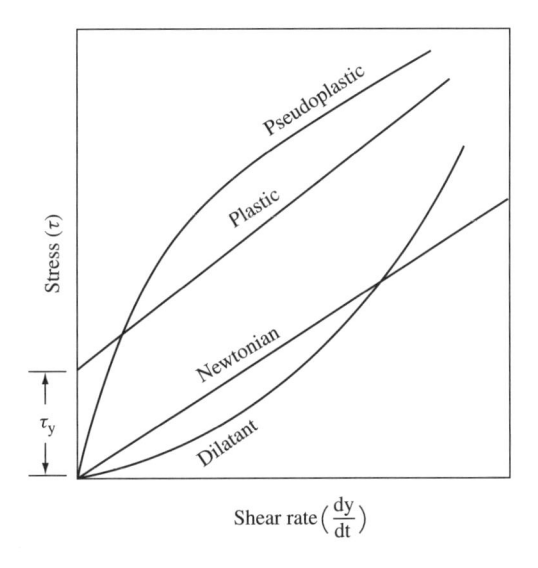

Figure 2.3 The four types of flow which may be exhibited by fluids: Newtonian, plastic, pseudoplastic and dilatant. (Reproduced from Becher [1].)

presumed to be due to the resistance of the particles themselves; it is often associated with instability.[†]

Although the concentration of an emulsion has a practical upper limit, it must not be assumed that a very dilute emulsion is necessarily more stable than a concentrated one. The word 'dilute' has been deliberately used in a non-quantitative manner. For a given proportion of stabiliser relative to the disperse phase, optimum stabilities are often achieved at 30–60 % of overall concentration. The phenomenon of 'creaming' by which attractions between the micelles crowd out the water molecules often makes the more concentrated emulsions the most stable.

2 SURFACTANTS AND STABILISERS

Surfactants, as already indicated, may be classified according to the charge of the effective emulsifying part of the molecule. Thus there are anionic, non-ionic and cationic types. In addition, some types vary in their charge with the pH of the solution, and are known as amphoteric. Thus a stearate ester of triethanolamine would be non-ionic at a pH of about 7 but cationic at a low pH, about 2, since it is only a very weak base. Its formulae, somewhat idealised, are

$$N \begin{cases} CH_2CH_2OOCC_{17}H_{35} \\ CH_2CH_2OOCC_{17}H_{35} \\ CH_2CH_2OOCC_{17}H_{35} \end{cases} \qquad Cl^- \; {}^+N \begin{cases} CH_2CH_2OOCC_{17}H_{35} \\ CH_2CH_2OOCC_{17}H_{35} \\ CH_2CH_2OOCC_{17}H_{35} \end{cases}$$

Triethanolamine stearate ester, non-ionic form 　 Triethanolamine stearate ester. hydrochloride, cationic form

Whilst most of the older types of ionised surfactants were carboxylate soaps, most modern types are either sulfates $XOSO_3M$ or sulfonates $Y.SO_3M$, where M represents an alkali metal or an organic base and X and Y are the entire balance of the molecules. In recent years, phosphate emulsifiers have been available.

Many modern non-ionic surfactants are based on a hydrophobic 'tail' attached to a chain of ethylene oxide molecules in a formula that could be represented as $X(OCH_2CH_2)_n OH$, where n can be as low as 2 or up to 100 or more, though most technical products have n below 50 and at least 4. The hydrophobic portion X may be a standard alkyl chain attached either to oxygen, giving an ether, or carboxyl, giving a salt, the former being the more usual and effective. The hydrophobic portion may be based on fatty alcohols,

---

[†] The term 'shear thickening' can also be used. BS 2015: 1992 gives a definition of dilatancy as the property of a paint or pigment paste that is manifested as a thickening or solidification on application of a shearing force. This definition applies more generally and is often referred to in describing the properties of some latices see (Chapter 3).

e.g. lauryl alcohol $C_{12}H_{25}OH$. It was often based on an alkyl phenol or an alkyl cresol, but the use of these are now discouraged because of the difficulty in biodegradation of these types.

The cationic surfactants consist almost invariably of amine salts, though in theory phosphonium or even sulphonium salts could be used. The simplest are the salts of alkyl amines of chain length $C_{12}-C_{18}$, the hydrochloride or acetate being the most common. In most cases cationic emulsifiers and the anionic type are unstable in each other's presence, and two oppositely charged emulsifiers will mutually precipitate. This is not invariable, however, and there are some exceptions where the product of the hydrophobic anion and the corresponding cation remain soluble, especially where both the anion and cation are modified with an ethylene oxide chain, thus imparting some additional properties of the non-ionic surfactants.

3 A SURVEY OF TECHNICAL SURFACTANTS

This section includes the principal classes and types of surfactants and a brief survey of their principal properties. In general trade names will not be mentioned as it is invidious to even attempt to make a selection from the large number of commercial products. Trade names and manufacturers may be included, however, when they represent the original patentee or manufacturer.

3.1 Anionic types—fatty acid soaps and allied salts

Soaps represent the earliest type of surfactant known, the origins being in early chemical history. Most modern soaps are the sodium or potassium salts of the straight chain fatty acids found in nature, and most commercial products represent a mixture of acids or a partially purified single acid. The principal acids are the straight chain alkyl fatty acids from lauric acid $C_{11}H_{23}COOH$ to stearic acid $C_{17}H_{35}COOH$. The soaps harden as the fatty acid chain increases in length, sodium soaps being harder than those of potassium. Occasionally a lithium soap may be used. The main unsaturated acid, in which the double bond is in the *cis*-9 : 10 position, is oleic acid $CH_3(CH_2)_7CH:CH(CH_2)_7COOH$. It produces soft, soluble soaps. These soaps are formed almost invariably from fatty acids in which the total number of carbon atoms is even, as with almost all natural products.

The higher chain alkyl fatty acid soaps, such as those derived from behenic acid $C_{21}H_{43}COOH$ and even higher fatty acids derived from waxes, are occasionally used for special purposes, although their soaps form micelles at low concentrations. Another fatty acid occasionally used is derived from tall oil, a by-product from paper manufacture, which contains rosin acids and a mixture of acids, $CH_3(CH_2)_4CH:CHCH_2CH:CH(CH_2)_7COOH$. Tall oil acids are normally low-cost products. The salts of rosin acids derived from abiotic acid, or various hydrogenated or disproportionated derivatives, are often included in emulsion polymerisation formulations of the monomers including butadiene.

The salts of naphthenic acids, which are saturated fatty acids containing cyclopentane rings are also occasionally included as emulsifiers. These naphthenic acids are petroleum by-products.

A soap is often prepared *in situ* by dissolving the fatty acid in the disperse phase and the alkali, e.g. sodium, potassium or ammonium hydroxides, in the water phase. This method usually produces fine particle size emulsions and is strongly recommended where possible. Whilst the use of ammonium hydroxide as the alkali has the advantage that on drying an emulsion no water-soluble materials remain, but only an insoluble fatty acid, the volatility of ammonia has certain disadvantages, especially when used with polymer emulsions, since loss of ammonia will cause a drop in pH and may lead to instability. With a coating on steel, 'flash' corrosion (to be discussed in a later volume) caused by a premature drop in pH may occur. In consequence, ammonia is often replaced by amines.

The simple amines, e.g. triethylamine $(C_2H_5)_3N$, are not often included in formulations because of the strong 'fishy' odour. However, the cyclic amine morpholine, $\overline{CH_2CH_2OCH_2CH_2N}H$ (b.p. 268.4 °C), is one of the most frequently used. Other amines in use are the alkanolamines, especially mono-ethanolamine $HOC_2H_4NH_2$ (b.p. 170.4 °C), diethanolamine $(HOC_2H_4)_2NH$, triethanolamine $(HOC_2H_4)_3N$ and diethylamino ethanol $(C_2H_5)_2(HOC_2H_4)N$. The amines find considerable use owing to their relatively innocuous physiological effects.

The acyl derivatives of sarcosine (methyl glycine) CH_3NHCH_2COOH and lauroyl sarcosine $C_{11}H_{23}CON(CH_3)CH_2COOH$ are available commercially and are rather interesting. The salts are more stable under slightly acid conditions than straight chain fatty acid salts and they are useful as a constituent in compositions for corrosion resistance since lauroyl sarcosine seems to displace water from a steel surface (see Chapter 10 in Volume 2).

Apart from pH stability, the major disadvantage of the soaps as emulsifiers is the insolubility of the calcium salts. This invariably creates difficulties when only hard water, which contains appreciable dissolved calcium salts, is available, and gives rise to emulsion instability. A strong tendency has therefore arisen for soaps to be replaced by the various sulfated and sulfonated emulsifiers described below.

3.2 Sulfates and sulfonates

The sulfates are in essence half esters, one of the hydrogens of sulfuric acid being esterified with a complex alcohol and the other being neutralised by a strong base, usually sodium hydroxide. They have the general advantage of stability over a wide pH range, although hydrolysis occurs both at high and low pH levels. Production is by a normal esterification process, using sulfuric acid, normal or fuming, or sometimes sulphur trioxide.

3.2.1 Sulfated fatty alcohols

The fatty alcohols such as lauryl alcohol $C_{12}H_{23}OH$, oleyl alcohol $C_{18}H_{35}OH$ and cetyl alcohol $C_{16}H_{33}OH$ are formed by controlled reduction of the natural fatty acids. The sulfation process usually implies including an excess of acid, with the result that technical neutralised products contain a certain quantity of sodium sulfate Na_2SO_4, although most are partially purified to at least 93 % active solids. Sulfated lauryl alcohol, or more properly sodium lauryl sulfate $C_{12}H_{23}OSO_3Na$, is often available as a readily soluble paste. The mixed sulfates of cetyl alcohol and stearyl alcohol, $C_{18}H_{37}OH$, are regular articles of commerce and are economic in cost. The potassium salts tend to be less soluble than the sodium salts in this series.

Certain of these products are based on synthetic alcohols, including secondary alcohols. They include Teepol® 610, a secondary alcohol sulfate $R_1R_2CHOSO_3Na$, where R_1 is a large alkyl group, molecular weight average 308, and R_2 is a relatively smaller one, and sodium 2-ethylhexyl sulfate, which has a C_8 branched chain and very good acid and alkali stability. It also has hydrotrope properties. A longer branched series, based on C_{14} and C_{17} secondary alcohols respectively, is available in solution form from Union Carbide as Tergitol 4 and 7 respectively.

3.2.2 Sulfates of ethoxylated alcohols

An alcohol may be reacted with several molecules of ethylene oxide before the sulfation process. Thus from lauryl alcohol it is possible to derive a product such as $C_{12}H_{23}(OC_2H_4)_n OSO_3Na$, where $n = 1-3$. These products are readily soluble, have good stability to salt solutions and hydrolysis, with efficient surfactants and stabilising properties. Their compatibility with other emulsifiers and stabilisers is generally good.

This class may be formed from oxyethylated alkyl phenols and the sodium salts are given a general formula such as $C_n H_{2n+1} C_6H_4(OC_2H_4)_m OSO_3Na$, where the alkyl group is substituted on the aromatic ring. If n and m are relatively large $\geqslant 4$, these sulfates have some non-ionic properties in addition, in particular better stability at low pH levels. Commercial products may be derived from an alkylated toluene rather than from an alkylated benzene.

3.2.3 Sulfated natural oils and esters

Historically this is the oldest class, but also the type with the least determinate formulae. The earliest example, introduced about 1875, is 'Turkey red oil', formed by the sulfonation of castor oil. The term 'sulfonation' which should be properly reserved for the formation of sulfonates (see below) is used here in a general sense. They are of relatively little use in emulsion polymerisation or in application as pigment dispersants, their main application being in textile treatment.

Other technical products useful as emulsifiers are formed by the sulfation of amyl or butyl oleates. They are available as solutions or gels of about 40 % concentration with a low sodium sulphate content. The reaction may be assumed to be an addition to the double bond of the oleate.

3.2.4 Sulfonates—general

Sulfonates form an exceptionally large class of surfactant and dispersant, almost every possible product which can react with sulfuric acid by electrophilic displacement having been subjected to sulfonation. The reaction of sulfonation is in general reversible, and an excess of reactant is necessary to ensure completion. In consequence, the neutralisation stage usually results in a small amount of sodium sulfate being present. The products are available as dried powders, pastes or slurries, usually cream to light brown in colour, and are nearly always supplied as the sodium salts, though the free acids, when stable, often have marked surface-active and emulsifying properties, even at their natural low pH.

3.2.5 Sulfonated aromatic and condensed ring compounds

To ensure an adequate balance of hydrophobe–hydrophile properties, the overall number of carbon atoms should be about 12–18. Benzene or toluene with a hydrocarbon side chain, which is in most cases octyl C_8H_{17} or nonyl (a branched chain compound) C_9H_{19}, is sulfonated to produce an emulsifier of general formulae:

Na alkylbenzene sulfonate   Na isopropylnaphthalene sulfonate

The indicated position of the side chains is a convention to indicate that their exact position has not been determined in most cases. The methyl CH_3 group of toluene need not be present. H atoms on the rings are not indicated here.

Because of the desirability for biodegradability, most modern sulfonates of this type have been manufactured with straight chain rather than branched alkyl groups, since the use in detergents represents a much greater volume of production than for specialised polymer emulsions. However, branched chain sulfonates in most cases remain available (see also Section 3.6.1).

In general the sulfonates formed from alkyl aromatic hydrocarbons are extremely versatile, giving good reduction of surface tension and excellent emulsifying properties. They are also soluble in many organic solvents, even in the form of sodium salts. The acids, however, have a marked catalytic

A survey of technical surfactants

activity for the hydrolysis of esters, even of polymeric esters such as polyvinyl acetate [6].

An alternative form of these products is obtained by adding a short alkyl chain to naphthalene, thence sulfonation. Technical products are sodium isopropyl naphthalene sulfonate (shown p. 62) or sodium butyl naphthalene sulfonate. Their properties resemble the alkyl benzene sulfonates, but they are more prone to darkening.

Other sulfonates based on aromatic hydrocarbons include the sodium salt of disulfonated dodecyldiphenyl oxide (Dow Chemical Co):

$$\begin{array}{c} \text{C=C} \\ \text{C} \quad \text{C-O-C} \quad \text{C} \\ \text{C=C} \quad \text{C=C} \\ \text{SO}_3\text{Na} \quad \text{SO}_3\text{Na} \ \text{C}_{12}\text{H}_{23} \end{array}$$

It has high water solubility and tends to solubilise non-ionic emulsifiers at high temperatures. Other sulfates and sulfonates are formed from an alkyl benzene which has been condensed with two molecules of ethylene oxide (see Section 3.6), with either sulfonation of the benzene ring or sulfation at the end of the oxyethylene units.

Petroleum sulfonates are formed as by-products of the refining of petroleum. The free acids have the generic formula $(C_n H_{2n-10} SO_3)_x H_x$, where n is >20. The hydrocarbon radical consists of three to six fused rings to which alkyl groups are attached. Oil-soluble petroleum sulfonates are of interest when it is desired to dissolve an emulsifier in the disperse rather than the water phase.

Lignin is a constituent of wood and is a by-product in the manufacture of cellulose. Its composition, as far as has been ascertained, is a complex condensed phenyl propane including phenolic and methoxy groups. It can be sulfonated to form lignosulfonates, which are low-cost commercial materials available as sodium salts. They are useful as pigment dispersants.

3.2.6 Aliphatic chain sulfonates

The direct formation of sulfonates from aliphatic hydrocarbons is accomplished by the joint action of sulfur dioxide and chlorine under the influence of actinic light to form a sulfonyl chloride, which is readily hydrolysed:

$$RH + SO_2 + Cl_2 \longrightarrow RSO_2Cl + HCl$$

This product is of considerable interest in Germany.

Two interesting classes of sulfate are based on acyl derivates of isethionic acid $HOC_2H_4SO_3H$ and taurine $NH_2C_2H_4SO_3H$ respectively. The derivatives are the Igepon® A series, $C_n H_{2n+1} COOC_2H_4SO_3Na$, and the Igepon® T series, $C_n H_{2n+1} CON(CH_3)C_2H_4SO_3Na$. Physical forms vary from gels and pastes to flakes or powders. These products have good surface-active properties and are efficient emulsifiers, but the solubility of the products with longer alkyl chains is low, sometimes less than 0.5 % at ambient temperatures. Some technical emulsifiers contain a considerable quantity of salt as diluent, or else unreacted fatty matter.

The isethionic derivatives are stable to hydrolysis at neutral or slightly acid pH levels and at alkaline pH ranges. The taurine derivatives are also stable at pH levels as low as 2, and are therefore useful under acid conditions.

3.2.7 The sulfosuccinates

A versatile range of products is a series derived from maleic acid, by addition of sodium bisulfite to the esters. These are of general formula

$$\begin{array}{c} ROOC.CH_2 \\ | \\ ROOC.CHSO_3Na \end{array}$$

The original products have become well known under the trade mark of Aerosol® (American Cyanamid Company, now Cytec Division), but there are now other manufacturers. Typical products have R = octyl, hexyl, amyl and tridecyl. Aerosol 102® is the sodium salt of a half ester and is

$$\begin{array}{c} CH_2COO(OC_2H_4)OC_{10-12}H_{21-25} \\ | \\ NaOOCCHSO_3Na \end{array}$$

The acid from which the series is derived,

$$\begin{array}{c} CH_2COOH \\ | \\ CH(SO_3H)COOH \end{array}$$

is usually known as sulfosuccinic acid.

More complex derivatives available include tetrasodium N-(1 : 2-dicarboxylethyl)-β-octadecylsulfosuccinamate:

$$\begin{array}{c} CH_2COONa \\ | \\ CHCOONa \\ | \\ CH_2CON-C_{18}H_{37} \\ | \\ NaO_3SCHCOONa \end{array}$$

which is known as Aerosol 22®. An allied surfactant is disodium N-octadecyl sulfosuccinamate (Aerosol® 18).

The series has the important advantage that most emulsifiers are available as 100 % active solids materials, possibly diluted with water or isopropanol.

A survey of technical surfactants 65

Solubility at 25 °C of sodium dioctyl sulfosuccinate, the most well-known member of the series, is restricted, being 1.5 % but the dihexyl sulfosuccinate has a solubility of 34.3 g in 100 ml of water, and the dibutyl compound a corresponding solubility of 39.2 %. Ambient solubility for bistridecyl sulfosuccinate is only 0.1 %, increasing with temperature to 5.5 g per 100 ml of water. However, it forms stable dispersions fairly readily in water. Equally, the sulfosuccinate can be diluted with water to form pourable gels, particularly if diluted with a polar solvent such as isopropanol.

The tolerance to electrolytes increases with reduction in length of the alkyl chain. The organic solvent solubility of the sulfosuccinates increases with the length of the alkyl chain, the dioctyl compound being soluble even in non-polar solvents such as benzene and kerosene. The bistridecyl ester is also soluble in many organic solvents, including vegetable oils and carnauba wax.

The surface tension falls most rapidly with concentration with the longer alkyl chains, as shown in Figure 2.4. In general, the presence of small percentages of a salt such as sodium sulfate helps to lower the surface tension. Aerosol TR has a surface tension of 29.6 dyn cm^{-1} at 0.1 % concentration by weight.

Aerosol 22® (see above) is of considerable significance in emulsion polymerisation. This may be because it contains several carboxyl groups as well as the sulfonate group. It is also of interest as having a low critical micelle concentration at 0.06 % of solids which may assist in obtaining unisize particles.

The technical product, a 35 % solution, is miscible with water in all proportions. It has high electrolyte tolerance, even to saturated sodium chloride solutions. Its reduction in surface tension is moderate, being about 42.5 dynes at

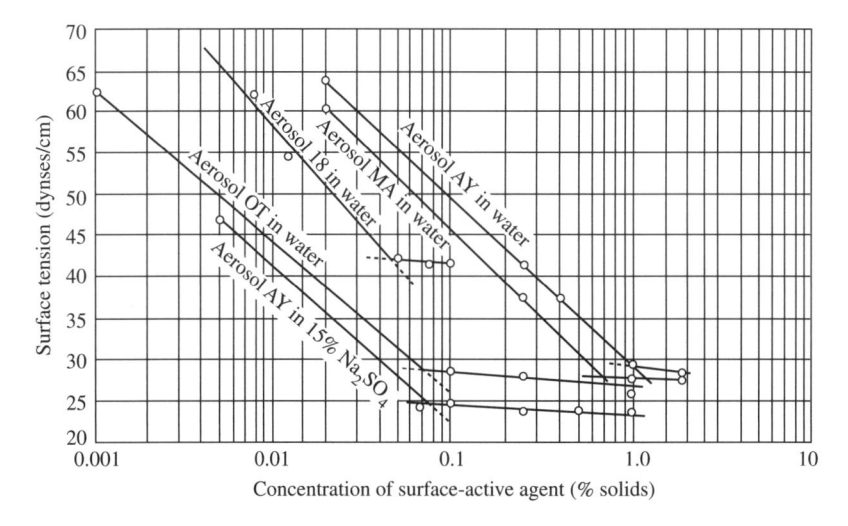

Figure 2.4 Surface tension of solutions of dioctyl sulfosuccinate esters: temperature 25 °C; Du Nouy tensiometer; Aerosol® OT, MA and Y are the dioctyl, dihexyl and diamyl esters respectively.

the critical micelle concentration. It is generally insoluble in organic solvents, although dilutable with the most water soluble such as ethanol.

The main interest of this sulfosuccinamate is as a solubilising agent, enabling slightly soluble organic compounds to have increased water solubility, a property it shares with other hydrophilic compounds such as sodium p-toluene sulphonate p-$CH_3C_6H_4SO_3Na$, and which may account for its use in emulsion polymerisation.

The sulfosuccinates are hydrolysed in acid or alkaline solutions, although stable at moderate pH levels.

3.3 Hydrotropes

Hydrotropes are compounds that will solubilise or increase the solubility of organic compounds in water or in aqueous salt solutions. They have a synergistic effect on solubility of organic compounds and will even solubilise compounds such as hydrocarbons and alcohols from long-chain hydrocarbons which are virtually insoluble in water. An interesting application is the extraction of lignin from wood pulp. The hydrotropes are allied chemically to the standard surfactants, in particular the sulfates and sulfonates, and most technical products consist of either shorter chain sulfates and sulfonates or short-chain alkylbenzene sulfonates.

Typical products are branched chain sodium octyl sulfate, $C_4H_9CH(C_2H_5)$-CH_2SO_4Na sodium p-toluene sulfonate p-$CH_3C_6H_4SO_3Na$ and sodium xylene sulfonate $(CH_3)_2C_6H_3SO_3Na$. They are available as powders or solutions, sometimes as the free acids or the ammonium salts.

Like the surfactants, the hydrotropes have the property of reducing surface tension of solutions, although to a limited extent, especially where there are only seven carbon atoms. Equally, the surfactants with a smaller number of carbon atoms, especially C_{12}, have an appreciable solubilising, as well as emulsifying, effect. Sulfates based on C_{10} are intermediate in properties. The property of solubilisation is important, as will be seen later, in emulsion polymerisation, and is often neglected in theoretical discussions.

3.4 Phosphates

Since phosphoric acid H_3PO_4 is a tribasic acid, it is possible to derive amphipathic molecules by substituting one or two hydrogen atoms by alkyl groups. In technical practice the phosphate-based surfactants are usually a mixture of mono- or dialkyl phosphates and are often available in the acid state, the products being neutralised as required.

In a modified version, the alcohols are partially ethoxylated. Typical formulae are $(RO)_2POOH$ and $ROPO(OH)_2$, where R is an alkyl group. If the molecules are partially ethoxylated the formulae become $[RO(C_2H_4O)_n]_2PO.OH$ and $RO(C_2H_4O)_nPO(OH)_2$, where R is an alkyl group and n is a small integer, normally 1–3. In a further modification, all or part of the acid groupings are modified to derivatives of phosphoramide $OPNH_2(OH)_2$. Some of the formulae are more complex and are derived from polyphosphoric acids.

The physical properties of these products vary according to the formulation. Most of the acids are brownish, fairly viscous liquids, although with increasing length of R they may be waxes. Water solubility varies with the hydrophile balance. Thus dodecyl hydrogen phosphate $(C_{10}H_{21}O)_2P(O)OH$ is a clear light-yellow liquid that is insoluble in water but soluble in most organic solvents. The disodium salt is water soluble, as is the disodium salt of di(2-ethylhexyl)sodium phosphate. These products have marked effects in reducing surface and interfacial tensions, a 1 % solution of di(2-ethylhexyl)phosphate having a surface tension of 32 dyn.

These phosphate derivatives tend to become insoluble in acid solutions, and should not be used below pH 5. A tendency also arises for precipitation to occur in the presence of calcium or magnesium ions, although this can be suppressed by additions of sodium tripolyphosphate. These considerations may not apply to phosphate-type surfactants in which a number of ethylene oxide groups are present. These phosphates may assist fire-resistant properties of compositions with which they are compounded.

3.5 Fluorochemical emulsifiers

Fluorochemical surfactants are a class in which the hydrogen content of the hydrophobic portion has been replaced by fluorine. They contain shorter chains than the more conventional types. Many are based on perfluorocaprylic acid $C_7F_{15}COOH$ and the perfluorosulfonic acid $C_8F_{12}SO_3H$, of corresponding chain length. While technical literature available from manufacturers does not always reveal the formulae of the various fluorocarbon surfactants, the general chemistry of these products and their surface tension reducing properties have been described [7].

The sulfonic acid can be modified via the sulfonamide to sulfonamido aliphatic acid $C_8F_{17}SO_2N(R)CH_2COOH$ and its salts. It can also be modified via the sulphonamide to a non-ionic surfactant $C_8F_{17}SO_2NR_2(C_2H_4O)_n$ H, which is extremely potent as a surface tension reductant, and both cationic surfactants and amphoteric derivatives such as $C_8F_{17}SO_2N(C_2H_5)CH_2COOK$, shown here as its potassium salt are technical products. The carboxylic acid can be subject to similar modifications, including a cationic derivative, $C_7F_{15}CONHC_2H_6N^+(CH)_3I^-$.

Fluorochemical surfactants are unique in reducing the surface tensions of water to 15–20 dyn at only 0.05 % of surfactant (Table 2.1). As the current cost is about 10 times that of more conventional surfactants, this is highly desirable. Solubilities in water are fairly restricted in most cases, being about 0.1 %, but are considerably increased by the addition of a cosolvent such as acetone or isopropanol. These surfactants show in general remarkable stability

Table 2.1 Surface tension of fluoroalkyl aqueous solutions

Concentration (%)	Zonyl S-11 (dyn cm^{-1})	Zonyl S-13 (dyn cm^{-1})
0.003	44.5	43.1
0.01	40.2	33.5
0.02	37.0	31.4
0.05	25.7	29.0
0.1	22.8	27.4
0.2	19.4	25.8
1.0	21.6	21.8

From data of E.I. Du Pont de Nemours.

in the presence of both strong alkalis and strong acids, some showing better reduction of surface tension in acid solution than in neutral solution.

Fluorocarbon surfactants have remarkably selective effects. In some cases they are preferentially absorbed on a surface and effectively decrease the wettability of that surface by the solution. Certain products will cause the rate of attack of hydrochloric acid on aluminium to increase greatly, while others will form a corrosion-resistant film. Fluorocarbon surfactants are used in polish formulations to assist in levelling and flow out, a result of the good surface-active properties.

Some interesting products combine the properties of the fluorochemical surfactants with those of the phosphates. An ammonium salt of a C_7 fluoroalcohol phosphate has been made available technically, as has the acid phosphate of mixed fluoroalcohols (Zonyl S-13 and Zonyl S-11 respectively; E.I. Du Pont de Nemours). The ammonium salt is readily soluble in water and partly soluble in the lower alcohols, because whilst the bifluoroalcohol ammonium salt is soluble, the monofluoroalcohol salt is insoluble in these alcohols. These products exhibit good stability to 10 % sulphuric acid at ambient temperature, but are slowly hydrolysed at the boil. Table 2.1 shows the surface tension of fluoroalkylphosphate solutions. Other fluorochemical emulsifiers are available from the 3M Comany (USA).

Allied to the above series is a phosphonate, γ-hydroperfluoroalkyl phosphonate $H(CF_2CF_2)_n PO(OH)_2$, where n is 4 or 6. It is a tan, semi-solid hygroscopic grease and a strong dibasic acid, whose sodium salt is very soluble in water. Like the above, it has powerful surface tension reducing quality.

These products may be of interest in emulsion polymerisation owing to the virtual absence of chain termination properties.

3.6 Non-ionic types

The solubility of ethylene oxide based surfactants increases with increasing length of the hydrophilic chain, which is usually a Gaussian blend of chains of a number of different lengths. The physical state is usually liquid for chains below about 20 ethylene oxide units and solid for chains where a substituted phenol is the hydrophobic portion. Where the hydrophobic portion is a long alkyl chain such as cetyl, the product is solid with fewer ethylene chains.

These non-ionic emulsifiers develop solubility at ambient temperature with increasing length of the ethylene oxide chain, but the shortest chains, with only about four ethylene oxide units, often have solvent solubility. These surfactants, except the most hydrophilic, tend to become turbid on raising the temperature, since micellar size increases with temperature until it becomes optically visible. Table 2.2 shows some variation in the properties of a technical range of non-ionic surfactants of this type. The 'nonyl' grouping quoted in these products is a branched chain surfactant. A frequently encountered alternative series is based on octyl cresol.

Table 2.2 indicates that about 16 moles of ethylene oxide are required in the octyl series and 17–18 moles in the nonylphenyl series to keep the cloud point above 100 °C. However, this series is often used with anionic emulsifiers, which apart from a marked synergistic effect as far as emulsification properties are concerned, tend to raise the cloud point. It should not be assumed that an emulsion stabilised, or partially stabilised, with non-ionic surfactant will automatically destabilize if heated above the cloud point.

Solubility in organic solvents follows inversely the solubility in water, but solubility in polar solvents such as the lower aliphatic alcohols is general, as

Table 2.2 Properties of technical non-ionic surfactants (Union Carbide)

Manufacturer and type	Active ingredient (%)	Average ethylene oxide	Average molecular weight	Surface tension[a] (0.1 % solution)	Cloud point (°C, 1 % solution)
Octylphenyl series					
Triton X-45	99	5	426	28	<0
Triton X-100	99	9–10	628	35	100
Nonylphenol series					
Triton N-57[b]	99	5	440	29	<0
Triton N-101	99	9–10	642	30	54
Triton N-150	70[a]	15	880	35	95
Nonylphenol series					
NP-14[b]	100	4	396	30	<0
NP-27	100	7	528	30	20
NPX	100	10–11	682	31	63
NP-40	100	20	1100	37	100

[a] Properties refer to 100 % material.
[b] not completely water soluble.

Note. The Triton series were formerly produced by Rohm & Haas Company. Some of the above data are quoted from earlier Rohm & Haas data.

is solubility in aromatic hydrocarbons. Compounds with up to five ethylene oxide units are also miscible to some extent with aliphatic hydrocarbons, and emulsification in mixed emulsifier systems is often achieved by adding the ethoxy derivative to the disperse phase. This is of interest in emulsion polymerisation, where a relatively hydrophobic non-ionic emulsifier may be dissolved in monomer and a more hydrophilic type dissolved in the water.

Because of steric considerations, especially the flat benzene ring which causes the emulsifier to lay flat on the surface of the disperse phase, the alkyl benzene ethoxylates make very good emulsifying agents. The types including a straight alkyl chain are also very useful.

Differences may occur between commercial products which are nominally of the same composition. This may be partly due to variations in the Gaussian scatter of the number of ethoxy groups, partly to polyethylene oxide units that have not been etherified and to variations in the hydrophobe.

Another interesting type is based on copolymers of ethylene oxide with propylene oxide of general formula $HO(C_2H_4O)_a(C_3H_6O)_b(C_2H_6O)_cH$ (a, b and c are integers). These products are based on the fact that as polypropylene oxide increases in molecular weight it becomes more hydrophobic and insoluble. The block copolymers of the type shown can thus be 'tailored' to give varying hydrophile–hydrophobe balances, and the average molecular weight of these products is much greater than most other surfactants, of the order of 2000–8000. The ethylene oxide content by molar proportions varies from 40 to 80 %. These block polymers have proved of great interest in emulsion polymerisation, and evidence suggests that monomer molecules may graft on to them rather readily. Some typical properties are shown in Table 2.3, which is based on the Pluronic® series of Badische Wyandotte (USA).

Table 2.3 Properties of non-ionic surfactants (Pluronic®), which are based on the formulation $HO(C_2H_4O)_a(C_3H_6)_b(C_2H_4O)_cH$. Reproduced by permission of BASF

	Manufacturers reference[a]					
	L31	L35	L44	L62D	L64	F68
Molecular weight	1100	1900	2200	2360	2900	8400
Surface tension[b]	41	49	45	43	43	50
Viscosity (cP at 25 °C)	175	375	440	385	850	1000
Melting/pour point (°C)	−32	7	16	−1	16	52
Cloud point (1 % aqueous solution, °C)	37	73	65	35	58	>100

[a]In the manufacturer's nomenclature L = liquid, P = paste, F = flakes. The first figure of the references refers to the base molecular weight of the propylene oxide unit and the second to the percentage of ethylene oxide (presumed molar) in the molecule, this being about 10 times the indicated digit. The above are a representative selection of the products available.
[b]In dym cm^{-1} at 25 °C Du Nony tensiometer.
From data of BASF.

Several other modifications of the non-ionic series are available commercially. In one case acetylenic alcohols are used as the hydrophobic group, i.e. the Surfonic® series of Air Products and Chemicals, Inc. (USA), which includes surfactants of the type of dimethyl octynediol and tetramethyl decynediol condensed with ethoxyl groups. In an unusual case a mercaptan is used rather than an alcohol or phenol in forming the molecule, thus giving a product of type $RS(C_2H_4O)_nH$.

A series of non-ionic surfactants of a different type have been obtained by dehydration of the polyhydric alcohol sorbitol in several stages and in a controlled reaction with a fatty acid. The formula of typical sorbitan esters is shown in the condensed form A and uncondensed forms B1 and B2:

Form A

$$\begin{array}{c} \text{HO-CH-CH} \quad \text{O} \quad \text{CH}_2 \\ \quad | \qquad | \qquad\qquad | \\ \text{CH}_2 \quad \text{CH-CH-OOCR} \\ \quad \diagdown \text{O} \diagup \end{array}$$

Form B

$$B1 \quad \begin{array}{c} \text{O} \\ \text{CH}_2 \quad \text{CHCH}_2\text{COOR} \\ | \qquad\qquad | \\ \text{HOCH} \quad \text{CHOH} \\ \diagdown \text{CHOH} \end{array} \qquad B2 \quad \begin{array}{c} \text{HOCH—CHOH} \\ | \qquad\qquad | \\ \text{H}_2\text{C} \quad \text{CHCHCH}_2\text{OOCR} \\ \diagdown \text{O} \diagup \end{array}$$

R = a typical alkyl chain from C12 to C18 in the various sorbitan ester derivativ

By oxyethylating the residual hydroxyl groups, a further series is produced. They are currently available from Seppic Industrial Division (France) and Honeywill and Stein (UK) with the trade marks Montane and Montanox respectively.

3.6.1 Biodegradable emulsifiers

It is considered desirable that for industrial and other applications emulsifiers should be biodegradable. The alkylphenol ethoxylates and their sulfonates are difficult to degrade biologically. Accordingly there is a tendency to replace them with alternatives. The non-ionic emulsifiers are in general replaced by ethoxylated fatty alcohols, not necessarily with straight chains [8–10].

3.7 Cationic types

The cationic types are based on compounds containing positive nitrogen, i.e. amines, although it is possible that some phosphonium or sulfonium compounds may be developed for the same purpose. The simplest types, as already indicated, are the simple long-chain amines and their salts. They are usually manufactured by reduction of the corresponding amide prepared from the fatty acid with the nitrile as an intermediary (RCN). Thus the technical products tend to have the composition of the alkyl groups of the original fatty acid, which is normally a mixture rather than a pure substance. A typical stearyl amine, Armeen® 18D (Akzo Chemie UK Limited, Armour Hess Division), contains 85 % of C_{18} alkyl chain (stearyl), 2.5 % of C_{17} (margaryl) and 12.5 % of C_{16} (palmityl). These amines vary from very dark liquids to pastes or waxy solids. The shorter chain amines have appreciable volatility; technical lauryl amine has a boiling range of 139–147 °C, whilst the octadecyl amines have boiling ranges up to 225 °C.

In a manner analogous to the fatty acids, emulsions may be formed by dissolving the amine in the disperse phase, with solubility in most organic liquids, both polar and non-polar being fairly good, and dissolving either a mineral acid, i.e. hydrochloric acid, or a weak organic acid, i.e. formic or acetic acids, in the water phase. The amine salt may be used directly in the water phase. Bivalent acids should be avoided as they may cause destabilization or a phase reversal. Because of the weakness of the long-chain alkyl amines, emulsions for which these are the sole stabilisers are unstable at neutral or high pH values. The surface tension reduction of these products is marked, especially for the amine salts of saturated alkyl chains.

The long-chain secondary of general formula RR^1NH in the salt form may be used as an emulsifier, as may the tertiary amines RR^1RN. The latter have the advantage of being strong bases and are therefore stable over a wide pH range. A cyclic compound available technically is formed by condensing oleic acid with 2-aminoethanol amine, forming an alkyl hydroxyethylimidazole:

$$C_{18}H_{35}.C{:}N{-}CH_2$$
$$\phantom{C_{18}H_{35}.C:}N{-}CH_2$$
$$\phantom{C_{18}H_{35}.C:N-}C_2H_4OH$$

The quaternary substituted ammonium salts are among the best known of the cationic series and are available in the bromide form. Typical products are lauryl trimethyl ammonium bromide and cetyl pyridinium bromide:

$$\begin{array}{c} CH \\ HCCH \\ \parallel\mid \\ HCCH \\ C_{16}H_{33}{-}N^+Br^- \end{array}$$

These products have moderate efficiency in reducing surface tension, but have a wide pH stability range. Surface-active cationic compounds with two amino groups are also possible, as in the alkylpropylene $RNHCH_2CH_2CH_2NH_2$, which can form water-soluble salts such as the diacetate.

The fatty acids and their modifications may be reacted with ethylene oxide to form derivatives of slightly weaker basicity, but combining the properties of

A survey of technical surfactants

both the cationic and anionic emulsifiers. They have better pH stability and in some cases have compatibility with anionic emulsifiers. A typical formula is

$$\underset{(CH_2CH_2O)_zH}{\overset{RNCH_2CH_2CH_2N(CH_2CH_2O)_xH}{|}} \quad (CH_2CH_2O)_yH$$

where R represents an alkyl group and x, y and z are small integers.

3.8 Amphoteric emulsifiers

Although a large range of amphoteric emulsifiers has been made available, their practical application for emulsion polymerisation and allied pigmented has been limited. The betaines can be represented as

$$(CH_3)_2{}^+N\!\!-\!\!C_xH_{2x+1}$$
$$\phantom{(CH_3)_2{}^+N\!\!-\!\!}CH_2COO^-$$

where x is usually 12–18. Cetyl betaine $[C_{16}H_{33}N^+(CH_3)_2(CH_2COO^-)]$ is a typical example, being normally available at pH 7–9. The reduction in surface tension is very efficient, and it is remarkable that at 0.025 % a typical product has a surface tension of 28.9 dyn cm^{-1}, but increases to 31.7 dyn cm^{-1} at 1 % concentration. These compounds with a dual charge are known as 'zwitterions', the overall charge depending on the pH. At a low pH, ionisation of the carboxyl groups is suppressed and the betaine is cationic. Near the isoelectric point, often about pH 5, the compounds act as if they are non-ionic, but at high pH levels they become anionic.

A further series includes N-lauryl β-aminopropionic acid $C_{12}H_{23}NHCH_2$-CH_2COOH and the corresponding N-lauryl β-aminobutyric acid $C_{12}H_{23}NHCH$-$(CH_3)CH_2COOH$. The free acids are liquids or gels, and the sodium salts are water soluble with some solubility in ethanol and isopropanol. A corresponding series with two carboxyl groups has the formula $C_{12}H_{23}N(CH_2CH_2COOH)_2$. These products are based on a technical 'coco' lauryl alcohol.

A cyclic compound of the series is an alkyl ethyl cycloimidinium-1-hydroxy-3-ethyl sodium alcoholate, 2-methyl sodium carboxylate:

$$\begin{array}{c} N\!\!\diagup\!\!\overset{CH_2}{}\!\!\diagdown\!\!CH_2 \\ R.C\!\!=\!\!N\!\!-\!\!CH_2CH_2ONa \\ OHCH_2COONa \end{array}$$

In these products both the anionic and cationic groups are weak, and the product does not tend to form insoluble internal salts.

Certain amphoteric betaine derivatives contain sulfonic acid groups.

3.9 Polymerisable surfactants

A polymerisable surfactant may be defined as a monomer capable of polymerisation which imparts to a resultant blend subjected to emulsion polymerisation a hydrophile–hydrophobe balance such that the resultant copolymer is self-emulsifiable, possibly with a greatly reduced content of additional surfactant on polymeric stabiliser. Usually quantities of 1–3 % on total monomers are included where there is a strongly hydrophilic group present, e.g. a sulfate or sulfonate, but up to 10 % for weaker hydrophiles such as the carboxyl group. The normal copolymerisation rules apply, as enunciated in Chapter 3, but the water solubility of these monomers may cause some complications with copolymerisation (see Chapter 3).

The principal classes are as follows:

(a) Compounds based on monomers containing sulfate, sulfonate, phosphate or phosphonate groups. Examples are sodium vinyl sulfonate, CH_2CHSO_3Na, sulfoethyl methacrylate $CH_2C(CH_3)COOCH_2CH_2SO_3Na$ and 2-acrylamido-2-methylpropane sulfonic acid $CH_2{:}CHCONHC(CH_3)_2CH_2SO_3H$, usually as the sodium salt.

(b) Monomers including carboxyl groups and their salts, e.g. acrylic acid and methacrylic acid. These have already been considered in Chapter 1.

(c) Monomers with multiple hydroxyl or ether groups, and including an unsaturated radical, e.g. a polyoxethylene methacrylate. An example of a mono-ester malcate of a 5 mol ethoxylated *p*-nonylphenol ethoxylate is $C_9H_{19}O(CH_2CH_2O)_5O.C({:}O).C{:}C.C(O)OH$. This compound is a slightly yellow waxy solid. It is 50 % soluble in water at 25 °C, has a pH of 3.3 and appreciable surface activity. There is slight solubility in most standard monomers [11].

A recent non-ionic polymerisable surfactant is the sequential addition of propylene oxide and ethylene oxide with diallylamine, which is used as the sole surfactant in emulsion polymerisation (Chapter 3) [12].

(d) Cationic types including monomers with various amino groups which may be quaternary. An example is diethylaminoethyl methacrylate $(CH_3)_2NC_2H_4OOCC(CH_3){:}CH_2$. This monomer produces cationically charged latices at low pH levels, below about 6, but non-ionic ones at higher pH 7.

For recent reviews, see references [13] to [17]. For supplementary references see the end of the chapter.

3.10 Dispersing agents

Many of the surfactants already described are also suitable as dispersants for solid pigments. For this function a dispersant should ensure that clusters of pigments are readily broken up into the ultimate discrete particles. Pigments

A survey of technical surfactants

consist essentially of fine solid particles often less then 1 μm average in diameter, but they are not necessarily spherical.

A dispersant should have an active group which is attracted to or absorbed on the surface of the pigment and also a water-soluble portion which assists indirectly by encouraging mutual repulsion of particles. Some mechanical energy is necessary in separating the particles, which enables an even pigmented composition to be obtained and assists in obtaining the full tinctorial strength of the pigment. Efficient dispersion is often accompanied by a marked decrease in the viscosity, which is partly due to the mechanical breakdown of agglomerates and partly due to the absorbed dispersant preventing 'bridging' of the pigment particles. Many polyphosphates are of use as dispersants. They include sodium hexametaphosphate $Na_4P_2O_7$, one of the most efficient, which in many cases displaces other dispersants already absorbed on to the particles.

Another well-known dispersant is a sulfonated condensate of naphthalene with formaldehyde. It is a mixture, but the principal component is

$NaSO_3$—[naphthalene]—CH_2—[naphthalene]—SO_3Na / [naphthalene]—CH_2—[naphthalene]—SO_3Na

This product, which in some preparations has increased condensation of the naphthalene ring, is available from several manufacturers and is water soluble but has negligible surface activity.

Several polymers and copolymers of acrylic and methacrylic acid are used both as dispersants and emulsifiers. They may vary in molecular weight and in the nature of the comonomer, e.g. methyl methacrylate. Salts of a copolymer of diisobutylene and maleic anhydride in the hydrolysed form as a maleic acid copolymer salt are well known as dispersants. Under some exceptional conditions dispersants for solids may also act as coemulsifiers in emulsion polymerisation. It may be assumed that they are absorbed on to the polymer particles in a manner similar to solid particles. Textbooks on organic coatings give more details on the nature of the surface stabilization of pigment dispersions [17–18].

3.11 Solid dispersants

Under some conditions a number of inorganic substances which have swelling properties in water may be used as auxiliary stabilisers and thickeners, imparting thixotropic properties. These include the bentonite and montmorillonite clays, a modification of which, Veegum®, is mined in the United States. These are complex magnesium aluminium silicates with some alkali. Some grades of colloidal alumina $Al_2O_3.xH_2O$ behave similarly. They swell in water to many times their own volume.

A 5 % dispersion has a viscosity of about 250 cP, increasing with age. Combinations with some cellulose ethers show synergistic increases in viscosity. These montmorillonite-type clays have some applications in polishes and in emulsion paints. Bentone® CT is a hectorite clay of this type, and Bentone® 4000 an organically modified hectorite (Rheox Inc. and Rheox Europe) [19].

Laponite, which is available in several grades in the UK from Laporte Industries, has a range of structured magnesium silicates. Some magnesium is substituted by lithium and some lattice positions are negatively charged. The charges are balanced by exchangeable sodium ions situated outside the lattice between the platelets. These platelets are approximately 1 nm thick and exist as aggregates held together by van der Waals forces. When Laponite is dispersed in water, the sodium ions diffuse away from the platelets. As hydration continues the clay swells, the platelets ultimately separating completely to give a clear colloidal dispersion. The platelets have a surface negative charge and the edges a small positive charge. As concentration increases, the surface-to-edge attraction causes gelling, but as the attractions are easily broken there is a strong thixotropic effect [20].

4 COLLOIDS AND HIGH MOLECULAR WEIGHT STABILISERS

Natural colloids have molecular weights of the order of 10^6. In general they consist of polysaccharides, often complex. Most of the formulae have been elucidated since about 1950. They have long found application in general emulsion technology, mainly in pharmaceuticals.

4.1 Starch and dextrin

Often in its degraded form dextrin, starch is frequently included as a stabiliser in emulsions. The empirical formula of both starch and dextrin is $(C_6H_{10}O_5)_n$, being based on anhydroglucose units. Variation is possible in starches and dextrins because starch has two possible chemical forms, dependent on the nature of the linkage between units. These are known as amylose, the straight chain type with entirely 1 : 4 glycosidic linkages:

and amylopectin, the branched chain variation with some 1 : 6 glycoside linkages:

Colloids and high molecular weight stabilisers

Whilst the two classes in any one natural starch may be separated by suitable solvents such as butanol/water mixtures, it is best to utilise purely botanical methods for production; e.g. sorghum starch is nearly pure amylopectin. Amylopectin has a number of advantages in emulsion manufacture, in particular emulsion polymerisation, as it does not 'revert', i.e. become insoluble, on standing due to crystallisation.

Amylopectin, which has a side chain every 20–30 anhydroglucose units, may have a molecular weight of up of 1 000 000, but the disorder caused by the branches ensures that solutions are both more readily obtainable and are of much lower viscosity than those using amylose. A 30 % solution of an amylopectin starch is possible at ambient temperature.

Another method of obtaining a starch for use as a colloid stabiliser or an emulsion thickener is by etherification of part of the CH_2OH groups. This increases molecular disorder, and thus aids solubility, as with the cellulose ethers, described in the following section.

4.2 Cellulose ethers

Cellulose is not water soluble and has the same empirical formula as starch, but varies in the nature of the linkage, which is α-glucosidic in starch and β-glucosidic in cellulose. The difference between the starch and cellulose molecules is shown below:

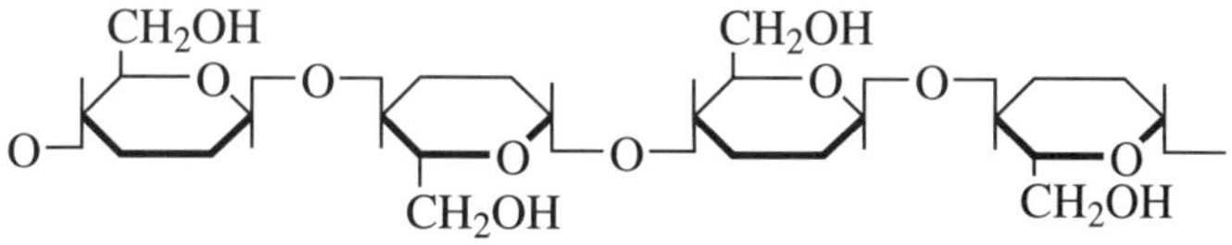

Cellulose

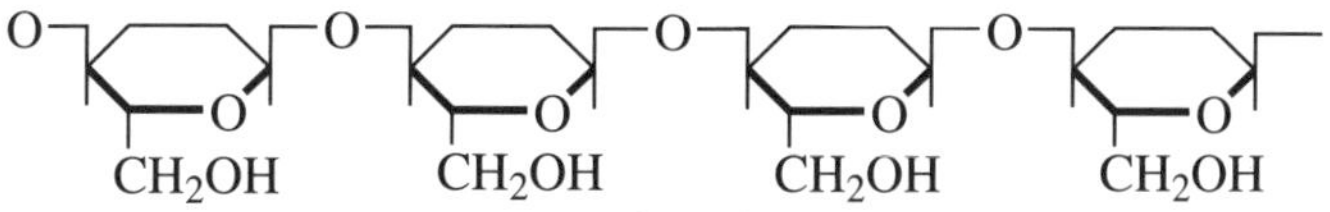

Starch

The unit of the cellulose molecule is

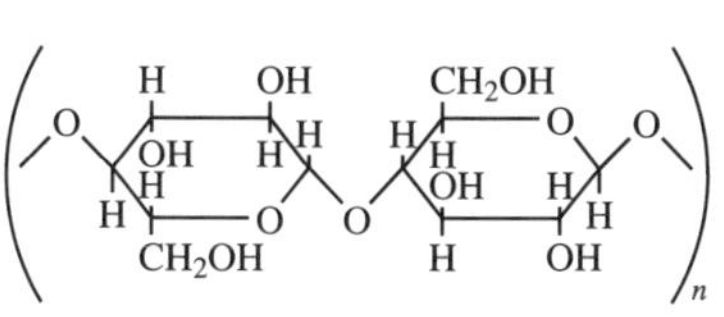

Whilst cellulose is insoluble in water, some derivatives of cellulose are water soluble and promote the formation of stable emulsions, particularly in conjunction with a low molecular weight emulsifier. The most useful of these in polymer stabilization is hydroxyethyl cellulose, in which between two and three of the available hydroxyls on each anhydroglucose ring have been ethoxylated, i.e. replaced by .OC_2H_4OH, a process analogous to the formation of the non-ionic surfactants previously discussed. In most cases only one ethylene oxide occurs per reacted hydroxyl group. These modified celluloses are of much lower molecular weights than cellulose proper.

Other derivatives are methyl cellulose, in which the methyl ethers of the hydroxyl groups are formed. These, however, have the disadvantage that they cease to be soluble above about 55 °C, coming out of solution in gel form. Methyl cellulose is thus more useful as a thickening agent than as an emulsifier, and is used mainly as an additive to pigment dispersions (see Chapter 4).

There are a number of other modifications. These include a mixed methyl-hydroxyethyl cellulose and hydroxypropyl cellulose, in both of which the tendency to gelation is below the boiling point of water. They could therefore only be used with caution during a polymerisation, but are useful as thickeners and with pigment dispersants. Sodium carboxymethyl cellulose, however, is soluble in water up to the boiling point. Most grades of water-soluble substituted cellulose derivatives require a preservative to prevent microbial breakdown.

4.2.1 Properties of technical cellulose derivatives

Whilst cellulose is insoluble in water, many derivatives of cellulose, in particular various ethers, are water soluble and promote the formation of stable polymer emulsions, particularly in conjunction with a low molecular weight emulsifier. They also have a major application in pigmented compositions including latices.

4.2.2 Methyl cellulose

The oldest of the commercial water-soluble derivatives is methyl cellulose, the simple ether of cellulose:

Methyl cellulose

It may be manufactured by the action of methyl chloride on the alkali derivative of cellulose, and the resultant salt is washed out with hot water in which methyl cellulose is insoluble. It is available technically as a powder, in granular form or as fibres. The degree of substitution (abbreviated DS) varies from 1.2 to 1.9. Commercial products have a degree of polymerisation from about 50,

Colloids and high molecular weight stabilisers

corresponding to a molecular weight of about 10 000 to 2000, which gives a maximum molecular weight of about 250 000. The viscosity of a 2 % aqueous solution varies from 17 to 100 000 cP. Solution is a simple matter provided that adequate stirring is available. The physical appearance, as in other cellulose derivatives, is a white-to-cream material, which may contain about 7 % of absorbed moisture. There may be about 1 % of residual salt content.

The solutions, particularly the high-viscosity variations, show some pseudoplastic properties. Little change in viscosity occurs with pH variation, but a moderate reduction in surface tension (ST) is caused, a 1 % solution having an ST of 54 dyn. Methyl cellulose precipitates from water when heated, the exact temperature varying according to the grade and averaging about 55 °C. Methyl cellulose is frequently included in emulsion paints and other pigmented compositions to provide viscosity, mechanical stability and good brushing properties. It is reasonably tolerant to small concentrations of salts.

4.2.3 Hydroxyethyl cellulose

This is the most useful derivative in emulsion polymerisation. An idealised formula is shown below:

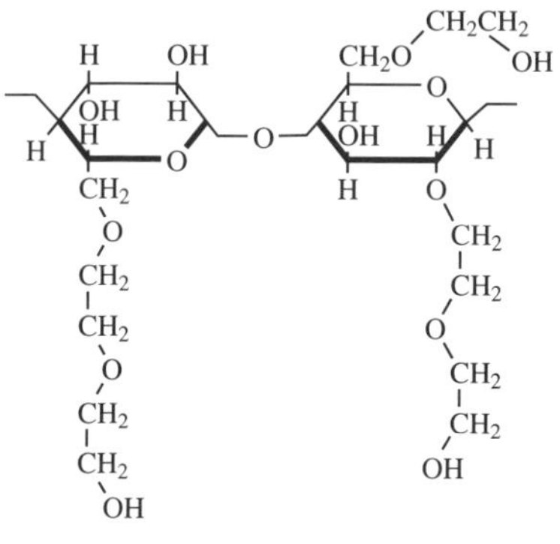

Hydroxyethyl cellulose

Hydroxyethyl cellulose is manufactured by the reaction of ethylene oxide C_2H_4O with sodium cellulose. The manufacturing process causes some degradation of natural cellulose, and molecular weights of all synthetic water-soluble cellulose derivatives are probably between 50 000 and 500 000. Technical products are available as fine granules passing a 40-mesh or 80-mesh sieve, and have a wide range of viscosities. The products usually contain up to 5 % of moisture, have a density of 1.38–1.40 and are white to pale cream in colour. A typical viscosity range of these products is shown in Table 2.4. Measurement is with the Brookfield viscometer, using a spindle and a speed suitable for the viscosity range.

It is possible for ethylene oxide to add on to both the hydroxyls of the cellulose units and also the hydroxyls at the end of the ethoxyethylene groups already added. Thus, a chemical description would define both the degree of substitution (DS), implying reaction with the anhydroglucose unit, and molar substitution (MS), which is the average number of C_2H_4O that have reacted with each anhydroglucose unit. Of two major products available to industry, one (Natrosol, Aqualon, USA and Netherlands) has an MS of 2.5, whilst the other (Cellosize, Union Carbide) has a DS of 1.9–2.0 and an MS of 1.6–2.0.

The manufacturing process causes some degradation of natural cellulose, and the molecular weights of synthetic water-soluble cellulose derivatives, as shown in Table 2.4, are between about 90 000 and 1 300 000. Technical products are available as fine granules, the bulk passing through either 40-mesh or 80-mesh sieves. The products usually contain up to 5 % of moisture, have an actual density of 1.38–1.40, a bulk density about 0.6, are neutral in pH and are white to pale cream in colour.

Technical hydroxyethyl cellulose is best dissolved, as with all cellulose derivatives, by adding the powder slowly to the vortex of rapidly stirring water, hot or cold. Some time is required to effect complete solution. Whilst a 10 % solution of the lowest viscosity grades can be achieved, the highest viscosity types will only form a 1 % solution without gelling. Solubility, however, is not lost, even at 100 °C.

Increased substitution of the anhydroglucose units gives increased resistance to degradation by the enzyme cellulase. Some technical enzyme-resistant grades are manufactured, e.g. Cellosize® ER HEC (Union Carbide) and the Natrosol® B series (Aqualon).

Figure 2.5 shows the effect of concentration on the viscosity of hydroxyethyl cellulose solutions. Figure 2.6 shows the effect of shear rate on the

Table 2.4 Viscosity of hydroxyethyl cellulose grades (Natrosol). Reproduced by permission of Hercules International United

		Viscosity (Brookfield 25 °C)		
Type	Molecular weight	1 % solution	2 % solution	5 % solution
250HHR	3400–5000	—	—	1.3×10^6
250HR	1500–2500	—	—	1.0×10^6
250MR	—	4500–6500	—	7.2×10^5
250GR	—	150–400	—	3.0×10^5
250JR	—	—	150–400	—
250LR	—	—	75–100	9.0×10^4

Note. The above are typical ranges of a technical product and are not necessarily specifications. Measurement is with the Brookfield viscometer series using a spindle and speed suitable for the viscosity range.
From data in technical literature of Aqualon Company (USA).

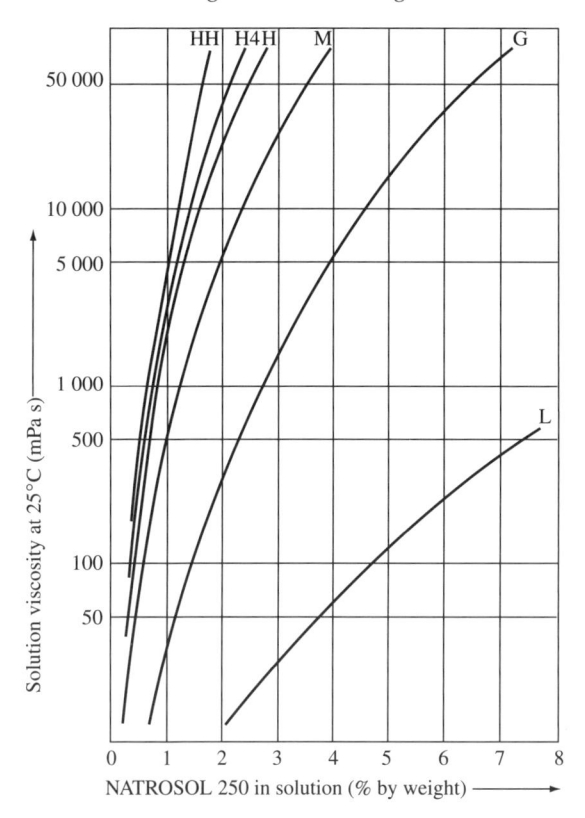

Figure 2.5 Effect of concentration on various grades of hydroxyethyl cellulose (Natrosol®). (Reproduced by permission of Hercules International Limited.)

viscosity of hydroxyethyl cellulose solution. Figure 2.6 illustrates the effect of shear rate on the viscosity of technical hydroxyethyl cellulose solutions. These solutions are pseudoplastic, the viscosity varying with the shear rate (Figure 2.6). A slight reduction of surface tension occurs, solutions being about 63–69 dyn against 72.4 dyn for pure water. Viscosity does not vary greatly with pH, and there is only a moderate temperature coefficient.

The lowest viscosity grades are normally used for emulsion polymerisation, a process that degrades higher molecular weight products to a fairly constant low viscosity, presumably through a peroxide attack on weak linkages in the cellulose chain. If various hydroxyethyl celluloses, or indeed other water-soluble cellulose derivatives are blended, a logarithmic function must be used to estimate the resultant viscosity.

A relatively recent innovation is a modified hydroxyethyl cellulose also containing some long-chain alkyl groups, e.g. Natrosol® Plus HMHEC, Grade 330 (Aqualon). The high viscosity of aqueous solutions is obtained by

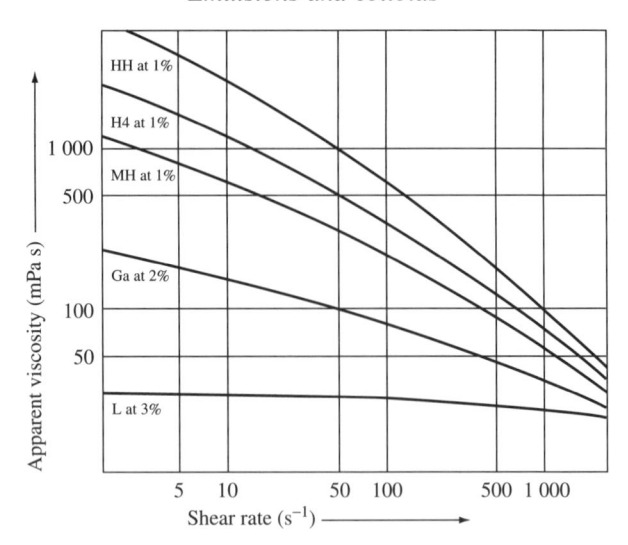

Figure 2.6 Effect of shear on various grades of hydroxyethyl cellulose (Natrosol®). (Reproduced Aqualon by permission of Hercules International Limited.)

association of hydrophobes as well as hydrogen bonding with water molecules. Mention of this type of product has already been mentioned in Section 3.9.

4.2.4 Hydroxypropyl cellulose

This is a variant of hydroxyethyl cellulose in which propylene oxide is used instead of ethylene oxide in the manufacturing process, resulting in a product whose side chains contain only secondary hydroxyl groups, as illustrated in the idealised formula on pg. 79, in which the MS is 3.0. The product physically resembles hydroxyethyl cellulose and has a limited moisture absorbency.

One major distinction between hydroxypropyl cellulose and hydroxyethyl cellulose is that the former tends to precipitate from water above 45 °C for the lowest molecular weights and 40 °C for the highest. This insolubility in hot water prevents the use of the product in emulsion polymerisation other than in exceptional circumstances. A further distinction is the strong surface activity of the product, the aqueous solution at 0.1 % having a surface tension of 45 dyn (Klucel®, Aqualon). Hydroxypropyl cellulose is also readily soluble in a variety of organic solvents, including methanol, ethanol and ethylene chlorohydrin, whilst acetone and chloroform tend to be borderline solvents.

Solution is by the same method as for hydroxyethyl cellulose, though other methods such as cooling a hot slurry with vigorous stirring are available. The product is recommended as a formulation additive in emulsion paints. It has the advantage of solubility in the glycols which are sometimes used as additives.

4.2.5 Carboxymethyl cellulose

This product, or more correctly sodium carboxymethyl cellulose, differs from all the above in being ionic in character. A simplified formula is

[Chemical structure diagram of sodium carboxymethyl cellulose]

It is manufactured by the reaction of sodium cellulose with chloroacetic acid, and is available in a wide degree of substitutions, products being specified as containing from 0.38 units of side chain per anhydroglucose unit up to 0.9 and exceptionally up to 1.4. The usual wide range of molecular weights, and hence viscosities, is available, with medium viscosity grades, those giving a viscosity of 100–1000 cP at 2 % aqueous solution, being preferred for most applications involving pigmented compositions. The aqueous solutions, which are easily prepared, are pseudoplastic with a tendency towards thixotropy. Technical products may have appreciable residual salt content.

Unlike methyl cellulose, there is no tendency to precipitation on heating, presumably because of the ionic nature of the product, but the solution viscosity falls markedly on heating. There is little effect on stability over the pH range 5–11. At low pH levels, about 2–3, the free acid of carboxymethyl cellulose tends to precipitate. Note that the abbreviation CMC should be avoided as it might be confused on occasion with 'critical micelle concentration'.

Sodium carboxymethyl cellulose shows the non-Newtonian behaviour of solutions of rigid particles. It tolerates most salts where the anion will also produce a soluble form of the product. In other cases there is a definite tolerance limit, in some cases above the saturation point of the added salt. There is a wide range of compatibilities with other cellulose products and many other colloids.

4.2.6 Other cellulose derivatives

A number of derivatives consist of combinations of the above modifications of cellulose, different side chains being present in the same molecule and probably in the same anhydroglucose unit. They include:

(a) Methyl ethyl cellulose, in which some of the methyl groups are replaced by ethyl. The properties are similar to those of methyl cellulose.

(b) Ethyl hydroxyethyl cellulose, which is markedly surface active, and soluble in dimethyl sulphoxide, $(CH_3)_2SO$, as well as in water. Trade names include Bermocoll.
(c) Hydroxypropyl methyl cellulose and hydroxyethyl methyl cellulose are further derivatives of commerce.

All of these cellulose water-soluble derivatives are liable to microbiological attack, which causes degradation and may lead to a more major breakdown. Hydroxypropyl cellulose is considered to be the most resistant to biological attack, followed by hydroxyethyl cellulose (but note special grades quoted previously), methyl cellulose and carboxymethyl cellulose, in that order. Various preservatives may be used in conjunction, depending on the use, but the organomercurials and chlorinated phenols are generally effective, as is *tert*-butyl tin oxide for uses where its toxicity is no handicap (see also Chapter 10).

The following is a list of some commercial suppliers:

Great Britain: Courtaulds
United States: Aqualon, Dow Chemical Company, Union Carbide
The Netherlands: Aqualon
Sweden: Berol, Svenska Cellulose AB, Uddeholms AB
Germany: Aqualon, Farbwerke Hoechst

4.3 Other semi-synthetic colloids

Several semi-synthetic products are not derived from starch or cellulose. These semi-synthetic products include alpha and gamma proteins, which are derived from soya bean. Casein, the principal protein of milk, is sometimes used as a stabiliser, but owing to its tendency to putrefy its use is limited.

The principal classes of synthetic stabilisers and colloids of interest in applications of polymer emulsions are now considered.

4.4 Polyvinyl alcohol

Polyvinyl alcohol is the polymer derived from the theoretical vinyl alcohol, $CH_2{:}CHOH$, which is unstable and isomerises to acetaldehyde CH_2CHO. The polymer can, however, be prepared by hydrolysis or alcoholysis with methanol from polyvinyl acetate, and may be virtually completely hydrolysed, i.e. $(CH_2CHOH)_n$, or available as a partially hydrolysed polymer with residual acetate groups which may be written empirically as $(-CH_2CHOH-)_n-(-CH_2CHOOCCH_3-)_m$. Water solubility exists until about 35 per cent of residual acetate groups by weight are included in the polymer. The most frequently encountered variations have about 12 % (molar) acetate groups or about 20 % by weight, i.e. a saponification value

(SV) of about 120. These partially hydrolysed species are the principal ones used in emulsion polymerisation (Chapter 3) because of their better hydrophile–hydrophobe balance [6].

Practically, there is a great deal of variation in technical polyvinyl alcohols, even with the same SV and viscosity specifications. This is partly due to variations in the statistical scatter of molecular weights of the molecular species present, and also in some cases to a statistical scatter of the degree of hydrolysis of the species present. Other variations may be the nature of the end group, differences in 'head-to-head' or 'head-to-tail' polymerisation, whether the residual acetate groups are stripped off in linear fashion during the hydrolysis:

$$HO-CH_2-CH_2-[CH_2-CH(OH)]_m-[CH_2-CHOOCH_3)]_n-$$

or whether the removal of acetate groups in random [6, 21]:

$$HO-CH_2-CH_2CHOH-CH_2-CH(OOCCH_3)-CH_2-$$

$$CH(OOCCH_3)-CH_2CHOH-CH_2-CHOH-CH_2-CHOH-$$

$$CH_2CHOH-CH_2CHOOCHCH_3-$$

In the above, end groups are shown for convenience as hydroxyl. The variations may have a profound effect on the success or otherwise in the use of a partially hydrolysed vinyl acetate polymer in the emulsion polymerisation of vinyl acetate (Chapter 3).

The aqueous solutions of polyvinyl alcohol generally have not marked thixotropy. Viscosity increases with molecular weight and decreases to approximately one-third in the range 15–60 °C. The low viscosity grades tend to give a gel at about 25 % concentration at ambient temperature. Films formed from the aqueous solutions are clear with good tensile strength. The tensile strength of fully hydrolysed polyvinyl alcohol increases with stretching, and this grade has been used for fibre production.

4.4.1 Properties of polyvinyl alcohol as emulsifier

Commercial grades differ in average molecular weight and molecular structure, as already indicated. The fully (>99 %) hydrolysed grades tend to be insoluble in cold water and are used for filming when water resistance is not required, or where subsequent insolubilization (Chapter 7) is a process. The medium–low viscosity grades (12–15 mPa s of 4 % aqueous solution) are used in paper and paperboard coating and sizing, and the highest (55–65 mPa s) as adhesives.

The grades suitable as emulsifiers, in particular emulsion polymerisation, are medium or medium–low (5–6 or 23–27 mPa s), with the medium–low grades being the most suitable for emulsion polymerisation. Some grades are tackified by boration (Chapter 8) giving a very high viscosity, and have the

86   Emulsions and colloids

Table 2.5 Partially hydrolysed grades of polyvinyl alcohol. The table indicates polyvinyl alcohol grades under the Gohsenol trade mark as supplied by the Nippon Synthetic Chemical Industry Company Limited

Grade	Viscosity (cp)^a	Hydrolysis(mol %)
GH-23	48–56	87–89
GH-17	27–33	87–89
GM-14L	16–20	87–89
GL-05	4.8–5.8	87–89
GL-03	3–4	87–89
KH-20	44–52	78.5–81.5
KH-17	32–38	78.5–81.5
KP-08	6–9	71–75

^{a}Viscosity in centipoises of a 4 % aqueous solution at 20 °C by the Hoeppler falling ball method. Volatiles are about 5 %, pH 5–7 and ash maximum 0.7 % in all cases.

principal application in adhesives Table 2.5 gives properties of some technical partially hydrolysed polyvinyl acetate. A publication describes the manufacture of polyvinyl alcohol [22], whilst two recent volumes cover many aspects of its production, properties and uses [23, 24].

4.4.2 Chemically modified polyvinyl alcohols

Some hydrolysed copolymers of vinyl acetate have been described. In general, the acetate groups are hydrolysed to hydroxyl, but the comonomer units withstand the method of hydrolysis. Typical comonomer units are vinyl esters of fatty acids with 10 carbon atoms, in which the carbon atom, α to the carboxyl group, is highly branched [25, 26]. Application, as discussed in the following subsection, is in emulsion polymerisation.

A claim has been made to hydrolyse copolymers in which the comonomers are a substituted acrylamide, e.g. diisobutyl acrylamide or N-t-octyl acrylamide, and monomethyl maleate. They are claimed to impart improved freeze thaw and storage stability when used in emulsion polymerisation (Chapter 4) [27].

Some polyvinyl alcohols have been prepared so that —SH groups are present by including a suitable chain terminator in the initial polymer. They are useful as they function in a redox polymerisation system (Chapter 1, Section 2.1.3) [28]. Acetoacetylated pv alcohol is described [28A].

4.5 Polyelectrolytes

Most polyelectrolytes are vinyl-type polymers prepared by a free radical process. The monomeric acids, e.g. acrylic and methacrylic acids, can

be polymerised in aqueous solution without difficulty. They are generally available as either water-soluble sodium or ammonium salts. The ions of polyacrylic acid and polymethacrylic acid can be represented as $[-CH_2CH-]_m$ or $[-CH_2C(CH_3)-]_n$ respectively.
            $|$ $|$
           COO^- COO^-

Some copolymers of these acids with unsaturated esters are also useful water-soluble colloids, but as the proportion of ester monomer increases, the insolubility of the acid form diminishes, with a result that polyelectrolytes of this type are not normally useful below a pH of 6 [29–34]. Some copolymers form soluble polyelectrolytes in the alkaline state only.

Amongst them are several copolymers including maleic anhydride which tends to alternate with other monomers, for instance the copolymer with styrene is alkali soluble. Copolymers of vinyl acetate and small quantities (about 5 % molar) of crotonic acid (Chapter 1, Section 5.5), available in emulsion form, are also water insoluble below a pH of about 5, but gradually dissolve to salts of a vinyl acetate–crotonic acid copolymer as the pH is increased. Copolymers of vinyl acetate and acrylic acid behave similarly, with the disadvantage that viscosity increases continuously with pH, although this can be overcome by a terpolymerisation with a maleate or fumarate ester [35].

The alternating copolymers of vinyl acetate and maleic anhydride are of interest since they are water soluble, with hydrolysis to the corresponding acid:

$$[-CH(OOCCH_3)-CH_2]-CH-CH-)$$
$$\quad\quad\quad\quad\quad\quad\quad\quad\quad | \quad |$$
$$\quad\quad\quad\quad\quad\quad\quad\quad O{:}C \quad C{:}O$$
$$\quad\quad\quad\quad\quad\quad\quad\quad | \quad\quad |$$
$$\quad\quad\quad\quad\quad\quad\quad\quad OH \quad OH$$

The acid form, however, is not stable in aqueous solution and tends to gel, with liberation of acetic acid. This is presumably due to an internal lactonization and partial cross-linking, forming vinyl maleate units by inter- or intramolecular reaction. Copolymers of ethylene and maleic anhydride, which have been technically available in a range of viscosities, are also water soluble and may be used as such or as salts.

4.6 Other synthetic water-soluble polymers

The polymer of vinyl methyl ether, $(-CH_2-CH-)_n$, is water soluble, but,
                                              $|$
                                           OCH_3
like methyl cellulose, gels from a solution when heated to 33–35 °C. This polymer is also soluble in many organic liquids, and aqueous solubility seems to be associated with hydration of the polymer, as indicated by heat evolution on solution.

88   *Emulsions and colloids*

Polyvinyl pyrrolidone as a monomer (see Chapter 1, Section 5.5) can be represented as

$$\begin{array}{c} H_2C\!-\!CH_2 \\ | \quad\quad | \\ H_2C\diagdown \quad \diagup CO \\ N \\ | \\ CH\!:\!CH_2 \end{array}$$

It is water soluble, as are some of the copolymers with small quantities of other monomers of vinyl acetate. The copolymers are applied both as protective colloids and as dispersing agents.

Cationic polyurethanes, neutralised with formic acid, have been claimed as suitable emulsifiers in emulsion polymerisation [36].

4.7 Associative thickeners

A new type of hydrophobically modified water-soluble polymer has been introduced in recent years. Typical products are a styrene–maleic anhydride copolymer modified with a hydrophobic monomer in small amounts and a hydrophobically modified urethane ethoxylate. The principle is that a water-soluble polymer containing hydrophobic segments will absorb on to hydrophobic surfaces of latex particles so as to form a dynamic network structure, which should provide unique rheology with special reference to emulsion paints (Chapters 9 and 10 in Volume 2).

Thickeners also associate with the dispersed components to inhibit flocculation. Osmotic stabilization of the dispersed components against volume excluded flocculation results in lower shear viscosity. The relative performance of these thickeners is not constant and varies with the nature of the thickeners and surfactants in formulations. Much fundamental work has been performed on the function of these thickeners [37–41].

Other references will be found in Chapter 10, Section 1.4.1 in Volume 2, with special reference to emulsion paints.

REFERENCES

1. P. Becher, *Emulsions, Theory and Practice*, Rheinhold, New York and Chapman & Hall, London, 1957
2. K. Kozo and S. Friberg, *Emulsions and Solubilisation*, Wiley, 1986
3. W.D. Harkins, R.W. Matoon and M.L. Corrin, *J. Colloid Sci.*, **1**, 105–26 (1946)
4. J.H. Schulman and E.C. Cockbain, *Trans. Far. Soc.*, **36**, 651–61 (1940)
5. K. Shinoda and S. Friberg, *Emulsions and Solubilisation*, Wiley, 1986
6. J.E.O. Mayne and H. Warson (Vinyl Products), BP 655,734, 1951; J.E.O. Mayne, H. Warson and R. Levine (Vinyl Products), BP 766,565, 1957; H. Warson, in *Polyvinyl Alcohol*, C.A. Finch (ed.), Soc. Chem. Ind. Monograph No. 30, 1968, pp. 46–76
7. M.C. Allison, *Spec. Chem.*, **4**(3), 23, 26 (1984)
8. C. Baumann and D. Feustel, *Polym., Paint Col. J.*, **186**(4380), 44–8 (1996)

9. A.M. Fernandez *et al.*, in *24th Waterborne, Higher Solids and Powder Coatings Symp.*, New Orleans, 1997, pp. 182–91
10. T. Akagi *et al.* (Kansai Paints), JP 96 253,515,1996; *Chem. Abstr.* **126**, 31830 (1997)
11. T. Sauer and B. Stutzel (Huels), Eur. P 565,825, 1993
12. M.J. Anchor *et al.* (BASF), Eur. P 737,693, 1995; USP 54 78 883, 1995
13. H. Warson, *Polymerisable Surfactants*, Solihull Chemical Services, 1989, 98 pp. + vii
14. J. Amalvy *et al.*, *Proc. PRA 16th Int. Conf. on Waterborne, High Solids, and Radcure Technologies*, Frankfurt, 1996, paper 17, 13 pp.
15. A. Bouvy, *PaintIndia*, **46**(8), 29–34, 1996; *Eur. Coatings J.*, **1996**(11), 3 pp. (1996)
16. T. Nagai, *Trends Polym. Sci.*, **4**(4), 122–7 (1996)
17. A. Guyot and K. Tauer, *Polym. Sci.*, **11**, 43–65 (1994); *Double Liaison*, 1994 (455/6), pp. 17–19(Fr.), and III–IV(Engl.); Z.W. Wicks, F.N. Jones and S.P. Pappas, *Organic Coatings, Science and Technology*, Vol. 2, Wiley–Interscience, 1994, Ch. 20, p. 29 *et seq.*
18. H.L. Jakubauskas, *Surface Coatings, Science and Technology*, Swaraj Paul, 1996, Ch. 5, p. 441 *et seq.*
19. *Rheology Handbook, and Rheological additives*, Rheox Inc., Hightstown, New Jersey, 1997
20. Laponite, brochure, Laporte Industries, Widnes, Cheshire
21. R.K. Tubbs, H.K. Inskip and P.M. Subramanian, in *Polyvinyl Alcohol*, C.A. Finch (ed.) Soc. Chem. Ind. Monograph No. 30, 1968, pp. 88–103
22. J. Dickstein, and R. Bouchard, Polyvinyl alcohol, in *Manufacture of Plastics*, Reinhold, 1964, pp. 256–85
23. C.A. Finch (ed.), *Polyvinyl Alcohol, Properties and Applications*, Wiley, 1973
24. C.A. Finch (ed.), *Polyvinyl Developments*, Wiley, 1992
25. H. Warson and D. Robertson (Dunlop), BP 1,194,097, 1969
26. H. Warson and D. Robertson (Dunlop), BP 1,174,914, 1969
27. K. Takahashi, (Nippon Synthetic Chemical Industry), JP 96 281,091–2, 1996; *Chem. Abstr.*, **126**, 48016–7 (1997)
28. H. Ida and T. Sato (Kuraray), JP 96 301,908, 1996; *Chem. Abstr.*, **126**, 104569 (1997). M. Nakamae (Kuraray), JP 2000; 63407; *Chem. Abstr.* **132**, 181468 (2000)
28A. M. Saito, and M. Shibaya, JP 2000; 6327; *Chem. Abstr.* **132**, 180154 (2000)
29. J.E.O. Mayne and H. Warson (Vinyl Products), BP 648,001, 1950
30. H. Warson and R. Parsons, *J. Polym. Sci.*, **34**, 251–69 (1959)
31. H. Warson, *Peintures, Pigments, Vernis*, **43**(7), 438–46 (1967)
32. C. Schade (BASF), Eur. P 736,546–7, 1996
33. B. Schlarb *et al.* (BASF), Eur. P 727,441, 1995
34. M. Antonietta and M.C. Weissenberger, *Macromol. Rapid Commun.*, **18**(4), 295–302 (1997)
35. H. Warson and G. Reed (Dunlop), BP 1,114,316, 1969
36. M.A. Schafheutle *et al.* (Hoechst), GP 4,446,439, 1996
37. J.E. Glass *et al.*, in *XVI Fatipec Congress*, 1982, pp. 119–29
38. Su.C. Tso, G.W. Beall and J. Gordon, *J. Waterborne Coat.*, **7** (August), 3–8 (1989)
39. M. Huldén, E. Sjoeblom and P. Bosroem, in *21st FATIPEC Congress*, Vol. 1, Amsterdam, 1992, pp. 216–32
40. J. Prideaux, *J.Surf. Coat. Int.*, **76**(4), 177–183 (1993)
41. P.M. Macdonald, *Polym. Mater. Sci. Eng.*, **76**, 27–8 (1997)

Supplementary references for Section 3.9

A. Zicmanis *et al.*, *Colloid Polym. Sci.*, **275**(1), 1–8 (1997)

S. Abele *et al.*, *Langmuir*, **13**(2), 176–81 (1997)

These describe reactive surfactants based on maleates.

A. Bouvy, in *PRA 4th Nurnberg Cong. on Creative Advances in Coatings Technology*; 1997, paper 21, 9 pp.

This paper also includes emulsification of air drying alkyd resins.

3

Emulsion Polymerisation

H. Warson

1 GENERAL PRINCIPLES

Amongst the advantages of water as a polymerisation medium is its use as a controller of the heat of reaction. Whilst the kinetics of polymerisation are often markedly different from that in bulk, water has virtually no chain transfer properties, except in exceptional circumstances, and this makes for the formation of polymers with high molecular weight. An elementary summary of the kinetics of emulsion polymerisation is given in Section 8 of this chapter.

Polymer latices (emulsions) may be used directly in many applications, film formation and impregnation being the major methods, often after being pigmented or otherwise compounded, as indicated in Chapter 4 and the chapters on individual applications. In some cases there may be disadvantages in employing emulsion polymers because of the presence of water-soluble stabilisers when extreme resistance to water is required, or for electrical insulation. Special procedures must be adopted to obtain films of desirable hardness when the temperature of drying is below the minimum film-forming temperature (see Chapter 4, Section 4.3). In addition, whilst the boiling point of water is 100 °C, less than that of many solvents, evaporation may be inordinately slow, because of the high latent heat of water and high saturation of the atmosphere with water vapour.

1.1 Chemistry and fundamentals

The technology of emulsion polymerisation has made major strides in the last four decades, and in the last thirty years there has been a marked increase in our knowledge of the processes involved, which often vary with the nature of the emulsifiers, initiators, method of addition of components to the reactor and other factors. There has also been much study of latices and their properties, the latter often coming under the heading 'polymer colloids' [1–6].

The problem of emulsion polymerisation is that of performing a chemical reaction on droplets in emulsion form whilst maintaining the mechanical stability of the system. The problem is complicated by the considerable exotherm of the process, about 14–22 kcal mol^{-1}, according to the monomer. In order to complete the reaction in the shortest possible time, efficient cooling of the reactors is essential. In general, reactions should be driven as far as possible to completion; at least 99.5 % of the starting monomers should be completely polymerised, and if possible the residual monomer should be 0.1 % or below. If traces of monomer are objectionable, they must be removed by distillation or other method and environmental protection regulations carefully observed.

The most unstable state of a polymerisation is an intermediate one in which part only of the monomers have polymerised, the polymer formed being normally soluble in its monomer. This is not always the case, e.g. with vinyl chloride. This so-called 'sticky state', in which the cohesive forces between the particles are particularly strong, requires both vigorous mechanical agitation and a high stabiliser content to prevent premature coagulation. This difficulty is often overcome by adding the monomers gradually to the polymerising system. This also assists in controlling the heat of reaction and enables better copolymerisation of monomers with very different reactivities, e.g. vinyl acetate with acrylate esters.

In addition to the monomers and water, which with some emulsifiers should be 'soft' water free from calcium and magnesium compounds, a number of other components are necessary. These include the surfactants or stabilisers and the initiators. Buffers such as sodium carbonate, monosodium dihydrogen phosphate and disodium hydrogen phosphate are often present, as well as molecular weight regulators such as mercaptans. Ammonia solution, or rarely acids, must sometimes be added during polymerisation to maintain a desired pH level.

A polymer emulsion, preferably referred to as a latex (plural latices or latexes), is usually at a solids concentration of 25–65 % of solids, the balance being water. The higher figure is about the practical limit of stability and viscosity, although there have been some claims to higher concentrations. Most of the standard monomers as described in Chapter 1 may be used in emulsion polymerisation. There are a few exceptions such as vinyl alkyl ethers, which can only be copolymerised, not homopolymerised, and require a pH above 6.5 to avoid hydrolysis of the monomer. Under controlled conditions, monomers that are water soluble, e.g. methacrylic acid, may be copolymerised with monomers that are substantially water insoluble, e.g. styrene. Gaseous monomers, e.g. butadiene or vinyl chloride, may be polymerised in emulsion under pressure.

1.1.1 Stabilisers

A large number of the surfactants and stabilisers described in Chapter 2 are suitable for emulsion polymerisation. Most emulsion polymers are anionically charged. Most latices made with non-ionic emulsifiers are weakly anionic, especially where persulfate initiators are used. Cationic latices are manufactured in fairly small quantities.

A colloid is frequently included in the polymerisation system, occasionally without any other emulsifier. The function of a colloid, which may be a natural one such as gum acacia or starch, a semi-synthetic one such as hydroxyethyl cellulose, a synthetic polymer such as polyvinyl alcohol or a polymethacrylic acid salt, is partly to increase stability of the latex and partly to improve the viscosity characteristics. Both polyvinyl alcohol, mainly partially hydrolysed, usually to about 88 % in molar (not weight) proportion, and copolymers and alkali-soluble copolymers of acrylic and methacrylic acids have stabilising properties usually including a limited reduction of surface tension. Unusually, emulsion polymerisation has been performed with polyelectrolytes in water–ethanol mixtures [7].

Amphiphilic block copolymers, e.g. poly(alkyl methacrylate–sulfonated glycidyl methacrylates), are stabilisers for latex formation, and are also efficient dispersing agents for pigments in aqueous media [8]. Polystyrene–polyethylene oxide block copolymers are also suitable stabilisers in the emulsion polymerisation of styrene and methyl methacrylate, The polystyrene block is 1000 g mole^{-1} and the polyethylene oxide block molecular weight should be >1000. Latices are sterically stabilised and stable to high electrolyte concentrations, but flocculate at high temperatures [9].

Non-ionic surfactants of the type of polyoxyethylene condensates with a hydrophobic end group, e.g. nonylphenyl, and the ethylene oxide–propylene oxide condensates may be included in formulations, often in conjunction with an anionic surfactant. These non-ionic surfactants may be present in high proportions, often 4–7 % of monomers, as they have less effect on water resistance than anionic types. The idea of HLB (hydrophile–lipophile balance) is useful in assessing the most suitable type of non-ionic surfactant for a particular monomer system, although it is empirical in nature. In many cases, blends of emulsifiers with high and low ethoxy content prove more satisfactory than a single product of intermediate composition. Note, however, that many non-ionic surfactants are themselves blends, usually with a statistical scatter of ethylene oxide units.

Cationic surfactants occasionally cause some difficulty due to the reactive amino · NH_2 or imino :NH groups, causing inhibition of polymerisation. This is especially so when polymerising vinyl acetate or allied monomers, which hydrolyse readily forming acetaldehyde. The latter tends to react with the amines producing nitrogen derivatives, which form stable radicals and inhibit the polymerisation. Cationic surfactants, except for most quaternary types, may also react with initiators, normally a persulfate or peroxide or hydroperoxide, which may lead to both loss of effective initiator and decomposition of the amino surfactant. In some cases a cationic latex may be prepared from an anionic type, e.g. polyvinyl alcohol, stabilised by controlled charge reversal.

General principles 93

The ratio of stabiliser, whether of high or low molecular weight, to monomers depends on the concentration of the emulsion, the nature of the monomers and sometimes the viscosity. The normal addition of sulfated or sulfonated surfactant is 0.5–4 % or 0.1–2 % if used with a non-ionic or polymeric stabiliser.

Under exceptional circumstances it may be possible to prepare a stable latex in the complete absence of added stabiliser (see Section 1.4.1). The polymerisable surfactants already described in Chapter 2, Section 3.9, may make the addition of emulsifier unnecessary or enable reduced quantities to be employed.

Under some conditions, e.g. when emulsification produces relatively large size initial particles, the copolymerisation of the highly insoluble vinyl stearate with vinyl acetate gives an obviously heterogeneous copolymer, although the r_1, r_2 values are almost unity. This is probably due to the solubility of vinyl acetate in water, supporting the theory that the aqueous phase is the principal locus of emulsion polymerisation of vinyl acetate.

Light-decomposable emulsifiers, alkyl-substituted azosulfonates, may be used as emulsifiers, e.g. of methyl methacrylate. The emulsifiers decompose when exposed to ultraviolet (UV) light, causing precipitation of the latex formed, and may be used as a method of coagulation when desired [10–11]. Many of the emulsifiers/stabilisers indicated are described in greater detail with their specific applications later in this chapter.

An unusual emulsifier is prepared from a cationic polyurethane reacted with the diamine $(CH_3)_2NC_3H_6NH_2$ and formic acid in glycol ethers. It may be used for the redox copolymerisation of styrene and hydroxyethyl methacrylate at pH 4.8. The resultant latex is the basis of the formation of cationic electrophoretic coatings [12] (see also Chapter 11, Section 10.4 in Volume 2).

1.1.2 Free radical initiators

Whilst many free radical initiators are included in emulsion polymerisation formulae, water-soluble types are the most extensively used. Hydrogen peroxide is of principal interest in preparing vinyl acetate polymer and copolymer emulsions, if polyvinyl alcohol is the stabiliser. Otherwise, sodium, potassium or ammonium persulfates are the most general initiators. Whilst potassium persulfate is the most readily available, its limited solubility may be a disadvantage, particularly where extra initiator has to be fed into a polymerisation system. Most initiators operate at about 50–80 °C. It is sometimes desirable to feed in a buffer during a reaction, particularly with a persulfate initiator. In addition, the tendency of certain monomers, e.g. vinyl acetate, to slight hydrolysis also makes it desirable for the pH to be controlled.

Organic peroxides may be used, e.g. dibenzoyl peroxide, which should be dissolved in the monomer. *tert*-Butyl hydroperoxide may be part of the water or monomer phases, depending on the relative solubilities and whether it is part of a redox system.

Azoisobutyronitrile (Chapter 1, Section 2.1) is occasionally used as an initiator. A study has been made of the efficiency of several water-soluble, non-ionic azo initiators which have a 10-hour half-life temperature from 80 to 88 °C. Their efficiency is relatively low [13].

Diazonium salts have been previously used as redox initiators for acrylonitrile aqueous solutions, both with hypophosphorous acid and a bisulfite [14]. This reaction may also be possible with monomer emulsions.

Water-based redox systems described earlier (Chapter 1, Section 2.1.2) are also widely used, especially as they tend to operate at lower temperatures than single initiators. These systems should be started after a nitrogen purge with a high purity grade ($\geq$99.9 %) of the gas, but in many cases the nitrogen stream may be discontinued after starting, as the monomer vapours themselves act as a gas seal. In commercial production performed under reflux conditions it is often not necessary to de-oxygenate, although this is often required in theoretical studies. If degassing is not performed there may be an appreciable increase in inhibition time in the case of styrene. Amongst redox systems which have been examined are malonic acid/Mn^{3+} for acrylonitrile and methyl methacrylate [15] and aqueous acrylamide initiated by an acidified bromate/thiourea redox system [16].

Russian work has shown that in the dibenzoyl peroxide–cetyl pyridinium bromide redox initiation, the latter combines the function of an emulsifier with that of a reductant, the rate of polymerisation being of 0.5 order with respect to peroxide and first order with respect to the emulsifier [17].

The adduct of a lower (C_{3-8}) ketone and a bisulfite is the reductant in a redox system and is effective in reducing free monomer content to <0.1 %. This reaction has been studied [18, 19].

Surfactants with transfer agent properties have a typical formula HS—C_{11} $H_{22}(OCH_2CH_2)_n$ OH. In a polymerisation with styrene and an azo-type initiator the amount of *trans*-surf incorporated was independent of the HLB balance or the process used (see Sections 2.2 and 2.3), and polymodal molecular weight distributions were obtained [20, 21]. Similarly, polyvinyl alcohol, prepared so that it is mercaptan (HS) ended, may take part in a redox initiation, e.g. with potassium bromate, as well as acting as a stabiliser for a latex [22].

A review of redox systems covers developments until 1975 [23]. Organic peroxides, e.g. *t*-butyl hydroperoxide and peroxyesters, are more effective than persulfates and hydrogen peroxide in reducing the residual monomer in a redox system, with special reference to acrylic monomers. However, the molecular weights of the resultant polymers are lower than those formed with persulfates [24].

Radiation-induced polymerisation provided by ^{60}Co of vinyl acetate, styrene, and styrene–butadiene copolymer has been described in a specially designed

General principles

laboratory apparatus [25]. Acrylonitrile has also been polymerised in emulsion aided by γ-radiation [26]. Electroinitiation has been found feasible with an emulsion of vinyl acetate containing an ionisable emulsifier [27].

Ultrasound has been found to initiate polymerisation of emulsions of butyl acrylate and of vinyl acetate at 15 °C as well as mixtures of the two. The polymerisation rate depends on the concentration of monomer dissolved in the aqueous phase and inversely with its vapour pressure. Ultrasound using a 20 Hz generator, also initiates the emulsion copolymerisation of methyl methacrylate/butyl acrylate at 18 °C, with sodium dodecyl sulfate as emulsifier and ammonium persulfate as initiator [28, 29].

1.1.3 Water-soluble monomers

Water-soluble monomers are occasionally included in formulae. Some of these, e.g. methacrylic acid, are also soluble in monomers, whether ester or hydrocarbon. These are preferably added to the monomers before polymerisation, especially during a delayed addition (semi-continuous) polymerisation. More hydrophilic monomers, such as the sodium salts of methacrylic or itaconic acids, are insoluble in most monomers. If water solubility of the principal monomer is appreciable, e.g. vinyl acetate is soluble to 2.3 % at 20 °C and methyl methacrylate is about 1.6 % soluble, difficulty rarely occurs in copolymerisation. However, the rates at which each monomer enters a copolymer may be different from that expected from their reactivities, depending on the locus of polymerisation.

Where a secondary crosslinking (see Chapter 5) is desired, for which the presence of a carboxyl group is a possible requirement, additions of monomers should be such that an even copolymer is obtained. This also applies to other monomers with reactive groups. If the composition is irregular, crosslinking may be partly excessive, giving brittleness and it may be partly incomplete [30].

Hydrophilic monomers such as itaconic acid give difficulty in copolymerisation with butadiene or styrene, since the acid is in the water phase. Successful copolymerisation depends on the correct choice of surfactant such that a very fine particle size with corresponding large surface area is obtained. In addition, the principal monomers may be solubilised to some extent by the emulsifiers. Solubilisation, as distinct from emulsification, is not considered in detail here, but it suffices to say that the more soluble surfactants formed from shorter hydrocarbon chains, e.g. C_{10-12}, tend to increase the solubility of monomers such as styrene by micellar action.

A new principle for the emulsion copolymerisation of monomers with low water solubility with those of high water solubility is by the occlusion of the low solubility monomer within a hydrophobic cavity, typically methyl-β-cyclodextrin. In a typical emulsion polymerisation the monomers, lauryl acrylate/methyl methacrylate/ methacrylic acid (40 : 59 : 1), were pre-emulsified and added gradually to the stirred reactor containing the methyl-β-cyclodextrin and an anionic surfactant in solution. Copolymerisation is facilitated [31].

2 TECHNOLOGY, LABORATORY AND LARGE SCALE

2.1 Monomers

2.1.1 Liquid monomers

A typical laboratory organic exotherm reaction type of apparatus consists of cylindrical 'split' flasks of 1–3 litre size with a ground glass joint between the reactor and inlets for a condenser, possibly a dropping funnel, a long immersion type thermometer, optionally an inert gas inlet and with a suitable gland, e.g. polytetrafluoroethylene (PTFE) for the stirrer. Heating is best via a constant-level water bath, which enables the exotherm to be controlled. No definite recommendations can be given for the shape of the stirrer, although as a rule the blade, normally of stainless steel, should be as wide as possible without fouling the thermometer. The shaft may include a small upper blade.

The main variations of polymer technology are:

(a) All ingredients of the emulsion are added at the start.
(b) Delayed addition ('semi-continuous') polymerisation takes place, in which all or most of the monomers are added during the polymerisation, preferably to maintain a constant or minimum percentage of monomer. There is a final polymerisation at the end of the addition, sometimes with additional initiator. Composition control by a closed-loop strategy is described [32].
(c) A pre-emulsion with liquid monomers is prepared, with part of the monomers and the emulsifying agents, this pre-emulsion being added gradually to the reactor. This may be lead to the formation of 'core–shell' particles or to a graft copolymerisation according to conditions.
(d) In (b) and (c) the monomer feed may be varied either throughout or an abrupt change may be made at one or more points of the composition. This may also lead to either a graft or to 'core–shell' particles. A further description of the latter is included in Section 2.4.3.
(e) Continuous processes (see Sections 2.3.1 and 2.3.2).

Some highly sophisticated apparatuses have now been manufactured. It is possible to measure the reaction rate by heat control and by reduction in volume during the reaction.

The advantage of the gradual addition of monomer is the improved control of the reaction temperature. Because of the lower quantity of disperse phase until the final stages, there is less tendency for coagulation to occur. Final particles, usually of the order of 0.1 µm, are formed at the start, gradually building up as monomer is absorbed, although in most cases there is fresh nucleation and therefore a considerable scatter of particle size, often bimodal.

On the other hand, an excess of extremely small particles, less than 0.2 µm, may lead to instability later due to incomplete coverage of the particles with emulsifier owing to the very large overall surface area. A theoretical study is available [33].

The treatment of redox initiators may vary, but they should be added to the reactor in separate streams and must not be mixed in advance, although one component may be added in a delayed addition of part of the water phase or a pre-emulsion. In a few cases a component may be miscible with the monomers. A system frequently found is a persulfate salt or *tert*-butyl hydroperoxide with sodium formaldehyde sulfoxylate. In some cases a redox system of the type of *tert*-butyl hydroperoxide with a reducing agent is added at the last stage only (>98.5 % polymerisation) in order to bring unreacted monomer to the lowest possible level.

On a laboratory scale the addition of monomers in controlled quantities often presents difficulties, necessitating careful control of glass taps. Microscale pumps are now available for metering monomers on a small scale, and it is possible to make a number of additions to differing reactors from the same pump (see Figure 3.1). On a large scale the problem is much less difficult.

If it is desired to removed the last traces of monomers, steam stripping is sometimes used, or a vigorous current of air or preferably an inert gas. Vapours should be condensed, not discharged to the atmosphere.

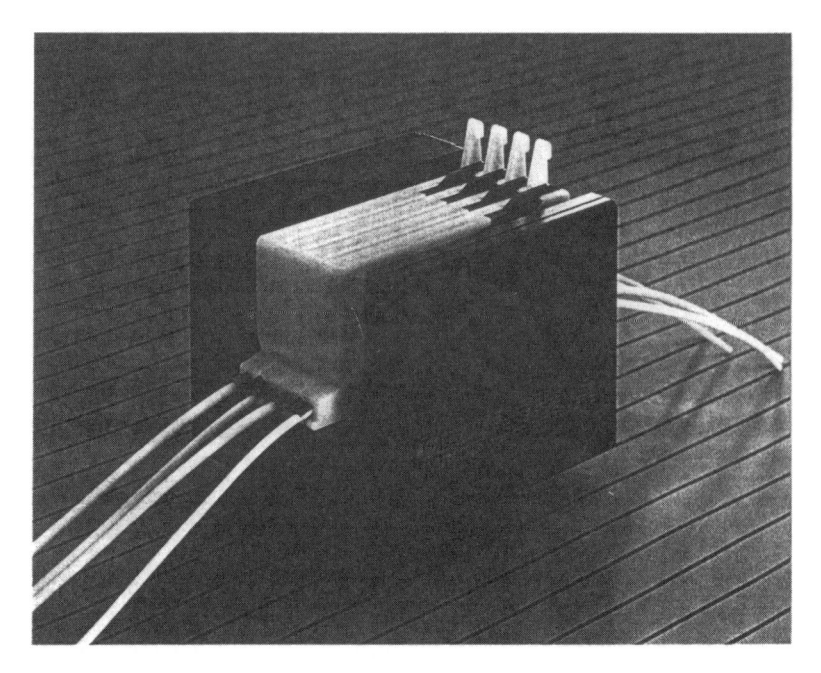

Figure 3.1 Micropump for adding monomers. (Reproduced by permission of Watson–Marlow Ltd.)

2.1.2 Volatile monomers

It is not possible to operate polymerisation in open reactors with a condenser for monomers that are gaseous at ambient temperature, e.g. ethylene, vinyl chloride and many fluorinated monomers. For monomers such as isoprene, b.p. 34.1 °C, or vinylidene chloride, b.p. 31.7 °, it may be possible to weigh them on a laboratory or plant scale at temperatures 0 °C and below, and to cool the condenser with brine from a refrigerated unit or by solid carbon dioxide. At the completion of polymerisation, surplus monomer may be stripped by raising the temperature, although complete removal of traces of unreacted monomer is not always easy. In accordance with safety laws in various countries surplus monomer should be condensed, and possibly recycled.

2.2 Pilot plants and large-scale reactors

For the smaller pilot plant polymerisations, from about 5 to 50 L, jacketed stainless steel vessels, which are essentially replicas of large-scale plants, are operated. These vessels are equipped for steam heating and cooling water and may be operated under reflux conditions or under pressure, provided that the reactor is suitably constructed. In the latter case a safety valve is desirable.

Liquid monomers may be polymerised under pressure, as this avoids problems due to foaming or sometimes the necessity for an inert gas stream. If a gradual addition of monomer is required in a pressure vessel, specially adapted inlets designed to operate under pressure are necessary. On the rare occasions when addition of a solid, often an initiator, is necessary, it is best added as a suitable slurry, generally aqueous.

Pilot plants should be equipped for all necessary additions, as with laboratory scale equipment, but are usually emptied through an aperture at the base. In general, the addition of monomer or other liquid components is controlled by a suitable gauge and pump. Problems such as airtight glands through which the stirrer operates and suitable clamps for the removable lid of the reactor are matters for chemical engineering design.

The design of full-scale equipment must take account of problems such as heat transfer, cooling and adequate, but not excessive, agitation. The latter is a complex problem and different agitators and agitator speeds may be desirable, depending on the final viscosity of the latex and viscosity changes during polymerisation. Vertical baffles are sometimes placed in reactors to improve mixing during manufacture, but they are sometimes a locus for the formation of coagulum. Many reaction kettles are of the order of 15 000–50 000 L. As the surface only increases to the two-thirds power of the increase in volume with the size of the reactor for constant shape, it is evident that heat transfer problems increase with reactor size. To overcome this a heat interchanger through which the polymerising emulsion is pumped is sometimes included. Auxiliary vessels such as header tanks for mixing water phases or pre-emulsions, and sometimes separate cooling vessels, are included in the design.

Safety storage and transfer of raw materials, in particular monomers, are also important. Tanks, shielded from direct sunlight and sometimes water-cooled, especially in hot climates, are desirable for storage of bulk monomers, although sometimes in smaller units monomer is pumped directly from drums of the order of 200 L. Adequate ventilation of sheds holding reactors is essential, although reactors are sometimes housed in semi-open conditions.

Accurate instrumentation is desirable on a large plant. Figure 3.2 illustrates typical commercial reactors, including a pressure plant. An early specification describes in detail an aluminium reaction kettle for the polymerisation of vinyl acetate (see Figure 3.3) [34]. Recent Japanese patents claim modified reactors in which the principal feature is the design of the stirrers [35]. Some specific problems in the industrial production of polyvinyl acetate are reviewed [36].

A technical-scale polymerisation normally takes 3–5 h to complete, excluding loading and emptying times. Exceptionally, polymerisations including butadiene may take up to 48 h, at a lower temperature than usual, circa 50 °C, for at higher temperatures secondary reactions occur which may form inhibitors. This time schedule is not necessary where butadiene

(a)

Figure 3.2 (a) Polymerisation plant and (b) polymerisation pressure plant. (Reproduced by permission of Vinamul Limited.)

(b)

Figure 3.2 (*continued*)

copolymers are formed for the production of synthetic rubber, since in this case polymerisations are short stopped.

The problem of residual monomers, even if only a fractional percentage, is of major importance since some monomers, especially acrylonitrile, are highly toxic. Methods of removal include sparging with air or nitrogen, and

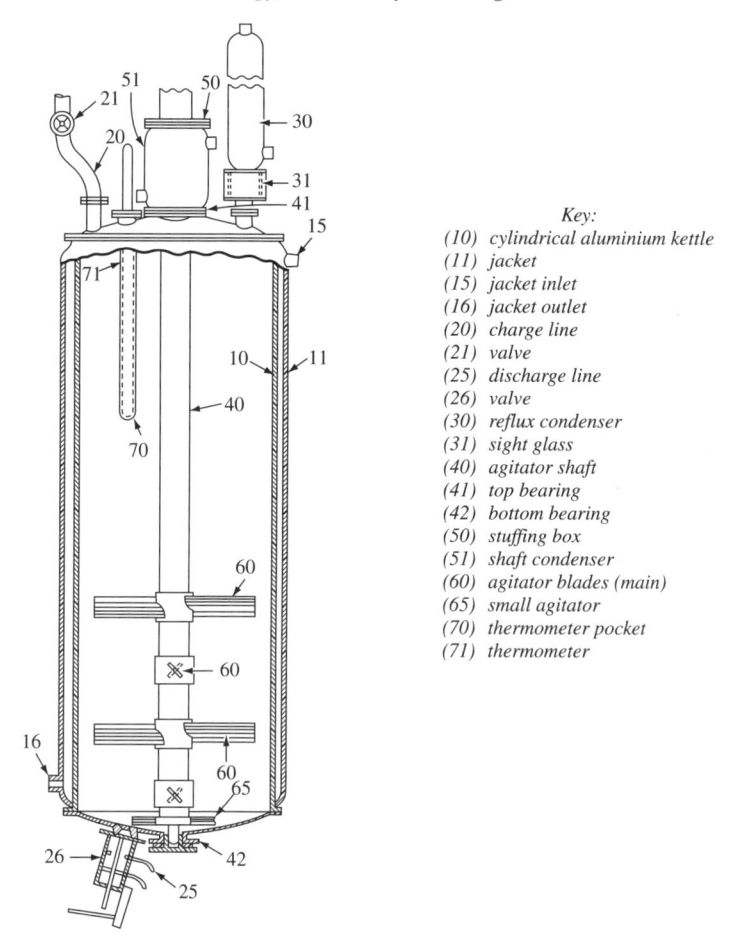

Key:
(10) cylindrical aluminium kettle
(11) jacket
(15) jacket inlet
(16) jacket outlet
(20) charge line
(21) valve
(25) discharge line
(26) valve
(30) reflux condenser
(31) sight glass
(40) agitator shaft
(41) top bearing
(42) bottom bearing
(50) stuffing box
(51) shaft condenser
(60) agitator blades (main)
(65) small agitator
(70) thermometer pocket
(71) thermometer

Figure 3.3 Emulsion polymerisation plant—diagrammatic. (Reproduced from Kiar [34].)

for vinyl acetate, vacuum has been applied, but foaming may be a more general problem [37].

Note that as in many cases residual monomer is absorbed by the polymer, removal may be difficult. One method of dealing with objectional monomers is to add another monomer which is less objectionable in small quantities towards the end of polymerisation and to continue the reaction, in this case a copolymerisation, adding more initiator if necessary. Thus residual styrene is sometimes polymerised to completion by adding about 1 % of methyl methacrylate towards the end of the reaction. A continuous process for the removal of residual monomer from latices employs a heat exchanger through which a latex is circulated [38].

Few latices are entirely free from coagulum, sometimes referred to as skinning, when it is caused by surface evaporation during transfer of hot latex. Sieving the latex before being drummed or tanked is necessary. A vibrating sieve with a fine nylon mesh, or alternatively a fine stainless steel mesh, is a convenient device (see Figure 3.4).

A general review of methods of manufacture of latices with special reference to engineering and plant safety is available [39].

2.2.1 On-line control of polymerisation

Techniques that have sometimes been used with volatile monomers, in particular vinyl acetate, such as control by reflux, are not valid, since considerable vinyl acetate is held in the formed polymer particles at temperatures considerably above its atmospheric boiling point (73 °C). At 80 °C, polymer particles will hold about 20 % of their weight of monomer without reflux.

Control by using cooling water is empirical, although it was the only possible method in the early days of emulsion polymerisation. Samples tested by gas chromatography and analysing for residual monomers gave a major improvement in control, although such methods are not suitable without special precautions for very volatile monomers.

A recent method employs calorimetry, requiring only on-line temperature measurements to calculate polymer conversion and composition, based on the heat involved during the reaction. The method has been used to produce homogeneous vinyl acetate–butyl acrylate copolymers of differing composition [40].

2.3 Special techniques; continuous processes

The batch or delayed addition processes that are standard have the disadvantage that heat is required for start-up of the reaction except for certain redox processes, and also for completion, whilst the exothermic stage may require strong cooling. Filling and emptying of vessels take a considerable time.

The optimum is a continuous process. Some early descriptions of suitable reactors have been given [41], but there are formidable difficulties in the design of a continuous reactor for polymerisation. Models for the steady state in a continuous stirred tank reactor have been deduced, but it is difficult to achieve in practice. In this case the emulsion in the reactor is not in equilibrium, but it polymerises as emulsion enters at the base and polymerised latex is stripped at the top. There is also a problem with oscillations in the reaction. Numerous highly mathematical theoretical calculations have been made to determine optimum conditions [42–54]. General problems of continuous reactors are considered in a review paper [55].

Otherwise, it is necessary to emulsify a stream of monomers at the entry to the reactor, and this must be followed by the emulsion proceeding slowly

Technology, laboratory and large scale 103

- Heavier gauge stainless steel contact parts.
- Demountable sieve assembly with built-in anti-splash guard.
- Heavy duty, easy-to-use clamps.
- Improved vortex action with increased throughput and mesh life.

- Low noise level.
- High frequency centrifugal vibrator.
- Adjustable amplitude and pitch for desired flow pattern.
- Low power consumption with standard or explosion proof motors.

Figure 3.4 Russell vibrating sieve, which is specially designed to handle high-viscosity dispersions such as vinyl emulsions and plastisols. (Reproduced by permission of Russell Finex Ltd.)

through a tube, possibly in several sections maintained at suitable temperatures. It may be necessary to add extra initiators or a second charge of monomers en route. The nature and quantity of initiator must also be carefully regulated in each process [56]. The final product must also be stripped of surplus monomer. Inhibitors in the monomer in a continuous feed stream may cause serious inhibition of the reaction under some conditions. This may occur in systems such as the polymerisation of methyl methacrylate where there is a strong Tromsdorf–Norrish gel effect, and there may be problems with the exotherm. Coagulum and coating of the reactor presents more difficulty than with a batch reactor. Entrained air may be a problem in a continuous process. It is claimed that using a pulsation source for a tubular reactor eliminates the reactor fouling and plugging known to have occurred previously [57].

Some possible designs for continuous reactors are described in the theoretical and patent literature [58]. Each process must be tailored to the monomers that are present. The nature of the stirring is also important as it may affect the regularity of a process, and also latex stability and particle size.

One of the difficulties of operating a truly continuous process is that conditions differ markedly from those prevailing in a batch reactor. It is therefore difficult to design a process for a specific product other than empirically on the reactor. It is also difficult to design a laboratory reactor of this type which would give significantly useful results. A tubular reactor has been used for the emulsion polymerisation of styrene [59–62]. A process for an ethyl acrylate–methacrylic copolymer was adjusted to resemble a batch product [63].

In one older process the outflow of emulsion is restricted so that the process operates under pressure. A small-scale continuous reactor is described having a length of 1.5 m and overall capacity of 10 L. This delivers 240 kg per day of polyvinyl acetate latex and after 1000 hours of operation, no precipitate forms on the sides of the reactor. In one example, an emulsion of vinyl acetate stabilised with 5.5 % of polyvinyl alcohol, saponification value 140, is initiated with t-butyl hydroperoxide and sodium formaldehyde sulfonate which is fed into the tube a short distance from the start. The reactor is heated in two sections, the lower half at 80 °C and the upper at 70 °C. The emulsion is pumped against gravity. After polymerisation the solids content is 51.5 % [64, 65].

The loop reactor (see Section 2.3.2) overcomes many of the problems of a continuous process. Continuous processes are reviewed, including methods of avoiding the oscillations, which are characteristic of one of them under operating conditions [66].

2.3.1 The cascade process

This is a modified continuous process, which is really a semi-continuous type and is probably operational with special reference to butadiene–styrene

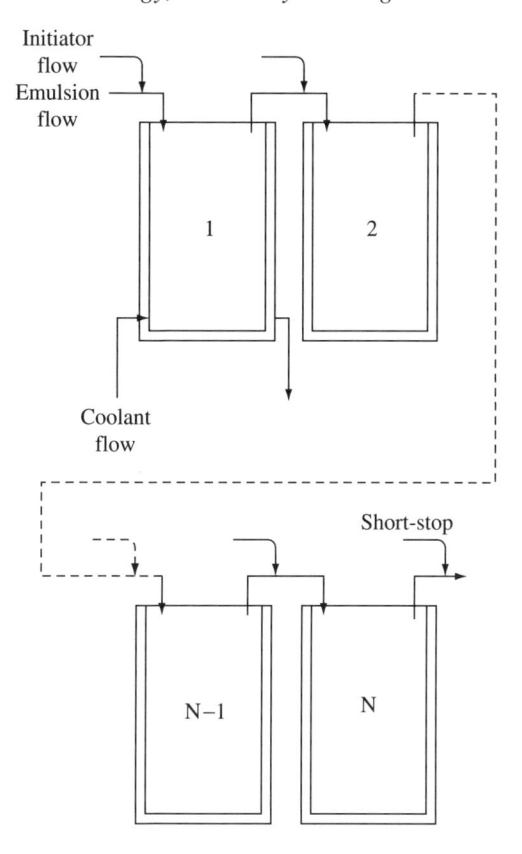

Figure 3.5 Cascade process.

latices; it has been claimed for vinyl chloride. In a 'cascade' system (see Figure 3.5) polymerisation is started in one reactor and taken to a level that may be as high as 80–90 % of the complete reaction, after which the reaction mixture is continuously pumped into a second, and possibly subsequent, reactor for substantially complete polymerisation. Simultaneously, the emulsion is renewed in the first reactor, either by a pre-emulsion addition or by the addition of individual components. Once the start-up reactor contents are polymerised to the desired degree, the reaction is continuous. If the main part of the reaction takes place in the first reactor, the second or subsequent reactors may be smaller. In this type of reaction the contents of the reactors are not evenly mixed, but polymerise as they pass through the vessel.

The polymerisation of vinyl acetate has been studied in a cascade process [67]. There have been a considerable number of disclosures on the polymerisation of butadiene, usually as a copolymer with styrene to form a synthetic rubber latex [68–75]. A typical process describes the controlled addition of butadiene and styrene with controlled addition of initiator solution in a rosin salt solution. The latter is usually a disproportionated rosin soap, but a naphthalene–formaldehyde condensate salt is sometimes included, which, although having negligible surface activity, assists stabilization of the latex. Stable latices of chloroprene and 2,3-dichlorobutadiene (19 : 1) have been prepared in a batch of seven reactors with a residence time of 25 min at 45 °C in each. The stabiliser is the disproportionated salt of abiotic acid [76].

A problem in a continuously stirred tank reactor is oscillations in the conversion. This and problems of continuous reactors are considered in a review paper [62].

2.3.2 The loop reactor

It seems fairly obvious that a continuous tubular reactor would be very desirable for a continuous polymerisation. A suitable design of tubular reactor is now available (Figure 3.6). A loop of stainless steel runs from the outlet to the inlet of a circulating pump. A multihead metering pump feeds monomers, water phase and initiators into the loop in the desired proportions. Sprays or jacket cooling are used to control the exotherm of the reactors. The contents of the loop are circulated independently at a rate considerably greater than the rate at which raw materials are fed to the loop. A pressure control valve releases completed emulsion from the loop at the same rate as the components enter.

The reactor shown in the diagram produces 20–35 tonnes of vinyl acetate/ Veova® latex every 24 hours. It occupies about 6 m² of floor space and is 2.5 m high. This reactor can be used for copolymers including ethylene with less of the pressure problems normally associated with this monomer. Careful adjustment of monomer flow is needed in the case where reactivities are very different. The products may have some marked differences from those produced in batch reactors and sometimes tend to have a fairly wide scatter of particle size.

The advantages of the loop reactor are its low installation costs, especially where pressure would otherwise be necessary, the small space it occupies and the fact that it is very simple to expand production by increasing the number of loops. In the event of operator error only a minimal quantity is lost; there is high conversion and less risk of hydrolysis with monomers such as vinyl acetate. The process has been developed technically by the Akzo Nobel Decorative Coatings, Darwen, Lancashire, England, following an earlier patent by Gulf Oil of Canada Limited, which formerly manufactured vinyl acetate

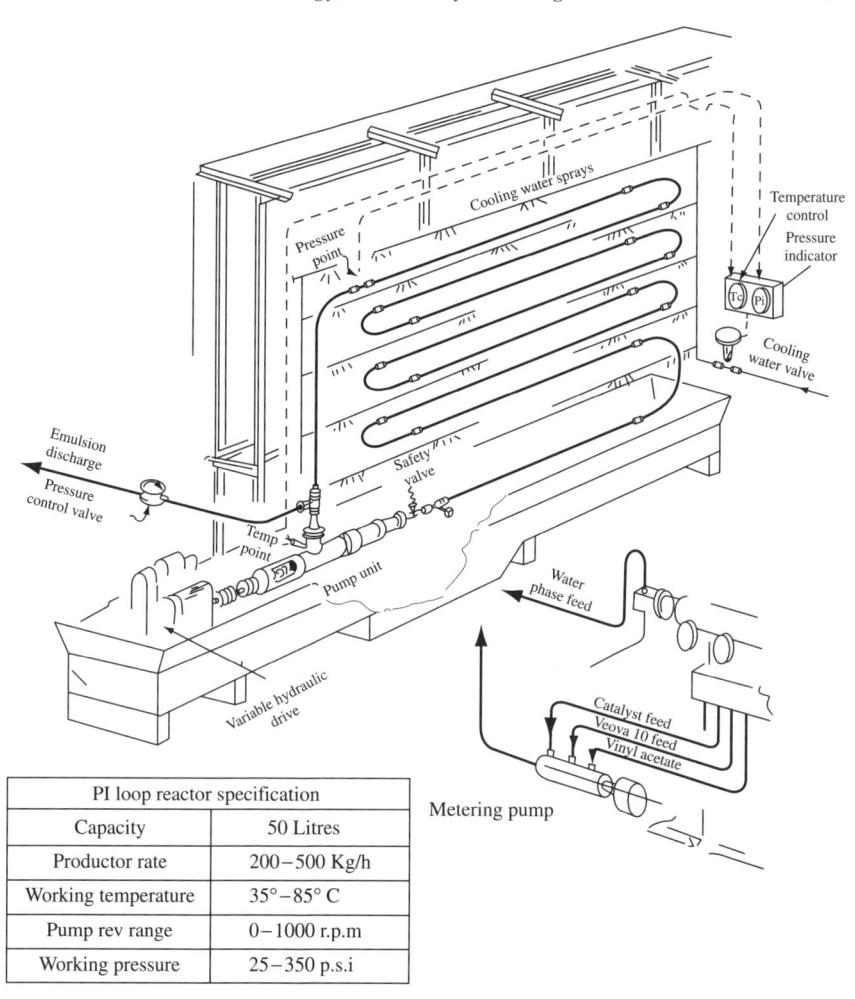

PI loop reactor specification	
Capacity	50 Litres
Productor rate	200–500 Kg/h
Working temperature	35°–85° C
Pump rev range	0–1000 r.p.m
Working pressure	25–350 p.s.i

Figure 3.6 Diagram of a typical loop reactor. Reproduced by permission of Akzo Nobel Decorative Coatings.

monomer and latices [77]. Pre-emulsification may be an advantage, as does a start-up with a completed previous run in the reactor [78].

Under operational conditions in which the heat removal capacity of a continuously stirred tank reactor is below the heat generation rate, it is claimed that there is no significant difference between this and a loop reactor. If heat generation is high there is a thermal runaway in the tank reactor [79].

There have been considerable developments, including fundamental research on particle formation and growth [80–88].

2.3.3 Special techniques; procedural variations

A number of other processes have been devised for continuous production, the nature of which depends on the monomers in use; e.g. a pressure type of reactor would be desirable for polymerisations including butadiene or vinyl chloride. Details of developments are regularly reported in a series of bulletins [89].

2.4 Varied polymerisation techniques

2.4.1 Seed latices

Other polymerisation variations are in the techniques operated rather than in the mechanical aspects. In some cases polymerisation takes place on a 'seed' of preformed emulsion. The process must be carefully operated, otherwise precipitation will occur. If, however, a monomer can be added to a preformed, fine particle size 'seed' formed by a conventional process, the monomer will become absorbed, possibly adsorbed only, by the 'seed' particles; it will then polymerise with an increase in particle size but without requiring further surfactant. Since the initial 'seed' may well have a surplus of surfactant, there may be some further initiation in the water phase. The process has the advantage of tending to produce 'monodisperse' latices (see the next subsection), since the particles tend to become even in size on growing. This is because absorption of monomer by the growing particles depends on their surface area and not on the volume.

2.4.2 Monodisperse latices

As will be mentioned later, the general pattern of particle size formation in latices is Gaussian in character. It is sometimes desirable to produce latices which are monodisperse in character with only a minimal variation in particle size.

One method of achieving this is by polymerisation on a seed latex, as shown in the previous subsection. The amount of seed latex included may be as low as 1 %, although it is usually higher, of the order of 4–10 %. The latex is prepared so that there is substantially no surplus emulsifier in the aqueous phase after polymerisation is complete, as distinct from that bound on the particle surfaces as stabilisers.

Alternative formulations are quoted in the literature, one possibility being a blend of anionic and non-ionic emulsifiers. These seem to function by producing large mixed micelles; i.e. the concentration is below the critical micelle concentration. This tends to restrict nucleation somewhat. The process may be combined with polymerisation on a seed [90–93].

It is also possible to obtain a substantially monodisperse latex by the controlled addition of a pre-emulsion. The addition of this pre-emulsion is under 'starved' conditions; i.e. addition is at such a rate that the amount of free monomer is always very low, thus ensuring reasonably even copolymerisation. A low proportion (see Tables 3.1 and 3.2) of sodium dodecylbenzene sulfonate is included in the formulation. This process operates successfully with seeded and unseeded reactions. It is successful with acrylic copolymers and may be used for copolymerisations involving monomers such as methacrylic acid which are both water soluble and soluble in ester and aromatic monomers. Monodisperse latices have been operated successfully in forming a vinyl acetate copolymer if new aqueous nucleation is avoided. The low proportion of emulsifier causes the resultant latices to have poor shear characteristics. Latex particle size was measured with an ICI–Joyce–Loebel disc centrifuge. In all experimental cases the ratio of the weight to number average was very close to 1, indicating a monodisperse product. There was minimal coagulum in each case [94] (see Tables 3.1 and 3.2).

Monodisperse latices of polystyrene have been formed with a persulfate initiator, giving stabilising sulfate end groups. They have been ion exchanged

Table 3.1 Monodisperse latices—vinyl acetate copolymer [94]. Reproduced by permission of Surfex Ltd

	Unseeded system A weight (g)	Seeded system weight (g)
Water phase		
Distilled water	226.46	226.46
Ammonium persulfate	0.35	0.35
Seed latex A	—	150.00
Monomer pre-emulsion		
Vinyl acetate	180.67	180.67
Butyl acrylate	31.88	31.88
Distilled water	70.00	70.00
Sodium dodecylbenzene sulfonate	0.50	0.50
Additional initiator solution		
Distilled water	20.00	—
Ammonium persulfate	1.00	—
Final initiator		
t-Butyl hydroperoxide, 70 %	2 ml	2 ml

Process:
  The water, initiator and seed latex, if included, are added to the reactor and maintained at 80 °C. Monomers are added at about 1 ml min^{-1} to 55 °C; then t-butyl hydroperoxide is added. Maintain for 15 min and then cool.

Table 3.2 Monodisperse latex—acrylic copolymer [70]. Reproduced by permission of Surfex Ltd

	Unseeded latex B weight (g)	Seeded latex weight (g)
Water phase		
Distilled water	267.50	—
Seed latex B, 27 % solids, $D_n = 788$ nm, $D_w = 798$ nm	—	230
Ammonium persulfate	1.35	1.08
Distilled water	—	60
Monomer emulsion		
Methyl methacrylate	105	84
Butyl acrylate	105	84
Methacrylic acid	2.55	2.55
Distilled water	70	56
Sodium dodecylbenzene sulfonate	0.16	0.19

Process:
As in Table 3.1, but there is no addition of persulfate solution during the reaction.

with purified ion exchange resins to remove adsorbed emulsifier and solute electrolyte. This enables the end groups to be determined by conductimetric titration. It results in a dispersion of monodisperse spheres with a constant and known surface charge due to chemically bound strong acid groups, which are ideal models for colloidal studies. They have varied applications, e.g. as calibration standards in electron microscopy and light scattering. Some further applications, including specialist medical ones, will be described in Chapter 17 in Volume 3 [95].

If styrene is polymerised in a water–ethanol mixture, including poly (ethylene glycol) methyl 11-(methacryloyl)undecyl ether, monodisperse particles are formed from the point where ~18 % of styrene has polymerised, and macromonomer–styrene copolymer accumulates on the surface [96].

Monodisperse latices of copolymers of styrene and divinyl benzene of unusually large particle size, up to 10 μm, based on a polystyrene seed latex have been prepared at about 20 % concentration. The formulation includes hexane, which is later extracted, enabling porous particles to be obtained. They have potential application interest [97].

An investigation into controllable factors in forming monodisperse latices considers that particle diameter distribution is solely a function of events occurring in the nucleation stage. High initiator concentrations lead to narrow size distribution as the rate of nucleation increases with initiator levels. Low emulsifier concentrations are favourable to monodisperse distributions because they shorten the time required for nucleation [98].

2.4.3 Core–shell latices

A further variation is known as a 'core–shell' copolymerisation and is intended to produce a mixed product by a sharp variation in the monomer composition during a delayed addition. This produces an inner sheath of one polymer or copolymer, known as the 'core', and an outer layer, or 'sheath', of another. There may be more than two stages. Carefully controlled conditions ensure that there is no excess diffusion of the added monomers into the core. The core–shell copolymers sometimes have the advantage that if the inner core gives a 'hard' and the outer shell a 'soft' copolymer, there may be a lower filming temperature than an even copolymer of the same overall composition [99]. The concept of core–shell polymerisation was pioneered by D.J. Williams and met with much opposition on thermodynamic grounds, but is now firmly established [100–105].

It has been claimed that thinner, softer shells on harder cores may require higher drying times than thicker shells with the same composition because the former are required to deform more to produce void-free films [106]. If a polymerisable acid is present, it is advised that this should be increased in the final stage for normal applications. In experiments with a polystyrene core and a butyl acrylate/methacrylic acid and with a monodisperse polystyrene latex as seed, it was found that the amount of acid in the shell controls the morphology. The shape of the particles is in general non-spherical [107].

A nuclear magnetic resonance study of a latex prepared by copolymerising butyl acrylate and methacrylic acid on to a polymethyl methacrylate seed indicates that some grafting occurs as well as core–shell particle formation [108]. Core–shell morphology is found in the copolymerisation of styrene and methyl methacrylate if styrene is added at a low feed rate [109]. Core–shell copolymers of vinyl acetate and methyl methacrylate were investigated by differential scanning calorimetry. Two distinct glass transition regions were found with positions unchanged regardless of the chemical composition [110].

Ionomeric core–shell latices have been prepared using a semi-continuous process, using a polystyrene core and styrene–n-butyl acrylate–methacrylic acid terpolymer shells. Analysis of the surface of the latex particles using X-ray photoelectron spectroscopy is preferred to conductiometric titration, which is time dependent [111].

In an interesting development it was noted that under some conditions there is an inversion of the core–shell structure. In general, where the first polymer is more hydrophilic than the shell, inversion of the core/shell may occur [112]. This is particularly the case where, if the second-stage monomer dissolves the first-stage polymer, the latter will tend to precipitate at the water–monomer surface, particularly in the presence of surfactant [113]. Thus, when methyl acrylate, which is relatively hydrophilic, is polymerised over a styrene latex at a 2 : 3 seed/monomer ratio with an oil-soluble initiator (AIBN) and a non-ionic surfactant, inversion occurs.

Isobutyl methacrylate, which is more hydrophobic, gives a high surface coverage of the polystyrene core particles. Interfacial tension between the polymer phases is a powerful factor in the morphology of the core–shell composites. Surface polarities and the viscosity of the monomer swollen particles also have a marked effect [112].

In a two-stage hydrophilic/hydrophobic system, low-temperature polymerisation reduces mobility of the phases, encouraging the formation of core–shell morphology [114] (see also Chapter 4, Figure 4.1). Methods for the study of the morphology of core–shell particles are given in Chapter 4, Section 2.1.

Core–shell particles, the core of which is formed of polybutyl methacrylate and the shell of poly butyl methacrylate-co-butyl acrylate-co-trifluoroethyl methacrylate have been prepared. The aim of these structured particles is to obtain as much copolymer as possible at the surface of the film to give it a more hydrophobic character and to use the minimum of the expensive fluorinated monomer. Full details are beyond the scope of this chapter, but it was shown that with minimal diffusion between the layers, the required shell was formed [115].

Core–shell copolymers with a hollow core have been prepared by the emulsion polymerisation of both vinyl ester and acrylic ester monomers. The process is the formation of an alkali swellable or soluble core that is encapsulated by a hard shell. Drying of the latex will provide hollow particles, as the alkali-sensitive material will diffuse through the hard shell [116].

Core–shell copolymers including Veova®-type monomers and acrylic acid with vinyl acetate have been prepared. The core in this case is hard and the shell soft, the core and shell having a T_g (Chapter 4, Section 4.3) >10 °C and <10 °C apart, with at least 5 °C between them. A formulation will be found later in Table 3.11 [117].

Monomers in both the core and shell of an acrylic copolymer may contain 4–5 % of a silyl group. The latices are used in a white enamel [118–119]. Some general references are given [120–135]. Details on film formation of core–shell latices are included in Chapter 4, Section 4.1.

2.4.4 Emulsifier-free latices

Notwithstanding the fact that an emulsifier is normally necessary to achieve stability, it is possible to prepare stable latices without the addition of any extraneous emulsifier. This sometimes occurs when a persulfate salt is used alone as an initiator or when a redox system, which may include a reducing acid of sulfur, is included (see Section 2.1). The end group of the polar sulfate or sulfonate group XSO_3Y or $XOSO_3Y$ (where X is the polymer chain and Y is H or a monovalent inorganic or organic alkali group) is sometimes sufficient to give a self-stabilising effect, particularly where the monomers themselves, e.g. vinyl acetate, are somewhat hydrophilic in character. In this case a rather higher percentage than usual (1–2 % by weight on monomers) of persulfate salt may be present, and concentrations of monomer tend to be lower than usual, about 25 %. The same principle applies to emulsifier-free latices prepared with redox emulsifiers [136]. These emulsions, without further stabilization, such as with non-ionic emulsifiers, tend to be rather unstable to salt additives and to compounding.

The self-emulsifying monomers (Chapter 1, Section 6.6) also have the property of forming stable latices without the addition of regular emulsifiers, whether of high or low molecular weight, although these are sometimes included as well. They generally promote the formation of stable latices [137–140].

Some studies have been made of the copolymerisation of styrene with acrylic acid or acrylamide and some of its derivatives under emulsifier-free conditions. With acrylic acid the reaction takes place in several stages, the water-soluble monomers first polymerising in the water phase, with a little styrene. In the second phase the main locus of polymerisation is in the monomer phase whilst in the third stage, due to depletion of styrene in the water phase, acrylic acid polymerises more rapidly with an increase in viscosity. Acrylamide and its hydrophilic derivatives are less effective in promoting stable latices than acrylic acid. N,N-Dimethylacrylamide copolymerises more readily, being less hydrophilic, but is less efficient in giving stable latices than acrylamide or N-(hydroxyethyl)acrylamide [141].

In the emulsifier-free copolymerisation of styrene and sodium styrene sulfonate, the latter causes a rapid increase in the rate of polymerisation with a potassium persulfate/sodium bisulfite initiator system, increasing rapidly with increasing sulfonate. This increase is attributed to the increased number of particles formed when sodium vinyl sulfonate is present and there is a gel effect enhanced by ion association. At low concentrations of functional monomer a monodisperse product is obtained, probably due to homogeneous nucleation (see the next section). At higher concentrations a bimodal size distribution is obtained, probably due to aqueous initiation of sodium vinyl sulfonate, which then acts as a locus of copolymerisation [142].

2.4.5 Mini-emulsions

It was known as far back as 1948 that the addition of a fatty alcohol improves the stability of methyl methacrylate latices and also improves the stability of vinyl acetate latices. Wool wax alcohols have also been used with the former [143–145].

In these polymerisations, hydrogen peroxide was the initiator and it is probable that a persulfate formed *in situ* with the sulfated lauryl alcohol emulsifier was the true initiator with the methacrylate and possibly with vinyl acetate. This system has been extended to the emulsion polymerisation of styrene, using sodium dodecylbenzene sulfonate and hexadecanol, a mixed system which produces porous monodisperse polymer particles, of size 200–400 nm, with increased latex stability. It was demonstrated that in these latices, with their very large particle surfaces, the principal locus of polymerisation is the polymer droplets.

This principle has been further extended to the inclusion of a long-chain alkane, such as hexadecane, instead of a fatty alcohol, which also leads to the formation of very stable latices [146]. Dodecyl methacrylate and stearyl methacrylate have also been used in a similar manner to hexadecane, these monomers probably being incorporated into the polymers [147].

In general, these very water-insoluble monomers, at 10 % of total monomer, produce droplet sizes in the mini-emulsion range [148]. In some cases a polymeric cosurfactant such as polymethyl methacrylate, added to 0.5–5 % of monomer, enables methyl methacrylate to give an average particle size of 10–500 nm with a polydispersity index of 1.05–1.40 [149]. A number of other studies have been published, with styrene or vinyl chloride, including the use of cationic emulsifier [150–155].

There have also been a number of allied studies of copolymerisation in emulsion. Mini-emulsion polymerisation of a 1 : 1 vinyl acetate–butyl acrylate mixture at 25 % concentration of monomers with the high proportion (13.33 %) relative to monomers of hexadecyl sulfate showed that the addition of hexadecane, added in varying proportions (up to 53.3 % relative to monomers), caused increased surfactant adsorption on the particles and a smaller initial droplet size. The morphology of the particles indicates that in the smaller particles containing the hexadecane (about 130 nm), the core of the particles is richer in polybutyl acrylate compared with the larger, more conventional particles that are formed in the absence of the hexadecane. The polymerisation rate was slower and the final average particle size was increased from 121–130 nm to 207–217 nm, but the latex stability was much greater [157]. Pentanol has been used as a coemulsifier in preparing mini-emulsions [158–160].

A batch process for a mini-emulsion polymerisation including a cosurfactant enables a 60 % high solids latex of a copolymer of styrene, 2-ethylhexyl acrylate and methacrylic acid to be attained. Without the cosurfactant the highest coagulum-free latex concentration was 50 % [161]. A semi-batch process has been disclosed [162].

Mini-emulsions can also be prepared by a continuous process [49]. Reviews for the preparation and properties of mini-emulsions are available [163–5].

2.4.6 Graft copolymerisation

The principles of graft polymerisation have been given in Chapter 1, Section 3.2. There are a number of useful applications for graft copolymers prepared in emulsion. These include the preparation of impact-resistant polystyrene in which styrene is grafted on to a latex of natural or synthetic rubber, the latter being a minor constituent.

Allied to this is the manufacture of acrylonitrile–butadiene–styrene (ABS), which includes a wide variety of materials and applications, depending on the method of production. Emulsion polymerisation gives the best method of giving a controlled formulation, which may include both graft and core–shell processes. The procedure commences with a polybutadiene latex, on which styrene and acrylonitrile are polymerised, with further initiator. The formed grafted latices are precipitated for use in the solid form, their major function being impact resistance.

There have been many developments in manufacture, chemistry and applications since the 1950s for which references should be consulted, as they are outside the scope of this volume and include more recent patent claims [166]. A modern reference quotes polymerising styrene(I) and butadiene(II) (40 : 760) to 35 % conversion, then adding I : II 120 : 680 and continuing polymerisation. Then acrylonitrile and I (300 : 700) are polymerised on 1000 parts latex solids [167]. Vinyl chloride has also been graft polymerised on to an acrylic latex including trimethylolpropane triacrylate [168]. Mixed acrylic–butadiene latices and acrylonitrile–divinyl benzene have been added to ABS latex [166–168].

3 VINYL ACETATE AND COPOLYMER LATICES

3.1 Historical development of latices

The first commercial synthetic resin emulsions (latices), as distinct from synthetic rubbers, were introduced in the early 1930s. The emulsifiers were simple soaps followed later by sulfates and sulfonates and the fatty acid esters of triethanolamine. The inclusion of water-soluble polymers, whether natural or synthetic, followed about a decade later.

Whilst the earliest solid contents of latices were only about 25 %, it was possible to raise this figure to 50 % with improvements in technology and many latices with over 60 % of solids prepared by emulsion polymerisation have been quoted in the literature. The mechanical stability of early products was poor as the prime purpose was a convenient method of production of solid polymer. Later products have been formulated to give good long-term mechanical stability and good stability to compounding.

The use of delayed addition techniques (Section 2.2.1) represented a major technical advance and enabled very large scale batch polymerisations of the order of 25 000 litres (25 tonnes) to be undertaken. The introduction of redox processes (see Chapter 1, Section 2.1.3), which are of particular interest in acrylic polymerisations, have also added to technical efficiency.

Commercially available latices are best considered under the major class headings, depending on the monomers included. The properties of a number of latices as marketed in various countries at the time of writing are included in Chapter 6 in more detail. This list is not claimed to be complete, and the inclusion of one manufacturer's product must not be interpreted as a recommendation to the exclusion of other products.

Whereas the nature of the monomers from which the polymers are formed is often disclosed, it is rare that any details are given about the stabiliser systems except in general terms such as 'anionic'. However, latices containing polyvinyl alcohol or cellulose derivatives are often described as such.

3.2 Polyvinyl acetate and related copolymers

The low cost of vinyl acetate monomer, about 40p per kg in Great Britain and $0.60 per kg in the United States at the time of writing (1999), has tended to make it one of the most popular starting points for latices intended for applications as emulsion paints and adhesives. The solid polymer has a second-order transition point (T_g) of 25–28 °C, increasing with molecular weight, and external plasticisation is simple. In addition, the weathering properties of films containing vinyl acetate polymer and corresponding copolymers are very good, especially with regard to ultraviolet light resistance and oxidative degradation. They are generally superior to copolymers of butadiene and styrene in this aspect.

Emulsion polymers have similar properties, with slight modifications due to the presence of emulsion stabilisers. Emulsifiers that have been used for polyvinyl acetate include phosphate ester salt surfactants and water-soluble polymers such as acrylate and methacrylate salts, including salts of alkali-soluble copolymers. Semi-natural products such as dextrins have also been successfully included in formulations. Many latices are prepared for captive use only. Tables 3.3 to 3.7 illustrate a range of formulations. Note that as polymerisation aspects have become well established, further improvements are only marginal, and most of those now quoted are from older publications and specifications.

Polyvinyl acetate and its copolymers have a marked tendency to 'creep' or 'cold flow', which causes a certain amount of difficulty in adhesive application. This has been overcome as far as possible by the use of polymers of the highest possible molecular weight. Films formed from a latex are generally tough and abrasion resistant, but have poor resilience and elasticity. The polymer is stable to a considerable temperature on heating, unlike polyvinyl chloride.

Polyvinyl acetate shows a slight sensitivity to water and is also relatively easily hydrolysed. The acetate units are replaced by hydroxyls under both acid and alkaline conditions. The salt of acetic acid is formed in the latter case. The unit equation is

$$H_2O + -CH_2CH(OOCCH_3) \longrightarrow -CH_2CHOH + CH_3COOH$$

Vinyl acetate and copolymer latices 117

Similarly, the polymer is fairly readily alcoholysed, especially with methanol. Therefore polyvinyl acetate does not have very strong water resistance. Moisture can be absorbed comparatively readily, especially when films have a high quantity of emulsifier. This is not necessarily objectionable as it enables films to 'breathe' and to transmit moisture when desirable. Typical examples are in paint used over brickwork or wood, in emulsion paint used externally in winter conditions in parts of the United States, and in paint in a kitchen, where there is a high humidity. Certain copolymers, such as those with branched chain fatty acids, have improved ester and alkali resistance.

Polyvinyl acetate is soluble in a wide range of solvents, including most ketones and esters with short hydrocarbon chains, methanol, 95 % ethyl alcohol, isopropanol, glycol ethers, diethylene glycol ethers and their esters, and most chlorinated solvents. It is soluble in benzene and in hot toluene, but xylol is a diluent for other solvents only. It solvents are water soluble, care should be taken when adding them to polyvinyl acetate latices, although most can be added slowly in small proportions with strong stirring. It is best to dilute the water-soluble solvents with water before addition to the latex.

Most technical latices tend to have comparatively large particle sizes, with diameters of about 0.5–1 µm, although some latices, particularly when prepared with ethylene oxide–propylene oxide emulsifiers or polymerisable surfactants, and some copolymers have somewhat smaller particle diameters.

An evaluation of suitable protective colloids for polyvinyl acetate latices showed that the highest latex viscosities were obtained with polyvinyl pyrrolidone and the lowest with carboxymethyl cellulose. The smallest particle sizes were obtained with hydroxyethyl cellulose and the lowest with gum acacia. All the latices were suitable for bonding porous materials such as paper, cloth and leather [169].

A number of modified polyvinyl alcohols have been claimed as protective colloids. These include a hydrolysed vinyl acetate–3-butenol 1-ol copolymer [169]. Acetoacetylated polyvinyl alcohol as a protective colloid gives a latex with temperature stability, especially if a urea of a derivative is added at about 1 % of latex. This also applies to copolymers with ethylene [170].

When used as stabilisers for the emulsion polymerisation of vinyl acetate with redox initiation of hydrogen peroxide and sodium formaldehyde sulfonate, 87 % hydrolysed copolymers of vinyl acetate and itaconic acid (99 : 1), with a degree of polymerisation of 1750, give a latex that forms films of high wet strength and good water resistance. Preferably, 1 % of a polyamide resin is added after polymerisation [171]. In some cases polyvinyl alcohols ending in -SH act as controllers of molecular weight [172] (see also Chapter 2, Section 4.4.2, and Reference [28]).

The polymerisation of vinyl acetate in the presence of polyvinyl alcohol has often been considered by those 'skilled in the art' to produce a partial graft of the monomer on to the polyvinyl alcohol. A fundamental study has shown that about 21.8 % of polyvinyl alcohol chains participate in grafting using Fourier transform infrared and nuclear magnetic resonance spectroscopies. Selective extraction with acetonitrile enables a quantitative separation of polyvinyl acetate from water-soluble polyvinyl alcohol and water-insoluble polyvinyl alcohol chains on which grafting occurred [173].

The plasticisation of polyvinyl acetate latices is described in Chapter 4, Section 5.2. The addition of plasticisers to a latex has the effect of slightly increasing the particle size. It is seldom necessary to prepare an independent plasticiser emulsion. It should be noted that some plasticisers, e.g. tritolyl phosphate (also known as tricresyl phosphate), enter the polymer particles much more slowly than, for example, dibutyl phthalate. Various proprietary plasticisers, usually of the polyester type, may be included in polyvinyl acetate compositions, especially where a particular property is required, e.g. the absence of migration into a substrate.

In order to obtain sufficient plasticisation for a continuous film at ambient temperature, about 5 % of plasticiser on polymer weight is desirable. Thus polyvinyl acetate emulsion adhesives are often plasticised to a slight degree to increase film strength and hence bond strength, although this might cause a slight increase in cold flow.

The following sections give a number of representative formulae. There are many variations in the literature, mainly described in patents.

3.2.1 Vinyl acetate, polyvinyl alcohol stabilised, delayed addition of monomer [174]

The following represents two standard types of formulation for these latices, which are of principal interest in adhesive formulation. In general the polyvinyl alcohol is an 88 % partially hydrolysed polyvinyl acetate, of low–medium molecular weight. Polyvinyl alcohols, in spite of similar specifications, vary widely in practice. Articles in several volumes indicate possible variations in the structure of polyvinyl alcohol [175–177]. A few modifications of the polyvinyl alcohol are described in recent patents. Examples of two typical latices are shown in Table 3.3.

3.2.2 Vinyl acetate, redox method with heel

In a modified process, vinyl acetate is polymerised on a 'heel' of about 20–25 % of a previously prepared latex of the same or similar type, whilst all the components, monomer, stabilisers and redox components are added simultaneously. This has the effect of producing a highly branched polymer, resulting in some limited crosslinking, and imparting the property that with a polyvinyl alcohol stabiliser the resultant latex may be diluted with ethyl alcohol, without the gelling that takes place with most polyvinyl acetate latices. The particle size of these latices tends to be relatively large, about 1 μm.

Vinyl acetate and copolymer latices

Table 3.3 Polyvinyl acetate latex and stabilised polyvinyl alcohol

	Latex 1	Latex 2
Water, potable grade	1000 kg	1000 kg
Formic acid (85 %)	to pH 3.4	to pH 3.4
Hydrogen peroxide, (35 %)	2.5 kg	2.5 kg
Polyvinyl alcohol, saponification value 100–120	40 kg	50 kg
Ferrous ammonium sulfate	to 7–10 ppm Fe^{2+}	
Vinyl acetate	1000 kg	1000 kg

Procedure:

(a) The water phase including polyvinyl alcohol and formic acid is heated to about 60 °C and then the hydrogen peroxide is added, followed by 10 % of the vinyl acetate. The reaction begins shortly with a rise in temperature and is allowed to rise to about 72–74 °C in about 30 min, cooling if necessary. At this stage the balance of monomer is added over about 3 h, allowing the temperature to rise slowly to about 80 °C at the end of the addition. Then the jacket temperature is raised with steam so that the reactor contents reach 92–96 °C, maintained for 30 min. Excess monomer can be removed by a careful sparge with nitrogen or air.

(b) The procedure is modified so that only 2 % of the monomer is added at the start, and the temperature is allowed to rise to 78–80 °C before the balance of monomer is added gradually; otherwise proceed as (a).

Final products:

The solids in each case are 50 % ±1 %. The pH should be raised to 4.0–4.5 by the cautious addition of sodium hydroxide solution; the viscosity of latex 1 is 40–60 cp and that of latex 2 is 5–8 P.

They have particular application in a procedure in the adhesives industry (see Chapter 8). A typical example is shown in Table 3.4.

3.2.3 Vinyl acetate and copolymers, mixed surfactant and colloid stabilisers; delayed addition of monomer

This is a standard method of production of latices with special reference to those suitable as media for emulsion paints, although they may be used for many other applications. Only a part of the monomer, 5–15 %, is added at the start of the reaction to a prepared water phase, usually including mixed non-ionic and anionic surfactants and a persulfate initiator. Polymerisation is commenced, and when the initial reaction is substantially complete, as shown by a slight fall in temperature, the balance of monomers is added, usually at a constant temperature of about 78 °C. In spite of the boiling point

Table 3.4 Polyvinyl acetate latex, prepared on 'heel' latex

	Weight (g)
Phase 1	
Polyvinyl acetate latex[a], stabilised polyvinyl alcohol 50 % of solids	100
Distilled water[b]	25
Sodium formaldehyde sulfoxylate	0.1
Phase 2	
Polyvinyl alcohol[c]	12.5
Hydrogen peroxide, 30 % aqueous	1
Sodium bicarbonate	0.25
Distilled water	225
Phase 3	
Vinyl acetate	215
Phase 4	
Sodium formaldehyde sulfoxylate	0.75
Distilled water	10

Procedure:

Add phase 1 to the reactor, warm to 70 °C and then add phases 2, 3 and 4 independently over about 3 h whilst the internal temperature rises to 80–82 °C. Maintain for about 30–60 min, raising the temperature to 92 °C. A non-ionic-type emulsifier may be included if desired. The final viscosity of the latex is rather sensitive to pH variations, which may be adjusted during the reaction by the addition of sodium bicarbonate solution.

[a] Most standard products are suitable, but optimally a latex of the same type as the final product is preferable. If a commercial product of the type in Table 3.3 is used, it will require three or four runs, using the final product of each run for the next to achieve the desired properties.

[b] Distilled water should be purged with pure nitrogen in all cases, but once the reaction has started, the nitrogen current may be discontinued.

[c] A medium viscosity polyvinyl alcohol, with a saponification value of about 120; with viscosity of a 4 % aqueous solution 20 cP is suitable. This is varied as required for the final latex viscosity.

of vinyl acetate being 73 °C, a polymerising latex will retain the monomer to a considerably higher temperature without any reflux taking place. The polymer, as emulsion particles, holds about 20 % of its weight of vinyl acetate at 80 °C without any appreciable reflux occurring. The absence of reflux is therefore not a guide to the completeness of polymerisation. The progress of polymerisation may be checked by taking samples and submitting them to

Vinyl acetate and copolymer latices

gas liquid chromatography. This testing procedure is essential with monomers such as the acrylate esters, which polymerise at a much faster rate than vinyl acetate.

A typical process is shown in Table 3.5 for a homopolymer of vinyl acetate. A copolymer in which 2-ethylhexyl acrylate is a comonomer, which can easily be prepared by the same method, is also shown. Most standard ester monomers such as di-n-butyl maleate or the vinyl esters of longer chain, including branched fatty acids, can be copolymerised in the same manner. As a general rule, in this type of formulation it is not necessary to use deionised water. This type of latex is prepared at about 55 % of solids. Viscosities may vary since an increase of hydrophilic-charged surfactant often tends to reduce viscosity. Specifications under the described conditions vary from 1 to 5 P.

3.2.4 Vinyl acetate polymers and copolymers, modified starch emulsifiers

Starch is probably the cheapest colloid and thickener available. Whilst natural starches are not in general suitable as colloids because of their high viscosity, the dextrins formed by degradation may be included in formulations, as may

Table 3.5 Vinyl acetate and copolymer latices (including hydroxyethyl cellulose)

	Parts by Weight (g)	
	Latex 1	Latex 2
Vinyl acetate	100	80
2-Ethylhexyl acrylate	—	20
Na dodecylbenzene sulfonate	0.5	—
Sulfated amyl/butyl oleate, ammonium salt, 50 % aqueous	—	0.5
Potassium persulfate	0.25	0.25
Sodium persulfate, 20 % aqueous*	0.5	0.5
Hydroxyethyl cellulose, low viscosity	2.5	2.5
Sodium bicarbonate	0.25	0.25
Water	80	80

Procedure:
> Prepare the water phase *except for the starred item*, adding the potassium persulfate last. Add 5 % of the monomers (mixed monomers in latex 2), heat with stirring for approximately 20–30 min and then add the balance over 3 h. If the rate of polymerisation falls, add the sodium persulfate solution as required; add monomer so as to maintain the internal temperature at 76–78 °C without appreciable reflux. Complete the polymerisation at 90 °C.

modified starches formed by addition of propylene oxide units. The high amylopectin starches, e.g. sorgum starch, which is almost 100 % amylopectin, are much more suitable than those based mainly on amylose. Dextrins derived from these starches, e.g. dextrins, whether formed by enzymic or acid degradation, are also suitable. Some examples of each type are given in Table 3.6 [178–180].

Table 3.6 Polyvinyl acetate latices and amylopectin stabiliser

	Parts by weight (g)		
	Reference [178]	*Reference [179]*	*Reference [180]*
Amylopectin starch	8.75	—	—
Sorghum starch, enzyme degraded	—	15.00	—
Sorghum dextrin, heat degraded	—	—	9.00
Sodium carbonate	1.00	—	—
Sodium bicarbonate	—	0.250	0.8
Cetyl alcohol (16 moles) ethylene oxide condensate	—	—	8.00
Ammonium salt, sulfated butyl oleate	—	—	1.00
Ammonium, persulfate	0.86 + 0.1	—	—
Potassium persulfate	—	0.50	0.80
Potassium hydroxide	—	to pH 4–4.5	—
Vinyl acetate	210	100	140
2-Ethylhexyl acrylate	—	—	60
Water	187	115	324

Procedures:

Reference [178]. A delayed addition process is used with about 10 % of the monomer and half of the initiator in most of the water phase, added at the start. The balance of monomer and initiator is added gradually at 80–85 °C; the supplementary initiator solution is added and the reaction is completed at 94 °C.

Reference [179]. The starch is slurried in about 2.6 times its weight of water and 0.35 % of a suitable enzyme (Nervanaise 10X) is added. The slurry is heated to 75 °C over 15 min and maintained for 30 min; 0.15 % of formic acid is added to destroy the enzyme. The latex is then prepared by a delayed addition process, the potassium persulfate and the potassium hydroxide as a 25 % solution, being added in three stages, the final being at the end of the delayed addition. The addition keeps the pH at 4.0–4.5.

Reference [180]. This entails a standard delayed addition process, with 5 % initial addition of monomer, the balance being added at 80 °C over 4.5 h. The temperature is finally raised to 90 °C. This type of emulsion gives a particle size of about 0.5–1.5 µm. The viscosity varies, being typically 2–7 p in the examples of reference [179] and about 0.2 p for emulsions prepared according to reference [180].

3.2.5 Vinyl acetate; ethylene oxide–propylene oxide block copolymer emulsifier

The block copolymers ethylene oxide–proplene oxide (EO–PO) (see Chapter 2, Section 3.6) are of considerable interest as emulsifiers for vinyl acetate and some copolymers since they seem to act as graft bases for the monomer. They form emulsions of much finer particle size than is usual with vinyl acetate, of the order of 0.1–0.3 µm in most of the above systems. The water resistance of the final latices is better than usual. Final emulsions, of the order of 50–55 % of solids, have a viscosity of the order of 15 p, with some 'false body', rapidly broken by shear. A suitable emulsifier is an ethylene oxide–propylene oxide block copolymer which is 80 % by molar ratio of ethylene oxide with an approximate molecular weight of 8000. A corresponding technical product is Pluronic ® F68 (EO–PO in Table 3.7). The process is rather critical in operation, and the delayed addition portion should be at least 90 % of the whole. The emulsions shown in Table 3.7 are taken from a specification [181].

3.2.6 Vinyl acetate, positively charged polymers

Positively charged cationic vinyl acetate latices may be obtained with a colloid such as a partially hydrolysed polyvinyl acetate or hydroxyethyl cellulose in conjunction with a cationic surface active agent, of which an example is given by the following general formula:

$$\text{R-CONH-}C_3H_6-\underset{\underset{HOCH_2-CH_2}{|}}{\overset{CH_3\diagdown \diagup CH_3}{N}}-H_2PO_4$$

where R is an alkyl radical with 12–18 carbon atoms. The phosphate salt is essential, a typical technical compound being γ-stearylaminopropyldimethyl-β-hydroxyethyl ammonium dihydrogen phosphate.

A development is the use of initiators which are themselves cationic in character, principally derivatives of azodiiso butyronitrile, e.g. azobis(iso butyramidine hydrochloride):

$$\underset{Cl^-H_2N^+\ \ \ CH_3}{\overset{CH_3}{\underset{|}{NH_2\diagdown_C-\overset{|}{\underset{\|}{C}}-N=N-\overset{CH_3}{\underset{|}{C}}-\underset{CH_3\ \ NH_2^+Cl^-}{C\diagup^{NH_2}}}}}$$

The advantage of these compounds is that they avoid possible destabilizing effects caused by the normal sulfate/sulfonate groups, especially as the oxidative character of persulfates may produce derivatives of amines that are

Table 3.7 Polyvinyl acetate latex and block copolymer emulsifier [181]

	Parts by weight (g)	
	Latex 1	Latex 2
Initial charge		
Water	34.01	14.36
Hydroxyethyl cellulose[a], 5 % aqueous	0.40	—
EO–PO	0.32	0.36
Sodium bicarbonate	0.13	0.13
Potassium persulfate	0.10	0.10
Vinyl acetate	2.55	0.51
Delayed addition		
EO–PO (24 % for latex 1; 22 % for latex 2)	13.18	12.80
Vinyl acetate	48.43	50.52
Potassium persulfate,		
4 % aqueous (late addition initiator)	0.52	0.52
	100	100

Procedure:
The initial addition is heated to 70 °C (5 % in latex 1; 1 % only in latex 2). The monomer is added at a rate so that the reaction is at incipient reflux. In latex 1 the emulsifier is added continuously from the time when 1.0 % of total monomer is added, and the addition is complete at 85–90 % addition of monomer. In latex 2 the emulsifier addition is complete when 50 % of the monomers are added. After completion of the addition, the terminal initiators are added slowly and the temperature is raised to 90–95 °C. The whole process takes 3.5 h.

[a]Unspecified grade.

inhibitory. Cationically modified colloids, e.g. guar gum, may also be included as in Table 3.8 [182].

3.3 Copolymers of vinyl acetate

Whilst many monomers have been successfully copolymerised with vinyl acetate, certain precautions have to be taken because of the wide difference in reactivity ratios which occurs in many cases. In particular, the acrylic series have a very wide diversity in reactivity $r_1 : r_2$ ratios ($r_2 =$ the acrylic monomer). Thus the $r_1 : r_2$ ratios for vinyl acetate and butyl acrylate are approximately 0.05 : 1. In practice in emulsion polymerisation, water solubility has a major effect, particularly where the initiators are water soluble, since compared with butyl acrylate or acrylates prepared from an alcohol with a larger carbon chain, vinyl acetate is much more water soluble. As this is

Table 3.8 Cationically charged latex [182]

	Weight (g)
Water phase	
Polyvinyl alcohol (d.p. 1400, 85 % saponified polyvinyl acetate), 6 % aqueous solution	160
Guar gum, cationic, 1–5 % N, 1 % aqueous, 43 cP	1.2
Sodium carbonate	0.5
Monomer phase (A)	
Vinyl acetate	10
Azobis(isobutyramidine HCl)	1
Monomer phase (B)	
Vinyl acetate	90

Procedure:
   Prepare the water phase; add the monomer phase A at 40 °C; heat slowly till 70 °C, then add monomer B over 2 h, commencing 10 min after the start of polymerisation and raise to 90 °C to give a 41 % latex.

a heterogeneous phase reaction, the greater water solubility of vinyl acetate tends to redress the balance of the $r_1 : r_2$ ratios.

However, if a fairly homogeneous copolymer is required, it is desirable to start a copolymerisation by a delayed addition, with either zero acrylate ester at the start or only about 2 % of the mixed monomers, and then add the balance slowly (sometimes referred to as 'starved' copolymerisation) so that the total monomer present at any one time is <2 %. Under these conditions a copolymer that is reasonably homogeneous is formed. The optimum conditions should be determined experimentally. Gas–liquid chromatography, used with samples taken from the reactor at various points in the reaction, is an efficient way of determining the course of the copolymer.

A kinetic study of the copolymerisation of vinyl acetate and acrylic acid, and a further one with the added inclusion of acrylamide, is available [183].

3.3.1 Vinyl acetate–ethyl acrylate copolymer, pre-emulsion feed, sulfosuccinate emulsifier

An example with 2-ethylhexyl acrylate as the comonomer is shown in Table 3.5. Further examples (from American Cyanamid Company) have a rather sophisticated process in which the components are added under controlled conditions as a pre-emulsion to a part of the water phase in the reactor. Note that the examples are redox processes. Note that this formulation also contains N-methylolacrylamide, which will be considered further in Chapter 5 as a monomer giving crosslinking properties (see Table 3.9).

3.3.2 Copolymers of vinyl acetate and ethylene

Copolymers of vinyl acetate with ethylene have been available for about two decades, and more recently terpolymers also include vinyl chloride. The manufacture requires high-pressure technology, and is most economically performed where ethylene can be taken directly from a pipeline, since any form of transport of ethylene invariably involves heavy losses due to its very low boiling point, -103.9 °C. Although the cost of ethylene has risen sharply in recent years, it is still the cheapest comonomer, being itself the major raw material for vinyl acetate.

The plasticising efficiency of ethylene is very high, about three times that of 2-ethylhexyl acrylate on a weight basis. Because of this, only low proportions of ethylene, about 5 % of vinyl acetate, are required to obtain flexible copolymer films at ambient temperatures. This is a disadvantage since pressure plant is much more expensive than normal reactors (see Figure 3.2) Copolymers including 10–15 % of ethylene are therefore of interest where there is a very high pigment/binder ratio, as in flat emulsion paints and especially in paper coating. Examples are shown in Table 3.10 [184, 185].

Experiments to obtain the maximum amount of ethylene that could be incorporated into a copolymer at reduced temperatures and pressures found that a 34 % ethylene content was obtained under optimum conditions at a pressure of 500 p.s.i. (35 kg cm^{-2}) and at 20 °C [186].

Latex terpolymers including vinyl acetate, ethylene and vinyl chloride have been developed. Because of the relative hardening influence of the vinyl chloride units it is possible to increase the relative proportion of ethylene. These terpolymers should be avoided under application conditions of strong light, since they have the typical tendency of vinyl chloride polymers to discolour slightly.

The analysis of copolymers and terpolymers including ethylene is difficult, probably because of considerable grafting during polymerisation and possibly in part on to the CH$_3$ of the acetyl radical. A number of other copolymers with three or more monomers have been disclosed and may be in production. Copolymers including groups capable of crosslinking are considered in Chapter 5.

3.3.3 Veova® copolymer

A seed latex process has been claimed for producing a vinyl acetate–Veova®–acrylic acid copolymer. This has already been briefly mentioned in Section 2.4.3. In this case the core has a $T_g > 13$ °C and the shell a $T_g < 13$ °C. Typical feed ratios are shown in Table 3.11. A second patent

Table 3.9 Vinyl acetate–ethyl acrylate copolymer latex and pre-emulsion feed

	Parts by weight (g)	
	Monomer A	Monomer B
Polymerisation vessel charge		
Water	62	62
Potassium persulfate, 5 %	20	20
Sodium bicarbonate	0.8	0.8
Sodium metabisulfite, 2 % aqueous	10	10
Aerosol A-102 (as 18 % solution)[a]	5.7	5.7
tert-Dodecyl mercaptan	—	0.2
Pre-emulsion		
Vinyl acetate	160	160
Ethyl acrylate	34	32
Sodium metabisulfite, 2 % aqueous	15	15
Water	44	95
Methanol	10	10
Aerosol® A-102[a] (as 18 % Solution)	5.7	—
Aerosol® OT, 37.5 % aqueous	—	1.33
tert-Dodecyl mercaptan	0.2	—
Delayed addition		
N-Methylolacrylamide, 60 % aqueous[b]	6.4	6.4
Water	10.0	10.4
Methacrylic acid	—	4
Diammonium phosphate	0.4	0.4
Surfonic N-95®, 50 % aqueous[c]	1	2
Separate addition		
Acrylamide	2	—
Water	20	—
Properties of resultant latices	Monomer A	Monomer B
Solids (%)	52.5	46
Particle size (μm)	0.35–0.6	0.15–0.19
Viscosity, Brookfield No.3 spindle 100 r.p.m. (cP)	596	140
Surface tension (dyn cm^{-1})	37	30
Film quality	Clear, continuous	Clear, continuous

[a]For formula see Chapter 2, Section 3.2.7.
[b]See Chapter 5 for further notes on this monomer.
[c]Surfonic N-95 is an alkylphenol polyoxyethylene and is a trade mark of the Jefferson Chemical Company.

Table 3.10 Vinyl acetate–ethylene copolymer latex [184, 185]

	Experiment number	
	1	2
Water (kg)	20	14.5
Polyoxyethylene nonyl phenyl ether (g) (cloud point 100 °C)	680	218
Polyoxyethylene nonylphenyl ether (g) (cloud point 55 °C)	340	151
Sodium vinyl sulfonate (g)	128	—
Sodium lauryl sulfate (g)	38	—
Hydroxyethyl cellulose (g)	—	157
Disodium hydrogen phosphate hydrate (g)	24	—
Citric acid (g)	56	—
Acetic acid (g)	—	9
Vinyl acetate (kg)	22.6	17
Potassium persulfate (g), 4 % aqueous (i)	300	500
Sodium formaldehyde sulfoxylate (g), 4 % aqueous (i)	25	Continuous addition as required
Potassium persulfate (g), 4 % aqueous (ii)	250	—
Sodium formaldehyde sulfoxylate (kg), 4 % aqueous (iii)	1.5	—

Procedure:

All ingredients except initiators and ethylene are added to the stirred reactor; the contents are purged with nitrogen and ethylene and then heated to 50 °C after addition of potassium persulfate (i). Ethylene is pressured at 36 atm, which takes about 11 min, and then the reaction is started by adding sodium formaldehyde solution (iii). The additional initiators (ii) are added as required, which is identical with experiment 4 of reference [185]. Experiment 1 takes 4.5 h and experiment 2 takes 10.5 h.

Properties. Experiment 1 has 48 % of solids, including 19 % of ethylene; the particle diameter is about 0.2 µm. Experiment 2 has 53 % of solids with an ethylene content of 18 %.

claims a core with the $T_g > 10$ °C and the shell <10 °C. In each case the difference between the core and shell must be at least 5 °C [117].

If alcohol ethoxylates are used for the environmentally less acceptable nonylphenyl ethoxylates in the formation of Veova® copolymers (see Chapter 1, Section 6.2), there is an increase in the molecular weight of the copolymer formed. This increases the water resistance and wet scrub resistance of derived coatings [187].

Table 3.11 Vinyl acetate–veova® copolymer for the core–shell process [117]

	Core	Shell
Vinyl acetate (g)	414	134
Veova 10® (g)	180	132
Butyl acrylate (g)	6	132
Acrylic acid	—	2

	Initial water phase	Pre-emulsion water phase
Water, demineralised (g)	245	600
Alkylaryl sulfonate, 10 % solution (g)	5	95
Non-ionic surfactant, 25 % solution (g)	—	80
Potassium persulfate (g)	—	4
Potassium carbonate (g)	—	4
Minimum film-forming temperature		0.5 °C
T_g measured by DSC		2.0/19.5 °C

Process:
  The pre-emulsion phase is formed from the monomers shown as the core and the pre-emulsion water phase. This is added at 80 °C over about 2 h and maintained for about 1/2 h. Then the shell monomers are added over about 2 h, and maintained for a further 1/2 h. Note that if the order of addition is reversed an inverted core–shell latex may result.

3.4 Alkali-soluble polymers

Vinyl acetate copolymerises well with crotonic acid, the $r_1 : r_2$ being 0.33 : 0. The zero implies that crotonic acid has no tendency to homopolymerise. About 5 % of crotonic acid produces a copolymer which is stable in emulsion form at pH 4, with the stabiliser formulation of anionic/non-ionic/hydroxyethyl cellulose operating quite well. Crotonic acid slows the copolymerisation somewhat, although it is uncertain whether this is due to radical transfer characteristics in the crotonic acid or to fortuitous impurities. It is possible to increase the crotonic acid to 15–20 %, but at the upper level complete polymerisation becomes difficult.

The crotonic acid must be dissolved in the monomer to give the necessary even copolymerisation (Experiment 1 in Table 3.12). The inclusion of acrylic acid with vinyl acetate has the disadvantage of a continuous thickening of the emulsion from a pH of about 3 until complete solution occurs. This disadvantage can be eliminated if an additional ester of maleic or fumaric acids is copolymerised with the monomers. Thickening does not take place until a pH of 5–5.5, with complete solubilisation [188].

The latices are of normal appearance, with the particle diameter averaging 0.5–1 μm. They dissolve to solutions which are virtually clear by adding a

Table 3.12 Alkali-soluble vinyl acetate copolymer latex [188]

	Parts by weight (g)		
	Experiment 1	Experiment 2	Experiment 3
Water	162	162	162
Hydroxyethyl cellulose	5	—	—
Sorghum dextrin	—	5.4	5.4
Sodium dodecylbenzene sulfonate	0.5	0.2	0.2
Cetyl alcohol/ethylene oxide (16 moles) condensate	2	6	6
Acrylic acid	—	7.4	7.4
Crotonic acid	10	—	—
Vinyl acetate	190	162.6	162.6
Dibutyl maleate	—	30	—
Dibutyl fumarate	—	—	30
Potassium persulfate	0.6	0.6	0.6
Potassium bicarbonate	—	0.3	0.3
Sodium persulfate, 20 % aqueous	3	—	—

Procedure:
  Of the total charge 5 % is added to the water phase in experiment 1 and 10 % in experiments 2 and 3. The balance of monomers is added at 76 °C, the temperature rising to 90 °C. In the case of experiment 1, the additional sodium persulfate solution is fed to the reactor during the second half of the polymerisation. Trace impurities in the crotonic acid have been known to slow the polymerisation considerably.

stoichiometric quantity of alkali, or ammonia solution. Dilution of the alkaline solution to 25 % is necessary for handling purposes.

The physical properties of the films resemble those of polyvinyl acetate. The major difference is the ready attack by dilute alkali. Plasticisation is possible, as with a vinyl acetate latex, although films from the alkali solubilised latex are not completely clear unless a plasticiser is included which has appreciable water solubility. Typical formulations are shown in Table 3.12 [188].

The molecular weight of commercial products is relatively low, probably with a viscosity weight average about 50 000. The latices are directly soluble in caustic alkalis, ammonia, borax, sodium carbonate and sodium bicarbonate solutions. Amines are not recommended as by-products and secondary reactions give rise to coloured products. Solubility starts at a pH between 6.5 and 7.0, and is complete at about pH 8.5, the viscosity rising considerably. In order to obtain a solution with good handling characteristics it is desirable that the final solids are reduced to 20–25 %.

Excess of alkali should be avoided, as this leads to hydrolysis of the vinyl acetate units, causing a further increase in viscosity. Excess of borax may also cause thickening, probably because of a reaction with vinyl alcohol units formed by hydrolysis. As a rule, it is desirable to use the alkali-solubilised solutions as soon as possible after preparation. The property of alkali solubilisation is not reversible. If the alkali-solubilised latices are made acid they coagulate into an insoluble mass. The modern developments, which involve 'core–shell' copolymers, are described in Section 2.4.1.

Terpolymers of crotonic acid with vinyl acetate, vinyl esters of fatty acids such as capric acid, and especially the vinyl esters of highly branched fatty acids, e.g. Veova 10®, including up to about 20 % of the latter, have the advantage that films from ammonia solution have considerable water and even alkali resistance. Terpolymers of the vinyl ester of Versatic® acids (the Veova® series) and crotonic acid are formed readily since the reactivities are similar. The terpolymers have similar alkali solubility characteristics, the main difference being increased film flexibility. Terpolymers have considerable water and even alkali resistance, but a somewhat greater excess of alkali is required to solubilise the latex [189]. On the other hand, it is difficult, if not impossible, to form terpolymers of reasonable homogeneity with acrylic monomers, vinyl acetate and crotonic acid because of the vastly greater reactivity of the acrylic ester series.

The major technical advantage of the alkali-soluble products is the increase in pigment binding power which makes them useful for applications requiring high loadings, as in binders for china clay in paper coating.

The property of thickening rather than dissolving on addition of alkali may be useful on some occasions. Some products, e.g. Acronal 500D (see Figure 3.7), have this property, the increase being of high order from a free-flowing latex almost to a gel by the addition of 2 % of a 25 % ammonia solution on emulsion weight. A latex with similar properties can be prepared containing approximately 48 % of vinyl acetate, 48 % of ethyl acrylate and 4 % of acrylic acid. The acid is copolymerised with part of the acrylic ester, and the balance of vinyl acetate and acrylic ester is then added by continuous addition. By this means, vinyl acetate is not directly copolymerised alone with the acrylic acid.

3.5 Equivalence of 'external' and 'internal' plasticisation

The plasticising efficiency of all the major ester monomers used for the internal plasticisation of vinyl acetate copolymers increases with the length of the alkyl chains. The upper limit has not been precisely defined, but is probably between 8 and 12 carbon atoms. Branched chains, as expected, are less efficient in producing flexibility than straight chains, although 2-ethylhexyl alcohol esters are popular because of the ready availability of the synthetic alcohol.

The most efficient plasticising monomers include n-heptyl acrylate, 2-ethylhexyl acrylate, 2-ethylhexyl fumarate and 2-ethylhexyl maleate, the efficiency being about 60–70 % of that of dibutyl phthalate, the standard for comparison. Ether alcohol esters are sometimes quoted, e.g. diethoxyethyl

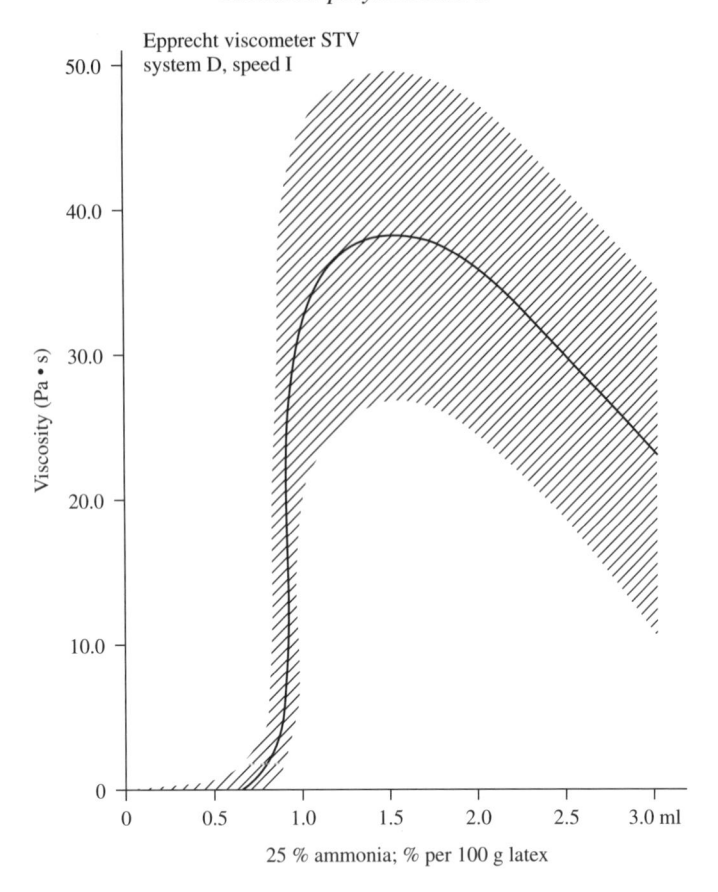

Figure 3.7 Change in viscosity of a polymer latex, Acronal 5000, leading to formation of an ammonium salt. (Reproduced from data of BASF.)

acrylate, and have high efficiency as internal plasticisers, although more expensive than the simple alkyl esters.

The vinyl esters of straight chain fatty acids are also used as plasticising monomers, efficiency increasing with the length of the alkyl chain in the fatty acid. The Veova® 10 type of monomer is less efficient, approximately 25–28 % of total monomer being required to be the equivalent of 10 % (on total film former) of dibutyl phthalate. This, however, is advantageous because a vinyl acetate–Veova® copolymer with 25 % of the latter is required to upgrade the water and alkali resistance of the copolymer to approximately its maximum extent, which tends to approach that of a 100 % acrylic copolymer.

The efficiency of ethylene as an internal plasticiser (Section 3.3.2) is probably due to its very low molecular weight and a weight percentage of ethylene gives a much higher molar percentage than any other comonomer. The alkali resistance of film from ethylene–vinyl acetate copolymers is of the same order as the Veova® copolymers mentioned previously. This may be due to a high degree of chain branching or grafting, causing steric hindrance, which improves resistance to hydrolysis [47].

In polymerisation formulations, a very similar system to vinyl acetate homopolymerisations can be adopted. A general rule, applicable particularly to the longer-chain acrylate esters, is that, above 15 % of comonomer, there should be a reduction of water-soluble polymer, e.g. hydroxylethyl cellulose, and a corresponding increase in non-ionic surfactant.

For special purposes the comonomer may be raised to 40 % of the total or even greater. In this case the stabiliser formation may need modification to approximate closer to that of acrylic esters. Care should be taken that the rate of monomer addition is controlled to prevent uneven copolymerisation because of the different monomer reactivities, unless an uneven copolymerisation is deliberately required.

The particle sizes of most copolymer emulsions tend to be lower than that of the corresponding homopolymer. For further details of copolymers with special reference to emulsion paints, see Chapter 9 of Volume 2. However, almost all comonomers reduce the ease of hydrolysis of vinyl acetate copolymers, the maleates less so than the others because of their relatively high polar character, the acids having two carboxyl groups on adjacent carbon atoms.

4 ACRYLIC LATICES

Latices based on acrylic monomers, usually in the form of copolymers, are generally more versatile than the vinyl ester series. A wide range of simple esters are available including many with extra functions, i.e. multiple double bonds or having other types of polymerisable groups, e.g. the amide group —$CONH_2$ and its derivatives or the epoxide $-CH-CH_2$ with O bridging, examples of which have been given in Chapter 1. A wide range of simple acrylic and methacrylic esters are available; the polymeric acrylic esters are less readily hydrolysed than polyvinyl acetate and the polymeric methacrylic esters are very difficult to hydrolyse. The relative properties have been described in Chapter 1.

In addition, both acrylic and methacrylic acids are readily available commercially and copolymerise readily with ester monomers, especially if dissolved in the monomers rather than the water phase. In general, methacrylic acid, being less hydrophilic, is more suitable for emulsion polymerisations, although many patent specifications quote acrylic acid as a comonomer. A variable feed rate addition has been used in the delayed addition emulsion polymerisation of a terpolymer latex prepared from methyl methacrylate, ethyl acrylate and methacrylic acid to give a polymer with almost constant composition [190].

Problems with emulsion stability vary with the nature of the monomers. Methyl acrylate, because of its relatively high water solubility and the insolubility of its polymer including absorbed monomer, tends to give difficulty in forming stable homopolymer latices, although there is little difficulty with copolymers. The polymers and copolymers of the higher acrylates also give some difficulty owing to the very high cohesion tendency of the polymer particles with absorbed monomer. Procedures giving hard strongly adsorbed shells of emulsifiers may avoid this problem of coagulum. Some examples that follow indicate suitable formulations.

The emulsion polymerisation of various copolymerisations of butyl acrylate/methyl methacrylate and vinyl acetate were studied using a 5 L stainless steel pilot plant, and data were used to establish useful formulations [191]. Most acrylic monomers copolymerise well with styrene, which is added as a 'hardener' instead of methyl methacrylate, which is rather more expensive. However, styrene as a comonomer tends to cause copolymers to yellow in sunlight. Nevertheless, styrene is included as a comonomer in many acrylic formulations, as will be obvious in the formulations that follow.

Besides acrylic and methacrylic acids, other derivatives are available for inclusion in emulsion copolymers. The ethoxyethyl esters $C_2H_5OC_2H_4OOCCX{:}CH_2$ ($X = H$ or CH_3) copolymerise very well, the O atom having a strong plasticising effect. The hydroxy esters $HOC_2H_4OOCCX{:}CH_2$ and $HOC_3H_6OOCCX{:}CH_2$ are available; they copolymerise readily, having formulations including the polymerisable acids without the hydroxy groups reacting with them, but giving extra film strength by van der Waals forces (see also Chapter 8).

Many practical formulations include monomers with multi-double bonds that have crosslinking properties. The most frequently described are the acrylic and methacrylic diesters. The simplest are the diesters of ethylene glycol, e.g. $CH_2{:}CXCOOC_2H_4OOCCX{:}CH_2$ ($X = H$ or CH_3). In practice, the diesters of longer-chain glycols are preferred as they have less reactivity than the ethylene glycol esters, and do not lead to premature or excess crosslinking which causes precipitation from the polymerising latex.

Other mild crosslinking monomers include allyl methacrylate $CH_2{:}CHCH_2OOCC(CH_3){:}CH_2$. Surprisingly, the divinyl esters of dibasic acids, such as divinyl adipate and divinyl sebacate, are not available in commercial quantities. See Chapter 5 for further details of crosslinking possibilities.

Occasionally a polymerisable siloxane is added as part of the monomers, an example being 3-methacryloyloxypropyltrimethoxysilane, added at 3 % of total acrylic monomers [192]. Vinyltrimethoxysilane at 5 % of additive on monomers is also quoted at 5 % of additive on total monomers [193].

4.1 Typical acrylic formulations

The formulation of a number of typical acrylic latices is now given together with modern and more complex developments in Section 4.2. Note that acrylic or methacrylic acid, usually at 1–2 % are included in most formulations. The acids give very considerable improvement in emulsion stability and also improve freeze–thaw resistance. Higher quantities of these acids are included for special purposes, e.g. when particle solubility in alkali is desirable. Acids also play a major part in the formulations of many latices formed on a seed, and in particular of 'core–shell' copolymers.

As methacrylic acid is less hydrophilic than acrylic acid due to the additional methyl group, this is the preferred acid for copolymerisation with acrylic ester monomers. This is well known, but has been confirmed in a theoretical investigation with butyl acrylate as the principal monomer. Methacrylic acid is also more evenly distributed in the latex particles than acrylic acid [194]. Hydroxyethyl acrylate, which is somewhat hydrophilic, also improves nucleation and stability during the polymerisation process [195].

In some cases, potentially crosslinking monomers are included. Many of these formulations include derivatives of acrylamide $CH_2{:}CHCONH_2$, especially methylolacrylamide $CH_2{:}CHCONH(CH_2OH)$ and its ethers. These are potential crosslinking agents and will be considered more fully in Chapter 5.

The technical monomers as shown in subsequent formulations normally contain traces of an inhibitor such as hydroquinone, usually <0.05 % and often <0.01 % to prevent premature polymerisation in storage. Experience shows that these traces of inhibitor have a negligible effect on the process of polymerisation.

4.1.1 Acrylic copolymer; delayed addition procedure

Table 3.13, with latices having identical water phases, shows three formulations of varied monomer proportions so as to vary the minimum film-forming temperature. All formulae give free flowing latices of fine particle size (circa 0.2 μm average). For most purposes the final latex will require neutralisation to a pH of 5.5–8.

4.1.2 Acrylic ester latex; two-stage redox process

Redox polymerisations are often performed in several stages because of premature exhaustion of the initiator system by side reactions, by chain termination and also because on a large scale the heat of reaction may cause difficulties in control, since there is a reduction in the relative ability of the surface area of the reactor walls to absorb the heat of reaction which depends on the volume. For a copolymer of butyl acrylate and methyl methacrylate by a two-stage process, see Table 3.14 [196].

4.1.3 Acrylic ester–styrene copolymer

This formulation includes an alkyldiphenyloxide disulfonate (Dowfax® 2A1) emulsifier. Note that many other sulfonate emulsifiers may be suitable, and also other acrylate esters, and ratios may be varied (see Table 3.15).

Table 3.13 Acrylic copolymer latex

	Weight (g)		
	Experiment 1	Experiment 2	Experiment 3
Monomer			
Ethyl acrylate	800	400	480
Methyl methacrylate	—	400	480
Methacrylic acid	—	—	8
Water phase			
Water initially in flask	200	200	200
Water	800	800	800
Sodium alkylaryl polyether sulfonate, 28 % solution	96	96	96
Ammonium persulfate	1.6	1.6	1.6

Procedure:
A pre-emulsion is formed with all components except the water, which is initially in the flask. To this latter in a 3 L flask is added 200 ml of the pre-emulsion. This is heated on a water-bath until the internal temperature reaches 82 °C, when reflux begins. The internal temperature rises to 90 °C in about 8–10 min when reflux subsides as the polymerisation nears completion. At this stage the pre-emulsion is added continuously for about 90 min, maintaining the temperature at 88–94 °C and finally at 97 °C to complete the polymerisation as far as possible. For most purposes the finished latex will require neutralisation to a pH of 5.5–8. Ammonia solution is normally used for this purpose.

Properties of final latices:

Solids content (%)	42.6–43.6		
pH	1.7–2.9		
MFT °C (minimum film-formation temperature—see Chapter 4)	5	21	0

4.1.4 Alkali-soluble polymer; acrylic polymer and seed latex

These two features are illustrated in Table 3.16. Section A follows normal delayed addition procedure, but the methacrylic acid has been raised to 15 %, at which the acid emulsion becomes easily soluble in alkali. In section B, 5 % of latex A (based on solids/monomers) is used as a seed on which to build the second emulsion. Although this results in a somewhat larger particle size, the resultant latex has a very low net quantity of water-soluble material, although it is alkali soluble. Note the high ratio of molecular weight controller in both sections. The final latex is itself alkali soluble, although it may be formulated with a lower proportion of methacrylic acid. The final formulation as shown in final latex B is intended for crosslinking operation with hexamethylolmethyl melamine (Chapter 5) [197, 198].

Table 3.14 Acrylic ester latex 2—stage process [196]

	Weight (g)	
	Stage 1	Stage 2
Deionised water	1000	—
Sodium alkylaryl polyether sulfonate, 28 % solution	31.6	—
Ethyl acrylate	253	283 g
Methyl methacrylate	168	188
Methacrylic acid	4	5
Ammonium persulfate	0.5	0.6[a]
Sodium hydrosulfate (in 6 ml of water)	0.6	0.5

Procedure:
The stage 1 ingredients, except initiators, are stirred at 15 °C. The two initiators are added individually under nitrogen. Polymerisation starts almost immediately and a peak temperature of about 65 °C is reached in about 30 min. After a further 5 min, the additional surfactant and mixed monomers are added, and, as previously, the initiators are added individually. The peak temperature of 65 °C is maintained for 1 h. The latex is cooled and is adjusted to a pH of 9.5 with 0.88 (28 %) ammonia solution. The solids content is about 46 %.

[a]In 1.5 ml of water.

Table 3.15 Acrylic ester–styrene copolymer

	Parts by weight
2-Ethylhexyl acrylate	140 g
Styrene	60 g
Dowfax® 2A1 (as 100 % active)	6.7 g
Ammonium persulfate	0.96 g
Water	390 g
Ferrous ion	6 ppm
Sodium *meta*-bisulfite	0.96 g
tert-Butyl hydroperoxide (70 %)	5 drops

Procedure:
This may be taken as a normal delayed addition process, about 5 % of monomer being present at the start; otherwise, in a two-stage (50 : 50) process or a three-stage (40 : 30 : 30) process, the initiators should be increased by about 50 %, the excess being added in the second or third stage. The starting temperature is 35 °C under nitrogen, rising to 82 °C, and is maintained at 85 °C for completion. If a stage process is used, allow the emulsion to cool somewhat before adding the extra stage, but do not attempt to complete each stage.

Table 3.16 Acrylic latices based an a 'seed' [197, 198]

	Parts by weight (g)	
	Seed latex A	Final latex B
Ethyl acrylate	65.0	70.0
Methyl methacrylate	20.0	20.0[a]
Methacrylic acid	15.0	10.0
tert-Dodecyl mercaptan	3.0	3.0
Seed latex A (40 % solids)	8.0	—
Potassium persulfate	0.25	0.5
Sodium persulfate	0.0625	0.2
Water (total)	140.0	140.0

Procedure:
  Both stages are prepared by normal delayed addition processes. The sodium persulfate is an addition at a late stage of polymerisation to ensure completeness of polymerisation. Then 13.6 g of diethylaminoethanol is added to latex B to give a clear solution, suitable for blending with the melamine aminoplast.

[a]Methyl methacrylate may be replaced by styrene.

4.1.5 Core–shell polymerisation

One example quotes itaconic acid or acrylic acid, but similar formulae may be used. Note that to produce a satisfactory film at ambient temperatures in this example the filming monomer premixture should contain a lower proportion of styrene (see Table 3.17) [199].

4.2 Developments in polymerisation of acrylics

There are a considerable number of patents and also a number of papers in theoretical journals giving various modified processes for the preparation of acrylic copolymers with varying properties; these are described in this section. Papers dealing with mechanistic and kinetic theories are considered in Section 8.9 of this chapter.

Whilst it is unusual for polyvinyl alcohol to be included as a stabiliser with an acrylic latex, an SH-terminated polyvinyl alcohol (see Chapter 2, Section 4.4.1) has been claimed in conjunction with a non-ionic surfactant [200]. Polyvinyl alcohol (type undisclosed) has been used with hydroxyethyl cellulose in the ratio of 1 : 0.1 % on monomers consisting of methyl methacrylate and butyl acrylate (1 : 1 %) with acrylamide and acrylic acid (0.1 : 0.5 %). A redox initiator system based on t-butyl hydroperoxide and sodium formaldehyde sulfoxylate was used for initiation [201].

Cellulose derivatives (Chapter 2, Section 4.2) are seldom used with acrylic monomers as they do not in general promote stability and give coarse particle

Table 3.17 Core–shell acrylic polymerisation

	Parts by weight (g)
Initiator–surfactant premixture	
Ammonium salt, alkylphenoxy polyoxyethylene sulfate (57 % solution)	13
Nonylphenoxy polyoxyethylene ether	15
Ammonium persulfate	4
Itaconic acid *or*	35
Acrylic acid (alternative to above)	30
Water	1028
Filming monomer premixture	
Ethyl acrylate	600
Styrene	200
Second emulsifier premixture	
Sodium salt, arylphenoxy polyoxyethylene sulfate	10
Water	30
Ammonium hydroxide (concentrated)	10
Non-filming monomer	
Styrene	200
Second initiator premixture	
Sodium persulfate	1
Water	20

Procedure:
  The film-forming monomers are added over 90 min to the initial water phase at 75–83 °C, with sodium persulfate added towards the end of polymerisation. A second emulsifier solution is added, followed by the entire styrene, and then additional initiator in two equal stages, the second when the reaction slows. The final film is clear and glossy with exceptional water resistance.

sizes. This is considered to be due to 'bridging' by the monomers between the cellulose derivative molecules. However, the use of carboxymethyl cellulose with a high level of carboxymethyl substitution, typically Ambergum ® 3021 (Hercules), has been described in the emulsion polymerisation of acrylic monomers. It is stated to give very small particle sizes and good shear stability [202]. Unsaturated siloxanes have also been copolymerised with various acrylic esters [203, 204].

A stable latex of ethyl acrylate is formed using a copolymer of vinylpyrrolidone and acrylic acid as emulsifier, with a persulfate initiator [205]. An amphoteric copolymer emulsifier formed from N,N-dimethyl-N(2-methacryloyloxyethyl)-N-(3-sulfopropyl) ammonium inner salt, with sodium styrene sulfonate (Chapter 2, Section 3.9) and butyl acrylate (60 : 35 : 5) at 5 % concentration, may be used in forming a methyl methacrylate latex with an average particle diameter of 0.20 μm. A feature is stability to coagulation when mixed with salt solutions, e.g. sodium sulfate and calcium chloride [206].

Disulfonated dodecyldiphenyl oxide sodium salt (Chapter 2, Section 3.2.4) as stabiliser enables a coagulum-free latex of methyl methacrylate and butyl acrylate at 45.2 % of solids to be formed, with a particle diameter of 40 μm and a weight average molecular weight of 1.22×10^6 [207].

Polybutyl methacrylate latices with solids of 61 %, and possibly up to 70 %, have been prepared by a two-stage process. In the first stage a seed latex is produced by a standard mini-emulsion polymerisation (Section 2.4.5). In the second stage, with the addition of 0.5 % of sodium lauryl sulfate on monomer, the solid content was raised to give a paste-like latex. This type of process, a gradual addition, tends to give latices tending towards the monodisperse state irrespective of initial particle distribution [208].

The emulsion polymerisation of higher alkyl (meth)acrylate esters C_{8-30} has been performed using two non-ionic surfactants, one with an HLB balance of 8–12 and the other 12–16. Typically the polymerisation of tetradecyl methacrylate with a persulfate initiator produces a stable latex at 99 % conversion after heating for 4 h at 70 °C [209].

Bicontinuous micro-emulsions containing methyl methacrylate, 2-hydroxyethyl acrylate and ethylene glycol dimethacrylate in which there are long- and short-chain cationic surfactants are precursors in the preparation of microporous polymeric materials [210]. Mini-emulsion polymerisation of acrylic monomers in the presence of an alkyd resin shows high stability and latices free from coagulum [211].

The average hydrolysis of glycidyl methacrylate in a copolymerisation with butyl acrylate is about 7 %, diminishing somewhat as the temperature is reduced. Carbon tetrabromide as the chain transfer agent reduces any internal crosslinking. Methacrylic acid may also be added without affecting the sol content of the particles [212].

Mixed ammonium persulfate and azobis-isobutyronitrile initiators enable high solid copolymers of methyl methacrylate–butyl acrylate to be formed [213]. The inclusion of 5 % of methacrylic acid in latices, optionally with 0.2 % of sodium bicarbonate, improves the chemical stability of the final latices, particularly at high pH [214].

5 STYRENE LATICES

5.1 Processes—general

Styrene as the sole monomer is relatively simply converted to polystyrene by emulsion polymerisation, even with single surfactants. Styrene latices may be prepared in reactors under autogenous pressure, or equipped with a reflux condenser. In the latter case it is good practice to purge the water phase with a current of pure nitrogen before commencing the reaction.

5.1.1 Soap-type surfactants

Table 3.18 gives an example of a polymerisation with a soap-type surfactant. Soft water should always be used with these emulsifiers. This formulation gives a very fine particle size latex, which finds application in emulsion polishes described in a later volume. These latices should be kept at pH 7.5 or above, although the non-ionic surfactants, as shown below, may give overall stability at a lower pH.

Latices of this type are not generally freeze–thaw stable. Freeze–thaw stability may be conferred by the addition of a non-ionic emulsifying agent of the alkylphenyl polyoxyethylene type at about 7 % of monomer, or alternatively by addition of about 2 % diethylethanolamine $(C_2H_5)_2NC_2H_4OH$. It is not recommended that the latter be added before polymerisation instead of ammonium hydroxide as it causes some inhibition or retardation of the reaction. However, the use of a redox system enables the reaction to be completed.

The very fine particle size, of diameter $\leqslant 0.1$ μm, is below the wavelength of light, and hence even in a concentrated form the latex appears translucent rather than opaque. The most satisfactory type depends not only on the exact acid soap used but also on external conditions such as the shape of the reactor and the nature and speed of the stirrers. It is essential that the product should produce no sediment of larger particles on standing or dilution. Viscosity is

Table 3.18 Styrene latex with a soap-type emulsifier [215]

	Parts by weight
Styrene 1	25
Styrene 2	975
Distilled coconut oil fatty acid	75
Water, softened	1115
Ammonia solution (0.88)	12
Potassium persulfate	9

Process:
    The potassium persulfate is dissolved in the water at about 50 °C followed by the ammonia and the fatty acid, which may be warmed to a fluid state. A current of pure nitrogen is bubbled slowly through the water phase. Styrene 1 is added. The temperature rises to 85 °C with a slight exotherm. After 15 min, the balance of monomer is added over about 4 h and held for about 1.2 h until polymerisation is substantially complete.

142   Emulsion polymerisation

difficult to control by the use of external agents, although apparent viscosity can be increased to some extent by the use of higher fatty acids, e.g. those based on a C_{18} fatty acid (Table 3.18).

Because of the very fine particle size and consequent large surface area, virtually all of the soap is absorbed on to the particles. In consequence, these latices have a relatively high surface tension, sometimes over 50 dyn cm^{-1}.

As with the latices based on vinyl acetate, a supplementary list is included in the Appendix to this chapter.

5.1.2 Vinyl acetate–maleic acid stabiliser

A copolymer of 1 : 1 vinyl acetate and maleic anhydride is prepared by copolymerisation of the monomers under nitrogen with a dibenzoyl peroxide initiator. The product is steam distilled, simultaneously hydrolysing the maleic anhydride, which is neutralised with sodium hydroxide or ammonium hydroxide solution to pH $\geqslant$ 7.

As this stabiliser solution is of low viscosity, it results in a polystyrene latex with a viscosity varying very sharply at constant non-volatile content with the

Table 3.19 Styrene latex–vinyl acetate–maleic acid copolymer emulsifier [215]

	Weight (g)			
	Formulation 1	*Formulation 2*	*Formulation 3*	*Formulation 4*
Styrene	150	144	250	187.5
Ethyl acrylate	—	—	—	62.5
Dibutyl phthalate	—	56	—	—
Non-ionic emulsifier[a]	9	12	—	5
Vinyl acetate–maleic anhydride copolymer[b]	7.5	6.7	41.6	15.7
Water	192	231	235	245
Potassium persulfate	1	1.3	2.5	2.5
Ammonia (28 % solution)	0.75	0.5	1.25	1.0
Final dry solids (%)	43.3	48.5	No record	No record
Final particle size (μm)	0.5–1	0.5	No record	No record

Procedure:
  Either bulk or delayed addition processes may be used. Formulations 1,2 and 4 require a temperature of about 60–65 °C for 5 h; for formulation 2 half of the persulfate is added at the start. The latex is heated for 6 h at 70 °C and then the balance of persulfate is added and polymerisation continued at 75 °C for 4 h. A nitrogen blanket at the start is preferred in all cases.

[a] Acetyl alcohol–ethylene oxide condensate; chain length about 13 units.
[b] The figures are a 30 % solution in all cases, neutralised to pH 7.5.

ratio of the copolymer emulsifier to monomer. With 1 % of the copolymer stabiliser, the final latex is free flowing. If the stabiliser is raised to 5 %, the final latex is paste-like.

Polystyrene is normally difficult to plasticise to give a flexible film. However, prepared with this copolymer stabiliser, sometimes including a non-ionic emulsifier, plasticisation is satisfactory, e.g. with dibutyl phthalate. This may be added either before or after polymerisation. Possible uses of these latices are as sizes for fibre glass before impregnation (see Table 3.19 [215]).

Evidence suggests that the styrene monomer grafts on to the vinyl acetate–maleic anhydride copolymer to some extent. This type of formulation is worthy of further development. It is also suitable for acrylic ester copolymers.

5.1.3 Cationic emulsifiers

Various cationic polymerisable emulsifiers give stable styrene latices and with a cationic initiator give monodisperse latices, but polymeric cationic surfactants with a low hydrophobe content do not give latex stability. Small amounts of polymeric surfactant are grafted on to the particles during the reaction [216].

6 DIENE POLYMERS AND COPOLYMERS

6.1 Introduction

Diene monomers differ from the vinyl and acrylic series in that there are two double bonds. The principal monomers are 1 : 3 butadiene $CH_2:CH.CH:CH_2$ and isoprene $CH_2:C(CH_3):CH:CH_2$. These monomers differ from other monomers with several double bonds, such as ethylene glycol diacrylate, in that only one double bond polymerises in the early stages of polymerisation, especially if temperatures are kept low, at about 50 °C. A reaction of unduly prolonged time or higher temperatures causes both crosslinking and dimerisation, which latter may cause further inhibition and other side reactions. Table 3.20 gives a typical formulation.

The main interest in diene copolymer latices, usually with styrene but occasionally including acrylonitrile, is in synthetic rubber. The term 'synthetic latex' is sometimes restricted to copolymer latices containing dienes. A current volume gives a survey of the emulsion polymerisation of styrene and butadiene for the purposes of the manufacture of synthetic rubber in which the ratio of these two monomers is approximately 1 : 3 respectively. The rubber is precipitated in a convenient crumb form. In addition, the polymerisation is usually 'short stopped' by an inhibitor. For this purpose it is not necessary in emulsion polymerisation to prepare a latex with a high degree of stability, whether to mechanical action, freezing and thawing or pigmentation. The manufacture of synthetic rubber will not be considered further here [217].

Table 3.20 Butadiene copolymers

	Parts by Weight (g)		
	Formulation 1	2 (initial addition)	3 (late addition)
Butadiene	33.9	12	24
Styrene	—	8.0	16
Acrylonitrile	19.7	—	—
Sodium lauryl sulfate	0.3	—	—
$C_9H_{19}C_6H_4(OC_2H_4)_7OCH_2COOH$	0.95	—	—
Dodecylbenzene-20-oxyethylene	0.45	—	—
Ethylenediamine tetra-acetic acid disodium salt, Na salt	0.45	—	—
tert-Dodecylmercaptan	0.58	0.1	0.2
Potassium persulfate	0.345	0.3	0.2
Sodium metabisulfite	0.25	—	—
Ferrous sulfate	0.05	—	—
Potassium carbonate	—	2	—
Oleic acid(1)	—	0.067	1.33
Oleic acid (final addition)	—	—	1.00
Water(deionised for B)	72.5	35	6
Non-volatile solids content (%)	41.5	—	59

Process:
1. Heat entire emulsion for 24 hours at 30 °C.
2. Polymerise at 75 °C in two stages. The gradual addition of monomers takes place over 4 h. This latex features highlight transmission.

6.2 Dienes: non-rubber applications

For non-rubber applications, one of the principal ones being the backing of carpets, a smaller proportion of diene, 60 % by weight of total monomers, is preferred and it is necessary to carry out the reaction to substantial completion under controlled conditions. It is desirable that polymerisation temperatures should be kept relatively low, preferably below 75 °C. Formulations include those of SB latices and butadiene–acrylonitrile copolymers (Table 3.20).

6.3 Carboxylated latex formulations

Amongst modified butadiene copolymer latices for non-synthetic rubber applications, those containing polymerisable acids require a mention. The acids concerned include itaconic acid, acrylic acid and methacrylic acid. The polymerisation process usually takes place at a pH of about 3.5–5.0 with emulsifier of the type used in acrylic polymerisations, rather than the slightly alkaline 'soap' type, which is favoured for the manufacture of synthetic rubber.

Diene polymers and copolymers

It is doubtful in the copolymerisation reaction with itaconic acid whether all of this acid actually copolymerises because of the lack of mutual solubility of the monomers. Adjustment of the emulsifiers may help to assist mutual solubilisation and thus to some extent assist copolymerisation. Copolymerisation may be somewhat easier with methacrylic acid. See Table 3.21 for an example.

A recent patent claims a formulation for a carboxylated latex in which a trace of an oil-soluble polymerisation inhibitor is added to the monomers before polymerisation (Table 3.22) [218]. The process can be further modified to give a core–shell structure by adding to a seed latex a preformed emulsion of 'hard' monomers also including acrylonitrile, glycidyl methacrylate and methacrylic acid [219].

In a seeded emulsion polymerisation of a styrene–butadiene rubber latex, a substituted unsaturated amide $N(C_2H_5)_3C_2H_4OOCC(CH_3):CH_2$ and a substituted acrylamide $(NHOCCH:CH_2)_2CH_2$ are included in the seed stage of a carboxylated styrene–butadiene latex. The polymer was quaternised with dimethyl sulfate [220].

Table 3.21 Carboxylated latex

	Relative parts by weight (g)
Butadiene	48
Styrene	50
Itaconic acid	2
Sodium dodecylbenzene sulfonate	1.5
Potassium carbonate	0.6
Sodium salt, ethylenediamine tetra-acetic acid	0.03
tert-Dodecylmercaptan	0.8
Ammonium persulfate	0.085
Water	102

Procedure:
This latex is prepared by a continuous two-reactor process in which all items except the persulfate are fed to a pump cooled to 25–30 °C. The reaction actually starts as a batch reaction, which has taken place after initiation in the first reactor, rising to 80–85 °C; the average residence time is 4.5 h in each reactor, the temperature of the second reactor being 85–90 °C. It then proceeds to a stripper where the pH is adjusted to 8.5 with ammonia solution. A typical product has a non-volatile solids content of 51–53 %. The Brookfield viscosity (spindle 2 at 20 r.p.m.) is 25–350 cP; the number average particle size is 1700–1900 Å. An uncured film is water dispersible. Note that all three cations in the above formulations are considered to be desirable.

Table 3.22 Carboxylated latex [218]

	Parts by weight (g)	
	Stage 1	Stage 2
Water	100	—
Sodium dodecylbenzene sulfonate	0.3	—
Sodium bicarbonate	0.2	—
Potassium persulfate	1.0	—
Butadiene	2.0	15.0
Styrene	7.0	63.0
Methyl methacrylate	1.0	9.0
Fumaric acid	1.0	—
Acrylic acid	1.0	—
Hydroxyethyl acrylate	1.0	—
p-Benzophenone	0.001	0.006
t-Dodecylmercaptan	0.05	0.05

Process:
   Stage 1 is polymerised at 65 °C to 70 % conversion; then stage 2 is added over 7 h, and continued for a further 3 h to give 98 % conversion.

7 VINYL HALIDES

7.1 Vinyl chloride polymers and copolymers

Vinyl chloride, and likewise vinylidene chloride, differ from most of the monomers already quoted in that the polymer is insoluble in its monomer, although there is some swelling of the polymer by its own monomer. This is slightly modified by copolymerisation, often with vinyl acetate. The major method of production is by suspension polymerisation, but new methods of direct bulk polymerisation have been devised.

For the polymerisation of vinyl chloride, emulsion techniques are often intended as a source of solid polymer only. Process improvements have been directed to efficiency of polymerisation and to polymer quality. These include better resistance to heat discoloration, improved electrical properties and higher molecular weights. Some new copolymers have been designed to ease processing problems. Various vinyl chloride and copolymer latices have been sold for compounding, and will be described in connection with the various applications.

The direct applications of latices based on vinyl chloride are much more limited than those of latices based on vinyl esters and acrylics. The principal reasons for the emulsion polymerisation of vinyl chloride is for preparation of polymer for specific purposes as distinct from preparation by suspension polymerisation. The direct use of latices of vinyl chloride and copolymers

Vinyl halides 147

is relatively limited, but copolymers including vinylidene chloride and an acrylic ester are now available which are claimed to have utility as the basis of corrosion-resisting emulsion paints. Even with preplasticisation it is not possible to obtain films with proper strength on ambient drying, and heating is necessary for the plasticiser to be absorbed into the polymer and to obtain full film strength.

7.1.1 Early processes

Because of the reasons indicated above there has been relatively little development of the batch preparation of emulsion polyvinyl chloride. Partly because the polymer is insoluble in its monomer, the emulsion particles that form have a very small particle size. The earliest patents include an application by Mark and Fikentscher in 1937 [221].

Another early specification is shown in Table 3.23 [222]. Ammonium persulfate has been used in allied processes, as have sodium or potassium persulfates. Redox systems have also been used. An alternative emulsifier is dioctyl sulfosuccinate, whilst complete latices have been produced on a 'seed' latex.

A summary account of the emulsion polymerisation of vinyl chloride and vinylidene chloride up to 1975 is available [223].

7.1.2 Graft copolymer

A graft copolymer has been disclosed in which vinyl chloride is fed to a prepared emulsion copolymer of butyl acrylate–acrylic acid–diallyl maleate. The separated graft copolymer is impact resistant and porous [224].

Table 3.23 Vinyl chloride latex formulation

	Parts by weight (g)
Water	200
Dibenzoyl peroxide	2
Sodium lauryl sulfate	3
Vinyl chloride (liquid)	200

Process:
> The vinyl chloride is forced into the reactor under carbon dioxide, hydrogen or nitrogen pressure and the temperature is adjusted to 45 °C at 5–6 atmospheres pressure. The pressurising gas removes considerable inhibition as well as improving colour and stability of the polymer. Polymerisation takes about 20 h for a 90 % yield. It is advised that the pH of a vinyl chloride latex be kept low, even as low as 2.5–3.

7.2 Vinylidene chloride polymers and copolymers

Vinylidene chloride $CH_2:CCl_2$, partly because of its low boiling point at 31.7 °C, its ease of polymerisation at low temperatures and the ease of direct oxidation of the monomer, causes practical problems. In addition, the pure polymer is highly crystalline and difficult to dissolve, except with high boiling chlorinated hydrocarbons. Most practical polymerisations are for copolymers, which are often acrylic esters and sometimes vinyl chloride (see the previous section). It is remarkable that for steric considerations the effect of the lower acrylic esters as comonomers is to raise the second-order transition of the copolymer rather than to lower it.

Vinylidene chloride (VDC) copolymer latices, however, are marketed as the basis of emulsion paints. Table 3.24 gives examples of latices with vinylidene chloride contents of 30 and 60 % respectively. Monomer addition is such that minimum free monomer is present in the latex at any one time to ensure even copolymerisation. The acrylic monomers also block 'unzipping', which results in the loss of HCl. A redox system is employed. The latices do not withstand freezing and thawing, but were part of a theoretical investigation [225, 226].

7.3 Monomers containing bromine

There has been little interest in monomers containing bromine. However, tribromostyrene (100 g) has been prepared in emulsion in the presence of heptane (34 g). The polymer is colorless and gives no odour during moulding [227].

7.4 Vinyl compounds containing fluorine

Simple monomers such as vinyl fluoride $CH_2:CHF$, vinylidene fluoride $CH_2:CF_2$ and tetrafluoroethylene $CF_2:CF_2$ must be treated as other gaseous monomers in designing suitable polymerisation apparatus. The preparation of latices from monomers such as vinyl fluoride $CH_2:CHF$ and vinylidene chloride $CH_2:CF_2$ is a comparatively recent development. The desirable emulsifiers usually contain fluorine (see Chapter 2, Section 3.5) Suitable newer monomers which may be submitted to emulsion polymerisation include 2-(N-ethylperfluoro-octane-sulfonamido-ethyl acrylate, a technical product FX-13 (3M Company), which is $C_nF_{2n+1}-SO_2N(C_2H_5)C_2H_4OC(O)CH:CH_3$, and the corresponding methacrylate, designated FX-14, where n averages 7.5 in each case. The densities of the solid monomers at 25 °C are 1.61 and 1.59 g/cm^{-3} respectively and the melting points range for FX-13 is 27–42 °C and for FX-14 is 32–52 °C. In a modified product FX-189 the side chain attached to nitrogen is C_4H_9 rather than C_2H_5. These monomers are generally soluble in aliphatic fluorinated hydrocarbons, acetone, toluene and ethyl acetate.

The polymerisation reactivity of these monomers is generally similar to the corresponding hydrocarbon acrylates. A water-soluble organic solvent such as

Vinyl halides

Table 3.24 Vinylidene chloride copolymer [225, 226]

	Parts by weight		
	30 % VDC	*60 % VDC*	
	Reference [225]	*Reference [225]*	*Reference [226]*
A. Water, deionised	19.90 g	19.90 g	—
Sodium persulfate	0.10	0.10	—
Sodium sulfite	0.10	0.10	—
Sodium dodecyldiphenyl oxide sulfonate, 50 % aqueous	0.50	0.50	—
B. Vinylidene chloride	14.90	19.80	79.1
2-Ethylhexyl acrylate	—	—	9.0
Butyl acrylate	24.35	16.40	—
Methyl methacrylate	9.44	2.50	9.0
Methacrylic acid	0.75	0.74	1.9
Acrylic acid	0.15	0.25	1.0
C. Water, deionised	178.70	178.70	67.0 (approx.)
Sodium lauryl sulfate	—	—	0.18
Na-1-Allyloxy-2-hydroxypropsulfonate sulfonate	—	—	0.5
Ammonium persulfate	—	—	0.5
Sodium persulfate	0.89	0.89	—
Sodium bisulfite	0.89	0.89	—
Sodium dodecyldiphenyloxide sulfonate, 50 % aqueous	4.47	4.47	—
D. Vinylidene chloride	133.40	268.10	—
Butyl acrylate	217.90	147.50	—
Methyl methacrylate	84.42	22.34	—
Methacrylic acid	6.70	6.70	—
Acrylic acid	2.22	2.22	—
Styrene/acrylate polymer seed particles	—	—	
Properties of latex [225]			Not available
Solids content (%)	circa 50	circa 50	
pH	1.19	1.8	
Average particle diameter (μm)	23	30	
Viscosity (cP)	40–55	40–55	

Procedure [134]:

Add phase A to reactor, add phase B at 70 °C under nitrogen and after 30 min add C and D continuously, introducing the monomers below the liquid level. Monomer addition takes 4 h and the surfactant/initiator phase 4.5 h. In the final $\frac{1}{2}$ hour raise the temperature to 90 °C and maintain for 1 hour. In reference [226] full details are not given; process is standard.

150   *Emulsion polymerisation*

Table 3.25 Polymerisation of vinylidene fluoride [135, 136]

	A	B
Water (L)	1.2	1.2
Sodium heptafluorobutyrate (g)	1.2	—
Ammonium perfluorooctanoate (g)	—	1.2
Methyl (or ethyl) acetate (g)	—	3.63
Ammonium persulfate, 0.5 % (1) (mL)	15	25
Vinyl fluoride, pressure (kg cm^{-2})	18–20	20
Temperature (°C)	90	90
Time (min)[a]	Not indicated	81
Ammonium persulfate, 0.5 %(2) (mL)[b]	—	7.5

[a] Time represents the time for a latex particle size of 0.21 μm.
[b] Several portions and added as required when pressure falls to 19 kg/cm.

acetone is often added at the start of an emulsion polymerisation. Many of these emulsion polymers such as fluoroalkyl surfactants are invariably used as emulsifiers. Table 3.25 shows two polymerisation formulations [228].

There have been a considerable number of patent disclosures, many describing copolymers. Micro-emulsions of tetrafluoroethylene and other fluorinated monomers have been claimed, based on stabilization with fluorinated emulsifiers [229–231]. Tetrafluoroethylene has also been copolymerised with propylene [232] and with various unsaturated perfluorinated monomers [233–236]. Chlorotrifluoro ethylene has been copolymerised with ethylene in the presence of 2,2,4-trimethylpentane [237].

Hexafluoroethylene has been copolymerised with vinylidene fluoride using persulfate and diisopropyl carbonate consecutively as initiators [238, 239]. Fluorochemical monomers have been polymerised in glycol or aqueous glycol using sodium dialkyl sulfosuccinate as emulsifier [240].

Other fluorinated acrylates have been claimed in copolymers with acrylic esters and styrene. Thus 2-(perfluoro-octyl) acrylate is included in a formulation including styrene, 2-ethylhexyl acrylate, hydroxyethyl acrylate, methacrylic acid and acrylamide (20 : 121 : 46 : 4 : 4 : 5). Neutralisation increases latex viscosity. The films from these latices are generally resistant to water and salt water and may have oil repellency [241, 242].

Chlorotrifluoroethylene is included in the core of latex particles consisting principally of ethylene and a highly branched vinyl nonanoate and crotonic acid, the shell omitting the fluoro compound. The neutralised latex is useful for weather-resistant coatings [243].

8 EMULSION POLYMERISATION—THEORETICAL CONSIDERATIONS

Numerous investigations have been made as to the mechanism and theory of the emulsion polymerisation of various monomers, either alone or in

Emulsion polymerisation—theoretical considerations 151

combination. The general development is indicated here with special reference to the kinetics of emulsion polymerisation, but a full description is outside the scope of a volume dealing with applications. There are a number of volumes dealing with specific aspects; these are mainly symposium volumes. A volume describes scientific methods for the study of polymer colloids [244]. A useful general symposium is summarised in the reference list at the end of this chapter. There is also a series of bulletins currently available [245].

Most fundamental work has been devoted to styrene, the polymerisation of which has represented the simplest system to investigate. It is almost insoluble in water, the polymer is soluble in the monomer and there is little tendency to chain transfer or branching. Sodium lauryl sulfate has frequently been included as the emulsifier, since it can be obtained in a relatively pure state and causes little chain transfer; also it has virtually no tendency to act as a graft base. A range of initiators have been examined, including persulfate salts, organic peroxides and hydroperoxides, and occasionally hydrogen peroxide with a trace of an iron salt. Azo-bis-isobutyronitrile or other dinitriles are used as alternatives, particularly where all oxidation reactions must be avoided.

Some latices may be stabilised by the use of un-ionised high molecular weight water-soluble polymers alone. In some cases this is aided by residues from a persulfate initiator. In others, principally when a polyvinyl acetate is stabilised by polyvinyl alcohol, stabilization is entirely steric in nature.

The presence of water-soluble monomers such as acrylic acid, methacrylic acid or itaconic acid creates further complications. In practice, they should be included in the monomer phase as long as there is mutual solubility. If dissolved in monomers, maleic anhydride will copolymerise before hydrolysing to the sluggishly polymerisable maleic acid. However, this type of copolymerisation is sometimes deliberately adjusted so that acid units are concentrated on the surface of the particles.

There have been a considerable number of investigations into the decomposition by the free radical mechanism of individual initiators, and also on the actual mechanism of a redox polymerisation as distinct from the method of initiation only. Early summaries are available of redox processes and theories including a conference report [246, 247].

The development of graft and 'core–shell' copolymers in emulsion form has resulted in a number of investigations as to the nature of the processes involved, which may take place on the surface of the particles. Alternatively, if there is rapid diffusion of the added monomers, there is unlikely to be a true 'core–shell' phenomenon. A modification is the polymerisation on a 'seed', in which case new nucleation may be avoided.

8.1 Types of mechanism

The stages in a polymerisation process are initiation, polymerisation and termination. The nature and behaviour of the free radicals formed is a major factor in understanding the process, whilst there are complications due to the fact that this is a heterophase reaction and similar processes may occur in either phase. There is a further complication in that several different reactions, or similar reactions but at different rates, may occur in the heterogeneous phase. The latter may at an early stage in the reaction consist of separate monomer and polymer particles, of which the latter may absorb monomer. A further variation is that the polymerisation may start with a 'seed' latex. In many processes most of these reactions continue simultaneously. With some monomers such as vinyl chloride, the polymer is insoluble in its monomer, although the polymer is swollen by the monomer, further affecting the kinetics. Crosslinking during the reaction (see Chapter 5) is a further complication.

Since Harkins first propounded a general theory of emulsion polymerisation in 1947, there have been many investigations that have given an insight into the complex processes involved [248]. Most of these have been concerned with the kinetics of polymerisation under emulsion conditions. Investigations have also been made as to the specific functions of various emulsifiers and stabilisers. Most patent references are concerned with empirical observations, but there has been considerable research on the nature of the absorption and protective function of emulsifiers on the surface of the particles and the morphology of the particles themselves. Much of the latter work is associated with J.W. Vanderhoff and his colleagues at Lehigh University (see Chapter 4 for further details).

It is now clear that the mechanisms involved in emulsion polymerisation are very different from those of polymerisation in bulk or from bead polymerisation (which tends to resemble a water-cooled bulk polymerisation) in which the beads are usually of the order of 1 mm in diameter. It is possible to divide emulsion polymerisation into several types, dependent on the monomers, which are sometimes overlapping, which may be tabulated as follows:

(a) Monomers highly insoluble in water (<0.1 %), of which examples are styrene, butadiene and vinyl stearate.
(b) Monomers with appreciable water solubility, but with polymers that are highly insoluble. Examples are methyl methacrylate and acrylonitrile.
(c) Monomers that are appreciably water soluble, and in which there is an appreciable hydrophilic character of the polymer. Vinyl acetate is the outstanding monomer of this class.
(d) Monomers that are highly water soluble, sometimes miscible in all proportions, and whose polymers are either completely soluble or highly swollen by water. Acids such as acrylic acid and methacrylic acid come into this category.

These factors are complicated by a number of practical conditions, e.g. copolymerisation which may involve two or more of the above classes.

Complications may include induced solubility; e.g. aqueous solutions of methacrylic acid tend to increase the solubility of methyl methacrylate. Where two phases are formed, partition coefficients of some monomers between water and an insoluble monomer come into play. This is further complicated by a partition of monomers between water and the polymer being formed. In addition, many emulsifiers, usually with a relatively low hydrocarbon content, appreciably increase the solubility of monomers in the aqueous phase; e.g. the sodium salt of *p*-toluene sulfonic acid increases the aqueous solubility of vinyl acetate, and in some formulations the aqueous solubility of butadiene and styrene is increased, for example, by the sodium salt of *p*-xylene sulfonic acid.

In most cases polymers are soluble in monomers as they are formed. There are exceptions, however, the principal cases being the polymers of vinyl chloride and acrylonitrile. This complicates polymerisation kinetics, especially in copolymerisation.

It is now convenient to consider the various theories of emulsion polymerisation in their historic context.

8.2 Theories of Harkins and Smith and Ewart

It had been observed from surfactant titrations that free monomer droplets rapidly disappeared during the emulsion polymerisation of styrene. Harkins considered that at the commencement of the reaction the principal loci of polymerisation are the solubilised monomer molecules inside the surfactant micelles. The hydrocarbon 'tails' of the emulsifier molecules can be considered as facing inwards towards the monomer, whilst the ionic polar groups, e.g. COO^-K^+, face outwards towards the water. As these polymer particles in the micelles grow, they absorb emulsifier and grow at the expense of the monomer in the water phase (Figure 3.8). Initiation of this type ceases at 13–20 per cent of polymer yield, after which the locus of polymerisation is the particles themselves. Harkins considered that at about 60 % polymerisation of monomer, all monomer has been absorbed by the polymer. This critical monomer/polymer ratio can be determined experimentally. This figure is based on 5 g of emulsifier per 100 g of monomer and decreases to 32 % when 10 g of emulsifier are present per 100 g of monomer When all monomer has been absorbed, the water phase ceases to be even a minor source of initiation.

From the Harkins theory it can be deduced that, with an increased quantity of surfactant, the amount of micellar material increases, and therefore the number of polymer particle nuclei formed in unit time is increased. This causes a reduction in particle size, and at the same time increases the reaction rate, a deduction that has been confirmed experimentally and supports the main theory of Smith and Ewart, indicated later, that there is only one growing chain per growing polymer particle. Harkins summarises the theory by stating that the principal function of the monomer droplets is not to serve as a primary locus of polymerisation but as a storehouse from which the monomer molecules

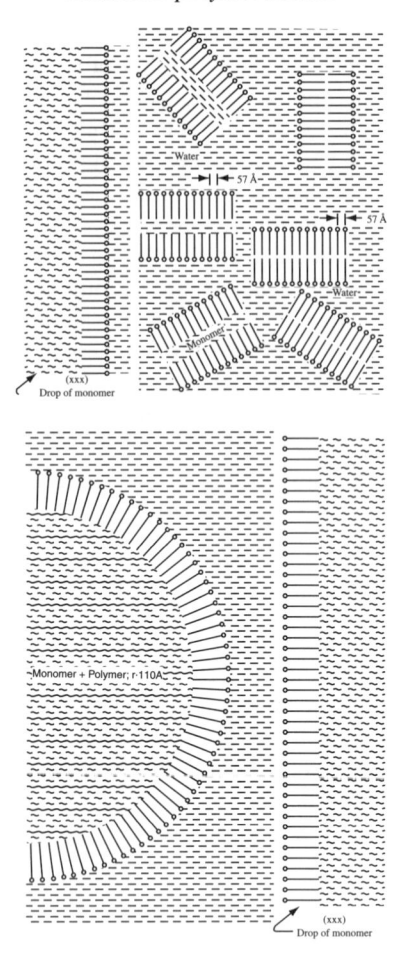

Figure 3.8 Highly idealised representation of diffusion from a monomer droplet through a monolayer of emulsifier (a) into micelles and (b) into a polymer–monomer droplet. All of the monolayers of emulsifier and also the emulsifier in the micelles exhibit the structure of a liquid with its irregularity and 'holes'. For simplicity the drawings show a regular structure. The micelles which contain monomer should be shown as containing many more emulsifier molecules than the others. (After Harkins [248].)

diffuse into the aqueous phase, where they may enter emulsifier micelles or polymer particles, in both of which they polymerise.

Later work by Harkins, utilising styrene and styrene–butadiene emulsions, showed that the surface tension of the emulsions increases at the point of disappearance of the emulsifier micelles. The micelles are considered to function, *inter alia*, by providing a much larger single target than a single monomer molecule by which a radical, either directly from the initiator or from a monomer free radical, may come into contact by diffusion. Each monomer molecule is 'interned' in 40–100 g of emulsifier in the micelles. The reaction of the radical with the monomer takes place in about 10^{-3} s, before any form of deactivation, i.e. mutual termination, can occur. In this time complete polymerisation has occurred. As there is a very large excess of micelles over polymer particles, the ultimate function of the emulsifier is to stabilise the polymer particles once they have formed. Polymerisation is not complete within each micelle, but there is a constant diffusion process; otherwise the molecular weight of the polymer formed would be much lower than is found in practice as there are only a limited number of active micelles. The idea of compartmentalisation of the termination reaction, i.e. that of polymer chains developing singly in a system of small drops, was also considered by Haward [249].

A complex mathematical treatment has been developed by Smith and Ewart [250, 251] which is not reproduced here. The kinetic treatment is based on consideration of the factors governing the polymerisation rate in a single swollen polymer particle and the problem of determining how each polymer particle forms. This has been based on the polymerisation of styrene and allied highly insoluble monomers.

Free radicals are generated at the rate of about 10^{13} s^{-1} mL^{-1} of water. It is presumed that all these radicals enter emulsifier micelles and that styrene commences to polymerise at the rate of about 10^3 molecules per second. Water saturated by styrene can provide monomer by diffusion to the growing polystyrene particle. This continues to grow until a free radical enters the growing polymer particle and terminates the chain in about 10^{-3} s. When all surfactant has been absorbed from the micelles, the number of particles formed in this time is 10^{14}–10^{15} cm^{-3} of aqueous solution. As free radicals are being formed at about 10^{13} s^{-1}, the period in which particles are being formed is from 10 to 100 s, assuming that all the radicals enter the emulsifier micelles.

Since one free radical enters a particle at 10–100 s intervals, and as the termination time is extremely short, it follows that on average a particle contains a growing radical half of the time in which polymerisation proceeds. The intervals of growth would thus govern the molecular weight of the polymer. It is also obvious that after an initial period in which new particles are formed, the rate of polymerisation will be virtually constant until the separate styrene monomer phase has disappeared. This assumes a constant generation of radicals.

Smith and Ewart showed that the number of particles N formed per mL of water up to the time that all of the emulsifier has been absorbed on to polymer particles is given by a law of the type:

$$N = k[P/\mu]^{0.4} [a^s S]^{0.6}$$
$$R_p = k_1 [e]^{0.6} [I]^{0.4}$$

where k is a constant between 0.37 and 0.53, P is the rate of formation of radicals, μ is the average rate of increase in particle size due to polymerisation, a^s is the interfacial area occupied by 1 g of surfactant, S is the amount of surfactant associated with 1 mL of water, $[e]$ is the concentration of emulsifier and $[I]$ of initiator and R_p is the rate of polymerisation.

The variation in k arises because the upper figure of 0.53 has been deduced assuming that only micelles can capture radicals. As polymer particles can also capture radicals the constant is too large. If one assumes that the radical diffusion flux, i.e. the number of radicals entering a particle per unit time/unit surface area is independent of the radius of the particle, the smaller factor, 0.37, has been deduced. However, since the smaller particles are relatively more efficient in capturing particles, this figure is too small.

The Smith–Ewart equation predicts that the number of particles produced per gram of emulsifier should be proportional to the three-fifths power of the emulsifier concentration. Equally, if the rate of formation of free radicals is proportional to the initiator concentration this law predicts that the number of particles should vary as the two-fifths power of the initiator concentration. Since the apparent size of particles depends on their diameter and not on their volume, the apparent particle size (diameter) will only vary slightly with the initiator concentration. Conclusions drawn from some experimental work must be considered doubtful since only very small errors in estimating numbers of particles may lead to major errors in theory. However, Bartholomé, Burnett and others have performed experiments which they consider prove the validity of the theories [252, 253].

The basis of Smith and Ewart theory, based on treating each particle as a separate unit, readily explains the high molecular weight of the polymers formed in emulsion polymerisation, since the process proceeds in each particle until the reaction is terminated by another radical entering the system. The larger the number of particles in a constant system, the fewer will be the termination reactions, provided that the rate of production of radicals is unchanged. This assumes that there is no chain termination with monomer and that terminations are due to disproportionation or mutual destruction of radicals The assumptions made in the simple Smith–Ewart theory may not be true in every case, and alternative equations have been deduced to cover exceptions. Thus, where the number of particles is small compared with the number of radicals, e.g. in the larger particle size emulsions, then every free radical tends to enter a particle in which several chains are growing simultaneously, and polymerisation in each particle tends to be continuous. Under these conditions the kinetics tend to approach a water-cooled bulk polymerisation.

Mixtures of anionic and non-ionic surfactants have been used to examine the generality of the Smith–Ewart theory. With styrene, the results were consistent with the theory when the weight % of the non-ionic surfactant (NP 40) was below 30 % of the total surfactant. There was considerable deviation when the non-ionic content was >50 %. The mixed system greatly improved the latex stability as the result of synergism between steric and ionic stabilization mechanisms, a 1 : 4 ratio of anionic to non-ionic surfactant giving the maximum polymerisation rate [254, 255].

8.2.1 Modifications by Gardon

The Smith–Ewart theory has been re-examined by Gardon to account for such factors as the escape of radicals from particles [256–259]. Gardon distinguishes between three stages in the polymerisation of a water-insoluble monomer. In the first stage the number of particles is increasing and monomer droplets are present; in the second the number of particles is constant, monomer still being present, whilst in the third the number of particles is constant, but there is no free monomer. It is also pointed out that for heterogeneous polymerisation, nucleation must be taken as a fourth process along with initiation, propagation and termination. Various calculations are made, based on a rate growth of particles, nucleation, molecular weight and particle size distribution. The equations are too complicated to be included here, but operating with both styrene and methyl methacrylate general agreement with the Smith–Ewart theory occurs in ideal cases. Theoretical predictions for the rate and molecular weight in the second stage are valid only for low initiator and high emulsifier formulae.

Gardon calculated the conditions when there is a slow termination rate, and thus several radicals coexist in the particle. It is found that $R_p = At^2 + Bt$ with very small particle latices, whilst on the other hand those of large particle size show a proportionality to t^2. Of many other equations derived by Gardon, only one will be quoted here, viz. the number average molecular weight M_n is given by

$$M_n = (2d_p N_A/R)(P/t)$$

where d_p is the polymer density, N_A is the Avogadro number, R is the number of radicals produced per mL of water per second and P is the volume of polymer formed in time t from the commencement of the reaction.

8.3 Theory of Medvedev

The Russian scientist Medvedev developed different theories as a quantitative explanation of emulsion polymerisation. There is one paper in the English language [260]. This theory ignores the presence of micelles, but considers the emulsifier present as one unit. The fundamentals of the theory, as based on experiments with styrene in the first place, is that polymerisation takes place on the surface of the particles. The water-soluble initiator reacts with an emulsifier molecule in the water phase, and the activated molecule enters the adsorption layer on the surface of the particles initiating the chain reaction.

From this assumption the following mathematical expression is derived:

$$R_p = kC_E^{0.5}C_I^{0.5}$$

where C_E is the concentration of emulsifier and C_I the concentration of the initiator. The closeness of this equation to one form of the Smith–Ewart equation should be noted, making quantitative proof rather difficult within the limits of experimental error.

An investigation of this theory showed that in many experimental series of polymerisation a threshold concentration occurred at which there was a sharp break in the polymerisation curve when plotted against emulsifier concentration. Above this minimal concentration, polymerisation is generally independent of emulsifier concentration. This is considered to be the point at which the particles are completely covered by emulsifier [261]. The results apply to methyl methacrylate and to chloroprene.

At high concentrations of initiator the dependence on initiator concentration falls, ultimately approaching zero. Medvedev considers that this is due to reactions between polymer and primary radicals. The maintenance of constant surface area during polymerisation inevitably means that there is some particle coalescence, It is surprising how easily this latter point has been overlooked, especially in view of the considerable skill that must be invoked in technological emulsification to prevent coalescence during the 'sticky state'. The nature and speed of stirring is known practically to alter the particle size, suggesting that under some vigorous stirring conditions the particles themselves may be broken up.

Whilst the general theory of Medvedev tends to ignore 'compartmentalisation' of growing radicals, it may give a more general explanation of emulsion polymerisation than the Harkins–Smith–Ewart theories. Many simple laboratory experiments tend to follow a square root law with regard to initiator concentration. In addition it is obviously necessary to account for the fact that with a persulfate, or redox initiation with salts of reducing sulfur acids, the growing polymer radical is at first water soluble.

8.4 Theoretical developments—Fitch and Ugelstadt

A quite different basis for the progress of emulsion polymerisation has been made by Fitch, who considered that the process is based on homogeneous nucleation and produced a model for particle formation based on free radical initiation in a homogeneous medium, with self-nucleation or particle capture being proposed. This theory was developed, often based on the work of earlier researchers, and was mainly based on studies of vinyl acetate and methyl methacrylate emulsion polymerisation. Stated simply, in emulsion polymerisation:

(a) There is homogeneous nucleation of single growing oligomeric free radicals in solution.
(b) A point is reached, varying with different monomers, at which the growing polymer chain is no longer soluble and nucleates to a primary particle. Thus there is aggregative nucleation of several oligomeric radicals from solution.
(c) The number of primary particles is approximately equivalent to the number of chains initiated in solution.
(d) The number of particles formed may be reduced by the combination of incompletely developed chains with polymer droplets before the former are adequately utilised.
(e) The number of particles may be reduced by coagulation. This depends on the nature and efficiency of the emulsifiers present. Where short chains are formed, e.g. with a sulfur acid end group, the emulsifier may be self-formed.
(f) The ultimate particle size in the case of persulfate-initiated latices is a function of the relative number of sulfur acid groups per particle.
(g) About 99.9 % of polymerisation occurs within the swollen polymer droplets and only $\leqslant 0.1$ % is identified with primary particles.

If hydrogen peroxide or an organic initiator is used the above may be slightly modified. This also applies to redox initiator systems. In the case of monomers with restricted solubility, initiation would be a slower process, but even a highly insoluble monomer such as styrene produces sufficient monomer in solution for initiation to take place. In addition, the emulsifiers have a strong solubilising effect on the monomer and thus assist in speeding the initiation process.

In all the above the effect of direct surface initiation of monomer particles must also be considered. The very complex reactions that take place, allowing for initiation, may depend on the surface of monomer exposed. Note that the surface varies as the square of the radius of the primary particles and there may be coagulation by collision, or by entry into monomer particles. Monomer is absorbed into the polymer particles, and at one stage in the reaction the number of monomer particles is zero. It is often forgotten that nature and speed of stirring may have a very marked effect on the speed and nature of the polymerisation. Ugelstadt has elucidated a mathematical solution for resolving the concentration of oligo radicals in the aqueous phase [262–265].

Many factors have to be resolved, including the efficiency in capture of radicals. A radical in solution may collide with 10 000–1 000 000 particles before it is captured. This may vary with the size of the particle and with the nature of the monomer. As particles grow the surface electrical potential is likely to increase as more charged radicals are captured, which leads to a reduction in capture efficiency, but as particles become more highly swollen this increases the possibility of irreversible radical capture.

Coagulation rates of particles may also affect the kinetics. The detailed mathematical consideration is beyond the scope of this volume, but readers should consult the references, in particular ACS Symposium 165 [266–273].

Ugelstad has expanded the understanding of the nature of nucleation by considering cases where very small particles are formed, the size being probably of the order of 0.1 µm. With particles of this size, the chance of collision due to the greater surface area, which varies as the square of the radius, is much greater, and in this instance nucleation takes place in the particles themselves. As a corollary to this, studies were made of the effect of simple emulsifiers such as styrene with the cationic octadecylpyridinium bromide and sodium hexadecyl sulfate. Whilst stability was poor with no further additives, the addition of hexadecyl alcohol gave increased stability, which further increased with the addition of hexadecane. To obtain stability with smaller particles obtained by vigorous stirring, the concentration of additive must be increased.

Some of these emulsions were polymerised with the aid of azo-di-isobutyryl nitrile and octadecyl pyridinium bromide as emulsifier. With equimolar proportions of hexadecane and the emulsifier, initiation takes place almost exclusively in the water phase, but where the molar ratio is 3 : 1, the particle sizes are bimodal, suggesting that initiation takes place in both the water and monomer phases. By comparison with an electron micrograph of the monomer particles, it was concluded that polymerisation takes place in both the monomer phase and water phase. Electron micrographs were also used to determine particle sizes.

This has been summarised in a short paper in which it is indicated that with very fine emulsions, the increased surface tends to absorb virtually all emulsifiers present. There is thus less emulsifier in the continuous water phase to facilitate nucleation and the formation of primary particles in the water phase [274].

Further studies by Ugelstad have developed the theory of homogeneous nucleation in emulsion, allowing for factors such as electrostatic repulsion and diffusion. The particle number theoretically goes through a maximum. Simultaneous nucleation and flocculation of primary particles may take place after interval 1 is completed in an emulsion polymerisation. This work was performed with seeded and unseeded emulsions, typically styrene with sodium dodecyl sulfate as emulsifer, but also in emulsifier-free systems. The rate of capture of oligomeric radicals by existing particles is considered to be governed by the absorption of oligomers with a chain length one less than the critical chain length for precipitation of the oligomer. The efficiency of particles in absorbing radicals may be calculated from available physical parameters such as diffusion constants and surface charge densities. This was also noted where the number of particles is large and of small size. For further details, the original papers should be consulted [264, 265, 275, 276].

Emulsion polymerisation—theoretical considerations 161

The ionic strength of the water phase affects the progress of polymerisation and the particle size in a number of cases, in particular where there are only ionic emulsifiers or an ionised polymerisable surfactant present. Thus if an emulsifier-free polystyrene latex is prepared in the presence of 1 % of sodium methallyl sulfonate with a persulfate initiator, the number of particles N_p varies as the 1.1 power of ionic strength [277].

As this volume is intended to give a general overview of the fundamentals of emulsion polymerisation as a prelude to understanding the nature of applications, only a summary has been given of developments in the theory of various aspects of the kinetics of emulsion polymerisation.

8.5 Copolymerisation—theoretical

Copolymerisation has already been quoted in previous sections as a practical art. It has been the subject of a number of fundamental studies that have been surveyed at some length [278]. Studies include the distribution of chemical composition and molecular weight (molar mass distribution). Various fractionation methods have been devised to obtain this information. Some theoretical investigations consider thermodynamic equilibrium, but the validity of this is limited since most of the latex systems are non-equilibrium ones.

Monomer partitioning plays a major part in the structure of copolymers. The apparent copolymerisation ratios may differ appreciably from the standard r_1, r_2 ratios as previously described (Chapter 1, Section 2.2) because of the heterogeneous nature of the system. Thus when styrene and either methyl acrylate or methyl methacrylate are copolymerised, studies show that in the first very early phase of polymerisation the polymer formed is rich in the two acrylic monomers because of their partial water solubility [279]. It has also been shown that where one monomer has some water solubility, there is general enrichment of the more water-soluble monomer in the earlier stages of polymerisation. However, other studies have indicated that there is an azeotropic composition at which the polymer composition is constant, at least up to 90 %. There is also a composition drift during copolymerisation. For further details the reference quoted should be consulted. The principles have been extended to terpolymerisation.

The control of composition of copolymers has been examined as a theoretical exercise, but as a practical art earlier sections of the current volume should be consulted (Sections 3.3 and subsequent sections in this chapter). Note also core–shell copolymerisation in Section 4.1.5 and Chapter 4, Section 7).

It has also been noted that the desorption of radicals from polymer particles may have a marked effect on the nature of copolymerisation. A growing radical, if still of low molecular weight with a very hydrophilic chain end, may diffuse away from the particle into the bulk of the aqueous phase. These rates have been calculated and the reader is referred to the general reference above. The termination rate in the particles is strongly diffusion controlled.

8.6 Specificity of surfactants and stabilisers

There have been relatively few references on the specific effects of various surfactants and colloids in the theoretical literature. The old adage that 'like emulsifies like' holds good to a considerable extent, e.g. the effectiveness of partially hydrolysed polyvinyl acetate as a stabiliser during the emulsion polymerisation of vinyl acetate. French has made some calculations involving the surface area of emulsifiers, but no suggestion has been made as to why only certain proportions of propylene oxide and ethylene oxide in the stabilisers used are satisfactory [280]. The concept of HLB has been used, particularly with non-ionic surfactants for obtaining suitable surfactants or a balance of surfactants for many emulsion systems. However, this does not state why these non-ionics are suitable and only gives an empirical approach to the subject.

An interesting feature of emulsifiers is that emulsifiers with more than one active group tend to give stable products of small particle size. A typical example is a dialkyldiphenyl sulfonate (Chapter 2, Section 3.2.5), available technically as Dowfax® 2A1. Some of the sulfosuccinates, which may include carboxyl as well as sulfonate groups, come into this category. This idea also applies to poly salts. Thus a copolymer of vinyl acetate and a maleic acid salt (Section 5.1.2) may be used as a stabiliser for an emulsion based on styrene, an acrylic ester or an acrylic ester–styrene copolymer, as already shown (Table 3.18). A polymethacrylic salt, although not often used without a coemulsifier, is in general a better emulsifier than a polyacrylic acid salt because of its better hydrophile–hydrophobe balance. It is difficult to formulate any general rule regarding the hydrophile–hydrophobe nature of emulsifiers and that of the various monomers. Steric factors may also play a part in assessing the suitability of surfactants and colloids, but relatively little has been worked out theoretically on this basis. The extent to which the emulsifiers solubilise the monomers and the solubility or otherwise of the polymer in its monomer all play a key part in determining the most suitable emulsifiers for the formation of a stable fine-particle latex.

The general issue of suitability is complicated by the fact that a suitable emulsifier for a monomer is not necessarily satisfactory for the corresponding polymer, for in most cases of emulsion polymerisation, as already discussed, polymerisation originates from the water phase. Studies by Schulman and Cockbain, already quoted (Chapter 2, reference [4]), on condensed surface films have a useful bearing on the stability of latices. Thus under some conditions a hydrophobic substance with a hydrophilic group, e.g. cetyl alcohol or cholesterol, can stabilise a polymethyl methacrylate latex or a latex of polyvinyl alcohol when used in conjunction with a water-soluble surfactant such as lauryl alcohol sulfate or sodium dioctyl sulfosuccinate [281, 282]. In some cases there is a condensed film on the surface of the particles due to two emulsifiers, the most hydrophilic being just on the particle surface and the more lyophilic just inside it, thus ensuring particle stability.

The phosphated emulsifiers (Chapter 2, Section 3.4) are occasionally encountered as stabilisers in emulsion polymerisation and are of particular interest where it is desired to impart some resistance to corrosion. However, the formulations of the emulsifiers themselves are more complex than sulfates. Thus there are mono- and di-esters, and a different balance of alcohol or alcohol ether chains may give the same overall empirical formula for different phosphated emulsifiers.

Furthermore, the manner of production of these ester salts may give products with varying quantities of the alcohols that have not been phosphated. There has been no systematic examination of phosphate-type ionised emulsifiers as far as has been ascertained, although they are sometimes quoted in patents.

Without doubt the suitability of the various phosphates is very dependent on the nature of the monomers, and this has been observed with various acrylic formulations. Probably the more hydrophilic monomers require a more hydrophilic phosphate emulsifier, but no further conclusions may reasonably be made.

A general review on the effect of emulsifiers in emulsion polymerisation is available. In a homologous series the general effect of increasing alkyl chain length is an increase in the rate of polymerisation to a limiting maximum because of an increase in the number of latex particles formed [283]. A survey of the problems involved in developing a theory of the function of emulsifiers is included in a conference reprint [284].

The significance of polymerisable surfactants, e.g. sodium vinyl sulfonate, has already been mentioned, as these compounds promote self-emulsifying properties with little or no supplementary emulsifier. It is also known that grafting of monomer to a stabiliser may improve, but exceptionally reduce, the function of an emulsifier. Since vinyl acetate does undoubtedly graft on to polyvinyl alcohol stabiliser, the modified polyvinyl alcohol may improve emulsifier properties, partly for the very simple reason that there is more of it and the modified product has equal stabilization efficiency. On the other hand, if there is excessive grafting, the modified product becomes less efficient and a rather unstable, thick latex is produced. It is interesting to note that an unstable polyvinyl acetate latex, sometimes produced inadvertently under large-scale conditions, may be restabilized by the addition of the emulsifier Aerosol®22 (Chapter 2, Section 3.2.7) which has several sulfonate groups.

Cationic emulsifiers have been quoted, both in theoretical studies and in a number of patents, but interest in them has been much more limited than with anionic emulsifiers. Under some conditions they may act as reductants in redox initiated polymerisation. On the other hand, they may tend to cause inhibition when used with vinyl acetate [285].

Water-soluble colloids probably function as protective envelopes around the particles, and if not charged, unlike the poly salts of methacrylic acid, have a purely physical action. In a manner similar to the theory above it may be found that a 'polyvinyl alcohol' (actually partially hydrolysed polyvinyl acetate) is more successful in emulsion polymerisation of vinyl acetate if the nominal saponification value, circa 120, is an average of a broad Gaussian scatter, rather than consisting of molecules whose degree of saponification varies little from the average [286].

The use of ionised polyelectrolytes, whether water soluble in their own right or alkali soluble, is quite general. Most are copolymers of a monomer and an acid. Various copolymers of acrylic and methacrylic acid with ethyl acrylate and methyl methacrylate are well known in emulsion form, and, except where the ratio of ester to acid monomer is less than about 1–2 : 20, they in general require alkali to solubilise them, usually with major increases in viscosity. Relatively low molecular weights, about 1000–5000, are usually the most suitable products. They are suitable for most acrylic-type monomers, usually with an auxiliary non-ionic or anionic emulsifier. Low molecular emulsion copolymers of acrylonitrile and acrylic acid are also suitable and are quoted for emulsion polymerisation of styrene [287].

A study has been made of the effect of the non-ionic surfactant Triton X-405 (octylphenoxy polyethoxyethanol) with several monomers. In the polymerisation of styrene two separate nucleation periods were noted, resulting in a bimodal final particle distribution. The first of the nucleation periods was attributed to homogeneous nucleation, the second to micellar nucleation.

With n-butyl acrylate the surfactant partitions primarily into the aqueous phase, leading to nucleation in the presence of micelles. There is limited aggregation in the early stages of the reaction, with in some cases new nucleation after 50 % conversion. High values of the average number of radicals per particle (>10) were observed at high potassium persulfate initiator concentrations.

In the copolymerisation of styrene and n-butyl acrylate, unimodal particle size distributions were found at the lowest and highest levels of emulsifier, but not at intermediate levels. The results were attributed to surfactant partitioning in the system. With the lowest levels of surfactant there is homogeneous/coagulative nucleation at the lowest levels and micellar nucleation at the highest [288–290].

Further information on the function of emulsifiers is given in the following section.

8.7 Latex particle stabilization

Some consideration has already been given to this subject Ottewill has reviewed the subject with special reference to styrene [291]. Stabilization has a number of different meanings. Thus it can refer to shear, freezing or heating, long-term storage and drying conditions. It may also refer to stability to additives such as pigments. Particles are stabilised by the electrical double layer and at a critical ionic concentration coagulation may occur. Theoretically a hydrated aluminium ion will neutralise a surface charge, causing almost instant precipitation, but an excess may impart a positive charge and under some conditions restabilize a latex. If a latex, e.g. of polyvinyl acetate, has only a weak negative surface charge with considerable steric stabilization, as with a polyvinyl alcohol stabilised latex, the charge may be reversed by the addition of a cationic surfactant, and this has been performed successfully commercially. The hydrophobic chain of an emulsifier must be compatible with the particle surface. Thus a $C_{11}H_{23}$ hydrocarbon chain shows little adsorption on to PTFE particles whilst a C_7H_{15} chain is strongly adsorbed.

In the case of styrene the surface-active oligomers that grow, associate and nucleate are not stable enough in the colloidal sense until enough charged groups have been formed on the surface to give the particle an adequate electrostatic surface potential. Hence they coagulate until this state is reached, forming large particles. Thus in the absence of surface-active agents, particle size may be controlled by controlling the ionic strength of the polymerising system. The assumptions have been verified qualitatively, although the particle size of the latices formed had particles of rather high coefficients of mean diameter in the range 0.5–0.97 μm [291].

In the polymerisation of styrene with two polymerisable surfactants, dodecyl sodium sulfopropyl maleate and tetradcyl sodium sulfomaleate, oligomers are formed with higher surfactant present in the early stages of polymerisation, only about 45 % of surfactant being on the particle surface [292].

Non-ionic surfactants such as dodecyl hexa-oxyethylene glycol monoether $C_{12}H_{25}(CH_2CH_2O)_6OH$ can adsorb on to polystyrene latex to give a monolayer on the surface in which the alkyl chains are adsorbed on to the surface and the ethylene oxide groups extended into the solution phase. This results in a heavily hydrated layer which prevents monomer entering into a primary particle.

Surface coagulation may occur at the water–air interface during filming due to desorption of the stabilising surfactant, partly due to their non-spherical shape in consequence of their crystallinity. This is particularly so with PTFE latices. The ionic surface groups at the chain ends may be concentrated irregularly at some points, thus depleting others of stabilising entities.

Surfactant interactions encountered in polyvinyl acetate and vinyl acetate–acrylic ester latices are:

(a) Simple saturation at the latex–water interface.
(b) Adsorption and penetration into the latex particle, leading to the formation of solubilised polyelectrolyte-type polymer–surfactant complex.
(c) Grafting to the polymer chains, especially with non-ionic surfactants.

Anionic surfactants of fairly low molecular weight and simple structure penetrate into the latex core, causing new adsorption sites to open up [293].

A number of references of interest are given [294, 295].

8.8 Further quantitative investigations

8.8.1 Styrene

Some authors have found that the number of particles increases with the square root of the amount of emulsifier, which is substantially micellar, other than the three-fifths index of Smith–Ewart [250, 251]. This discrepancy is sufficiently small not to be considered as very serious in the light of the approximations made, although the Medvedev theory would fit in exactly with the increasing number of particles in these results.

A study of critical micelle concentration of mixed sodium dodecyl sulfate and ethoxylated nonylphenol in the emulsion polymerisation of styrene concludes that the polymerisation is not in accord with the Smith–Ewart theory when the concentration of non-ionic surfactant is high [296].

An equation was developed by Van der Hoff, who connected the rate of polymerisation R_p over the range of particle concentrations $N = 5 \times 10^{15}$–5×10^{14}:

$$R_p = cN^{-0.83}$$

where the constant c depends on the concentration and nature of the emulsifier. It was also observed that at 55 % polymerisation a reduction occurred in the polymerisation rate corresponding with the point of disappearance of independent monomer droplets. Van der Hoff has also verified to some extent the Smith–Ewart theory by counting the absolute number of particles using potassium laurate as emulsifier [297]. The above empirical equation, however, is not in accord with the Smith–Ewart theory. Strong deviations from Smith–Ewart kinetics for the reaction rate of the seeded emulsion polymerisation of styrene have been found [298].

It is worth noting that dilute solutions of styrene in pure water and in dilute aqueous solutions of various surfactants have been found to polymerise at a rate dependent on the square of the styrene concentration. Since surfactants tend to solubilise the monomer appreciably, this might well cause a complication in the overall kinetics [299]. The number of particles observed has been found to differ significantly from that estimated by the Smith–Ewart theory if the emulsion particles are of 1–5 μm in size [300].

There was also a tendency to autocatalytic growth. Amongst other variations, a 'gel effect' has been noted with an emulsion of polystyrene [301] (see Section 8.9.1 for methyl methacrylate). This effect is a speed-up of polymerisation due to trapped radicals in a highly viscous medium. Thus a polymethyl methacrylate particle swollen with a limited amount of monomer causes an acceleration of polymerisation due to lack of a termination reaction.

At a low level, the presence of oxygen accelerates the polymerisation of styrene with sodium dodecyl sulfate as emulsifier, probably by forming traces of persulfate, but beyond this level it causes inhibition [302]. Styrene may be polymerised in emulsion in the absence of emulsifier, or at any emulsifier concentration below the critical micelle concentration, provided that an initiator with charged end groups is present. This indicates that latex particles are formed by nucleation of oligomeric particles formed in aqueous solution. This is not possible with *p-tert*-butylstyrene as it has only one-tenth of the water solubility of styrene [266, 303].

The inclusion of sodium styrene sulfonate with styrene resulted in very dramatic effects in polymerisation in the absence of emulsifiers, and preferably with a persulfate/bisulfite initiator system at constant temperature. The water-soluble monomer acts as a locus of polymerisation [304].

Thiol-terminated non-ionic surfactants, with *t*-butyl hydroperoxide, initiate the emulsion polymerisation of styrene and control molecular weight by chain transfer [305].

8.9 Emulsion polymerisation, acrylic and methacrylic esters

The control of surfactant level in a starved-feed emulsion of methyl methacrylate and butyl acrylate has been studied by measuring the instantaneous free surfactant level by surface tension measurements. If the surface tension of the latex was maintained at 45–57 dyn cm^{-1}, monodisperse particles with diameters in the range 500 nm–3 μm were prepared [306].

The radical polymerisation of methyl, ethyl, butyl, hexyl and 2-ethylhexyl acrylates in micro-emulsions (Section 2.4.5) was studied with an ammonium persulfate initiator. The maximum rates for methyl, ethyl, butyl and 2-ethylhexyl acrylates were proportional to the 0.53, 0.17, 0.46 and 0.59 powers of the initiator concentrations respectively and, except for ethyl acrylate, were proportional to the monomer concentration. This irregularity of ethyl acrylate, which has a reaction order of unity only at very low concentrations, is surprising [307]. See also Sections 8.9.1, 8.9.2 and 8.9.3 immediately following.

8.9.1 Emulsion polymerisation of methyl methacrylate

The major feature of the emulsion polymerisation of methyl methacrylate is the 'gel effect' already mentioned [301]. This effect due to trapped radicals was first observed by Norrish and Smith [308], but is often referred to as the 'Tromsdorff effect' [309].

Zimmt put the Harkins–Smith–Ewart theory in the form

$$R_p = K_p N (M)/(2N_A)$$

where R_p is the rate of propagation of mole L^{-1} s^{-1}, K_p is the propagation constant in L mole^{-1} s^{-1}, N is the number of particles per litre of water, N_A

is the Avogadro number and (M) is the monomer concentration in mole L^{-1} in monomer–polymer particles [310].

Methyl methacrylate was polymerised in a standard laboratory apparatus with a branched chain alkyl sulfate as surfactant and potassium persulfate as initiator, polymerising under nitrogen (Figure 3.9). The particle size seems to have been varied only by alterations in the stirrer speed. N was determined from particle diameters, both absolutely by electron microscopy and relatively by light transmission (see Chapter 4, Section 1.1.1). Figure 3.10 shows that only for particle sizes averaging 0.11–0.19 μm is the rate of polymerisation normally linear against time. For all larger particle sizes, polymerisation proceeds more slowly in accordance with the theoretical prediction. There is no other retarding effect, but there is a gradual acceleration of polymerisation from a point between 10 and 20 % conversion due to the gel effect, which can cause a considerable number of growing radicals to be present in a particle at the same time.

A generalised model has been developed to estimate the variation of particle concentration during the entire course of the emulsifier free emulsion polymerisation of methyl methacrylate [311].

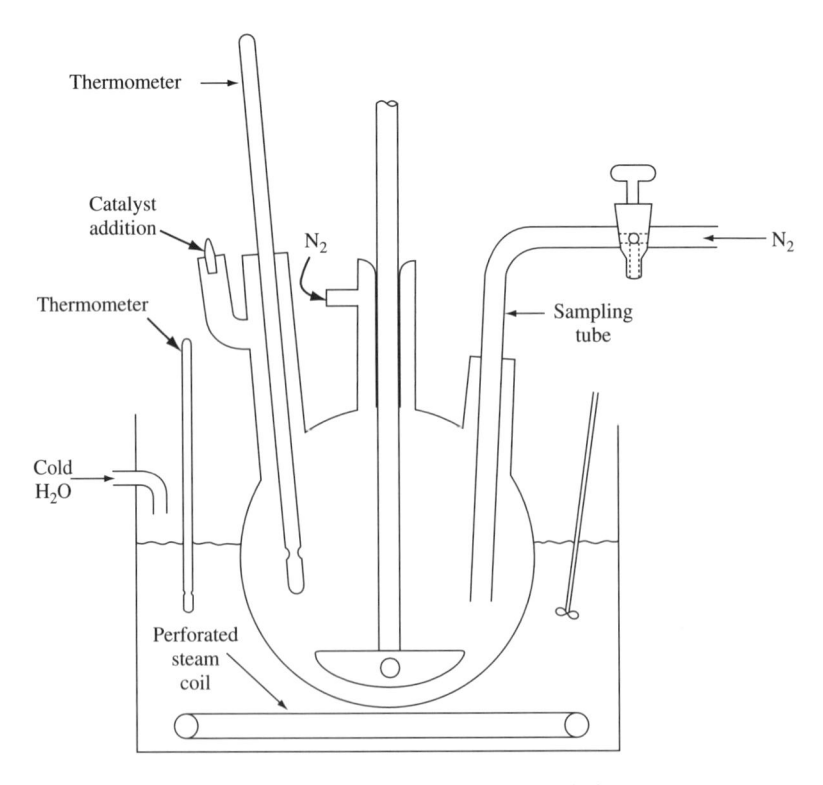

Figure 3.9 Schematic diagram of the experimental zimmt apparatus.

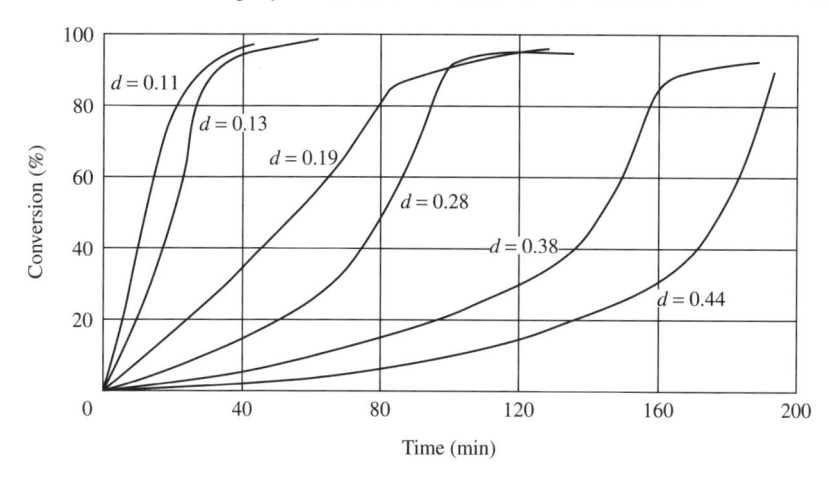

Figure 3.10 Conversion versus time curve for the emulsion polymerisation of methyl methacrylate, showing the final particle diameters (in μm) of various experiments. (Reproduced from reference [310].)

8.9.2 Emulsion polymerisation of butyl acrylate

Investigations into the effect of initiator on the emulsion polymerisation of butyl acrylate, with sodium dodecyl sulfate and a sorbitan-type non-ionic emulsifier, show that there is a typical sigmoidal curve of polymerisation versus time, except where the ammonium persulfate initiator was high (>0.005 mol dm^{-3}), in which case the steady state (interval 2) starts almost immediately. The rate of polymerisation versus the initiator concentration was 0.43, not far removed from the Smith–Ewart theory, which suggests that this figure is 0.4. If the oil-soluble initiator 2,2-azodiisobutyronitrile (AIBN) is included, the R_p is dependent only on the 0.25 concentration of ammonium persulfate. AIBN causes both a decrease in molecular weight and a reduced polymerisation rate. Results using dibenzoyl peroxide were similar, but suggestions as to the cause of these phenomena are beyond the scope of this volume [312].

8.9.3 Copolymers of butyl acrylate

The copolymerisation of methyl methacrylate with butyl acrylate in the presence of a non-ionic emulsifier (nonyl phenol, 24 ethylene oxide) follows the classical theory with a micellar mechanism for particle nucleation. With sodium dodecyl sulfate homogeneous nucleation also takes place, the balance between the two methods being dependent on emulsifier concentration and the monomer-to-water ratio [313].

In a general comparison between the polymerisation of butyl methacrylate and methyl methacrylate, Brodnyan et al. [314] concluded that water solubility was unimportant in determining the rate dependence and particle number dependence. The rate of polymerisation R_p obeys the equation

$$R_p = kC_E^{0.5}C_I^{0.5}$$

where C_E and C_I are the emulsifier and initiator concentrations. Note that this equation is identical with that of Medvedev.

The number of particles N is given by

$$N = kC_E \sim^{3.0} C_I \sim^0 C_M \sim^0$$

where C_M is the concentration of monomer over most of the range of emulsifiers investigated. Note that there is a resemblance between these results and those of French (next section) as far as dependence of the particle number on the cube of the emulsifier concentration is concerned, although the dependence on monomers differs.

The rate of polymerisation per unit surface area seems to be fairly constant, but of course the rate per particle changes dramatically as the emulsifier concentration is varied. The authors conclude that there is evidence that the particle surface may be the locus of polymerisation as predicted by Medvedev, but variations in the particle size distribution seem to indicate that more than one mechanism is operating. However, experimental variance is such that it is not possible to decide on the relative validity of the two theories, bearing in mind that their effective mathematical relationships are very close [314].

Further studies of the kinetic behaviour in the copolymerisation of methyl methacrylate and butyl acrylate have been made [315]. Studies have also been made of the effect of anionic and non-ionic emulsifiers or mixtures on the copolymerisation of styrene and n-butyl acrylate, with special reference to particle size and size distribution. Laser light-scattering techniques and transmission electron microscopy were employed [316].

8.10 Emulsion polymerisation of vinyl acetate

Vinyl acetate shows major deviations from the general theory. This is due to the fact that not merely is the monomer appreciably water soluble (about 2.5 % with little temperature variation), but also that the point at which polymer separates from solution to form emulsion particles occurs only after an appreciable amount has polymerised. Some solubility measurements of an aqueous solution of polyvinyl acetate dissolved in its monomer have been made [317]. In addition, a certain amount of fairly low molecular weight polymer with sulfated end groups is formed, due to residues from the persulfate initiators. This fact was clearly recognised by Priest who, working with very dilute emulsions, nevertheless produced stable latices. He considered that not more than 0.1 % of polymerisation is identified with primary particles, the remainder being polymerised in the swollen polymer droplets [318].

However, Stannett and coworkers, who extended this work by polymerisation of vinyl acetate on to a seed latex, concluded that the rate of polymerisation is proportional to the 0.2 power of the number of particles. This is not in accord with Harkins' theory and its development. The authors conclude that most of the polymerisation takes place in the water phase [319]. It has also been shown that termination is essentially an aqueous reaction.

The work of French focuses some attention on the emulsifier as well as the mechanism of polymerisation [320]. The block copolymers of ethylene oxide and propylene oxide were used in this case at variable concentrations which range from 2 to 7 %, whilst the overall monomer was present at 55 %, a typical industrial concentration. The diffusion of monomer into polymer particles was followed by short stopping at various concentrations, centrifuging the polymer and determining the ratio of monomer to polymer in the polymer layer. It was concluded that the monomer has largely diffused into polymer droplets by the time that 13.5 % conversion is reached. This implies that particles have reached the maximum size at about 13.5 % conversion, which was determined experimentally.

The surface area of the particles was approximately a linear function of the surfactant used, and the logarithm of the viscosity of a polyvinyl acetate emulsion was proportional to the total surface of the total surface concentration, and hence to the total polymer surface area. Mathematically, French found the following:

$$N = anD_s^2$$

which can be alternatively expressed as $N = d^2 (an)^3 (600F_m)^2$, where N is the number of particles per 100 g of latex. D_s is the diameter of a polymer particle in cm, a is the cross-sectional area of a surfactant molecule in cm^2, n is the number of surfactant molecules per 100 g of latex, F_m is the weight fraction of monomer initially present and d is the density of the polymer particles. Thus, in this case the number of particles varies as the cube of the surfactant concentration.

It is interesting to note that some of the ethylene oxide–propylene oxide emulsifiers are very effective for vinyl acetate, preferably with a colloid stabiliser, producing a very fine particle size latex at 55 % concentration. This may be due to some grafting of the monomer.

By combining the results of French with those of Mesrobian and coworkers [321], who found that using polyvinyl alcohol as an emulsifier R_p was proportional to the 0.6 power of the emulsifier concentration (as in Smith–Ewart kinetics), we arrive at Stannett's conclusion, that the rate of polymerisation is proportional to the 0.2 power of the number of particles, showing consistency of data, irrespective of the emulsifier type. Other early researches have indicated that, with the use of a persulfate initiator, no additional emulsifier is necessary [322, 323].

Emulsion polymerisation

The general correlation between the rate of polymerisation emulsifier concentration (S), the original concentration of initiator (C_0) and the phase ratio P has been given by Gerschberg as

$$R_p = K(S)^{x \pm 0.07} (C_0)^{0.45 \pm 0.05} (P)^0$$

For monomers whose water solubility is below 0.04–0.07 %, $x = 0.6$, as predicted by the Smith–Ewart theory. For the more water soluble monomers, x equals a gradually diminishing in quantity, as shown from Fig. 3.11 [324].

There have been a considerable number of later papers attempting to explain the behaviour of vinyl acetate polymerisation under specific conditions of concentration, emulsifier concentration, initiator concentration and ionic strength. Without doubt the formation of oligomeric species ending with a sulfate or sulfonate group from the initiator has a profound effect on the reaction since particle growth is in the first instance in the water phase. The escape of radicals from particles, which also initiate further polymerisation, is another complicating factor.

A comprehensive study by Litt and Stannett and coworkers in 1970 gives a model with the following form:

$$R_p = C_1[I](M_p N_p)^{0.5} (1 + C_2 M_p/r^2[M_{aq}])^{-0.5}$$

where C_1 and C_2 are constants, the latter allowing for the sweep-up of aqueous phase radicals by the polymer particles; $[I]$ is the initiator concentration (in mole L^{-1} of water), r is the radius of the particle and M_p is the monomer concentration in the polymer phase and M_{aq} the monomer concentration, both in mole L^{-1} [325].

Friis [326] disagreed with some aspects of this work. He found that under varied conditions at 50 °C the rate of polymerisation is proportional to the 0.5 power of the initiator concentration and the 0.25 power of the number of particles. The number of particles was proportional to the 0.5 ± 0.05 power of the

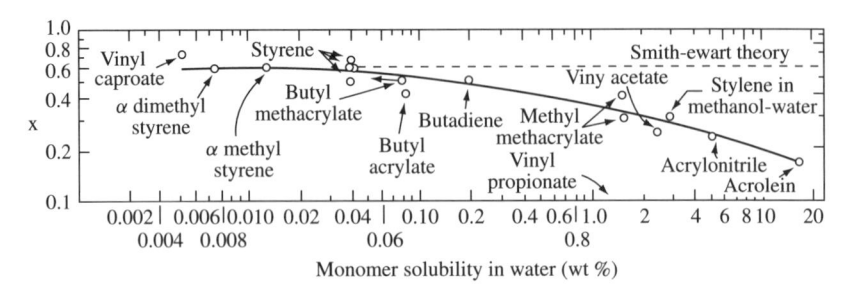

Figure 3.11 Emulsion polymerisation kinetics showing the dependence of the polymerisation rate on emulsifier concentration plotted against monomer solubility in water, $R_p \propto (S)^x$.

emulsifier concentration, but is independent of the initiator (potassium persulfate) concentration. However, Zollars comments that the changing dependence of the particle number on the initiator level may be the cause of wide variations in the reaction rate order with respect to the initiator level reported by other investigators [327, 328].

A paper by Litt and Chang noted the surprising result that polyvinyl acetate is appreciably solubilised in a solution of sodium tridecyl sulfate [329]. A seeded emulsion polymerisation was used as a basis for the investigation, sulfate or sulfonate emulsifiers which were purified as far as possible being included. The conclusions were that the rate of polymerisation is dependent on the 0.6 power of the initiator concentration, the 0.12 power of the particle concentration and the vinyl acetate volume to the 0.39 power. It is independent of the type of emulsifier and its concentration and ionic strength in the emulsion. The R_p is almost independent of monomer concentration until 85–90 % concentration. A summary of the results is:

(a) The persulfate ion enters the particle where it either initiates or terminates the polymerisation, depending on whether the particle contains a radical or not. As only 1–2 % of particles contain radicals, polymerisation is usually initiated in this case.
(b) The radical in a particle can chain-transfer to the monomer, generating a monomer radical. The kinetically important chain transfer is probably on vinyl hydrogen. This is an aqueous radical.
(c) The aqueous radical may be swept up by a dead particle. It may react with the initiator forming a sulfate ion radical which then enters a particle. It may also be swept into a particle containing a radical, and may occasionally escape without terminating, but will diffuse back until it reinitiates or terminates [329].

Polyvinyl acetate with a high molecular weight and a low degree of branching is obtained by low-temperature emulsion polymerisation of vinyl acetate with a potassium persulfate/dimethylethanolamine initiator system. Polyvinyl alcohol formed by saponification indicated linearity and a high molecular weight [330].

A symposium volume already quoted above gives a further selection of papers on vinyl acetate polymerisation [331].

8.10.1 Copolymerisation—mainly of vinyl acetate

In general, copolymerisations in emulsion follow a rather similar course kinetically to polymerisation of single monomers, but there are marked discrepancies, such as those caused by modifications of the gel effect and also where one of the monomers has some water solubility (see the next section). There are numerous claims in the patent literature to various copolymerisations.

In copolymerisations where all monomers are substantially water-insoluble, the copolymerisation rate depends on the normal reactivity ratios. If the two monomers are added at the start of the reaction, the composition will vary, the most reactive monomer forming the major part of the polymer first formed, the latter gradually reverting to the less reactive monomer as the reaction proceeds. If both reactivity ratios r_1 and r_2 are below unity, all of the monomers will be included in the copolymers. If one of the monomers has a reactivity above unity, which implies that it has a tendency for its radicals to add preferably to its own monomer, the least reactive monomer will ultimately polymerise on its own, which may mean either that the particles will be heterogeneous or, where there is new nucleation, there will be particles of varying composition. Where a monomer has zero reactivity other than by copolymerisation, part may remain unpolymerised. This also applies to monomers which tend to alternate; if a monomer with zero reactivity towards its own radicals is in excess in molar proportion, part of this monomer will remain unpolymerised. It is to be appreciated that the above consists only of generalisations and there are numerous variations in specific cases, especially where several monomers are involved.

Where there is appreciable water solubility of one or more of the monomers, there is a more marked tendency for initiation, and at least part of the propagation to take place in the aqueous phase. This has already been considered in connection with the general kinetics of the polymerisation reaction. Thus, with the copolymerisation of vinyl acetate with butyl or 2-ethylhexyl acrylate, of which the reactivity ratios are highly in favour of the acrylic monomer, the copolymerisation may produce a higher ratio of vinyl acetate than expected because of the prior initiation of the vinyl acetate. If vinyl acetate is copolymerised with a vinyl monomer of approximately the same reactivity, e.g. Veova10® (see Section 1.1), it is found at the apparent end of the reaction, which is normally a minimum of 99.5 % complete, and there is always a slight excess of Veova 10 even though it is the smaller constituent. This is because its much lower aqueous solubility reduces slightly its apparent reactivity. The nature of the initiator may have an appreciable effect on the nature of copolymerisation. This applies particularly to the contrast between a water solubility initiator, usually a persulfate or a redox system including a persulfate, and an organic initiator such as dibenzoyl peroxide [332].

Constant composition copolymers are prepared industrially by the delayed addition process, the addition of mixed monomers being added at such a rate that polymerisation is as fast or faster than the rate of addition of monomers. Under these conditions, free monomer at any time is minimal, a factor that balances the differences in reactivity of the monomers. The conditions are determined empirically, based on taking samples of the polymerising emulsion, inhibiting and examining for residual monomers. A satisfactory rate is determined experimentally and is used as a basis for a standard formulation. Some complications occur with very volatile monomers, and suitable pressure apparatus for addition must be used for a delayed addition (semi-continuous) process with monomers such as vinyl chloride or butadiene if they are components in an emulsion copolymer.

A technique has been described by which non-uniform emulsion polymers can be produced in a controlled manner. It involves the continuous addition of a monomer mixture into a stirred tank containing another monomer mixture. This continuously changing mixture is then fed into a reactor producing polymers whose instantaneous monomer composition varies. It may produce valuable results for industry, e.g. a broad transition range [333]. A design for an apparatus to ensure constant composition of copolymers has been described [334].

8.10.2 Vinyl acetate and acrylic copolymers

Many of the studies of vinyl acetate copolymers include acrylic esters, especially butyl acrylate. As expected, a semi-continuous process in which the monomers are added under starved conditions with sulfosuccinate emulsifiers provides a much greater degree of homogeneity in copolymer composition than in a batch process in which all the monomer is added in a single stage. This is well known in industry [335–339]. However, if an 88 % hydrolysed polyvinyl acetate (technically 'polyvinyl alcohol') is used as the emulsifier, vinyl acetate polymerises more rapidly than butyl acrylate. This is attributed to hydrogen abstraction from the polyvinyl alcohol chains by vinyl acetate monomer, which then proceeds to add on forming branched chains. This occurs until the average degree of hydrolysis is 82 %, at which level precipitation tends to occur [340].

8.11 Styrene copolymers (except with dienes)

Copolymers of styrene, especially with acrylic ester monomers, have been the subject of most researches into the nature of copolymerisation. A considerable amount of this research has been undertaken by Guillot and Pichot at the CNRS Laboratories in Vernaison, France. Results can only be summarised here. If styrene is copolymerised with methyl acrylate, with sodium dodecyl sulfonate as the emulsifier and either potassium persulfate or a redox system also including sodium hydrogen sulfite at 50 °C, both initial and final copolymers as formed are richer in the more hydrophilic monomer. Varying ratios of monomers were used. The former is due to the water solubility of methyl acrylate, the latter to its relatively lower activity compared to styrene. The particle number (N_p) tends to increase with monomer mixtures rich in methyl acrylate, contrary to styrene polymerisations in which N_p remains constant [341].

In the copolymerisation of styrene and butyl acrylate, 1 : 1 by weight with potassium sulfopropyl methacrylate with a persulfate initiator, end groups are sulfate and sulfonate. Thus 30 % of the sulfo- monomer is fixed on the surface compared with 10 % when it is replaced by methacrylic acid. Most of the functional monomer is incorporated into the particle surface during the last 30 % conversion. It is considered that there are two loci of polymerisation, both in the aqueous phase and the monomer swollen particles [342, 343]. This work was extended to terpolymers also including methyl methacrylate. In a batch polymerisation this was consumed at the fastest rate, followed by butyl acrylate and styrene. Gas liquid chromatography (GLC) was used to study reaction kinetics and polymer compositions were verified by nuclear magnetic resonance (NMR) [344]. Ethyl acrylate copolymerisation with styrene has also been investigated in a batch emulsion polymerisation. The general study is rather similar to that of styrene with methyl methacrylate, but ethyl acrylate is somewhat less active [345]. The polymerisation rate increases with increasing butyl acrylate ratio in copolymerisation with styrene [346].

Further studies of the copolymerisation of methyl methacrylate with styrene are available [347–349]. Earlier studies have shown that even when styrene is copolymerised with methacrylic acid under standard conditions with a persulfate initiator there are about six times as many sulfate groups on the surface of the particles as carboxyl groups [350].

8.12 Butadiene copolymers

Both radiation and persulfate methods have been used to graft styrene to a polybutadiene latex average particle diameter of 2400 Å. Conversion curves were found to be complex functions of the initial monomer concentration and the number of particles and temperature. The simple Smith–Ewart theory did not in general apply to these systems. Radiation gave somewhat lower graft efficiencies, but this was more than 80 % under the best conditions, and radiation was considered to be viable for industrial use [351].

The reactivity ratios of styrene and butadiene have been reported as 0.78 : 1.4. Whilst this indicates a slight tendency of butadiene to copolymerise with itself, the closeness of the quotient $r_1 r_2$ to 1 indicates a more or less random copolymerisation. These monomers are usually polymerised in a single stage, even in large reactors. Whilst for use in synthetic rubber the copolymerisation is short-stopped to obtain a desirable product, when the latices are used for coating purposes polymerisation takes place slowly at a temperature of about 50 °C and may take up to 48 h for completion. These latices usually contain higher proportions of styrene than the synthetic rubbers (35–40 % compared with 25 % for synthetic rubbers). Acrylonitrile is sometimes used as an additional monomer or as an alternative to styrene in copolymerisation with butadiene, often in a grafting process by addition across the residual double bonds.

Thus an azeotropic mixture of styrene and acrylonitrile (80/20 weight/weight) was copolymerised under various conditions with a polybutadiene seed, the additive monomers being equal in weight to the seed, both batch and semi-continuous processes being used, a suitable emulsifier at 35 °C being 3 % of Igepal CO-990, a nonylphenol–ethylene oxide adduct with 100 moles of ethylene oxide and 6 % at 50 or 70 °C. This gave no secondary particle initiation. The redox initiators used were either cumene or *t*-butyl hydroperoxide, the latter being preferred, with sodium formaldehyde sulfoxylate and ETDA and chelated Fe^{2+}. Slower monomer addition and higher temperatures gave higher grafting rates, although coagulum increased. Grafting also increased with decreasing particle size of the seed latex. Poly(styrene-acrylonitrile) inclusions were formed inside the polybutadiene particles at higher monomer-to-polymer ratios [352].

Experiments with the copolymerisation of styrene and butadiene (approximately 2 : 1), sometimes with the addition of acrylic acid or acrylamide with a high persulfate ratio, an emulsifier system consisting mainly of a non-ionic emulsifier but with a lower proportion of anionic emulsifier, showed that particle size was governed by the small amount of third comonomer added. No results corresponded to the Smith–Ewart theory for particle number. It was assumed that primary particles originated in aqueous solution and that they consecutively agglomerate. Particle size is also governed by the persulfate concentration. Nucleation is complete at 1–2 % conversion and particle growth between 15 and 30 % depending on the non-ionic stabiliser used. There is a gel effect at about 35 % [353].

8.13 Copolymerisation including a water-soluble monomer

The modelling of emulsion polymerisations that include monomers with significant aqueous solubility and their influence on the radical transport out of the water-swollen polymer particles and the characterisation of oligomers formed in the aqueous phase is presented [354]. The pH has a strong influence on the polymerisation of acids such as acrylic acid and methacrylic acid, both as homopolymers and in copolymerisation [355].

If acrylonitrile, which has an aqueous solubility of about 7 %, is copolymerised with styrene, there are appreciably less acrylonitrile units in the copolymer than would be predicted by theory. If a correction factor is applied, allowing for this solubility, the composition of the copolymer is that expected theoretically. This also seems to indicate that this copolymerisation takes place in the monomer droplets [356].

Methyl methacrylate and sodium methacrylate copolymerise in a curious manner. Whilst the consumption of the ester monomer follows zero-order kinetics, that of the salt follows first-order kinetics. The phenomenon is explained by assuming that there is aqueous phase initiation [357]. The copolymerisation of methacrylic acid with either styrene or methyl methacrylate and ethyl acrylate with which it is blended and added in a delayed addition without any buffering is quite complex. The acid copolymerises as fast as it is added. Under standard conditions there are varying quantities of residual monomer which may depend on whether a surfactant or a seed latex is present and also whether *tert*-dodecyl mercaptan is included in a formulation [358].

If styrene is copolymerised with acrylic acid in a delayed addition reaction, then regardless of monomer addition rate, the unreacted acrylic acid increases rapidly at first. Extended addition time gives more copolymer formation and less styrene homopolymer [359]. If styrene and itaconic acid are copolymerised in dioxan, in which they are both soluble, a ready copolymerisation results. On the other hand, an aqueous phase polymerisation results in almost pure polystyrene only being formed [360].

Many emulsion formulations, principally intended for carpet backing, however, claim to copolymerise itaconic acid successfully with butadiene and styrene, as judged by the final properties of the copolymers formed. This acid (h. 1, Section 6.5), with m.p. of 167–168 °C, does not homopolymerise readily under normal conditions and is quite water soluble (9.5 %) at 25 °C, as would be expected of an acid with two carboxyl groups. The solubility in styrene is less than 0.01 %. Although the copolymerisation ratios, on which there has been limited research, suggest that copolymerisation is very much in favour of the main monomers, there is undoubtedly some copolymerisation. This seems to a large extent to be a function of the surfactants present, which may solubilise styrene and butadiene sufficiently for a reasonable copolymerisation, or alternatively present a very large surface due to formation of very minute monomer emulsion particles. There is little published evidence on the extent of copolymerisation, or whether there is any free unchanged itaconic acid after completion of technical copolymerisations.

If styrene is polymerised in emulsifier-free aqueous media with one of acrylamide, methacrylamide or N,N-dimethylacrylamide at various ratios with a potassium persulfate initiator, there are three stages of polymerisation. At first acrylamides polymerise preferentially in the aqueous phase. After particle formation styrene polymerises exclusively. The decrease in the concentration of styrene droplets causes the transfer of the main reaction locus to the aqueous phase, leading to the formation of appreciable amounts of polymer dissolved in the aqueous phase. With N-(hydroxymethyl)acrylamide, monodisperse latices with a particle diameter of 350–1100 µm are formed that one dependent on the manner of monomer charge [361].

Much relatively recent work is published in collective volumes, resulting from symposia organised by the CNRS, France (see references [362] and [363]).

9 EMULSION POLYMERISATION IN SUPERCRITICAL CARBON DIOXIDE

The newest development in emulsion polymerisation is the use of supercritical carbon dioxide as the medium. An early successful polymerisation was performed using polymers of relatively low molecular weight of 1,1-dihydroperfluorooctyl acrylate as stabilisers. These fluoropolymers are very soluble in supercritical carbon dioxide, as are some silicones, these being termed CO_2-philic, most others being CO_2-phobic [364]. The carbon dioxide is vented at the end of the reaction, leaving the polymer as a fine powder [365]. Researchers from the same laboratory in the University of North Carolina have also prepared polyacrylic acid by the same method with an AIBN initiator. Pressures used are from 110 to 340 bar [366].

A study of the stabiliser indicated that the efficiency of the stabiliser is dependent both on the nature of the anchor group and soluble group on the stabilising polymer, and also on the distribution of the soluble group in the backbone [367]. The process has been extended to the preparation of polystyrene, using siloxane-based block copolymers [368]. Copolymers of methyl methacrylate and acrylic acid with core–shell, inverted core–shell and concentric shell morphologies have been prepared by his method, using a dibenzoyl peroxide–N,N-dimethylaniline redox initiator [369].

Polymerisation in supercritical CO_2 is of its nature restricted to obtaining polymers as solid powders, where they are substantially above the second-order transition point (T_g).

10 INVERSE EMULSIONS

A number of polymer emulsions have been prepared based on water-soluble polymers, sometimes as aqueous solutions, which are dispersed in a solvent in which the polymer is insoluble. These are referred to as inverse emulsions. There is some variation in nomenclature. The term 'dispersions' is sometimes used, but this term has alternative meanings including polymerisations in which a monomer is soluble in a solvent but the corresponding polymer is insoluble, precipitating in a fine powder.

Typical examples of invert emulsions include polymers and copolymers of acrylamide, which have many practical uses including tertiary oil recovery and as flocculants in many industries. An advantage of inverse emulsions is, like aqueous latices, that their viscosity is independent of the molecular weight of the polymer.

There are a number of important differences between inverse emulsions and aqueous latices. Thus the monomers are often solid and are dissolved in water, the solution being emulsified. Stabilization is necessary as in aqueous emulsions. Block copolymers are used as stabilisers, as are some non-ionic emulsifiers. Typical block copolymers are of the ABA type, where A equals a polyester based on 12-hydroxystearic acid and B is a polyoxyethylene. Note that the particles formed tend to be polymers swollen by the water originally present. The final solid is usually about 35 % of the total emulsion.

Free radical initiators may be water soluble, added before emulsification or soluble in the continuous medium. Redox initiation is possible and is preferred, as a reaction below about 50 °C is more controllable and is less likely to cause precipitation. In this case the two components may be in different phases. Control methods are rather similar to that in normal emulsion polymerisation. The stability of these emulsions is usually less than that of aqueous emulsions, but in many cases occurs after the polymer has settled, with stirring restoring the homogeneity.

Early work on this subject is described in a review by Vanderhoff and his colleagues at Lehigh University. This included the inverse emulsion polymerisation of a solution of sodium p-vinylbenzene sulfonate dispersed in o-xylene with dibenzoyl peroxide as the initiator. The nature of the reaction with special reference to the action of the initiator is discussed in the original paper [370]. This system has been used to form particles with a hydrophilic core and a hydrophobic shell. Acrylamide and N,N'-methylenebismaleimide are formed into an inverse emulsion in hexane with Span 80, a sorbitan-type emulsifier (see Chapter 2, Section 3.6). The paste-like product (after redox initiation with ammonium persulfate/sodium metabisulfite) is diluted with a hexane solution of a vinyl monomer, e.g. styrene, with an additional initiator, e.g. cumene hydroperoxide or vinyl chloride, which apparently does not need further initiator and is polymerised to give a core–shell structure. Note that neither polystyrene or polyvinyl chloride is soluble in hexane [371].

Monodisperse polymer particles have been prepared from 4-vinylpyridine and also copolymers with methacrylic acid by dispersion polymerisation with a block copolymer of styrene and butadiene as stabiliser and a mixture of dimethylformamide and toluene as solvent (5–20 : 95 : 80). Monodisperse particles of about 1 μm are formed [372].

A fuller description of the chemistry of inverse emulsions is given in a recent volume [373].

11 INTERPENETRATING NETWORKS (IPN)

Interpenetrating networks (IPN) are an intimate mixture of two or more polymers that are not identical, usually crosslinked and normally not miscible. Whilst they are not held by covalent bonds, they do not behave as arbitrary mixtures, but tend to have unique properties.

IPN polymers may be produced by sequential emulsion polymerisation. The first polymer is normally crosslinked *in situ*, e.g. by the inclusion of a crosslinking agent such as divinylbenzene before the second monomer, or a blend of monomers is added and polymerisation completed. A short time should be allowed for equilibriation.

There are many complications in the procedure. Thus some graft polymerisation may occur on the first polymer after further initiator has been added. New nucleation must be avoided with the second polymer.

Unusual physical properties are that the IPN may have T_g of both polymers involved. Because the first and second networks are intertwined on a molecular scale, higher tensile strengths have been observed as well as higher moduli and higher impact strengths at temperatures intermediate between the two T_g levels.

There have been a number of patent applications in respect of IPNs. One describes the use of an interpenetrating network as a binder for fibres, especially fiberfill. Fiberfill gives thickness and also insulation to a fabric. It requires a binder for which an IPN is suitable [374]. In this case the monomers, prepared with typical initiators and emulsifiers, were vinyl acetate, monoethyl maleate, triallyl cyanurate, acrylamide and N-methylolacrylamide (238 : 0.75 : 0.31 : 0.31 : 2 : 9.5). The second addition involves a preprepared emulsion which includes styrene and N-methylolacrylamide (100 : 2.45), which is added directly to the first latex, equilibrated for 30 min at 60 °C and initiated by a redox initiator. After completing polymerisation and cooling, two final additions to complete crosslinking are added. These are water, 50 % aqueous zinc nitrate, phosphoric acid (4 : 0.88 : 0.44), followed by water, biocide, 37 % formaldehyde solution (2 : 0.01 : 0.01). The IPN latex is diluted to 22 % and coated on to a polyester staple web prebonded with a polyvinyl acetate latex. A textbook, published in 1981, is available [375].

12 CONCLUSIONS

A survey has been presented of the principal papers that seek to explain some of the facets of emulsion polymerisation in more precise terms. This has been done mainly as an aid to sufficient understanding of the problems involved for the reader who is fundamentally interested in the applied side of these products. It is by no means exhaustive, and includes much of the fundamental earlier work.

There has been much development in our understanding of the various processes that take place in emulsion polymerisation since the 1972 edition of this volume, and also of the nature of the particles formed and the function of the emulsifiers. However, no one theory is universally applicable. In the words of one paper, albeit thirty years old at the time of writing [376]:

> All this indicates that the present theory remains inadequate, for although the reaction period when the rate of polymerisation is constant is to date considered interpretable, questions such as what will this rate be or how quickly it will be reached, still pose themselves. The answers to these problems lie in the understanding of the processes by which latex particles originate, and the dependence of these processes on initiator and emulsifier concentrations, and adsorption phenomena on the particle surfaces. This latter has in fact been one of main sectors of our knowledge, as will be shown in later chapters. This has been aided by the major advances in instrumentation in the past two decades. A list of further references of interest is included as references [377] to [386].

REFERENCES

1. Polymer Colloids Preprints, NATO Advanced Study Institute, University of Trondheim, 1975
2. D.R. Bassett and A.E. Hamielec (eds.), *Emulsion Polymers and Emulsion Polymerisation*, ACS Symposium Series 165, 1981
3. R. Buscall, T. Corner and J.F. Stageman (eds.), *Polymer Colloids*, Elsevier Applied Science Publishers, 1985
4. Emulsion polymerisation (emulsion copolymers), *Makromol. Chem.*, Supplementary Vol. 10/11, 1985
5. E.S. Daniels, E.D. Sudol and M.S. El-Aasser, *Polymer Latexes*, ACS Symposium Ser. 492, 1992 (see also reference [386])
6. *Radical Copolymers in Dispersed Media*, Hüthig & Wepf, 1995
7. T. Corner, *Colloids Surf.*, **3**, 119–29 (1981)
8. P. Teyssie et al., In *23rd FATIPEC Congress*, vol. A, 1996, pp. A 100–11
9. N. Berger et al., *Polym. Bull.*, **33**(5), 521–8 (1994)
10. T. Mezger et al., *Proc. XXI Int. Conf. Org. Coat. Sci. Technol.*, Athens, 1995, pp. 313–27
11. T. Metzger *Polym. Mater. Sci. Engng (ACS)*, **73**, 153–5 (1995)
12. M.A. Schafheutel et al. (Hoechst), GP 4,446,439, 1996
13. H. Toreii et al., *J. Polym. Sci.*, **34**(7), 1237–43 (1996)
14. H. Warson, *Makromol. Chem.*, **105**, 228–45 and 246–50 (1967)
15. N. Ganga Sevi and V. Mahadevan, *Makromol. Chem.*, **152**, 177–86 (1972)
16. G.S. Misra and C.V. Gupta, *Makromol. Chem.*, **168**, 105–18 (1973)
17. S.N. Trubitsyna et al., *Uzb. Khim. Zh.*, **19**(6), 50–3 (1975) (in Russian); *Chem. Abstr.,; CA*, 1976 **84**, 106–142; S.N. Trubitsyna et al., *Tr. Politekh. Inst.*, **132**, 15–23 (1974) (in Russian); *Chem. Abstr.*, **85**, 78468 (1976)
18. B. Stanger (BASF), Eur. P 771,328, 1997; PCT 33775,1995
19. K.A. Shaffie et al., *J. Polym. Sci. A, Polym. Chem.*, **35**(15), 31 (1997)
20. F. Vidal et al., *Collold Polym. Sci.*, **273**, 1199–207 (1995)
21. A. Guyot and F. Vidal, *Polym. Bull. (Berlin)*, **34**(5/6), 569–76 (1995)
22. M. Nakamae et al. (Kuraray), JP 278,212 1995; *Chem. Abstr.*, **124**, 119742 (1996)
23. H. Warson, in *Emulsion Polymerisation*, I. Piirma and J.L. Gardon (eds.), ACS Symp. Series No. 24, 1976, pp 330–40
24. V.A. Stannett et al., *J. Appl. Polym. Sci.*, **13**, 1175–80 (1969)
25. V.A. Stannett et al., *J. Macromol. Sci., Chem.*, **A8**(2), 353–72 (1974)
26. V.V. Polikarpov et al., *Vysokomolek. Soyed*, **A16**, 2207–13 (1974); *Polymer Sci. USSR*, **16**(10), 2559–66 (1974) (Engl. trans.)
27. B.M. Tidswell and A.W. Rain, *Br. Polym. J.*, 409–16 and 407–28 (1975)
28. G. Cooper et al., *J. Colloid Interface Sci.*, **184**(1), 52–63 (1996)
29. G. Brown, H. Chou and J. Stoffer, *Prepr. Polym. Mater. Sci. Engng*, **74**, 300–1 (1996)
30. H. Warson, *Peinture, Pigments, Vernis*, **43**, 538–46 (1967)
31. W. Lau (Rohm & Haas), Eur. P 710,675, 1996.
32. J.M. Asua et al., *Polym. Reactor Engng*, **1**(4), 461–98 (1992–3)
33. H. Gerrens and Z. Koll, *u.Z. Polymere*, **227**, 92 et seq. (1968)
34. M. Kiar (Shawinigan Chemicals), BP 568,886, 1945
35. H. Hayashida et al. (Mitsui Toatsu Chemicals), JP 278,210, 1995; *Chem. Abstr.*, **124**, 88290 (1996); K. Nagasaki and M. Taukiide (Asahi Chemical Industry), JP 292,002, 1995; *Chem. Abstr.*, **124**, 88291 (1996)
36. H. Warson in *Emulsion Polymerisation of Vinyl Acetate*, M.S. El-Aasser and J.W. Vanderhoff (eds.), Applied Science Publishers, 1981, pp. 1–10
37. S.J. Baum and R.D. Dunlop, *B.I.O.S. 1102*, HM Stationery Office, 1947, pp. 5–6
38. J.-P. Bindelle (Solvay), Eur. P 551,681, 1993
39. N. Kondekare, *PaintIndia*, **45**(4), 47–51

40. J.M. Asua et al., *DECHEMA Monograph.*, **131**, 655–71 (1995)
41. Consortium für Electrochemische Industrie, BP 1,003,290, 1965; Consortium für Electrochemische Industrie, BP 1,068,086, 1967
42. B.W. Brooks, *Rev. Chem. Engng*, **2**, 403–43 (1983)
43. B.W. Brooks and E.E. Badder *Chem. Engng Sci.*, **39**, 1499–1509 (1984)
44. B.W. Brooks and G. Raman, *Chem. Engng Sci.*, **42**, 1439–49 (1987)
45. B.W. Brooks, *Chem. Engng Sci.*, **44**, 857–61 (1989)
46. R.A. Wessling, *J. Appl. Polym. Sci.*, **12**, 309–19 (1968)
47. A.W. DeGraff and G.W. Poehlein, *J. Polym. Sci.*, **A-29**, 1955–76 (1971)
48. G. Poehlein in *Polymerisation Reactors and Processes*, J.N. Henderson and T.C. Bouton (eds.), ACS Symposium Series No. 104, 1979, pp. 1–14
49. G.W. Poehlein and J. Schork, *Trends in Polym. Sci.*, **1**(10), 00–00 (1993)
50. M. Namura et al., *J. Appl. Polym. Sci.*, **15**, 675–91 (1971)
51. N. Namura and M. Harada in *Emulsion Polymers and Emulsion Polymerisation*, D.R. Bassett and A.E. Hamielec (eds.), ACS Symposium Series No. 165, 1981, pp. 121–43
52. G.E. Meira, A.F. Johnson and J. Ramsay, in *Polymerisation Reactors and Processes*, J.N. Henderson and T.C. Bouton (eds.), ACS Symposium Series No. 104, 1979, pp. 253–66
53. R.K. Greene, R.A. Gonzalez and G.W. Poehlein, ACS Symposium Series No. 24, 1976, pp. 341–66
54. J.B. Rawlings and W.H. Ray, *Am. Inst. Chem. Engng J.*, **33**, 1663–77 (1987)
55. G.R. Meira et al., *Polym. Reactor Engng*, **3**(3), 201–33 (1995)
56. K.H. Lee and J.P. Marano, in *Polymerisation Reactors and Processes*, J.N. Henderson and T.C. Bouton (eds.), ACS Symposium Series No. 104, 1979, pp. 221–51; J. Meuldijk, F.H.A.M. Van Den Boomen and D. Thoenes *DECHEMA Monogr.*, **131**, 45–54 (1995) (in English)
57. W. Herrig and H.J. Kessler (Bayer), BP 2,243,613, 1991; equiv. GP 4,013,715, 1991
58. E. Pettelkau (Bayer), GP 2,520,891, 1976; M. Ghosh and T.H. Forsyth, in *Emulsion Polymerisation*, I. Piirma and J.L. Gardon (eds.), ACS Symposium Series No. 24, 1976, pp. 367–78
59. B.W. Brooks, *Ing. Chim. Ital.*, **18**, 26–8 (1982)
60. B.W. Brooks, *Rev. Chem. Engng*, **1**, 403–30 (1983)
61. J.B. Rawlings and W.H. Ray, *Am. Inst. Chem. Engrs. J.*, **33**, 1633–77 (1987)
62. A.L. Rollin et al., in *Polymerisation Reactors and Processes*, J.N. Henderson and T.C. Bouton, ACS Symposium Series No. 104, 1979, pp. 113–36
63. A. Penlidis, G.L. Rempel and P.J. Scott, *Polym. Reactor Engng*, **3**(2), 93–130 (1995)
64. J. Marco, BP 889,088, 1962
65. F.E. Tirik and F. Castenada (Kativo), USP 3,296, 168 1967
66. G.W. Poehlein and F.J. Schork, *Trends in Polym. Sci.*, **4**(6), 173–6 (1996)
67. T.V. Kreitser et al., *Zh. Prikl. Khim. (Leningrad)*, **55**(7), 1647–51 (1982). *Chem. Abstr.*, **97**, 128188 (1992)
68. Mitsubishi Chemical, JP 116501, 1981; *Chem. Abstr.*, **95**, 44251 (1981)
69. J. Stöckel et al. (VEB), East GP 148,225, 1981
70. A.A. Abdullaev and T.T. Ismailov USSRP 954,395, 1992; *Chem. Abstr.*, **98**, 17883 (1983)
71. A.A. Abdullaev and F.M. Abdullaev, USSRP 994,471, 1983; *Chem. Abstr.*, **98**, 162213 (1983)
72. T.T. Ismailov and M.A. Mekhtiev, USSRP 1,030,369, 1983; *Chem. Abstr.*, **99**, 195609 (1983)
73. O. Chechik et al., *Kauch. Rezina*, **1984**(1), 18–21 (1984); *Chem. Abstr.*, **100**, 104820 (1984)
74. A.E. Hamielec et al., *Chem. Engng Commun.*, **35**(1–6), 123–40 (1985)
75. F. Steffers et al. (VEB C.W. Buna), East GP 238,236, 1986
76. R. Musch et al. (Bayer), GP 3,002,711, 1981

77. R. Lanthier (Gulf Oil of Canada), USP 3,551,396, 1970; BP 1,220,777, 1971
78. C. Abad et al., *DECHEMA Monogr.*, **131**, 87–94 (1995); *Polymer*, **36**(22), 4293–9 (1995)
79. C. Abad et al., *J. Appl. Polym. Sci.*, **56**(4), 419–24 (1995)
80. K.R. Geddes, *Chem. Ind.*, 21 March, 223–7 (1983)
81. K.R. Geddes, *2nd Asia-Pacific Conference*, prepr. paper 181, 1975
82. K.R. Geddes, *Br. Polym. J.*, **21**, 433–41 (1989)
83. K.R. Geddes, in *Handbook of Adhesive Technology*, A. Pizzi and K.L. Mittai (eds.), 1985, pp. 431–41
84. K.R. Geddes and M. Wilkinson, *Chem. in Britain*, December, **29**(12), 1050–2 (1993)
85. K.R. Geddes, in *Surface Coatings*, Vol. 3, A.D. Wilson, J.W. Nicholson and H.J. Prosser (eds.), Elsevier, new york, 1990, pp. 199–228
86. M. Bowtwill, *Mod. Paint Coat.*, **83**(7), 46,49 (1993)
87. D.Y. Lee et al., *Polym. Engng Sci.*, **32**(3), 198–205 (1992)
88. G. Scheider, *Processing*, 21 July 1991; A. Labbadene and P. Baille, *J. Appl. Polym. Sci.*, **51**(3), 503–11 (1995)
89. *Emulsion Polymerisation and Applications of Polymer Emulsions*, a series of bulletins issued by Paint Research Association, Teddington, Middlesex, UK
90. I.M. Krieger et al., *J. Paint Technol.*, **40**, 541 *et seq.* (1968)
91. J.S. Dodge, M.E. Woods and I.M. Krieger, *J. Paint Technol.*, **42**, 71 *et seq.* (1970)
92. Y.S. Papir, M.E. Woods and I.M. Krieger, *J. Paint Technol.*, **42**, 571 *et seq.* (1970)
93. J. Ugelstad et al., *Proc. Int. Conf. Org. Coat. Sci. Technol.*, State University of New York at New Paltz, 1978, pp. 415–45
94. S.T. Eckersley, G. Vandezande and A. Rudin, *J. Oil. Col. Chem. Ass.*, **72**(7), 273–5 (1989)
95. J.W. Vanderhoff, *J. Coll. Interface Sci.*, **28**, 336 *et seq.* (1968)
96. J. Liu et al., *J. Polym. Sci. A., Polym. Chem.*, **35**(16), 3575–3583 (1997)
97. J.W. Vanderhoff, M.S. El-Aasser and C.M.J. Cheng, *Polym. Sci.*, **30**, 235–44, 245–56 (1992)
98. D.A. Cauley and R.W. Thompson, *J. Appl. Polym. Sci.*, **27**, 363–79 (1982)
99. C. Tongyu et al., *J. Appl. Polym. Sci.*, **41**(9–10), 1965–72 (1990)
100. M.R. Grancio and D.J. Williams, *J. Polym. Sci.*, **A-1**, 8, 2617–29 (1970)
101. M.R. Grancio and D.J. Williams, *J. Polym. Sci.*, **A-1**, 8, 2733–45 (1970)
102. M.R. Grancio and D.J. Williams, in *New Concepts In Emulsion Polymerisation, J. Polymer Sci.*, **C7**, 139 *et seq.*
103. D.J. Williams, P. Keusch and J. Prince, *J. Macromol. Sci.-Chem.* **A7**(3), 623–46
104. D.J. Williams and P.J. Keusch, *Polym. Sci. Polym. Chem. Ed.*, **11**, 143–62 (1973)
105. D.J. Williams, *J. Polym. Sci., Polym. Chem.*, **11**, 301–3 (1973)
106. A. Rudin et al., *Coll. Polym. Sci., Proc. Int. Colloq. Copolymerisation and Copolymers*, CNRS, Lyon, France, 1989, pp. 103–4
107. L. Rios et al., *Coll. Polym. Sci.*, **269**(8), 812–24 (1981)
108. V.I. Eliseeva et al., *Vysokomolek. Soedin A*, **31**(2), 263–8 (1989); *Chem. Abstr.*, **111**, 40194 (1989)
109. J.-E. Jonsson et al., *Macromol.*, **27**(7), 1932–7 (1994)
110. W.-D. Hergeth et al., *Coll. Polym. Sci.*, **268**(11), 991–4 (1990)
111. A. Arora et al., *J. Appl. Polym. Sci.*, **58**(2), 301–11, 313–22 (1995)
112. V.L. Dimonie, M.S. El-Aaser and J.W. Vanderhoff, *Polym. Mater. Engng. Sci.*, **58**, 821–5 (1988)
113. S. Lee and A. Rudin, *J. Polym. Sci., Polym. Chem.*, **30**(5), 865–7 (1992)
114. S. Lee and A.J. Rudin, *J. Polym. Sci., Polym. Chem.*, **30**, 2211–6 (1992)
115. D. Jouhué et al., *Forum de la Connaisance*, 1996 AFTA/CORI (18 pp. French, 10 pp English)
116. S.B. Choi et al. (Lucky), Eur. P 674,673, 199; PCT, 11265, 1995; *Chem. Abstr.*, **123**, 199730 (1995)
117. D.M.C. Heymans (Shell International Research), Eur. P 614,919–20, 1994

References 185

118. Y.-Y. Lu (Minnesota Mining Manufacturing), PCT 25,499, 1994; *Chem. Abstr.*, **123**, 200581 (1995)
119. N. Ando *et al.* (Kanegafuchi Chemical Industry), Eur. P 621,325, 1997
120. S. Nelson *et al.* (GAF), USP 4,419,492, 1983
121. S.A.F. Bon *et al.*, *J. Appl. Polym. Sci.*, **58**(1), 19–29 (1995)
123. F. Dobler *et al.*, *J. Appl. Polym. Sci.*, **44**(6), 1075–86 (1992)
124. Y. Xu *et al.*, *Huogong Xuebao*, **42**(6), 683–9 (1991)
125. R. Buter and P.M. Postma (Akzo), Eur. P 555,903, 1993
126. M. Okubo and H. Ahmed, *J. Polym. Sci., Polym. Chem.*, **34**(15), 3147–53 (1996)
127. M. Nakano and F. Hoshino (Mitsui Toatsu Chemicals), JP 188,913, 1995; *Chem. Abstr.*, **124**, 11050 (1996)
128. S. Rimmer *et al.*, *Polymer*, **37**(18), 4135–9 (1996)
129. R.H. Ottewill *et al.*, *Coll. Polym. Sci.*, **64**(6), 1123–34 (1997)
130. K. Yanagiuchi and K. Tamazawa (Taisei Kako), JP 316,242, 1995; *Chem. Abstr.*, **124**, 263552 (1996)
131. R.M. Blankenship *et al.* (Rohm & Haas), Eur. P 696,602, 1994
132. M.E.J. Dekkers and M.F. Adams (GEC), USP 5,356,955, 1994
133. T. Tomita *et al.* (Mitsubishi Yuka Badische), Eur. P 619,341–2, 1994
134. J. Pavinlec and M.J. Laxar, *Macromol. Sci.*, **A31**(10), 1469–79 (1994)
135. O. Karlsson *et al.*, *J. Appl. Polym. Sci.*, **63**(12), 1543–55 (1997)
136. A. Renfrew and W.E.F. Gates, USP 2,296,403, 1943
137. H. Warson, *Polymerisable Surfactants and Their Applications*, Solihull Chemical Services, 1989
138. T.R. Aslamazova, *Prog. Org. Coat.*, **25**, 109–67 (1995)
139. R.M. Fitch *et al.* (Cook Paint & Varnish), USP 3,501,432, 1970
140. J. Reimers and F.J. Schork, *Polym. Reactor Engng*, **3**(4), 361–95 (1995)
141. H. Kawaguchi, T. Sugi and Y. Ohtsuka, in *Emulsion Polymers and Emulsion Polymerisation*, D.R. Bassett and A.E. Hamielec, ACS Symposium Series No. 165, 1981, pp. 145–56
142. J.W. Vanderhoff, M.S. El-Aaser *et al.*, *J. Polym. Sci. A, Polym. Chem.*, **30**, 171–83 (1992)
143. J.E.O. Mayne, H. Reichard and H. Warson (Vinyl Products), BP 607,704, 1948
144. J.E.O. Mayne and H. Warson (Vinyl Products), BP 627,612, 1949
145. J.E.O. Mayne and H. Warson (Vinyl Products), BP 648,001, 1950
146. J. Ugelstad, M.S. El-Aasser and J.W. Vanderhoff, *J. Polym. Sci., Polym. Lett.*, **11**, 503 *et seq.* (1973)
147. C.S. Chern and T.J. Chen, *Coll. Polym. Sci.*, **275**(6), 546–54 (1997)
148. J.L. Reimers and F.J. Schork, *Polym. Prepr. (ACS)*, **38**(1), 461–2 (1997)
149. K.J. Fontenot *et al.* (Georgia Tech. USA), USP 5,686,518, 1997
150. A.R.M. Azad, J. Ugelstad, R.M. Fitch and F.K. Hansen, in *Emulsion Polymerisation*, I. Piirma and J.L. Gardon (eds.), ACS Symposium Series No. 24, 1976, pp. 1–23
151. J. Ugelstad, F.K. Hansen and S. Lange, *Makromol. Chem.*, **175**, 507 *et seq.* (1974)
152. A.R.M. Azad, R.M. Fitch and J. Ugelstad, ACS Symposium Series No. 9, 1975, pp. 135 *et seq.*
153. E. Haque and S. Qutubuddin, *J. Polym. Sci. C., Polym. Lett.*, **26**, 429–32 (1988)
154. G.W. Hallworth and J.E. Carless, in *Theory and Practice of Emulsion Technology*, Symp. Brunel University, UK, 10–18 September 1974, prepr. 265 *et seq.*
155. M. El-Aaser *et al.*, in *Polymer Latexes*, E.D. Sudol and M.S. El-Aaser (eds.), ACS Symposium Series No. 492, 1992, pp. 72–98; J.W. Vanderhoff, M.S. El-Aaser *et al.*, in *polymer Latexes*, E.D. Sudol and M.S. El-aasser (eds.), ACS Symposium Series No. 492, 1992, pp. 99–113.
156. J.W. Vanderhoff, M.S. El-Aasser and J.J. Delgado, *Polym. Sci. A, Polym. Chem.*, **24**, 861–74 (1986)
157. J.W. Vanderhoff and J.J. Van den Hul, *J. Macromol. Sci.-Chem.*, **A7**(3), 677–707 (1973)

158. J.W. Vanderhoff, in *Emulsion Polymers and Emulsion Polymerisation*, D.R. Bassett and A.E. Hamielec (eds.), ACS Symposium Series No. 165, 1981, pp. 61–83
159. J.W. Vanderhoff, M.S. El Aasser *et al.*, *J. Polym. Sci. A, Polym. Chem.*, **30**, 691–702 (1992)
160. J. Santhanalakshmi and K. Anandhi, *J. Appl. Polym. Sci.*, **60**(3), 293–304 (1996)
161. L.L. de Arbina and J.A. Asua, *Polymer*, **33** (22), 4832–7 (1992)
162. S. Ghosh and N. Krishnamurti, *Paint Ink Int.*, **9** (1), 19–21 (1996)
163. H.S. Wu and E. Kaler (W.L. Gore & Associates), Eur. P 730,610; PCT 14715, 1995
164. I. Aizopura *et al.*, *Cidepint Anales*, 53–62 (1996) (in English)
165. M. Antonietti *et al.*, *Macromol. Chem. Phys.*, **195**, 441–66 (1995)
166. P.A. Lovell and P. Dorian, in *Emulsion Polymerisation and Emulsion Polymers*, P.A. Lovell and M.S. El-Aasser (eds.), Wiley, 1997, pp. 667 *et seq.*; D.M. Kulich, P.D. Kelley and J.E. Pace, in Encyclopedia of Polymer Science and Engineering, Vol. 1, H.F. Mark *et al.* (eds.), New york, 1985, pp. 385 *et seq.*
167. Y. Li and Y. Nakai (Mitsubishi Rayon), JP 286,833, 1997; *Chem. Abstr.*, **128**, 23818 (1998)
168. T. Omura *et al.* (Sekisui Chemical), JP 316,139, 1997; (1998) **128**, 75832
169. T. Kawahara *et al.* (Kuraray), JP 228,625, 1995; **124**, 30656 (1996)
170. Y. Mitsui *et al.* (Toho Rika), JP 279,644, 1994; **122**, 162618 (1995); H. Yamamoto *et al.* (Denki Kagaku), JP 03,102, 1995; **123**, 115601 (1995)
171. M. Nakamae *et al.* (Kuraray), Eur. P 587,114, 1994
172. M. Nakamae *et al.* (Kuraray), JP 295,010, 1993; **120**, 166876 1994; M. Nakamae and T. Sato (Kuraray), JP 263806, 1994; **122**, 56845 (1995)
173. G.S. Magallanes-Gonzales *et al.*, *J. Polym. Sci., Polym. Chem.*, **34**(5), 849–62 (1996)
174. R.D. Dunlop, F.I.A.T. 1102, Office of Military Government for Germany (US), 1947, p. 4
175. C.A. French (ed.), *Properties and Applications of Polyvinyl Alcohol.*, Society of Chemical Industry, 1968
176. C.A. Finch (ed.), *Polyvinyl Alcohol, Properties and Applications*, Wiley, 1973
177. C.A. Finch (ed.), *Polyvinyl Alcohol Developments*, Wiley, 1992
178. D.J. Guest (ICI), BP 870,287, 1961
179. J.E.O. Mayne, H. Warson and R.M. Levine (Vinyl Products), BP 932,389, 1963
180. H. Warson and D. Dargan (Dunlop), BP 1,038,623, 1966
181. A.E. Cory *et al.* (Shawinigan Chemical), BP 940,200, 1963
182. R.H. Pelton, *Polymer Colloids Reprints* (Nato Advanced Study Institute, University of Trondheim), 1975 (pages are unnumbered; see contents page for approximate position)
183. L.-G. Tang *et al.*, *Eur. Pol. J.*, **32**(9), 1139–43 (1996)
184. Cumberland Chemical, BP 1,068,976, 1967
185. M.K. Lindemann and R.P. Volpe (Air Reduction), BP 1,117,711, 1968
186. G.A. Vandezande and A. Rudin, in *Polymer Latices*, ACS Symposium Series No. 492, 1992, pp. 114–33
187. F. Decocq *et al.*, *Proc. XXIII Fatipec Congress*, Brussels, 1996, Vol. B, pp. 232–44
188. H. Warson and D.S.W. Dargan (Dunlop), BP 1,088,634, 1967
189. H. Warson and G.A. Reed (Dunlop), BP 1,144,316, 1969
190. J.I. Amalvy, *J. Appl. Polym. Sci.*, **59**(2), 339–44 (1996)
191. M.A. Dube and A. Penlidis, *Polym. Int.*, **37**(4), 235–48 (1995)
192. T. Uele and H. Iwai (Kansai Paint), JP 190,504, 1986; *Chem. Abstr.*, **106**, 6551 (1987); H. Ohata and H. Saga (Nisshim Chemical Industry), JP 73,543, 1996; *Chem. Abstr.*, **125**, 35530 (1996)
193. M. Sagara *et al.* (Nippon Shokubai), JP 158,767, 1987; *Chem. Abstr.*, **107**, 238648 (1987)
194. C.-S. Chern *et al.*, *J. Appl. Polym. Sci.*, **62**(3), 585–94 (1996)
195. C.S. Chern *et al.*, *J. Macromol. Sci.*, **A33**(8), 1077–96 (1996)
196. V.A. Demarest *et al.* (Rohm & Haas), Eur. P 623,659, 1995
197. J.L. Smith and R.D. Singer (Dunlop), BP 1,003,318, 1965
198. H. Warson, *Peintures, Pigments, Vernis*, **43**(7), 438–46 (1967)

References

199. A.E. Staley Manufacturing, BP 928,251, 1963
200. M. Nakamae and T. Sato (Kuraray), JP 179,705, 1994; *Chem. Abstr.*, **121**, 182415 (1994)
201. T. Sauer (Huels), Eur. P 671,420, 1995
202. W. Salomon, *PRA Proc. 3rd Nürnberg Congress on New Technologies for Coatings and Inks*, 1995, paper 21, 9 pp.
203. G. Huang and K. Li, *Hecheng Xiangjiao Gongye*, **17**(4), 224–7 (1994) (in Chinese); *Chem. Abstr.*, **122**, 188921 (1995)
204. T. Yamauchi and Y. Kamiyama (Asahi Kasei), PCT 29,196, 1995; *Chem. Abstr.*, **124**, 120312 (1996)
205. D.I. Jon and J.C. Hornby (ISP Investments), USP 5,225,474, 1993
206. K. Ogura (Toyo Boseki), JP 109,221, 1996; *Chem. Abstr.*, **125**, 115524 (1996)
207. S. Roy and S. Devi, *Polymer*, **38**(13), 3325–31 (1997)
208. J.R. Leiza, *J. Appl. Polym. Sci.*, **64**(9), 1797–809 (1997)
209. Y. Kobayashi (Sanyo Chemical Industry), JP 170,805, 1993; *Chem. Abstr.*, **120**, 77928 (1994)
210. T.H. Chieng *et al.*, *Polymer*, **37**(21), 4823–31 (1996)
211. S.T. Wang *et al.*, *J. Appl. Polym. Sci.*, **60**(12), 2069–76 (1996)
212. J.M. Geurts, *J. Appl. Polym. Sci.*, **61**(1), 9–19 (1996)
213. K.J. O'Callaghan *et al.*, *J. Appl. Polymer. Sci.*, **58**(11), 2047–55 (1995)
214. C.-S. Chern and C.-H. Li, *Polym. J. (Tokyo)*, **28**(4), 343–51 (1996) (in English)
215. J.E.O. Mayne, H. Warson and G.E.J. Reynolds (Vinyl Products), BP 887,356, 1962
216. D. Cochin *et al.*, *Macromol*, **30**(8), 2278–87 (1997)
217. D.C. Blackley, *Polymer Latices, Science and Technology*, 2nd edn, Vol. 2, Chapman & Hall, 1997, Chs, 10 and 11, p.137 *et seq.*,
218. W. Fujiwara *et al.* (Sumitomo Dau), JP 17,511, 1993; *Chem. Abstr.*, **119**, 50130 (1993)
219. K. Sato and T. Sugimura (Nippon Zeon), JP 255,457, 1993; *Chem. Abstr.*, **120**, 79560 (1994)
220. H. Itoda *et al.* (Mitsui Cyanamid), USP 5,250,602, 1993
221. H. Mark and H. Fikentscher (I.G. Farben), USP 2,068,424, 1937 (Priority 1931)
222. F.K. Schoenfeld (B.F. Goodrich), USP 2,168,808, 1939
223. H. Warson, *Polym. Paint Col. J.*, **178**(4220), 625–7, 645; (4226), 865–6, 871, 878 (1988)
224. I. Zschach *et al.* (Buna), GP 4,330,238, 1995
225. H.R. Friedli and C.M. Keillor, *J. Coat. Technol.*, **59**(748), 66–73 (1987)
226. T.A. Morgan *et al.* (Goodrich, B.F.), US 5,344,867, 1994
227. A. Mukai (Teijin Chemicals), Eur. P 565,075, 1993; *Chem. Abstr.*, **120**, 246040 (1994)
228. J.P. Daikin, JP 34,814, 1983; *Chem. Abstr.*, **99**, 140608 (1983); J.P. Daikin, JP 63,711 1983; *Chem. Abstr.*, **99**, 159044 (1983)
229. J.A. Abusieme and A. Chittofrati (Ausimont), Eur. P 771,823, 1997
230. H.S. Wu (Gore & Associates), PCT 22,928, 1994
231. H.S. Wu (Gore & Associates), Eur. P 764,173, 1996; PCT, 34583, 1995
232. G.K. Kostov and P.C. Petrov, *J. Polym. Sci., Polym. Chem.*, **32**(12), 229–34, 2235–9 (1994)
233. P. Colaianna *et al.* (Ausimont), Eur. P 648,787, 1995
234. A. Funaki and T. Takakura (Asahi Glass), JP 340,716, 1994; *Chem. Abstr.*, **123**, 33897 (1995)
235. L. Mayer and G. Lohr (Hoechst), Eur. P 612,770, 1994
236. J.A. Abusleme and P. Maccone (Ausimont), Eur. P 625,526, 1994
237. J.A. Abusleme and G. Gregorio (Ausimont), Eur. P 612,767, 1994
238. H. Aihara *et al.* (Daikin Kogyo), PCT 17,876, 1996; *Chem. Abstr.*, **125**, 117082 (1996)
239. J.A. Abusleme and M. Albano (Ausmont), Eur. P 739,911, 1996; *Chem. Abstr.*, **125**, 331252 (1996)
240. M. Ogura and S. Chiba (Du Pont–Mitsui Fluorochemicals), Eur. P 718,363–4, 1996
241. W. Shinokawa and Y. Fukazawa (Hoechst Gosei), JP 17,538, 1993; *Chem. Abstr.*, **119**, 141317 (1993)

242. H. Hokonohara and M. Nagata (Mitsui Toatsu Chemicals), JP 140,237, 1993; *Chem. Abstr.*, **119**, 252235 (1993)
243. H. Tanaka *et al.* (DaiNippon Ink & Chemicals), JP 56,942, 1994; *Chem. Abstr.*, **121**, 111596 (1994)
244. F. Candau and R.H. Ottewill (eds.), *Scientific Methods for the Study of Polymer Colloids and Their Applications*, Kluwer Academic Publishers, 1990
245. *Emulsion Polymer Technologies*, Bulletins, Paint Research Association, Teddington, Middlesex; monthly series
246. R.G.W. Bacon, *Q. Rev.*, **9**, 287 *et seq.* (1955)
247. H. Warson, *Polymer Colloids Preprints*, Nato Advanced Study Institute, University of Trondheim, 1975, 5 pp. (pages not numbered)
248. W.D. Harkins, *J. Am. Chem. Soc.*, **69**, 1428–44 (1947); W.D. Harkins, *J. Polym. Sci.*, **5**, 217–51 (1950)
249. R.N. Haward, *J. Polym. Sci.*, **4**, 273–88 (1949)
250. W.V. Smith and R.H. Ewart, *J. Phys. Chem.*, **16**, 592–9 (1948)
251. W.V. Smith, *J. Am. Chem. Soc.*, **70**, 3695–702 (1948)
252. E. Bartholomé, H. Gerrens *et al.*, *Zeit. für Elektrochemie*, **60**, 334–48 (1956)
253. G.M. Burnett and R.S. Lehrle, *Proc. R. Soc. (Lond.)*, **A253**, 331 *et seq.* (1959)
254. L.-J. Chen *et al.*, *Coll. Surf.*, **129**, 1–30, 161–8 (1997)
255. C.S. Chern *et al.*, *Polymer*, **38**(8), 1977–84 (1997)
256. J.L. Gardon, *J. Polym. Sci.*, **6**, 623–41 (1968)
257. J.L. Gardon, *J. Polym. Sci.*, **6**, 643–64 (1968)
258. J.L. Gardon, *J. Polym. Sci.*, **6**, 665–85 (1968)
259. J.L. Gardon, *J. Polym. Sci.*, **6**, 687–710 (1968)
260. S.S. Medvedev, *Int. Symp. on Macromol. Chem.*, 147–90 (1959)
261. T. Krishan and M. Margaritova, *J. Polym. Sci.*, **52**, 139–52 (1961)
262. J. Ugelstad and F.K. Hansen, *Rubber Chem. Technol.*, **49**(3), 536 *et seq.* (1976)
263. J. Ugelstad and F.K. Hansen, *J. Polym. Sci., Polym. Chem.*, **16**, 1953 *et seq.* (1978)
264. J. Ugelstad and F.K. Hansen, *J. Polym. Sci., Polym. Chem.*, **17**, 3033–45 (1979)
265. J. Ugelstad and F.K. Hansen, *J. Polym. Sci., Polym. Chem.*, **17**, 3047–67 (1979)
266. A.S. Dunn, *Eur. Polym. J.*, **25**, 691–4 (1989)
267. R.M. Fitch and T.-J. Chen, *Polym. Prepr., Am. Chem. Soc. Div. Polym. Chem.*, **10**(1), 424–32 (1969)
268. R.M. Fitch, *Polym. Prepr., Am. Chem. Soc. Div. Polym. Chem.*, **11**(2), 807–10 (1970)
269. R.M. Fitch, *Br. Polym. J.*, **5**(6), 457–65 (1973)
270. R.M. Fitch, in *Emulsion Polymers and Emulsion Polymerisation*, D.R. Bassett and A.E. Hamielec (eds.), ACS Symposium Series No. 165, 1981, pp. 1–29
271. R.M. Fitch, in *Polymer Colloids Preprints*, Nato Advanced Study Institute, University of Trondheim, 1975, 25 pp. (pages not numbered)
272. J. Ugelstad and E.K. Hansen, in *Polymer Colloids Preprints*, hato advanced Study Institute University of Trondheim, 1975, 70 pp.
273. J. Ugelstad and R.M. Fitch *et al.*, in *Emulsion Polymerisation*, I. Piirma and J.L. Gardon (eds.), ACS Symposium Series No. 24, 1975, pp. 1–23
274. J. Ugelstad, Emulsification of monomer, initiation in monomer droplets, in *Polymer Colloid Preprints*, Nato Advanced Study Institute, University of Trondheim, 1975 (pages not numbered)
275. J. Ugelstad and F.K. Hansen, *J. Polym. Sci.*, **17**, 1953–78 (1979)
276. J. Ugelstad and F.K. Hansen, *J. Polym. Sci.*, **17**, 3068–82 (1979)
277. H.-S. Chang and S.-A. Chen, *J. Polym. Sci. Chem. Ed.*, **26**(4), 1207–29 (1988)
278. G.H.J. Doremmaele and H.A.S. Schoonbrood and A.L. German, in *Comprehensive Polymer Science*, J.C. Bevington (general ed.), G.C. Eastmond, A. Ledwith, S. Russo and P. Sigwalt (volume eds.), Vol. 4, Part 2, Pergamon Press, 1989, Ch. 3, pp. 41–65
279. R.M. Fitch, *Makromol. Chem., Makromol. Symp.*, **35/36**, 549 *et seq.* (1990)

References

280. D.M. French, *J. Polym. Sci.*, **32**, 395–411 (1958)
281. J.E.O. Mayne, H. Reichard and H. Warson, BP 607,704, 1948
282. J.E.O. Mayne and H. Warson, BP 627,612, 1949
283. A.S. Dunn, in *Emulsion Polymerisation*, Academic Press, 1982, Ch. 6, pp. 221–45
284. A.S. Dunn, in *Polymer Colloids Preprints*, Nato Advanced Study Institute, University of Trondheim, 1975, 7 pp. (includes extensive bibliography)
285. A.S. Dunn, *Chem. Industry*, December, 1406–12 (1971)
286. H. Warson, unpublished
287. E.V. Gulbekian, *J. Soc. Cosmet. Chem.*, **21**, 471–82 (1970); **30**, 171–83 (1992)
288. M.S. El-Aasser et al., *J. Polym. Sci. A, Polym. Chem.*, **35**(17), 3813–25 (1997)
289. M.S. El-Aasser et al., *J. Polym. Sci. A, Polym. Chem.*, **35**(17), 3827–35 (1997)
290. M.S. El-Aasser et al., *J. Polym. Sci. A, Polym. Chem.*, **35**(17), 3837–46 (1997)
291. R.H. Ottewill et al., *Br. Polym. J.*, **5**, 347–62 (1973)
292. H.A.S. Schoonbrood et al., *Ciidepint Anales*, **1996**, 197–208 (1996) (in English)
293. B.R. Vijayendran and T. Bone and C. Gajria, in *Emulsion Polymerisation and Emulsion Polymers*, D.R. Bassett and A.E. Hamielec, ACS Symposium Series No. 165, 1981, pp. 225–238
294. R.H. Ottewill, in *Emulsion Polymers and Emulsion Polymerisation*, D.R. Bassett and A.E. Hamielec, ACS Symposium Series No. 165, 1981, Ch. 2, pp. 31–59
295. C.P. Roe, *J. Coll. Interface Sci.*, **37**, 93–101 (1971)
296. L.-J. Chen et al., *Coll. Surf.*, **122**(1/3), 161–8 (1997)
297. B.M.E. Van der Hoff, *J. Polym. Sci.*, **33**, 487–90 (1958)
298. T. Stockhausen et al., *DECHEMA Monogr.*, **131**, 137–46 (1995)
299. F.A. Kolthoff and I.M. Bovey, *Zeit für Elektrochemie*, **60**, 400–4 (1956)
300. C.P. Roe and F.D. Brass, *J. Polym. Sci.*, **24**, 401 et seq. (1957)
301. H. Gerrens, *Zeit für Elecktrochemie*, **60**, 400 et seq., (1956)
302. S.M. Hasan, *J. Polym. Sci., Polym. Chem.*, **20**, 3031–68 (1982)
303. A.S. Dunn, *Polym. Int.*, **30**, 547–50 (1993)
304. M.S. El-Aasser, J.W. Vanderhoff et al., *J. Polym. Sci. A, Polym. Chem.*, **30**, 171–83 (1992)
305. A. Guyot and F. Vidal, *Polym. Bull. (Berin), Polym. (Berlin)*, **34**(5/6), 569–76 (1995)
306. Z. Wang et al., *J. Coll. Interface Sci.*, **177**(2), 602–12 (1966)
307. I. Capek et al., *Polym. Int.*, **43**(1), 1–7 (1997)
308. R.G.W. Norrish and R.R. Smith, *Nature*, **150**, 336 et seq. (1942)
309. E. Tromsdorff H. Kohle and P. Lagally, *Makromol. Chem.*, **1**, 169 et seq. (1948)
310. W.S. Zimmt, *J. Appl. Polym. Sci.*, **1**, 323–8 (1959)
311. Y.-C. Chen et al., *J. Polym. Sci., Polym. Chem.*, **34**(13), 2633–491, 2651–64 (1996)
312. I. Capek, *Macromol. Chem. Phys.*, **195**, 1137–46 (1994); I. Capek and P. Potisk, *Angewandte Makromol. Chemie* (1994)
313. C. Pichot J. Guillot and B. Emelie, *Makromol. Chemie. Suppl.*, **10/11**, 43–57 (1985)
314. J.G. Brodnyan et al., *J. Coll. Sci.*, **18**, 73–90 (1963)
315. E. Unzueta and J. Fircada, *Polymer*, **36**(22), 4301–8 (1995)
315A. M.-G. Weng, Z.-X.,Z.-R. Pan, Gaodeny Xuexiao, Huaxue Xueabo (**21**(1), 148–51 (2000)) in Chinese. *Chem. abstr.*, **132**, 194717 (2000) (152–5, 156–9).
316. A.M. Santos and F.M.B. Coutinho, *Polym. Bull. (Berlin)*, **30**(4), 407–14 (1993) (in English)
317. V.T. Stannett, R.H. Marchessault and E. Vanzo, *Diss. Abstr.*, **25**(3) (1964)
318. W.J. Priest, *J. Phys. Chem.*, **56**, 1077–82 (1952)
319. R. Patsiga, M. Litt and V. Stannett, *J. Am. Chem. Soc.*, **1960**, 801–4 (1960)
320. D.M.J. French, *J. Polym. Sci.*, **32**, 395–411 (1958)
321. J.T. O'Donnell, R.B. Mesrobian and A.E. Woodward, *J. Polym. Sci.*, **28**, 171 et seq. (1958)

322. D.H. Napper and A.G. Parts, *J. Polym. Sci.*, **61**, 113–26 (1962); D.H. Napper and A.E. Alexander, *J. Polym. Sci.*, **61**, 127–33 (1962)
323. A.S. Dunn and P.A. Taylor, *Makromol. Chem.*, **83**, 207–19 (1966)
324. D. Gerschberg, *Inst. Chem. Engin, Symp. Ser.*, **3**, 3–4 (1965)
325. V. Stannett, M. Litt and R. Patsiga, *J. Polym. Sci.*, **A-1**(8), 3607–49 (1970)
326. N. Friis and L. Nyhagen, *J. Appl. Polym. Sci.*, **17**, 2311–27 (1973)
327. R.L. Zollars, *J. Appl. Polym. Sci.*, **24**, 1353–70 (1979)
328. M. El-Aasser, J.W. Vanderhoff and J. Delgardo, *J. Polym. Sci. A, Polym. Chem.*, **24**, 861–74 (1986)
329. M.H. Litt and K.H.S. Chang, in *Emulsion Polymers and Emulsion Polymerisation*, D.R. Bassett and A.E. Hamielec (eds.), ACS Symposium Series No. 165, 1981, pp. 455–70
330. R.P. Mkhitaryan *et al.*, *Arm. Khim. Zh.*, **40**(11), 719–23 (1987); *Chem. Abstr.*, **108**, 132389 (1988)
331. M.S. El-Aasser and J.W. Vanderhoff (eds.), *Emulsion Polymerisation of Vinyl Acetate*, Applied Science Publishers, 1981
332. J. Barton, in *Copolymerisation and Copolymers in Dispersed Media*, Hüttig & Wepf. Basel, 1990, pp. 41–58
333. D.R. Bassett and K.L. Hoy, in *Emulsion Polymers and Emulsion Polymerisation*, D.R. Bassett and A.E. Hamielec (eds.), ACS Symposium Series No 165, 1981, pp. 371–87
334. A. Guyot *et al.*, in *Emulsion Polymers and Emulsion Polymerisation*, D.R. Bassett and A.E. Hamielec (eds.), ACS Symposium Series No. 165 1981, pp. 415–36
335. J.W. Vanderhoff, C. Pichot *et al.*, in *Emulsion Polymerisation of Vinyl Acetate*, M.S. El-Aasser and J.W. Vanderhoff (eds.), Applied Science Publishers, 1981, pp. 215–52
336. C. Pichot, J. Guillot and X.Z. Kong, *2nd Colloque Internat. sur Milieu Disperse*, 1989, pp. 152–4
337. B.W. Brooks and S. Sajjadi, *Institute of Chemical Engineers, Res. Event, 2nd Euro Conf. Young Res. Chem. Engng*, 1996, Vol. 1, pp. 494–6, 527–9
338. X. Kong *et al.*, *Gaofenzi Xuebao*, **1995**(3), 308–14 (1995) (in Chinese); *Chem Abstr.*, **123**, 112918 (1995)
339. G.A. Vandeande and A. Rubin, *Polymer Latices*, ACS Symposium Series No. 492, 1992, pp. 114–33
340. J.W. Vanderhoff, M.S. El-Aasser *et al.*, *Characterisation and Copolymers in Dispersed Media*, Hütig & Wepf, Basel, 1990, pp. 477–98
341. J. Guillot and W. Ramirez-Marquez, *Makromol. Chem.*, **189**, 379–97 (1988)
342. J. Guillot, C. Pichot and J.L. Guillaume, *J. Polym. Sci. A*, **26**, 19377–459 (1988)
343. C. Pichot, A. Revillon and J.L. Guillaume, *Makromol., Suppl.*, **10/11**, 69–86 (1985)
344. J. Guillot *et al.*, *Prepr. 2nd Int. Colloque, Les Copolymerisations et les Copolymères en Milieu Disperse*, Lyon, 1989, pp. 25–7
345. J. Guillot, *et al.*, pp. 156–9
346. E.M.A. Cruz *et al.*, *Makromol. Chem., Suppl.*, **10/11**, 87–103 (1985)
347. M.S. El-Aasser *et al.*, *Makromol. Chem., Macromol. Symp.*, **35/36**, 59–86 (1990)
348. M. Nomura, K. Takahashi and K. Fujita, *Makromol. Chem., Macromol. Symp.*, 13–22 (1990)
349. M. Nomura *et al.*, *International Symposium on Radical Copolymers in Dispersed Media*, Hüthig & Wepf, Zug, 1995, pp. 233–42
350. H. Ono, E. Jidai and K. Shibayama, *Br. Polymer J.*, **7**, 109–17 (1975)
351. V. Stannett *et al.*, *J. Macromol. Sci. Chem.*, **A14**(5), 739–62 (1980)
352. M.S. El-Aasser, J.W. Vanderhoff *et al.*, *J. Appl. Polym. Sci.*, **41**, 2463–77 (1990)
353. U. Türck, *Angewandte Makromol. Chemie*, **46**, 109–33 (1975)
354. G.W. Poehlein, *Macromol. Symp.*, **92**, 179–94 (1995)
355. J. Guillot, *2nd Colloque International sur Les Copolymerisations et les Copolymères en Milieu Disperse*, Lyon, 1989, pp. 163–5
356. H.G. Fordyce and E.C. Chapin, *J. Am. Chem. Soc.*, **69**, 695 *et seq.* (1947)

References

357. H. Warson and R.J. Parsons, *J. Polym. Sci.*, **34**, 251–69 (1959)
358. H. Warson, *Peintures, Pigments, Vernis*, **43**, 438–46 (1967)
359. L.W. Morgan and D.P. Jensen, *Makromol. Suppl.*, **10/11**, 59–67 (1985)
360. H.G. Fordyce and E.C. Ham, *J. Am. Chem. Soc.*, **69**, 695 *et seq.* (1947)
361. H. Kawaguchi, Y. Sugi and Y. Ohtsuka, in *Emulsion Polymers and Emulsion Polymerisation*, D.R. Basset and A.E. Hamielec, ACS Symposium Series, No. 165, 1981, pp. 145–56
362. *Makromoleculare Chemie, Supplementary Volume*, **10/11** (1985)
363. *Radical Copolymers in Dispersed Media*, Hüthig & Wepf, 1995
364. J.M. DeSimone and E.E. Maury, *Polym. Prepr. (Am. Chem. Soc.)*, **35**(2), 868–9 (1994)
365. J.M. DeSimone, E.E. Maury and T.J. Romack, *Polymer Prepr. (Am. Chem. Soc.)*, **35**(2), 741–2 (1994)
366. J.M. DeSimone and D. Betts, *Prepr. Polym. Mater. Sci. Engng. (Am. Chem. Soc.)*, **74**, 400 (1996) (there are several other useful references in this volume)
367. C. Lepilleur and E.J. Beckman, *Macromol.*, **30**(4), 745–56 (1970)
368. J.M. DeSimone and D.A. Canelas, *Polym. Prepr. (Am. Chem. Soc.)*, **37**(2), 542–3 (1996)
369. J.M. DeSimone, J.L. Young and Y.-L. Hsiao, *Polymer Prepr. (Am. Chem. Soc.)*, **38**(1), 506–7 (1997)
370. J.W. Vanderhoff *et al.*, *Polymerisation and Polycondensation Reactions*, Advances in Chemical Series, American Chemical Society No. 34, pp. 32–51
371. L. Hangquan and E.J. Ruckenstein, *Appl. Polym. Sci.*, **61**(12), 2129–36 (1996)
372. K. Takahashi *et al.*, *Macromol. Rapid Commun.*, **18**(6), 471–5 (1997)
373. F. Candau, in *Emulsion Polymerisation and Emulsion Polymers*, P.A. Lovell and M.S. El-Aasser (eds.), Wiley, 1997, Ch. 21, pp. 723–41
374. M.K. Lindemann and K. Deakon (Sun Chemicals), USP 4,616,057, 1986
375. H. Sperling, *Interpenetrating Networks and Related Materials*, Plenum Press, New York, 1989
376. A.G. Parts, D.E. Moore and J.G. Watterson, *Makromol. Chem.*, **89**, 156 *et seq.* (1965)
377. R.M. Fitch and T.-J. Chen, *Polym. Prepr. (Am. Chem. Soc.)*, **10**(10), 424–32 (1969)
378. R.M. Fitch, *Polym. Prepr. (Am. Chem. Soc.)*, **11**(2), 807–10 (1970)
379. R.M. Fitch, *Br. Polym. J.*, **5**(6), 457–65 (1973); *Chem. Abstr.*, **81**, 37825 (1974)
380. R.M. Fitch, in *Emulsion Polymers and Emulsion Polymerisation*, D.R. Bassett and A.E. Hamielec (eds.), ACS Symposium Series No. 165, 1981 pp. 1–29
381. R.M. Fitch, *Polymer Colloids Preprints*, Nato Advanced Study Institute, University of Trondheim, Norway, 1975, 25 pp. (pages not numbered)
382. J. Ugelstad and E.K. Hansen, *Polymer Colloids Preprints*, Nato Advanced Study Institute, University of Trondheim, Norway, 1975, 70 pp. (pages not numbered)
383. J. Ugelstad and R.M. Fitch *et al.*, in *Emulsion Polymerisation*, I. Piirma and J.L. Gardon (eds.), American Chemical Society, 1976, pp. 1–23
384. A.S. Dunn, *Eur. Polym. J.*, **25**, 691–4 (1989)
385. R.G. Gilbert, *Emulsion Polymerisation, A Mechanistic Approach*, Academic Press, 1995
386. P.A. Lovell and M.S. El-Aasser (eds.), *Emulsion Polymerisation and Emulsion Polymers*, Wiley, 1997

Supplementary references

Section 2.3

D.A. Paquet and W.H. Ray, Comparisons, batch, continuously stirred tank reactors, tubular reactors, *Am. Inst. Chem. Engrs J.*, **40**(1), 73–87, 87–96 (1994)

K. Katoka *et al.*, Continuous emulsion polymerisation of styrene, *Chem. Engng. Sci.*, **50**(9), 1409–16 (1995)

Section 4.1

J.A. Brinkley and T.E. Say (Rhone Poulenc Speciality Chemicals), Eur. P 644,205, 1995

4

Latex Properties Relative to Applications

H. Warson

1 INTRODUCTION

This chapter is concerned with properties that apply to latices as a class. The specific properties of some latices are included as examples, whether by virtue of the polymer content and the method of the inclusion of its monomers or because of the stabiliser system used. Also included is a general account of compounding, both with liquids such as plasticisers and solvents, including transient plasticisers, and with solids, usually pigments and extenders.

The general stability properties of a latex include mechanical stability to shear, stability to various additives and pH stability. The first named is largely a function of the stabiliser system, as considered in Chapter 3. Mechanical stability depends on the efficiency of the stabilisers in providing a tight 'envelope' around the particles with special reference to steric stabilisers, whilst ionic stabilisers enable the particles to repel each other. Other major factors are the density of the polymer relative to that of the medium and the latex viscosity. However, some latices are prepared with a formulation that enables the polymer particles to settle readily, either because this is desirable in a particular application or, as in the case of most polyvinyl chloride emulsions and some butadiene copolymer rubbers, it is a convenient method of preparing the polymer in a fine particle size.

In the case of a polymer in which the second-order transition point T_g (Chapter 1, Section 4.3) is below ambient temperature, it is essential that the repulsive charges on the polymer particles are greater than the forces that tend to cause coalescence or else that the steric stabilization is adequate.

Many latices tend to 'cream', either upward or downward, depending on the relative density of the emulsified polymer. This is not usually a serious defect, and a creamed latex is often homogenised by simple stirring. Sometimes a harder 'cake' or solid may form at the base of a container, particularly on long standing or when the product has been exposed to severe temperature fluctuations. It may be possible to disperse this sediment by vigorous stirring if coalescence is still incomplete. In this latter case the sedimented product may be looked on as a tightly packed mass of polymer particles in which

Introduction

the stabilisers are acting as a 'cement' that can be dispersed by vigorous mechanical effort, a phenomenon similar to that considered later in pigmented systems (Section 6).

If a latex is highly diluted, the tendency to settlement will increase, due mainly to the gravitational forces at the low viscosities employed and also to the fact that the concentration of stabiliser in the water phase has been reduced. Normally there is an equilibrium between the absorbed emulsifier on the particles and that which remains in solution or in micellar form. On dilution this equilibrium may be disturbed, with the result that there may be some destabilization. Special formulations are required for latices which are required to be stable at low concentrations, e.g. 1–5 %. This may apply to some impregnation processes in the textile and paper industries. In general under these conditions, the overall concentration of stabiliser must be increased to give adequate latex stability.

In some cases the phenomenon of creaming may be induced, particularly where an upward creaming is required in the case of a polymer, the density of which is below that of the aqueous phase. This is well known in rubber latex technology, and may be extended generally to latices in other applications. It is particularly noticeable when polyelectrolytes, e.g. sodium polymethacrylate, are added to a latex of relatively low concentration. Creaming up to a concentration of 60 % is known. The cause of this phenomenon may be a repulsion due to the charge on the polyelectrolyte ions which tend to crowd the charged latex particles together.

There is an increasing tendency to refer to the subject of emulsion polymerisation under the more general title of 'polymer colloids'. The major features have been listed as follows:

Particle size
Particle size distribution
Chemical nature of surface groups
Concentration of surface groups, e.g. surface charge density
Location of surface groups, uniform or irregular
Presence of residual monomer
Composition of copolymer latices
Morphology of the particle, e.g. 'core–shell'
Concentration of absorbed emulsifier
Concentration of adsorbed polymeric stabiliser
Thickness of steric stabilising layer
Reaction of the particle to the environment, e.g. swelling of polyelectrolyte latices on uptake of organic solvent
Nature of interaction between the particles
Surface potential of the articles in dilute and concentrated latices
Particle surface morphology, rough or smooth [1]

A review with 41 references indicates practical methods of controlling latex morphology [2]. A number of tests for mechanical stability will be indicated in Chapter 6.

2 VARIOUS PROPERTIES AND DESCRIPTIONS

2.1 Emulsion particles and resultant properties

Whilst earlier research on emulsion polymerisation has concentrated mainly on the kinetics of the reaction, during the past two decades attention has been focused mainly on the nature of the particles formed, e.g. the core–shell type [2], (Section 2.3) and the nature of phenomena such as adsorptions and film formation. This has been aided by a variety of sophisticated test apparatus, some of which will be quoted in subsequent pages. The results have assisted in designing products for various applications.

2.1.1 Particle size distribution

Turbimetry at several wavelengths and elastic light scattering (ELS) may be used to determine the particle size distribution of latex particles, using Mie scattering theory. Computer-generated results using the two methods simultaneously with a polystyrene latex have proved useful [3].

Further experiments have indicated that field flow fractionation (FFF) is useful for latices with a broad particle size distribution as shown by the elution curve. Photocorrelation spectroscopy (PCS) has to be repeated at more than one scattering angle to obtain an accurate mean diameter for a latex with a broad size distribution. Scanning electron microscopy (SEM) yields values 20 % lower than the alternative methods, attributed to shrinkage because of sample drying and high vacuum measurements [4].

2.2 Particle characterization

There have been many studies on the morphology of individual latex particles. Atomic force microscopy (see Section 4.1.1) has been used in studying homogeneous and composite particles prepared from butyl acrylate–methyl methacrylate copolymerisations. During adsorption and drying, the packing of the particles was determined by capillary forces; deformation is caused by interfacial forces [5].

There has also been considerable examination of the structure of particles formed from core–shell copolymers, usually referred to as 'structured particles' (see also Chapter 3, Section 2.4.3) Details are often too theoretical for this volume, but they have some significance in film formation, and in particular the minimum film-forming temperature (Section 4).

Figure 4.1 gives a general indication of the process in the reaction medium. Figure 4.2 indicates the principal types of structured particles [6].

The three methods of initiation, viz. micellar, homogeneous and in the particles, affect their morphology. Homogeneous initiation takes place with monomers such as vinyl acetate and methyl acrylate, whilst some functional monomers such as methylolacrylamide are also initiated in the same way. The ideas of Grancio and Williams on core–shell polymerisation are now generally accepted [7, 8].

Various methods, including electron microscopy with selective marking of one constituent, light diffusion, neutron diffusion and estimation of active surface groups, have been used to elucidate particle structure. Indirect methods include rheology, colloidal stability, mechanical properties, coalescence of films and permeability.

If vinyl acetate and butyl acrylate are submitted to a batch emulsion polymerisation, core–shell particles are formed, but with variable composition throughout the reaction, except in the last stages The core is rich in acrylate

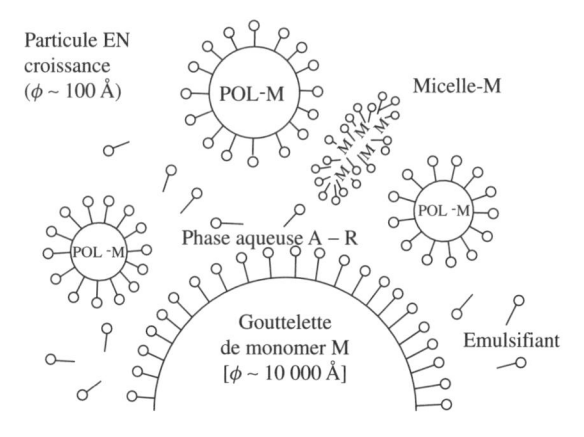

Figure 4.1 Reaction medium in emulsion polymerisation.

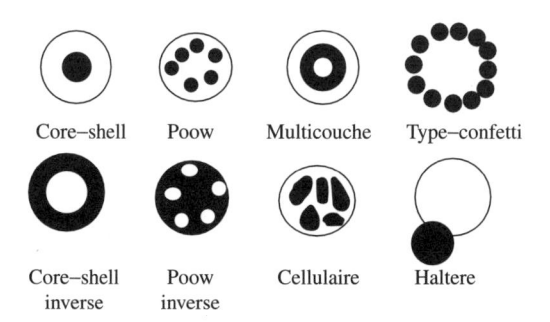

Figure 4.2 Principal types of structured particles. (Reproduced from reference [6].)

and the shell may be almost pure polyvinyl acetate, dependent on monomer ratios. The more usual method of performing core–shell processes is by a stage addition of the various monomers. The actual addition is modified by a number of factors, including the partition of monomers between the particles and the aqueous phase, mutual compatibility of the chains, viscosity in the interior of the particles, the nature of seed particles and the process, whether by stage or gradual addition of monomers.

Whether new particles are formed may depend on the surface area of a seed latex and on monomer solubility. Grafting, caused by chain transfer, may play a part. An inverted core–shell structure is favored by hydrophobic initiators such as AIBN, whilst more complex structures, i.e. a sandwich type, are formed by water-soluble initiators such as potassium persulfate.

Most hydrophilic monomers, which includes many functional monomers, tend to partition into the water phase. Thus polymer chains are initiated in the water phase from which they are precipitated, but hydrophilic end groups stabilise the particles. It is preferable to conduct emulsion polymerisation such that functional monomers tend to be localised on the particle surfaces.

The surface groups introduced, which require characterization, may be due to functional monomers, and also to initiator radicals or surfactant molecules introduced to the surface by chain transfer, often non-ionic emulsifier of low HLB. Conductimetry and potentiometry are simple methods for the analysis of acid groups, both strong and weak. The surface area of an emulsifier at the interface (A^s) and the polarity of the polymer (X_p) are related by the expression:

$$\log A^s = k + f(X_p)$$

where k is a constant.

The charge density on the latex particles may be measured by methods based on the electrophilic mobility. Many instruments are able to measure the zeta potential, but there are difficulties with very small particles (0.01 μm) or if the macromolecules have non-ionic emulsifiers absorbed on the surface. Methods for the analysis of serum include conductimetry and gel permeation chromatography. Spectrometric methods are available for mixtures of anionic and non-ionic emulsifiers. Functional groups in the interior of particles may be measured by non-aqueous titration, e.g. in pyridine, titrating with a quaternary amine, or under some conditions it may be possible to titrate so that sodium hydroxide solution diffuses through the particles.

2.3 Core–shell copolymers—surface and buried groups

The location of the groups, interior or surface, depends on the nature of the monomer, the emulsifier and the initiator. Semi-continuous (gradual addition) methods and a stage addition by varying the composition facilitate a functional group being on the surface of the particles. Neutralization of acids has a marked effect on the location of carboxylic groups and tends to give more water-soluble polymer or to have carboxylic groups concentrated on the surface [6].

A test with a semi-continuous three-stage emulsion polymerisation, in which methacrylic acid was added in the last stage only, was performed using potentiometric titration and conductometric titration in 1,4-dioxane. It was concluded that about 27 % of ionisable groups were buried in the particle interior [9].

Studies including NMR on latex particles with a polybutyl acrylate core and a polymethyl methacrylate shell concluded that to characterize interphase structure, the temperature of measurement should be optimised [10, 11]. General reviews are available [12–17] (see also reference [175]).

The formation of emulsions based on core–shell particles has produced a wide variety of changes compared with latices in which the particles are substantially homogeneous in composition, Studies have indicated that if the core is more hydrophilic than the shell, under some conditions inversions will occur and the core becomes the shell [18–20].

The surface morphology of polybutyl acrylate–polymethyl acrylate core–shell latices has been studied by atomic force microscopy (see Section 4.1.1), giving roughness and height distributions [21]. Whilst it is probable that the second-order transition point or minimum film-forming temperature (see Section 4.3) is lowered by the formation of a core–shell latex, compared with a homogeneous latex of the same overall composition, there have been contradictory publications on this, and it may depend on whether the outer shell is 'hard' or 'soft' [22–25].

Complications may occur by absorption of the second or later monomers into the shell, or an intermediate layer, or by grafting. If there is a crosslinking monomer, interpenetrating networks may be formed [26]. It is advised that optimum conditions for grafting include a high seed particle number, a water-soluble initiator, addition of monomer under starved conditions, the core in general should be more hydrophobic than the shell and the core and shell polymers should be incompatible [27]. See also Section 7 for a discussion of the production of organic pigment extenders by core–shell techniques.

2.4 Adsorption of surfactants on latex particles

The adsorption of sodium dodecyl sulfate and the competitive adsorption of the non-ionic surfactant Triton X-405 on a polystyrene latex has been studied by small-angle X-ray scattering, since that test has good sensitivity. After complete coverage is attained the sulfate forms free micelles. Competitive adsorption is an equilibrium process, but the non-ionic surfactant is more strongly adsorbed [28].

2.5 Emulsion viscosity

2.5.1 Definitions and theory

Viscosity is a branch of rheology, the science of the way in which materials flow and deform. It applies mainly to liquids of pseudoliquids, and in specific units is defined as the shear stress divided by the shear strain. A Newtonian fluid is one in which the shear stress is proportional to the shear strain. The current unit of viscosity is the pascal-second (Pa s), which is related to the former unit poise (P) as follows:

$$1 \text{ Pa s} = 10 \text{ P}; \quad 1 \text{ mPa s} = 1 \text{ cP}$$

Viscosity is usually denoted by η.

In most cases if a surfactant of low molecular weight is the sole latex stabiliser, the final product, at the usual concentrations of up to 55 %, will have a comparatively low viscosity, usually 0.1 Pa s or less. This rule, however, is not entirely universal, particularly if a surfactant with strong micelle formation tendencies has, in addition, low water solubility. This sometimes occurs with latices containing comparatively high concentrations of the sodium salt of dioctyl sulfosuccinic acid. A latex of very fine particle size, of the order of 0.1 μm or less, may also have a pseudoplastic flow (Chapter 2, Section 2.4), or thixotropy. This may be due to the formation of surface layers by the high quantity of stabilisers that are present, or it may be a frictional effect cause by the high surface areas involved.

Another feature of viscosity, especially of a thixotropic type, is the phenomenon known as 'bridging' by which an emulsifier, particularly of high molecular weight or containing at least two active groups, bridges several emulsion particles simultaneously, thus creating a barrier to stirring and increasing apparent viscosity. This bridging may be broken by the energy imparted by high agitation [29].

As a general rule, the inclusion of a colloid in the latex formulation tends to increase latex viscosity, although this may only be obvious at higher solids contents, above about 50 %. An exception is when the colloid itself gives a viscous solution at low concentrations. Viscosity in this case depends on the aqueous concentration of the colloid and not on its concentration relative to polymer.

As the particle size of latices of somewhat similar formulations decreases, the viscosity of the latices tends to increase. This may be due to the mutual charge repulsion. It is also possible that surplus surfactant forms an associated layer around the stabilised particles, often in association with excess water molecules. This phenomenon is often referred to as 'hydration of the particles'. The increase in viscosity of a latex on adding polyvinyl alcohol is explained by the excluded volume effect of an absorbed monolayer of polyvinyl alcohol on the polymer particle surface [30].

Extreme conditions of shear provide a source of energy which may shear this outer layer, thus reducing the mutual obstruction of particles relative to each other, thus reducing viscosity. The sudden onset of this phenomenon may cause the existence of a 'yield point' in a stress–strain curve, which is the minimum value at which an applied stress causes a viscosity change.

It must be stressed, as already indicated (Chapter 2, Section 1.2.3), that whilst dilute latices behave almost as low-viscosity Newtonian fluids, viscosity is not a properly defined quantity when a non-Newtonian fluid is under consideration. A three-dimensional structure is formed when the mean separation between particles is reduced to a similar dimension to the range of the interparticle forces. Viscoelasticity is then observed. Thus viscosity can only be defined under very specific conditions of test, which must not only include the method of test and the units employed but also the prehistory of the sample.

2.5.2 The measurement of viscosity

Table 4.1, which gives the viscosity in poises of a series of Mowilith® latices, illustrates clearly the differences caused by the use of different viscometers and the wide range of viscosities that is technically acceptable for the same product. The Brookfield viscometers give readings based on the torque of a revolving cylinder in a relatively wide outer cylinder. These are about double the results given by viscometers which are based on a concentric moving cylinder with a narrow gap between this and the outer fixed cylinder. The rotational type of viscometer is the most suitable type for routine viscometry measurements. An alternative type is the cone and plate viscometer. Cone and plate geometry is the fixation of a conical vertex perpendicular to and in point

Table 4.1 Viscosity range of Mowilith dispersions

Product	Stabiliser	Total solids (%)	Viscosity (P)		
			Brookfield RVT 20 r.p.m.	Epprecht STV	Höppler DIN 53015
Mowilith D	PV alcohol	50	160–280 (Sp.6)	100–220 (MS-C1)	200–600
Mowilith DN	PV alcohol	50	25–55 (Sp. 4)	10–40 (MS-Cll)	10–50
Mowilith DN	PV alcohol	60	75–135 (Sp. 6)	60–120 (MS-Cl)	80–120
Mowilith DH	Colloid	50	250–450 (Sp. 6)	60–200 (MS-Dlll)	300–700

[a]Dispersion is the term used in Germany for latex or polymer emulsion.

contact with a flat plate. When the cone is made very obtuse (δ less than 4°) and rotated at constant speed (ω), precise viscosity measurements are obtained at absolute and uniform values of shearing rate and stress.

In order to measure strain against a constantly changing stress more accurate instruments are needed. These advanced instruments, of principal interest for research purposes, are called rheometers. The Weissenberg Rheogoniometer, now manufactured by Carri-med (Dorking, UK), is a typical advanced instrument. Many current rheometers give computerised results. The following is a list of manufacturers of suitable instruments for measuring latex viscosity at the time of publication, but is not exhaustive:

Bohlin Instruments Division, Cirencester, UK, Cranbury, New Jersey, USA and Mühlacker, Germany
Brave Instruments, Liege, Belgium
Brookfield Engineering Laboratories, Inc., Broughton, Massachuaitts, USA
Brookfield Viscometers Limited, Harlow, Essex, UK
Carri-med, Dorking, Surrey, UK
Contraves AG, Zurich, Switzerland
Contraves Industrial Products, Ruislip, Middlesex, UK
Engaln Rheometrics, Piscataway, New Jersey, USA and Frankfurt, Germany
ICI Cone and Plate Viscometer, Research Equipment (London) Limited
Mettler-Toledo, Schwerzenbach, Switzerland and Beaumont Leys, Leicester, UK
Nametre Company, Edison, New Jersey, USA
Paar Scientific, London, UK
Physica, Stuttgart, Germany
Ravensfield Designs, Heywood, Lancashire, UK
Sheen Instruments, Teddington, Middlesex, UK

Many of the above companies produce excellent technical literature on rheology and its measurement. See Chapter 10 in Volume 2 for further details of the effects of different viscometers on pigmented systems.

It is noteworthy that Mowilith DN, shown in Table 4.1, is of identical character to Mowilith D and of substantially similar formulation. Mowilith DN is probably formed from Mowilith D by strong shearing action [31]. Figure 4.3 is an example of the variation in viscosity caused by shear. Figures 4.3(a) and (b) indicate the variation of Vinamul® 9900 over high and low shear rates using the Ferranti–Shirley viscometer. The relatively small increase in shear stress with shear rate is equivalent to a fall in viscosity with increasing shear rate.

2.5.3 Viscosity, dilatancy and rheopexy phenomena

A study of the kinetics of thixotropic recovery shows that the concentration of the disperse phase is the main factor affecting the recovery. The effect of

Various properties and descriptions

Shear rate
= 172.8 s^{-1}
per division

Shear rate
= 19.9 s^{-1}
per division

Shear stress = 4314 dyn cm^{-2} per division

(a) Wide shear range

Shear stress = 9205 dyn cm^{-2} per division

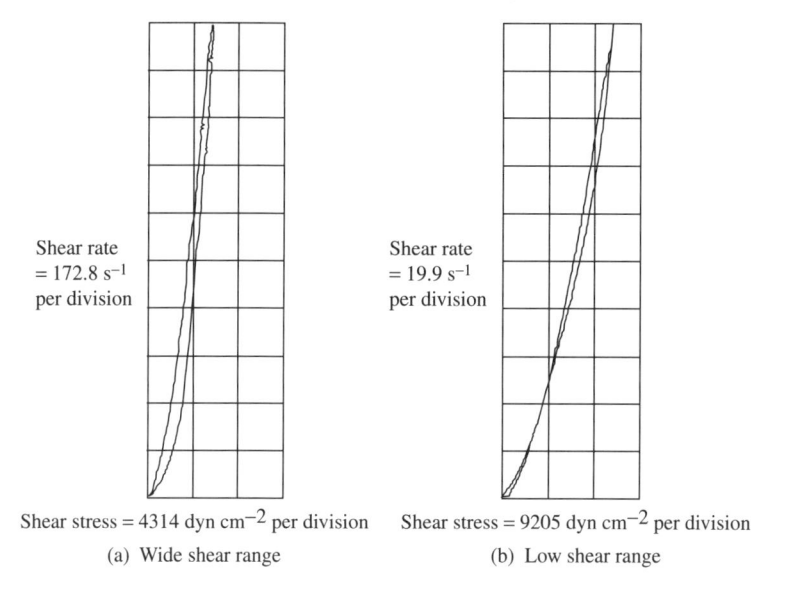

(b) Low shear range

Figure 4.3 Stress–strain curves for Vinamul 9900. Minor fluctuations in the graph are an instrumental artefact. (Reproduced from data of Vinamul Limited.)

dilution on the decrease of concentration of the disperse and dispersing phase of the system is negligible [32].

In a few cases an increase in shear stress of a non-Newtonian liquid produces no increase in the velocity of deformation after an optimum. In other words, the apparent viscosity rises with increased shear. This is known as dilatancy.[†] The differences between pseudoplastic flow, thixotropy and dilatancy may be summarised by stating that in pseudoplastic flow, the particles of the system or their surrounding sheaths, the attraction between the particles in a system is broken by a finite stress.

This reduction in mutual attraction does not take place spontaneously in the case of a thixotropic product. With dilatancy there is a purely mechanical hindrance, which is especially noticeable in concentrated systems. Hence a large deformation force will tend to increase the packing of the particles, and ultimately cause coalescence in a manner similar to that which causes a drying film to coalesce (Section 4). Thus dilatancy is often associated with mechanical instability.

An allied phenomenon is rheopexy, which is characterized by the sudden increase in viscosity of 2–3 orders of magnitude within a narrow range of shear rate. This is due to groups being present at the surface of the latex particles which are able to form hydrogen bonds. At a definite shear rate these

---

[†] According to BS 2015 : 1992, dilatancy is a more specialised phenomenon where thickening is accompanied by an increase in volume.

202   *Latex properties relative to applications*

groups interact to give a network structure of latex particles. This has been demonstrated by electron microscopy [33].

2.5.4 Viscosity phenomena with latices

Pseudoplasticity, thixotropy and dilatancy have a considerable bearing on application properties. This is specially the case with thixotropy, which will be considered in further detail (Chapter 10 in Volume 2) when considering emulsion paints.

Whilst most of the above phenomena apply to emulsions generally, they have particular reference to latices (polymer emulsions). There are some major differences with emulsions of one liquid in another. Thus polymer particles, except those with a very low T_g, have strong resistance to deformation. In most cases, once coalescence has taken place, it is almost impossible to reconstitute the products as latices. There are some exceptions to this, e.g. self-dispersible latices that include water-soluble monomers or latices with high concentrations of dispersants such as polyvinyl alcohol. In the latter case films, apparently coherent, consist of particles of polyvinyl acetate in which the polyvinyl alcohol acts as an adhesive.

In an examination of saturation adsorption and penetration of anionic surfactants in a model vinyl acetate–acrylic ester latex, it was noted that the penetration of certain surfactants such as sodium lauryl sulfate leads to the formation of polymer–surfactant complexes and considerable thickening of the latices [34, 35].

Recent work has often been designed to give specific viscosities under differing conditions of shear. The associative thickeners (Chapter 2,

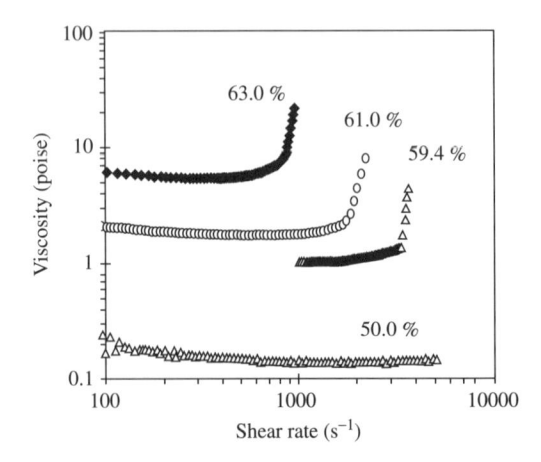

Figure 4.4 Shear Thickening of Hycar latex H5 (Hycar®) at different solids concentration.

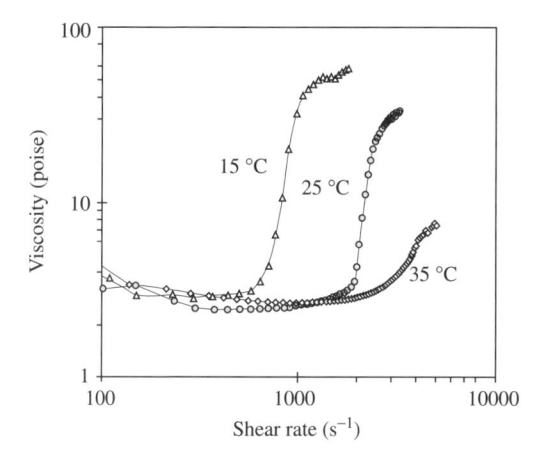

Figure 4.5 Effect of shear thickening behaviour of H5 with 61 % of solids. (Reproduced from data of PMSE Division, American Chemical Society.)

Section 4.7) are often added for this purpose, especially with regard to controlling viscosity under differing conditions of shear [36, 37].

The effect of shear stress on a technical anionic acrylic latex gives rise to critical shear rates that correspond to a sudden and rapid reversible increase in viscosity. As the latex concentration increases from 59.4 % of solids to 63.0 %, the critical shear rate of shear thickening drops from 3400 to 800 s^{-1}, but at 50 % of solids there was no shear thickening at an accessible shear rate. There is hysteresis in the return loop. As temperature is increased from 15 to 35 °C, the critical shear rate is higher and the increase in shear rate greatly reduced [38].

Titanium chelates have been added to latices to improve viscosity characteristics at low shear. Further details will be included in Chapter 10 in Volume 2, as these modifications apply very largely to emulsion paints. Viscosity phenomena of pigmented latices are also included in Chapter 10, Section 1. Viscosity phenomena regarding viscosity problems of latices with crosslinking possibilities are included in Chapter 5. (see also references [39] to [41]).

3 FREEZE–THAW AND TROPICAL STABILITY

3.1 Freeze–thaw stability—general principles

The stability of a latex to freezing is partly a function of the polymer, optionally plasticised, and partly of the medium. With regard to the polymer, freeze–thaw stability can be considered in the light of the various transition points, of which, especially in copolymers, there may be several. The one that would correspond to the true melting point is often above the decomposition temperature of the polymers, and it is the second-order transition point (T_g)

that is of principal interest (Chapter 1, Section 4.3). This is the temperature above which particles of the polymer tend to flow together spontaneously. As will be shown later, below this temperature, continuous films of polymer are not readily formed on drying a film of latex, but see Section 4.3 on minimum film-forming temperature.

On freezing a latex, it is obvious that if the freezing point is below the T_g, there will be less chance of coalescence of the particles and therefore a better chance of a latex containing the polymer to be resistant to freezing. As the latex freezes, ice tends to separate, and the latex gradually becomes more concentrated. A level will soon occur at which the entire latex no longer has sufficient liquid present for the latter to be a continuous phase. The particles may have sufficient charge around them for the mutual repulsion of like charges to enable the particles to stay as discrete entities, and thus the latex to withstand freezing and re-thawing. This condition occurs where there is an excess of certain anionic emulsifiers, although it does not occur in every case. Dissolved materials also depress the freezing point of water, thus assisting stability to freezing in another way. Colloids may assist stability to freezing by forming a rigid envelope around polymer particles. Steric and hydration effects are major causes of stability or otherwise [42]. One theory which has been advanced is that 'bound water' in the particles may impart stability to freezing. Rochow and Mason made a general study of the breaking of latices by freezing [43].

Amongst other general factors causing coagulation on freezing are the concentration of salt impurities, establishment of true contact between adjacent films of emulsifier, with loss of the orienting influence of water, and subsequent diffusion of emulsifier away from these regions. In consequence, globules tend to coalesce as soon as the thawing of the ice permits them to change shape. The mechanism of coagulation during freezing may be considered to be an accumulation of disperse phase particles which are rejected by the growing ice crystal lattice and which ultimately force the primary particles into contact with each other and so promote coalescence.

Barb and Mickucki, who worked mainly with polystyrene latices, also found that the particles which have coagulated tend to be the smallest. The more rapid the freezing and the more concentrated the latex, the larger are the final particles. In one series of experiments the coarsest latex particles, of about 1 µm diameter, produced by far the smallest size particles after coagulation. They concluded that with very fine particles of the order of 0.1 µm diameter, a significant portion of the polymer chains in each latex particle are free from intersegmented interference. In such a case the concept of the second-order transition may not apply, and surface interaction will permit a weld of finite mechanical strength [44].

The method of freezing and subsequent thawing may have a secondary effect on freeze–thaw stability. Thus a very quick freeze at low temperatures may not allow sufficient time for an equilibrium to be established. On the other hand, partial freezing of drums of latex, e.g. in cold weather, is normally very slow, allowing ample time for equilibrium to take place. This is more likely to cause gelation than a rapid freeze.

Tensiometric examination of the freezing process in thin layers of film from a Russian latex (SK25) shows that three stages are discernible, differing in the rate of particle agglomeration. The stages are:

(a) Freezing in the presence of the liquid layer
(b) The point of disappearance of all liquid
(c) The ageing of the completely frozen layer at the operable temperature [45]

It is considered that the reduction of the aqueous phase during freezing immobilizes the latex particles and brings them into stationary contact. The continued decrease in temperature freezes the hydration layers on the particles and the stresses created cause agglomeration [45].

When latices are stabilised with soaps, the longer chains of fatty acid soap molecules, e.g. stearic acid, in the adsorption layers lose their mobility first on freezing, and thus latices stabilised in this way are less stable than latices prepared with shorter chain fatty acid soaps, e.g. based on lauric acid. Aggregation increases with time. The phenomena on freezing are caused by dehydration of the adsorption layers, the effect being rather similar to adding an electrolyte [46].

Some surfactant-type additives to a latex are incorporated into the absorbed surface layer of emulsifiers on the surface of the particles, and both may reduce the freezing point of a latex and improve freeze–thaw resistance [47].

Deuterium nuclear magnetic resonance has been used to give a correlation between the aqueous fraction of butadiene–styrene latex absorbed by emulsifying agent and the aggregate stability of the latex during freezing and thawing. Addition of large amounts of emulsifier to the latex led to the formation of a hydrated mantle around the latex particles. An increased fraction of this mantle led to increased wedging pressure between latex particles [48].

Whilst a latex, irrespective of the polymer base, which contains sufficient acrylic or other acid groups to render it completely soluble in ammonia solution or an amine, would be stable in all conditions of frost, compositions prepared from this type of latex soften when washed with mild alkali. In addition, the molecular weight of such a copolymer must be restricted, as it is not possible to obtain a reasonably concentrated solution at a pourable viscosity at high solids content.

3.2 Freeze–thaw stability—technical methods of control

Possibly the simplest method of ensuring stability below 0 °C is the addition of a water-soluble solvent which depresses the freezing point of water. Ethylene glycol and propylene glycol are frequently added to latices, and also a number of other water-soluble solvents such as ethylene glycol monoethyl ether. These latter are also used as coalescing solvents (Section 5). However, their low volatility may be objectionable under some conditions of application. In some cases, if a latex is destabilised to some extent by freezing and subsequent thawing, purely mechanical action such as vigorous stirring may restore a homogeneous and stable product, although viscosity may be affected.

Latices prepared from acrylic and methacrylic esters, and also from styrene, butadiene and other monomers, often obtain freeze–thaw stability by the copolymerisation of about 2 % of acrylic or methacrylic acids, or occasionally other polymerisable acids such as itaconic acid. The effect of these acids as comonomers is probably due to the charge on the particles causing sufficient repulsion to prevent coalescence. Stability to freezing is often retained, even in the acid state. Very high proportions of a copolymerised acid may, however, have the reverse effect, probably due to the particles swelling because of their increased hydrophilic character.

A patent quotes terpolymers of esters including blends of 'soft' monomers, e.g. acrylic esters or the longer alkyl chain methacrylic esters, with 'hard' monomers such as methyl methacrylate or cyclohexyl acrylate with 0.5–2.5 % of polymerisable acid based on the carboxylic groups present. In the alkaline state both mechanical and freeze–thaw stability is greatly improved [49].

Russian researchers indicate that the addition of non-ionic surfactants at 7–10 % improved the freeze–thaw resistance of a local S-B latex, whereas other surfactants, presumably ionized, lowers it [50].

For freeze–thaw stable latices of styrene, the presence of a polymerisable acid is considered essential. Suitable surfactants include the blend of a water-soluble anionic emulsifier, e.g. the sodium salt of an alkylarylpolyoxyethanol sulfate, a potentially oil-soluble emulsifier such as sodium dioctyl sulfosuccinate and an octylphenol–ethylene oxide condensate as a non-ionic surfactant. The process described is based on preparing a pre-emulsion, adding about one-third of this to the reactor and the gradual addition of the rest after polymerisation has started. The initiator system is a redox type based on t-butyl hydroperoxide and a reducing sulfur acid [51].

An alternative scheme includes a diene monomer together with styrene, an acrylic ester and a carboxylic acid. This is prepared in a similar manner to the above, but about 1 % of carboxylic acid is added towards the end of the reaction [52]. A somewhat similar specification claims that a 1 : 2 butadiene–styrene latex is freeze–thaw stable if an alkylaryl polyoxyethylene is included such that the ethylene oxide chain has from two to five ethylene oxide groups only [53].

Monomers including other hydrophilic groups may be included to ensure stability to freezing and subsequent thawing. Thus, a composition prepared with non-ionic stabilisers containing 280 parts of vinyl acetate, 120 parts of dibutyl maleate and 6.6 parts of acrylamide, and initiated with a redox system of t-butyl hydroperoxide and the salt of a reducing sulfur acid, is stable to freezing and thawing [54].

In a rather interesting chemical modification of an emulsion polymer, based on a 37 : 61.5 : 1.5 monomer ratio of methyl methacrylate–ethyl acrylate–methacrylic acid, the resultant copolymer is reacted with aqueous propyleneimine

$$\mathrm{CH_3\underset{\underset{NH}{\backslash\;/}}{CHCH_2}}$$

to esterify about half of the carboxylic groups present. The modified copolymer, which contains units of aminoester $COOCH(CH_3)CH_2NH_2$, has been found to impart a high degree of freeze–thaw stability without damaging other properties such as film adhesion and film resistance to water. This specification also discloses the addition of either N-methylformamide, N-methylacetamide, N-methyl-propionamide or formamide as an additive from 0.1–5 % of polymer, the optimum being about 3 % of additive. Ethylene–vinyl acetate copolymer latices also containing carboxylic groups are iminated to give improved freeze–thaw and mechanical properties. The latex stabilisers are polyvinyl alcohol or hydroxyethyl cellulose [55].

Russian work indicates that the addition of 0.4 % of calcium chloride $CaCl_2$, based on the initial weight of monomer in a polyvinyl alcohol stabilised polyvinyl acetate latex, enables freeze–thaw stability to be achieved down to $-70\ °C$ [56].

A freeze–thaw stable vinyl acetate latex is described which claims a system containing 5–50 % by weight on polymer of post-added plasticisers, surface-active agents containing 20–75 % by weight of plasticisers and protective colloid containing 0.3–6 % by weight of latex. These latices are claimed to be stable down to $-40\ °F$ [57]. The inclusion of sodium vinyl sulfonate as a monomer is also claimed to give improved freeze–thaw resistance [58].

Polyvinyl acetate latices stabilised with polyvinyl alcohol tend to be stable to freezing followed by thawing. This may be connected with the grafting of monomer on to the stabiliser, which takes place to varying extents.

Some emulsifier systems that sometimes improve freeze–thaw stability are of the sulfate and sulfonate types, irrespective of the nature of the polymers. A number of amphoteric reagents have also been claimed. The inclusion of mixtures of emulsifiers based on copolymers of ethylene oxide and propylene oxide is recommended. A typical type is the Pluronic® series, which according to composition may be water soluble or water dispersible only. A typical example, with special reference to vinyl acetate polymers and copolymers includes 0.2–0.5 % of a water dispersible Pluronic, 0.1–1.0 % of a water miscible type, whilst an anionic surfactant such as an alkylbenzene sulfonic acid sodium salt is present from 0.05–0.5 %. It is claimed that freeze–thaw stable pigmented systems can be prepared with a pigment volume concentration of over 55 % using standard pigments, both white and colored [59].

3.3 Freeze–thaw stability—tests

A simple test is to place a filled sample bottle of known size in a refrigerator at, say, $-10\ °C$ for a standard time, about 1 h, and then thaw at 40 °C for 1 h and examine. If the latex remains stable the process is repeated. In most cases, if a latex remains stable to one cycle, it will remain stable almost indefinitely.

A compounded latex will not necessarily have the same stability as the latex from which it is derived. In the case of a normal pigmented latex, e.g. an emulsion paint, it is rather unlikely that the paint will be stable if the original latex is unstable to freezing. However, some paints with considerable additives, and more complex compositions such as emulsion polishes, are stable to freezing and re-thawing, but this is not necessarily the case.

3.4 Tropical stability

Tropical stability is bound up with the fact that a latex is fundamentally a metastable system. With increased temperature, viscosity tends to become lowered, enabling gravitational effects to be more pronounced. In most cases polymers have a density greater than the aqueous solution, and will tend to sediment. A few, especially some synthetic rubber latices, may tend to cream. Certain chemical changes are also accelerated by heat, even at the experimental 'tropical' temperature of 40 °C. Thus several emulsifying agents of the sulfonate type tend to dissociate:

$$C_{18}H_{37}C_6H_4SO_3Na + H_2O = C_{18}H_{37}C_6H_5 + NaHSO_4$$

This dissociation is accompanied by a rise in surface tension of the latex, which can be measured readily. It is, however, only when dissociation reaches such a level that the protective 'envelope' itself begins to decompose that instability is likely to occur. On the other hand, once a stable latex is formed, the protective 'envelope' may remain as a stabiliser, irrespective of diminished solubility of the stabiliser at higher temperatures, although several surfactants are rather less soluble hot than cold, whilst a number of polyoxyethylene derivatives have 'cloud points' at varying temperatures. This latter property is also associated with a tendency to precipitation during polymerisation, which sometimes occurs at about 90 °C when some polyethylene derivatives of short-chain ethylene oxide are used as surfactants. This is also true of the less frequently used diesters of polyethylene oxides. In addition, a number of colloids precipitate almost completely at elevated temperatures. Methyl cellulose begins to precipitate at about 55 °C and polyvinyl methyl ether at 30 °C.

Some latices, mainly of vinyl acetate homopolymers and copolymers, which are stabilised with various polyvinyl alcohols tend to be less stable at elevated temperatures. Polyvinyl alcohols with a relatively high residual acetate content (saponification value over 200) tend to precipitate from solution when heated, and the equivalent may be formed by grafting during the polymerisation of vinyl acetate with a polyvinyl alcohol stabiliser.

A purely mechanical effect of storage of latices at high temperatures is the ready formation of skin due to the difficulty of suppressing evaporation completely, especially as, in practice, the latices may have to be transferred from reactors at temperatures of 35–40 °C. The addition of propylene glycol may tend to reduce the skinning effect, this glycol being useful against both freezing and excessive heat on storage.

Storage of drums of latex in the open in sunny tropical climates may also have a deleterious effect on some latices, since the internal temperature may reach 60 °C. Prolonged storage under these conditions is known to cause slight precipitation or 'stringing' in vinyl acetate copolymer latices. The problem is overcome by a careful balance of surfactants, in particular of the non-ionic type. Sometimes a slight increase in water-soluble polymeric stabiliser, e.g. hydroxyethyl cellulose, may be necessary.

4 EMULSION FILMS

4.1 Film formation

The formation of a film from a solution, usually concentrated, of a polymer is a comparatively simple process. Evaporation of the solvent causes the liquid to concentrate, viscosity gradually increasing until all the solvent has evaporated, after which the characteristic properties of the polymer are imparted Even in this case some complications may arise. Thus there may be undesirable flow during the drying process if the drying solution is not in a horizontal plane. This results in a film of uneven thickness. Absorption of the solvent into a porous substrate may cause some difficulty. If a solution is used as an adhesive, shrinkage may pull the adhesive film from one of the adherents, or alternatively the stress during drying may cause cracking.

Film formation from a latex is a much more complex process and has resulted in much research. The system is essentially a heterogeneous one, and therefore evaporation of water alters the overall characteristics. In addition, there is much greater heterogeneity in the drying liquid than a film deposited from a solution. The term 'film' as understood in this context refers essentially to a coating or layer which in most cases will be less than 10 thou (0.25 mm) thick, although this is not a rigid limit. The term 'film' is also intended to cover films of adhesives, i.e. one that adheres to two surfaces rather than one, and also to latices used for impregnation, which forms a coating on the inner interstices of a substrate, e.g. in paper impregnation.

When a latex film dries, water evaporates from the surface, and this is in most cases the only method by which moisture can escape. The special case of a porous substrate will be treated separately. The water that is lost from the surface is replaced by diffusion from within the film itself, which is a process that takes place in a finite time. A drying film is not an equilibrium system.

The drying of a film can be divided into three stages, an initial stage operating until 60–70 % of polymer by volume remains. In this stage the particles have Brownian movement and evaporation is substantially the same as from a dilute solution of the stabilisers. Next there is an intermediate stage in which the particles come into contact and tend to coalesce irreversibly. In this stage the rate of evaporation decreases with decreasing water–air interface. In the final stage, water in capillaries escapes by diffusion, and later slowly through the coalesced film. Experiments have been made with three copolymer latices: vinylidene chloride–n-butyl acrylate (7 : 3), ethyl acrylate–methyl methacrylate (2 : 1) and styrene–butadiene (2 : 1). It should be noted that sodium 2-sulfoethyl methacrylate was used as a polymerisable surfactant. The final latices, of particle diameter 0.1–0.2 µm, were steam stripped to remove surplus monomer. The rate of loss of water approaches asymptotically that of water transmission through the film itself [60–66].

If a continuous film is formed, even as a micro-layer on the surface, it will reduce continuously the amount of evaporation that can take place. If the vapour transmission through the film is below the rate of evaporation from the surface, a skin will form on the surface which effectively seals off the liquid inside from further evaporation. Under these conditions it is extremely difficult to form a completely dried film.

If transmission of moisture through the upper surface is sufficient to replace moisture lost by evaporation and prevent a micro-film forming or, as in the case of many polyvinyl acetate products, where the rate of moisture transmission through a film is high, no skinning will occur. Skinning should not occur where the polymer has a minimum film-forming temperature which is above the temperature of evaporation.

Skinning can be avoided by slowing down the rate of evaporation from the surface, e.g. under conditions of relatively high humidity, or by evaporation at low temperatures, since the vapour pressure of water falls many times more rapidly than the diffusion rate. Equally, skinning is made worse by increasing the temperature of evaporation. This causes difficulty, sometimes in estimating the solids content of a latex. In this case evaporation under high vacuum may be used, as it has the effect of lowering the boiling point of water to below that of the minimum film-forming temperature. Unless raising the temperature above 100 °C is undesirable, the problem is usually solved more effectively by drying at 105–120 °C for a shorter period, when the combined effect of the high vapour pressure of water and the softening effect of temperature on the the polymer enables water vapour to escape, giving a cratering effect. This also occurs sometimes when pigmented latex finishes are stoved.

Edge effects may cause some irregularity, generally a star-shaped cracking due to additional shrinkage tensions on the edge. These edge effects are important in certain applications, such as in the formation of gloss or very clear films, when it is highly desirable that edge effects should not occur between application of a brush or roller. This problem will be discussed in further detail in chapter 10 of volume 2, Section 3.4. The addition of a small quantity of a glycol, often 1 : 2 propylene glycol, assists in keeping the edges 'open', since their rate of evaporation is much slower than that of water.

Even with the simplest latex the existence of stabilisers has some effect on the drying and filming properties. In general, the time for complete fusion of the particles will be finite, and as an interim stage the film can be regarded as particles that are to some degree adhered together by the stabilisers. It may be several days, particularly with vinyl acetate type latices, including those with either internal or external plasticisation, before full film strength is gained. The distribution of emulsifiers between the surface of latex particles has been studied by adsorption titration and equilibrium dialysis and centrifugation, with special reference to emulsifiers with a high critical micelle concentration. Note that the distribution of emulsifiers in a film is often irregular [67]. Nevertheless, the fine structure of films from emulsion polymers shows the fine structure of particles, even when solvents have been added [68–71].

The segregation of film formation has been studied using energy filtering transmission electron microscopy. When the samples were prepared below the MFT, separate domains of surfactant in detachment replicas and ultra-thin sections were detected. The mobility of emulsifier at the surface of the particles is increased upon deformation and during coalescence in the process of film formation [72].

The migration of sodium dodecyl sulfate in latex films was studied using infrared (IR) and attenuated total reflectance. Fourier transform infrared photography (see Section 4.1.1) showed that in a latex copolymer of methyl methacrylate, ethyl acrylate and methacrylic acid, there is surface enrichment of surfactant, with a parabolic distribution of surfactant throughout the thickness of the film [73, 74]. Similarly, sodium dioctyl sulfosuccinate was found at the highest concentrations at a film–air interface. The hydrophilic groups of the surfactant are oriented parallel to the surface [75–77].

The presence of emulsifiers in a latex film often has a marked effect on adhesion, some tests having been made on glass with a 2-ethylhexyl methacrylate latex [78, 79]. The presence of colloids, especially hydroxyethyl cellulose, may in some cases cause a 'bridging' of particles, resulting in an apparent coarsening of particle size if added late to the latex. This is also the case with most acrylic latices if hydroxyethyl cellulose is a constituent of the original emulsion. This bridging causes clustering whilst drying, reducing gloss and clarity due to the surface becoming more irregular. Large particle size latices (circa 1 μm), when drying, leave clusters of emulsifier in 'holes' in the film causing a matte appearance. However, these can be readily washed out [80].

Temperature and humidity have a very strong influence on film formation. Low temperature and/or low humidity may prevent film formation by precluding the compaction process involved in particle coalescence [81].

Another complication that may affect film formation is the porosity of the substrate, causing the phenomenon known as 'wicking'. Limited wicking, especially in the case of a pigmented composition, may assist the drying process, but it is generally undesirable. Wicking as a rule will absorb the water phase, but few polymer or pigment particles, and so may cause the latex to break before the various filming forces have had any opportunity to act. Wicking is normally overcome by precoating the porous substrate with a dilute latex, occasionally a diluted emulsion paint, or by precoating with a solution of a material, e.g. a silicone that renders the substrate repellent to water [82].

Dynamic thickness measurements have been made on a latex film during the drying process using a laser profilometer, which showed that film thickness decreases more quickly than is expected from weight loss profiles [83].

Acrylic ester copolymers tend to form films much more readily than vinyl ester copolymers, and the addition of coalescing solvent is usually unnecessary. Vinyl chloride and vinylidene chloride copolymers are exceptional cases (see Section 6.)

In general with core–shell type polymer particles, there is little difference in the coalescing temperature whether the particles are inner soft, outer hard or vice versa, although under some conditions the reverse has been claimed [84]. Fluorescence non-radiative energy transfer (see Section 4.1.1) has been used to study the structure of core–shell latex particles. The studies included fluorinated polymer [85–87]. Core–shell acrylic ester latices have been prepared with various amounts of methacrylic acid in the shell to alter surface characteristics of the particles. Coalescence in water occurs even below the T_g of the latex [88].

In some cases a core–shell structure may eliminate the necessity for the addition of a coalescing solvent (Section 5.4). This has been illustrated by a latex comprising a polybutyl methacrylate core and a butyl methacrylate/butyl acrylate shell which did not require a filming aid, compared with a polybutyl methacrylate homopolymer to which filming aids were added [89, 90] (see also Section 4.3).

The addition of a latex with T_g of about 60 °C when added to a soft film-forming latex with T_g of about 0 °C, increased the block resistance without affecting the film properties, even in considerable ratios [91]. A review includes studies of latices stabilised by polyacids which it is claimed never truly coalesce, but nevertheless have high cohesion [92].

Critical relative humidity has been applied to the drying of water-borne coatings. Thus the relative air humidity of a binary organic solvent/water system remains constant until one of the solvents has completely evaporated [93]. During evaporation, very hydrophilic solvents such as ethylene glycol monoether acetate, which, although having some aqueous solubility, tend to migrate towards the polymer phase. Thus, for two solvents having similar molar size, the more polar one evaporates faster from the coalescing film, although skinning may cause a complication [94].

The rate of drying of non-woven fibrous networks could not be predicted, except where the networks are saturated. Water transmission in a fibrous network may be greater than that on a solid substrate [95].

Micro-foam is a problem that may occur during the spray application of latices. The retention of bubbles of about 10 µm radius in a film adversely affects the clarity, gloss and tensile properties of the film. The ability of a coating to dissipate foam is influenced by the nature of the cosolvent which may be present, the initial pH and any base that may be added, as these affect the viscosity rise in the film during drying [96].

General reviews of film formation from latices are available [97–99].

4.1.1 Methods of study—film formation

Scanning electron microscopy is quoted in many papers on the mechanisms of film formation. Nuclear magnetic resonance (NMR) is a well-established method of analysing the macromolecular architecture of emulsion polymers. A referenced review is available [100].

Small angle neutron scattering (SANS) has been used to study particle coalescence with deuterated polystyrene after removal of the aqueous phase and surfactants [101, 102]. Dielectric measurements at microwave frequencies have also been used to study latex coalescence. It is possible to discriminate whether a latex forms a continuous film [103, 104].

A non-destructive method for the measurement of properties during film formation is based on surface amplitude decay, i.e. the decrease in surface variation that occurs as levelling of the film between application and coalescence. The method used two-beam optical interference which neither disturbs the film nor affects evaporation from the surface [105].

Fourier transform infrared spectroscopy (FTIS) has been used to examine surfactant desorption. Thus when hexadecylpyridinium chloride was used as a stabiliser for a poly(2-ethylhexylmethacrylate) latex stabilised with hexadecylpyridinium chloride, the cationic emulsifer underwent crystallisation after film formation and desorption. Other emulsifiers also showed desorption after filming [78].

A steady state fluorescence technique has been used to examine the annealing of films formed from high T_g particles above their T_g. In conjunction, scanning electron microscopy was used to elucidate data. Films were prepared from a mixture of naphthalene- and pyrene-labelled polymethyl methacrylate latex particles [106–109].

Atomic force microscopy (AFM) is a very recent addition used for the investigation of latex films. It is basically a probe system (Figure 4.6). A sharp point mounted on a spring cantilever is brought into close contact with a surface until an up-force due to the atomic forces are experienced as the probe is then pulled across the sample. The forces deflect the cantilever, and the movement is detected by a reflected laser beam trained on a photodetector. AFM studies have concentrated on the examination of surface topography as a means of following film formation. The particulate nature of even well-formed films is already known using electron microscopy. Figure 4.7 shows an example. As an illustration of the difference between a film from an acrylic–styrene latex, including a coalescing solvent, the water vapour transmission(WVT) was 140 g m^{-2} every 24 h, whilst the comparable solvent-borne film had a WVT of 20 g m^{-2} every 24 h.

AFM has the ability to give data in the vertical axis, allowing the peak-to-valley height over a series of particles to be determined. AFM has shown

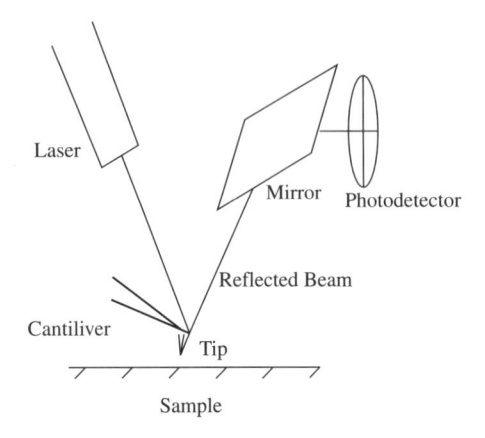

Figure 4.6 The basis of the AFM is the movement of a probe mounted on a cantilever as it moves across a sample surface.

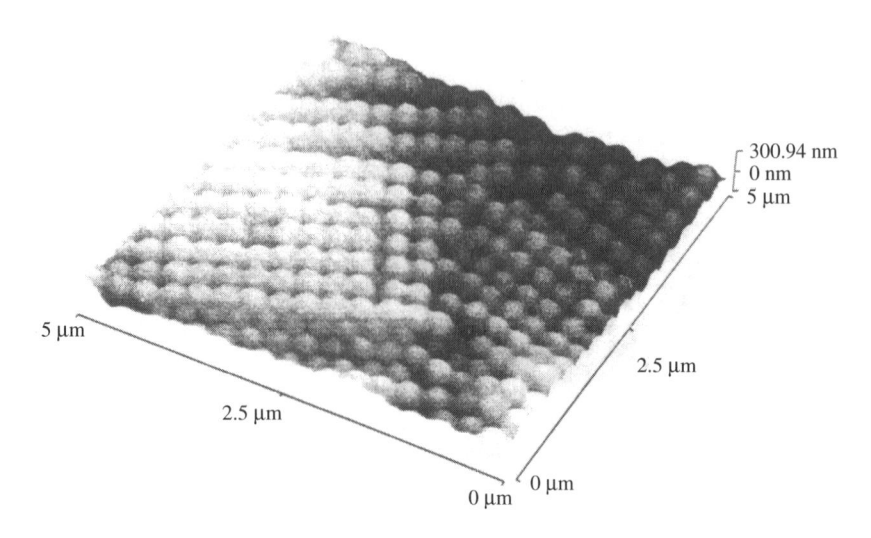

Figure 4.7 AFM scan of a well-ordered latex film.

that annealing an n-butyl acrylate latex at 60 °C for 0, 1 and 2 h respectively, the average line roughnesses were 12, 7.2 and 2.3 nm respectively. Heating to 80 °C eliminated the particulate nature of the films, leaving a residual line roughness of 0.53 nm line roughness. These tests were extended to paint films and also to the effect on surfaces of surfactants.

Atomic force microscopy (AFM) has been used to study exudation of sodium dodecylbenzene emulsifier from a butyl methacrylate latex. The addition of Texanol® coalescing agent increases surface exudation. Latex films dry from the edges of a film inward, and the mechanism of latex film formation may differ from one end of the drying front to the other [110].

AFM results seem to confirm the fact that capillary pressure is the dominant factor in latex film formation by investigating the kinetics of film formation of wet and dry latex systems, using polybutyl methacrylate as the model latex, which gives a hydrophobic polymer and may be prepared surfactant-free [111]. Atomic force microscopy has also been used to examine compatible and incompatible latex blends [112].

Further investigations using AFM are now available [113–121A].

4.2 Theoretical developments

One of the earliest suggestions made in postulating a mechanism for particle coalescence during drying was that the main contribution to the energy required for particle coalescence is produced by surface tension forces as the total surface energy decreases. This theory used Frenkel's equation:

$$\theta^2 = 3\gamma rt/2\pi\eta$$

where θ is the half-angle of incidence, as in Figure 4.8, η is the particle viscosity, r is the radius the particle and t the elapsed time. As θ and r are the only variables in any one latex, θ^2 was determined as a function of r by electron microscopy, and it was confirmed that it is proportional to $1/r$ (Figure 4.8) [122].

By this theory the pressure on the particles varies inversely with the latex particle diameter, and it has been calculated that where the particle diameter is 500 Å, the pressure during drying with $\gamma = 30$ dyn cm^{-1} is not less than 1740 p.s.i. (122 kg cm^{-2}). An alternative suggestion is that particles will coalesce if the driving force provided by capillary pressure is sufficient to overcome the resistance of the polymer particles to deformation, and some approximate calculations were made [123]. It was calculated that for a particle of radius 1 μm (10 000 Å) and assuming $\gamma = 30$ dyn cm^{-1}, the maximum modulus G for film formation $= 1.1 \times 10$ kg cm^{-2}, or 2.5×10 kg cm^{-2} if $g = 70$ dyn cm^{-1}. This figure alters inversely with the particle diameter (see Figure 4.9). Optical microscopy and motion pictures, used with latex particles

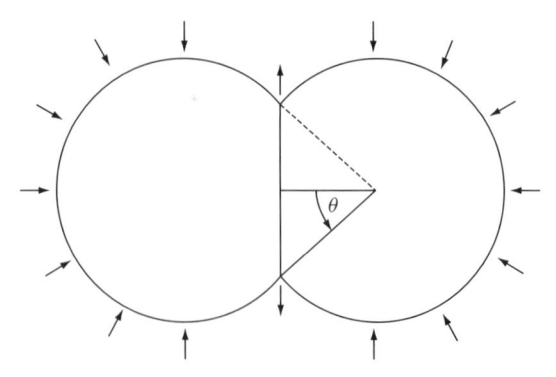

Figure 4.8 Coalescence of spheres by viscous flow caused by surface tension forces [122].

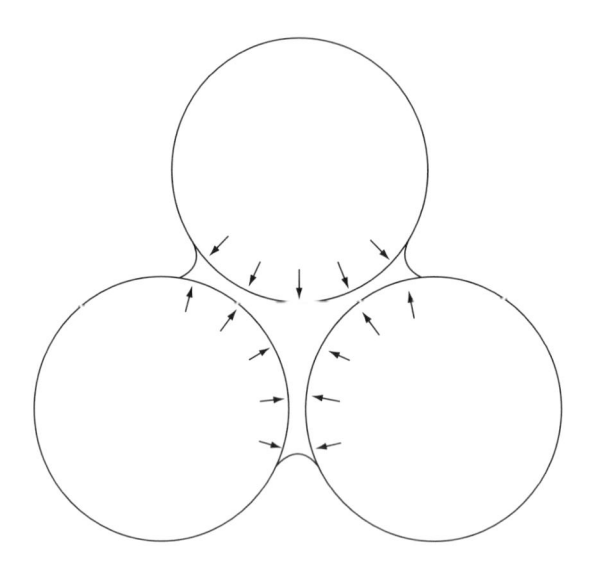

Figure 4.9 Coalescence of spheres caused by capillary forces [123].

1.2 µm in diameter, showed Brownian movement, the convection currents in a drying drop of latex and the formation of latex particle crystallites.

When a drop of dilute latex dries a crust appears around the edge. The evaporation of the water gives rise to convection currents, which transport particles from the centre to the edge, streaming just above the lower surface of the drop. These currents then carry the particles not deposited at the edge back towards the centre, streaming just under the upper surface of the drop [124].

A further suggestion which has been made is that there is an additional force called autohesion, which is the mutual interdiffusion of free polymer chain ends across the particle–particle interface in the coalesced film [125].

A rather useful general picture of the drying mechanism has been given by Sheetz [126]. Water evaporates until the system becomes so concentrated that there is 60–70 % of polymer by volume. At this stage evaporation is substantially the same as a dilute solution of emulsifier and the particles have Brownian movement. Thence the repulsive energy between the particles is overcome and flocculation begins. The rate of evaporation decreases with a decreasing water–air interface. At about this point the spheres in the surface begin to emerge from the liquid, and are consequently subject to the forces of capillarity G_n, normal to the surface, and G_p, parallel to the surface. G_n tends to push the particle in a direction normal to the surface, exerting a compression on the matrix of particles below the surface, which contract or deform, forcing water to the surface. Simultaneously, G_p tends to distort the surface layers of particles so that 'holes' or gaps are closed. In addition, if the surface free energy of the polymer–water interface is significant, wet sintering will contribute to the forces, fusing the particles in the early part of the compaction process.

During this time, whilst water is being evaporated from the air–water interface, water is also diffusing through the particles in the surface and evaporating. Ultimately the only way in which water can leave is by diffusion through the particles, as the compressive pressure compacts the film with no void spaces. This is considered to be the case when compaction is about half complete. Allowance should be made for the fact that an appreciable amount of evaporation from the surface will occur before compaction starts, the local concentration of solids being at least 74 % if the latex consists theoretically of monosize spheres.

After compression the surfaces remain substantially intact, and it is several days before final equilibrium is reached as either the absorbed surfactant groups congregate together, as they will if there is a large surplus of surfactant, or become dispersed evenly through the film where there is little surplus surfactant or where a minimal quantity has been used. Unless the polymer is particularly free flowing, i.e. far above the T_g, the boundaries are never entirely obliterated. Thus when a film is rewetted, blanching occurs more rapidly with a large particle size latex [127, 128].

A 1995 paper also using AFM shows a direct correlation between the kinetics of film formation and the rheological properties of the latex polymer [129]. It is also surprising that whilst latices of 'soft' polymers dry more slowly than those of high T_g, blends dry even more slowly. The rate of drying is considered to be controlled not by the wet part of the film but by the boundary between the wet and dry parts [130]. Alternating current (a.c.) conductance has been used to elucidate the three stages in drying [131]. Recent work confirms the views of Sheetz [132].

Heterogeneous copolymers with a multiphase particle morphology may have a complex method of film formation depending on the nature of the heterogeneity. Slow interdiffusion may take place depending on the components, even leading to an inversion of the morphology. On the other hand, polymer segments with a high T_g may prevent further coalescence [133–136].

In many cases films from latices, especially nominal copolymers, display heterogeneity. Many test methods have been used including analytical ultra-centrifugation, mechanical spectroscopy, tensile tests, electron microscopy and atomic force microscopy [137, 138].

Latex films have Youngs modulus systematically higher than the corresponding films from solution. This is due to the fact that the hydrophobic shells of the latex particles form a continuous phase, which increases the modulus due to polar interactions [139].

The complex viscoelastic behaviour of polymer latex film formation has been recently described. With sterically stabilised particles, e.g. with partially hydrolysed polyvinyl alcohol shells, annealing has a marked effect on the viscoelastic properties of a film, a transition at ~ 75 °C being noted, in addition to the normal one at ~ 35 °C [140].

In a butyl methacrylate latex (PBMA) stabilised by sodium dodecylbenzene sulfonate, a Texanol® coalescing agent (see Section 5.4) promoted surface exudation of the sulfate (see Figure 4.10) [141]. In further tests with butyl methacrylate latex the addition of a non-ionic surfactant enhances the interdiffusion rate in film formation, but a typical anionic surfactant inhibits interdiffusion. Whilst the former tends to plasticise the polymer and increase interparticle diffusion, the anionic emulsifier remains in particle boundaries and interstitial regions and acts as a barrier to interdiffusion [142].

Thermodynamic considerations also control surfactant behaviour during drying, as does surfactant–polymer compatibility [143]. The inclusion of acid molecules in a copolymer decreases internal stress during filming and gives improved mechanical properties due to increased molecular interaction. When copolymers of ethyl acrylate and methyl methacrylate (36 : 54 and 51 : 39, each with 10 parts of methacrylic acid), were copolymerised with 10 % of methacrylic acid and then neutralized with 10 % of methacrylic acid and to various degrees with dimethylaminoethanol, the MFT decreased as swelling with increased base decreased [144–146].

The rheology of acrylic ester monomers including methacrylic acid varies according to the point of addition of the acid during polymerisation. Addition of acid in the second half at pH 5 gives the highest viscosity. This is due to the higher acid concentration which causes particle swelling (see also Chapter 3, Section 4.1.4) [147]. See also Chapter 10 in Volume 2, as these modifications apply very largely to emulsion paints.

The hydrophobically modified thickeners (HEUR and HMHEC; see Chapter 2, Section 4.7) have been studied with regard to their 'in can' behaviour and also the effects of coalescing solvents [148].

A steady state fluorescence technique has been used to examine the annealing of films formed from pyrene-labelled polymethyl methacrylate particles above their high T_g. Scanning electron microscope data were used in

Emulsion films 219

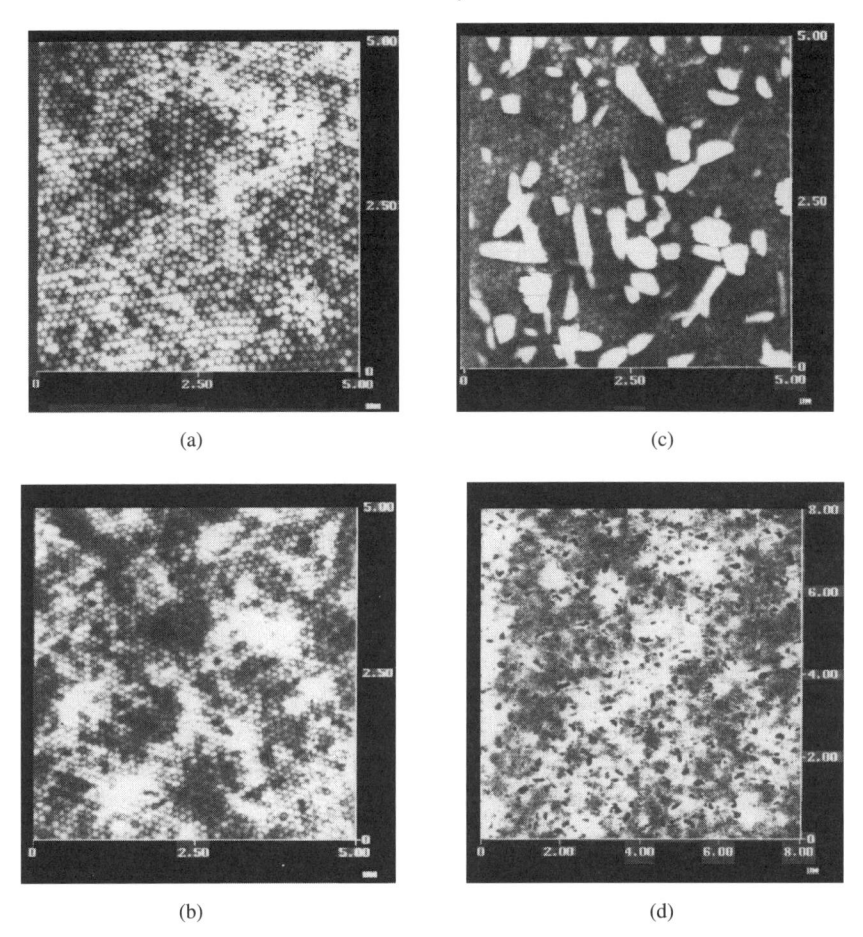

Figure 4.10 (a) Nacent PBMA latex film. (b) PBMA latex film annealed for 5 min at 70 °C. (c) PBMA latex containing 10 wt % of Texanol® annealed for 5 min at 70 °C. (d) PBMA latex containing 10 wt % of Texanol® annealed for 5 min at 70 °C, washed with distilled water. (Reproduced from Johui *et al.* [141].)

conjunction to interpret results. Polyamyl methacrylate has also been studied by this technique [149–151].

A recent study considers that the nature of the latex, with regard to particle size, viscoelasticity of the polymer and the kinetics of drying and deformation, affect the process of film coalescence. Film formation does not occur under unique conditions and environmental conditions may affect the process. As the character of the hydrophilic shell at the particle surface is poorly defined, the true nature of the surface region is very important in the study of film formation [152].

Silica, both untreated and functionalised by treatment with methacryloxypropyl-trimethoxysilane (see Chapter 5) has been encapsulated in a latex of ethyl acrylate. Small angle neutron scattering (SANS) indicated that the functionalised silica was well dispersed, and very high elongations at break were obtained with this treated silica. There may be grafting of the ethyl acrylate emulsion polymer on the silica [153].

For further studies see references [154] to [160].

4.3 Minimum film-forming temperature

At this stage it is desirable to consider a temperature known as the minimum film-formation temperature (MFT). The name is self-explanatory and refers specifically to films from latices. The MFT value is often quoted as being synonymous with the second-order transition point T_g (Chapter 1, Section 4.3). This is not the case, since the strong cohesive forces already mentioned tend to cause film cohesion a few degrees below the T_g [161]. The MFT and the T_g are major factors in film formation [162]. Hydrogen bonding may have a marked effect on results [163].

A number of descriptions of suitable apparatus appear in the literature. This usually consists of an aluminium or a copper plate which has a 'dry ice' (solid carbon dioxide) tank at one end, a heater at the far end and a number of thermocouples located along the plate. The top is protected by a glass plate and the sides are protected by insulating material. A current of dry air can be passed under controlled conditions over the plate to prevent condensation at the cold end. A technical apparatus is available (Sheen Instruments, Teddington, UK). The latex to be tested is spread over the plate with a standard film spreader and allowed to dry. The MFT is found by visible inspection at the point where the film becomes white and discontinuous. It is found in practice to be relatively sharp [164–168].

Some variations in MFT may be expected, even with two latices of an identical polymer with similar molecular weight ranges if the particle sizes are very different. Thus copolymers of butyl acrylate, methyl methacrylate and methacrylic acid, otherwise similar, have appreciable differences in MFT. The difference between the smallest particle size (150 nm) and the largest (1235 nm) shows a variation between 11.5 and 16.4 °C [169, 170]. This may also occur if the stabiliser systems vary. In extended tests in an alternative investigation, it was found that variations with particle size could not be proven and seemed independent of it. It was found, however, that the quantity MFT minus T_g varied with the polarity of the sample, being 10 °C for a copolymer of 75 % of styrene and 25 % of n-butyl acrylate, and, surprisingly, -3 °C for a 1 : 1 ethyl acrylate–methyl methacrylate copolymer [171]. Substituting 3–5 % of 2-ethylhexyl acrylate for styrene in a latex lowers the MFT by 10–15 °C. Other substitutions such as butyl acrylate or the addition of external plasticisers cause the expected lowering of the MFT [172].

Emulsion films

A range of bimodal particle size systems, including large and small particles, was tested, and at a 4 : 1 ratio of large to small particles there is a minimum MFT [173]. The same authors have found that the maximum volume packing fraction of bimodal latices is greater than that of monomodal latices, reducing shrinkage on drying and generally giving better film properties [174].

A study compares the MFT of methyl methacrylate (MMA)/butyl acrylate (BA) and styrene (ST)/ethyl acrylate (EA) systems, in each case comparing a copolymer with a core–shell copolymer and a blend of independent latices (see Figures 4.11(a) to (c)). It will be noted that in the former case a particle shell of methyl methacrylate produced much higher MFT results than a copolymer. Figure 4.11(b) indicates differences between internally hard–externally soft and externally hard–internally soft particles. The results are as expected: the latex with an externally soft polymer gives a much lower MFT, which is generally useful technically. Figure 4.11(c) gives a very sharp contrast where the curves for core–shell copolymers superimpose, irrespective of which phase is added last in copolymerisation. This seems to be caused by an inversion phenomenon, suggesting that polystyrene is always the internal phase (see Section 1.1). Figure 4.12 shows corresponding electron micrographs. Figure 4.13 gives electron micrographs of latex particles showing deformation of polybutyl acrylate particles during preparation, but only limited deformation of a core–shell copolymer with polymethyl methacrylate as the external phase [175].

The general principles of film formation hold for pigmented systems as well as for the latices alone.

4.4 Specific studies—film formation

4.4.1 Vinyl acetate and copolymer latices

Latex particles of polyvinyl acetate during drying have been observed by electron microscopy (the type of emulsifier is not stated). It took 30 days for a homogeneous state to be reached. Rather surprisingly, the rate of film formation from vinyl ester latices decreases with increasing length of the hydrocarbon chain in the acid unit, e.g. polyvinyl acetate forms films faster than polyvinyl butyrate [176].

Tests on a film of polyvinyl acetate have given the rather surprising information that most of the film has not coalesced until the percentage of water lost from the film corresponds to a film composition of 95 % of non-volatiles and 5 % of water. At this stage film coalescence occurs rapidly at 21 °C, the temperature of the test. Films from polyvinyl acetate latices stabilised with hydroxyethyl cellulose and polyoxyethylene ethers undergo further coalescence after normal film formation, except for crosslinked vinyl acetate copolymers. The effects are due to differing interdiffusion of molecular segments [177].

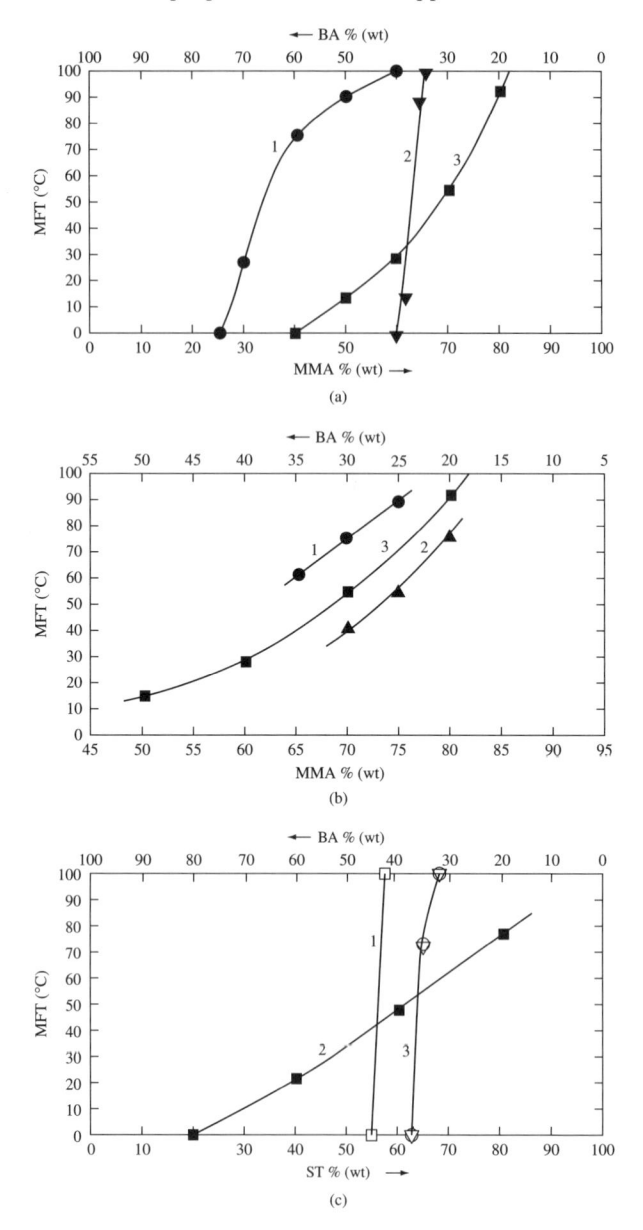

Figure 4.11 (a) Curves of polymer compositions in different morphology particles versus MFTs of latices for the MMA/BA system: ● core BA/shell MMA; ▼ blend; ■ copolymer. (b) Curves of polymer compositions in gradient latex particles versus MFTs for the MMA/BA system: ● from internally soft to externally hard; ▲ from internally hard to externally soft; ■ homogeneous copolymer latex. (c) MFTs of latices for the ST/EA system: □ blend; □ copolymer; △ core EA/shell ST; ○ internal ST/external EA. Reproduced by permission of John Wiley & Sons, Inc.

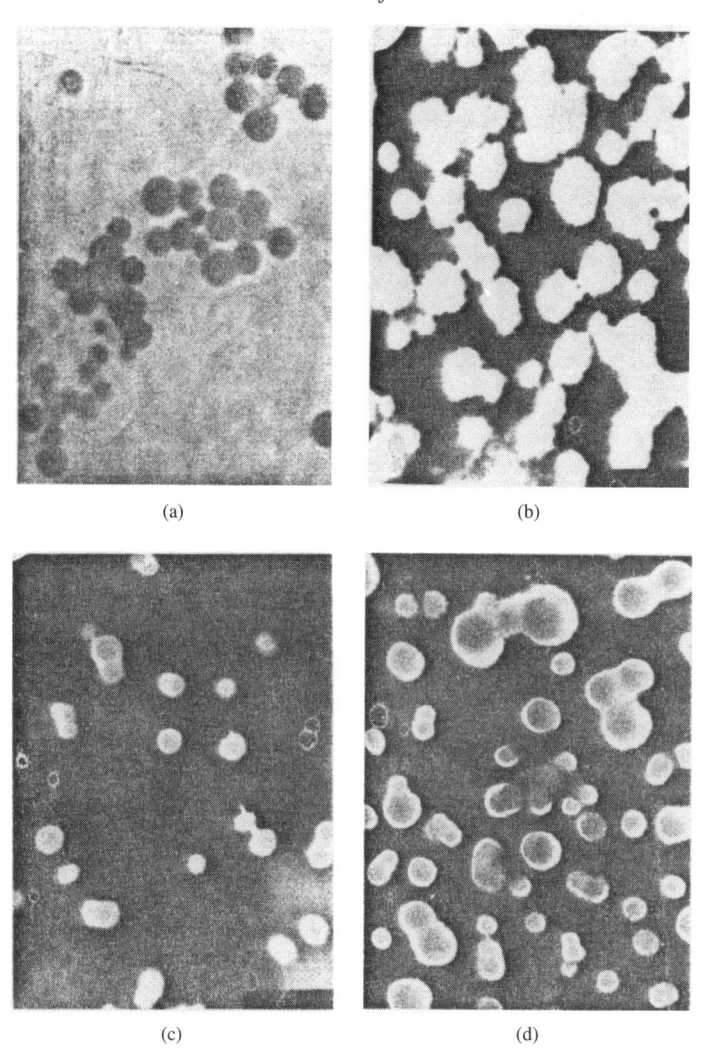

Figure 4.12 Electron microscope images of latex particles for the ST/EA system: (a) PST homopolymer latex particles: (b) PBA homopolymer latex particles: (c) IST/IIEA latex particles: (d) IEA/IIST latex particles, I where I and II are as in the original Tong you *et al.* paper [175]. (Reproduced by permission of John Wiley & Sons, Inc.)

Scanning electron microscope studies of polyvinyl acetate, ethylene–vinyl acetate and ethylene–vinyl acetate–vinyl chloride terpolymers, using films that have been fused to various degrees, indicate that fusion depends on conditions of film formation as well as the temperature [178]. If films are formed from vinyl acetate–Veova® copolymers above the T_g, the water absorption is

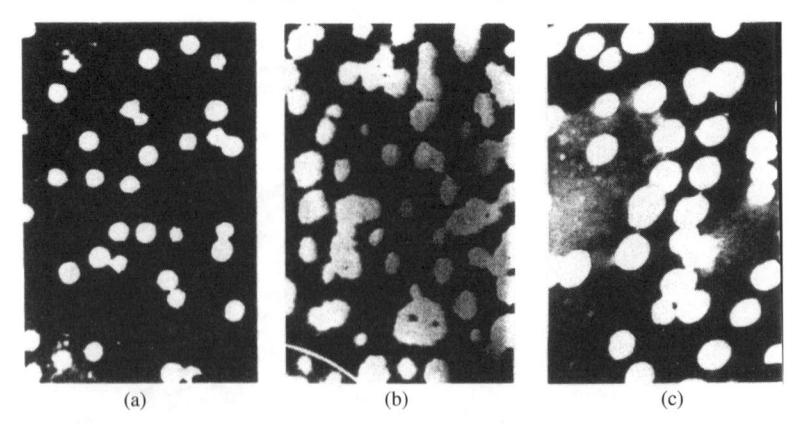

Figure 4.13 Electron microscope images of latex particles for the MMA/BA system: (a) PMM homopolymer latex particles; (b) PBA homopolymer latex particles; (c) core BA/shell MMA(1 : 1 wt) latex particles. (Reproduced from Tongyu *et al.* [175] by permission of Wiley, USA.)

relatively high, but the tendency to whitening is low since the extractables are evenly distributed in the film. If an attempt is made to form a film below the T_g the opposite effects are noted. This is an extension of earlier work in which small particle size latices of the same copolymer behave in a similar fashion. Larger particle size latices (0.5–1.0 μm diameter) show rapid whitening and a relatively rapid early absorption of water. However, because of the rapid leaching out of emulsifier, in contrast to the films formed from latices of fine particle size, there is greater resistance to water when the film is redried [179].

In general, the balancing of a film from a latex occurs because of its heterogeneous structure; emulsifier and initiator fragments on oligopolymers tend to concentrate on the surface. Technical nonylphenol polyoxyethylene emulsifiers with 20 and 50 ethylene oxide units respectively and sodium lauryl sulfate have been examined as surfactants in the preparation of vinyl acetate and vinyl acetate–butyl acrylate latices and subsequent film formation. Water plasticises a 85 : 15 copolymer; this should be considered in film-formation theories. The surfactants with shorter ethylene oxide chains are better plasticisers, but poorer steric stabilisers. The 50-unit ethylene oxide surfactant has a very marked effect on film formation, and in particular is the most compatible with the latex film surface. Sodium lauryl sulfate, popular with academic researchers but rarely used in industry, shows incompatible regions on the latex surface [180, 181].

4.4.2 Acrylic latices

Scanning electron microscopy has been used in defining the four stages of film formation of acrylic latices. These are the initial latex, a close-packed array with water-filled interstices, a densely packed array of deformed particles and the continuous coating with no internal solid–solid interfaces. An intermediate stage between stages 2 and 3 coincides with optical clarity. The kinetics of particle coalescence in the third stage is a function of the T_g of the polymer [182].

A blend of acrylic latices consisting of a uniform low glass transition temperature (T_g) and high T_g polymer particles, which may be of a non-homogenous composition, may have a lower T_g and probably a lower MFT than a uniform particle size having the same average composition. Figure 4.14 shows the latex MFT as a function of the percentage hard phase. The latex types soft: hard and hard: soft refer to sequential polymerisation of monomers. The blend in the diagram (Figure 4.14) is a mixture of varying percentages of the soft phase (T_g of 5 °C) and the hard phase (T_g of 60°) [183A].

Studies have shown that if a latex of acrylonitrile–ethyl acrylate copolymer contains added sodium alginate (3 phr) or sodium lauryl sulfate, the initial latex being emulsifier free, filming at 30 °C causes the sodium lauryl sulfate to be localised at the air interface, whilst the alginate is localised at both interfaces. Localization is reduced at 50–70 °C, but increases again at 100 °C, due to polymer coalescence. Increased acrylate ester content of the copolymer reduces localization by reducing the second-order transition point [184].

In allied investigations with films formed on a polyester base from a persulfate–initiated butyl acrylate–acrylonitrile (97 : 3) copolymer, localised distribution of the emulsifier was measured by electric surface reactivity. Results

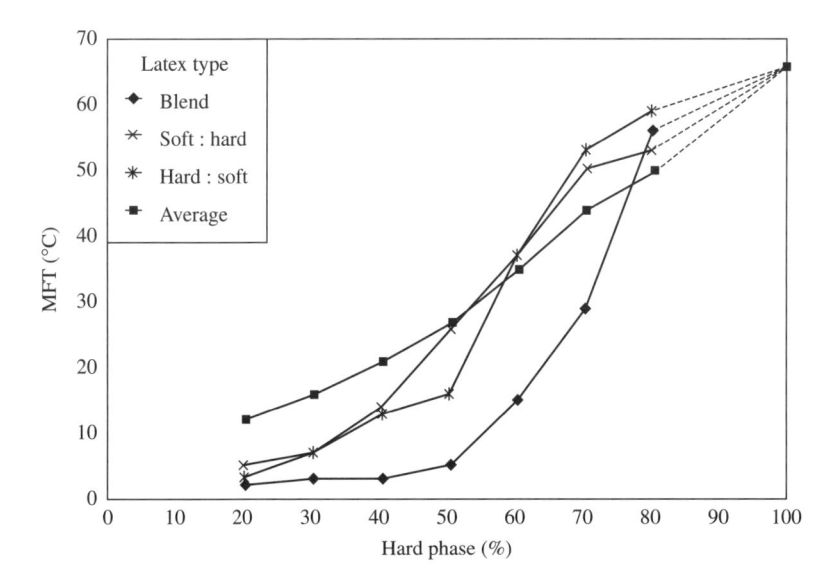

Figure 4.14 Latex MFT as a function of the percentage hard phase [183].

indicate that an emulsifier with poor compatibility and relatively high molecular weight migrates towards the interface and reduces peel resistance [185].

Similar investigations at 40–50 °C with an emulsifier-free ethyl acrylate–methyl methacrylate emulsifier–free latex, optionally with added dodecylbenzene sodium sulfate, confirm the general non-homogenous nature of film drying, including skinning and dry polymer forming from the edges, experiments being conducted at 40–50 % relative humidity [186].

A turbimetric study has been made of emulsion copolymers of methyl methacrylate and butyl methacrylate [187]. Surfactant adsorption on the surface has a marked influence on the film formation of acrylic latices, *inter alia*, and complete saturation of the acrylic surfaces may cause difficulty in film formation due to charge repulsion [188]. Consolidation of an acrylic polymer latex film depends on the carboxyl content and its dissociation, particle size and electrolyte concentration [189]. Films from butyl methacrylate latices have been examined by a freeze fracture transmission electron microscope technique. Film preparation time and temperature and ageing time and temperature influenced the state of coalescence of their films and their permeabilities to gases and aqueous solutes. Postadditives did not affect film quality [190, 191].

As will be shown in the following section, coalescing solvents are often considered to be desirable for good film formation. A new strategy for zero volatile organic solvents in film-forming latices is a blend of soft and hard particles. The soft latex particles form the continuous phase and fill the voids between the hard particles, leading to transparent films without the volatile organic compounds (VOC). The hard particles impart good mechanical properties to the films. The techniques for study include scanning electron microscopy and atomic force microscopy. The latices chosen for study were based on polymethyl methacrylate and poly(butyl acrylate–butyl methacrylate). Polymer compatibility is necessary to obtain clear films [192].

Experiments have been made to form copolymers of vinyl acetate and butyl acrylate (3 : 1) so that the MFT is much lower than the T_g, in order to avoid completely the addition of volatile organic compounds. Experiments were made by a standard delayed addition process using a pre-emulsion feed which included sodium dodecylbenzene sulfonate as the emulsifier and a redox system based on ammonium persulfate and sodium formaldehyde sulfoxylate. Full results and variations were made available, but preliminary results show that whereas a standard latex had a T_g of 18 °C and an MFT of 10 °C, variations in procedure produced in one case a T_g of 35 °C, and an MFT of 1 °C only. Further details and characteristics of the resultant paints will be presented by the authors in further papers [193].

In a series of tests on acrylic latices with particle sizes of 63–458 nm, it was confirmed that the latices with smaller particles have lower MFTs than those with larger particles [194]. Under some conditions the formation of a skin during drying of an acrylic-based latex does not hinder subsequent evaporation of the water [195].

A well-referenced review of film formation from latices, including pigmented latices, is available and includes a mathematical treatment [196]. Details of film formation where crosslinking occurs will be found in Chapter 5. Further information on pigmented films is included in Chapter 10, in Volume 2.

4.4.3 Styrene–butadiene latices

Investigations on a 67 : 33 styrene–butadiene (SB) copolymer have shown that when a copolymer film was aged for 3 h the contours of individual particles could be discerned, although the film was dry and transparent. Complete coalescence took about two weeks, but after some further time an exudate appeared, which was found to be the stabiliser originally present. Results are shown in Figure 4.14 [197A].

In the evaporation of water from a film of an SB latex, the concentration of particles increases on the surface due to their low diffusion coefficient until about 60 %, when the apparent density of the cream increases spontaneously, followed by surface coagulum formation [198]. The viscoelastic properties of a series of carboxylated styrene–butadiene latex films depend on the pH of the latex and the nature of the ions used for neutralization [199].

For further studies on film formation of latices see references [200] to [203].

4.5 Summary

To summarise, surface tension forces, capillary forces and autohesion all play a major part in film formation, with temperature, humidity, film thickness and wicking playing a major role. The nature of the polymers present is of course a major feature, particularly when mixed compositions are present. Some modifications will be considered in Chapter 10 and 13 in Volumes 2 and 3.

A review of the literature on the dynamic viscoelastic properties of latex polymer films and their measurement is available [204]. A recent symposium gives studies on film formation by various spectroscopic methods, including the effects of acid groups on film stratification [205].

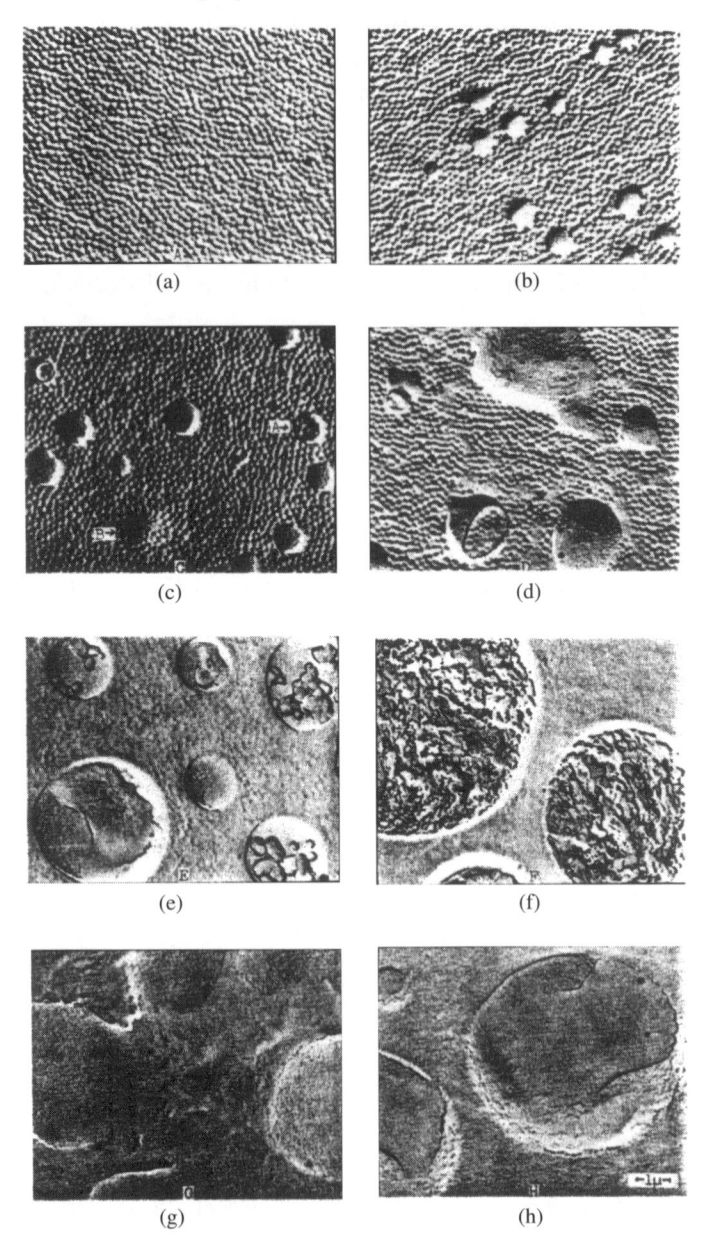

Figure 4.15 Effect of ageing on the surface structure of a 67 : 33 styrene–butadiene latex film: (a) 2.5 h, (b) 1 day, (c) 3 days, (d) 6 days, (e) 9 days, (f) 14 days, (g) 35 days, (h) 40 days [197]. (Reproduced by Permission Marcel Dekker.)

5 ADDITION OF PLASTICISERS AND COALESCING SOLVENTS

5.1 Introduction

An organic liquid which is added to a polymer in order to make it more flexible is known as a plasticiser. A plasticiser is essentially non-volatile at ambient temperatures and should have some mutual solubility, or compatibility, with the polymer concerned. The compatibility need not be complete. Thus 2-ethylhexyl phthalate has only limited compatibility with polyvinyl acetate. Since in nearly all cases the polymer is in the major proportion, compatibility may be interpreted as solubility of the plasticiser in the polymer rather than vice versa.

In some cases an organic liquid may be slowly volatile, even at ambient temperatures. These compounds are referred to as 'transient plasticisers' or coalescing solvents. Examples are ethoxyethyl acetate $C_2H_5OC_2H_4OOCCH_3$, b.p. 156.3 °C, and dimethyl phthalate $o\text{-}C_6H_4(OOCCH_3)_2$, b.p. 282 °C. The former is a limiting case of volatility, whilst the latter is a limiting case between a normal and a transient plasticiser. Dimethyl phthalate is sometimes deliberately added to a formulation, e.g. for an emulsion paint, because of its insect-repelling properties, The normal readily volatile solvents such as ethyl acetate and toluene are not usually classed as coalescing solvents.

At the other extreme, solid plasticisers are added to polymers, often dissolved in another plasticiser or in a solvent. An example is dicyclohexyl phthalate, m.p. 62–65 °C. Solid plasticisers are, as a rule, only effective above their melting points, but there is often a time lag before their plasticisation is effective. An interesting application is in the formulation of delayed-action heat-seal adhesives (Chapter 8, Section 5.3).

Plasticisers which have a particularly wide application with the emulsion polymers described in these volumes are the various alkyl phthalate esters, perhaps the most widely used plasticisers, including various phosphate esters, both aryl and alkyl, of which tritolyl phosphate (also known as tricresyl phosphate) is often encountered. Other plasticisers include the diesters of adipic acid $HOOCC_4H_8COOH$ and sebacic acid $HOOCC_8H_{16}COOH$, which have special interest where low-temperature flexibility is required, as well as some esters of citric acid $HOOCCH_2C(OH)(COOH)CH_2COOH$ and tartaric acid $HOOC(CHOH)_2COOH$. A number of plasticisers are based on esters of polyethylene glycols $XOOCCH_2(CH_2OCH_2)_nCH_2OOCX$, where X is a fatty acid residue. Esters of p-toluenesulfonic acid, p-toluenesulfonamide and mixed o- and p-N-ethyl-toluenesulfonamides are occasionally encountered. p-Toluenesulfonamide has the rather interesting property of being an 'anti-plasticiser' for some polymers, although it functions more conventionally with condensation resins such as melamine–formaldehyde and polyamides.

Many polymers, broadly referred to as 'polyvinyls', experience plasticiser migration dependent on the surface with which they are in contact, but the phenomenon is particularly noticeable when another 'vinyl' film is coated, e.g.

polyvinyl chloride, or on a substrate such as leather, and even some rubbers. Constant contact with water may cause a plasticiser to be gradually leached out if it has even slight aqueous solubility. This occurs with some short-chain dialkyl phthalates. In addition, contact with cleaning solvents may leach plasticiser, even if the polymer proper is unattacked.

In consequence, a series of plasticisers of much higher molecular weight than usual (upwards of 1000) have been developed for specialised applications. Most of these are polyesters such as polyethylene glycol adipates and polyethylene glycol sebacates or similar compounds, sometimes with a different end group. The general formula is $H(C_2H_4O(OCC_nH_{2n}COOC_2H_4O)_mH$, where $n = 4$ or 8, although the end group need not be hydroxyl. These products should not be confused with polyoxyethylene glycol derivatives, which are often surfactants. These plasticisers, although not as efficient as those of lower molecular weight, are much more difficult to leach from polymers than lower molecular weight types and they do not tend to migrate.

The acrylic ester polymers as a class, in distinction to those based mainly on methacrylic acid, do not require plasticisation. The term 'internal plasticisation' is used when the desired flexibility, softness or lower T_g is obtained by copolymerisation of a monomer such as styrene or methyl methacrylate, which yield hard polymers with a monomer such as an acrylic ester, which imparts the desired properties by direct copolymerisation. However, in many cases the addition of an external plasticiser is desirable, either for consideration of cost, or possibly because internal plasticisation may cause some other objectionable property, such as loss of resistance to certain solvents.

Tests on latices of polybutyl methacrylate latex and core–shell latices of polybutyl methacrylate cores and 80 : 20 butyl methacrylate/butyl acrylate shells with 2,2,4-trimethyl-1,3-pentanediol and diethylene glycol monobutyl ether coalescing agents indicate that the latter is located mainly at the particle surface, increasing its efficiency. The former has a more lasting effect [206].

5.2 The mechanism of platicisation in emulsion

When plasticiser is first added to a latex and emulsified directly by the surplus emulsifier present in the system, it can be considered as forming discrete particles which are stabilised independently from the polymer particles. Similar conditions apply where a plasticiser has been submitted to pre-emulsification.

Ultimately there will be a small increase in the overall size of the particles. Since, however, only the diameters of particles are measured, this will be difficult to observe, except with an electron microscope or a very accurate sedimentation device (Section 8.7). Absorption may be faster with a fine particle size (circa 0.1 µm) latex, but the repulsive forces in a highly charged system may slow absorption because of the inability of the plasticiser to penetrate the surface layer on the polymer particles. There is some evidence that systems which are based substantially on non-ionic emulsifiers only are easier to plasticise than those which are strongly anionically charged.

5.3 Plasticisation of principal polymers

The principal polymers used in latex form which require plasticisation for effective film formation or other application properties are polystyrene, polymethyl methacrylate, polyvinyl acetate, polyvinyl chloride, polyvinylidene chloride and a number of other copolymers based on these monomers. Polyacrylonitrile is not used as a homopolymer in latex form, but mainly as a copolymer with butadiene, this being more properly included in synthetic rubber latex technology. Acrylonitrile is occasionally disclosed in patents as a comonomer with acrylic esters.

5.3.1 Plasticisation of polystyrene

Polystyrene, although it can be plasticised with dibutyl phthalate and other plasticisers, is rarely used on its own. It is either a constituent of a copolymer, often with butadiene, or with an acrylic ester, or it is plasticised by being part of a more complex formulation, as in emulsion polishes (Chapter 14 in Volume 3). In these compositions, tributoxyethyl phosphate is often included as a plasticiser, in quantities up to about 5 % of the polystyrene present. The particular function of tributoxyethyl phosphate is that of a levelling agent during film formation, probably because it is soluble in water to the extent of about 7.3 % at 25 °C, and thus helps to promote more even film drying.

5.3.2 Plasticisation of polyvinyl acetate

Polyvinyl acetate, which has a T_g of about 25 °C, depending on molecular weight and to some extent on the stabiliser system, requires plasticisation in many applications. Small quantities of dibutyl phthalate or di-isobutyl phthalate, or of blends of a phthalate ester with tritolyl phosphate, about 3–6 % of the total weight of the solids, are sufficient to ensure adequate film formation at ambient temperatures; hence these are often added in adhesive compositions (Chapter 8, Section 4.1.2). These small additions are sufficient to ensure adequate film formation at ambient temperatures, without appreciably altering properties such as cohesion and tensile strength. Suitable plasticisers should be of relatively high polarity; e.g. 2-ethylhexyl phthalate is not a good plasticiser for polyvinyl acetate and tends to exude from films in additions of more than 5 %, but it may be blended with other plasticisers. Many non-ionic surfactants of the alkylphenol–polyoxyethylene and block copolymers of ethylene and propylene oxides also function as plasticisers for polyvinyl acetate.

In the plasticisation of polyvinyl acetate in latex form some plasticisers such as dibutyl phthalate are absorbed quickly. A simple test is to dye the plasticiser with an oil-soluble dye which is also soluble in the polymer, dispersing the dyed plasticiser in a latex, and then examine it microscopically. In a few hours no distinction is observed between any of the particles. This equilibrium is hastened by stirring and heating. Most plasticisers can be dispersed directly in a vinyl acetate latex by stirring, with optional use of heat, since there is normally a large excess of stabiliser present because of the relatively high particle size (0.5–1.0 μm). Plasticisation of a polyvinyl acetate latex is normally accompanied by a slight increase in latex viscosity [207].

The effect of plasticisation of polyvinyl acetate emulsion films have been demonstrated by the Paint Research Laboratory, Teddington, England. Electron micrographs of films of vinyl acetate latex, unplasticised and plasticised by 20 % with dibutyl phthalate, are shown (Figures 4.16 and 4.17), and also

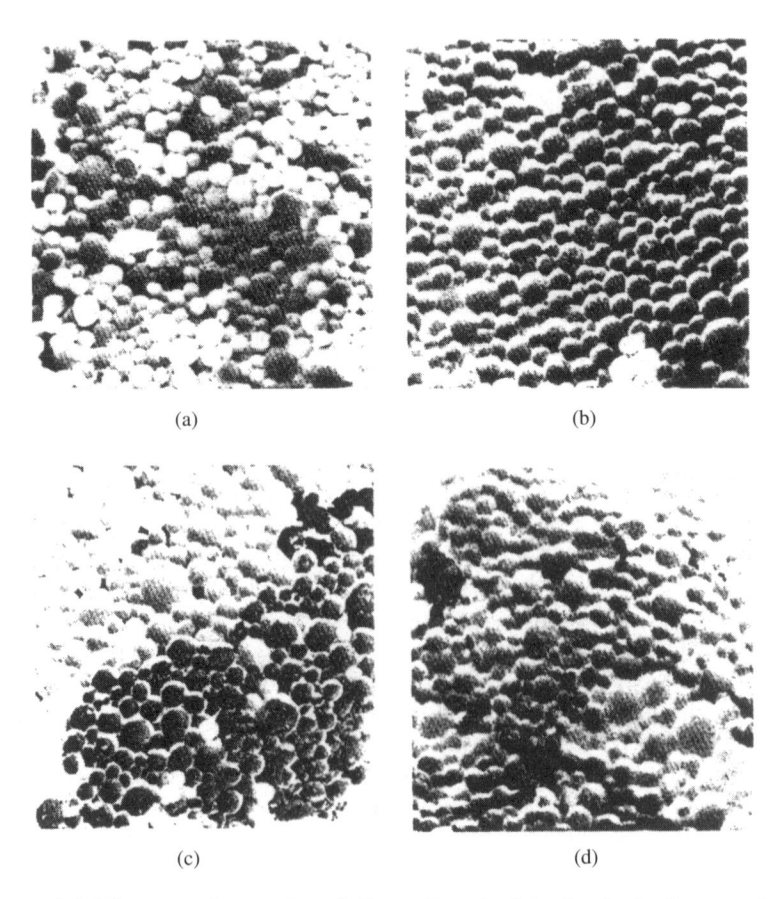

Figure 4.16 Electron micrographs of films of unplasticised polyvinyl acetate latex. (×5000 magnification and shadowing angle of 45°): (a) sample A—touch dry; (b) sample B—touch dry; (c) sample C—touch dry; (d) sample B—10 days after application. (Reproduced by permission of the Paint Research Association.)

Addition of plasticisers and coalescing solvents 233

(a) (b)

(c) (d)

(e)

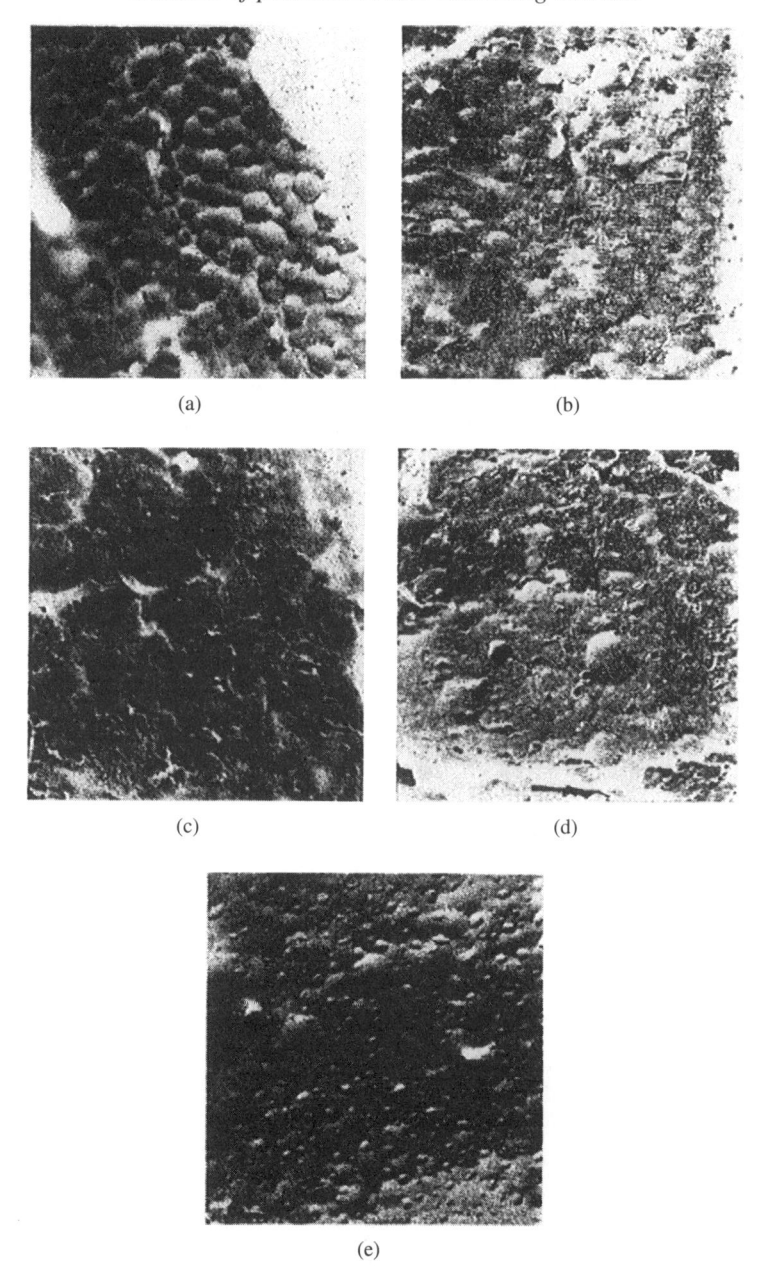

Figure 4.17 Electron micrographs of films of plasticised 29 % dibutyl phthalate) polyvinyl acetate latex (×5000 magnification and shadowing angle of 15°): (a) touch dry; (b) 1 day after application; (c) 4 days after application; (d) 8 days after application; (e) another sample, 9 days after application. (Reproduced by permission of the Paint Research Association.)

photomicrographs of changes in the latices themselves with time (Figures 4.18 and 4.19) [208].

If a polyvinyl acetate is plasticised with dibutyl phthalate, microphotography has shown that the plasticiser is absorbed into the polymer after not more than 30 minutes of stirring. This process takes much longer if tritolyl phosphate is the preferred plasticiser. Note that dibutyl phthalate and sometimes di-isobutyl phthalate are the normal preferred plasticisers for polyvinyl acetate.

With certain plasticisers such as tritolyl phosphate, dibutyl sebacate or polyesters such as the polyglycol adipates, migration of the plasticiser is very much slower. In these cases, if a film of plasticised polymer is cast, it is rather weak and sometimes opaque. After the latex has dried thoroughly, the plasticiser is very gradually absorbed into the film, which gradually takes on normal characteristics even if the minimum filming temperature of the polymer alone is above the temperature of film drying (see Figures 4.18 and 4.19) [209].

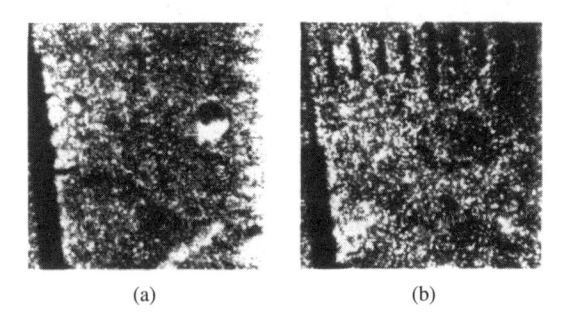

Figure 4.18 Photomicrographs of latex plasticised with dibutyl phthalate: (a) after 5 min of stirring; (b) after 30 min of stirring (×300 magnification). (Reproduced by permission of the Paint Research Association.)

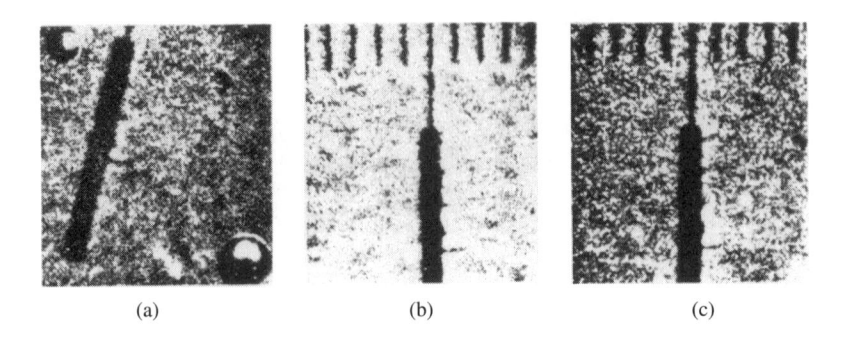

Figure 4.19 Photomicrographs of latex plasticised with tricresyl phosphate: (a) after 5 min of stirring; (b) after 30 min of stirring; (c) after overnight standing (×300 magnification). (Reproduced by permission of the Paint Research Association.)

Addition of plasticisers and coalescing solvents

Experiments performed on a 20 % postplasticised technical vinyl acetate latice also show the difference between a plasticiser with rapid action and those that remain as discrete particles in the latex and plasticise the films very slowly [208]. Plasticiser was added to the emulsion for 35 min at 15 °C, with rapid stirring. After ageing for standard periods, the turbidity of the films was measured with a photoelectric cell and a galvanometer. The results are shown in Table 4.2. The scale is arbitrary, but a sheet of plate glass gave a reading of 3–5 and a frosted glass plate 200. Under experimental conditions, an unplasticised latex did not produce a coherent film. A figure under 25 may be interpreted as a clear film, 25–50 as translucent and above 50 as cloudy (Table 4.2) (see also Section 5.4).

Table 4.2 may be interpreted as indicating that the effect of plasticisation with dibutyl phthalate is almost immediate and, that tritolyl phosphate has a gradual effect on plasticisation, a rise in temperature, as expected, increasing the rapidity of absorption. Table 4.3 shows the loss of plasticiser of films from a latex that had been plasticised to 20 % of solids weight. Films of 75 μm on aluminium were prepared. Natural weathering was under summer conditions. The water extraction test involved a continuous spray equivalent to the amount of water involved in 500 of accelerated weathering [208].

The rate loss of film fusion agents are plotted in Figures 4.20 to 4.22 [209]. They show that, although ethoxyethyl acetate has a vapour pressure many times that of dipropylene glycol, a greater part of it is retained in the polyvinyl acetate film. Reproduced by permission of the Paint Research Association.

5.3.3 Plasticisation of polyvinyl chloride

Polyvinyl chloride is unique in that, at ambient temperatures, plasticisers are virtually unabsorbed into the polymer, but tend to form pastes. Nevertheless, there is some absorption on standing, especially with lower molecular weight plasticisers such as dibutyl phthalate, with the result that gelation occurs on ageing. This is less pronounced with di-2-ethylhexyl phthalate, and also with phosphate and polyester plasticisers. In the manufacture of polyvinyl chloride sheet, temperatures of the order of 145 °C must be applied in order that the

Table 4.2 Turbidities of polyvinyl acetate latex films containing 20 % of plasticiser after ageing [208]. Reproduced by permission of the Paint Research Association

Plasticiser in mixture	Temperature of film application (°C)	Turbidity of film from mixture aged for						
		0 h	5 h	1 day	2 days	4 days	10 days	30 days	60 days
Dibutyl phthalate	15	9	—	—	9	—	6	—	9
Tritolyl phosphate	15	171	81	43	23	26	—	—	—
Tritolyl phosphate	25	98	47	20	21	15	—	—	—

Latex properties relative to applications

Table 4.3 Plasticiser contents of polyvinyl acetate films after weathering [208]

Plasticiser	Accelerated weathering for			Plasticiser content (% of original plasticiser content) after			
	250 h	500 h	1000 h	Natural weathering for 3 months	Heat ageing at 45° for		Water extraction
					250 h	1000 h	
20 % dibutyl phthalate	71	—	38	88	82	70	95
20 % tritolyl phosphate	88	93	85	95	94	94	—

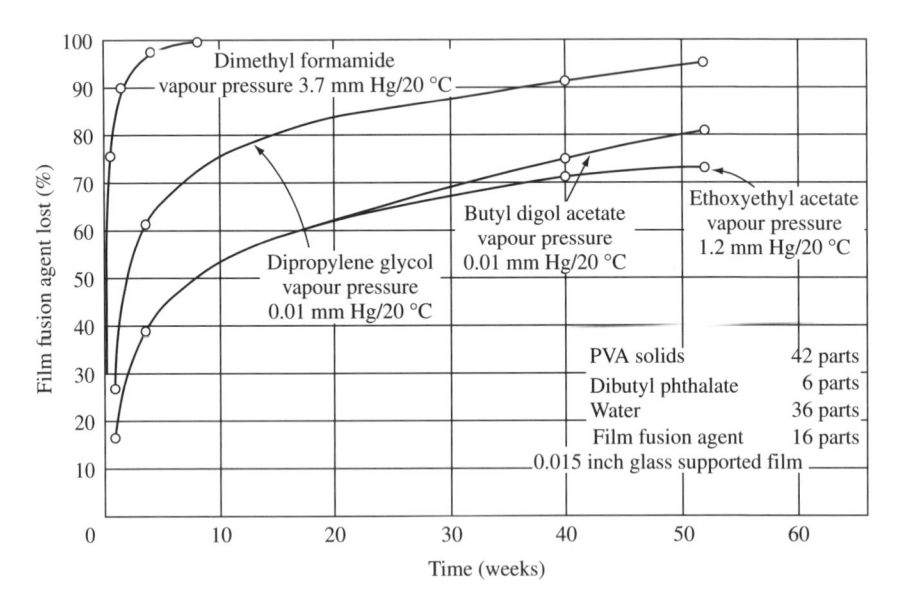

Figure 4.20 Rate loss at 23 °C of film fusion agents from plasticised PVA films; butyl digol equals dibutyl glycol. (Reproduced by permission of the Paint Research Association.)

full strength of the plasticised composition is obtained; i.e. the plasticisers have been fully absorbed. This process is irreversible, and the plasticiser in not exuded on cooling.

In the case of a polyvinyl chloride latex and some vinyl chloride copolymer latices, special considerations apply. With most other polymers, the added plasticisers will gradually coalesce with the individual polymer particles by a process of diffusion, principally through the water phase and to a very limited

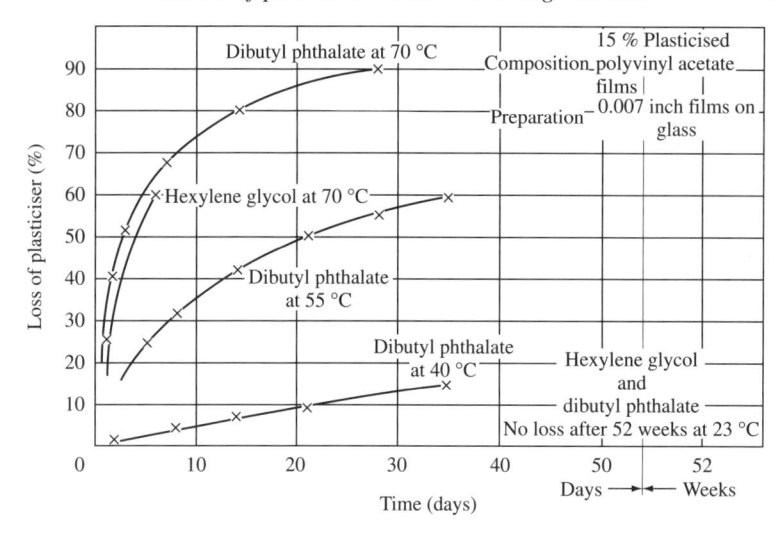

Figure 4.21 Plasticiser loss of polyvinyl acetate films with increasing time. (Reproduced by permission of the Paint Research Association.)

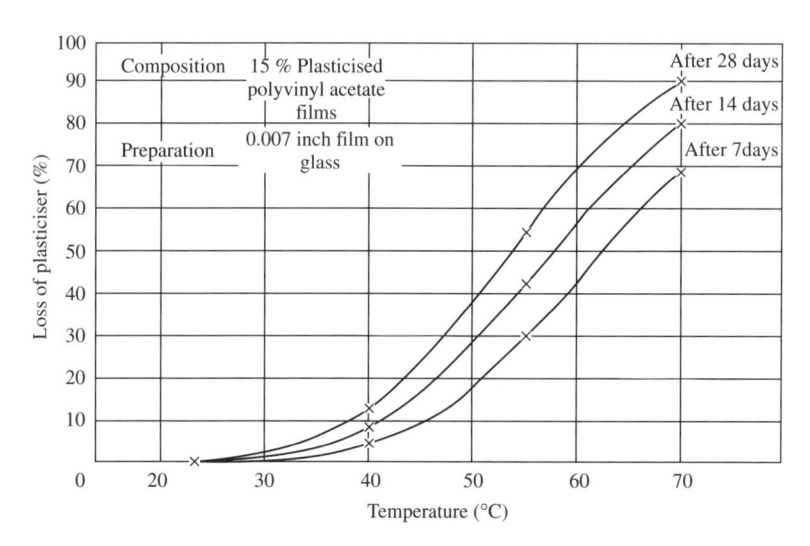

Figure 4.22 Dibutyl phthalate loss from polyvinyl acetate film at different temperatures [209].

extent by particle–particle contact. It has been suggested that absorption of plasticiser into a polymer particle may under some conditions produce an intermediate state in which only the exterior layers of the polymer particles are swollen by the plasticiser.

Plasticised films of polyvinyl chloride do not in general give satisfactory films when dried, even at temperatures approaching 100 °C. Experiments to test relative plasticiser absorption in a vinyl chloride–vinylidene chloride (25 : 75) copolymer have shown a wide variation in absorption times (Table 4.4). The tests were performed by partial centrifugation. The addition of a solvent such as toluene was ineffective in reducing the absorption time of butyl stearate, but toluene was very effective, especially when present at over 10 %, in increasing the absorption rate of 2-biphenylyl diphenyl phosphate [210].

However, if an emulsion is preplasticised before polymerisation, the plasticiser is absorbed to some extent into the polymer particles during the polymerisation and a lower temperature, e.g. 50–90 °C, is required to obtain film strength, although even in this case a final treatment of 130 °C is advised.

5.4 Transient plasticisers (coalescing solvents)

It is often required to form a film of a polymer for which the minimum filming temperature is above ambient temperature, but for which normal permanent plasticisation is undesirable. Equally it may be desirous that a film, particularly of a polyvinyl acetate copolymer, should attain its full integral strength soon after drying. For this purpose transient plasticisers (coalescing solvents) are added.

The principle of their addition is identical with that of plasticisation, a modification being that many of the transient plasticisers have appreciable water solubility, even amounting to complete miscibility. The solubility of water in the transient plasticiser may further assist film formation. The coalescing solvent, must, of course, be a true solvent for the polymer or, to use the normal terminology, it must be compatible with it.

Many of the liquids that come in this category contain either hydroxyl groups in which the volatility is restricted by hydrogen bonding, ether groups

Table 4.4 Absorption of 15 % of plasticiser in vinyl chloride copolymer latex [210]. (Reproduced by permission of American Chemical Society)

Plasticiser or solvent	Time for absorption (day)
Toluene	0
Triethylene glycol 2-ethylbutyrate	0
Triethylene glycol dipelargonate[a]	0
2-Ethylhexyl adipate	4
2-Biphenylyl diphenyl phosphate	14
Dioctyl phthalate[b]	23
Butyl stearate	48

[a]Synonymous with caprylic or octoic acid.
[b]Probably 2-ethylhexyl phthalate.

or carboxylic ester groups, all of which are relatively less volatile than hydrocarbons, relative to their boiling points. Solvents such as the most well-known aromatics, e.g. toluene and xylene, are sometimes added to latices, but as their volatility is equal to that of water, or somewhat greater, they cannot be included in the category of transient plasticisers.

All of these medium boiling materials enter the polymer particles very rapidly, and do not require ageing. Apart from the property of improving film formation, they sometimes improve freeze–thaw stability (Section 3). In some cases esters used as coalescing solvents, e.g. ethylene glycol monoethyl ether acetate $C_2H_5OC_2H_4OOCCH_3$, are objectionable on the grounds that they are liable to hydrolysis, generating free acetic acid if the pH is on the acid side or an acetate salt if the pH is > 8. In the former case, the liberation of acetic or another organic acid may tend to destabilise the whole composition. The formation of salts may also have the same effect, and the effect is even more objectionable when the latex is part of a pigmented formulation.

The addition of a water-soluble solvent to a latex requires care, as in many cases direct addition will cause precipitation of the polymer particles. This is akin to adding ethyl alcohol or acetone to a latex, where the affinity of the solvent for water can be considered as similar to a dehydrating action. Equally, direct addition will create temporary areas of high concentration, where the particles will swell or dissolve partially in the solvent–aqueous phase blend, and more thorough mixing will cause precipitation since the swollen particles no longer have a protective envelope. The usual recommended method for the addition of a water-soluble coalescing solvent is that it should be diluted with at least its own weight of water, and added to the latex in a thin stream with good stirring or agitation. This precaution is not normally necessary where the solubility of the solvent is restricted.

An alternative method by which coalescing solvents may be introduced applies in the case of pigmented latices. In this case the coalescing solvent may be introduced into the pigment paste rather than the latex, so avoiding the necessity of diluting with extra water. It should be noted that the uptake of solvents, e.g. toluene, hexane and cyclohexane, into latex particles may vary by three orders of magnitude, dependent on the water solubility of the solvent and the viscosity of the polymer [211].

The distribution of a water-soluble plasticiser, whether transient or otherwise, is probably dependent on both the nature of the solvent and the polymer, with an appropriate partition coefficient which may not, however, follow the simple Nernst law of partition for low molecular weight compounds. As water evaporates, the relative concentration of the transient plasticiser becomes higher, and where the solubility of water in this compound is appreciable, water may ultimately tend to act as a cosolvent, improving flow properties and hence film integration. This is of great interest in the formulation of gloss paints (Chapter 10 in volume 2). Coalescing solvents also tend to improve pigment binding efficiency (Section 6) [212].

240   *Latex properties relative to applications*

Table 4.5 gives a list of some common coalescing solvents. Note that derivatives of propylene glycol are now often used instead of ethylene glycol derivatives, principally because of some doubts on the toxicity of ethylene glycol derivatives (see later).

Certain additives such as hexylene glycol and 1 : 3 propylene glycol, already mentioned in connection with the improvement of resistance to cycles of freezing and thawing, are rather on the boundary of being true transient plasticisers, with special reference to vinyl acetate polymers and copolymers, since they are cosolvents rather than true solvents. Glycols tend to concentrate as water evaporates without breaking the emulsion nature of the drying film. Since they evaporate slowly they give good 'wet edge' properties, which are desirable in applications such as emulsion paints (Chapter 10 in volume 2). These glycols, however, may be objectionable on porous substrates since they tend to become absorbed rapidly by 'wicking' (Section 4.2). This would also tend to pull some polymer into the substrate. This results in a matt, irregular film, and if the latex is pigmented, it will be poorly bound.

Some tests with Texanol (see Table 4.5) have been described. It functions in a similar manner to a plasticiser [213]. In addition to the above there have been investigations into the inclusion of alkyl acetates, based on relatively low water miscibility and 2-(1-cyclohexyl)cyclohexanone as coalescing solvents [214].

If a copolymer which should impart the same degree of flexibility is tested, the time for coalescence is slow, although it will ultimately occur. It may be deduced, therefore, that there is a ready migration of some liquid plasticiser between particles. If a coalescing solvent is used instead of a quick-acting plasticiser, film formation is equally rapid. Conditions during drying are somewhat different, since most coalescing solvents are either appreciably soluble in water or more usually completely miscible. In the latter case the solubility of the polymer particles in the medium tends to increase during drying and conditions for coalescence are improved, and in the latter stages of drying approach that of a polymer evaporating from solvent. Hence coalescing solvents are often included in formulations with vinyl acetate copolymers in order that film strength may be attained rapidly.

New ester-type coalescing agents of general formula RCO(CH2)n(COOR") have been disclosed where $R' = 1\text{--}4C$ alkyl, $R'' = 2\text{--}6$ alkyl and n is >2. A typical example is butyl 5-oxocaproate [215].

Butyl benzoate, b.p. 250 °C, is hydrolytically stable and has no appreciable effect on latex viscosity or pH. It is recommended as a transient plasticiser for latex polymers with $T_g > 0$ °C, e.g. copolymers of acrylic esters with styrene. Good hardness and block resistance are rapidly developed [216]. Isobenzyl benzoate has been evaluated in a range of latex paints and found to be comparable with standard ether alcohol agents [217].

1-Phenoxy-2-propanol $CH_3CHOHCH_2OC_6H_5$ is an effective coalescing agent for acrylic latices over a wide range of temperatures. It has low water

Table 4.5 Typical coalescing solvents

Chemical name	Formula	Molecular weight	Specific gravity 20:20	Boiling point at 760 mm Hg (°C)	Solubility	
					In water (%)	Water in (%)
1:2 Propylene glycol	$CH_3CH(OH)CH_2OH$	76	1.0381	187.3	Infinite	Infinite
1:3 Butanediol	$CH_3CH(OH)CH_2CH_2OH$	90	1.006	207.5	Infinite	Infinite
1:4 Butanediol	$HOCH_2CH_2CH_2CH_2OH$	90	1.0154	228	Infinite	Infinite
2-Methylpentane 2,4-diol	$CH_3CHOHCH_2C(CH_3)OHCH_3$	118	0.9234	197.1	Infinite	Infinite
Methoxyethylene glycol	$CH_3OC_2H_4OH$	76	0.9663	124.5	Infinite	Infinite
Ethoxyethylene glycol	$C_2H_5OC_2H_4OH$	90	0.9311	135.6	Infinite	Infinite
n-Butoxyethylene glycol	$n-C_4H_9OC_2H_4OH$	118	0.9019	171.2	Infinite	Infinite
Diethylene glycol methyl ether	$CH_3OC_2H_4OC_2H_4OH$	120	1.0211	194	Infinite	Infinite
Diethylene glycol ethyl ether	$C_2H_5OC_2H_4OC_2H_4OH$	134	1.0273	195	Infinite	Infinite
Propylene glycol methyl ether	$CH_3OCH_2CH_2CH_2OH$	90	0.963	124.6	Infinite	Infinite
Propylene glycol ethyl ether	$C_2H_5OCH_2CH_2CH_2OH$	104	0.902	133	Infinite	Infinite
Propylene glycol t-butyl ether	$(CH_3)_3COCH_2CH_2CH_3)H$	132	0.87	153	17.5	Approx. 6
Dipropylene glycol methyl ether	$CH_3[OCH_2CH_2CH_2]_2OH$	148	1.020	188	Infinite	Infinite
Dipropylene glycol n-propyl ether	$C_3H_7[OCH_2CH_2CH_2]_2OH$	176	0.922	212	19	20.5
3-Hydroxy-2,2,4-trimethyl-O-pentyl-isobutyrate (Texanol®)		216	0.96	244–7	Insoluble	0.9

solubility with minimised loss into a porous substrate and does not retard evaporation [218].

In an unusual process, trimethylolpropane triacrylate is used as a coalescent for radiation-curable coating systems. In this case, while in its monomeric state it functions to give film coalescence, volatility is not necessary, as the film is finally cured after film formation [219].

It should be noted that aqueous amines sometimes have the effect of coalescing solvents, especially in polymers containing polymerisable acids. A coalescing solvent may fail at high humidities as it retards film formation [220]. For the effect of coalescing solvents on vinyl acetate copolymers films in paints, see Chapter 9, Section 3.4 in volume 2.

In general, transient plasticisers possess no fire hazard because of their relatively high boiling points and tendency to water solubility. However, prolonged repeated inhalation of the glycol ether solvents is undesirable. Some undesirable effects of ethylene glycol monoethyl ether have been reported in the United States and several manufacturers have discontinued production, generally in favour of the derivatives of propylene glycol, which show little signs of toxicity. In general, most paper and paper board products containing them may be used for food coatings [221].

The loss of coalescing solvents from latex coatings is controlled by the surface resistance to evaporation. Plots of residual volatiles versus time are given and comparisons made with glycols and coalescing solvents. Solvent loss from the void precursers in pittmentised (microvoid) coatings (see Section 7) is given. Clearly marked breaks indicate the substantial removal of water [221A].

Transient plasticisers when included in a formulation migitated to some extent the advantage of having an aqueous medium only, although they are rarely present at more than 5 % of polymer weight, and often at 2–3 % only. These additives are not a part of the true non-volatile solids content, and to that extent represent an additional item in the cost of the final product.

For general reviews see references [222] and [223].

6 PIGMENTATION

6.1 Introduction

The addition of pigments and extenders to latices represents one of the principal methods of compounding. A pigment may be defined as a solid product which imparts both opacity and colour, treating both black and white as colours. An extender is an inert mineral, or exceptionally an organic compound with the correct refractive and dispersing properties for visible light, and is used to extend the principal pigment. An extender has weak pacifying powers compared with the principal pigment. The inclusion of extenders is partly on cost considerations, since extenders such as china clay and chalk are cheap materials compared with titanium dioxide TiO_2, the principal pigment in use.

Pigmentation

A principal application of pigmented latices is in emulsion paints, which will be a topic to be discussed in Chapters 9 and 10 in volume 2. Many other processes require pigment and extender additions to latices. These include paper coatings, leather coatings, certain textile processes and some compounded adhesives. In this chapter the problems of pigmentation will be considered in general terms.

The major problem of obtaining an even and homogeneous dispersion of the pigments (a term often used comprehensively) to include extenders is due to the fact that the polymer is not in a continuous phase, as it would be in an oil paint. It is first essential to obtain a good pigment dispersion in an aqueous medium that remains stable when it is compounded with the latex and any other desired additives such as microbiocides. Secondly, it is also important that the finished pigmented product, which will be referred to as a paint for simplicity, fulfills all desired requirements for the application. These may depend on the method chosen. The simplest is by brushing, but other methods include dipping, spraying, hand rollers, machine rollers of various types and possibly electrodeposition.

The nature of application of pigmented latices governs to a considerable extent the properties that are built into it. Thus, products intended for brushing must not drag under the brush. Brush drag is related to dilatancy (Section 2.5.3) since there is a considerable shear force, of the order of 5000 s^{-1}, which must be applied by an ordinary paint brush. The suitability of a paint for brushing is best ascertained by examining the viscosity behaviour of an instrument such as the Ferranti–Shirley viscometer, which measures stress against strain over a wide range of speeds. Any sudden or irregular increase of stress with increasing strain (speed or rotation) indicates instability at high shear. Another suitable instrument is the Weissenberg rheogoniometer, which also measures the vertical component of strain. For further suitable viscometers see Section 2.5.

Latices of very fine particle size, of the order of 0.2 μm particle diameter or less, tend to give poor brushing properties, whether used alone or pigmented, although there are some exceptions. On the other hand, most latices of particle size 0.5–1.5 μm tend to 'slip' on brushing, whether or not they are pigmented.

The presence of water-soluble polymers tends to improve brushability; a natural colloid, gum arabic (acacia), is one of the most efficient, especially when it is a component part during the polymerisation. Unfortunately, natural products of this type are subject to a wide variation in properties depending on origin and on the season of the year in which the trees have been tapped.

The semi-synthetic types, in particular hydroxyethyl cellulose and carboxymethyl cellulose, are often included as pigment paste additives, improving dispersion and ultimate flow. The former is often a component in the formulation of latices intended for emulsion paints. Methyl cellulose is also sometimes used as an additive for pigment pastes.

Where application of a latex-based composition is by spraying, some dilution is necessary for optimum viscosity. The use of a coalescing solvent is particularly desirable to prevent such spraying defects as 'cobwebbing', or of premature precipitation. Exceptional mechanical stability is required in latices that are applied in the pigmented state by an airless spray, where the shear forces are greater than usual.

Whilst roller applications are unlikely to produce major problems, machine coating, which is useful in paper coating and in some textile processes, requires a very careful balance of properties, both rheological and in the general balance of the composition and its constituents. Absorption into the substrate must be considered, and drying properties carefully balanced in the light of its application.

Occasionally, a pigmented latex is used for impregnation purposes where a fine particle-type latex may be essential. At the other extreme, a composition may be required in a semi-solid paste form, e.g. as a crack filler.

6.1.1 The dried composition

Pigmented latices are required for various types of coating, and they must have the required properties for the application under consideration. Thus, a pigmented film should have the correct opacity and thickness. It may be 'egg-shell' (semi-gloss) or in some cases it may approach a complete gloss.

The coatings must be able to withstand extreme operating conditions, such as extremes of temperature, differential expansion and contraction when on an exterior substrate; they may have to withstand corrosive atmospheres or chemical attack, and textile finishes should withstand dry-cleaning solvents. A special feature of pigmented systems is that composition must either not 'chalk' or have uncontrollable chalking. Chalking is a breakdown of the surface film, causing the pigment to powder, which is induced by light under the catalytic influence of the pigment itself. Anatase titanium dioxide, TiO_2, is one of the principal pigments inducing chalking. Limited chalking is sometimes desirable, as it enables a surface to be self-cleaning. More often it is objectionable, as it reduces the life of a pigmented coating. Rutile titanium dioxide is now almost entirely the only titanium dioxide included in coating compositions.

Various modifications are described in Section 6.3. There must be no chemical reaction between the latex or any of its components. Chalk, a typical extender, is sensitive to low pH values, and if zinc oxide is included the pH should be as near neutral as possible. Certain pigment dyestuffs are reactive, either physically or chemically, and irregularities with Hansa Yellow may be due to the fact that it is soluble in certain plasticisers and coalescing solvents from which it crystallises, losing much of its pigmentary power.

6.2 Pigmentary power

The optical properties of white pigments depend on the reflectivity, tinting strength and hiding power. It is considered that these can be reduced to two parameters, the scattering coefficients S and the absorption coefficient K. S is the proportion of incident light that is scattered by the pigment and reflected. It is defined by $S = (dR_0/dX)_{x \to 0}$, where R_0 is the reflectance of a pigmented film over a black substrate and X is the film thickness. The units are the reciprocal of length. S depends on the physical properties of the chemical compound used as a pigment, on the particle distribution, the volume concentration and on the state of distribution. K and S are related by the equation

$$K/S = f(R_\infty = (1 - R_\infty)^2 2R_\infty$$

where R_∞, the reflectivity, is that reflectance which, on increasing film thickness over a substrate, does not change further, a theory that has been developed quantitatively [224].

The smaller the particle size of a pigment, the greater the number of interfaces that interfere with the linear transmission of light, and therefore the greater the hiding power. This ceases to be true when the particles become less than the wavelength of the shortest wavelength of visible light, about 4000 Å. A film containing particles between 0.4 and 1 µm diameter may therefore appear yellow-white because of its scattering of the longer wavelengths of light, whilst one of 0.2 µm has a bluish-white tone. In very thin films blue light is reflected off the surface more efficiently than that of the longer yellow and red wavelengths; hence in very thin films bluish-white pigments seem to have greater hiding power.

Whilst hiding power varies inversely with particle size, the effects are highly complicated by mutual interferences and the distribution in a binder. Agglomeration, or the clustering of particles, results in loss of tintorial strength. The function of grinding the pigment is to break down clusters, not to grind the ultimate particles into something smaller. A fineness of grind gauge is a suitable tool for estimating the efficiency of dispersion of a pigment.

Some calculations have been made of the efficiency of hiding power in a pigment dispersion which is rather similar to that of close-packed spheres in a latex [225]. At very high ratios of pigment to the overall composition (pigment–volume–concentration), the efficiency of a pigment is reduced as the particles become progressively closer together. If a coarse extender is added, it has only a diluting effect. A fine particle size extender acts as if it partly replaces the vehicles in which the principal pigment is contained, and often results in an appreciable increase in hiding power.

As 'oil absorption' of a pigment increases, determined by a simple rub-out test, the efficiency of hiding power increases. Hiding power may be measured quantitatively in square metres per gram.

The above has been quoted in order to give an elementary idea of the principles of pigmentation. There are a number of standard works on the subject [226–229].

6.3 Some common pigments and extenders

6.3.1 Titanium dioxide

The principal white pigment currently in use is titanium dioxide, TiO_2, prepared technically from the mineral ilmenite. Other white pigments such as zinc oxide, which is rather reactive chemically, and lithopone are found occasionally in some formulations for emulsion paints, but they are rare. With its combination of high opacity and chemical inertness, titanium dioxide is generally the sole white pigment, apart from extenders, in almost every application. Two forms are available with different crystal structures. They are anatase, with a somewhat open structure, and rutile, which is more compact. Figure 4.23 illustrates the crystalline structure of the two pigments. Table 4.6 shows the difference in properties of the two pigments. Rutile, which is a later introduction than anatase, is the more stable form. The older sulfate process for the manufacture of titanium dioxide has to some extent given way to the more modern chloride process. Table 4.6 gives the properties of anatase and rutile.

Rutile, the grade used in water-based systems, invariably contains a number of chemical additives, some of which are on the surface of the particles. These improve its performance, e.g. to reduce the amount of dispersing solution required to obtain a workable viscosity of a pigment paste and to ease dispersion by wetting agents. Some of these additives, such as zinc oxide, are always present in small quantities because of the nature of the manufacturing process. The other main additives are alumina Al_2O_3, silica SiO_2 and sometimes cerium or zirconium compounds. The overall quantity of additives is considerable, and in most cases the effective quantity of titanium dioxide is of the order of 92–95 %. Suitable grades that are most easily wetted by water-based systems are recommended by the manufacturers.

Rutile is normally of a very fine particle size. Figure 4.24 shows the particles under high magnification and most are usually between 0.4 and 1nm in average diameter, although the term 'diameter' is rather a misnomer, as will be seen from the plate. Rutile, because of its particle size, has a slight yellowish cast which may be corrected by a trace of blue pigment. The specific surface is 6–7 m^2 g^{-1} for the purest grades, but rises to 30 m^2 g^{-1} in some cases. Further information on the various grades of rutile suitable for paint manufacture will be included in Chapter 9 in volume 2.

A survey of the production of titanium dioxide and its properties is available [230].

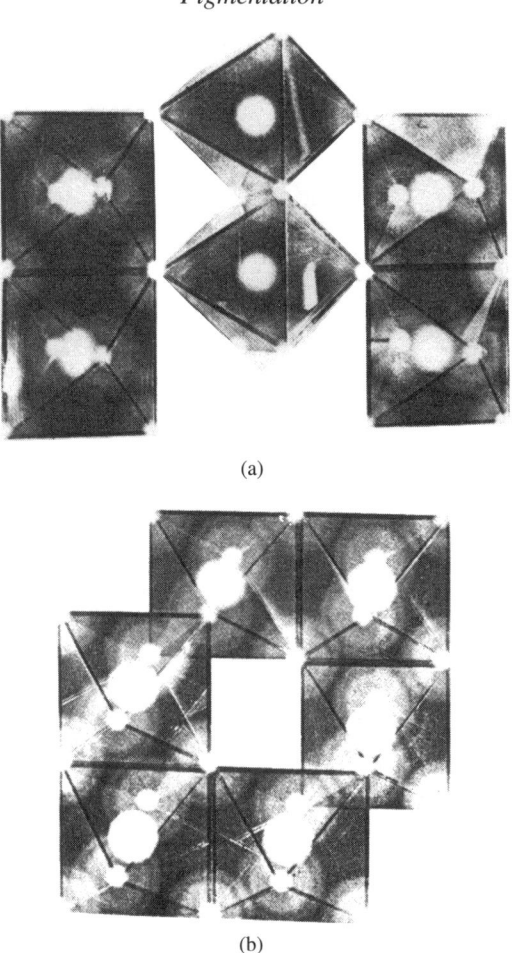

Figure 4.23 (a) Crystalline structure of rutile and (b) crystalline structure of anatase. (Titangesellschaft GmbH, Leverkusen, Germany.)

Table 4.6 Properties of rutile and anatase, titanium dioxide

Property	Rutile	Anatase
Specific gravity	4.2	3.9
Refractive index	2.71	2.53
Dielectric constant	114	31
Moh's hardness	6.7	5.5–6
Melting point (°C)	1825	Converts to rutile at high temperature

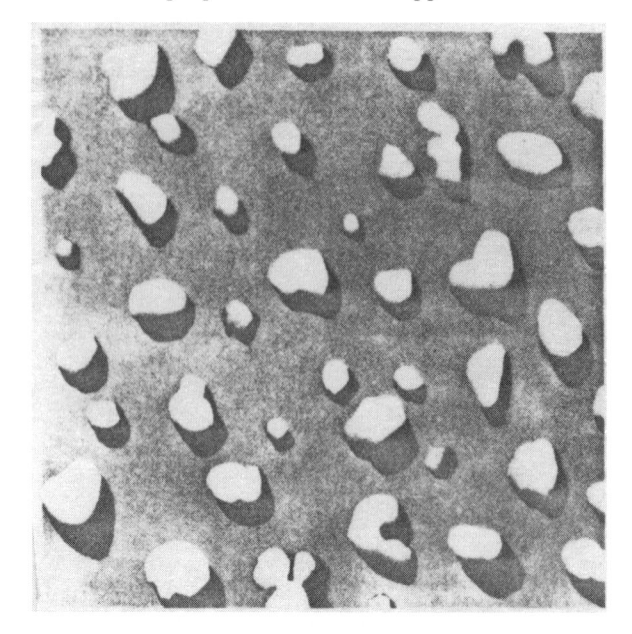

Figure 4.24 Electron micrograph of titanium dioxide pigment particles (enlarged 30 000 times). (Titangesellschaft GmbH, Leverkusen, Germany.)

6.3.2 Other white pigments

Zinc oxide and lithopone, which have about equal tinting strengths, are occasionally encountered, mainly in older formulations. Antimony oxide is included in some formulations intended for fire resistance.

6.3.3 Extenders

Of the numerous extenders, the majority are of interest in emulsion paints, but several are the major constituents of paper coatings. The following is a list of some principal classes of extenders:

Calcium carbonate (chalk), whether synthetic or natural
Clays, especially china clays
Various natural and semi-synthetic silicates
Silica
Talcs
Barytes

Calcium carbonate is available in a ground form, of diameter 2–20 µm. The best natural products are derived from calcite, which is finely ground and undergoes a suitable surface treatment, e.g. with 1 % of a stearate. Most grades are suitable as extenders for emulsion paints and, contrary to expectation, they give no trouble due to 'gassing' over a wide pH range.

China clay, an extender widely available for general purposes, and especially as a filler for paper coating, has the formula Al_2O_3, $2SiO_2$, $2H_2O$ and is believed to have originated by the decomposition of feldspar in granite rock. It is available in various forms from fine powders from 0.2 to 10 μm in size to pellets and lumps. The rather interesting rheological properties are explained in Section 6.4.1. A typical grade of china clay has a specific gravity of 2.6, refractive index of 1.60, surface area of 13.0 $m^2 g^{-1}$, a moisture content of 1–2 % and a loss of weight on ignition of 13.1 %.

Most grades of china clay have some water absorption and are therefore not efficient in giving water-repellent films. The largest size has the lowest water absorption. Modified clays, e.g. the Polestar® series of EEC International Europe, have low water absorption and assist film brightness. Lightly calcined china clay has improved hiding properties relative to the regular material, this being partially due to entrapped air during the process. These calcined china clays are also available coated with an aminosilane coupling agent (Polarise® of EEC) and treated with a vinyl functional silane for use with free radical crosslinking agents.

With the exception of barytes $BaSO_4$ (barium sulfate, natural or synthetic) and silica SiO_2 itself, all other extenders are mineral silicates. A number of synthetic aluminium silicates finds application as extenders. They are easily wetted out by water, have some buffering action and may replace titanium dioxide to 15–30 % in a paint. They are more stable to phosphate ions than extenders which contain calcium ions, and their chemical formula is substantially that of calcined china clay.

Several other silicates resemble china clay. Thus colloidal forms of attapulgite, $3MgO$, $1.5Al_2O_3$, $8SiO_2$, $9H_2O$ (Attagel series, Mineral and Phillip Corp, New Jersey, USA), are available. The particle size of this product is about 13 μm and the surface area is 210 $m^2 g^{-1}$. Whilst forming thixotropic gels in water, these silicates are not true water-soluble or water-swellable types, and are best considered as pigment extenders.

Mica is a potassium aluminium silicate, occasionally used as an extender when ground. Its use detracts from weathering in external applications. Further information on the application of extenders will be included in Chapter 9 (raw materials for the paint industry) and chapter 13 (the paper industry) in Volume 2.

6.4 Pigmentation of latices

Unlike paints based on the solution of a resin in a solvent, which depend on a medium wetting, the pigments as part of the dispersing process, forming a homogeneous composition with a latex, depend on the pigments being ground independently with suitable dispersing agents, followed by subsequent mixing with the latex and minor additives.

The term 'pigment' has been used here in a general sense, and to some extent the method of dispersion will depend on the balance of true pigments and extenders. The principal extenders, such as china clay, talc and barytes, are often blended with the pigment, especially for emulsion paints. In paper coatings, china clay is the main solid component, whilst in leather coatings, a relatively minor use, iron oxides are the main pigment constituent.

Technical pigments are available in the form of agglomerates requiring grinding in order to separate them into ultimate particles which may be as low as 0.2 μm in diameter or corresponding size depending on the structure. The ultimate particle size is the lowest beyond which they cannot be readily comminuted. Extenders are usually of the order of 1 μm and many common extenders are greater than 5 μm. Many pigments such as iron oxide, orange chrome and barytes are highly irregular in shape.

As a general rule the pigment blend is ground and wetted out in water in an independent operation. A conventional wetting agent, such as has been described in Chapter 2, or specific dispersing agents may be used. The latter have the property of being good dispersants for the pigments without having specific surface-active properties. They include sodium hexametaphosphate, the sodium salts of sulfonated condensate of formaldehyde and naphthalene. A number of polyelectrolytes, such as the sodium or ammonium salts of polyacrylic acid, polymethacrylic acid and salts of a 1 : 1 copolymer of di-isobutylene and maleic acid, obtained by copolymerisation with maleic anhydride, are used as pigment dispersants. A small quantity of a thickener such as carboxymethyl cellulose may be added optionally. Other additives such as transient plasticisers, anti-foams and microbiocides may be added at this stage.

The ratios of pigment to water phase may vary, but would normally be of the order 55 pigment : 45 water to 50 : 50. This may depend to some extent on the water demand of the pigment. Detailed formulations will not be included here as they may vary with the final application.

Conventional paint grinding machinery may be used to prepare pigment dispersions. A simple but messy type is the ball mill, which can be filled and allowed to grind for 16–24 hours if required. Modern developments, now available in Great Britain, include the 'Torrance Cavitation Disperser' and the Pilamec 'Vibracon' (Fig. 4.25) which is the nearest to a 'continuous' ball mill that has yet been devised. Many other types are in use, the main disadvantage, as in the triple roll mill, is the loss of water by evaporation, making exact calculations difficult.

In some cases it is possible to use the latex itself as all or part of the grinding medium. This is particularly true of latices from vinyl acetate and its copolymers. These are of large particle size of the order of 0.5–1.0 μm, and therefore include a large amount of 'surplus' emulsifier, which may be used

Pigmentation 251

Figure 4.25 The Pilamec 'Vibracon'.

to wet out the pigment blend. This is exceptional since the heavy grinding required for dispersing a pigment may be sufficient to cause a latex to coagulate. Some technical latices including a colloid, especially a polyelectrolyte, are suitable for the direct addition of pigment, and were popular in the early years of polyvinyl acetate emulsion paints because of the great ease of manufacture [231].

Coloured mineral pigments such as the red oxides (iron oxides) used in leather finishing are ground independently, often in alkaline casein, and are added to a latex, usually an acrylic type, as required (Chapter 14 in Volume 3). Organic lake pigments are often available pre-ground as concentrates, and are added to the base pigment paste of titanium dioxide, with the exception of some paints based on deep shades. The only major use for these lake pigments in latex systems is for emulsion paints.

Non-ionic surfactants are, in general, not the most satisfactory sole dispersing agents for the majority of pigments such as titanium dioxide. The use of cationic surfactants is very rare and, though possible, is not recommended since the natural charge on most pigments is negative.

6.4.1 Treatment of china clay

There are some negative charges on the faces of the lamellae of china clay due to irregularities in the silica surfaces. This leads to a natural slightly acid pH with some positive charges on the crystal edges, leading to some natural flocculation, the particles as a whole having a slightly negative charge. At a pH of 10.5, the positive charges are neutralized by adsorption of anions. This is particularly noticeable with polymeric anions, both those of Calgon®

(sodium hexametaphosphate) and organic polyanions, which greatly aid dispersion since both edges and faces become negatively charged. Polyanions tend to form a cage-like structure with the finest particle sizes of china clay. The hydroxyl ion is also strongly absorbed and tends to keep the clay on dispersion.

Studies have been made of adsorption by china clay of a number of surface-active agents and the behaviour with synthetic latices. Sodium polyacrylate effectively prevents adsorption of anionic emulsifiers on to china clay, which probably accounts for the stabilising effect of the polyelectrolyte on acrylic latices to which china clay has been added. The deflocculating action is very efficient. With a vinyl acetate latex of unspecified formulation the best deflocculant was a mixture of sodium polyacrylate and tetra-sodium pyrophosphate ($Na_4P_2O_7$), another useful dispersing agent. Thus a mixture of a polyacrylate salt and a polyphosphate seems the best overall deflocculant in systems where china clay is included [232].

6.5 General properties of pigmented systems

A fully pigmented polymer latex is a highly complex system, but in fundamental behaviour it depends on the medium, the polymer present and the pigments and extenders. There is little doubt that some redistribution of the stabilisers takes place between the various components, a matter that has not been appreciated sufficiently. Absorptions for a pigment such as titanium dioxide will be quite different from that of an extender such as china clay, which contains a number of highly hydrophilic groups and different charges on the faces and edges of each crystallite.

Each particle, whether pigment or polymer, remains distinct until the system is dried. On drying, there is a fundamental change in the relationship between the pigment and the binder. The binder attempts to form a continuous film, as explained previously in Section 4.1. The polymer tends to coat the pigment particles and may become the continuous phase. For this it is necessary that the composition should be reasonably homogeneous; i.e. if there are multicomponents they should all be compatible. It is also necessary that the critical pigment volume concentration, as explained in the next paragraph, is not exceeded.

6.5.1 Critical pigment volume concentration

The pigment volume concentration (PVC) is the ratio of pigment to the total composition by volume. It should be obvious that this volume ratio is more critical than a weight ratio in film formation. As long as the film of polymer is substantially a continuous phase, there is relatively little change in the properties of the film from the composition. There is a point, however, at which there is an abrupt change, and this corresponds to the pigment volume concentration at which the binder just ceases to become a continuous film and functions as a type of 'spot weld' between the particles of pigment. At this stage the film becomes more porous, tensile strength and scrub resistance decrease sharply and gloss or reflectance will undergo a sharp decrease. This point is known as the critical pigment volume concentration (CPVC).

Efficiency of pigment binding depends on a number of factors. Firstly, there is the polymer itself. In general acrylic polymers have greater pigment binding capacity than polymers based on vinyl acetate. This is possibly a function of the higher elasticity of acrylic polymers as a class. Particle size is also very critical, and large particle size (circa 1 μm) latices have, in general, a poorer pigment binding capacity than small particles (circa 0.2 μm), which are closely similar in formulation [233]. The efficiency of binding can in part be measured by a scrub test, but principally by the level of the CPVC. The higher the CPVC, the more efficient is the pigment binding.

For guidance in formulation, it is desirable to find the point at which the CPVC occurs. This may be determined graphically by a number of methods, including scrub tests against gradually rising PVC levels. Relative elongation and tensile strength may also be used graphically. Alternatively, a theoretical calculation may be made, based on the water absorption test. This is the amount of water plus a standard dispersing agent used to wet out a pigment to a level that the pigment as dispersed ceases to be dilatant [234].

There is obviously a theoretical basis to this since the minimum amount of liquid required to wet the pigment completely and to fill the voids between the particles must be related volumetrically to the amount of polymer latex required for the same function. Some tests using both elongation and tensile tests on an Instron tester have been described [235]. In this test, films 280–200 μm thick were prepared on rigid polythene and were aged at 23 °C and 50 % relative humidity (RH) for seven days before testing.

Several pigments were tested with three latices. A is a plasticised polyvinyl acetate with hydroxyethyl cellulose as the protective colloid, B is as A, but with a polyvinyl alcohol stabiliser, and C is a copolymer stabilised with hydroxyethyl cellulose. The particle size of latex A and latex B was about 1 μm. The MFT of latex A was −2 °C and latex B −0.5 °C. Results in which the two methods were applied show that they agree with each other in almost all cases, as shown in Table 4.7. It is noteworthy that poor grinding causes the apparent CPVC to rise, presumably as agglomerates can be more readily coated with a smaller quantity of polymer. Of course, this would lead to other defects in a final film.

In a different series of tests, details of formation of monolayers of dispersant have been studied to find the optimum surfactant concentration on pigment/latex systems. The surfactant concentration for monolayer formation seems to be half of the critical micelle concentration, apparently irrespective of the nature of the pigment. Adsorption isotherms of sodium oleate on titanium dioxide are included [236].

Amongst other methods that have been used to determine CPVC, a graph of the contrast ratios may be mentioned. This is obtained by painting over a

Table 4.7 Critical pigment concentration values [235]

Pigment	CPVC (%)			
	Latex A		Latex B	
	Elongation	Tensile	Elongation	Tensile
Rutile (type A)[a]	40	40	36.5	36.5
Rutile (type B)	40	38	36	32
Lithopone	50	50	53	55
Barytes[a]	55	57	59	61
Kaolin (type A)	33	33	36	36
Kaolin (type B)	36	38	42	40
Calcium carbonate (type A[a])	48	50	51	53
Calcium carbonate (type B)	51	51	55	55
Mica (type A)	18	18	20	22
Mica (type B)	21	21	26	26
Talc (type A)	32	30	35	33
Talc (type B)	35	35	36	38
Green chromium oxide	43	43	46	48

[a]With latex C, rutile (type A) results were 40 and 40 %, barytes 60 and 62 % and calcium carbonate 51 and 53 % respectively.

black base and using a suitable instrument to estimate the covering power. There is a very sharp increase in contrast ratio at the CPVC since above this level the voids between particles of pigment contain air instead of medium, and the bigger difference in refractive index increases the hiding power of the extender. The method is less sensitive with titanium dioxide. With a medium sized calcium carbonate, this method gave a CPVC with an acrylic latex of 48 %, as against 49 %. In Table 4.7 the original included trade marks have not been reproduced.

Small particle size latices give better CPVC values than corresponding latices with higher particle sizes, since the smaller particles enable packing efficiency to be better. In addition, a coalescing solvent such as ethylene glycol, up to about 10 % for an acrylic latex with a T_g of 20 °C, will raise the CPVC of a calcium carbonate dispersion from 39 to 54 % [237].

6.6 Pigment encapsulation

A number of efforts have been made to encapsulate pigment particles so that they can be used in this form in various coatings. A process is described in which titanium dioxide is first made hydrophobic by treatment with titanates. Styrene or methyl methacrylate was polymerised in emulsion in the presence of the treated pigment. With styrene there was substantial coagulation at the latter stage of polymerisation. The nature of the reaction, in particular the position of the emulsifiers, was studied, and also the possibility of creating multilayers around the pigment particles, but details are too complex for this text [238].

7 ORGANIC EXTENDERS

There are other ways of obtaining appreciable hiding power other than by the direct use of pigments and extenders. As has once been pointed out, the white cliffs of Dover consist of chalk, with little direct tinting power. Fundamentally differences in refractive indices, in particular where air is one constituent, can contribute to appreciable hiding power, even when the compound concerned has no appreciable hiding power of its own right.

Microvoid pigments may be based on solutions of polymers such as cellulose nitrate or an acrylic copolymer which are carefully precipitated by a less volatile non-solvent [239]. The formation of microvoid pigments was shortly extended to types based on polymers first formed in emulsion, and this type of microvoid is now readily available commercially. Much of the pioneering work was performed in the laboratories of the Rohm and Haas Company (USA). The principle of one of the earliest patents is that a water-insoluble polymer is prepared by sequential emulsion polymerisation in which an acid polymer core is partially encased in a sheath polymer. This sheath is permeable to volatile bases such as ammonia to permit swelling of the core on at least partial neutralization. The polymer may serve as all on part of the binder, but on drying it forms microvoids in the particles which can act as pigments by virtue of the light scatter caused by the microvoids [240].

The process has to be carefully controlled to have the desired effect. The core copolymer is usually formed on a seed with a minimum amount of added emulsifier to keep the system stable. The core contains sufficient carboxylic acid groups to be swellable with alkali. Earlier patents restrict this to a volatile alkali such as ammonia, or volatile amines, but fixed alkalis such as sodium hydroxide are also quoted in later specifications. The core monomers should preferably contain a small quantity of a crosslinking monomer so as to restrict diffusion of the shell monomers into the core. The outer shell is based on monomers producing a relatively hard copolymer, with a T_g of at least 50 °C, and preferably at least 70 °C, but which should contain 1–5 % of a polymerisable acid. This enables diffusion of the alkali to take place into the particles, causing swelling. On carefully drying, the particles, which remain distinct, have considerable hiding power because the microvoids cause scattering of light.

The optimum diameter for the seed particles is 0.03–0.2 μm and that of the core particles is 0.1–0.5 μm. A small addition of emulsifier, 0.05–0.5 % by weight, may be desirable to avoid the formation of any coagulum, but without other undesirable effects. The sheath may include up to about 5 % by weight of an acid monomer. The amount of polymer deposited to form sheath polymer is such that the multistage polymer particles are of diameter 0.2–2 μm in the unswollen conditions. There is wide variation between claimed ratios between the core and the shell, with ratios of 1 : 8 and 1 : 50. The core–shell ratio is preferably 1 : 8 to 1 : 49.

The crosslinking of the sheath serves to render it more stable structurally. Therefore, on drying the swollen particles to remove swellant, the shrinkage of the swollen core produces microvoids therein, but the sheath resists collapse so that microvoids are retained within the particles, the shape of which remains spherical [240].

The swelling stage is best carried out at the T_g or above to favour rapid swelling of the alkaline agent through the sheath to the core. By softening the sheath, the core is allowed greater freedom of expansion against the limitations of the sheath. After expansion the swelling agent is removed by drying below the T_g, which should be as rapid as possible to assure efficient formation of microvoids.

A number of variations are possible. A second outer sheath may be added, preferably with a T_g at or below ambient temperature. This is useful where the products described above are included as part of the binder in an emulsion paint. Table 4.8 gives an example of the preparation of the seed polymer on which the core is formed. Table 4.9 gives a formulation for a core polymer and Table 4.10 gives several versions of the shell (sheath) [240].

Titration with 0.5N potassium hydroxide of latex B shown in Table 4.7 shows all of the theoretical carboxylic acid (3.5 meq g^{-1} of solid polymer). After completion of the core–shell latex, as shown in Table 4.10, the acid titration is zero, showing that the core is well encapsulated. Adding 0.16 meq g^{-1} of ammonium hydroxide shows no swelling after 24 h at ambient temperature. After heating this ammoniacal latex for 24 h at 95 °C, the particle diameter averaged 1.15 μm, indicating that there had been swelling. A similar phenomenon occurs with neutralization with triethylamine. Examination of swollen and unswollen particles by oil immersion with an optical microscope shows dark circles of 0.8 μm because of the voids inside the swollen particles, but unswollen particles were virtually invisible because of the small difference in refractive indices between the polymer and the oil. Electron microscopy also shows major differences. A test on blending the ammonia-swollen latex with a commercial acrylic latex (Rhoplex®AC 54) at 1 : 3 on solids showed considerable light scattering with the swollen film, but the unswollen film gave a clear film [240].

As a variation, it is possible to form micropores by swelling the cores with a solvent. The partition coefficient of the non-aqueous solvent between the core and the shell should preferably be 1.5. The quantity of iso-octane added in a typical example is of the order of the total core by volume, and it may be added before or after the shell is polymerised, preferably the former [241, 242].

A specification claims that a double shell is considered to give better results than a single shell in that undesired rupture during the swelling stage is

Table 4.8 Seed latex for microvoid latex [240]

A. First addition to reactor at 78 °C:	
Deionized water	2900 g
Sodium dodecylbenzene sulfonate	5.5 g
Heat to 78 °C with stirring; then prepare B/u_i:	
B. Deionized water	266 g
Sodium dodecylbenzene sulfonate	0.4 g
Butyl acrylate	416 g
Methyl methacrylate	374 g
Methacrylic acid	10.4 g
C. Add 50 g of B to the flask; then add:	
Ammonium persulfate (predissolved)	3 g
Water	10 g
After 15 min add the balance of emulsion B to the flask, allowing the temperature to rise to 85 °C. Maintain 15 min when addition is complete.	
D. At 55 °C, add the following to complete polymerisation:	
t-Butyl hydroperoxide (70 %)	1 g
Sodium formaldehyde sulfoxylate	0.5 g
Water, deoxygenated	20 g
E. At 20 °C add the following, followed by filtration through a fine mesh:	
Aqueous ammonia, 28 %	10 g
Final product:	
Solids content	19.5 %
pH	9.4
Average particle diameter	0.6 μm
If the sodium dodecylbenzene sulfonate was reduced to 2, then the average particle diameter is 0.095 μm.	

avoided and the swelling, which produces vesicles, is more regular. The formation of vesicles is emphasised. This variation of the core–shell process also enables emulsion paints with improved gloss to be obtained. The core should contain an unsaturated monomer, typically styrene with at least 15 % of polar monomers. An illustration of the patent is given in Tables 4.11 and 4.12 [243; see also 244].

A further example quotes an acid-free core formed from ethyl acrylate, methyl methacrylate and allyl methacrylate, an intermediate shell from methyl methacrylate, styrene, methacrylic acid and allyl methacrylate, and an outer rigid shell based on the same monomers with acrylic acid. The expansion stage is achieved by a solution of sodium hydroxide to give an organic pigment suitable for gloss enhancement in paper and paperboard coatings [245]. Unrelated specifications are rather similar, and 1,3-butadiene may be an alternative monomer [246, 247].

Table 4.9 Core latex [240]

A. Add the following to the reactor at 84 °C, under nitrogen:	
Water, deionised	2115 g
Sodium persulfate (as solution)	4.2 g
Water	25 g
Acrylic seed latex A of Table 4.7, 19.5 % solids, PD 0.095 μm	62 g
B. Prepare monomer emulsion as follows, and add to seed over 3 h at 85 °C:	
Water, deionised	235 g
Sodium dodecylbenzene sulfonate	0.8 g
Methyl methacrylate	490 g
Methacrylic acid	210 g
Ethylene glycol diacrylate	3.6 g
To complete, hold at 85 °C for 30 min, cool to 25 °C and filter.	
Properties:	
pH	2.3
Solids content	22.4 %
Average particle diameter	0.35 μm
C. A diluted sample of latex is neutralized to pH 10 with ammonia solution. On examination with an optical microscope, the average particle diameter is 0.8 μm. This corresponds to a swelling of about 12 by volume.	

Table 4.10 Formation of core–shell copolymer [240]

A. Add the following at 85 °C to reactor:	
Water	80 g
Sodium persulfate	0.07 g
Latex B of Table 4.7 (equivalent to 1 g solid polymer)	4.5 g
B. Add over one hour at 85 °C:	
Methyl methacrylate	18 g
Maintain at 85 °C until polymerisation is substantially complete.	
Average particle diameter	1 μm

A modification to the general process adjusts the reaction so that in the multi-shell structure pores penetrate to the surface. In this case the inner core contains 1–5 parts of a polymerisable acid, preferably acrylic or methacrylic acids, per 10 parts of the principal vinyl monomer. A functional monomer may be included in limited amounts. The first shell may also contain a limited quantity of acid up to 2 %, to ensure latex stability, and also a crosslinking monomer. There is a second shell which may contain up to 3 % of polymerisable monomer and at least 0.1 %, and less than 3 %, of a crosslinking monomer. It is added after alkaline treatment. A seed polymer may be used to form the core. The first shell is preferably 2–6 times the weight of the core polymer and the outer shell is 2–10 times the weight of the combined core and inner shell. The final particles are 0.1–1.2 μm in diameter.

Table 4.11 Core–double shell latex [243]

	Seed latex (g)	Core latex (g)
A1. Initial water phase in reactor:		
Water	478.0	2402
Sodium dodecylbenzene sulfonate	2.8	—
Ammonium persulfate	1.6	5.6
t-Butyl hydroperoxide	0.15	—
Sodium formaldehyde sulfoxylate	0.1	—
Seed polymer latex	—	138.9
A2. Monomer pre-emulsion:		
Water	140.1	405
Butyl acrylate	219.2	144.5
Methyl methacrylate	197.0	639.5
Methacrylic acid	5.6	422.3
Sodium dodecylbenzene sulfonate	0.2	3.2
Ammonium persulfate (see Procedure below)	—	1.1

Procedure:

Seed. Add monomer pre-emulsion at 85 °C at 1.2 g min^{-1}. Maintain for 30 min after completion. Cool and filter through a 200 mesh screen.

Core. Add monomer at 5 g min^{-1} at 85 °C. Then 30 min after monomer addition add the persulfate. At the end hold for 40 min and filter through a 200 mesh screen.

	First shell (g)	Second shell (g)
Initial water phase in reactor:		
Water	1714.0	—
Ammonium persulfate	2.3	—
Core latex above	549.8	—
Monomer pre-emulsion:		
Water	287.3	255.0
Methyl methacrylate	412.5	—
Butyl acrylate	59.3	24.0
Styrene	675.0	480.0
Sodium dodecylbenzene sulfonate	3.8	2.3
Nonylphenyl (40) polyoxyethylene condensate	15.0	9.0

Procedure:

To the core water phase above, add the first and second shell emulsions consecutively at 5 g min^{-1}. One hour after complete addition, cool to 40 °C and filter.

Table 4.12 Alkali treatment of double shell layer latex [243]

Core–shell latex as in Table 4.8	150.0 g
Sodium hydroxide, 10 % aqueous	9.4 g
Sodium mondodecyl diphenyl oxide disulfonate	0.2 g

Procedure:
    Allow the latex plus alkali to stand at ambient temperature for 22 h, and then add surfactant. Heat to 90 °C and maintain for 30 min. Cool to ambient temperature and filter through a 200 mesh screen. The number average particle size is 495 nm.

It is believed that the inner shell does not completely cover the pore under the limiting conditions of the claims, and that, as a result, penetrating pores are formed, improving the oil absorption properties and gas permeability of the particles. Tables 4.13 and 4.14 illustrate the process [248].

In the same way as most of these core–shell microvoids are prepared with an acid copolymer which can be swollen, it is also possible to use alkaline monomers such as dimethylaminoethyl methacrylate and an acid, a range of dilute acids being possible. An unusual feature is that the core and shell are grafted by addition of a polyfunctional compound. The formulations tend to be complex [249].

Table 4.13 Multishell emulsion particle–core emulsion

	Seed emulsion X (g)	Core emulsion Y (g)
A. Addition to reactor:		
Water	2500	Total X
Potassium persulfate	1.5	—
B. Pre-emulsion:		
Styrene	16.0	—
Methacrylic acid	0.3	40.
Methyl metacrylate		55.0
Butyl acrylate	—	5.0
Divinylbenzene	0.2	—
Sodium dodecylbenzene sulfonate (as solution in water)	0.2	0.3
Water	7.0	40.0

Process:
    Polymerise seed at 80 °C, adding pre-emulsion gradually; particle diameter is 0.6 μm. The pre-emulsion was added and polymerised over 1 h and then aged for 1 h.
    Particle size   0.12 μm

See Table 4.12 for the procedure.

Table 4.14 Formation of shells

	Inner shell (g)	Outer shell (g)	
		Stage 1	Stage 2
Preliminary addition			
Core latex	Total (Table 4.11)	—	—
Inner shell latex	—	490 g	—
Ammonium persulfate (as solution)	3.0	—	0.5 g
Water	30.0	78.0	5.0
Ammonia, 28 % aqueous	—	8.5	—
Second pre-emulsion:			
Methyl methacrylate	468.0	—	—
Butyl acrylate	120.0	—	—
Methacrylic acid	12.0	—	—
Sodium dodecyl sulfate (as solution)	1.2	—	0.6
Water	240.0	—	120.0
Styrene	—	—	297.0
Acrylonitrile	—	—	3.0

Process:
  Inner shell. Add pre-emulsion at 80 °C over 1 h to preliminary addition and age for 2 h.
  Outer shell. Stage 1: heat at 85 °C for 30 min to give swelling. Stage 2: add emulsion over 30 min and age for 90 min.

Further examples of applications are given in chapter 10 in volume 2, where emulsion paints, paper coatings and heat-sensitive recording materials are considered.

REFERENCES

1. M.S. El-Aasser and R.M. Fitch (eds.), *Position Paper, Future Directions in Polymer Colloids*, Martinus Hijoff Publishers, 1987, pp. 107–17
2. A. Rudin, *Macromol. Symp.*, **92**, 53–70 (1995)
3. G. Elicabe and G. Frontini, *J. Coll. Interface Sci.*, **181**(2), 669–72 (1996)
4. S. Lee et al., *Anal. Chem.*, **68**(9), 1545–9
5. H.-J. Butt and B. Gerharz *Langmuir*, **11**(12), 4735–41 (1995)
6. C. Pichot, *Bull. Soc. Chim. France*, **1987**(4), 725–33 (1987)
7. D.J. Williams, and M.R. Grancio, *J. Polym. Sci.*, **A-1**(8), 2617 (1970)
8. D.J. Williams, *J. Elastoplastics*, **3**, 187–200 (1971)
9. S. Kawaguchi et al., *J. Coll. Interface Sci.*, **176**(2), 362–9 (1995)
10. K. Landfester and H.W. Spiess, *Macromol. Rapid Commun.*, **17**(12), 875–83 (1996)
11. K. Landfester and H.W. Spiess, *Macromol.*, **29**(18), 5972–80 (1996)
12. J.V. Vanderhoff, *2nd Colloque International sur les Copolymerisations et Copolymères en Emulsion*, CNRS Lyon, 1989, pp. 81–3
13. J. Guillot et al., *NATO ASI Series, Mathematical and Physical Sciences*, Vol. C303, Kluwer Academic Publishers, 1990, pp. 97–127

14. W.D. Hergeth et al., Coll. Polym. Sci., **268**(11), 991–4 (1990)
15. W.D. Hergeth and J. Lange, J. Macromol. Chem., Macromol. Symp., **52**, 283 (1991)
16. L. Rios et al., Coll. Polym. Sci., **269**(8), 812–24 (1991)
17. D. Charmot et al., Polymer, **37**(23), 5237–45 (1996)
18. V.L. Dimonie, M.S. El-Aasser, and J.W. Vanderhoff, Polym. Mater. Engng. Sci., **58**, 821–5 (1988)
19. L.W. Morgan et al. (Johnson & Sons), Eur. P 338,486, 1990
20. S. Lee and A. Rudin, J. Polym. Sci., Polym. Chem., **30**(5), 865–71 (1992)
21. F. Sommer et al., Langmuir, **11**(2), 440–8 (1995)
22. V.I. Eliseeva et al., Vysokomol. Soedin. Ser. A, **31**(2), 263–8 1989; Chem. Abstr., **111**, 40194 (1989)
23. A. Zosel et al., 2nd Colloque International sur les Copolymerisations et les Copolymères en Emulsion, CNRS, Lyon, 1989, pp. 109–11
24. S. Xia et al., Xiangtan, Daxue Ziran Kexue Xuebao, **11**(2), 57–62 (1989); Chem. Abstr., **112**, 99813 (1990)
25. C. Tongyu et al., J. Appl. Polym. Sci., **41**(9/10), 1865–72 (1990)
26. D.A. Greenhill and D.J. Hourston, Polym. Mater. Sci. Engng, **60**, 643–5 (1989)
27. W.D. Hergeth et al., Polymer, **60**(10), 1913–7 (1989)
28. J. Bolze et al., Coll. Polym. Sci., **274**(12), 1099–108 (1996)
29. M.A. Cohen Stuart, in Future Direction in Polymer Colloids, M.S. El-Aasser and R.M. Fitch (eds.), Martinus Nijhoff, 1987, pp. 229–39
30. D. Eagland and G.C. Wardlaw, Coll. Polym. Sci., **256**(11), 1079–94 (1978)
31. S.J. Baum and R.D. Dunlop, FIAT 1102, HM Stationery Office, 1947
32. R. Lapasin, F. Sturgzi, and A. Alexandrini, Ind. Vernici, **35**(12), 3–9 (1981)
33. K. Heine et al., Ang. Makromolek Chemie, **27**, 37–56 (1972)
34. B.R. Vijayendran, T. Bone and J. Gajriam, J. Appl. Polym. Sci., **26**(4), 1351–9 (1981)
35. B.R. Vijayendran, and R.M. Fitch (eds.) Polymer Colloids, Vol. 2, Pergamon Press, 1980, pp. 209–14
36. J.C. Kennedy, J. Meadows and P.A. Williams, Polym. Mater. Sci. Engng, **73**, 202–3 (1995)
37. W. Wetzel, Polym. Mater. Sci. Engng, **73**, 329–30 (1995)
38. A.M. Jamieson et al., Polym. Mater. Sci. Eng., **73**, 440–1 (1995)
39. D.C.H. Cheng and R.A. Richmond, Rheol. Proc. Int. 8th. Congr., **2**, 575–80 (1980) (comparison of viscometers)
40. I.M. Krieger, in Rheology of Polymer Colloids, R. Bus (ed.),
41. C.K. Schoff, Encycl. Polym. Sci., **14**, 445–501 (1988) (review).
42. D.C. Blackley, Polym. Latex Int. Conf., prepr. paper 9, Coll. J., (1979)
43. T.G. Rochow and C.W. Mason, Ind. Eng. Chem., **28**, 1296 et seq. (1936)
44. W.G. Barb and W. Mickucki, J. Polym. Sci., **37**, 499 et seq. (1959)
45. L.N. Afas'eva et al., Kolloid Zh., **35**(4), 735–8; (1973) Coll. J., USSR, **35**(4), 682–4 (1973) (English trans. of previous paper)
46. R.E. Neiman et al., Coll. J., USSR, **38**(1), 109–110 (1976) (English trans.)
47. O.G. Kiseleva et al., Kolloid Zh., **53**(5), 858–60 (1991); Chem. Abstr., **115**, 234389 (1991)
48. V.Yu. Grigorev and B.P. Nikolaev A.M. Shlyakov, Kolloid Zh., **53**(3), 453–7 (1991) Chem. Abstr., **115**, 94226 (1991)
49. W.R. Conn, B.B. Kine and W.O. Prentiss (Rohm & Haas), USP 2,795,634, 1957
50. I.I. Eliseeva, T.B. Gonsovskaya and R.E. Neiman, Kolloid Zh., **32**(6), 856–9 (1970)
51. Shell International Research, BP 940,366, 1963
52. Shell International Research, BP 951,172, 1964
53. J.H. Musch (Firestone Tire & Rubber), USP 3,075,246, 1963
54. C.E. Blades and A.J. Buselli (Cumberland Chemical), USP 3,231,534, 1965
55. Kaplan and R.A. Ripley (Du Pont de Nemours), USP 3,553,116, 1971

References

56. A.E. Akopyan, E.A. Enfiadzyan and R. Rostombekyan, *Arm. Khim. Zh.*, **20**(11), 926 *et seq.* (1967); *Chem. Abstr.*, **69**, 11057 (1968)
57. D. Schoenholz and C.O. Morrison (Shawinigan), BP 676,763, 1952
58. Hoechst, Austrian P 221,810, 221,812, 1960
59. Celanese, BP 834,900, 1960
60. G.W. Poehlein, J.W. Vanderhoff and W.J. Witmeyer, *Polym. Prepr. Am. Chem. Soc.*, **16**(1), 268–73 (1975)
61. W.A. Gate, *et al.*, *J. Coll. Interface Sci.*, **27**(1), 32–7 (1968)
62. J.W. Vanderhoff, E.B. Bradford, and W.K. Carrington, *J. Polym. Sci. Symp.*, **41**, 155–74 (1973)
63. J.W. Vanderhoff, *Polym. News*, **3**(4), 194–204 (1977)
64. N. Pramojaney, G.W. Poehlein and J.W. Vanderhoff, *Org. Coat. Plast. Chem. Prepr.*, **42**, 55–61 (1980)
65. D. Sullivan, *Proc. 4th Water-borne Higher Solids Coat. Symp.*, University of South Mississipi, 1977, Vol. 2, pp. 196–204
66. G.P. Bierwagen *et al.*, *Org. Coat. Plast. Chem. Prepr.*, **42**, 51–4 (1980)
67. R.E. Neiman and G.A. Gorenkova, *Coll. J., USSR*, **37**(2), 274–8 (1975) (English trans.)
68. D. Distler and G. Kanig, *Coll. Polym. Sci.*, **256**(11), 1052–60 (1978)
69. G.A. Vadezande and A. Rudin, *Polym. Mater. Sci. Engng. (ACS)*, **73**, 149–50 (1995)
70. D. Juhue *et al.*, *Polym. Mater. Sci. Engng (ACS)*, **73**, 86–8 (1995)
71. J. Richard, *Polym. Adv. Technol.*, **6**(5), 270–5 (1995)
72. A. Du Chesne *et al.*, *Polym. Int.*, **43**(2), 187–96 (1997)
73. J.I. Amalvy and B. del Amo, *CIDEPINT Anales*, **1995**, 59–67 (1995)
74. J.I. Amalvy and D.B. Soria, *Prog. Org. Coat.*, **28**(4), 279–83 (1996)
75. M.W. Urban and B.-J. Niu, *J. Appl. Polym. Sci.*, **56**(3), 377–85 (1995)
76. M.W. Urban and L.K. Tebelius, *J. Appl. Polym. Sci.*, **56**(3), 387–95 (1995)
77. M.W. Urban and L.R. Martin, *J. appl. Polym. Sci.*, **62**(11), 1887–92 (1996)
78. E. Kientz *et al.*, *Polym. Int.*, **34**(2), 125–34 (1994)
79. E. Kientz *et al.*, *J. Adhes. Sci., Technol. Sci.*, **10**(8), 745–59 (1996)
80. H.A. Oosterhof, *J. Oil Col. Chem. Ass.*, **48**, 256–81 (1965)
81. B.S. Snyder *et al.*, *Polym. Prepr. Am. Chem. Soc.*, **35**(2), 299–300 (1994)
82. R.E. Zdanowski and G.L. Brown, *33rd Mid-year Proc.*, Chemical Specialities Manufactures Association, New York, 1958.
83. J. Mulvihill *et al.*, *Prog. Org. Coat.*, **30**(3), 127–39 (1997)
84. A. Rudin *et al.*, *2nd Colloque Les Copolymerisations et les Copolymères en Milieu Disperse*, CNRS, Lyon, 1989
85. E. Perez, *Langmuir*, **12**(13), 3180–7 (1996)
86. P. Marion *et al.*, Proc. XXIII FATPEC Congr. Brussels, 1996, Part A, pp. 73–9
87. P. Marion *et al.*, *J. Appl. Polym. Sci.*, **64**(12), 2409–19 (1997)
88. F. Dobler *et al.*, *J. Coll. Interface Sci.*, **152**(1), 1–11,12–21 (1992)
89. D. Johue and J. Lang, *Double Liaison*, **42**, 464–5 (in French), III–X (in English) (1994); *Macromolecules*, **28**(4), 1306–9 (1995)
90. G.C. Overbeek *et al.*, *Polym. Mater. Sci. Engng (ACS)*, **73**, 1450–1 (1995)
91. S.T. Eckersley and B.J. Helmer, *J. Coat. Technol.*, **69**(684), 97–107 (1997)
92. M. Joanicot *et al.*, *Double Liaison*, **40**, 452–3 (in French), V–XIV (in English) (1993)
93. P.W. Dillon, *J. Coat. Technol.*, **49**(634), 38–49 (1977)
94. D. Sullivan, *Proc. 4th Water-borne Higher Solids Coat Symp.*, University of South Mississipi, 1977, Vol. 2, pp. 196–204
95. N. Pramojamey, G.W. Poehlein and J. Vanderhoff, *Pap. Synth. Conf. Proc.*, 1980, pp. 253–61
96. M. Geghart and F. Loflath, Proc. XXI Conf. Org. Coat. Sci. Tech., Athens, 1995, pp. 209–20
97. S.T. Eckersley, and J. Rudin, *Appl. Polym. Sci.*, **53**, 1139–47 (1994)

98. S.T. Eckersley and J. Rudin, *J. Coat. Technol.*, **62**(780), 89–99 (1990)
99. T. Imoto, *Prog. Organic Coat.*, **2**(3), 194–201 (1974)
100. M.F. Llauro *et al.*, *Macromol. Symp.*, **92**, 117–32 (1995)
101. J.N. Yoo *et al.*, *Macromolecules*, **23**(16), 3962–7 (1990)
102. Y. Chevalier *et al.*, *Polym. Mater. Sci. Engng*, **73**, 95–6 (1995)
103. Y. Chevalier *et al.*, *Trends in Polym. Sci.*, **4**(5), 145–51 (1996)
104. F. Henry *et al.*, *Coll. Polym. Sci.*, **267**(2), 167–78 (1990)
105. M. Klaskov, *Farg. Lack*, **36**(3), 53–8 (1990)
106. M. Canpolat and O. Pekcan, *J. Appl. Polym. Sci.*, **59**(11), 1699–1707
107. M. Canpolat and O. Pekcan, *J. Polym. Sci., Polym. Phys.*, **34**(4), 691–8 (1996)
108. M. Canpolet and O. Pekcan, *Polymer*, **36**(23), 4433–8 (1995)
109. M. Canpolat and O. Pekcan, *Polymer*, **34**(15), 3319–21 (1993)
110. D. Juhue, *et al.*, *Polym. Mater. Sci. Engng (ACS)*, **73**, 86–7 (1995)
111. D.J. Meier and F. Lin, *Polym. Mater. Sci. Engng*, **73**, 993–4 (1995)
112. D.J. Meier and Y. Ming, *Polym. Mater. Sci. Engng (ACS)*, **76**, 25–6 (1997)
113. J. Mulvihill, Forum de la Connaissance AFTPA/CORI, prepr., 1996, 8 pp.
114. B. Gerharz *et al.*, Proc 22nd Int. Org. Coat. Athens, 1996, pp. 133–4 1998
115. J. Lin and D.J. Meier, *Polymer Prepr. (ACS)*, **37**(2), 591–2
116. J. Lin and D.J. Meier, Proc. XXI Int. Conf. Org. Coat. Sci., Athens, 1995, pp. 297–311
117. J. Lin and D.J. Meier, *Langmuir*, **12**(11), 2774–80 (1996)
118. A.-C. Hellgren and T.-Q. Li Proc. PRA 4th Nurnberg Congr. on Creative Advances in Coatings Technology, 1997, paper 49, 22 pp.
119. H.-J. Butt *et al.*, *Coll. Polym. Sci.*, **272**(10), 1218–23 (1994)
120. A.G. Gilleinski and C.R. Hegedus, *Polym. Mater. Sci. Engng*, **73**, 142–3 (1995)
121. L. Sarrazin, *Double Liaison*, **41**(463), XI–XVI in (English), 19–24 (in French) (1994); L.-L. Xing, J.E. Glass and R.H. Fernando, *Polym. Mater. Sci. Engng* (ACS), **76** (1997)
122. J. Frenkel, *J. Phys., USSR*, **9**, 385 *et seq.* (1945)
123. G.L. Brown, *J. Polym. Sci.*, **22**, 423 *et seq.* (1956)
124. J.W. Vanderhoff, World Paper Trade Review, 24 August 1967
125. S.S. Voyutskii, *J. Polym. Sci.*, **32**, 528 *et seq.* (1958)
126. D.P. Sheetz, *J. Appl. Polym. Sci.*, **9**, 3759–73 (1965)
127. J.W. Vanderhoff, E.B. Bradford and W.K. Carrington, *J. Polym. Sci. Symp.*, **41**, 155–74 (1973)
128. L.T. Bakhshieva and V.M. Chesunov, *Izv. Yyssh. Ucheb. Zaved. Tekhnol. Legk. Prom.*, **1973**(6), 29–33; *Chem. Abstr.*, **81**, 97048 (1974)
129. D.J. Meier and F. Lin, *Polym. Mater. Sci. Engng (ACS)*, **73**, 84–5 (1995)
130. J. Feng and M.A. Winnik, *Polym. Mater. Sci. Engng (ACS)*, **73**, 90–1 (1995)
131. L.A. Dissado, *et al.*, *J. Phys., Appl. Phys.*, **22**(5), 713–16 (1989)
132. F. Dobler *et al.*, *J. Phys., Appl. Phys.*, **22**(5), 1–2 (1989)
133. H. Kast, *Makromol. Chem. Suppl.*, **10/11**, 447–61 (1985)
134. L.H. Sperling, M. Sambasivam and A. Klein, *Polym. Mater. Sci. Engng (ACS)*, **73**, 45–6 (1995)
135. A. Zosel and G. Ley, *Polym. Mater. Sci. Engng (ACS)*, **73**, 43–4 (1995)
136. A. Guyot *et al.*, *Polym. Mater. Sci. Engng (ACS)*, **73**, 50–1 (1995)
137. A. Zosel *et al.*, *Makromol. Chem., Macromol. Symp.*, **35**(6), 4223–6 (1990)
138. D.J. Meier and F. Lin, *Polym. Mater. Sci. Engng (ACS)*, **73**, 93–4 (1995)
139. J.Y. Charmeau, E. Kientz, and V. Holl, *Polym. Mater. Sci. Engng (ACS)*, **73**, 48–9 (1995)
140. J. Richard, *Polym. Mater. Sci. Engng (ACS)*, **73**, 41–2 (1995)
141. D. Johué *et al.*, *Polym. Mater. Sci. Engng (ACS)*, **73**, 86–7 (1995)
142. E.M. Boczar *et al.*, *Polym. Mater. Sci. Engng (ACS)*, **73**, 8–9 (1995)
143. M.W. Urban, *Polym. Mater. Sci. Engng (ACS)*, **73**, 137–8 (1995)
144. L.A. Bukhareva and B.L. Khavkina, *Acta Polym.*, **34**(9), 559–6 (1983) *Chem. Abstr.*, **99**, 123525 (1983)

145. M. Okubo et al., *Nippon Setchaku Kyokaishi*, **25**(3), 95–100 (1989); *Chem. Abstr.*, **111**, 116119 (1989)
146. F. Dobler et al., *2nd Colloque International sur les Copolymerisations et les Copolymères*, CNRS, Lyon, 1989, pp. 91–3
147. J.I. Amalvy and B. del Amo, *Surf. Coat. Int.*, **80**(2), 78–81 (1997); J.I. Amalvy, *Pigment Resin Technol.*, **26**(6), 363–9 (1997); **27**(1), 20–7 (1998)
148. F. Molenaar et al., *Prog. Org. Coat.*, **30**(3), 141–58 (1997)
149. M. Campolet and O. Pekcan, *J. Polym. Sci., Polym. Phys.*, **34**(4), 691–8 (1996)
150. M. Campolet and O. Pekcan, *Polymer*, **36**, 4433–8 (1995)
151. Z. Fu et al., *Polym. Mater. Sci. Engng. (ACS)*, **73**, 362–3 (1995)
152. S.T. Eckersley, *Polym. Mater. Sci. Engng (ACS)*, **73**, 3–4 (1995)
153. Ph. Espiard et al., *Polymer*, **36**(23), 4385–9,4391–5 (1995)
154. T. Imoto, *Prog. Org. Coat.*, **2**(3), 194–201 (1974)
155. S.T. Eckersley and A. Rudin, *J. Appl. Polym. Sci.*, **53**(9), 1139–47 (1994)
156. A.G. Gillicinski and C.R. Hegedus, *Polym. Mater. Sci. Engng (ACS)*, **73**, 142–3 (1975)
157. J.L. Keddie et al., *Polym. Mater. Sci. Engng (ACS)*, **73**, 144–5 (1995)
158. S. Mazur and C. Argento, *Polym. Mater. Sci. Engng (ACS)*, **73**, 47 (1995)
159. E.F. Meyer, *Polym. Mater. Sci. Engng (ACS)*, **73**, 360–1 (1995)
160. F. Zhenwen et al., *Polym. Mater. Sci. Engng (ACS)*, **73**, 362–3 (1995)
161. K. Weinmann *Farbe u. Lack*, **59**(1), 34–44 (1973)
162. G. Mason, *Br. Polym. J.*, **1973**(5), 101–8 (1973)
163. K.L. Hoy, *J. Paint Technol.*, **45**(578), 51–8 (1973)
164. T.F. Protzman and G.L. Brown, *J. Appl. Polym. Sci.*, **4**(10), 81–4 (1960)
165. H.G. Cogan, *Off. Dig.*, **33**(434), 465 et seq. (1961)
166. M. Jendraszek et al., *Lakokras. Mater. Ikh. Primen.*, **1981**(2), 10–11 (1981); *Chem. Abstr.*, **95**, 63771 (1981)
167. P. Hartmut et al., *Plaste Kautsch.*, **28**(7), 380–3
168. M. Chainey, M.C. Wilkinson and J. Hearn *J. Appl. Polym. Sci.*, **30**(11), 4273–85 (1985)
169. S. Eckersley and A. Rudin, *Polym. Mater. Sci. Engng*, **58**, 1115–9 (1988)
170. S. Eckersley and A. Rudin, *J. Coat. Technol.*, **62**,(780), 89–99 (1990)
171. J.G. Brodnyan and T.J. Konen, *J. Appl. Polym. Sci.*, **8**, 687 (1964)
172. V.V. Kondretuk et al., *Naucho-Tekh. Sb*, **4**(9–11), 241 (1978); T. Cao et al., *J. Appl. Polym. Sci.*, **41**(9/10), 1965–72 (1990)
173. F. Buckmann and F. Bakker, *Eur. Coat. J.*, **1995**(12), 922–7 (1995)
174. F. Buckma and F. Bakker, *Double Liaison*, **1995**(475), 12–26 (in French), III–XI (in English) (1995)
175. C. Tongyu et al., *J. Polym. Sci.*, **41**, 1965–72 (1990)
176. M.A. Eneiadzhyan and E.T. Pogosyan, *Arm. Khim. Zh.*, **27**(3), 263 et seq. (1974)
177. H. Wiest, 14th FATIPEC Congress Book, pp. 705–10
178. H. Wiest, *Defazet*, **329**(10), 378–81 (1978)
179. T.C. Aten, *Emulsion Polymers Symp., Prepr. Conf.*, Plastics and Rubber Institute, 1982, pp. 4/1–4/11; W.C. Aten and T.C. Wassenburg, *Plast Rubber Process Applic.*, **3**(2), 99 (1983)
180. B.R. Vijayendran, T. Bone and L.C. Sawyer, *J. Dispersion Sci. Technol.*, **3910**, 81–97 (1982)
181. G.A. Vandezande, *J. Coat. Technol.*, **68**(860), 63–73 (1996)
182. J.L. Keddie et al., *Macromol.*, **28**(8), 2673–82 (1995)
183. G.C. Overbeek et al., *Polym. Mater. Sci. Engng (ACS)*, **73**, 140–1 (1995)
184. N. Okubo et al., *Nippon Setchakun Kyokaishi*, **17**(7), 264–9 (1981); *Chem. Abstr.*, **98**, 7439 (1983)
185. N. Okubo et al., *Nippon Setchakun Kyokaishi*, **17**(7), 185–91 (1981); *Chem. Abstr.*, **95**, 116343 (1981)
186. N. Okubo et al., *J. Polym. Sci. Chem.*, **19**(1), 1–8 (1981)

187. van Tent and K. te Nijenhuis, *Verfkroniek*, **64**(7/8), 296–301 (in Dutch)
188. J. Snuparek, *Farbe u. Lack*, **1978**(6), 394–8 (1978)
189. V.I. Eliseeva, *Kobunshi Kako*, **27**(6), 205–12 (1978); *Chem. Abstr.*, **89**, 164271 (1978)
190. J. Hearn et al., *J. Oil. Col. Chem. Ass.*, **73**(11), 467–70 (1990)
191. B.J. Roulstone et al., *Polym. Int.*, **24**(2), 87–94 (1991)
192. J. Feng and M.A. Winnik, *Polym. Mater. Sci. Engng (ACS)*, **73**, 252–3 (1995)
193. S.F. Thames and R. Hariharan, *Polym. Mater. Sci. Engng (ACS)*, **73**, 366–7 (1995)
194. M.W. Urban and B.-J. Nio, *Polym. Mater. Sci. Engng (ACS)*, **73**, 358–9 (1995)
195. S.T. Eckersley and A. Rudin, *Prog. Org. Coat.*, **23**, 387–402 (1994)
196. K. Weinmann, *Farbe u. Lack*, **59**(1), 34–44 (1973)
197. J.W. Vanderhoff and E.B. Bradford, *J. Macromol. Chem.*, **1**, 335–60 (1960); J.W. Vanderhoff, E.B. Bradford, H.L. Tarkowski and M.C. Jenkins, *J. Macromol. Chem.*, **1**, 361–97 (1966)
198. A.V. Lebedev, *Kauch Rezina*, **32**(2), 14–16 (1973)
199. C.S. Kan and J.H. Blackston, *Macromol.*, **29**(21), 6853–64 (1996)
200. N.A. Senatova, *Ref. Zh. Khm.*, Abstr. 17S871 (1973); *Chem. Abstr.*, **81**, 153192 (1974)
201. I. Langfelder, *Zb Pr. Chemickotehnol. Fak. SVST*, 171–80 (1974) in (Sloval); *Chem. Abstr.*, **82**, 98920 (1975)
202. F. Henry et al., *2nd Colloque International sur les Copolymerisations et les Copolymeres*, CNRS, Lyon, 1989, pp. 100–2
203. K.M. O'Connor and S.-L. Tsaur, *J. Appl. Polym. Sci.*, **33**, 2008–27 (1987)
204. J. Richards, *Trends in Polym. Sci.*, **4**(8), 272–9 (1996)
205. M.W. Urbain et al., *Polym. Mater. Sci. Engng. (ACS)*, **82**, 375–87 (2000)
206. D. Juhue and J. Lang, *Proc. PRA 14th Int. Conf.*, Copenhagen, 1994, paper 12, 11 pp.
207. Y. Wang and M.A. Winnik, *Macromol.*, **20**(21), 4731–2 (1990)
208. Research Memoranda 204 and 212, Paint Research Association, Teddington, Middlesex, 1954
209. Research Memorandum 265, Paint Research Association, Teddington, Middlesex 1958
210. R.E. Dillon, E.B. Bradford and R.D. Andrews, *Ind. Engng. Chem.*, **45**, 728 (1953)
211. G. Loehr, in *Polymer Colloids Symposium*, R.M. Fitch (ed.), Pergamon Press, 1980, pp. 71–81
212. A.J. De Fusco, *Proc. 15th Water-borne and Higher Solids Symp.*, University of Southern Mississippi, 1988, pp. 297–329; D.H. Guthrie, *Proc. 15th Water-borne and Higher Solids Symp.*, University of Southern Mississippi, 1988, pp. 208–11
213. W.E. Gunter, *Proc. 15th Water-borne and Higher Solids Symp.*, University of Southern Mississippi, 1988, pp. 265–96
214. J. Mieziva et al., *Farbe u. Lack*, **95**(1), 9–12; J.R. Kelsey (BP Chemicals), Eur. P. 501,614
215. B. Hudson, (BP Chemical), Eur. P. 119,026, 1984
216. J. Friel (Rohm & Haas), Eur. P 26,982, 1981
217. W.D. Arendt, Proc. 16th Water-borne and Higher Solids Symp., New Orleans, 1989, pp. 149–61
218. J.S. Kelyman, D.H. Guthrie and H.F. Hussey, *Mod. Paint Coat.*, **76**(10), 155–60 (1986)
219. J.M. Loutz et al., *Polym. Paint. Col. J.*, **178**(4219), 571–3 (1988)
220. V. Calder, *Proc. Water-borne and High Solids Coating Symp.*, University of South Dakota, 1977, pp. 27–47
221. R. Quartermaster and J. Swinselman, *FATIPEC XXI Congress Book*, Vol. 3, 1992, pp. 109–23
222. E.P. Cruz, *Eurocoat. J.*, **1991**(4) 204–15 (1991)
223. A.J. De Fusco, *Proc. 15th Water-borne and Higher Solids Symp.*, University of Southern Mississippi, 1988, pp. 297–330
224. E. Lund and F. Vial, *Farbe u. Lack*, **65**, 695 (1959)
225. F.B. Stieg, *The Geometry of White Hiding Power*, Kronos Information Bulletin 11E, Kronos, UK and Germany

References

226. P. Swaraj, *Surface Coatings, Science and Technology*, Wiley, Chs. 3 and 5, pp. 2nd Ed. 1996 Ch: 3 + 5.
227. W.M. Morgams, *Outlines of Paint Technology*, 3rd edn, Edward Arnold, 1990, Chs. 2 to 7 and 9, p. 133 and Ch. 15, pp. 266–78
228. Surface Coatings Association, Australia, *Surface Coatings*, Vol. 1, 3rd edn., Chapman & Hall, 1993, Chs. 25 to 30, pp. 409–538
229. Z.W. Wicks, F.N. Jones and S.P. Pappas, *Organic Coatings, Science and Technology*, Wiley, Vol. 1, 1992, Ch. 16, pp. 267–95, Ch. 18, pp. 306–22; Vol. 2, Ch. 20, 29–54, Ch. 23, pp. 65–104
230. J.H. Braun, *J. Coat. Technol.*, **69**(68), 59–72 (1997)
231. H. Warson (Vinyl Products), BP 697,452, 1953
232. H. Warson (Vinyl Products), BP 751,622, 1956
233. R.F. Avery and W.F. Scheufele, *Franl. V.S. TAPPI*, **45**(2), 105 *et seq.* (1962)
234. F.P. Liberti and R.C. Pierehumbert, *Off. Dig.*, **31**, 252 *et seq.* (1959)
235. B. Beradi, 6th FATIPEC Congress Book, **67** et seq. (1962)
236. I.V. Sutareva, V.V. Verkholantsev and S.L. Tantsman, *Lakokras Mater. Ikh. Primen*, **1976**(2), 46–8 (1976)
237. B. Schaller, *J. Paint Technol.*, **40**, 433 (1968)
238. R.Q.F. Janssen *et al.*, *R. Soc. Chem.*, **138** (Encapsulation and Controlled Release), 102–16 (1993)
239. S.Y. Yakubovitch *et al.*, *XXI FATIPEC Congress Book*, 1976, pp. 206–9
240. A.M. Kowalski and R.M. Blankenship (Rohm & Haas), Eur. P. 22,633, 1981
241. A. Kowalski, M. Vogel, and D. Scott (Rohm & Haas), Eur. P, 342,944, 1989
242. H. Touda and Y. Takagashi (Nippon Zeon), Eur. P. 404,184, 1990
243. H. Gharapetian, G.W. Chip and A. Rudin (Tioxide), BP 2,242,437, 1991
244. G.K. Chip and A. Rudin (Tioxide), USP 4,985,469, 1991
245. D.I. Lee *et al.*, (Dow Chemical), USP 5,521,223, 1996
246. D.I. Lee *et al.*, (Dow Chemical), PCT Patent 04603, 1994
247. H. Toda (Nippon Zeon), JP 102,042, 1995; *Chem. Abstr.*, **123**, 1445 (1995)
248. K. Someta *et al.*, (Mitsui Toatsu Chemicals), Eur. P. 467,646, 1991
249. D.P. Lorah and M.P. Frazza, (Rohm & Haas), Eur. P 478,193, 1992

5
Crosslinking and Curing

H. Warson

1 INTRODUCTION

The formation of polymers is fundamentally one of linear addition. Whilst there may be appreciable chain branching, in general the polymer molecules formed are of finite size. This is because the unsaturated molecules that take place in addition reactions are bifunctional in nature: the double bond is considered as bifunctional [1]. If monomers have more than two functional groups, they may become multifunctional. Whilst this is difficult to represent in a planar diagram, it will be readily seen that a three-dimensional or crosslinked network is rapidly built up. This will produce a product of the same type as cured alkyd resins, usually formed by the condensation of glycerol or pentaerythritol with phthalic anhydride, or phenoplasts, formed by condensation of phenol with formaldehyde.

The term 'curing' is generally synonymous with crosslinking, but it has reference to a specific industrial process, e.g. a stoving paint. 'Thermosetting' is often used, but the term should be restricted to reactions where heat is necessary to effect cure.

In practice, because of differing reactivity of double bonds, the presence of two unsaturated groups in a molecule does not imply that both will polymerise at the same rate and produce a crosslinked three-dimensional structure. The obvious examples are butadiene $CH_2{:}CH.CH.CH_2$ and isoprene $CH_2{:}C(CH_3).CH.CH_2$. When these molecules are polymerised under controlled conditions, substantially only one double bond in each molecule takes part in the polymerisation (see Chapter 1, Section 6.8). Similarly, molecules containing two allyl groups, e.g. diallyl phthalate $C_6H_4(COOCH_2.CH.CH_2)_2$, polymerise to an appreciable extent as if they were bifunctional only, with a single unsaturated group taking part in the polymerisation, although ultimately this monomer will form an insoluble structure, either alone or with a comonomer such as styrene. Another phenomenon, cyclopolymerisation, which occurs with methacrylic anhydride $[CH_2{:}C(CH_3)CO]_2O$, prevents crosslinking taking place, but will not be discussed further.

A monomer of the type of p-divinylbenzene $C_6H_4(CH{:}CH_2)_2$ is one of the most efficient comonomers in producing crosslinked structures, and will do so appreciably in a concentration of 0.15 % relative to another monomer,

Introduction

such as styrene, if the overall molecular weight is high enough. This type of copolymerisation may be shown diagramatically as

$$-CH(C_6H_5).CH_2.CH.CH_2.CH(C_6H_5)-$$
$$| $$
$$C_6H_4$$
$$|$$
$$-CH(C_6H_5).CH_2\overline{CH.CH_2}.CH(C_6H_5)-$$

For all degrees of crosslinking except the very lowest, a three-dimensional structure is built up. It may be readily seen that the higher the molecular weight of the polymer, the more effective will be the crosslinking. Thus, for a polymer consisting of 100 units, e.g. a low molecular weight polystyrene, an addition of 0.1 % of a divinylbenzene, taken as a molar proportion, which links two polymer units together, is capable, on average, of linking one-fifth of the polymer units. On the other hand, with a high molecular weight polymer of 10 000 units, i.e. $M_w \sim 10^6$, each molecule of divinylbenzene on average provides at least five crosslinking sites per molecule. This provides a very rapid network formation, particularly if the normal Gaussian distribution of crosslinks is allowed for. This particular aspect of crosslinking, on the effect of polymer molecular weight, is significant. It is possible, in a suitably controlled polymerisation, to include *p*-divinylbenzene up to 2 % w/w of total monomers without causing gelation [2]. Controlled increase of molecular weight of this type is used in alkali-soluble and alkali-thickenable latices.

Crosslinking in a polymer produces several advantages. Solubility, whether solvent or aqueous, is heavily reduced, usually to a 'swelling index', and possibly completely eliminated. Some physical properties of a polymer, e.g. abrasion resistance and the property generally referred to as 'cohesive strength', are much improved if it is modified by a crosslinking process. On the other hand, the flexibility of a polymer is reduced by being modified. The general balance of altered properties thus makes a crosslinked polymer highly desirable in many applications, e.g. in industrial finishes where solvent attack is likely, where abrasive dusts may be present, in textile finishes that are required to withstand dry cleaning and in removable adhesives, where high cohesive strength is desirable. Many of these are discussed further in later chapters.

In industrial practice, crosslinking is usually effected by a two-stage process. It is possible to have a monomer which will first polymerise by a normal free radical addition process and then the polymer formed may crosslink in a second stage; an additional catalyst is often required. Methylolacrylamide (Section 8.2) is in this category, a trace of acid being necessary for crosslinking. Similarly, glycidyl methacrylate (Section 6.1) will first polymerise by normal vinyl addition, producing $[-CH_2.C(CH_3).C(CH_3)COOCH_2\overline{CHCH_2O}]_n$. Heating with amines or anhydrides causes a second polymerisation of the oxirane group ($-CH.CH_2O$). Since these groups can easily react with corresponding groups of any chain if steric conditions permit, to build an oxyethylene CHX.CH$_2$O- chain, where X is itself a polymer 'tail', which may become part of a chain, it will be seen that a network structure may develop rapidly.

In many cases crosslinking is effected by the external addition of another polymer of relatively low molecular weight which also has crosslinking properties. It may also be effected by the external addition of a relatively simple polyfunctional molecule. Thus copolymers which include carboxyl groups will react with various polymers containing hydroxyl groups, epoxides or aminoplasts which contain methylol —CH$_2$OH groups, the special feature being that this group is attached to nitrogen, giving enhanced reactivity. This will be considered further in Section 8.1. Other types of crosslinking are based on sulfur vulcanisation, which is used with natural and synthetic rubbers, and oxidative crosslinking, as with drying oils.

The crosslinking of a vinyl polymer, which may also take place with a latex-based product, in an uncontrolled reaction might lead to film shrinking, cracking or excessive embrittlement. If a composition is pigmented (Chapter 4), the wetting of the pigment or extender by the polymer may be disturbed, resulting in 'cratering' or the formation of 'fish-eyes'. In some cases adhesion to a substrate may be reduced by excessive crosslinking. Equally there may also be film yellowing or discoloration.

Many crosslinking polymers of industry are based on acrylic monomers of various types, so much so that crosslinking or thermosetting addition polymers have become almost synonymous in some cases with thermosetting acrylics. This is because many addition polymers capable of subsequent cure are acrylic based, since the acrylic and methacrylic acid units or the methylolamide groups in the polymer molecules are well adapted to a crosslinking reaction because of the reactivity of these groups. These reactive monomers have few complications in addition polymerisation in emulsion. There are some exceptions to the inclusion of reactive acrylic monomers, e.g. monomer molecules containing diene groups.

The degree of crosslinking is discussed in the following sections. It may be mentioned at this stage that the degree of swelling in standard solvents and measurements of the stress–strain characteristics of the swollen polymer gels have been used. In theory, but not in practice, a giant molecule of almost infinite molecular weight may be formed.

1.1 Problems of crosslinking in latices

The stability of a latex depends on the individual particles being stabilised, either electrostatically or sterically. If indefinite crosslinking takes place, it becomes extremely difficult for individual particles to remain stable, as a crosslinking process forms extremely large molecules, the molecular weight of which is difficult to ascertain because of lack of solubility, although modern techniques of measurement overcome this. It is therefore impractical to obtain a highly crosslinked polymer in latex form, although partial crosslinking is possible, as is shown by vulcanised synthetic rubber latex. The ideal case would be to prepare a latex which was not crosslinked during the polymerisation process, but will crosslink on drying, preferably at ambient temperature, but in some cases by the application of heat, a catalyst, radiation, including gamma ray and electron beam, or by pH control. An example is a latex containing a polymerised acid group which is neutralised with ammonia or a volatile amine, which are lost on drying. If there is present in the formulation a monomer such as acrylamide or one of its derivatives, the lower pH on drying will encourage a crosslinking reaction by the active group. Further details are included in Section 8.

Whilst the crosslinking of a vinyl resin prepared in latex form imparts many useful properties, the reaction must be carefully controlled in order that it has adequate utility. An uncontrolled reaction may lead to film shrinkage or cracking or excessive embrittlement. In addition, if a latex is pigmented, the wetting of the pigment by the polymer or dispersants might be disturbed, resulting in cratering or the formation of 'fish-eyes'. There may also be film yellowing or discoloration.

To obtain a successfully cured product which is not excessively swollen by solvents or attacked by other reagents, the crosslinking groups must be present at regular intervals in substantially all individual polymer molecules present, allowing for steric hindrance. Many practical experiments have failed through this being ignored, particularly where acrylic or methacrylic acids have been copolymerised with water-insoluble monomers in emulsion. It is usually found that the acids have completely polymerised whilst there is considerable water-insoluble monomer remaining in emulsion. The latter will ultimately polymerise, but with few, if any, crosslinking groups available [3].

A theoretical study stresses the inhomogeneity which occurs in the early stages of crosslinking. It is considered that soon after the start of the crosslinking reaction, 'microgel' particles are formed which are highly crosslinked and which have pendant double bonds, which are shielded sterically from further reaction [4]. A simulated model for free radical polymerisation accompanied by crosslinking performed with microgel particles gives qualitative agreement with experiments on the dependence of conversion on the number of pendant double bonds [5].

The importance of the second-order transition point T_g and the corresponding minimum film formation temperature MFT have been mentioned previously. Latex films must be prepared above the MFT to obtain a coherent film, and thus to have adequate film strength, but with resistance to abrasion and blocking when coated objects are stacked or other objects placed on the film. If limited crosslinking readily takes place on filming, this problem is overcome. Excessive crosslinking, however, may cause loss of flexibility and elasticity.

In some cases it is possible to produce a 'pseudo-crosslinking' effect. An example is the inclusion of a hydroxyl group in one monomer, e.g.

hydroxyethyl acrylate, and an acid group in another molecule, e.g. methacrylic acid. These are usually included at about 1.5 % each by weight with various acrylic ester monomers in a polymerisation. They do not react under conditions of emulsion polymerisation, but on film formation they give added strength to the resultant films, e.g. in adhesives. In other cases, e.g. when an epoxide-containing monomer such as glycidyl methacrylate is present with an acid-containing monomer, crosslinking may take place via the epoxide group, particularly on heating. Other more complex monomers with reactive groups are discussed later.

1.2 A survey of crosslinking

For the purposes of this section, crosslinking will be discussed under the headings of the principal group in the vinyl polymers, including double bonds. A summary will also be given of the active compounds which, in conjunction with the vinyl polymers, give rise to crosslinking. The types often referred to as 'vulcanisation groups' and closely associated with rubber will not be discussed.

Crosslinking may take place on the polymer alone, aided by an external agent such as a catalyst, often an acid, but sometimes an alkali, a metal compound or sometimes heat alone. It is highly desirable, in many cases, that an emulsion polymer, after coating as a film, should crosslink at ambient temperature, without any problem of instability arising in the latex, even if standing for a considerable time. If an acid catalyst is required, the problem of 'cold cure' is overcome by including a catalytic quantity of an ammonium salt, sometimes a volatile amine salt, of the acid in the latex.

A review considers general aspects of crosslinking reactions derived from linear polymers [6]. A series of surveys is available of the technology, properties and applications of crosslinked latices, including epoxides, epoxyesters, polyurethanes, silicones and phenolic emulsions [7–11]. A published symposium describes many aspect of the processes that occur during crosslinking [12].

Hydrophilic functional groups, as in N-methylolacrylamide, tend to orient on the surface during emulsion polymerisation, whilst the more hydrophilic groups, e.g. glycidyl methacrylate, are randomly distributed. Emulsion polymers including the first type have improved stress deformation and thermomechanical properties, and crosslinking reactions as well as hydrogen bonding are maximised. However, in the former type crosslinking on the surface by an added reagent tends to prevent migration inside the particle. Reactions of the second group, if slower, are more homogeneous [13].

The modelling of crosslinking kinetics in emulsion polymerisation has been examined and extended to include systems that employ high mole fractions of divinyl monomer. Under special, rather limiting conditions, the present kinetic models reduce to Flory's theory of network formation in which an equilibrium system is assumed. These kinetic approaches provide a great insight into the phenomenon of network formation and are readily applied to emulsion polymerisation processes [14].

Various reactions are described in the literature involving vinyl-type polymers with condensation resins of such types as aminoplasts and phenoplasts, and even with stabilised polyurethanes, as described in Chapter 6. These are usually not added in catalytic quantities but in more substantial quantities, sometimes as much as 50 % by weight. They normally need heat for the reaction to take place. External epoxides have sometimes been added to induce crosslinking.

In cases where an external agent is required, it may be blended as a technical product with the latex, provided the latter remains stable. It may be necessary to include the additive just before application and it may be available commercially as a 'two-part' product. In this chapter the various reactions are considered in the light of the active groups giving rise to crosslinking, whether a polymer is used alone or with a further additive. The latter may itself be polymeric, although usually often of relatively low molecular weight, and is often a polymer precursor rather than a polymer itself, e.g. dimethylolurea, the simplest compound formed from urea and formaldehyde, or a simple diglycidyl ether. The chapter is therefore divided into the sections indicated below. It must be appreciated that there is often a major overlap between these sections.

2 ESTIMATION OF CROSSLINKING

2.1 Cure studies

Some studies quoted here refer to crosslinking in water-soluble products in aqueous solution which are crosslinkable, but they are considered to be useful in elucidating the nature of crosslinking in emulsion-type products. Some may refer to non-vinyl macromolecules, which are described in more detail in Chapter 7. A rheological and chemical overview of thermoset curing, useful for process automation, is available [15].

In simultaneous crosslinking with emulsion polymerisation, the crosslinking density tends to be fairly high, even in the early stages of the reaction. Crosslinking density development is substantially different from that in bulk and solution polymerisations in which the crosslink density increases with time, especially in the early stages. The monomer concentration in the polymer particles is lower than that for linear emulsion polymerisations as long as the monomer droplets exist, resulting in an enhanced tendency towards crosslinking reactions. This research concluded that homogeneous networks cannot be formed for vinyl/divinyl emulsion polymerisation. However, a kinetic method which was developed could be used to identify semi-batch methods to control the structure of the copolymers [16].

Time–temperature transformation cure diagrams of thermosetting polymeric systems include the onset of phase separation, gelation, vitrification and full cure, the results being considered in the light of the Flory theory [17, 18]. Dielectric cure analysis is a modern method for characterizing cure properties [19].

The crosslink density is a function of the density of the functional groups, effective number average functionality, ratio of the number of crosslinking functional groups to prepolymer reactive groups, the gel fraction of the crosslinked system and the extent of crosslinking in the gel phase [20].

2.1.1 Measurement of the degree of swelling

The degree of swelling of a 99 : 1 acrylic acid–ethylene dimethacrylate salt network shows a minimum at 15 % neutralization due to hydrogen bond formation between ionized and non-ionized carboxyl groups. Data relating to photoelastic behaviour are also included [21, 22].

The swelling ratios of copolymers of styrene with multifunctional acrylic and methacrylic esters, divinylsulfone and divinylketone are given as a function of the crosslinker content, and have been compared with crosslinking with divinylbenzene and isopropylbenzene. Crosslinking is quantified as an excess percentage vinyl group in the monomer mixtures [23]. Swelling characteristics, which are used in conjunction with thermoelastic properties to characterise networks formed by either peroxide or photocrosslinking, indicate that the latter gives a more homogeneous distribution of crossover points than peroxide cured samples [24].

2.1.2 Viscosity characteristics

A general analytical model is given for determining chemoviscosity variations during the cure of thermosetting resins. It is assumed that R_p is diffusion controlled and linearly inversely proportional to the viscosity of the medium over the entire cure cycle. This results in a non-linear diffusion equation which may be solved numerically. Predictions agree with experimental results using the epoxide precursor Epon 828 with a rheometer. A model predicts chemoviscosity over six orders of magnitude [25].

The viscosity–time curves of thermosetting coatings prior to gelation have been measured on an automated Ferranti–Shirley viscometer. They have been analysed to obtain data relating to cure rheology and results have been used to derive physical and chemical activation parameters to predict cure performance [26, 27].

A system including a technical epoxide, methacrylic acid, methyl methacrylate and an oligourethane acrylate has been examined for viscoelastic properties and network structure. The first radical stage starts as a microgel and ends as a homogeneous polymer network with an increase in viscosity and reaction rate. The second stage is a polyaddition in which the rate of reaction of epoxide and carboxyl groups depends on the relative contribution of heteroatom and carbon chains in the first stage [28].

2.1.3 Gel permeation chromatography

Gel pemeation chromatography has been used in the determination of cross-linked structure of polymers, an example being 1,2-polybutadiene cured with *tert*-butyl styrene [29].

2.2 Cure tests

Tests on elasticity modulus are considered to be the most efficient for estimating crosslinking and swelling [30]. Torsion braid analysis (TBA) is a technique for characterizing the cure and properties of thermosetting systems. Time–temperature–transformation cure diagrams have been produced, and the T_g is also an index for monitoring cure [31, 32]. A further analysis concludes that it is not possible to detect a gelation point by TBA, but the second, vitrification, peak is correct [33].

A combination of TBA and tensile mode dynamic analysis allows for the observation of curing reactions over a wide range of temperatures and times, including gel time and long-term curing in one measurement. It applies to different crosslinker types and catalysts [34].

Spectrophotometric methods provide the best method for determining the degree of cure of interior coatings for metal cans. This involves the analysis of material extracted under standard conditions with tetrahydrofuran and measurement of absorption of the extract at 280 nm [35].

Dielectric cure analysis (DEA) is a technique for investigating the processing behaviour and physical and chemical structures of polymers through the measurement of their dielectric properties. It is useful in estimating both the degree of polymerisation and the degree of crosslinking. Its original use was in the aerospace industry, but is now used more widely in thermosetting polymers [36].

3 MULTIPLE DOUBLE BONDS

3.1 Polymers including divinylbenzene and allied monomers

Divinylbenzene $CH_2{:}CH{-}C_6H_4CH{:}CH_2$ (see Chapter 1, p. 30) contains two vinyl double bonds, both of which polymerise and copolymerise readily, and is very useful in increasing molecular weight if it is included in restricted ratios or to impart insolubility in solvents in larger ratios, which cannot be defined quantitatively. This is irrespective of whether the vinyl groups are in the *ortho meta* or *para* positions. Another feature of divinylbenzene, unlike dienes, is that the two double bonds are equal and thus can readily polymerise simultaneously. Other features have already been mentioned in Section 1.

Microgels are formed by the emulsion polymerisation of 1,4-divinylbenzene in the presence of sodium lauryl sulfate and potassium persulfate. The product is not homogeneous; there is some coagulum which decreases with an increase in emulsifier. The particle sizes are in two ranges from 0.1 to 0.3 μm and from 0.5 to 10 μm with primary and secondary agglomerates. The phenomena are explained by the sulfate anion radicals formed from potassium persulfate adding to the pendant vinyl groups of the microgels, thereby rendering them soluble in methanol. These hydrophilic groups at the particle surfaces also prevent adsorption of the hydrophobic parts of emulsifier molecules. The number of sulfate groups increases with time and are included in about three-quarters of the polymers as a whole [37].

Crosslinked latices of fine particle size, including 1–5 % by weight of divinyl benzene, are prepared using sodium styrenesulfonate as a polymerisable surfactant. Polymerisation takes place at 91 °C with a persulfate initiator. This is preferred to redox initiation at 40–50 °C, which gives larger, more disperse particles [38].

A multilayer process utilises a copolymerisation in from four to seven stages (Table 5.1). The monomer addition, which includes small ratios of divinylbenzene, is balanced so that here is increasing softness in successive layers, except for the outer layer. There is probably some simultaneous grafting [39].

A two-stage latex, vaguely described as an 'overpolymerisation', has a core of a copolymer of 2-ethylhexyl acrylate, vinyl acetate and acrylic acid, followed by a shell at about one-tenth of the original weight of 2-ethylhexyl acrylate, acrylic acid and divinylbenzene (29.7 : 1 : 1.5). The final latex has a lower viscosity (13 versus 33 P) compared with a similar latex omitting the divinylbenzene. This is probably due to a lower hydration level of the particles due to the crosslinking agent [40].

As an other example, 100 parts of a divinylbenzene of 55 % purity, the balance being monounsaturated styrenic monomers, is heated with 8 parts of a low molecular weight 98 : 2 styrene–methacrylic acid copolymer, of number average molecular weight 3100 and weight average 5000, to give highly crosslinked polymer particles, of average diameter 0.38 μm, featuring good heat stability, even at 300 °C [41].

Table 5.1 Multistage copolymerisation including divinylbenzene [39]

	Stage 1	Stage 2	Stage 3	Stages 4–6	Stage 7
Potassium salt, fatty acid	1.2	—	—	—	—
$FeSO_4 \cdot 7H_2O$	0.003	—	—	—	—
Sodium pyrophosphate, $Na_4P_2O_7$	0.3	—	—	—	—
Dextrose	0.4	—	—	—	—
Water	100.0	—	—	—	—
Styrene	10.0	9	6	—	15
Divinylbenzene	0.3	0.01	0.01	0.01	0.01
Allyl methacrylate	0.1	—	—	—	—
Butadiene	—	6	9	15	—
n-Octyl mercaptan	—	—	—	—	0.03

The above are typical patent examples of the effects of the addition of small quantities of divinylbenzene to the monomer basis of latices. There are many examples in the patent literature of the use of technical divinylbenzene.

3.2 Polymers derived from acrylic and allied esters—multiple double bonds

A new kinetic model for network structure during the emulsion polymerisation of vinyl and divinyl monomers has been applied to the copolymerisation of methyl methacrylate and ethylene glycol dimethacrylate. The crosslink density is high, even in the early stages of polymerisation, and the networks are very heterogeneous. The kinetics of crosslinking and the network structure differ from those in bulk and solution polymerisation [42].

Many glycol esters have been included in acrylic and methacrylic ester formulations to obtain controlled crosslinking. A large number of glycol esters of both acids are available commercially. The simple diesters are based on ethylene glycol acrylate or methacrylate $C_2H_4(OOCCX:CH_2)_2$, where X is H or CH_3. In crosslinking they are very reactive and are liable to produce undesired gels in emulsion, even in small quantities. In consequence, glycol esters with longer hydrocarbon chains are preferred. However, cautious addition of ethylene glycol dimethacrylate has been claimed during a copolymerisation of styrene with methyl methacrylate [43].

Neopentyl diacrylate with methacrylic acid, 2-hydroxyethyl acrylate, styrene and ethyl acrylate (0.75 : 15 : 15 : 60 : 210) are included in a copolymerisation in which 1.5 parts of butyl mercaptan are added halfway through the reaction. The mercaptan has the property of controlling both the molecular weight of the polymer and particle morphology [44]. Trimethylolpropane trimethacrylate has also been disclosed as a controlled crosslinker in an acrylic latex, increasing film hardness and resistance to acetic acid [45].

Crosslinked latices with antistatic properties are formed by the emulsion polymerisation of butyl acrylate, methacrylic acid and tetraethylene dimethacrylate with sodium lauryl sulfate as emulsifier and potassium persulfate as initiator. After polymerising for 6 h at 85 °C, vinyl chloride and further initiator solution are added and the reaction continued for 3 h at 65 °C. After neutralization, a T_g of -5 °C is observed. These latices have antistatic utility [46].

A two-stage emulsion process is used to provide crosslinked acrylic polymer particles suitable as matting agents in thermoplastic polymers. The first stage is the polymerisation of styrene–butyl acrylate–acrylonitrile (60 : 30 : 10); the second stage is similar, with the addition of hexanediol diacrylate and N-methylolacrylamide (Section 10). The latex, partially crosslinked, is coagulated with dilute sulfuric acid [47].

A range of microgel latices prepared by a core–shell technique has been converted to emulsion paints. Microgel latices prepared with trimethylolpropane trimethacrylate showed slightly inferior performance with regard to slight precipitate formation, monomer conversion, shear and heat stability and electrolyte tolerance compared with latices including ethylene glycol dimethacrylate. However, the former latices showed better water resistance, water absorption and vapour transmission properties, and provided superior emulsion paints [48].

The viscosities of microgels of ethyl acrylate, methacrylic acid and 1,4-butanediol diacrylate have been studied. The low shear viscosity at 0.95 % by weight when neutralised with sodium hydroxide is about five orders greater in magnitude than when unneutralised ($\sim 10^{-2}$ P), due to microgel swelling [49].

Ethylene glycol dimethacrylate may be included in emulsion compositions intended as polymer pigments (see Chapter 4, Section 7, especially reference [246]. A copolymer based on styrene and methyl methacrylate with 6 % on monomers of acrylic acid and 1 % of the dimethacrylate is described. The final procedure involves neutralization and then acidification to produce porous particles useful as pigmentary materials [50]. An allied process is based on a two-stage emulsion polymerisation in which the glycol dimethacrylate is included in the first stage, which may include methacrylic acid [51, 52].

There are other claims to a core-shell morphology of polymers, usually of the core containing monomers with two or more double bonds, with a softer shell, with various claimed advantages. Systems including an acrylamide derivative are described in Section 8.1. An example of a crosslinked core is based on butyl acrylate with methacrylic acid, 1,4-butylene diacrylate and allyl methacrylate followed by methyl methacrylate-ethyl acrylate as a shell. The polymer formed is applied as a high impact strength modifier for nylon-6. Various claims to a core of butylene dimethacrylate and butyl acrylate have been disclosed, with a possible addition of methacrylate (see Section 6.1) [53-58].

A different type of dimethacrylate used as a comonomer is formed by reacting a polycaprolactone with acryloyl chloride to form an acrylic-terminated polycaprolactone. This is copolymerised in emulsion, producing separated products ranging from elastomers to tough plastic materials [59].

3.3 Allyl polymers—multiple double bonds

The allyl group —CH_2—CH:CH_2 is only weakly polymerisable or copolymerisable by free radical methods. This is explained simply by ready chain transfer. The reactivity may vary somewhat depending on the attached groups. It should be noted that all olefinic hydrocarbons with the double bond are in the 1 : 2 position, except for the vinyl CH_2:CH., which are allylic in character. However, allyl compounds with two or more allyl groups can be polymerised in a controlled fashion so that they can be partially polymerised and, by more rigorous treatment, finally form a crosslinked compound. Typical amongst these allyl compounds is o-diallyl phthalate o-$C_6H_4(COOCH_2CH:CH_2)_2$, which is occasionally included in emulsion polymerisation formulations as a monomer which may have subsequent crosslinking properties. Occasionally the allyl group is found in a compound such as allyl methacrylate $CH_2{:}CHCH_2OOCC(CH_3){:}CH_2$. This section will describe a number of allyl compounds, especially from the patent literature, which are included in emulsion formulations. Whilst some references may not refer specifically to latices, they indicate the nature of the reaction [60].

Diallyl esters of linear dicarboxylic acids with carbon chain lengths from two to ten have been prepared and used as difunctional crosslinking agents for vinyl polymers. This enables the reactivity of the copolymers to be determined in terms of chemical structures. Swelling in chloroform was used to determine the average molecular weight between crosslinks and thus its efficiency [61].

A polymerised diallyl phthalate latex, optionally including allyl esters, is disclosed. About half of the monomer with the sodium salt of an alkylphenyl polyoxyethylene phosphate as emulsifier and about 0.75 % on the monomer of potassium persulfate is heated at 80 °C to 50 % conversion; then the balance of monomers is added over 4 h. This gives a 43 % solids latex of low viscosity and good chemical stability. It is assumed that substantially only one double bond is polymerised and that curing is capable at a later stage [62].

Diallyl phthalate is occasionally included at 2–6 % in polymerisations including an acrylic ester, styrene and acrylic acid [63].

A suitable latex for impregnation comprises a diallyl phthalate prepolymer, diallyl maleate and styrene (15 : 1 : 4) and is formed with dibenzoyl peroxide and methylethyl ketone peroxide initiators. This emulsion is unusual in that polymerisation takes place after impregnation, e.g. of an 0.08 mm paper [64].

Stable latices have been described which are prepared from allyl methacrylate (1), diallyl phthalate (II), vinyl chloride, vinylidene chloride, 2-ethylhexyl acrylate and methacrylic acid (0.5 : 5.0 : 20.4 : 71.3 : 6.3 : 2.0). The allyl methacrylate is included in the first half of conversion to give a copolymer with a gel content of 60 %; diallyl phthalate is added in the conversion range 70–90 % to provide the residual unsaturation for autoxidative crosslinking. Films from this latex have a good modulus and solvent resistance and good elongation before break [65].

A rather complex multistage process has been described in which preemulsified monomer feeds are added to the reactor sequentially. These include potential crosslinkers such as allyl methacrylate and diallyl phthalate. A photosensitizer (see Section 10), typically benzophenone, is added to the final composition to catalyse the crosslinking reaction. The degree of crosslinking enables a high degree of extensibility to be maintained [66].

In an emulsion polymerisation of ethylene and vinyl acetate, an allyl perester $(CH_3)_3COOCO_2CH_2CH{:}CH_2$ is included. This enables a film to crosslink if heated to 150 °C. The allyl ester is unusually stable at polymerisation temperatures [67].

Allyl ethyl malonate $CH_2{:}CHCH_2OOCCH_2COOC_2H_5$ (5 % of monomers) is copolymerised with ethyl acrylate using anionic–non-ionic emulsification and persulfate/bisulfite redox initiation. After precipitation the polymer can be vulcanised as a rubber [68].

Allyl alcohol may be reacted, using a catalyst with carbon monoxide under high pressure, to form allyl vinyl acetate. This is a derivative of vinyl acetic acid with formula $CH_2{:}CHCH_2COOCCH_2CH{:}CH_2$. This rather interesting compound has two points of polymerisation, although the condensation causes both double bonds to be allylic in character. Copolymerisation with possible subsequent weak crosslinking is of interest [69].

An allyl acrylate–butyl acrylate (1 : 99) copolymer, prepared as a latex at 32.6 % of solids, is blended 3 : 1 with a styrene–butyl acrylate–acrylonitrile copolymer with 32.6 % of solids. This produces a thermoplastic elastomer when precipitated with aluminium sulfate. The few crosslinks provided by the allyl groups, or possibly pendant acrylic groups, provide the required properties in the copolymer [70].

A two-stage process includes the emulsion polymerisation of ethyl acrylate with triallyl cyanurate to form a crosslinked polymer followed by the further addition of methyl methacrylate and N-cyclohexylmaleimide, which, after polymerisation, form an elastomer [71].

For further references to allyl crosslinked polymers see references [111] to [111C] in Section 6.1.2 reference and [270] in Section 13.2.

3.4 Diene polymers

An elementary account of the polymerisation of dienes, in particular butadiene, has been included in Chapter 3, Section 6. It is sufficient to add here that under controlled conditions only one double bond at a time will polymerise and that two types of addition occur, both 1 : 2, the addition of which may be illustrated as $-CH_2.\underset{\underset{CH:CH_2}{|}}{CH}-CH_2.\underset{\underset{CH:CH_2}{|}}{CH}-$, and 1 : 4, which is
$-CH_2.CH{:}CH-CH_2-CH_2.CHCH{:}CH.CH_2-$.

Controlled polymerisation or copolymerisation, usually with styrene or acrylonitrile, ensures that the polymer as formed is substantially linear, but crosslinking is possible as a secondary reaction. These copolymers are of course the basis of synthetic rubbers, which are usually crosslinked by the process of vulcanisation, which involves the use of sulfur or sulfur compounds, often with an accelerator to reduce cure time. Organic peroxides and polyurethanes have also been used for the same purpose.

The crosslinking of butadiene compounds is facilitated by the inclusion of carboxyl groups (see Chapter 3, Section 6.2) The inclusion of an acid such as itaconic acid or methacrylic acid usually has the effect of increasing the 'gel content', i.e. the amount of insoluble matter obtained when a film is treated with a selected solvent. The latices are invariably made alkaline with a volatile base such as ammonia or an amine, the removal of which seems essential for crosslinking to proceed.

Some reactions which indicate these effects include a polymerisation recipe (not indicated) using azobisisobutyronitrile as initiator and 0.7 g/100 g of *t*-dodecyl mercaptan on monomers as the chain transfer agent, giving a copolymer of butadiene and acrylonitrile which at 90 % conversion gives only 2 % of gel. This increases rapidly with ageing unless hydroquinone is added to stop further free radical processes leading to crosslinking [72].

In some cases where a latex is pigmented, the pigments themselves are suitable catalysts for crosslinking. This applies to a compound of a metal in Groups V, VI and VII, or the iron vertical subgroup of Group VIII. The weight ratio of pigment to binder should be at least 0.5 : 1 and preferably 1 : 1. Technical products which have been found successful are an ochre and red oxide (ferric oxide Fe_2O_3), chromic oxide Cr_2O_3, scarlet chrome $PbMoO_4$ and Prussian Blue $KFe(CN)_4Fe(CN)_2$. The procedure is especially useful in stoved paints. If a pigment is not included, 0.23 % based on latex solids of ferric alum, manganese naphthenate, an iron naphthenate and ferrous ammonium sulfate all induce crosslinking. Cobalt salts were ineffective [73, 74]. As crosslinked products of this type are often included in industrial metal finishes, further examples will be given in a later chapter.

Copolymers of acrylonitrile, styrene, isoprene and maleic anhydride, obtained by emulsion polymerisation, show high thermal stability, as acrylonitrile causes intramolecular cycles during polymerisation. On thermal treatment of the copolymer, crosslinking occurs due to the unsaturated C:C bonds in the polymer which give rise to crosslinking, aided by the acid groups present [75].

The crosslinking reaction in these carboxylated latices is probably an addition of the carboxyl group across the residual double bonds, a single unit, based on a 1 : 4 addition of methacrylic acid across two chains:

$$-CH_2-CH:CH_2-CH_2-\underset{\underset{\underset{-CH_2-CH_2.CH.CH_2.CH_2-}{|}}{\underset{O}{|}}}{\underset{C:O}{|}}{C(CH_3)}-CH_2-CH$$

A partially crosslinked rubber-like copolymer is formed by emulsion polymerisation of chloroprene (Chapter 1, Section 6.3) with $\leqslant 2$ % of N,N'-dimethylaminoethyl methacrylate (Chapter 1, Section 6.4.4). The polymerisation rate and average molecular weight of the copolymer increases as N,N'-dimethylaminoethyl methacrylate is increased, which also promotes better physicochemical properties than pure chloroprene rubber [76].

The ammonium salt of a carboxylated latex has been used to emulsify butyl acrylate in a polymerisation, cobalt naphthenate being added. The initial polymer has no gel content, but on standing at ambient temperature the gel content becomes 30 and 70 % after 1 and 10 days respectively [77].

3.5 Other vinyl-type polymers with residual unsaturation

The diene monomers described in the previous section are a unique type, since by resonance both double bonds are polymerisable. In many other vinyl monomers, unsaturation, usually in the alcohol group of an ester of a polymeric acid, gives rise to polymerisation of a different type. When these groups are incorporated as side chains in a polymer, they crosslink by oxidation in a manner somewhat similar to that of the drying oils. One group of monomers which is frequently encountered is that of derivatives of cyclopentadiene, which is allylic in character. Cyclopentadienyl acrylate (CPDA) is one example. Dicyclopentadienyl methacrylate (DCPDM), chemical name (5,2,1,0)-3 : 8-decadiene, is frequently quoted in patent specifications, and so occasionally is the corresponding acrylate (DCPDA) [78]. CPDA methacrylate as a film cures in air at ambient temperature in 1 day and may be part of a more complex adhesive composition [79].

$$\begin{array}{c} \diagup CH_2 \diagdown \\ HC CH \\ \| \vdots\!\!\!-OOCCH:CH_2 \quad CPDA \\ HC\!\!-\!\!-\!\!-CH \end{array}$$

$$\begin{array}{c} \diagup CH \diagdown \diagup CH_2 \diagdown \\ HC CH_2 CH CH \quad DCPDM \\ CH_2C(CH_3)COO\!\!-\!\!\!+\!\!\!\!\!\!\!   \| \\ HC\!-\!CH\!-\!CH\!-\!-\!CH \end{array}$$

In the above formulae, the dotted lines indicate the probable location of the acrylate and methacrylate groups according to published formulae, indicating that they could be on either of the two carbons on the left-hand side of the formulae.

DCPDA may be copolymerised by emulsion polymerisation together with methyl methacrylate, butyl acrylate and methacrylic acid. The emulsifier is a t-octylphenol-40 ethylene oxide condensate and the initiator is 0.12 % of ammonium persulfate with 0.16 % of sodium hydrosulfite, polymerisation taking place at 62 °C with a final addition of t-butyl hydroperoxide of 0.12 %; cobalt naphthenate of 0.06 % is added as an accelerator with methylethyl ketone oxime of 0.25 % as a stabiliser. For a good ambient cure in the presence of air it is recommended that extra reactive monomer equal to one-third of the total polymer weight be added. The ring double bonds are probably the reactive sites in the cure reaction [80].

Specifications describe copolymers of DCPDA with hydroxyethyl acrylate, methyl methacrylate, butyl acrylate and acrylic acid (5 : 5 : 39 : 50 : 1), which is mixed with 5 % of additive with a 1 : 1 adduct of trimethylolpropane–triallyl ether, and coated on steel with the addition of cobalt and lead naphthenates and cyclohexanone peroxide. This coating hardens to a 2B level after 7 days, probably by atmospheric oxidation and crosslinking [81, 82].

A latex formed from styrene, butyl acrylate, 2-ethylhexyl acrylate, acrylic acid and DCPDA (9 : 75.5 : 10 : 2.7 : 4) also contains acrylamide and 2-(acryloyloxy)ethyl 4-(4-chlorobenzyl) benzoate (0.5 : 1). The emulsifier is the polymerisable surfactant sodium sulfoethyl methacrylate (0.5). There is limited crosslinking [83]. The related compound, dicyclopentadienyloxyethyl methacrylate (DPOMA), is prepared by the addition of ethylene glycol to dicyclopentadiene, reducing the number of double bonds to one, and then transesterifying with methyl methacrylate. This monomer (MW = 262) has the advantage of much less odour than dicyclopentadienyl methacrylate. It is cured in a similar fashion by the addition of cobalt naphthenate [84]. The structure of DPOMA is

DPOMA has also been used in a core–shell emulsion polymerisation (Chapter 3, Section 2.4.3), as given in Table 5.2. It is useful, *inter alia*, as an overprint varnish with exceptional detergent resistance. There is probably some autoxidation with possible grafting of the shell to the core [85].

The terpene derivative 5-ethylidene-2-norbornene, a bridged ring compound allied to cyclopentadienyl compounds, is reacted with methacrylic acid at 120 : 84 to form a methacrylate ester, and then forms a copolymer, as given in Table 5.3. The elastomer formed is separated and cured as shown in Table 5.3 [86].

Cyclopentadienyl and dicyclopentadienyl maleate and fumarate have been prepared and crosslinked with cobalt naphthenate accelerator, although emulsion polymerisation has not been described [87].

Unsaturated air-drying binders for coatings of a different type are formed by the reaction of glycidyl methacrylate (I) (see Section 6) with an unsaturated drying oil fatty acid. An example quotes a two-stage process of polymerisation to form a latex, as in Table 5.4. The latex formed, which includes some coalescing solvent, is useful as a binder for coatings and cures as drying oils [88].

The acrylic ester of cardanol, 3-(8,11,14-pentadecatrienyl) phenol, copolymerises with styrene or methyl methacrylate to form crosslinked beads. Cardanol is a naturally occurring material derived from the plant *Anacardium occidentale* L. It is possible that the unsaturated esters of cardanol may also operate in emulsion polymerisation, under controlled conditions [89].

Table 5.2 Core–shell copolymer including dicyclopentyloxyethyl methacrylate [85]

	Parts by weight	
	Core	Shell
Water, deionised (I)	140.0	140.0
Amphoteric surfactant (42 % aqueous)	5.0	10.1
Methyl methacrylate	127.5	255.0
Butyl acrylate	297.5	—
Styrene	—	42.5
Diethylaminoethyl methacrylate	—	106.3
Dicyclopentenyloxyethyl methacrylate	—	21.2
Octane thiol	—	21.3
The following are added incrementally:		
Ammonium persulfate	0.85	2.38
Water deionised (II)	80.0	100.0
Latex properties:		
Solids content		42.4 %
Viscosity, Brookfield at pH 4.8		940 cP

Table 5.3 Latex including norbornene unsaturated ester [86]

	Parts by weight
Norbornene methacrylate	2.0
Ethyl acrylate	98.0
Lauryl mercaptan	0.02
Polyoxyethylene dodecyl ether	2.0
Sodium dodecyl sulfate	2.0
Ammonium persulfate	0.05
Sodium bisulfite	0.05
Water as required	(optimally 100–150)
Treatment of the separated polymer is as follows:	
Elastomer above	100
Carbon black	50
Stearic acid	1
Dicumyl peroxide	4

Knead at 60 °C for 30 min and postcure for 6 h at 150 °C.

4 THE CARBOXYL FUNCTION IN POLYMERS

4.1 Introduction

The principal carboxylic acids that take part in secondary reactions are acrylic and methacrylic acids in the form of copolymers and more rarely as homopolymers. Crotonic acid and itaconic acid also take part in these secondary reactions.

Table 5.4 Latex including reaction product—methacrylate and unsaturated fatty acid [88]

	Weight (g)
Stage 1	
Glycidyl methacrylate	131.5
Soya bean oil fatty acids	250.0
Form adduct of above and then proceed to:	
Stage 2	
Above adduct	322.6
Methacrylic acid	52.2
Butoxy ether of ethylene glycol	430.0
Azobisisobutyronitrile	Unspecified
Polymerise the above polymer solution at 120 °C and then proceed to:	
Stage 3A	
Polymer solution (Stage 2)	146.0
Water	560.0
Sodium lauryl sulfate	4.4
Ammonia, 25 % solution	9.3 mL
Ammonium persulfate	0.86
Stir above and then add with stirring:	
Stage 3B	
Methyl methacrylate	68.0
Butyl methacrylate	250.0
Heat the above at 70 °C for 7 h to produce a latex with an average particle size of 220 nm.	

Crotonic acid mainly occurs in copolymers with vinyl acetate and other vinyl esters since it copolymerises well with vinyl acetate, even in emulsion form, provided that it is dissolved in the monomer and not in the water phase. Copolymerisation with about 4 % in emulsion gives a latex that is irreversibly alkali soluble. Above about 20 % of total monomer, the polymerisation in emulsion becomes too slow to be practicable. Technical alkali-soluble latices contain about 5 % of crotonic acid. These are readily soluble in ammonia solution, which facilitates blending with water-soluble resins which react with acid groupings.

Itaconic acid copolymerises to some extent with butadiene and styrene to form a terpolymer, useful *inter alia* for carpet backing. This depends on emulsion conditions, probably some solubilisation of the major monomers by the emulsifiers, since styrene and itaconic acid will normally only polymerise in a mutual solvent such as dimethyl sulfoxide. The resulting latex is not alkali soluble, about 1–2 % of acid being the normal percentage of total monomers [90].

Water-based polymers containing acid (carboxylic) groups are usually employed in the alkaline state, with ammonia or a volatile amine such as morpholine $\overline{CH_2CH_2OCH_2NH}$ being the solubilising agent. This is irrespective of whether the polymer is alkali soluble, an alkali-thickenable latex, which often occurs, or a latex containing carboxyl groups, the viscosity properties of which are apparently unaffected by alteration of pH.

Under some conditions it is possible to form a copolymer with a carboxylic anhydride, in particular maleic anhydride, if it is dissolved in monomer, as its rate of polymerisation is much faster than that of hydration to maleic acid.

For reactions of carboxylated polymers containing a diene as monomer see Section 3.4. Some reactions of polymers containing carboxyl groups under the influence of various types of radiation are included in Section 10. A number of other reactions will be included in Chapter 9 (industrial finishing) in Volume 2.

4.2 Combinations with hydroxyl and epoxide groups

Carboxylic groups in polymers may react with hydroxyl and epoxide groups at elevated temperatures, usually about 150 °C. In a simple case, the reaction of polyacrylic acid with a polyhydroxylic alcohol, e.g. glycerol, forms a thermoset product by the formation of ester groups between the carboxyl group and the hydroxyl groups, the high joint functionality ensuring ready crosslinking.

A reaction of an acid polymer is also possible with alkyd resins containing an excess of hydroxyl groups and may also occur with a hydroxy-containing vinyl polymer. Polyvinyl alcohol can be made to react with polymers containing acrylic acid, although with some difficulty. Limited intermolecular as well as intramolecular reactions may be possible where both hydroxyl and carboxyl groups are in the same copolymer, as, for example, a latex including about 1.5 % of monomers of methacrylic acid and 1.5 % of hydroxyethyl or hydroxypropyl acrylate or methacrylate, which will give a pseudo-crosslinking effect, probably by hydrogen bonding rather than a direct reaction [90A]. This may be represented as

$$X-OH-O \atop HO-CY$$

Tetraisopropyl titanate and tetra-n-butyl titanate have been used as catalysts for crosslinking dual function acrylic polymers containing hydroxyl and carboxyl groups at temperatures of only 140 °C. Whilst this refers to polymers containing acid groups prepared in solvent, but water-solubilised by an amine, it probably applies also to latex films.

An epoxide may react with the carboxyl group. Conventional types of epoxide resin (condensates of diphenyolpropane and epichlorhydrin) may be employed, or any other type with a plurality of epoxide groups. In this case the inclusion of a catalyst such as an amine might encourage self-crosslinking of the epoxide. The fundamental simple reaction is

$$R-COOH + \underset{CH_2CHR}{\overset{O}{\triangle}} = RCOOCH_2\underset{OH}{\overset{|}{C}}HR$$

The hydroxyl group may react further with carboxyl, but if there is a plurality of carboxyl groups and epoxide groups, it will be obvious that a crosslinked network will develop.

In the formula below, n is about 24 for acrylic acid and 31 for methacrylic acid. With acrylic acid, the equivalent of the copolymer is about 10.6 that of the epoxide if acrylic acid is copolymerised and 12.3 for methacrylic acid. Approximately between 1 in 17 and 1 in 20 carboxyl groups react with epoxide. Thus, for an acid copolymer of molecular weight 100 000 at least 5 to 6 carboxyls will react per polymer molecule—more than an adequate number for crosslinking to occur. Curing is at 125–130 °C [91].

The reaction may be represented as

$$R^1-CH_2-CH-(C_2H_4)_n-R^2$$

[reaction scheme showing carboxyl group reacting with diglycidyl epoxide and second carboxyl-containing polymer to form crosslinked product with R^1, R^2, R^3, R^4 polymer residues]

where R^1, R^2, R^3 and R^4 are polymer residues. The glycidyl and the carboxyl groups may be included in the same molecule under controlled conditions. Only about 7 % of the glycidyl group is hydrolysed to a dihydroxide, and mutual reaction with the carboxyl is negligible under normal emulsion polymerisation conditions.

Many further examples will be included in Section 6 of this chapter and in the application chapters.

4.3 Metal oxides

It seems obvious that the reaction of a di- or tri-metal oxide or hydroxide with a carboxyl group on a polymer would produce a network effect. In practice there is little evidence that this proves useful, largely because the neutralised acid polymers using heavy metals rarely have useful properties and lose desirable ones such as adhesion. There are a few claims to crosslinking by metal oxides [92].

Claims have been made that carboxyl-terminated liquid polybutadiene and nitrile rubber latices are ionically crosslinked by metal oxides, increasing the tensile strength and many other properties [92A, 92B]. Many other metal compounds form complexes with polymers, and these are considered in Section 13. Some crosslinked derivatives of polyvinyl acetate are included in Chapter 8 on adhesives (Section 4), in particular, some chromium compounds.

A water-soluble zirconium salt, e.g. zirconium ammonium carbonate, is added to a neutralised butadiene copolymer latex, including itaconic acid, acrylic acid and hydroxyethyl acrylate. The addition is about 1 % of zirconium as metal to the polymer. Zinc salts may also be used. The process is for paper coating [93].

5 THE HYDROXYL FUNCTION

An obvious reaction for polymers containing the hydroxyl group is with the carboxyl group. This has already been described in the previous section. Reactions involving the hydroxyl and epoxide group are included in Section 6. The reactions of the hydroxyl group are significant, partly because of residual hydroxyl groups in vinyl ester polymers, i.e. polyvinyl alcohol units which are usually present by slight hydrolysis of polyvinyl acetate or by grafting of a polyvinyl alcohol emulsifier. The principal basis for crosslinking of the hydroxyl group is the inclusion of the monoesters of ethylene glycol and 1 : 2 propylene glycol, principally by acrylic and methacrylic acids.

Polymers containing hydroxyl groups are not normally regarded as capable of self crosslinking, but it has been claimed that copolymers of styrene and/or acrylic esters including β-hydroxyethyl acrylate are cured by heat using a zinc perchlorate catalyst. This may possibly be due to etherification [94].

Reactions with various methylolamide groups, whether internal or external, are discussed in Section 8. Reactions with aldehydes and the formation of metal complexes are considered here.

5.1 Reactions with aldehydes

One of the simplest methods of crosslinking, with particular reference to hydroxyl groups in polyvinyl acetate, is the addition of a solution of glyoxal CHO.CHO, a dialdehyde that is only stable at pH < 7. This will react only in the dried film at ambient temperature and not with the emulsion particles. Glyoxal also reacts to some extent with other hydroxyl-containing colloids such as hydroxyethyl cellulose. The crosslinking reaction can be represented as

$$\begin{array}{c} CH_2 \\ | \\ CHOH \\ | \\ CH_2 \\ | \\ CHOH \end{array} + \begin{array}{c} H\ H \\ |\ | \\ OC.CO \end{array} + \begin{array}{c} CH_2 \\ | \\ HOCH \\ | \\ CH_2 \\ | \\ HOCH \end{array} \longrightarrow \begin{array}{c} CH_2 \\ | \\ CHO \\ | \\ CH_2 \\ | \\ CHO \end{array} \begin{array}{c} H\ H \\ \backslash |\ | / \\ C.C \end{array} \begin{array}{c} CH_2 \\ | \\ OCH \\ | \\ CH_2 \\ | \\ OCH \end{array} + 2\, H_2O$$

A blend of two latices, one of which contains 20 % of propylene glycol monoacrylate with acrylic esters and with sodium lauryl sulfate emulsifier, has a 40.3 % solids content latex. Another 40 % latex contains ethyl acrylate and acrolein CH_2:CH.CHO (95 : 5). Films from a blend tend to cure, as measured by acetate swelling, even at ambient temperature. The cure is accelerated by heating a film at 150 °C. The reaction is presumably a standard acetalisation with the multiple CHO groups on the polymer itself. Acrolein is a very reactive material, with a very pungent odour. It both oxidizes and polymerises in air and is difficult to handle.

The difficulties of handling acrolein have been partially circumvented by using an acetal derived from acrolein. One possibility is the product formed by condensation of acrolein with pentaerythritol, forming 3,9-divinyl-bi-(metadioxane), which contains two heterocyclic six-membered rings:

$$CH_2{:}CHCH \underset{OCH_2}{\overset{OCH_2}{\diagup\!\!\!\diagdown}} C \underset{CH_2O}{\overset{CH_2O}{\diagup\!\!\!\diagdown}} CHCH{:}CH_2$$

If hydroxyl groups are present, this compound acts by transacetalation. It can be added to various formulations for a vinyl acetate latex, but because of the risk of precipitation it should not be added at more than 0.3 % of total monomers, and preferably at 0.1 %, at which low level it is effective in improving water resistance [95].

If 100 g of a polyvinyl acetate latex is acidified to pH 1 with hydrochloric acid, it will hydrolyse slowly at ambient temperature. Then 1.3 g of 37 % formaldehyde solution is added. Under these conditions a crosslinked polymer, not a formal, is formed. The reaction is probably that of a molecule of formaldehyde reacting with two hydroxyl groups formed by hydrolysis from two different polyvinyl acetate molecules. A repeat of this process along various chains induces crosslinking [96].

5.1.1 Dioxalane derivatives

The similarity between the acetals and derivatives of 1,3-dioxalane has given it an interest in allied polymers, and therefore they may be described here. Dioxalanes with 2-alkyl substitution (the numbers are markers to the carbon atoms),

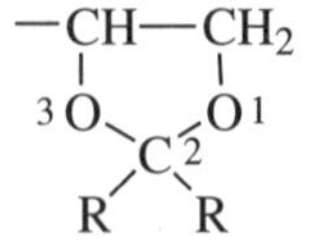

$$-CH-CH_2$$
$$||$$
$$^3OO^1$$
$$\diagdown C^2\diagup$$
$$RR$$

which are cyclic ketals, crosslink readily when exposed to ultraviolet light, the 1,3-dioxalane group acting as an internal ultraviolet light sensitiser [97].

An alternative is given by derivatives of 2-oxo-1,3-dioxalane

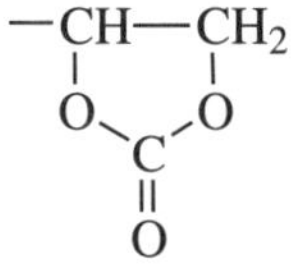

$$-CH-CH_2$$
$$||$$
$$OO$$
$$\diagdown C\diagup$$
$$\|$$
$$O$$

which are cyclic carbonates, readily hydrolysed by bases. 2-Oxo-1,3-dioxalan-4-yl) methyl acrylate and the corresponding methyl maleate and methyl itaconate have been prepared. They have been polymerised in solvents and will crosslink thermally with elimination of carbon dioxide, but do not crosslink by ultraviolet light. Although these derivatives have been polymerised in solvent, they may well be sufficiently stable at acid pH levels for stable latices to be obtained, which can then be thermally treated [97, 98].

6 THE EPOXIDE FUNCTION

The epoxide function (sometimes referred to as oxirane)

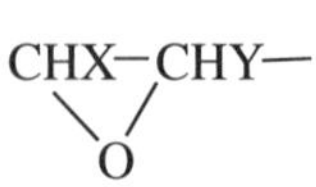

$$CHX-CHY-$$
$$\diagdown\diagup$$
$$O$$

may be reacted both with carboxyl groups and hydroxyl groups, as indicated above. It is, in general, surprisingly stable in water, depending on the nature of the X and Y groups and also on the overall molecular weight. Electron-withdrawing groups, such as the aromatic ring, particularly when it carries phenolic groups and is of high molecular weight, tend to reduce aqueous hydrolysis towards zero. When hydrolysis takes places a 1 : 2 glycol is formed, with the general reactivity of hydroxyl groups. Even a monomer such as glycidyl methacrylate

$$CH_2{:}CH(CH_3)COOCH_2CH.CH\underset{O}{\diagdown\diagup}$$

when copolymerised in emulsion, is only hydrolysed to about 7 % of the epoxide groups.

There are usually residual hydroxyl groups left after the reaction of an epoxide with a carboxyl group, unless there is an excess of reacting carboxyls. These hydroxyl groups are insufficient to make the cured polymer hydrophilic in character, but are sufficient to impart some polarity, which helps adhesion to wood, plaster, brickwork, cement and metals. A difficulty that must be resolved in each individual case is the tendency of an epoxide to condense with itself rather than with a reactive vinyl or acrylic polymer. It should also be noted that as the molecular weight of the epoxide (sometimes incorrectly referred to as a 'resin') increases, a larger proportion of it may be required, since many 'epoxies' of industry have only two epoxides, irrespective of molecular weight.

The crosslinking of styrene–butyl acrylate copolymer latices functionalised by the inclusion of glycidyl methacrylate has been studied. The location of the epoxide groups on the surface of the latex particles was determined by surface analysis, using soap adsorption methods and by the analysis of kinetic data based on the reaction of the epoxide group with hydrochloric acid. It was thus possible to determine the distribution of epoxide groups between the interior and the surface of the particles. Hexamethylenediamine as the crosslinking agent was added to the latex immediately before the casting of films. These were either heated for 12 h at 100 °C or maintained at ambient temperature for 6 weeks. Crosslinking was followed by swelling experiments and by measurement of mechanical properties, the latter being dependent on the radial distribution of the groups. It was concluded that the cured film is a highly crosslinked network with soft, poorly crosslinked inclusions [99].

The reactions of epoxides with carboxylated polymers have already been described (Section 4.2). The optimum ratios for the two reactants should be stochiometric with respect to each other, taking the epoxide as equivalent to a single carboxyl group. However, as the overall molecular weight of the bisphenol type of epoxide increases, compatibility with most vinyl polymers begins to decrease. For most acrylic copolymers, this point containing carboxyl is where the equivalent weight with respect to epoxide groups is 500–525. A basic catalyst should be used if these systems are stoved.

Monomers including epoxide groups, usually the glycidyl group as in glycidyl methacrylate, may be crosslinked with conventional catalysts of the type of polyamines or anhydrides (see Chapter 7). A polymer containing a monomer such as glycidyl methacrylate may react with an external compound containing a hydroxyl or carboxyl group, heat and sometimes a catalyst. Equally, a polymer with a reactive group such as hydroxyl or carboxyl may react with an external epoxide. The latter is more common in commercial practice because of cost considerations. The group may be catalysed to react with itself to give chain polymers, and where there is a multiplicity of groups, it may give rise to crosslinking. The formation of epoxide polymers in emulsion is described in more detail in Chapter 7. The simplest and most frequently encountered epoxide with high reactivity is bisphenol A diglycidyl ether

$$CH_2\overset{O}{-}CH.CH_2-O-\underset{CH_3}{\overset{CH_3}{\underset{|}{C}}}-O-CH_2.CH\overset{O}{-}CH_2$$

which may be used to crosslink emulsion polymers including reactive groups. The general formula below represents more complex derivatives of bisphenol A:

$$CH_2\overset{O}{-}CH.CH_2-\{O-\underset{CH_3}{\overset{CH_3}{\underset{|}{C}}}-OCH_2CHOH.CH_2\}_{\overline{n}}-CH\overset{O}{-}CH_2$$

Alternatively, the epoxide group may be an integral part of a polymer, the group being included with an unsaturated acid. Glycidyl acrylate and glycidyl methacrylate are the simplest monomers (Chapter 1). Section 6.1 considers latex polymers in which one or more of the monomers contains the glycidyl or other epoxide groups. Section 6.2 describes polymers where the added epoxide-containing polymer is external.

A number of technical epoxies have an aliphatic basis, e.g. vinyl cyclohexene dioxide. The higher molecular weight epoxies are viscous oils or even solids, which can cause some difficulty in emulsification.

6.1 Internal epoxide groups in unsaturated monomer

The principal compound of interest in this series is glycidyl methacrylate. Whilst it polymerises very readily, it is difficult to retain in the monomeric condition for long periods, In addition, the cost is rather high. It copolymerises readily with standard monomers, the reactivities r_1 and r_2 for glycidyl methacrylate (M_1) and styrene (M_2) being 0.63 and 0.34 respectively (see Table 5.5). For glycidyl methacrylate and methyl methacrylate these ratios become $r_1 = 0.75$ and $r_2 = 0.69$ [100, 100A]. The corresponding acrylate is also available. The corresponding compounds of other polymerisable acids,

Table 5.5 Copolymer latices including glycidyl methacrylate [100]

	Parts by weight			
	1	2	3	4
Ethyl acrylate	97	97	97	94
Glycidyl methacrylate	3	3	3	6
Water	210	210	210	210
Sodium lauryl sulfate	1.5	1.5	1.5	1.5
α,α-azodi-isobutyramidine dihydrochloride (initiator)	0.05	0.05	0.05	0.05
Dodecyl mercaptan	0.025	0.05	0.10	0.05
Final pH	7.1	6.8	6.0	6.6
Solids %	31.5	31.9	31.3	31.6
Polymer: inherent viscosity[a]	2.74	1.76	1.00	2.03

[a] Inherent viscosity is the relative viscosity (0.5 % solution of polymer in ethyl acetate at 25 °C) of the solution to solvent -1.

such as crotonic, itaconic, maleic and fumaric acids, are occasionally encountered in the literature. One problem with glycidyl methacrylate is the presence of 100–200 parts of dimethacrylates in technical monomer, which may lead to premature crosslinking [100B].

In general, an epoxide-containing monomer such as glycidyl methacrylate does not react with a polymerisable acid to any appreciable extent during an aqueous polymerisation, even in solution when there are no substantially insoluble ester monomers present [100C].

Glycidyl methacrylate copolymers tend to polymerise further via the epoxide group in much the same way as low molecular weight epoxides, using dibasic acids, anhydrides and amines with similar conditions and stoichiometry (see Chapter 7). Temperatures of 100–150 °C are required. Stability of the compositions is the usual problem, which is somewhat accentuated with emulsion products. A copolymerisation formulation is shown in Table 5.5 [100].

A copolymer, prepared by semi-continuous seeded emulsion copolymerisation of methyl methacrylate, butyl acrylate, acrylic acid and varying quantities of glycidyl methacrylate, has been characterized by swelling experiments and electron microscopy. The self-crosslinking of thin films was investigated by Fourier transform infrared spectroscopy and differential scanning calorimetry. The polymer particles had a core–shell structure. The swelling capacity of the copolymer diminished with increasing epoxide content as a result of ester formation when heating a copolymer film at 140 °C for 45 min [101].

In general, delay in adding glycidyl methacrylate during emulsion polymerisation is advised [102]. To ensure good control of copolymerisation of a terpolymer of ethyl acrylate, methyl methacrylate and glycidyl methacrylate in a three-stage process proceed as follows. In the first stage of polymerisation only 10 % of the ester monomers are added and 73 % in the second stage.

Then 6.5 % of glycidyl methacrylate (on initial monomers) is added with the balance of the simple ester monomers. The dried films have improved water resistance [103].

It is preferable to form copolymers including both glycidyl and acid groups, as in Table 5.6 [91]. These are generally stable if maintained at a pH close to 7 by the addition of ammonia solution. In some cases, e.g. example 2 of the table, the addition of a coalescing solvent to ensure good filming is desirable. The latices may be pigmented, and generally show extremely good adhesion. They will be referred to in Chapter 11.

An early specification describes a copolymer of vinyl acetate with acrylic esters, together with about 3 % of acrylic or itaconic acids and glycidyl methacrylate [104]. A terpolymer of vinyl acetate, glycidyl methacrylate and trimethylolpropane trimethacrylate (75 : 20 : 1) is also disclosed [105]. A further terpolymer of vinyl acetate, 2-ethylhexyl acrylate, acrylic acid, glycidyl methacrylate with *tert*-dodecyl mercaptan is also described [106].

A formulation uses blends of vinyl acetate and acrylic monomers, together with itaconic or acrylic acids and glycidyl methacrylate. Polymerisation is by a redox system, rising to 80 °C in about 3 1/4 h. Typical formulations are shown in Table 5.7 [107].

Table 5.6 Copolymer latices including a glycidyl ester and an acid [91]

	Parts by weight	
Ethyl acrylate	86	—
Butyl acrylate	—	27
Acrylonitrile	—	63
Styrene	10	—
Glycidyl methacrylate	—	5
Glycidyl acrylate	2	—
Acrylic acid	2	—
Methacrylic acid	—	5
Tergitol NP-35	2	—
Sodium lauryl sulfate	—	0.5
Sodium metabisulfite	0.50	0.10
Potassium persulfate	0.50	0.30
Water (total)	148.5	200
Dodecyl mercaptan	—	0.5
Solids (%)	40	33

The procedures are convential ones for redox reactions; polymerisations take place under nitrogen. In 1 above the sodium meta-bisulfite is added in two parts. The final pH is 3–3.5 and may be adjusted with ammonia solution to 7–7.5 depending on further procedures.

The epoxide function

Table 5.7 Crosslinking composition: acid/epoxide type, includes vinyl acetate [109]

	Parts by weight			
Vinyl acetate	105	225	105	225
Ethyl acrylate	375	—	375	—
n-Butyl acrylate	—	257.7	—	257.7
Itaconic acid	10	1.5	—	—
Acrylic acid	—	—	1.0	1.5
Glycidyl methacrylate	10	2.0	10	2.0
Nonylphenol/5 ethylene oxide adduct	25	25	25	25
Water, total	450	450	450	450
Sodium *meta*-bisulfite	1	1	1	1
Ammonium persulfate	0.5	0.5	0.5	0.5

A cationic emulsifier may be used in terpolymer acrylic latices with glycidyl methacrylate and 2,2′-azobis(2-amidinopropane)dihydrochloride as initiator [108]. It is possible to add basic catalysts to expedite crosslinking, but these should not be included until just before application. Thus a suggested formulation for emulsion polymerisation contains 0.9–2.7 mol % of glycidyl acrylate, 2.9–15.4 mol % of styrene, 80–94 mol % of ethyl acrylate and 0.8–5.5 mol % of acrylic acid. Temperatures of 100–110 °C are required for cure. This cured copolymer is useful for backing rugs [109].

Included in an acrylic monomer blend is 4 % by weight of glycidyl methacrylate as well as 0.3 % of 4-methacryloyloxy-2,2,6,6-tetramethylpiperidine. The emulsifier is the ammonium salt of poly(acrylic acid)dodecylthioether, with a degree of polymerisation of only 14. This probably acts as a molecular weight controller [110].

A 50 % latex including a polymerisable acid (Table 5.8) is grafted with glycidyl methacrylate, inhibitors as shown in the formulation preventing direct polymerisation, and modified by the addition of metal naphthenates which have a catalytic effect. Films on paper or aluminium show properties similar to hardened alkyds after 30 days [110A].

Further examples illustrating the crosslinking of the carboxyl group with the aid of epoxides include the blending of a copolymer of ethylene and acrylic or methacrylic acid, containing 10 % by weight of acid, with vinylcyclohexene diepoxide and Portland cement, the latter presumably functioning partly as an alkaline catalyst as well as a filler. A typical ratio quoted is 50 parts of acid copolymer to 0.25 parts of epoxide and 16.7 of Portland cement [110B].

6.1.1 Two-pack latex system in single vessel

A system in which an epoxide emulsion and a curing agent could be included in a single container would serve a useful practical purpose. A latex containing

Table 5.8 Grafted latex with air drying films [110A]

	Parts by weight
1. Latex, monomer phase only shown (50 %):	
Butyl acrylate	25.64
Methyl methacrylate	17.25
Methacrylic acid	3.73
2. Graft stage:	
To latex as above add and maintain at 80 °C for 30 min.	
Glycidyl methacrylate	4.62
Tetraethyl ammonium bromide	0.09
Hydroquinone	0.05
The reaction is substantially an esterification.	
3. Final composition:	
Graft latex as above	100
Nonylphenol polyoxyethylene	1.5
Conaphthenate, 6 % solution in aliphatic solvent	1

an amino group would be useful for this purpose. A copolymer of aminopropyl methacrylate is useful for use with a methacrylate copolymer and an acrylic monomer. This should be prepared at a pH > 11. There are several problems to ensure a successful crosslinking reaction. It is advised that in order to obtain an even reaction, the particle sizes of each latex should be similar, to avoid problems with the coalescence and adhesion of the film, because the packing of the matrix determines the contact area between the complementary reactive latices. About 10 % of reactive monomer is optimum, preferably with the same backbone monomer, butyl methacrylate, in a series of experiments.

6.1.2 Allyl glycidyl ethers

The only other common epoxide-containing monomer which has been disclosed is allyl glycidyl ether $CH_2{:}CHCH_2OCH_2CH\overset{O}{\overset{/\backslash}{C}}H_2$ [111]. Addition polymerisation with this monomer is less ready, and there is a strong tendency for polymer molecular weight to be limited by chain transfer through the allyl part of the molecule (Chapter 1, Section 6.9). It is quoted in a number of specifications. An example is an emulsion copolymer of vinyl acetate, ethyl acrylate, allyl ether and crotonic acid (86 : 10 : 3 : 10). Part of the monomers, excluding ethyl acrylate, are added in the first stage of polymerisation, which includes only 0.05 % (on total monomers) of crotonic acid [111A].

A latex polymer, stabilised with a non-ionic emulsifier and sodium lauryl sulfate utilising a redox initiator system, is formed from ethyl acrylate, butyl acrylate, ethoxyethyl acrylate, diallyl glycidyl ether and glycidyl methacrylate (47 : 25 : 2.5 : 2.5). Both monomers are essential for good results. Curing in

The epoxide function 297

this case is performed as a solid mass after precipitation of the latex, using standard methods [111B]. A general account is available [111C].

A less usual epoxide, allyl β,β-dimethyl glycidate $CH_2{:}CHCH_2OOCCHC(CH_3)_2$
$$\underset{O}{\diagdown\diagup}$$
has been disclosed, preparation being by Darzen's reaction [111D]. It forms copolymers with vinyl acetate in the weight ratio of 1 : 9.

6.1.3 Miscellaneous formulations including glycidyl esters

A storage stable latex which cures slowly at ambient temperature in film form is prepared from a latex containing methyl methacrylate, butyl acrylate and glycidyl methacrylate (159 : 95.2 : 63.4) with a proprietary acrylic polymer including amine groups. A formulation is given in Table 5.9 [112].

A latex is formed with the monomer content shown in Table 5.10 in the presence of a seed of a polystyrene latex including a carboxylic acid monomer. The resultant latex is made alkaline with dimethylamine, residual monomers, steam stripped and the pH adjusted to 9 with sodium hydroxide solution. Suitably modified, it finds application as a paper coating. There is probably some crosslinking between epoxide and acid groups [112A].

A technical epoxide prepolymer DER 383 is esterified with methacrylic acid and then copolymerised in emulsion with glycidyl methacrylate and

Table 5.9 Crosslinking latex including a glycidyl group and an amino compound [112]

	Parts by weight
Acrylic latex, including glycidyl methacrylate(as solids)	28.52
Emulsifier	0.5
Coalescing agent	8.0
Wetting agent	1.0
Plasticiser	0.5
Aquaplus 100M (amine containing acrylic polymer)	25.36

Table 5.10 Latex containing monomers with glycidyl and carboxyl groups [112]

	Parts by weight
Butadiene	38
Styrene	42
Methyl methacrylate	10
Acrylonitrile	5
Fumaric acid	2
Glycidyl methacrylate	3
Dimethylamine(added only after completion)	4.8

2-ethylhexyl acylate (187.5 : 44.4). The latex formed is further heated with methacrylic acid, styrene and miscellaneous additives (48.33 : 16.52 : 35 : 0.15). Grafting takes place to some extent forming a product described as a rubber-modified polymer. This copolymer is blended with a cobalt salt (methylethyl ketone peroxide) and cured to give a hard coating of high tensile strength with a heat distortion temperature of 97.5 °C [113].

A curable composition may be formed from a monomer mixture including a glycidyl group and a hydroxyalkyl acrylate or methacrylate. A typical example is styrene, butyl acrylate, hydroxypropyl methacrylate and glycidyl methacrylate in the weight ratio 3 : 3 : 3 : 2 : 2. The only catalysts found useful in giving required hardness and solvent resistance were tetrabutyl ammonium hydroxide and 1,4-diazabicyclo[2.2.2.]octane [114].

Latex copolymers of vinyl chloride with 3–30 % of glycidyl methacrylate and acrylic acid have been disclosed, boron trifluoride being added as a curing agent. Curing is with boron trifluoride BF_3 [115].

6.2 External epoxide compounds

Groups of hydroxyl, carboxyl, amine derivatives and the epoxide group itself within polymers may react with an external compound or even a polymer containing an epoxide group. Some of these are described in earlier sections of this chapter. If several epoxide groups are available in a single compound, crosslinking is enhanced.

It is necessary to emulsify external epoxide compounds before blending with the vinyl-type latex. This usually causes some difficulty since conditions which tend to self-condensation must be avoided. The formulation in Table 5.11 gives a suggested formulation for the emulsification of Ethoxyline Resin 190, a product once available in the United States [116].

Useful catalysts are 2,4,6-tri(dimethylaminomethyl)phenol and benzyldimethylamine. Useful ratios of copolymer to epoxide compound are from about 3 : 2 to 1 : 1, with catalyst at about 0.6–2.5 %. These products will be referred to again in Chapter 11 in Volume 2 on industrial finishes.

Many epoxide compounds other than those with polymerisable vinyl groups come into the category of compounds, based on $-CH_2-CH-CH_2$. An example is diglycidyl phthalate $O\text{-}C_6H_4(OOCCH_2CH\underset{O}{\overset{}{\diagdown\diagup}}CH_2)_2$. In this case an acrylic copolymer containing acrylic acid is first formed in isopropanol solution, which when neutralised and diluted with water forms a spontaneous emulsion. After the addition of diglycidyl phthalate under controlled ratios there is a good pot life [117].

A curing adhesive is formed from a latex of a copolymer of acrylic acid, butyl acrylate, ethyl acrylate and vinyl acetate, which is combined with a quarter of its weight of an emulsion of Epon 828®, which is mainly bisphenol

Table 5.11 Emulsification of epoxide resin precursor [116]. Reprinted from progress in organic coatings, vol 20, no 1, 1992, pages 73–80, with permission from Elsevier Science

	Parts by weight			
	1	2	3	4
Ethoxyline 'resin'	1800	1800	1800	1800
Distilled water	1197	1197	1197	1197
Hydroxyethyl cellulose	9	9	—	—
Sodium salt, carboxymethyl cellulose	—	—	—	8.9
Gum acacia	—	—	9.4	—
Octylphenol ethylene oxide (1 : 30) condensate (70 % aq.)	63.4	90.6	—	—
Sodium alkylaryl polyether sulfate	—	—	181.2	—
Sodium alkylaryl polyether sulfonate	—	—	—	226.8

Emulsification is effected by pouring the surfactant solution slowly into the epoxide, which is vigorously stirred at 90 °C until inversion takes place; then surfactant is added more rapidly. Finally, the solution is cooled and passed through a colloid mill.

A diglycidyl ether. A solution of tetraethylenepentamine is added prior to application [118]. A high molecular weight epoxide resin may be combined with an acrylic latex including epoxide groups [119].

An acrylic latex including carboxyl groups (Neocryl A-622), which also contains a coalescing agent, is blended with a cycloaliphatic epoxide (UVR 610) and aluminium acetoacetonate as catalyst. Film formation takes place normally and crosslinking takes place gradually in the film. After 10 days at ambient temperature the film is resistant to more than 200 rubs with methylethyl ketone [120].

In a complex process, butadiene, acrylonitrile, divinylbenzene and methacrylic acid (71 : 20 : 3 : 6 by weight) were polymerised in emulsion to 90 % conversion to produce a partially crosslinked rubbery polymer. This latex is mixed with Epikote 828® (identical with Epon 828) and a tetrafunctional glycidylamine-based epoxide resin (10 : 72 : 18) for 30 min to obtain a modified epoxide. This is heated at 90 °C for 4 h in the presence of triphenylphosphine and vacuum dried to form a modified epoxide resin precursor. This is further cured with diaminodiphenyl methane (27 parts) at 80 °C for 2 h and then at 150 °C for 3 h to produce a product with T_g of 198 °C and fracture toughness of 3.2 kg cm$^{-1.5}$ [121].

An epoxide derivative of bisphenol A (Epiclon® 850) is included in a two-stage polymerisation, as shown in Table 5.12, in both stages of which potassium persulfate is the initiator (not shown in the table). A coating from the latex dried at 25 °C for 48 h is insoluble in toluene, a reaction apparently taking place between the epoxide and the acid, probably catalysed by the diethylaminoethyl methacrylate [122, 123, 123A, 124].

Table 5.12 Acrylic latex with added reactive external epoxide [122]

	Parts by weight	
	Stage 1 latex	Stage 2 latex
Water	324	120
Polyoxyethylene nonylphenyl ether	16	—
Sodium dodecylbenzene sulfonate	4	—
Epiclon 850 (epoxide derivative)	40	—
Butyl acrylate	200	—
Ethyl acrylate	—	120
Methyl methacrylate	192	74
Acrylic acid	8	—
Stage 1 latex	—	452
Dimethylaminoethyl methacrylate	—	4
Solids content (%)	50.3	50.2

A terpolymer of styrene, butyl acrylate and acrylic acid, prepared in ethylene glycol monopropyl ether solution, is inverted to an oil in water (o/w) emulsion by the addition of an ethylene oxide–propylene oxide block copolymer (Pluronic F68) and water. It is blended further with the diglycidyl ether of bisphenol A and an amine catalyst [125].

A specification discloses a latex of ethyl acrylate, 2-hydroxyethyl acrylate and methacrylic acid which is used with glyceryl diglycidyl ether as a stoving crosslinked coating for iron, but which is water swellable [126].

An alternative type of aliphatic epoxide, which may be used as a cure for a latex including carboxyl groups, is an epoxidized linseed oil, in which the epoxide groups are formed in the double bonds of the oil. Even if there is only a single epoxide per acid chain, this will ensure that there are at least three epoxide groups per molecule to ensure ready and efficient crosslinking. A typical composition contains 9.5 % of additive of epoxidized linseed oil on an acrylic latex including carboxyl groups, which with 1 % of a catalyst, e.g. ammonium acetate, enables cure to take place in 5–30 min at 75–150 °C [127].

In a theoretical investigation, the crosslinking of a copolymer of acrylic esters including N-methylolacrylamide (Ch. 5.8) with a water-soluble epoxide, in equimolar amounts with the amide, was examined. The resultant film was examined by IR spectra. Orientation of the hydrophilic groups on the particle surface has some effect on the optimum ratios [128].

7 DI-ISOCYANATES AND POLYURETHANES IN VINYL-TYPE CURES

The chemistry of polyurethanes is described in Chapter 7, and this section is restricted to those examples where an isocyanate or polyurethane is used to modify a monomer with a reactive group or as an external crosslinking agent. Blocked di-isocyanates are usually required for these reactions. Because of stability considerations in emulsion, there are only a limited number of relevant examples [129].

The reaction product of a phenol-modified di-isocyanate with hydroxyethyl acrylate, which thus contains latent isocyanate groups, is copolymerised in emulsion with ethyl acrylate and styrene in the presence of an anionic surfactant and a redox initiator. On filming and baking at 175 °C a glossy coating is obtained [130].

A technical polyisocyanate, Tolanate HCT, is reacted with methylethyl ketoxime and then with 2-hydroxyethyl acrylate to form a monomer which is copolymerised in emulsion with styrene and acrylic acid to give a self-crosslinkable copolymer latex [131]. A copolymer of vinyl acetate, butyl acrylate and 2-hydroxyethyl acrylate (16 : 4 : 1) is treated similarly with diphenylmethanedi-isocyanate [132].

An emulsion copolymer of ethyl acrylate, methyl methacrylate and diacetoneacrylamide may be reacted with the reaction product of 2-hydroxyethyl acrylate and isophorone diisocyanate (1 : 1 molar) to give crosslinked products useful in coatings and adhesives [133].

A latex in which one of the monomers is 2-hydroxyethyl acrylate and another is the reaction product of propylene glycol monomethacrylate with hexamethylene diisocyanate–water adduct forms a curing composition with hexamethoxymethylmelamine (Section 8). A film has 3H pencil hardness, with good gloss and impact resistance [134]. Thick coatings, e.g. for undersealing motor vehicles, are prepared from a water thinnable urethane/polyester resin and an acrylic latex [135].

A modified di-isocyanate is included in a latex together with aminopropylmethyldiethoxysilane (see Section 11) [136].

8 POLYMERS CONTAINING (METH)ACRYLAMIDE

8.1 Chemistry of the reactions

Polymers containing amide groups can be insolubilised by the use of aldehydes, acids, ionising radiation and polyvalent cations. The principal amide-containing monomer is acrylamide $CH_2:CHCONH_2$. On stoving, acids tend to join two acrylamide units by an imide group formation with loss of ammonia. Aluminium salts tend to form complexes, whilst aldehydes, especially formaldehyde, add on at pH 8–10 to the NH_2 groups to form N-methylol groups .$NHCH_2OH$. Other aldehydes add on to acrylamide in a similar manner. The methylol groups condense further at 100 °C or above, with loss of formaldehyde and water. This reaction is described further below.

The most useful member of this series is N-methylolacrylamide, alternatively known as N-(hydroxymethyl)acrylamide $CH_2:CHCONHCH_2OH$ (NMAM), formed by the direct reaction of formaldehyde on acrylamide, and is normally supplied as an aqueous 60 % clear solution at pH 7–7.5. N-Methylolacrylamide may be polymerised in aqueous solution by the normal addition mechanism using a persulfate initiator. The polymer, however, still possesses the reactive methylol group which, because of its attachment to the NH. group, gives it high activity characteristics.

The N-methylol group also appears in the precursors of aminoplasts, which are also based on the N-methylol group. The principal initial compounds in this group are dimethylolurea $O:C(NHCH_2OH)_2$, the compound formed by the condensate of urea with formaldehyde, and also compounds formed by the condensation of melamine with formaldehyde (see Section 8.2 and Chapter 7, Section 2).

The methylol group takes part in crosslinking reactions, including those with partly condensed aminoplasts and with copolymers including N-methylolacrylamide as one of its monomers. The condensation is fundamentally as follows:

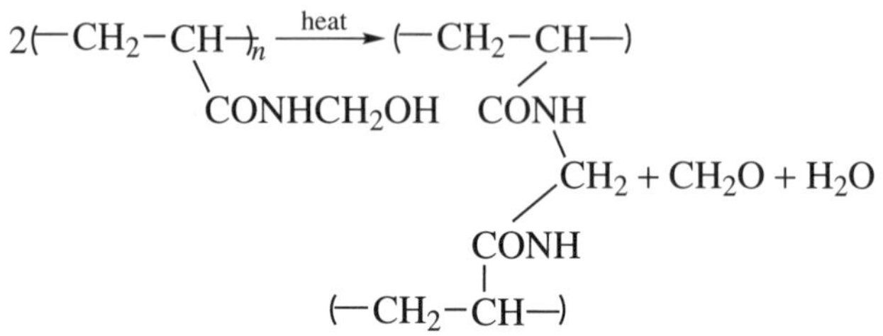

$$2(-CH_2-CH-)_n \xrightarrow{heat} \begin{array}{c} (-CH_2-CH-) \\ | \\ CONHCH_2OH \quad CONH \\ \backslash \\ CH_2 + CH_2O + H_2O \\ / \\ CONH \\ | \\ (-CH_2-CH-) \end{array}$$

A very useful modification is the further etherification of the methylol groups to form ethers, often the methyl ether $CH_2{:}CHCONHCH_2OCH_3$ or the butyl or isobutyl ethers, $CH_2{:}CHCONH_2CH_2O(iso)C_4H_9$. These ethers are easily prepared since etherification of a methylol group attached to nitrogen is virtually as easy as a normal esterification. They are of particular interest where it is desired that the compound should take part in a reaction with another active group, e.g. carboxyl, which may be on the same or on another molecule.

Note that there is a difference between these ethers and the product formed by direct substitution of one of the hydrogen atoms on the amino nitrogen. It is not always clear in the literature as to which products are referred to, and there may be some errors in consequence.

A study of the crosslinking of a copolymer of butyl acrylate and n-butoxy or isobutoxy methacrylamide concludes that complete crosslinking of the acrylamide derivatives occurs at 120–140 °C. It proceeds with the elimination of the formal $CH_2(OC_4H_9)_2$ and the formation of $CONHCH_2NHCO$ crosslinks. This is a slight variation on the suggested reaction as shown above [137]. For the reaction if a melamine aminoplast is included, see Section 8.2.

In general methylol derivatives crosslink more readily than types hitherto mentioned, and will crosslink at ambient temperature provided that the pH is reduced to ∼4. The reaction is rather slow and may be estimated by the resistance of the cured film to perchlorethylene C_2Cl_4 or another solvent. At ambient temperature, times $\geq$ 24 hours or above are generally necessary. At 60 °C this is reduced to 1 h, at 149 °C to 5 min and seconds only are sufficient at 205 °C [137A]. The slow ambient reaction is an advantage since it is found that emulsions and latices remain perfectly stable for long periods, and only begin to crosslink when they are dried to form films.

The rate of cure may be increased by the addition of weak acids such as oxalic acid and citric acid, which can be included in a latex, making a 'one-pack' system possible. Alternative latent acid catalysts such as diammonium phosphate, ammonium chloride and magnesium chloride may be included. About 3 % of catalyst on polymer weight is a suitable quantity.

It is also possible for an N-methylol group to react with polycarboxylic acids directly, the product being a urethane type of linkage:

$$-CH_2-CHCONHCH_2OC_4H_9 + HOOC-X$$
$$= -CH_2-CHCONHCH_2OOC-X + C_4H_9OH$$

where X is a polymer residue. This reaction is of interest since some sources mention the polymerisation of an N-methylol derivative and an acid such as methacrylic acid or itaconic acid in the same process to produce a stable latex.

In the latex copolymerisation of NMAM with butyl acrylate and either styrene or methyl methacrylate, it was noted that the NMAM content of the particles is low, whatever the method of its introduction. Some water-soluble polymers, mainly NMAM, were separated by centrifugation. The NMAM in the particles were located as a 'hairy' shell of grafted or strongly absorbed polymers. Films were obtained in the pH range 2–8 [137B].

The crosslinking characteristics of emulsion polymers containing N-isobutoxymethyl)acrylamide (iso-$C_4H_9OCH_2NHOCCH:CH_2$) was studied by thermal evolution analysis and by thermal gas chromatography. Isobutanol evolution as a function of temperature was used to determine kinetic parameters. Crosslinking is promoted by the presence of copolymerised carboxyl groups, but not by monomers that contain hydroxyl groups. Triethylamine catalyst is retained to some extent by films, even if heated above 200 °C, but nevertheless crosslinking starts about 120 °C, in accordance with other studies [138].

The N-methylolethers are also reactive with hydroxyl and epoxide groups. Several alternative condensations are possible, such as an ether link between methylol or methylol ether groups:

$$-CH_2-CHCONHCH_2OC_4H_9 + HOCH_2NHOCCH-CH_2-$$
$$= -CH_2-CHCONHCH_2-O-CH_2NHOCCH-CH_2 + C_4H_9OH$$

With an epoxide the condensation is

$$R-CH-\overset{O}{\overset{/\ \backslash}{C}H_2} + C_4H_9OCH_2NHOCH-CH_2-$$

$$= R-\underset{OC_4H_9}{\overset{|}{C}H}-CH_2OCH_2NHOCCH-CH_2-$$

where R represents the balance of an epoxide molecule which, in some cases, may be a short-chain epoxide, or alternatively a polymer which may contain methacrylate units.

The kinetics of the EP of N-methylolacrylamide, N-methylolmethacrylamide and N-(2-hydroxyethyl)methacrylamide with acrylonitrile, methyl methacrylate and vinyl acetate at 50 °C have been calculated. The addition of up to 8 % of alkylolamide with vinyl acetate or acrylonitrile increases the viscosity of the copolymer formed, but further increases sharply lowers it. The acrylonitrile copolymer, after spinning to a fibre, may be insolubilised by crosslinking at 150 °C [139].

The location of N-isobutoxymethylacrylamide segments at the latex particle surface in a copolymer promotes crosslinking, which takes place at temperatures below 121 °C, especially where acid groups are also present. This system was compared with one in which crosslinking was achieved by an external melamine aminoplast. The N-isobutoxyacrylamide system develops better water resistance at a lower temperature, but there is poorer chemical and stain resistance [140].

In the subsequent subsections the cure of latex polymers containing an acrylamide derivative, the cure of other polymers with reactive groups utilising aminoplasts and the more complex derivatives will be discussed separately.

8.2 Simple derivatives of acrylamide

Some of the investigations described below are not necessarily from emulsion copolymers, but as curing takes place in the films, usually above 100 °C, the reactions during curing from latices should be substantially the same.

Most latices contain N-methylolacrylamide (NMAM) rather than acrylamide. It is available as a 60 % solution, which is rather easier and safer to handle, and it is slightly less reactive than acrylamide, which makes for easier copolymerisation, especially with vinyl acetate. Methacrylic acid is often present in addition. NMAM solution is usually added at a controlled rate during a copolymerisation, sometimes with acrylic monomers, resulting in an even copolymerisation and therefore an even copolymer. This causes the final curing process to be more efficient [141].

Derivatives of methacrylamide are sometimes encountered and less frequently the amides of crotonic acid; the half amides of maleic acid HOOC:CHCONH$_2$ and of other dibasic acids such as fumaric acid and itaconic acid are quoted. Methanol is often added to an emulsion polymerisation as it is considered to prevent any self-condensation of methylolamide groups during the process, but some formulae have been disclosed without methanol.

A number of disclosures have been published in which N-isobutoxyacrylamide or similar derivatives are claimed. This is formed presumably by the action of isobutyraldehyde on acrylamide in a manner similar to that of formaldehyde. Thus a mixture of butyl acrylate, acrylic acid and N-isobutoxyacrylamide (584 : 16 : 20) is added to a water phase using an anionic emulsifier, Fenopon® EP110, and an ammonium persulfate/sodium bisulfite redox system and then polymerised. If 0.2 % of p-toluenesulfonic acid is added to the latex before dry films are prepared, these films are readily crosslinked, even to some extent at ambient temperatures. The effect of trichlorethylene on the film is minimal if curing takes place at 120 °C. Infrared spectroscopy has been used to examine crosslinking.

A somewhat more complex process takes in a two-stage process, as indicated in Table 5.13. Curing by heating is used as the standard method [142].

8.2.1 Formulations mainly based on acrylic esters and/or styrene

A formulation includes 2-ethylhexyl acrylate, acrylonitrile, styrene, NMAM, 2-hydroxyethyl methacrylate and itaconic acid (39 : 5 : 32 : 2 : 2). Note that there are three potential crosslinking monomers in this formulation and also a titanium compound as the chelating agent. A relatively low curing temperature is possible [142A].

An allied formulation includes ethyl acrylate, trimethylolpropane triacrylate, acrylic acid, NMAM, styrene and a methallyl sulfonate salt which assists emulsification (103 : 5 : 8 : 100 : 146 : 1.5). The latter also includes the ammonium salt of a sulfate alkylphenoxy polyoxyethylene. Initiation and polymerisation takes place at 25 °C with potassium persulfate, ascorbic acid and ferrous sulfate as a redox system, the addition being gradual. Final neutralization is made with dimethylaminoethanol to a pH of 7.1 at 35.3 % of solids [143].

Increasing the cure temperature and the inclusion of two initiators of different activities increases the yield and also the thermal stability of acrylamide–acrylonitrile–trimethylol trimethacrylate–styrene copolymers. Final unreacted monomers are also decreased [144].

A specification describes a two-stage process in which the first includes styrene, butyl acrylate, methyl methacrylate, acrylic acid and acrylamide (427 :

Table 5.13 Copolymer including isobutoxymethacrylamide [142]

	Parts by weight	
	1	2
Stage 1 latex	—	250
Water	145	755
2-Ethylhexyl acrylate	26	160
Ethyl acrylate	—	32
Methyl methacrylate	29.1	—
Hydroxyethyl acrylate	4.4	—
Acrylic acid	7.7	—
Isobutoxymethacrylamide	—	48
Styrene	26	160
Polyethylene glycol nonophenyl ether ammonium persulfate	a	—
Butoxyethanol	—	122
Water(with butoxyethanol)	—	122
tert-Butyl hydroperoxide	a	—
Ascorbic acid	a	—
Diethanolamine	a	—
Process: (Heat for 4 h at 75 °C (stage 2).		
Final solids (%)	—	30
Final pH	—	8.4

^aNot indicated in abstract.

258.4 : 163.2 : 21.8 : 13.2) and the second, styrene and butyl acrylate only, with dodecyl mercaptan (227 : 228 : 1). Potassium persulfate is the initiator, each stage being pre-emulsified. The two-stage process probably assists film formation [145].

A further specification quotes a pre-emulsion in which the (proprietary) emulsifier concentration is below the critical micelle concentration in the early part of the reaction. The monomers are ethyl acrylate, butyl acrylate, methyl methacrylate, acrylic acid and N-methylolacrylamide (60 : 15 : 10 : 10 : 5) [146].

8.2.2 Formulations with vinyl acetate as the principal monomer

Vinyl acetate copolymers with NMAM are prepared in a polyvinyl alcohol (>91 % hydrolysed polyvinyl acetate) solution with a minor proportion of an anionic surfactant. The vinyl acetate is added initially, and after polymerisation starts NMAM is added at a rate such that 85 % or more enters the reactor at the rate of 0.01–0.03 parts per minute (on the total monomer). Ammonium persulfate is the initiator and the main reaction takes place at 68 °C [147]. A variation includes ethylene [147A].

An emulsion copolymer of vinyl acetate, (isobutoxymethyl)acrylamide and acrylic acid (37.5 : 0.8 : 0.8) with a standard anionic/non-ionic emulsifier system and a bicarbonate buffer has been prepared under an ethylene pressure of 20 kg cm^{-2}. As expected, the resulting films are self-crosslinking [148].

Another process describes the addition of vinyl acetate and NMAM at a controlled rate to an aqueous solution of Aerosol® 22, a sulfosuccinate with carboxyl groups (see Chapter 2, Section 3.2.7) and fully hydrolysed polyvinyl alcohol. This forms a latex with a good shelf life, suitable, *inter alia*, as an adhesive [149].

A variant also includes diallyl maleate as a comonomer, which acts synergistically with NMAM, and also a polymerisable acid [149A]. In a somewhat more complex procedure a copolymer is formed of vinyl acetate as the principal monomer with butyl acrylate as a comonomer, also with 1 % on monomers of diallyl maleate as a synergistic crosslinking agent. The emulsifiers are the disodium salt of the sulfosuccinic ester of an ethylene glycol mono C_{10-12} ether with potassium persulfate as initiator. NMAM solution is added continuously with part of the surfactant at 65–71 °C, followed by a final initiator addition. The film from the latex is highly crosslinked in 3 min at 132 °C [150].

Ethylene may be a comonomer instead of the acrylic ester [151, 152]. Itaconic acid may also be included [153].

8.2.3 Formulations including vinyl chloride

A stable copolymer latex of ethyl acrylate, vinyl acetate, vinyl chloride and NMAM (25 : 28 : 45 : 2) includes sodium dodecyl diphenyl ether disulfonate as stabiliser and disodium hydrogen phosphate as a buffer. The solids content is 45 % [154].

A copolymer including ethylene is based on vinyl chloride as the principal monomer with acrylamide and *N*-(isobutoxymethyl)acrylamide (27 : 66.5 : 3.4 : 0.3). The ethylene is under high pressure at 127 kg cm^{-2}; 1800 psi). Ammonium persulfate is the initiator. The final polymer has a T_g of -7 °C and is 71 % soluble in tetrahydrofuran. Film formation at the pH of 2.9 probably encourages film formation. The product is an impregnated fibre [155].

A latex based on butyl acrylate, NMAM and vinyl chloride with sodium vinyl sulfonate as polymerisable surfactant (55 : 6 : 44 : 1) has a 44.8 % solids content and an average particle diameter of 113 nm. On drying, sheets with minimum film temperature at 9 °C are formed and the latex is coated on to corona-treated polyester film, which is suggested to be suitable as a receptor sheet for thermal transfer printing [156].

A latex, of which the film is crosslinkable even at ambient temperature, includes vinyl chloride as the principal monomer and also includes a carboxylic acid and NMAM, each up to 2 % of total monomers. An acrylic acid may be present; bis(β-chloroethyl) vinyl phosphonate, which imparts fire-resistant properties, is an optional monomer. A range of emulsion stabilisers may be utilised, a redox initiator being preferred. The rate of the crosslinking reaction is increased by heating, without the necessity of approaching the discoloration point [157].

8.2.4 Formulations including isoprogyl-, butyl- and isobutyl, isobutoxy acrylamide

Isopropylacrylamide $CH_2:CHCONHC_3H_7$ and methylene bisacrylamide $(CH_2:CHCONH)_2$, which includes two double bonds, have been copolymerised together to form a latex using sodium dodecylbenzene sulfate as emulsifier. Below 32 °C the particles are highly swollen, with average particle diameters of 400–800 nm. Above 32 °C, diameters decrease to half. Swelling showed the expected trends in crosslink density, with the presence of both methanol and urea [158].

A copolymer latex is formed from vinyl acetate, dibutyl maleate and N-isobutoxyacrylamide (900 : 500 : 28). The latter has the advantage of being monomer soluble. The aqueous phase consists of a seed latex, sodium benzoate buffer and polyvinyl alcohol and water (74 : 0.4 : 53.3 : 1100) and a redox initiator system including sodium formaldehyde sulfoxylate and ferrous sulfate (5 : 0.02). Polymerisation takes place at pH 4.8 at 69–74 °C. Films are water resistant and 33.1 % is insoluble in trichlorethylene compared with 4.3 % of insoluble material if the N-isobutoxymethacrylamide is omitted. The water-resistant films are useful as adhesives [159].

N-Butoxymethacrylamide is also included in a monomer formulation consisting of butyl acrylate, styrene and acrylic acid (3 : 59 : 27 : 1) with 10 parts of acrylamide in the water phase. The emulsifier is nonylphenyl pentaethylene glycol, with a persulfate–bisulfite initiation with the incremental addition of both redox components. The final latex, rendered acid, is stable for > 6 months, and films are crosslinkable above 80 °C [160].

N-Butoxymethacrylamide is also described in a number of specifications for acrylic ester copolymers, with methacrylic acid to assist crosslinking. The amide derivative is included at 5–8 % and the acid at 10–2 %. Typically, cure on an aluminium plate is for 30 s at 260 °C [161–163].

8.3 Diacetoneacrylamide (DAAM)

This and the following subsections describe some more complex derivatives of acrylamide.

Diacetoneacrylamide (DAAM), otherwise shown as N-(1,1-dimethyl-3-oxo-butyl)-acrylamide $CH_2:CHC(O)NHC(CH_3)_2CH_2C(O)CH_3$, is formed by the condensation of acrylonitrile and acetone in the presence of sulfuric acid. It is neutralised with ammonia solution, and on a laboratory scale may be recrystallised from petroleum ether. It is a solid with m.p. 57 °C, b.p. 120 °C, at 8 mm pressure; the specific gravity, molten, is 0.998 at 60 °C. It is soluble in water and most organic solvents. The reaction is general, between a hydroxy-ketone, e.g. diacetone alcohol and acrylonitrile, which occurs at 70 °C in the presence of sulfuric acid [164, 165]. One such homologue is N-(1,1-dimethyl-5-oxohexyl)acrylamide CH_2:$CHC(O)NHC(CH_2)_3C(O)CH_3$, formed by the condensation of 6-methyl-5-hepten-2-one with acrylonitrile [166]. The monomer, DAAM, has some surface activity.

The polymer of diacetoneacrylamide, unlike the monomer, is not water soluble, although it is hydrophilic. It is soluble in many organic solvents, but not in aliphatic hydrocarbons. The polymer is potentially reactive, and hence may be used in crosslinking reactions. The carbon atoms α to the keto group can take part in such reactions as the aldol condensation and the Mannich reaction [166A]. The amide hydrogen is relatively inert due to steric hindrance, but the inclusion of DAAM in copolymers improves wetting properties and improves adhesion to aluminium, glass, concrete and polyvinyl chloride. The inclusion of the monomer in polymers increases water vapour and gas transmission, which may prevent a paint from blistering under some conditions.

The copolymerisation of DAAM occurs readily with most monomers. The reactivity of DAAM (r_1) with acrylamide (r_2) is 0.075:1, with styrene (r_1) is 1.77:0.49 and with methyl methacrylate (r_1) is 1.68:0.57 [167].

Both monomer and polymer may be methylolated under alkaline conditions, e.g. with butanolic potassium hydroxide or a quaternary ammonium hydroxide. It is important to note that this takes place at pH 9–10 on the carbon atoms adjacent to the ketonic groups and not on the amide group. Each hydrogen may be substituted by a methylol group. A technical product is about 50 % methylolated.

In a number of systems 3 % of DAAM has proved the optimum where crosslinking is required, and hexamethoxymethylmelamine is advised as the most suitable agent. A pH near 3 gives the optimum cure, but 85–95 % of acetone insoluble material is obtained with an all-acrylic latex after cure over a wide pH range without the addition of an external crosslinking additive. At ambient temperature, crosslinking occurs in 4 to 16 h.

For satisfactory crosslinking, oven curing is required with a film from a vinyl acetate–DAAM latex, to which a melamine aminoplast has been added to obtain the maximum content of acetone-insoluble matter at an optimum pH of 3–4. Variations with latex copolymers of this type may be due to the difficulties in obtaining an even copolymerisation, although a delayed addition (semi-continuous) process is advised.

Adipic acid dihydrazide as a crosslinking agent for copolymers of 2-ethyl-hexyl acrylate, acrylic acid and DAAM (21.6 : 98 : 21 : 21) has been disclosed. Styrene may also be a comonomer [168]. An adipic acid hydrazide crosslinker is also included in a mixed acrylic ester–styrene–diacetoneacrylamide latex, together with titanium dioxide in a paint which cures on filming at ambient temperatures [169].

The photochemical crosslinker of polyDAAM or an acrylic acid copolymer, either in water or in water/acetone (1 : 3) in the presence of solutions of ammonium dichromate, is incomplete [170] (see also Section 10).

8.4 Methylene bis acrylamide (MBAM)

Crosslinking effects are derived from the inclusion of methylene bisacrylamide (MBAM), formed by the reaction of formaldehyde and acrylamide in the presence of sulfuric acid or hydrochloric acid and an inhibitor. The reaction involves the elimination of water:

$$2CH_2{:}CHCONH_2 \xrightarrow{H^+} CH_2[NHC(O)CH{:}CH_2]_2 + H_2O$$

The kinetics of the polymerisation of MBAM has been studied with redox initiation with acidic permanganate–thiourea:

$$R_p = k[MBAM]^{1.0}[thiourea]^{0.5}[KMnO_4]^0[H^+]^{-0.5}$$

Propagation is believed to be by cyclopolymerisation. If Ce^{IV} is used as the oxidant, with thiourea, $R_p = [thiourea]^{0.5}[Ce(IV)]^{1.5}[MBAM]^{1.5}$ [171, 172].

A study of the copolymerisation of acrylamide with MBAM indicates that the reactivity ratios (MBAM = r_2) are 2.20 : 1. The molecular weights of the copolymer decrease under corresponding conditions with increasing MBAM from 5 to 20 % [173].

MBAM is often included in polymerisation formulations to ensure some crosslinking without imparting solubility in water or solvents. In one example a terpolymer of butyl methacrylate, styrene and acrylic acid (2 %) is neutralised with sodium hydroxide to pH 9. This is mixed with 2 % of acrylamide and 0.12 % of MBAM (on polymer) and polymerised with ammonium persulfate and β-dimethylaminopropionitrile to give a uniform gel. The latter additive is probably a molecular weight controller; it is possible that some grafting takes place [174]. N,N'-Methylenebisacrylamide is included with 1 % of diethylaminoethyl methacrylate in a latex (which may be quaternised), with a redox initiator system. The quaternised polymer, mildly crosslinked, gives a dry 0.55 mm film of the latex on glass, which had a water absorption factor of 120 [175].

8.5 The aminimides[†]

The aminimides have the general formula for unsaturated derivatives $CH_2{:}C(R).C({:}O).N(R_1).(CH_2)_n C({:}O).N^{-+}NR_2R_3R_4$, where R = H or CH_3

[†] The aminimides were available from the Ashland Chemical Company, USA, in the 1970s. They are not currently available from this source.

(i.e. acrylate or methacrylate), R_1 = H, alkyl, alkaryl or cyclohydrocarbyl, R_2, R_3, R_4 are alkyl or hydroxyalkyl and $n = 1–4$.[‡]

The original preparation of aminimides was by the action of an acid chloride with an unsymmetrical alkyl hydrazine to produce a hydrazide. Alkylation of the hydrazide forms hydrazonium salts, analagous to quaternary ammonium salts. The action of a base removes the elements of HCl to form the aminimide; see the reactions below [176]:

$$RCOCl + H_2NN(CH_3)_2 \longrightarrow RCONHN(CH_3)_2$$

$$RCONHN(CH_3)_2 + CH_3X \longrightarrow RCONHN^+(CH_3)_3X^-$$

$$RCONHN + (CH_3)_3X^- + OH \longrightarrow RCON{-}N^+(CH_3)_3 + H_2O + X^-$$

A development of the general method has been the reaction of the more nucleophillic nitrogen of unsymmetrical dimethylhydrazine with an epoxide in the presence of a carboxylic acid ester [177]. This gives much better yields than the classical synthesis and is represented as[§]

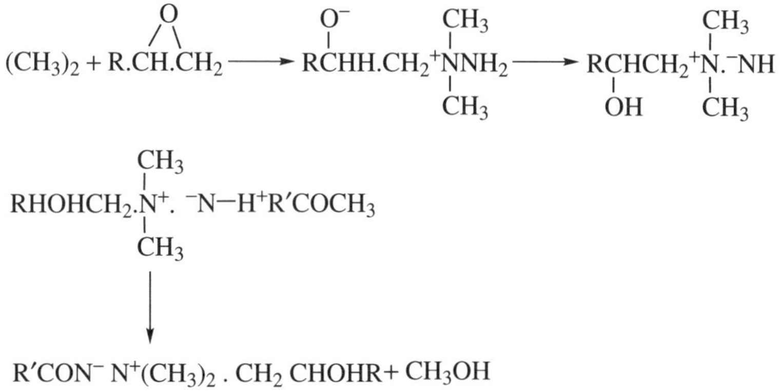

$$(CH_3)_2 + R.\overset{O}{\overset{/\backslash}{CH.CH_2}} \longrightarrow RCHH.CH_2{}^+\underset{CH_3}{\overset{O^-}{\underset{|}{\overset{|}{N}}\underset{|}{NH_2}}} \longrightarrow RCHCH_2{}^+\underset{OH}{\overset{CH_3}{\underset{|}{\overset{|}{N}}.{}^-NH}}\underset{CH_3}{\overset{}{}}$$

$$RHOHCH_2.\overset{CH_3}{\underset{CH_3}{\overset{|}{N^+}}}. {}^-N{-}H^+R'COCH_3$$

↓

$$R'CON^- N^+(CH_3)_2 . CH_2 CHOHR + CH_3OH$$

(Note that in the final stage the aminimide formula has been reversed for clarity.)

The aminimides rearrange to form isocyanates above 120 °C, with elimination of a tertiary amine:

$$R.C(O)N^-N^+R'_3 \longrightarrow R.N{:}C{:}O + R'_3N.$$

where R is an unsaturated group. It is possible to produce unsaturated and potentially polymerisable isocyanates.

[‡] This is more complicated than the formulae shown later, and is actually a generalised formula. Derivatives of this formulation may be referred to as acrylaminoalkanaminimides.

[§] In the following equations, + represents a positive charge and − represents a negative charge.

The principal interest in latex technology is the formation of aminimides of acrylic and other unsaturated acids. One class is represented by the equations above. Thus glycine methyl ether hydrochloride is reacted with methacryloyl chloride in the presence of potassium carbonate, forming methyl N-methacryoylglycinate. This is reacted with *asym*-dimethylhydrazine and propylene oxide, forming 1,1-dimethyl-1–1-(2-hydroxypropyl)amine N-methacryloylglycinimide [177A]. Another example which has been claimed is 1,1-dimethyl-1-(2-hydroxyethyl)amine hydrochloride [178, 179].

Several acrylic aminimides, including groups with longer alkyl chains than methyl on the positive nitrogen, are soluble in aliphatic hydrocarbons. Some Q and e values have been deduced and are given in Table 5.14 [179A].

Many polymers and copolymers of aminimides have been prepared [176]. These include water-soluble types and emulsion polymers. Peroxide initiation cannot be used as the functional groups in aminimides are susceptible to oxidation, but standard redox systems are suitable. In non-aqueous solution azobisisobutyronitrile is a useful initiator. Thus a 20 % copolymer of N-methacryloyl aminimide with methyl methacrylate has been described. Infrared spectra indicate the presence of absorption bands due to amide, ester and aminimide groups. The product is thermoset on heating for 30 min at 160 °C.

Under curing conditions, besides the rearrangement to an isocyanate, there is a reaction with an active hydrogen atom, usually from hydroxyl or amino along the chain. The polymer may represent a combination of hardness and flexibility which is usually lacking in crosslinked polymers, whilst the urethane type of crosslink tends to provide both solvent and chemical resistance. It is possible to prepare monomers with various secondary sites suitable for crosslinking.

Aminimide monomers do not copolymerise with styrene, but their hydrochlorides are easily prepared, and these copolymerise with styrene. The copolymers may be thermolysed directly, in solution or in the solid phase, to copolymers with pendant isocyanate groups [176].

Unsaturated aminimides are formed by the action of unsymmetrical dimethylhydrazine and the chloride of a lipophilic radical, e.g. decyl chloride and maleic anhydride, to form an aminimide, *cis*-$HOOCCH{:}CHCON^{-+}N(CH_3)_2C_{10}H_{21}$. After copolymerisation in emulsion

Table 5.14 Q and e values of aminimides and some comparisons [179A]

	Q	e
Methyl methacrylate	0.74	0.40
Methacrylamide	1.46	1.24
1,1,1-Trimethylamine methacrylimide	0.183	−0.60
1,1-Dimethyl-(2,3-dihydroxypropyl) methacrylimide	0.24	−1.24
1,1-Dimethyl-1-(2-hydroxypropyl)amine methacrylimide	0.12	−2.45
Acrylonitrile	0.60	1.20
Vinyl acetate	0.026	−0.22

with ethyl acrylate and 2-ethylhexyl acrylate, this latex may be coated on to poly(ethylene terephthalate) film to give an adhesive tape [180].

A typical emulsion polymerisation includes acrylic esters, hydroxypropyl methacrylate and about 4 % of trimethylamine methacrylimide with an anionic emulsifier and with redox initiation [181].

Speciality monomers containing the aminimide functional group have been used to prepare a thermosetting acrylic solution, emulsion and powder coatings with the monomers functioning as urethane precursors. Suitable solution polymers contain about 15 % of hydroxyethyl acrylate and 10 % of 3-(1,1-dimethyl-1-(2-hydroxypropyl)amine with styrene and butyl acrylate (43 : 32 %) in 2 : 1 xylene/butanol. Azodi-isobutyronitrile is the initiator with 0.15 %, on monomers, of dibutyltin dilaurate added after polymerisation. The product has a typical solids content of 50 %.

Dimethyl-1,2-dihydroxypropylamine methacrylimide, which is synthesized using glycidol in the epoxide route (see reference [177]), is an alternative monomer. It has the benefit that 1,2-dihydroxypropylmethylamine, which is non-volatile, is released on conversion to isocyanate during stoving:

$$CH_2{:}C(CH_3)CON^{-+}N(CH_3)_2CH_2CHOHCH_2OH \longrightarrow$$

$$CH_2{:}C(CH_3)NCO + (CH_3)_2NCH_2CHOHCH_2OH$$

Note that isopropenyl units are formed.

The non-volatile hydroxyamine can itself function in crosslinking, and care must be taken in determining the overall NCO:CH ratio. For paints, the polymers are precipitated and ground or spray dried. They may be pigmented with titanium dioxide, a typical pigment, the binder ratio being 1 : 0.6.

After application, films from this type of composition show resistance to chemicals and stains and are unaffected by overbake. Detergent and bleach resistance is excellent and likewise resistance to salt spray, which does not induce blister formation. However, under some conditions, particularly with the highest molecular weight polymers, there may be a poor flow rating on application and some trouble with gas blistering when a volatile hydroxyamine is released. This applies particularly to powder coatings [182].

Water-based paints may be formed from a conventionally polymerised emulsion copolymer also containing hydroxyethyl acrylate. The aminimide may be added at the beginning or end of the polymerisation, optionally in the latter case as an aqueous solution [182].

Hydroxy-containing acrylic copolymers may be crosslinked with bisaminimides of general formula $(CH_3)_3N^{+-}NOC(CH_2)_nCON^{-+}N(CH_3)_3$. The composition has indefinite ambient stability. A temperature of 163 °C is necessary for cure by the release of isocyanate. A typical example is bis(trimethylamine)adipimide. Note that the acrylic group is not in this case in the aminimide function. In the main polymer, because of relatively low reactivity, a high hydroxyl number, circa 110, is desirable and there should be equivalence of hydroxyl and NCO functions. Attenuated total reflectance techniques have been used to determine the course of the reaction [182].

1,1-Dimethyl-1-(2-hydroxypropylamine)methacrylimide reacts with succinic anhydride to form a suitable crosslinking compound. This rearranges in the stoving cycle by a similar reaction to that shown previously:

$$CH_2:C(CH_3)CON^{-+}N(CH_3)_2CH_2CH(CH_3)OC(O)C_2H_4COOH \longrightarrow$$

$$CH_2:C(CH_3)NCO + (CH_3)_2NCH_2CH(CH_3)OC(O)C_2H_4COOH \longrightarrow$$

Crosslinking takes place in the simultaneous presence of a hydroxyl-containing acrylic polymer solution. The isocyanate liberated crosslinks with the hydroxyl groups of the acrylic copolymer, whilst the amino acid which is formed also contributes to the crosslinking of both the epoxide and the hydroxyacrylate copolymer [182].

It is possible to utilise an acrylic aminimide to crosslink an epoxide resin, typically the liquid Epon® 812. It is also possible to crosslink an epoxide as Epikote 828® (Shell) with an aminimide in the absence of an acrylate copolymer with active hydroxyl groups, especially in the presence of hexahydrophthalic anhydride. The products have a long pot life and are useful in adhesives and coatings [183].

A practical synthesis of trimethylamine acrylimide starts from 1,1,1-trimethylhydrazinium p-toluenesulfonate and 3-chloropropionyl chloride. The intermediate trimethylamine 3-chloropropionimide is easily transformed into trimethylamine acrylimide through a single-step dehydrohalogenation. The monomer $CH_2:CRCON^-N^+(CH_3)_3$ is also described [184].

8.6 Acrylamidoglycollates

An acrylamidoglycollate enables cures to take place without amines or the liberation of formaldehyde. An active hydrogen donor is necessary. The films formed are non-yellowing and acid resistant, useful in both binders and adhesives. The formula for the methyl ester (MAGME) is $CH_2:CH.C(O)NHCH(OCH_3)COOCH_3$.

The copolymerisation of a batch including 5 % of MAGME, styrene, butyl acrylate (5 : 25 : 75) has been studied in an emulsion with sodium dodecylbenzene sulfonate as emulsifier at 5.8 g L^{-1}, in which the ratio of monomer to water was 0.43 : 1. Conversions of monomers are shown in Figures 5.1 (a) and (b) and also the cumulative fraction of MAGME in the copolymer against conversion. The results are explained by the water solubility of the monomer, starting in the water phase, but transferring to the particles as they became insoluble. Styrene completely polymerises before the butyl acrylate and the largely aqueous MAGME increases towards the end of the reaction [185].

A typical latex is formed from ethyl acrylate, methyl methacrylate, methyl acrylamidoglycollate, acrylamide and p-toluenesulfonic acid (52.5 : 42 : 3 : 2.5 : 2) by conventional methods, forming a 40 % latex. Films from this may

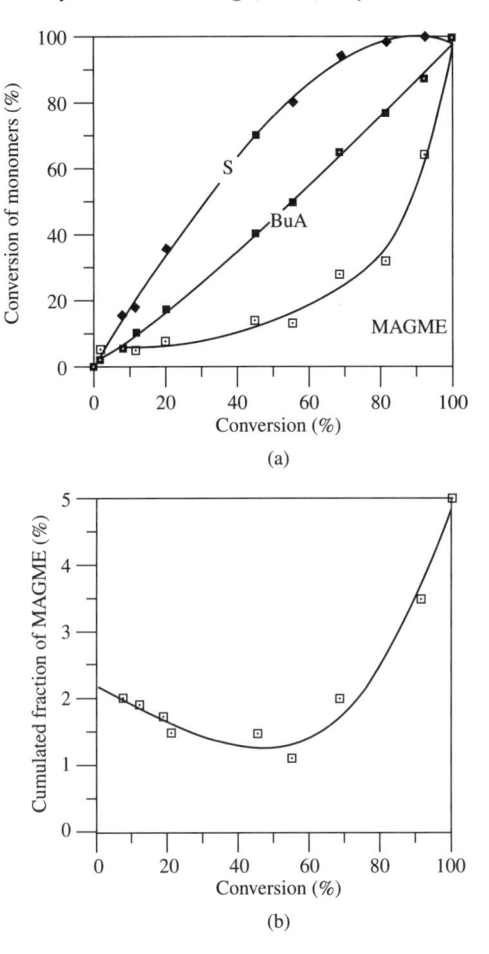

Figure 5.1 (a) Conversion of monomers versus total conversion. (b) Cumulated fraction of MAGME in the polymer versus conversion.

cure at ambient temperatures, the catalyst being desirable, or at 150 °C for 5 min without the catalyst. Films formed have swelling ratios (swollen film area/initial film area) in methylethyl ketone between 4.0 and 5.8 [186, 187].

8.7 Other amide derivatives with unsaturation

Crosslinkable copolymers have been prepared from NMAM and a methylenediamine derivative of acrylamide, e.g. N-formyl-N'-methacryloylmethylenediamine $HCONHCH_2NHCOC(CH_3):CH_2$, the latter at $\sim 3.5\%$ of total monomers. Polymerisation is by a conventional redox process [188].

The derivative $CH_2:CHCONHCH(OCH_3)CH(OCH_3)_2$, which can be considered to be derived from an acetal of glyoxal, was copolymerised with butyl acrylate, methyl methacrylate and acrylic acid (1.96 : 41.27 : 56.27 : 0.5) to obtain a latex, with 45.1 % of solids and viscosity of 44.5 mPa s. This is cured for 10 min at 150 °C to give a film with a or T_g 25 °C and 140 % swelling in chloroform. There is no detectable formaldehyde emission [189].

A novel crosslinker, tetramethylolglycouril at 1–5 % (on vinyl acetate), was included with NMAM in a latex stabilised by 3–5 % of polyvinyl alcohol. The latices have a good shelf life after adding the catalyst and good viscosity stability, and may be used with radio-frequency curing [190] (see Section 10).

Alternatively, 2.5 % of allyl N-methylolcarbamate CH_2:$CHCH_2$.$OCONH$-OCH_2OH (formed from allyl carbamate CH_2:$CHCH_2OCONH_2$ and formaldehyde) is included in a vinyl acetate copolymer latex. It has approximately the same reactivity as vinyl acetate and thus copolymerises well and provides a site for crosslinking. A similar percentage of acrylamide is also added cautiously during the polymerisation. A film from the latex when heated becomes insoluble in trichlorethylene [191].

8.8 Cure with aminoplasts

It is not always possible to indicate from an abstract, and sometimes from an original paper, whether specific reference is made to emulsion polymers. If they are likely to be useful in aqueous conditions many papers and specifications will be included here. In some cases the direct use of a latex with an aminoplast solution is claimed in patent specifications.

Dimethylolurea $OC(NHCH_2OH)_2$ has already been mentioned as a potential curing agent. The principal aminoplasts of interest in cure systems for acrylic latices are condensates of melamine with formaldehyde:

$$\text{Melamine} \qquad \text{Trimethylolmelamine}$$

The trimethylolmelamine is for most purposes further reacted so that the second H atom on the NH_2 is reacted with formaldehyde and then etherified to form hexamethoxymethylmelamine:

Polymers containing (meth)acrylamide

In practice the etherification is not quite complete and the actual number of etherified groups in technical products is 5.5–6. Hexamethoxymethylmelamine is a white crystalline solid, soluble in many common solvents including water if there is a small quantity of ethyl alcohol or isopropyl alcohol present. Whilst most solvent-based systems of this type use a mixture such as xylene–butanol, comment here will be restricted to water-based compositions.

For the crosslinking reaction it is normal for a latex, usually acrylic ester based, to contain sufficient carboxyl groups to be soluble in ammonia solution, or preferably an amine, and then blended with a solution of hexamethoxymethylmelamine. Studies of the crosslinking reaction have shown that, following the loss of volatile alkali, the free carboxyl groups which are generated react with the methylolamine ester, forming methylene esters, which because of their high functionality crosslink rapidly. A simplified unit reaction may be as follows [192]:

$$\begin{array}{c} CH_3OCH_2-N-C \\ {|} \\ [CH_3O]CH_2 \\ {|} \\ (CH_3)C-COO{H} \\ {|} \\ CH_2 \end{array} \longrightarrow \begin{array}{c} CH_3OCH_2-N-C \\ \diagdown\diagup CH_2 \\ O \\ (CH_3)C-C=O \\ {|} \\ CH_2 \end{array}$$

The above is based on units of methacrylic acid.

Typical copolymers taking part in this reaction contain >8 %, preferably >10 %, of polymerisable acid. Methacrylic acid is the most useful with emulsion polymers. Molecular weights of polymers should be low, ~10 000, to avoid excessive functionality which would lead to formation of brittle films. To ensure this, a chain transfer agent such as *tert*-dodecyl mercaptan is added to the initial emulsion before polymerisation, preferably in the water phase. It is also highly desirable that the copolymerisation should be even, most easily accomplished by adding all the monomers at a controlled rate, with acid blended with other monomers.

By careful blending, the reaction between the two components is virtually quantitative, as shown by titrations of residual acid, treating the crosslinked polymer as an ion exchange resin. An acid catalyst is desirable to obtain a reasonable reaction rate; *p*-toluenesulfonic acid is amongst the most practical.

The disadvantage of latices is that they include water-soluble substances, which generally do not take part in the crosslinking reaction. This may be overcome to some extent by preparing the original latex using a 'seed latex'. About 5 % of a previously prepared latex is used as a seed. In the second stage only monomers, and a further addition of initiator, are required, no further emulsifier being necessary. Water-solubles are thus reduced to below 1 % of the polymer. Typical formulations are shown in Table 5.15 [193].

In order to solubilise the latices in Table 5.15, trialkylamines or monohydroxyalkylamines with b.p. <175 °C and heat of solution above

Table 5.15 Seed latex formulation and latex for crosslinking [193]

	Parts by weight	
	Seed latex	Latex for crosslinking
Ethyl acrylate	65.0	60.0
2-Ethylhexyl acrylate	—	10.0
Styrene	25.0	20.0
Methacrylic acid	10.0	10.0
Seed Latex (40 % aq.)	—	12.5
Sodium lauryl sulfate	2.5	—
tert-Dodecyl mercaptan	3.0	3.0
Potassium persulfate	0.25	0.4
Sodium persulfate	0.0625	0.2
Hydrochloric acid (17.5 % w/w)	—	0.1
Water	150.0	150.0

The sodium persulfate, in part of the water, is gradually added during each of the two stages at 82–83 °C.
Both stages employ the normal process of gradual addition of monomers. 13.6 g of diethylaminoethanol added to the latex for crosslinking gives a clear solution.

5500 calmole^{-1} at infinite dilution are advised, suitable examples being triethylamine and diethylaminoethanol. Addition of these amines enable reasonable viscosity stability to be achieved. This is approximately 3–6 P at 50 % solids content, hardly changing for at least 8 weeks. Addition of 45 parts of the dimethyl ether of trimethylolmelamine to 100 parts of copolymer (w/w on solids) forms a stable composition which can form films by baking on to a mild steel panel after air drying. There should be no 'flash' corrosion in the drying stage due to pH fall. Even after immersion in water at 38 °C for 24 h, the films showed no blistering, softening or loss of adhesion [194]. The addition of an alcohol such as isopropanol increases the stability of the mixed acrylic latex with the melamine derivative [195].

In considering this type of latex, finish stability considerations, both chemical, mechanical and of viscosity, are paramount. This applies to both the initial latex and to the blend with the aminoplast. The composition must remain homogeneous during the drying of the film. This is one of the reasons for the volatility restriction on the alkali used. Hydroxyalkylamines may also have both a solubilising effect on the polymer resin blend and an indirect one by forming the salt. An excessively high polymer molecular weight may cause irregularities, e.g. matt or 'cissing' films due to excessive functionality of the polymer, and lack of flow properties. The types quoted here are also used as a basis for stoving coatings, the subject of a later chapter.

It may be stressed that the use of the ethers rather than the simple methylolmelamines enables a much smoother reaction to take place, and produces conditions that discourage self-condensation of the melamine derivative. In preparing satisfactory crosslinked acrylics and vinyls it is in principle necessary to choose conditions that will discourage self-condensation and promote the desired interaction.

The chemistry of acrylic copolymers including hydroxyl and carboxyl groups or hydroxyl alone with a melamine aminoplast has been studied, utilising infrared spectroscopy. An almost fully methylated product (Cymel® 300) and partially methylated or butylated methoxymelamines were used. An equation is

$$C_{\text{eff}} = \left(\frac{P}{E}\right)\frac{\text{PA}_{\text{eff}} + \text{PA/PB}_{\text{eff}} + M\ \text{PC}_{\text{eff}}}{2}$$

where e = effective, E = the equivalent weight of the acrylic functional groups, P is the fraction of total solids that is acrylic polymer, PA is the extent of the reaction of acrylic functional groups, PC is the extent of the reaction of the melamine methylol groups, M is the ratio of melamine methylol groups to acrylic functional groups, C = crosslink density, PA_{eff} is the probability that the acrylic functional group has formed a crosslink leading to an infinite network and PB_{eff} and PC_{eff} are similar parameters for the melamine alkoxy and the melamine methylol respectively. The equation allows for the fact that not all of the crosslinks are elastically effective. With Cymel® 300 the rate of the reaction at low temperatures depends on the concentration, but the latter only slightly affects the extent of reaction. Even with excess, the reaction does not go to completion.

With a partially methylated methoxymelamine the reaction level increases with temperature, but reaches a plateau. Melamine–melamine crosslinks contribute to the network in the latter cases, taking place at 130 °C compared with the 180 °C required for a fully methylated product. Acid catalysts have a marked effect in reducing the cure temperature with Cymel 300 and an active acrylic polymer, by as much as 50–60 °C in some cases. *p*-Toluene sulfonic acid is a good catalyst [196].

These investigations also show that the major reactions are those of the functional groups on the acrylic polymer with the methoxylated hydroxymelamine alkoxy groups. There is some self-condensation of the hydroxymelamine groups; hence the recommendation is justified that the fully methoxylated melamine derivative is desirable. The reaction of methoxy groups of the melamine derivative with the reactive acrylic increases with temperature, as does at first the self-condensation of the melamine hydroxyls [197, 198].

A computerised program has been initiated to find optimum conditions for the MF resin/active acrylic reaction [199–201].

A latex prepared by a standard method with acrylonitrile, butyl acrylate, acrylic acid and β-hydroxyethyl acrylate (42 : 46 : 2 : 10) can be crosslinked with a methoxylated methylurea resin, even at 80 °C, the optimum amino content on a solids basis being 28.5 %. Figures 5.2 (a) to (e) give an indication

Crosslinking and curing

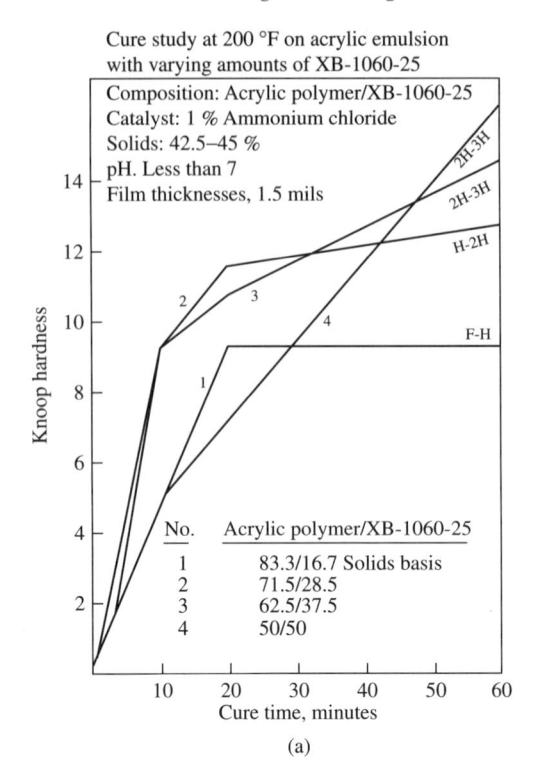

Figure 5.2 Degree of cure of acrylic latex/methylated urea resin [203].

Cure study at 200 °F on acrylic emulsion with varying amounts of XB-1060-25

Composition: Acrylic polymer/XB-1060-25
Catalyst: 1 % Ammonium chloride
Solids: 42.5–45 %
pH. Less than 7
Film thicknesses, 1.5 mils

No.	Acrylic polymer/XB-1060-25
1	83.3/16.7 Solids basis
2	71.5/28.5
3	62.5/37.5
4	50/50

(a)

of the degree of cure by the Knoop hardness test of the cured film. Figure 5.2 gives a suggested mechanism for cure. The corresponding melamine derivative would give little reaction at this temperature. However, whilst films from the blend have good hardness, elasticity and adhesion combined with good gloss and chemical resistance, the shelf life of the mixed latex and resin is only 6 days at a maximum, much less than when the melamine derivative Cymel® 300 is used as an alternative crosslinker [202].

N-Butoxymethacrylamide occurs in a number of compositions, typically at 5 % of this compound with 10 % of methacrylic acid, and is part of a hard acrylic copolymer which is solubilised by ammonia to an aqueous composition [203]. One example (Table 5.16) gives a formulation (emulsifiers and water are not indicated) for a coating for aluminium, pigmented before application on to aluminium. It has good adhesion and water resistance. Note the lower-than-usual proportion of melamine aminoplast [204]. A self-crosslinking acrylic copolymer including amide and/or acid groups may be blended with dispersible or soluble melamine or urea resins polyols or amines and an acid catalyst as a hardener [205].

Polymers containing halogens 321

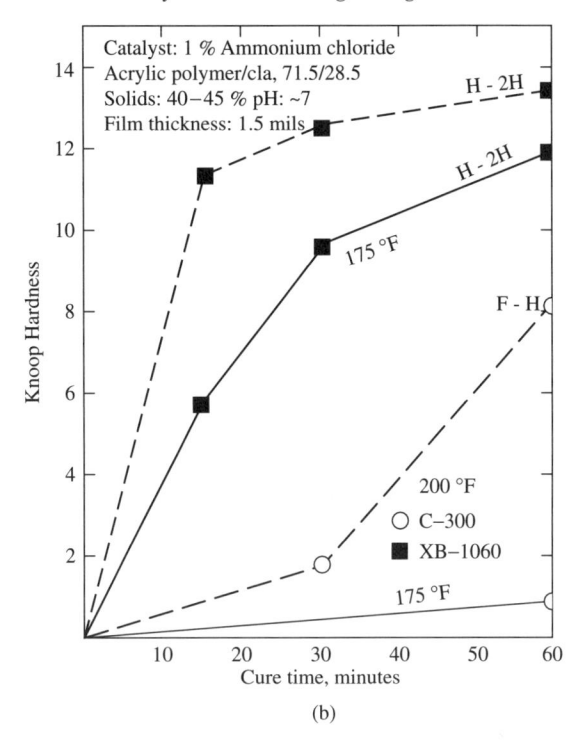

Figure 5.2 (*continued*)

Unusually, butoxymethylmelamine is dissolved in ethyl acrylate and styrene (20 : 152 : 38) whilst *N*-methylolmelamine is dissolved in an aqueous phase including an ethoxylated nonylphenol and a buffer. The additives seem to have no major effect on polymerisation, although possibly there may be some chain transfer. The resultant latex gives a polymer that is thermoset on heating [206].

9 POLYMERS CONTAINING HALOGENS

This section refers to those monomers and polymers in which crosslinking takes place by reaction of labile halogen atoms, usually chlorine. In many cases, however, the chlorine atoms in one group are activated by adjacent groups, making the loss of the atom relatively easy. This may involve the replacement of the chlorine by another group such as hydroxyl, but the loss of chlorine may be turned to advantage by allowing different molecules of polymer to combine, thus giving rise to crosslinking. This usually occurs under alkaline conditions, rather than the low pH normally employed, but may have the disadvantage of the production of a salt as a by-product.

A polymer such as polyvinyl chloride is fairly stable, although there is some discoloration on exposure to light, causing the formation of alternate

322   *Crosslinking and curing*

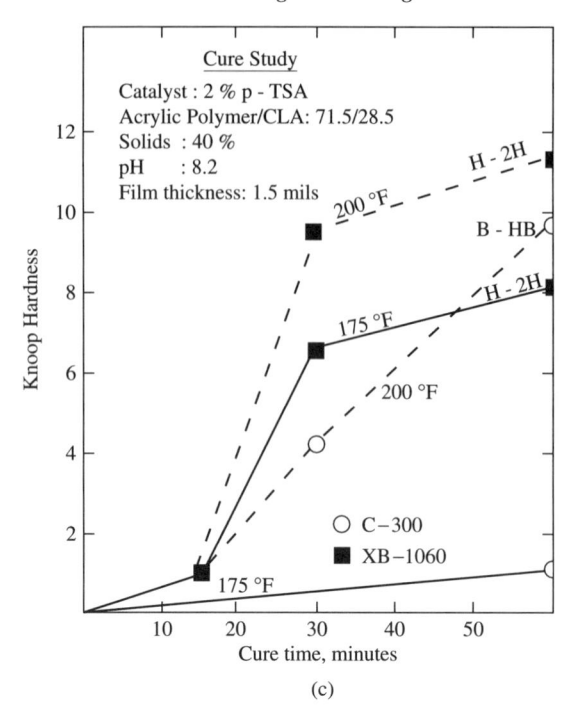

Figure 5.2 (*continued*)

double bonds by limited loss of hydrogen chloride. This is overcome by the inclusion of stabilisers. Polymers and copolymers of vinyl chloride are not readily crosslinked unless groups with more labile chloride atoms are included.

Vinyl chloroacetate is an example of a compound where the chlorine atom is labile. Thus an emulsion copolymer containing, *inter alia*, 0.1–5 % of hydroxyethyl acrylate and 1–10 % of vinyl chloroacetate is vulcanisable after isolation of the copolymer. The method of vulcanisation is not indicated. A latex formulation is shown in Table 5.17 [207]. The above is from Japanese sources, but an alternative European patent has been disclosed, which claims that aluminium hydroxide may also be used to effect crosslinking [208].

Chloroacetoxyethyl acrylates and methacrylates and the butyl and propyl equivalents in copolymers, which also include acrylamide and hydroxypropyl methacrylate, have been quoted in a copolymer formulation in which is included a polyamide resin. This is especially useful in the preparation of water-based gloss paints [209].

A vulcanisable acrylic latex includes 15 % of chloromethylstyrene and 0.4 % of methacrylic acid, the polymerisation being of a conventional redox type. The separated solid is vulcanised as a rubber with bis[4-(α-α-dimethylbenzyl) phenyl] amine and 2-aminopyridine. A phosphonium salt, $C_6H_5CH_2$-$P(PH_3)^+Cl^-$, may also be present in the curing agents [210, 211].

Cure study
Acrylic emulsion with XB-1060-25

Composition: Acrylic Polymer/XB-1060-25 71.5/28.5
Catalyst: 1 % Ammonium Chloride on total solids
Solids: 45 %
pH. 6.7
Viscosity, 85 cps
Film thickness, 1.5 miles

(d)

Possible curing mechanism

R = CH₃ or H

(e)

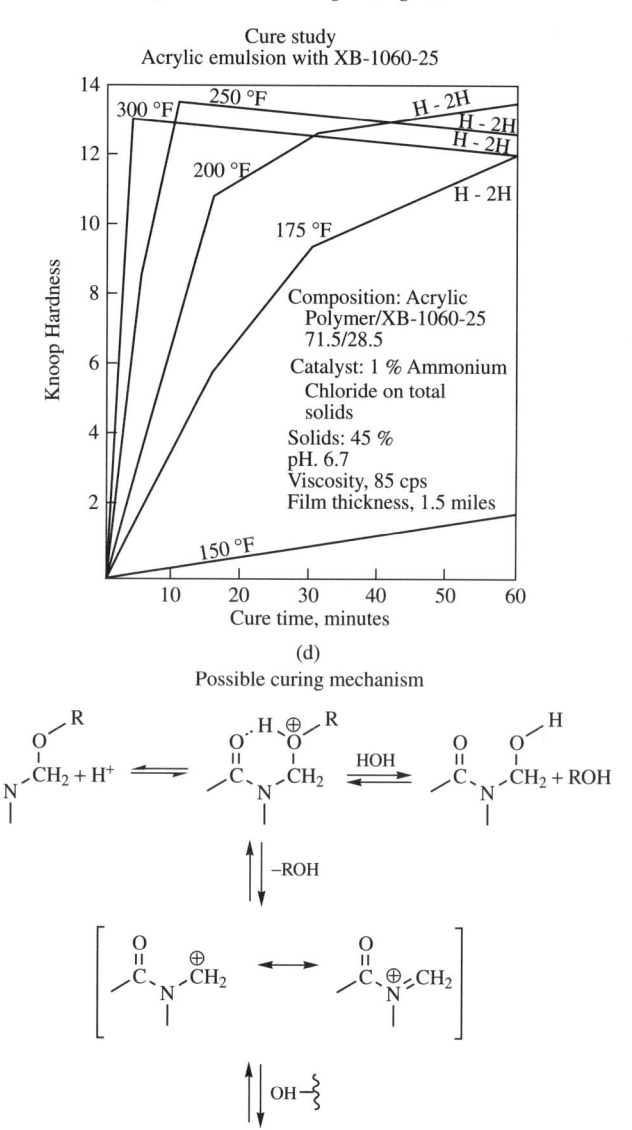

Figure 5.2 (*continued*)

Table 5.16 Latex monomers and cure formulation [204]

	Parts by weight
A. Monomers in latex	
Butyl acrylate	328.5
Methyl methacrylate	342.5
Diethylaminoethyl methacrylate	16.9
B. Formulation for cure	
Latex A above, 40 % solids	101.9
Melamine aminoplast (Cymel® 1141)	6.88
Titanium dioxide pigment paste	62.41
Butoxy diglycol	8.51
Water	5

Table 5.17 Vulcanisable copolymer containing vinyl chloroacetate [207]

	Parts by weight
Prepare the following water phase:	
Polyoxyethylene dodecyl ether	2
Sodium dodecyl sulfate	1
Water	200
Heat to 45 °C under nitrogen and then add in separate streams:	
Sodium persulfate	0.04
Sodium hydrogen sulfite NaHSO$_3$	0.04
Prepare the following monomer mixture:	
A. Ethyl acrylate	48
B. Butyl acrylate	25
C. Methoxyethyl acrylate	24
D. 2-Hydroxypropyl acrylate	0.5
A–E are mixed and then the following are added to the reactor:	
Vinyl chloroacetate	2.5
A–D mixture	6.15
Dodecyl mercaptan	0.00125
Polymerisation commences and then the balance of comonomer mixture A–D is added at 60 °C and maintained for 1 h.	

Dimethyldiallyl ammonium chloride, 3-chloro-2-hydroxypropyl acrylate and dimethylaminopropyl methacrylamide hydrochloride (1.5 : 0.3 : 0.2 mol) are polymerised in aqueous solution with a persulfate initiator. The pH is adjusted to 5–5.5 with sodium hydroxide to give an aqueous solution, which is stable for 6 months at ambient temperature. Gelling takes place within 165 min if the pH is raised to 9.5 with sodium hydroxide solution. No doubt this could be adapted to normal emulsion polymerisation with a suitable adjustment of monomers. The gelling is evidently due to the labile nature of the chloride atom on the acrylate ester [212].

Water-borne crosslinkable fluorochemical coatings are of considerable interest in 'non-stick' surfaces and monomers ending in CF_3 have the lowest surface energy possible. Water-borne crosslinkable fluorochemical (WXF) coatings are formed from a polymer including a fluoroaliphatic group and at least one anionic group. A second polymeric component has at least one oxazoline group. The reaction is shown below:

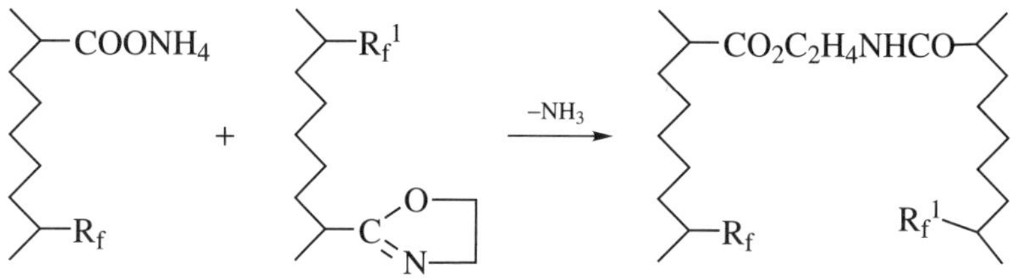

$$\{-COONH_4 \quad + \quad \{-R_f^1 \quad \xrightarrow{-NH_3} \quad \{-CO_2C_2H_4NHCO-\} $$
$$\{-R_f \qquad \{-C\underset{N}{\overset{O}{\diagdown}}\quad\quad\quad \{-R_f \qquad R_f^1-\}$$

Typical fluorochemical monomers with a terminal CF_3 group include C_8F_{17}-$SO_2N(C_2H_5)CH_2CH_3OC(O)CH:CH_2$. The fluorochemical 'tails' tend to orient towards the air interface with the 'tails' out. The extent of crosslinking can be controlled by adjusting the carboxylate to oxazoline ratio. An interesting application is an anti-graffiti coating [213].

A copolymer of fluorinated unsaturated hydrocarbons $CH_2:CF_2.C_3F_6$ and the brominated ether $BrC_2H_4OCF:CF_2$ (50 : 24.8 : 24 : 0.4 mol %) with CBr_2-F_2 as the chain transfer agent, formed by emulsion polymerisation, forms a rubber after vulcanisation, the product having good oil resistance, an elongation of 210 %, tensile strength of 17 MPa and a compression set of 38 % [214].

10 RADIATION CURING

The two principal methods of radiation cure are ultraviolet and infrared, the former being the principal one in practical use. Electron beam cure, radio-frequency heating, γ-rays and microwave have also been used, and in some cases natural light alone is sufficient. The detailed theory of the reactions involved, as well as a full description of the necessary apparatus, are beyond the scope of this volume, but a number of texts and reviews are available. The technical demand for radiation-curable raw materials has increased considerably in recent years [215–220A].

There are some limitations to the use of radiation cure. These include the cost of curing equipment and the limitations depending on the size and nature of the object being subject to radiation cure. Infrared (IR), microwave and radio-frequency cure tend to resemble standard oven heating. Ultraviolet (UV) cure and electron beam (EB) heating depend on a different type of reaction with a high-energy input, resulting in the breaking of chemical bonds and a subsequent chain reaction of the radical formed, involving crosslinking, occasionally by an ionic mechanism. When latices are involved, radiation cure will take place after forming a film or impregnation depending on the ultimate application.

General advantages of radiation cure include high productivity, because of the very fast curing reaction, the ability to work with heat-sensitive substrates and energy and space savings, Limitations are a possible health hazard as radiation-curable materials, especially monomers, can cause skin irritation and sensitization, with UV-restricted film thickness, limited adhesion in some cases to substrates, handling problems with high-viscosity formulations and some deinking problems if recycling is desired. In addition, equipment costs are high compared to methods involving direct heating or solvent removal [221].

10.1 Ultraviolet curing

UV radiation has the advantage that there is only little heating of the object being treated, which is of advantage in heat-sensitive substrates, although some infrared heat is also evolved. The successful use of UV radiation depends on the availability of a low cost and practical source of radiation. The principal commercial sources are medium-pressure mercury vapour lamps, which are up to 2 m long, typically with power outputs of 80 W cm^{-1}, although lamps with higher power outputs are available. The radiation emitted consists of a continuous wavelength distribution of radiation with peaks at 254, 313, 366 and 405 nm (Figure 5.3) [222].

Electrodeless lamps powered by microwaves are in commercial use. They are of particular advantage when the lamps are doped with elements other than mercury, e.g. to increase radiation in the near-visible region. For efficient operation the distance between the lamp and the object being treated must be reasonably uniform, and is therefore most efficient with flat surfaces that can be moved or cylindrical objects that can be turned. It should be noted that UV radiation is hazardous to the operator, leading to severe burns, and must be shielded. Ozone is also liberated so appropriate ventilation is desirable. Note that pigments may affect the efficiency of the UV cure and that oxygen inhibition is prominent.

Water as a diluent in UV-curable formulations can either be used directly with a solvent or as part of an emulsion system without a loss in performance of coating properties such as hardness, elasticity and reactivity. As shrinkage during polymerisation leads to excess free volume, there may be a microphase-separated morphology of the crosslinked coatings, thus evenly distributing excess free volume generated during polymerisation. This leads to continuous and homogeneous network formation, uninterrupted by stress cracks due to shrinkage [223].

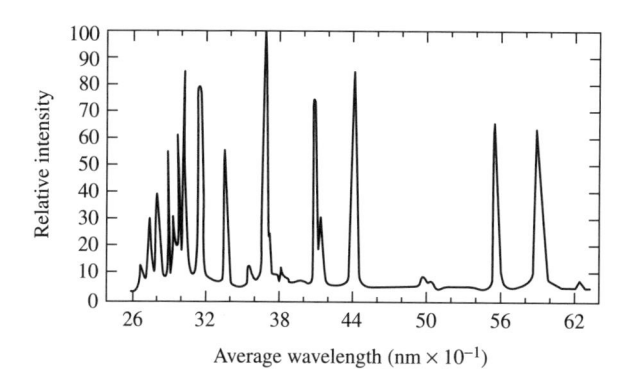

Figure 5.3 Energy distribution of radiation from a medium pressure mercury vapour lamp. (With acknowledgement to Plenum, New York.)

10.1.1 Photoinitiators

The operation of UV cure depends on the co-existence of a photoinitiator, which is often used directly in photopolymerisation, but the principle is similar to conventional crosslinking. UV radiation dissociates the photoinitiator into radicals, or occasionally into ionic species. These react with the polymer being cured. e.g. by hydrogen abstraction, forming a radical or ionic species, and commence the curing reaction. Under some conditions, e.g. on a film, photoinitiators may retard cure since a photoinitiator has also the effect of light screening. There is therefore an optimal photoinitator concentration that decreases with film thickness.

In many photochemical processes, a photosensitizer is used. These additives undergo hydrogen or electron transfer and can generate radicals from molecules with low excitation energies. This enables photoinitiation to be extended to the visible part of the spectrum. The curing (crosslinking) of a polymer under the influence of UV requires initiation in the same way as a monomer. In the process, the absorbed energy is transferred to an energy acceptor which produces an initiator radical by hydrogen cleavage. The photosensitizer undergoes no net change. Photocoinitiators only accept energy from the photosensitizer. They are usually dyestuffs, typical examples being eosin, chlorophyll, methylene blue and thioxanthone. The last named causes photocleavage of quinoline-8-sulfonyl chloride in a bimolecular process.

Natural rubber latex may be vulcanised by addition of 2-ethylhexyl acrylate (5 g) and 1 % ammonia solution (39 mL) to 166 g of latex. After standing for 16 h, it is irradiated with γ-rays from a Co^{60} source for 5 h at 1 Mrad h^{-1}. The treated rubber shows improved elongation, tear and tensile strength. There is reduced permanent elongation compared with treating rubber latex with the more hydrophilic 1,6-hexane glycol diacrylate [224].

10.1.2 Photoinitiators for radical cure

Benzoin ethers were the first to be used commercially. The dissociation may be represented as

$$\underset{H}{\underset{|}{C_6H_5-\overset{O}{\overset{||}{C}}-\overset{|}{\underset{|}{C}}-C_6H_5}} \xrightarrow{h\nu} C_6H_5-\overset{O}{\overset{||}{C}}\cdot + \underset{H}{\underset{|}{C_6H_5-\overset{OR}{\overset{|}{C}}\cdot}}$$

Both of the radicals formed take part in polymerisation and curing reactions. However, they exhibit rather low stability in reactive coatings, which is attributed to the labile benzylic hydrogen atom. Package stability is improved if the benzylic carbon atom is fully substituted.

A latex, prepared as in Table 5.18, to which a polyol methacrylate and benzoin methyl ether have been added, has been embossed on to a decorative sheet. After drying at 95 °C, it is overlaid coat side down on a stainless steel plate, irradiated with UV light for 5 s, and freed from the polyethylene sheet to give a decorative protective embossed surface [225]. A review on industrial photoinitiators is available [226].

Acetophenone derivatives, e.g. the ketal 2,2-dimethoxy-2-phenylacetophenone, which do not have the problem of a labile hydrogen atom, are effective photoinitiators with package stability. Dissociation can be explained by the fact that the sluggish dimethoxybenzyl radical initiator undergoes further dissociation to the very reactive methyl radical, while increases with temperature:

$$C_6H_4-\overset{O}{\overset{||}{C}}-\underset{OCH_3}{\overset{OCH_3}{\overset{|}{C}}}-C_6H_5 \xrightarrow{h\nu} C_6H_5-\overset{O}{\overset{||}{C}}\cdot + C_6H_5-\underset{OCH_3}{\overset{OCH_3}{\overset{|}{C}}}\cdot$$

$$C_6H_5-\underset{(OCH_3)_2}{\overset{OCH_3}{\overset{|}{C}}}\cdot \longrightarrow C_6H_5-C-OCH_3 + CH_3\cdot$$

For further initiators of this type and for cationic photoinitiators, the volumes already quoted should be studied. Water and alcohols act as terminators for these photoinitiators, but their main interest is in non-aqueous polymerisation. Some varied examples from the literature which are quoted may not necessarily refer to latices, but should be easily adaptable to products prepared from latices.

10.1.3 Monomers and polymers in UV cured systems

Because of viscosity problems, the polymers submitted to crosslinking are of relatively low molecular weight, often referred to as oligomers or prepolymers.

Table 5.18 Latex for UV light cure [225]

Stage 1: latex preparation	
2-Ethylhexyl acrylate	53 kg
Methyl methacrylate	40 kg
Acrylic acid	5 kg
Polyethylene glycol dodecyl ether	3 kg
Ammonium persulfate	0.5 kg
Water	100 kg
Stage 2: modification for UV treatment	
Latex above	60 kg
Trimethylolpropane triacrylate	30 kg
Benzoin methyl ether	5 kg

Oligomers must contain a reactive centre which can undergo further polymerisation. As mentioned previously, it is often immaterial whether the oligomers have been prepared by emulsion or other methods. The fastest cure rates are obtained with acrylic functionality, which is faster than methacrylic functionality. Methacrylic functionality cures faster than allyl functionality, which in turns cures faster than vinyl functionality.

Differential scanning calorimetry (DSC) is recommended as the best analytical technique for the estimation of the degree of cure in UV-curable clear coats [227].

Suitable acrylic oligomers may be derived from epoxy acrylates which, in general, are prepared by reacting 1 mole of the diglycidyl ether of bisphenol A with 2 moles of acrylic acid Section 6.1 although other types, in particular aliphatic rather than aromatic ethers, are used. Urethane acrylics, obtained by end capping a polyester or polyether polyol with a di-isocyanate, followed by reaction with a hydroxyacrylate, are unlikely to be of interest as emulsion derivatives. This also applies to polyester acrylates. Epoxides may be cured with cationic photoinitiators. There has been a recent surge of interest in acrylic monomers with other functional groups including phosphoric acid and some silanes.

10.1.4 Practical examples of UV cure

Radiation cure of acrylic and methacrylic esters with multiple double bonds, often in copolymers, takes place under UV cure. In addition, methoxyether acrylates show a good response [228].

A formulation for a 50 % latex, prepared conventionally with a sulfonate/non-ionic emulsifier system as in Table 5.19, is modified as shown in Table 5.20, coated on to a polythene sheet to form a 10 µm layer, heated at 85 °C, bonded to stainless steel and irradiated by a high-pressure mercury lamp. The polythene sheet is stripped to form an adherent, water-resistant surface layer [229].

Table 5.19 Acrylic latex for crosslinking [229]

The following formulation gives the monomers only:	
2-Ethylhexyl acrylate	60 kg
Ethyl acrylate	35 kg
Acrylic acid	2 kg
2-Acryloyloxyethyl phosphate	3 kg

Table 5.20 UV curing composition [229]

Latex as in Table 5.20	100 kg
Tetraethylene glycol dimmethacrylate	50 kg
Benzoin isobutyl ether	3 kg
Cure (see text)	

A similar acrylic latex with $T^g \sim 28$ °C consists typically of methyl methacrylate, ethyl acrylate and acrylic acid (63 : 35 : 2). It is blended with trimethylolpropane triacrylate, applied to phosphated steel panels, dried at first ambient and then at 66 °C and cured with medium-pressure mercury lamps with a proprietary photoinitiator [230].

Related patents describe photopolymerisable water-soluble polymers such as polypyrrolidinoethyl methacrylate in admixture with polyvinylpyrrolidone, gelatin and lithium acrylate. Methylene blue is used as a sensitizer; these compositions are designed to cure under white light and are used to prepare holograms [231, 232].

A carboxylated styrene–butadiene latex and a water-dispersible sensitizer have been used in conjunction as a coating on a lithographic printing plate. After exposure to light, the unexposed parts are removed with water, but the crosslinked portion forms a rigid matrix structure [233].

An SB latex including acrylic acid monomer units is crosslinked in the presence of benzophenone tetracarboxylic anhydride and methyl diethanolamine by the application of radiation. Ionic bonds are formed *inter alia* [234].

A series of photosensitive acrylates and methacrylates have been prepared, derived from the cyclic alcohol 1-aza-5-hydroxymethyl-3,7-dioxabicyclo[3,3,0] octane:

$$\begin{array}{c} 7 \\ 6\ H_2C \diagup^{O} \diagdown CH_2\ 8 \\ HOCH_2 - \underset{|}{C}\,^5 - \underset{|}{N}\ 1 \\ 4\ H_2C \diagdown_{O} \diagup CH_2\ 2 \\ 3 \end{array}$$

The intermediate alcohols are prepared by reaction of *tris*-hydroxymethylaminomethane with two aldehydes, which may be added consecutively, modifying the CH_2 groups on carbons 2 and 8. The esters are formed by transesterification

Radiation curing   331

with ethyl acrylate or ethyl methacrylate with a dibutyltin oxide catalyst. The monomers may be converted to salts. Whilst polymers were prepared in solution, these cyclic ether derivatives should be stable for copolymerisation in emulsion. When the α substitution is a halogenated phenyl replacing hydrogen, cure rates are very fast with UV radiation. Halogenated photosensitizers otherwise increase UV cure. The bicyclic moiety seems to be responsible for the observed UV cure. Suggested applications include coatings, inks, printing plates and photoresists *inter alia*, since masked areas do not crosslink [235].

An acrylic copolymer including dicyclopentenyloxyethyl methacrylate (see Section 3.2) to which a photoinitiator is added crosslinks as a result of photopolymerisation on exposure to light. The compositions are suitable for the formulation of anticorrosive lacquers and paints [236].

Photosensitive polymers are formed by condensation of poly(2-hydroxyalkyl) acrylate with cinnamoyl chloride. Crosslinking is due to cyclobutane formation during irradiation, and possibilities for forming acrylic emulsion polymers, the films of which are capable of crosslinking, are present [237].

10.2 Cure by electron beam (EB)

EB polymerisation, whether initiation or curing, is by the action of high-speed penetrating electrons which form what is described as excited molecules. High-energy electrons, generated by charging a tungsten filament at high negative potential, at 150–300 keV, are directed by magnets in a curtain through a metal window to the coatings to be cured. The polymerisation or cure reactions are oxygen inhibited.

The original EB generators, operating at 500 V, required massive shielding to protect operators, and were very bulky and expensive to run. Modern types are more energy efficient and operate at 150–300 kV. For further details of electron beam generators and diagrams a reference is available [238]. The electrons formed in these systems have much greater penetrating power than the photons formed by UV, giving much greater latitude with thick films and with additives, and the inclusion of pigments in films does not reduce its effectiveness.

Cure by electron beam does not require photoinitiators and sensitizers, which are expensive, and may leave undesirable fragments of molecules. However, electron beam curing limitations are the necessity for an inert atmosphere and the high initial outlay required for high-vacuum and high-voltage technology.

The reactions caused by electron beams are mainly free radical in type. The monomers and oligomers used are similar to those in UV curable systems. EB has been used to form low surface energy coatings on substrates, such as are formed by the cure of perfluoroacrylates, e.g. 2-(*N*-butylperfluorooctane sulfonamido)ethyl acrylate or diacrylates such as perfluoroether diacrylate [239, 240].

Electron beam doses of up to 100 kilogray are used to enhance the performance of emulsion pressure-sensitive adhesives which have a Tg at least 20 °C below the temperature of use. Multifunctional monomers, e.g. pentaerythritol triacrylate or 1 : 6 pentaerythritol diacrylate, are added before irradiation [241].

Copolymers containing glycidyl methacrylate and an alkylbenzene polyoxyalkylene methacrylate, e.g. 2-ethylhexyl acrylate–glycidyl methacrylate–nonylphenyloxyethylene methacrylate (98 : 2 : 2), are curable by irradiation, typically a 6 Mrad electron beam. This may be applied after coating on polyester, giving an adhesive for aluminium [242].

11 INCORPORATION OF SILOXANES

The formation of emulsions containing siloxanes will be considered more fully in Chapter 7. There has been a rather limited number of disclosures in which siloxanes have been included in curable latices. Emulsion copolymerisation with vinyl monomers is possible with some silanes, e.g. γ-methacryloxy propyl trimethoxysilane $CH_2=C(CH_3)-C(O)-OC_3H_6Si(OCH_3)_3$, giving possibilities for crosslinking by ultimate alcoholysis or hydrolysis, or the action of peroxides under more drastic conditions [243, 244]. Crosslinking may be attained cationically [245].

Examples of this procedure include hydrogels of 2-hydroxypropyl acrylate and alkyl (meth)acrylate monomers which exhibit poor mechanical properties, but can be improved by the inclusion of bis(2-methacryloyloxy) propoxy dimethylsilane as the crosslinker. This was prepared by reacting 2 moles of 2-hydroxypropyl methacrylate with 1 mole of dimethyldichlorsilane in the presence of triethylamine in benzene [246].

A two-stage copolymerisation, including two silane derivatives, is described in a Japanese patent as shown in Table 5.21 [247].

Aqueous silicone emulsions that dry to clear elastomers contain crosslinked polydiorganosilane particles as a dispersed phase with <10 % of the particles

Table 5.21 Latex including silanes [247]

	Parts by weight
Stage 1	
Butyl acrylate	216
Methyl methacrylate	216
Methacrylic acid	18
γ-Methacryloxypropyltrimethoxysilane	6.9
trimethoxymethylsilane	101.5
Stage 2	
Butyl acrylate	245
Methyl methacrylate	245
Methacrylic acid	10

having a diameter >1000 nm. To obtain a clear film the emulsions are prepared with the alkali or ammonium salt of a C_{6-20} alkyl sulfate [248].

Tetraethylsilicon $Si(C_2H_5)_4$ is added at 17–17.5 % to a 40 % polyvinyl acetate latex including 15 % of dibutyl phthalate, with 7–8 % of phosphoric acid, which is an accelerator, both hydrolysing the polymer and catalysing the crosslinking while improving wet rub resistance, hardness and fire resistance [249].

Copolymers of standard alkyl methacrylates, alkylsiloxane monomers and comonomers selected from acrylamide, with sodium styrene sulfonate and/or acrylic acid, give thin films that cure at ambient temperatures. A typical 45 % latex is formed from butyl methacrylate, γ-methacryloxypropyl trimethoxysilane and acrylamide (80 : 20 : 2.5). A related specification does not contain acrylamide [250–257].

In another example a blend of butyl methacrylate, styrene and γ-methacryloxytriethoxysilane (240 : 30 : 30 parts by weight) is pre-emulsified and then fed into further emulsifiers buffered to pH 7.5 with sodium bicarbonate and polymerised. The product is self-curing, typically 40 % of solids and 0.13 µm average particle diameter. The filtered latex does not develop grit [258].

Trisodium vinylsiliconate $H_2C:CHSi(ONa)_3$ is included with butyl acrylate, methyl methacrylate and butyl methacrylate (5 : 30 : 40 : 25), which are polymerised with ammonium persulfate/sodium bisulfite redox initiation at 60 °C and pH 7 to provide a 40 % latex. A resultant film cures at ambient temperature or slightly above [259].

Under some conditions a low-density polythene is cured by the addition of a vinyl unsaturated silane by treating with an electron beam, which causes the gel content to increase [260].

12 CURE OF EVA AND SATURATED POLYMERS

It is well-known that saturated polymers of the type of polyethylene and especially ethylene–vinyl acetate (EVA) polymers can be cured by peroxidic compounds at relatively high temperatures. *tert*-Butyl hydroperoxides and dicumyl peroxide are among the suitable peroxides. The reaction probably operates by hydrogen abstraction, forming multiradicals which combine, giving a three-dimensional structure. Whilst polymers of this type need not be prepared by emulsion, or even suspension polymerisation, the principles of the cure are similar, irrespective of the method of preparation of the polymers.

There is little published information referring to polymers specifically prepared by emulsion. Amongst general examples an EVA (3 : 7) copolymer is crosslinked by immersion in a polyvalent metal alkoxide, e.g. titanium tetraisopropoxide, and held for 24 h at 50 °C. Crosslinking may be a combined hydrolysis/alcoholysis with titanium chelate crosslinking [261].

13 OTHER CROSSLINKING SYSTEMS

13.1 Acid and alkali induced crosslinking

A latex copolymer of vinyl acetate and butyl acrylate (7 : 3) with $\sim$1 % of potassium hydroxide is crosslinked by heating at 140 °C for 8 min, imparting resistance of the polymer to dry cleaning solvents. There is probably internal crosslinking with elimination of butyl acetate [262, 263].

13.2 Crosslinking via β-ketonic acid esters

Esters of acetoacetic acid, e.g. the ethyl ester $CH_3.CO.CH_2.COOC_2H_5$, are well known in organic synthesis and it is not surprising that they have been adapted to crosslinking reactions with reactive groups on polymers. Acetoacetoxyethyl methacrylate is commercially available. Reactivity leading to crosslinking occurs with conventional melamine and di-isocyanates. A summary of the possible reactions is given in Figure 5.4 (a) to (h). Note that Figure 5.1(e) indicates keto-enol tautomerism, increasing the versatility of these compounds [264].

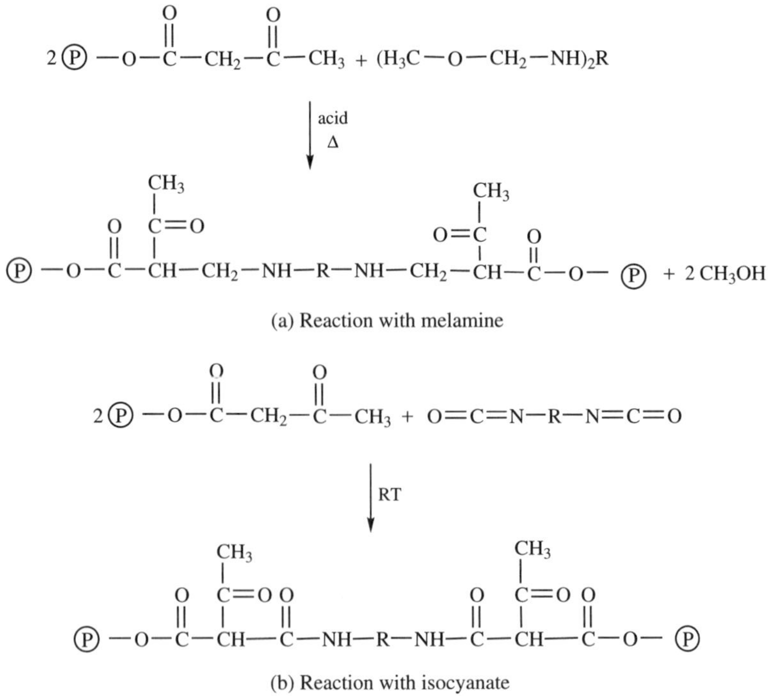

Figure 5.4 Crosslinking of diketones. (From reference [264].)

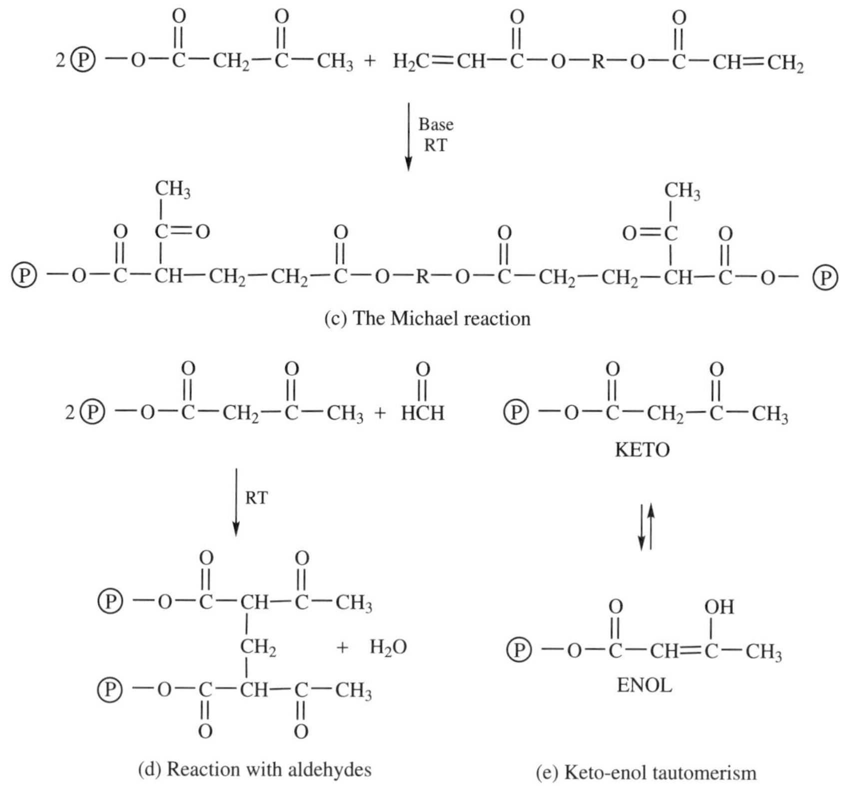

Other crosslinking systems 335

$$2 \text{P} - \text{O} - \overset{\text{O}}{\overset{\|}{\text{C}}} - \text{CH}_2 - \overset{\text{O}}{\overset{\|}{\text{C}}} - \text{CH}_3 + \text{H}_2\text{C} = \text{CH} - \overset{\text{O}}{\overset{\|}{\text{C}}} - \text{O} - \text{R} - \text{O} - \overset{\text{O}}{\overset{\|}{\text{C}}} - \text{CH} = \text{CH}_2$$

↓ Base, RT

$$\text{P} - \text{O} - \overset{\text{O}}{\overset{\|}{\text{C}}} - \overset{\overset{\text{CH}_3}{|}}{\underset{|}{\text{CH}}} - \text{CH}_2 - \text{CH}_2 - \overset{\text{O}}{\overset{\|}{\text{C}}} - \text{O} - \text{R} - \text{O} - \overset{\text{O}}{\overset{\|}{\text{C}}} - \text{CH}_2 - \text{CH}_2 - \overset{\overset{\text{CH}_3}{|}}{\underset{|}{\text{CH}}} - \overset{\text{O}}{\overset{\|}{\text{C}}} - \text{O} - \text{P}$$
(with pendant C=O groups on the CH carbons)

(c) The Michael reaction

$$2 \text{P} - \text{O} - \overset{\text{O}}{\overset{\|}{\text{C}}} - \text{CH}_2 - \overset{\text{O}}{\overset{\|}{\text{C}}} - \text{CH}_3 + \text{HCH}$$

↓ RT

$$\text{P} - \text{O} - \overset{\text{O}}{\overset{\|}{\text{C}}} - \overset{|}{\underset{|}{\text{CH}}} - \overset{\text{O}}{\overset{\|}{\text{C}}} - \text{CH}_3$$
$$\overset{|}{\underset{|}{\text{CH}_2}} \quad + \text{H}_2\text{O}$$
$$\text{P} - \text{O} - \overset{\|}{\underset{\text{O}}{\text{C}}} - \overset{|}{\underset{|}{\text{CH}}} - \overset{\|}{\underset{\text{O}}{\text{C}}} - \text{CH}_3$$

(d) Reaction with aldehydes

$$\text{P} - \text{O} - \overset{\text{O}}{\overset{\|}{\text{C}}} - \text{CH}_2 - \overset{\text{O}}{\overset{\|}{\text{C}}} - \text{CH}_3$$
KETO

⇅

$$\text{P} - \text{O} - \overset{\text{O}}{\overset{\|}{\text{C}}} - \text{CH} = \overset{\overset{\text{OH}}{|}}{\underset{|}{\text{C}}} - \text{CH}_3$$
ENOL

(e) Keto-enol tautomerism

Figure 5.4 (*continued*)

A review of the use of acetoacetate functional compounds to crosslink hydroxyfunctional resins, particularly in the role of reactive diluents, indicates that the degree of modification controls the hydrophilic/hydrophobic balance, the reactivity, reduced viscosity and ultimate film properties. The acetoacetate moieties improve adhesion, salt spray resistance and water/humidity resistance in coatings [265].

In a two-pack crosslinking system, suitable for automobile refinishes, one pack contains an acrylic copolymer with pendant acetoacetate groups and the other a component with aromatic aldimine groups. As an aromatic aldimine, e.g. prepared from diethylenetriamine and benzaldehyde, has relatively low reactivity with water compared with the acetoacetate group, the components may be mixed with water immediately before use. A quoted copolymer is formed from acetoacetoxymethacrylate, glycidyl methacrylate, maleic anhydride, n-butyl methacrylate and isobutyl methacrylate [266].

A somewhat similar system contains a copolymer of ethyl acrylate, butyl acrylate, butyl methacrylate, styrene, methacrylic acid and acetoacetoxyethyl

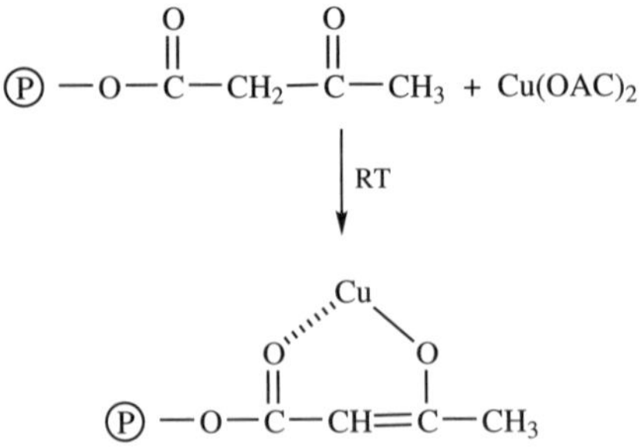

336   Crosslinking and curing

$$\text{(P)}-O-\overset{O}{\underset{\|}{C}}-CH_2-\overset{O}{\underset{\|}{C}}-CH_3 + Cu(OAC)_2$$

$$\downarrow RT$$

$$\text{(P)}-O-\overset{O\cdots}{\underset{\|}{C}}-CH=\overset{O}{\underset{|}{C}}-CH_3 \quad \text{(with Cu bridging)}$$

(f) Chelation

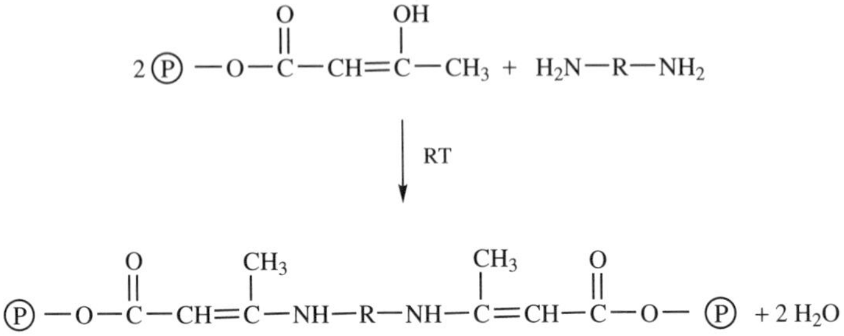

$$2\,\text{(P)}-O-\overset{O}{\underset{\|}{C}}-CH=\overset{OH}{\underset{|}{C}}-CH_3 + H_2N-R-NH_2$$

$$\downarrow RT$$

$$\text{(P)}-O-\overset{O}{\underset{\|}{C}}-CH=\overset{CH_3}{\underset{|}{C}}-NH-R-NH-\overset{CH_3}{\underset{|}{C}}=CH-\overset{O}{\underset{\|}{C}}-O-\text{(P)} + 2\,H_2O$$

(g) Reaction with diamines

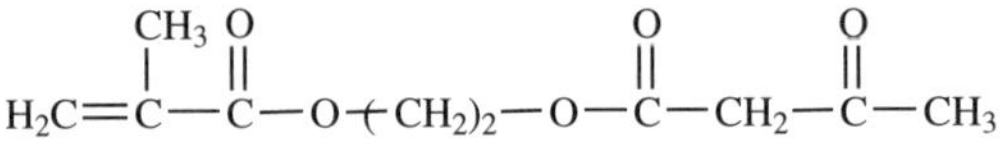

$$H_2C=\overset{CH_3}{\underset{|}{C}}-\overset{O}{\underset{\|}{C}}-O-(CH_2)_2-O-\overset{O}{\underset{\|}{C}}-CH_2-\overset{O}{\underset{\|}{C}}-CH_3$$

(h) Acetoacetoxyethyl methacrylate (AAEM)

Figure 5.4 (*continued*)

methacrylate (25 : 20 : 20 : 20 : 20 : 7 : 220). The ratio of carboxylic groups to acetoacetate groups is 0.87. The preferred ratio of the acetoxy compound to total polymers is 0.5–1.8 mol kg^{-1} [267–268A].

Table 5.22 gives the formulation of a heat-resistant coating that may be crosslinked with adipic acid dihydrazide. The composition may be pigmented and is strippable from steel plates [269].

Allyl acetoacetate is included in a latex, mainly based on vinyl acetate as the principal monomer and a polyvinyl alcohol (Poval 205) as emulsifier. The formulation is Poval 205, vinyl acetate, allyl acetoacetate, ethylene and acetic acid (100 : 1940 : 19 : 200 : 200). Polymerisation is by redox with hydroxide and a reducing sulfur compound [270].

2-Acetoacetoxypropyl methacrylate, b.p. 115–122 °C at 0.5 mm, prepared by the action of diketene on hydroxypropyl methacrylate, has been polymerised in emulsion by standard methods with anionic/non-ionic emulsifiers, and has been crosslinked by chelation with multivalent metal ions, an example

Other crosslinking systems

Table 5.22 Latex including acetoacetoxyethyl methacrylate [269]

A. Initial water phase	
Polyvinyl alcohol (PVA-205)	8 g
Water	413 g
B. Pre-emulsion (continuously stirred)	
Butyl acrylate	360 g
Acrylonitrile	204 g
Acetoacetoxyethyl methacrylate	36 g
Polyoxyethylene nonylphenyl ether	20 g
Sodium dodecybenzene sulfonate	3 g
Water	130 g
Initiator (type and quantity not included in the abstract)	
Polymerisation times and temperatures are not indicated.	
C. Final modification	
Latex above	100 g
Adipic acid dihydrazide	0.5 g
Pigment and thickeners were also added to give a steel coating.	

being an EDTA metal complex, which may be illustrated as follows:

$$\begin{array}{c} O \\ \parallel \\ O \diagdown C \cdot CH_2 \\ \\ M \cdots N - CH_2 - CH_2 - N \cdots M \\ \\ O - C \cdot CH_2 \\ \parallel \\ O \end{array} \quad \begin{array}{c} O \\ \parallel \\ CH_2 C \diagdown \\ O \\ \\ CH_2 C = O \\ \parallel \\ O \end{array}$$

The crosslinking ability of the polymer was substantiated by torsional braid analysis which indicated that coatings containing zinc chloride $ZnCl_2$ were cured both at ambient temperature and on heating. Other salts, including aluminium sulfate $Al_2(SO_4)_3$, magnesium chloride $MgCl_2$ and stannous chloride $SnCl_2$, were also effective. Zinc acetate $Zn(OOCCH_3)_2$ at a stochiometric level of 20 % gave the best results at ambient temperature, yielding coatings from latex with very good hardness, gloss with water and solvent resistance. Crosslink densities were high, and films from the latex were superior to those from similar solvent cast systems [271].

A low molecular weight acetoacetate has been used in the crosslinking of acrylic copolymers including reactive groups. Thus a copolymer of glycidyl methacrylate and methyl methacrylate is converted to a ketimine-functional polymer by reaction with a ketimine derived from 3-amino-1-methylaminopropane. This copolymer is crosslinked with trimethylolpropane tris(acetoacetate). Curing takes place at ambient temperature [272, 273].

The β-ketonic group may be used indirectly. Thus a latex is prepared from butyl acrylate, methyl methacrylate and acetoacetoxyethyl methacrylate with n-dodecyl mercaptan as the chain transfer agent. The latex is treated with ammonia solution to convert all of the acetoacetoxy ethyl groups to the corresponding enamine. The resultant product is mixed with polyallyl glycidyl ether, a water-soluble cobalt drier, and methyl ethyl ketoxime to form a curable coating composition. The final reaction is an auto-oxidizable one, tending to resemble the drying oils [274].

A seeded copolymerisation of butyl acrylate, methacrylic acid and acetoacetoxyethyl methacrylate including n-dodecyl mercaptan as the chain transfer agent and with a persulfate initiator is reacted with ammonium hydroxide to produce an enamine by reaction with the acetoacetate groups. The treatment enables shear strength of a resultant film to be drastically increased without loss of peel strength or tack when used as an adhesive [275].

Copolymers of butadiene with 0.5 % of methacryloylacetone CH_3COCH_2-$COOC(CH_3):CH_2$ or methacrylolylacetylacetone $CH_2:C(CH_3)COOCH_2$-$COCH_2COCH_3$, formed in emulsion with sodium oleate stabiliser, may be crosslinked by metal chelation [276].

13.3 Crosslinking with reactive nitrogen compounds

The formation of a 97 % gel in the latex formed by the emulsion copolymerisation of C_2F_4, $CH_2:CF_2$ and C_3F_6 is achieved by the addition of 1 % of 1,3,5 tris-(3,3-difluoro-2-propenyl)-S-triazine-2,4,6-) 1H, 3H,5H)-trione. This latter is formed by the reaction of 1–3-dibromo-3-difluoropropane with cyanuric acid in the presence of an alkaline earth base. Cyanuric acid is a simple triazine of formula

$$\begin{array}{c} N \\ HO.C \diagup \diagdown C.OH \\ | \| \\ N \diagdown \diagup N \\ C.OH \end{array}$$

Note the resemblance of cyanuric acid to the formula for melamine [277].

Vinyl-substituted cyclic hemiamidals and their interconvertible acetal precursors, e.g. acrylamidobutyraldehyde dimethyl acetal $CH_2CH.C(O)NCH_2$-$CH_2CH_2CH(OCH_3)_2$, have been included as latent crosslinkers and substrate reactive functional comonomers in solution and emulsion copolymers. They show low-energy cure, a potential long shelf-life and stability even when catalysed. They have good adhesive properties, water and solvent resistance and do not release formaldehyde (Figure 5.5) [278–280].

Self-crosslinking latex blends comprise acrylic latices with hydrazine residues and latices with polymers containing carbonyl groups. An example is shown in Table 5.23. On mixing A and B (2 : 1), a blend consisting of latices

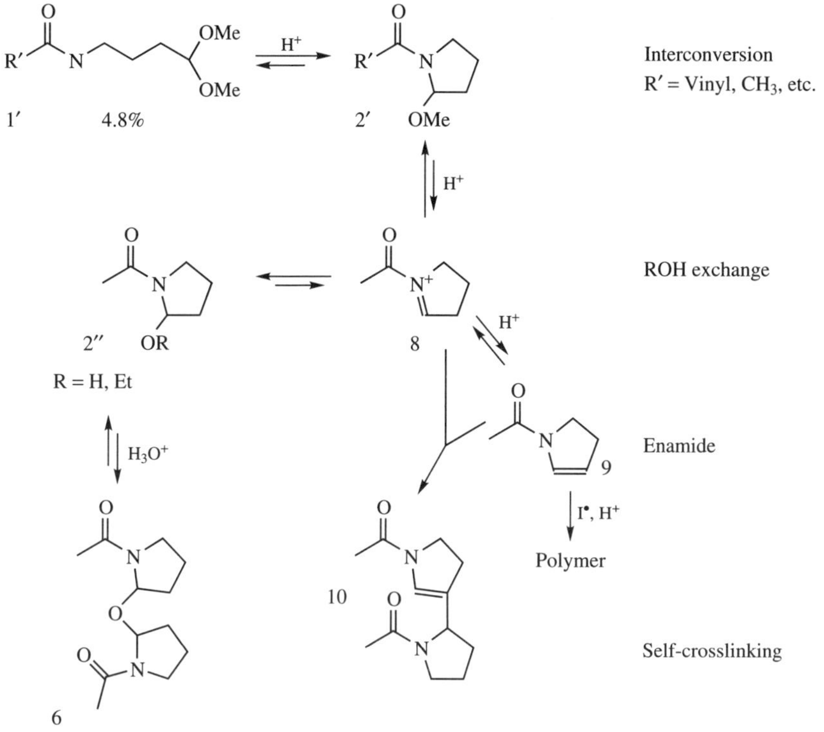

Figure 5.5 Curing reactions of vinyl-substituted cyclic hemiamidals. (From reference [278]). (Reproduced by permission of American Chemical Society.)

with films of different hardness, anti-corrosive and water-resistant crosslinked coatings, which are suitable *inter alia* for paper, wood and metal, are produced [281].

Technical water-soluble polyhydrazides have been employed to form a crosslinking coating of a proprietary latex. Coatings crosslink within 1 min at ambient temperature [282].

Tetramethoxymethylglycouril is reported as a crosslinker for powder coatings, but may be suitable for emulsion coatings. This compounds reacts with hydroxyl groups on polymers under the influence of strong acid catalysts, liberating methanol. The compound resists hydrolysis to release free formaldehyde. The formula is [283]

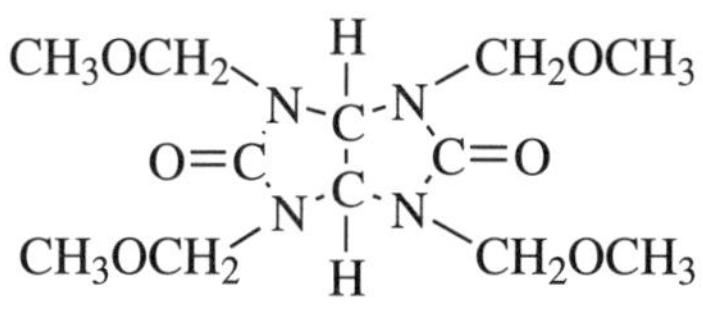

Table 5.23 Self-crosslinking latex blend [281]

	X (g)	Y (g)
A. Water	180	180
Na salt, p-nonylphenyl monosulfonate–ethylene oxide (20) condensate	5	5
p-Nonylphenol-ethylene oxide (1 : 25) condensate	20	20
B. To the above add stepwise:		
Water	200	200
Emulsifier[a]	25	25
Styrene	242	242
n-Butyl acrylate	—	215
Acrolein	11	11
Acrylamide	10	10
Acrylic acid	—	10
Hydrazine hydrate	—	15
Potassium persulfate (as solution)	2.5	2.5
Water	85	85

[a] Acrylamidobutyraldehyde dimethyl acetal.

Polyvinylpyrrolidones may be activated by pretreatment with ammonium persulfate to form active centres, the exact nature of these not being described, followed by addition of styrene and methylvinyl ketone, the latter at 1–7 % of total monomers. Further initiator and a non-ionic surfactant is added to the emulsion first formed. Polymerisation results in a graft on to the water-soluble polymer. The active keto groups in the polymer may be crosslinked, e.g. by diamines [284].

13.3.1 Carbodi-imides

Polycarbodi-imides of low molecular weight are viscous oily oligomers which crosslink polymers that contain carboxyl groups. They react with carboxylic acids to form N-acyl ureas at elevated temperatures in the presence of amines or in polar environments. Their formula is

$$R-\bigcirc-[N=C=N-\bigcirc]_n-R$$

These crosslinkers can be made to be self-emulsifying in water. Water-dispersible polycarbodiimides can crosslink water-borne polymers containing carboxylic groups under ambient conditions. These water-soluble carbodi-imides may hydrolyse with time. Coatings modified with aromatic polycarbodi-imides have a pot life about 12 h but cure in 2–6 h. Coatings modified with aromatic polycarbodiimides show increases in stain, salt spray and solvent resistance. As alternatives, various alkyl, phenyl and cyclohexyl carbodi-imides were prepared and characterized [285]. Because carbodiimides react with acids it is possible to prepare alkyl carbodi-imide ethyl methacrylate [286, 287].

An emulsion polymer of methyl methacrylate, butyl methacrylate, methacrylic acid with mercaptoacetic acid as controller (9200 : 250 : 100 : 17.2) was prepared by a standard method at pH 5 with mixed anionic/non-ionic surfactants. To 150 g of this 40.66 % latex a preprepared cyclohexyl carbodi-imide ethyl methacrylate (17.2 g as monomer) emulsion was added. When the latex was heated at 80 °C the precipitated polymer showed 95 % of the theoretical amount of unsaturation. There are reactive sites that can be cured thermally using t-butylperoxybenzoate or photochemically (Section 10). Coatings from the modified latices were cured at ambient temperature [288]. Aromatic carbodi-imides can be self-emulsifying, with a pot life of 12 h, but cure in 2–6 h [289].

Oxime ethers, including modifications which are polymerisable (see diagrams), are formed by reacting oxime ether alcohols with specific isocyanates. They may be used in place of polyols for crosslinking radical copolymers or condensation polymers containing keto- or aldehyde groups [290]:

$$\begin{matrix}CH_3\\C_2H_5\end{matrix}\!\!>\!\!C\!=\!N\!-\!O\!-\!CH_2\!-\!CH(OH)\!-\!O\!-\!(CH_2)_4\!-\!OCH(OH)\!-\!CH_2\!-\!O\!-\!N\!=\!C\!<\!\begin{matrix}CH_3\\C_2H_5\end{matrix}$$

$$CH_2\!=\!C\!-\!\underset{\underset{CH_3}{|}}{C}(:O)\!-\!NH_2\!-\!\underset{\underset{O}{\overset{\|}{}}}{C}\!-\!O\!-\!-\!(CH_2)_3\!-\!O\!-\!N\!=\!C(CH_3)_2$$

A ketimine formed from diethylenetriamine, methyl isobutyl ketone and glycidyl methacrylate (393 : 900 : 450) is polymerised in emulsion with butyl methacrylate and methyl methacrylate (40 : 220–135.3) at pH 8.1–9.0. On mixing with melamine acrylate at 10 : 1 ratio, acidifying to pH 4 and forming a film on a substrate, cure takes place at ambient temperatures in a few minutes [291].

13.4 Intra- and inter-molecular reactions

Partially hydrolysed isobutene–maleic anhydride–maleimide copolymer (240 g of 10 % solution) is used as a stabiliser for 150 g of vinyl acetate with 0.8 g of ammonium persulfate. Heating films of the resultant latex at 150 °C for 1 min ensure a great reduction in swelling on water immersion. This is probably due to an effective crosslinking between the polymer and the emulsifier with the elimination of acetic acid, and possibly inter- or intramolecular lactone formation. Other copolymers of this type, particularly in neutralization, with ammonia or a volatile amine, will probably react in the same way. There is some evidence that a vinyl acetate–maleic anhydride copolymer (as salt) will act in the same way with some ester monomers [292].

13.5 Unspecified

Whilst glyoxal is known as a crosslinker for hydroxyl groups (Section 5), it is also claimed as a crosslinker for a latex of butyl acrylate–methyl methacrylate–methacrylic acid–acrylamide–N–[2-(methacryloyloxy) ethyl] ethyleneurea (350 : 150 : 13.5 : 7.5 : 10). 100 ml of this 48.9 % latex is mixed with 0.65 parts by weight of glyoxal and cast as a film, 500nm thick. This had a tensile strength of 4 N nm^{-2} and elongation at break of 304 % compared with corresponding values for the untreated latex of 2.7 N mm^{-2} and 545 %. Swelling with tetrahydrofuran was also reduced [293].

Crosslinking agents based on the 2-acetylglutarate esters of polyols have the following general formula.

$$(CH_2-C(:O)-\underset{\underset{CH_2-CH_2-C(:O)-O)_xR''}{|}}{CH}-C(O)-OR)$$

Where R = t-butyl, t-amyl or phenyl, R is a di-, tri- or tetrafunctional polyol residue and x is an integer from 1 to 5. They may be used to crosslink a range of polymers including functional groups such as polyepoxides, polyisocyanates, amino resins and polyunsaturated compounds. Curing takes place generally at 150–200 °C [294].

A very unusual claim is that of a latex in which the emulsifiers have a terminal alkylthio group and an acid value ⩾200. The emulsifiers are prepared by polymerising standard monomers with an unsaturated acid, e.g. acrylic acid, in the presence of an alkyl mercaptan (6–18 carbon atoms). The emulsifier apparently reacts at the surface of polymer particles during and subsequently to coalescence, even at ambient temperatures [295]. It is possible to crosslink latices based on acrylic monomers by the use of a peroxide which is directly polymerisable because of included unsaturation, e.g. dimethylvinylethynyl-methyl $tert$-butyl peroxide [296].

Another peroxy-functional copolymer is 1,1-dimethyl-pent-4-en-2-ynyl $tert$-C_4H_9OOH, the product itself being formed in emulsion. It may be reacted with styrene in a redox system to form a styrene copolymer, which may be further grafted with acrylonitrile in the presence of Fe^{2+}, the latter product crosslinking on heating. Recourse to the original Russian paper is necessary for further elucidation [297].

Bis(1,3,2-dioxothioan-2-oxido-4-methyl)malonate, a cyclic compound including a sulfur atom, and which is free from formaldehyde, is used to crosslink a methyl methacrylate–methacrylic acid–styrene latex. Another patent by the same inventors claims bicycloamide acetals; a typical example is 5-ethyl-3-phenyl-1-aza-4,6-dioxabicyclo[3,3,0]octane [298].

An epithiopropyl methacrylate is analagous to a glycidyl methacrylate in that a sulfur atom is attached to two carbon atoms. The formula is

$$CH_2-CH-CH_2OOCCH(CH_3)C:CH_2$$
$$\diagdown S \diagup$$

It is polymerisable in emulsion, but crosslinks very readily, and the amount included should be restricted to 6 % of the total monomers to avoid premature crosslinking [299].

REFERENCES

1. P.J. Flory, *Principles of Polymer Chemistry*, Cornell University Press, 1953, p. 31
2. E.J. Carlson (B.F. Goodrich), USP 2,726,320, 1955, Ex. 8
3. H. Warson, *Peintures, Pigments, Vernis*, **43**, 438–46 (1967)
4. H.M.J. Boots and R.B. Pandey, *Polym. Bull.*, **11**, 415–20 (1984)
5. K. Hümmel, *Angew. Makromol. Chem.*, **76/77**, 25–38 (1979)
6. J.R. Crawe and B.C. Bufkin, *J. Paint Technol.*, **50** (641), 41–55 (1978)
7. J.R. Crawe and B.C. Bufkin, *J. Paint Technol.*, **50** (643), 67–83 (1978)
8. J.R. Crawe and B.C. Bufkin, *J. Paint Technol.*, **50** (644), 83–109 (1978)
9. J.R. Crawe and B.C. Bufkin, *J. Paint Technol.*, **50** (645), 70–100 (1978)
10. J.R. Crawe and B.C. Bufkin, *J. Paint Technol.*, **50** (647), 65–96 (1978)
11. J.R. Crawe and B.C. Bufkin, *J. Paint Technol.*, **50** (649), 34–67 (1978)
12. *Polym. Mater. Sci. Engng*, **56** (1987) (several papers)
13. V.I. Eliseeva, *Br. Polym. J.*, **7**(1), 33–49 (1975)
14. H. Tobita and A.E. Hamielec, *Makromol. Chem., Macromol. Symp.*, **35/36**, 213–29 (1990)
15. C.A. May, *ACS Symp. Ser.*, **227**, 1–24 (1983)
16. A.E. Hammielec and H. Tobita, *Polym. Int.*, **30**(2), 177–83 (1993)
17. J.K. Gilham, *Polym. Mater. Sci. Engng*, **54**, 8–11 (1986)
18. H.M.J. Boots et al., *Br. Polym. J.*, **17**(2), 219–23 (1985)
19. *Eur. Coat. J.*, **1995**(12), 930–5 (1995)
20. K. Luo, *Gaofenzi Tingxun*, **1986**(4), 204–8 (in Chinese) (1986); *Chem. Abstr.*, **105**, 209879 (1986)
21. M. Havsky, J. Mikes and K. Dusek, *Polym. Bull. (Berlin)*, **3**(8–9), 481–7 (1980)
22. A. Arcozzi, S. Faffumi and R. Arletti, *Pitture e Vernici Europe*, **48**(11), 51–61 (1992)
23. J.A. Dale and J.R. Millar, *Macromolecules*, **14**(5), 1515–8 (1981)
24. B. Haidar, A. Vidal and J.B. Donnet, *Br. Polym. J.*, **15**(2), 120–4 (1983)
25. T.H. How, Nasa Contract CR-172443-NASI26:1724433, 1984
26. R.R. Ley, *J. Coat. Technol.*, **56**(718), 49–56 (1984)
27. M.B. Roller, *Polym. Engng Sci.*, **15**, 406–14 (1975)
28. V. Lipatov et al., *Polimery (Warsaw)*, **29**(4–5), 159–64 (1984)
29. A.J. Ayorinde et al., *ACS Symp. Ser.*, **18**, **245**, 321–31 (1984)
30. W. De Winter and R. Van Haute, *Polym. Bull. (Berlin)*, **4**(1–2), 133–9 (1980)
31. J.K. Gillham, *Prepr. Polym. Mater. Sci. Engng (ACS)*, **63**, 742–52 (1990)
32. S.M. Hill and W.W. Wright, *Br. Polym. J.*, **22**(3), 189–93 (1989)
33. H. Stutz and J. Mertes, *J. Appl. Polym. Sci.*, **38**(5), 781–7 (1989)

34. Th. Frey et al., 20th Int. Conf. on Organic Coating Science Technology, Athens, 1994, pp. 125–41
35. P.P. Winner, *Prepr. Polym. Mater. Sci. Engng (ACS)*, **1994**, 125–41 (1994)
36. *Eur. Coat. J.*, **1995**(12), 930–5 (1995)
37. W. Obrecht and W. Seitz, *Makromol. Chem.*, **177**(8), 1877–88, 2235–40 (1976)
38. H.B. Sunkara et al., *J. Polym. Sci. Polym. Chem.*, **32**(8), 1431–5 (1994)
39. K. Kishida et al. (Mitsubishi Rayon), GP 2,728,618, 1977
40. S. Hashimoto et al. (Toyo Ink), JP 78 84,092 *Chem. Abstr.*, **89**, 164375 (1978)
41. K. Kasai et al. (Japan Synthetic Rubber), Eur. P 335,029
42. H. Tobita et al., *Polymer*, **34**(12), 2569–74 (1993)
43. T. Takeochi (Soken Chemical Engineering), JP 241,310, 1986
44. D.R.A. Bassett and K.L. Hoy (Union Carbide), GP 2,647,593, 1975
45. K. Hering et al. (Lonza), BP 1,668,142, 1976
46. S. Saito et al. (Tokuyama Soda), JP 87 13,414; *Chem. Abstr.*, **107**, 60010 (1987)
47. J.N. Yoo et al. (Lucky), JP 05,715, 1994; *Chem. Abstr.*, **122**, 56819 (1995)
48. C. Samant and R. Pal, Proc. PRA 3rd Asia-Pacific Conference on Advances in Coatings, Inks and Adhesive Technology, Singapore, 1993, paper 18, 21 pp.
49. B.E. Rodrigues-Douglas and M.S. Wolfe, *Prepr. Polym. Mater. Sci. Engng*, **73**, 198–9 (1995)
50. H. Toda et al. (Nippon Zeon), JP 91 279,493; *Chem. Abstr.*, **116**, 196510 (1992)
51. J.W. Vanderhoff, M.S. El-Aasser, and J.M. Park, *ACS Symposium Series No. 192*, 1991, Ch. 17, pp. 272–81
52. E. Schwarzenbach et al. (BASF), GP 4,013,724, 1991
53. J. Oshima et al. (Takeda Chemical Industry), Can. P 2,028,160, 1991
54. J.-W. Kim, *J. Appl. Polym. Sci.*, **63** (12), 1589–600 (1996)
55. R.A. Dickie and S.S. Labana (Ford-Werke), GP 2,163,461, 1972
56. R.A. Dickie and S.S. Labana, GP 2,163,504, 1972
57. R.A. Dickie and S.S. Labana, GP 2,163,464, 1972
58. R.A. Dickie and S. Neuman (Ford-Werke), GP 2,167,478, 1972
59. H. Li, and E. Ruckenstein, *Polymer*, **36**(11), 2281–7 (1995)
60. C.E. Schildknecht, Allyl Compounds and Their Polymers, Wiley–Interscience, 1973
61. A.G. Andreopulos, *J. Appl. Polym. Sci.*, **34**(7), 2389–97 (1987)
62. A. Suzuki et al. (Osaka Soda), GP 2,504,104, 1976
63. M. Fujii and N. Hirai (Yamata Iron & Steel), FP 1,460,751, 1966; *Chem. Abstr.*, **67**, 12644 (1967)
64. F.J. Dany (Hoechst), GP 2,450,949, 1976
65. J.C. Padget, Eur. P 297,781, 1989
66. L.S. Frankel et al. (Rohm & Haas), Eur. P 522,789, 1993
67. Kuraray, JP 139,514, 1981; *Chem. Abstr.*, **96**, 105248 (1982)
68. M. Kuraya (Toyo Steel), JP 82891, 1978; *Chem. Abstr.*, **89**, 164742 (1978)
69. J.F. Knifton (Texaco), USP 4,025,547, 1977
70. Y. Isobe et al. (Toa Gosei), JP 76 17,247; *Chem. Abstr.*, **85**, 22581 (1976)
71. T. Kyo and M. Baba (Mitsubishi Petrochemical), JP 93 25,225; *Chem. Abstr.*, **119**, 51134 (1993)
72. I. Piirma and S. Laferty, *Polym. Mater. Sci. Engng (ACS)*, **64**, 280 (1991)
73. G.R. Brown (Dunlop), BP 1,189,302, 1970
74. F. Heins and Matner (Bayer), GP 2,830,455, 1980
75. S.N. Trubitsyna and M.A. Askarov, *Izv. Uchebn. Zaved. Khim. Khim. Tekhnol.*, **26**(1), 86–9 (1983); *Chem. Abstr.*, **98**, 73689 (1983)
76. R.S. Arutyunyan et al., *Arm. Khim. Zh.*, **43**(9), 604–8 (1990) (in Russian); *Chem. Abstr.*, **114**, 187274 (1991)
77. Sekisui, JP 81 00,808; *Chem. Abstr.*, **94**, 105002 (1981)
78. Denki, JP 84 8,773; *Chem. Abstr.*, **101**, 73928 (1984)

References

79. Denki, JP 84 25,855; *Chem. Abstr.*, **101**, 24507 (1984)
80. W.D. Emmons (Rohm & Haas), USP 4,100,133, 1978
81. Hitachi, JP 32,518, 1981; *Chem. Abstr.*, **95**, 44862 (1981)
82. R. Ukita (Hitachi), USP 4,309,330, 1982
83. R. Reeb (Rhone-Poulenc), FP 2,697,530, 1994
84. J.A. Levelle, *Prepr. Polym. Mater. Sci. Engng*, **55**, 138–42 (1986)
85. H. Lorah and M.S. Frazza (Rohm & Haas), Eur. P 478,193, 1992; USP 5,212,251, 1992
86. O. Yamada (Nippon Mectron), JP 87 138,511, *Chem. Abstr.*, **108**, 206399 (1988)
87. T. Lizuka (Dainippon Ink & Chemicals), JP 87 263,209; *Chem. Abstr.*, **108**, 206399 (1988)
88. K.H. Weinert *et al.* (Lankwitze Lackfabrik), PCT Int. Applic. 14,763, 1992
89. G. John and C.K.S. Pillai, *Makromol. Chem. Rapid Commun.*, **13**(5), 255–91 (1992)
90. H.G. Fordyce and G.E. Ham, *J. Am. Chem. Soc.*, **69**, 695–6 (1947)
90A. H. Warson and D. Kinsler (Dunlop), BP 1,174,914, 1969; equivalent to FP 1,535,014, 1968; S. Afr. P 05,372, 1967
91. Dupont, BP 1,087,286, 1967
92. W. Machtle *et al.*, *Coll. Polym. Sci.*, **273**(7), 708–16 (1995)
92A. H. Matsuda and Minoura, *J. Appl. Polym. Sci.*, **24**(3), 811–26 (1979)
92B. J.W. Vanderhoff *et al.*, *J. Appl. Polym. Sci.*, **24** (3), 704–7 (1979)
93. S. Hayano *et al.* (Misui Toatsu Chemicals), JP 192,997, 1994; *Chem. Abstr.*, **122**, 12399 (1995)
94. S.N. Lewis and A. Mercurio (Rohm & Haas), USP 3,272,788, 1966
95. H. Warson and D.S.W. Dargan (Dunlop), BP 1,044,527, 1966
96. A.S. Michaels (W.R. Grace), USP 4,385,155, 1983
97. G.F. D'Alellio and J. Caiola, *J. Polym. Sci. A-1*, **5**, 285 *et seq.* (1967)
98. G.F. D'Alellio and T. Huemmer, *J. Polym. Sci. A-1*, **5**, 307–21 (1967)
99. S. Magnet *et al.*, *Prog. Org. Coat.*, **20**(1), 73–80 (1992)
100. J.A. Simms, *J. Polymer Sci., J. Appl. Polym. Sci.*, **5**, 58 *et seq.* (1961)
100A. N.F. Sorokin, L.G. Shode and R.A. Dudakova, *Lakokras Mat.*, **1964**, 112 *et seq.* (1964)
100B. J.M. Geurts, Forum de la Connaissance AFTPA/CORI, 1996 (not numbered)
100C. H. Feicke *et al.* (VEB C.W. Buna), East GP 223,600, 1985
101. Z. Xu *et al.*, *J. Appl. Polym. Sci.*, **56**(5), 575–80 (1995)
102. S.H. Ganslaw and J.L. Walker (National Starch), GP 2,365,789, 1976
103. Y. Matsukawa *et al.* (Nippon Oils), JP 75 34,076
104. I. Ono *et al.* (Toyo Soda), JP 74 13,871
105. K. Araki *et al.* (Japan Atomic Energy Research Institute; Kansai Paint), JP 78 11,927; *Chem. Abstr.*, **89**, 112721 (1978)
106. A. Tsunemi and A. Kimura, JP 78 41,335; *Chem. Abstr.*, **89**, 130614 (1978)
107. P. Cantor *et al.* (Celanese), USP 3,223,670, 1965
108. M. Izubayashi (Nippon Shukubai Chemical Industry), Eur. P 320,594, 1987; *Chem. Abstr.*, **11**, 21544 (1989)
109. H. Mayfield and W.F. Hill (Union Carbide), BP 1,072,042, 1967
110. T. Nikaya *et al.* (Nippon Shokubai), PCT 10,184, 1993; *Chem. Abstr.*, **120**, 10436 (1994)
110A. S.G. Mylonakis (De Soto), USP 4,244,850, 1981
110B. Dupont, BP 1,087,286, 1967
110C. P.F. Sanders (Du Pont deNemours), Can. P 551,589, 1958
111. C.E. Schilknecht, Allyl Compounds and Their Polymers, Wiley–Interscience, 1973, pp. 406–12
111A. A.J. Casey and C.E. Blades (Cumberland Chemical), USP 3,247,151, 1966
111B. K.K. Uchiyama, JP 85 120,708; *Chem. Abstr.*, **104**, 7004 (1986)
111C. *Encyclop. Polym. Sci.*
111D. H. Warson *et al.* (Vinyl Products), BP 995,726, 1965
112. P.R. Sampath and D.J. Grosse (S.C. Johnson), PCT 14,715, 1991; *Chem. Abstr.*, **116**, 153940 (1992)

112A. M. Tsurumi and T. Sato (Asahi Chemical Industry), JP 146,905, 1989; *Chem. Abstr.*, **112**, 38272 (1990)
113. D.K. Hoffman *et al.* (Dow), USP 4,690,988, 1987
114. E.H. Jonker and M.J. Weber (Akzo), Eur. P 508,536, 1991
115. M. Sakai *et al.* (Nippon Oils & Fats), JP 73 07,506; *Chem. Abstr.*, **80**, 15657 (1974)
116. S. Magnet *et al.*, *Prog. Org. Coat.*, **20**(1), 73–80 (1992)
117. A. Yamazaki and F. Tateyama (Nihon Yunyakku), JP 79 34,390; *Chem. Abstr.*, **91**, 58867 (1979)
118. F. Davis (National Starch), Eur. P 56,452, 1982
119. G.P. Craude and D. Bode (Glidden), Eur. P 577,958, 1994
120. D.E. McGee (Rohm & Haas), Eur. P 537,910, 1993
121. Y. Igarashi *et al.* (Japan Synthetic Rubber), JP 92 202,451; *Chem. Abstr.*, **118**, 148720 (1993)
122. F. Yoshino *et al.* (Dainippon Ink and Chemicals), USP 4,973,614, 1991
123. G.L. Brown and M.A. Tobias (Mobil Oil), USP 4,028,294, 1977
123A. L. Hsueh-Chi (Air Products & Chemicals), Eur. P 647,664
124. L.C. Graziano *et al.* (Rohm & Haas), Eur. P 398,577, 1991.
125. M.M. Miller (Hi-Tek Polymers), Eur. P 459,651, 1991
126. J.R. Gross and R.T. McFadden (Dow), USP 3,926,891, 1975
127. L.Z. Pillon and H.S.G. Slooten (BASF), Eur. P 329,027, 1989
128. I.S. Pinskaya and V.I. Eliseeva, *Probl. Sin. Issled. Svoistv. Pererab. Lateksov.*, **1971**, 74–9 (1971); *Chem. Abstr.*, **78**, 72994 (1973)
129. M. Colucci (OECE Industrie Chimiche), Eur. P 562,282, 1995
130. Y. Chikazoe and Y. Yamomoto (Dainippon), JP 75 113,593
131. M. Desbois *et al.* (Rhone-Poulenc Chimie), PCT 13,731, 1994; *Chem. Abstr.*, **122**, 106738 (1995)
132. T. Juki *et al.* (Kuraray), JP 93 011,270; *Chem. Abstr.*, **119**, 97594 (1993)
133. K. Haeberle *et al.* (BASF), Eur. P 596,291, 1995
134. M. Susuki (Asahai Chemical), JP 79 138,025; *Chem. Abstr.*, **92**, 148667 (1980)
135. W. Staritzbichler (Vianova Kunstharz), Eur. P 663,928; PCT 07,932, 1994
136. E.R.-H. Chang *et al.* (PPG Industries), Eur. P 608,274, 1994
137. E. Krejcar and J. Snuparek, *Chem. Prum.*, **29**(10), 534–8 (1979); *Chem. Abstr.*, **92**, 23433 (1980)
137A. American Cyanamid Company (now Cytec), Technical Literature
137B. C. Bonardi *et al.*, *New Polym. Mater.*, **2**(4), 295–314 (1991)
138. D.R. Basset, M.A. Sherwin and S.L. Hager, *J. Coat. Technol.*, **51**(657), 65–72 (1979)
139. M.A. Askarov, T.G. Kulagina and Z.M. Niyazova, *Vysokomol. Soed. B*, **15**(5), 326–9 (1973)
140. D.R. Bassett and M.A. Sherwin, Proc. 4th Water-borne and Higher Solids Coat. Symp., University of South Mississipi, 1977, pp. 157–90
141. T. Inoe (Nippon Zeon), JP 95 149,808; *Chem. Abstr.*, **123**, 289924 (1995)
142. M & T Chemicals, JP 78 28,687; *Chem. Abstr.*, **89**, 112755 (1978)
142A. A. Yamamoto (Toyo Ink), JP 76 131,554; *Chem. Abstr.*, **86**, 74609 (1977)
143. Lonza, JP 77 73,930 (Swiss Original)
144. S.N. Zeldin *et al.*, *J. Appl. Polym. Sci.*, **24**(2), 455–64 (1979)
145. A. Pffaffenschlager *et al.* (Glasurit do Brasil), PCT 07,191, 1993
146. S. Tamaoka and I. Namikata (Kao), JP 93 178,916; *Chem. Abstr.*, **119**, 150760 (1993)
147. P.F. Stehle *et al.* (Borden), GP 2,139,262, 1973
147A. P.F. Stehle *et al.* (Borden), GP 2,301,099, 1974; BP 1,407,864, 1974
148. T. Tsuchihara and S. Kubo, *Nippon Setchaku Kyokaishi*, **17**(11), 455–61 (1981); *Chem. Abstr.*, **96**, 123739 (1982)
149. P.F. Stehle and J. Dickstein (Borden), USP 3,730,933, 1973
149A. Union Oil, California, USP 3,714,099, 1973

150. G. Biale (Union Oil, California), USP 3,714,096, 1973
151. G. Biale and R.L. Pilling (Union Oil, California), USP 3,714,100, 1973
152. W. Haas et al. (Wacker-Chemie), Eur. P 621,289, 1994
153. G. Biale (Union Oil, California), USP 3,714,099, 1973
154. I. Ito, et al. (Sumitomo), JP 76 74,081; Chem. Abstr., **85**, 124978 (1976)
155. H.M. Andersen (Monsanto), GP 2,325,730, 1973
156. T. Wirth, G. Oetter and K.H. Etzbach (BASF), GP 4,104,463, 1992
157. B.K. Mikofalvy and D.P. Knecht (B.F. Goodrich), BP 1,309,513, 1973
158. W. McPhee et al., J. Coll. Interface Sci., **156**(1), 24–30 (1993)
159. M.K. Lindemann (G.S. Tanner), USP 4,001,160, 1977
160. J. Snuparek, Cz. P 165,265, 1976
161. Du Pont de Nemours, Neth. P Applic. 11,330, 1978
162. B.G. Dicklesteel and D.F. Anders (De Soto), Eur. P 15,077, 1980
163. K. Yamaguchi and T. Goto (Dainippon Ink & Chemicals), JP 145,344, 1995; Chem Abstr., **123**, 289828 (1995)
164. L.E. Coleman et al., J. Polym. Sci. A, **3**, 1601–8 (1965)
165. L.E. Coleman (Lubrizol), USP 3,277,056, 1966
166. D.I. Hoke, J. Polym. Sci., Polym. Chem. Ed., **11**(10), 2711–2 (1973)
166A. G. Jalics (SCM), USP 3,663,573, 1972
167. C.L. McCormick, G.S. Chen, J. Polym. Sci., Polym. Chem. Ed., **22**(12), 3633–47 (1984)
168. R. Blum et al. (BASF), GP 3,536,261–2, 1987
169. M. Sugishima et al. (Kansai Paint), JP 249,987, 1992; Chem. Abstr. **118**, 171128 (1993)
170. H. Yoshida et al., Nagoya-Shi Kogyo Kenkyu Hokuko, **1972** (48), 6–10; Chem. Abstr., **78**, 148560 (1973)
171. L. Chindrella et al., Polym. Commun., **28**(4), 25–8 (1987)
172. S. Paulajan et al., Polymer, **24**(7), 906–8 (1983)
173. C.L. McCormick and K. Blackmon, Angew. Makromol. Chem., **141**, 73–86 (1986)
174. T. Sunamori and N. Nishii (Mitsubishi Rayon), JP 75 140,545
175. N. Watanabe et al. (Mitsui-Cyanamid), USP 5,132,358, 199
176. Ashland Oil Inc., Technical Bulletin Aminimides, 1972
177. J.A. Hartlage and W.J. McKillip (Ashland Oil)
177A. Kuraray, JP 144,110, 1983; Chem. Abstr. **100**, 87156 (1984)
178. H.J. Langer and N.A. Randen, Polym. prepr. (ACS), **16**(1), 490–7 (1975)
179. B.M. Culbertson and H.J. Langer, Polym. Prepr. (ACS), 498–505, 506–13 (1975)
179A. Ashland Chemical Company, Columbus, Technical Literature, 1974
180. R.C. Gasman (Kendall), USP 3,803,220, 1972
181. R.C. Slagel (Ashland), USP 3,527,802, 1970
182. W.J. McKillip, B.M. Culbertson, G.M. Gynn and P.J. Menardi, Ind. Eng. Chem. Prod. Res. Develop., **13**(3), 197–201 (1974)
183. K. Natueda et al. (Permachem Asia), GP 2,357,121, 1974
184. L.D. Taylor et al., J. Polym. Sci., Polym. Chem., **21**(4), 1159–64 (1983)
185. J. Guillot and S. Magnet, Prepr. 2nd Colloque International sur les Copolymerisations et les Copolymères en Milieu Disperse, CNRS, Lyon, 1989, pp. 28–31
186. H.R. Lucas et al. (American Cyanamid), Eur. P 224,736, 1987
187. R.G. Lee et al., Proc. Water-borne and Higher Solids Coat. Symp., New Orleans, 1989, pp. 10
188. J. Bibka et al. (Casella), GP 2,251,922, 1974
189. F. Cuirassier, W. Didier and A. Blanc (Soc. Francaise Hoechst), Eur. P 337,873, 1989
190. J.G. Iacoveillo and D.W. Horwat (Air Products & Chemicals), USP 5,182,328, 1993
191. C.S. Lindemann (C.H. Tanner), GP 2,222,730, 1972
192. R. Saxon and J.H. Daniel, Div. Org. Coat. Plast. Chem. (ACS), **22**(1), 277–289 (1962)
193. R. Singer and J.L. Smith (Dunlop), BP 1,003,318, 1966
193A. E.J. Carlson (B.F. Goodrich), USP 2,726,230, 1955, Ex. 8

194. R.D. Singer and J.L. Smith (Dunlop), BP 1,006,042, 1967
195. D.H. Klein and W.J. Elms, *J. Paint Technol.*, **45**(576), 68–75 (1973)
196. D.R. Bauer and R.A. Dickie, *Prepr. Org. Coat. Plast. Chem.*, **42**, 451–62, 463–8 (1979)
197. D.R. Bauer and R.A. Dickie, *J. Polym. Sci., Polym. Phys.*, **18**(10), 1997–2014 (1980)
198. D.R. Bauer and R.A. Dickie, *J. Coat. Technol.*, **54**(685), 57 (1982)
199. D.R. Bauer and R.A. Dickie, *Prepr. Polym. Mater. Sci. Engng*, **52**, 550–4 (1985)
200. D.R. Bauer and R.A. Dickie, ACS Symposium Series no. 313, 1986, pp. 26–74
201. H.J. Spinelli, *Prepr. Org. Coat. Appl. Polym. Sci.*, **47**, 529–34 (1982)
202. L.J. Calbo, *Prepr. Org. Coat. Plast. Chem.*, **32**(2), 359–64 (1972)
203. Du Pont de Nemours, Neth. P Applic, 11,330, 1978
204. S.K. Das and C.M. Kania (PPG), USP 4,476,286, 1984
205. K. Dickerhof (BASF), Eur. P 279,441, 1988
206. S. Horiguchi *et al.* (Dainichesika Colour), JP 75 24,382
207. Nippon Oil Seal, JP 112,212, 1980; *Chem. Abstr.*, **93**, 240972 (1980)
208. Hoechst, BP 1,424,724, 1976
209. D. Pirck and G. Füchs (Deutsche Texaco), USP 4,110,2282, 1982
210. Nippon Mektron, JP 115,443, 1982; *Chem. Abstr.*, **98**, 35880 (1983)
211. Nippon Mektron, JP 128,746, 1982; *Chem. Abstr.*, **98**, 73688 (1983)
212. National Starch & Chemical, JP 87 36,409; *Chem. Abstr.*, **107**, 97294 (1987)
213. G. Moore *et al.*, *Surf. Coat Int.*, **78**(9), 377–9 (1995)
214. V. Arella *et al.* (Ausimont), Eur. P 211,251, 1987
215. R. Holma and P. Oldring (eds.), *UV and EB Curing Formulations for Printing Inks*, SITA Technology, 1988
216. Z.W. Wicks, N.F. Jones and S.P. Pappas, Organic Coatings, Vol. 2, Ch. 11, pp. 253–72 (note references quoted)
217. P. Swaraj (ed.), *Surface Coatings, Science and Technology*, Wiley, 1996, Sec. 9.2, pp. 714–86
218. C. Becker, *Polym. Paint Col. J.*, Suppl., 81–92 (1988)
219. P.G. Garratt and K.F. Klimesch, *Polym. Paint Col. J.*, **184**(4343), 30–2 (1994)
220. K. Elsen, *Surf. Coat. Int.*, **1994**, 234–42 (1994)
220A. J.F.G.A. Jansen *et al.*, New developments in Radcure, in *Forum de la Connaissance*, AFTPA Paris/CORI, 1999, 12 pp. (pages not numbered)
221. M. Uminski and L.M. Saija, *Surf. Coat. Int.*, **78**(6), 244–49 (1995)
222. Patent Reference not available
223. L. Haussling *et al.*, *Prepr. Polym. Mater. Sci. Engng*, **1995**, 72581–2 (1995)
224. K. Kakuchi *et al.* (Japan Atomic Energy Institute), JP 85 181,132; *Chem. Abstr.*, **104**, 51898 (1986)
225. Nitto Electric, JP 69,179, 1985; *Chem. Abstr.*, **103**, 38417 (1985)
226. C. Armstrong, *Eur. Coat. J.*, **1994**(4), 178–90 (1994)
227. P. Swaraj, *Surf. Coat. Int.*, **77**(8), 336–8 (1994)
228. F.S. Stowe and R. Lieberman, *Surf. Coat. Aust.*, **23**(8), 8–11 (1986)
229. S. Yamata *et al.* (Nitto Electric), JP 86 103,574; *Chem. Abstr.*, **106**, 6493 (1987)
230. A.B. Brown *et al.* (Rohm & Haas), Eur. P 486,278, 1990
231. P.J. Harris *et al.* (Ciba-Geigy), Eur. P 297,050, 1990
232. Cray Valley, Eur. P 639,626, 1990
233. D.C. Thomas (Richardson Graphics), USP 4,186,069, 1980
234. J.R. Slocombe (Monsanto), USP 4,074,670, 1983
235. R.J. Himics, *Prepr. Org. Coat. Plast. Chem.*, **33**(1), 274–80 (1973)
236. X. Coqueret *et al.* (Elf-Atochem), PCT 20,016, 1995
237. T. Nishikubo *et al.*, *Nippon Kagaku Kaishi*, **1972**(9), 1626–30 (1972)
238. M.R. Cleland, *Proc. of Radtech 92 North America*, Boston, 1992, pp. 765–9
239. S. Paul, in *Surface Coatings, Science and Technology*, P. Swaraj (ed.), Wiley, 1996, pp. 775–9

240. J. Pacansky and R.J. Waltman, *Prog. Org. Coat.*, **18**, 79 et seq. (1990)
241. P. Mallya et al. (Avery Dennison), Eur. P 536,146, 1990; equivalent to USP 5,011,867, 1990
242. Toa Gosei, JP 123,665, 1980; *Chem. Abstr.*, **94**, 85274 (1981)
243. T. Masuda et al. (Kanegafuchi Chemical), JP 94 25,502; *Chem. Abstr.*, **121**, 11845 (1994)
244. T. Masuda et al. (Kanegafuchi Chemical), Eur. P 578,229, 1994
245. M. Maegawa (National Starch), Eur. P 600,074, 1993; PCT, 25,625, 1993
246. A. Despande et al., *Org. Coat. Appl. Polym. Sci. Proc.*, **46**, 353–5 (1982)
247. T. Yamauchi (Asahi Chemical Industry), JP 93 93,071; *Chem. Abstr.*, **119**, 51501 (1993)
248. B.E. Craig et al. (Dow Corning), Eur. P 542,498, 1993
249. N.I. Li et al., *Lakokras Mater. Ikh. Primen.*, **1979**, 57–9 (1979); *Chem. Abstr.*, **91**, 212667 (1979)
250. W. Shimokawa and T. Koshio (Hoechst Gosei), JP 227,312, 1991; *Chem. Abstr.*, **116**, 42654 (1992)
251. J. Barwich et al. (BASF), Eur. P 624,633, 1990; equivalent to USP 5,532,321, 1990
252. K. Nakayama et al. (Kanegafuchi Chemical Industry), JP 95 03,115; *Chem. Abstr.*, **123**, 86129 (1995)
253. N. Ando et al. (Shinetsu Chemical), Eur. P 598,294, 1994
254. N. Ando et al., JP 94 322,313; *Chem. Abstr.*, **122**, 293577 (1995)
255. N. Ando, et al., Eur. P 621,325 1994
256. T. Yamazaki et al. (Shinetsu Chemical), Eur. P 653,446, 1995
257. K. Hahn and K. Doren (Huels), Eur. P 640,630, 1995
258. T. Hatstori and M. Hasegawa (Toa Gosei Chemical), JP 149,840, 1995; *Chem. Abstr.*, **123**, 314893 (1996)
259. T. Masuda et al. (Kanegafuchi Chemical), JP 166,790, 1994; *Chem. Abstr.*, **121**, 303209 (1994)
260. W. Kleeberg et al. (Siemens), GP 2,823,820 1979
261. M. Oki and T. Tsukada (Nippon Unicar), JP 79 61,268; *Chem. Abstr.*, **91**, 141611 (1979)
262. J. Kabela et al., Cz. P 179,471, 1979; *Chem. Abstr.*, **92**, 59774 (1980)
263. H. Warson in *Properties and Applications of Polyvinyl Alcohol*, C.A. Finch (ed.), Society of Chemical. Industry, 1968, pp. 46–76
264. F. Del Rector et al., *J. Coat. Technol.*, **61**(771), 31–7 (1989)
265. D.R. Eslinger, *Am. Paint Coat. J.*, **78**(30), 54–65 (1994)
266. K. Kyo-Jun and R.C. Williams (Reichhold Chemicals), Eur. P 522,469, 1990
267. I. Kriessmann et al. (Vianova Kunstharz), Eur. P 555,774, 1993
268. G. Brindopke et al. (Hoechst), Eur. P 481,335, 1990
268A. B. Klumperman, J. Geurts and J. Verstegen, *Polym. Mater. Sci. Engng*, **76**, 177 (1997)
269. N. Nakagawa et al. (Nippon Carbide), JP 95 126,572; *Chem. Abstr.*, **123**, 289826 (1995)
270. J. Yoshii et al. (Sumitomo Chemical Industry), JP 93 287,248; *Chem. Abstr.*, **120**, 272713 (1994)
271. J.R. Graw and B.G. Bufkin, *J. Coat. Technol.*, **52**(661), 73–87 (1980)
272. L.J. Molhoek and H.J. Wories (DSM), Eur. P 483,915, 1992
273. H. Tucker (Tremco), Eur. P 603,716, 1994
274. D.A. Bors et al. (Rohm & Haas), Eur. P 492,847, 1992
275. B.S. Snyder and D.A. Bors (Rohm & Haas), Eur. P 573,142, 1994
276. M. Iwata et al. (Ouchi Shinko Chemical), JP 73 28,089
277. J.P. Erdman (Du Pont de Nemours), USP 4,211,868, 1980
278. R.K. Pinschmidt et al., Crosslinked Polymers, ACS Symposium Series No 367, pp. 454–78
279. R.K. Pinschmidt et al., *Prepr. Polym. Mater. Sci. Engng*, **56**, 785–9 (1987) (two papers)
280. F.L. Marten (Air Products), USP 4,360,632, 1982
281. S. Abe et al. (Mitsubishi Yuka Badische), JP 87 72,742; *Chem. Abstr.*, **108**, 57951 (1988)
282. K. Kayanuma and K. Hiral (Ajimoto), Eur. P 629,657, 1994

283. W. Jacobs *et al.*, Proc. XVI Int. Conf. on Organic Coatings Technology, Athens, 1990, pp. 509–24
284. E.S. Barabas and M.M. Fein (GAF), USP 3,635,868, 1972
285. C. Palomo and R. Mestres, *Synthesis*, **1981**, 373 *et seq.* (1981)
286. W. Brown, *Surf. Coat. Int.*, **78**(6), 238–42 (1995)
287. J.W. Taylor, M.J. Collins and D.R. Bassett, *Prepr. Polym. Mater. Sci. Engng*, **73**, 102–33 (1995)
288. W.T. Brown, *Proc. 21st Water-borne and Higher Powde Coat. Symp.* New Orleans, 1994, Vol. 1, pp. 40–53
289. J.W. Taylor, M.J. Collins and D.R. Bassett, *Prep. Polym. Mater Sci. Engng*, **73**, 102–3 (1995)
290. G. Bauer *et al.* (BASF), Eur. P 617,012–3, 1994
291. J.T.K. Woo (SCM), USP 4,328,144, 1982
292. T. Tanaka, O. Ohhara and K. Moritami (Kuraray), JP 102,389, 1977; *Chem. Abstr.*, **88**, 74920 (1978)
293. R. Baumstark and M. Portugall (BASF), GP 4,334,178
294. P.C. Heldt *et al.*, *J. Appl. Polym. Sci.*, **56**, 1161–7 (1995)
295. M. Izubayashi *et al.* (Oji Paper), Eur. P 320,954, 1989
296. S.A. Voronov *et al.*, USSRP 614,416, 1978; *Chem. Abstr.*, **89**, 111278 (1978)
297. V.A. Puchin *et al.*, *Sint. Fiz-Khim Polim.*, **14**, 39–43, 1974; *Chem. Abstr.*, **82**, 31527 (1975)
298. Y. Takao *et al.* (Nippon Zeon), JP 345,977–8, 1994; *Chem. Abstr.*, **122**, 316106,316108 (1995)
299. B.G. Bufkin, *J. Coat. Technol.*, **53**(673), 549–58 (1981)

6
Technical Polymer Latices

H. Warson

1 INTRODUCTION

The first commercial polymer latices (synthetic resin emulsions), as distinct from synthetic rubber latices, were introduced in the 1930s. One of the first patents described the polymerisation of butyl methacrylate, ethyl acrylate and a copolymer of acrylonitrile and methyl acrylate as 25 % solids latices [1]. The first emulsifiers were simple soaps, followed by various sulfates and sulfonates, and the fatty acid esters of triethanolamine. The inclusion of water-soluble polymers, whether natural or synthetic, as stabilisers followed about a decade later.

With improvements in technology, it was possible to raise the solids content first to 50 %, and some latices with solids content as high as 65 % have been quoted in the technical literature. The mechanical properties of early products were poor, the purpose of polymerisation being a convenient preparation of the solid polymer. Later products have been formulated to give good long-term mechanical stability and stability to compounding, already considered in Chapter 4. Practical developments have been of a semi-empirical nature, being in parallel with the intense academic interest previously described.

In the case of vinyl chloride, the use of emulsion techniques is still often used only to provide a source of solid polymer. Process improvements have been made to improve the quality of the polymer and the general efficiency of polymerisation. The former includes higher molecular weights, improved resistance to discoloration by heat, improved electrical properties and the modification of the polymers, sometimes by copolymerisation, to ease processing problems. Some polyvinyl chloride latices are sold for the purpose of compounding and will be described later in connection with various applications. The earliest patents include one in which Mark and Fikentscher are the inventors [2].

The use of delayed addition techniques (Chapter 3, Section 2.1.1) represented a major technical advance and enabled very large scale batch polymerisation, of the order of 25 000 litres (25 tonnes) to be undertaken. A modification is to place part of the emulsifier solution into the reactor and to add a pre-emulsion from an overhead tank, this usually requiring stirring to keep the very concentrated monomer emulsion reasonably stable. In this case the initiator may be in either or both phases. The introduction of redox processes (Chapter 1, Section 2.1.3 and Chapter 3, various examples in Sections 3 to 7), which are of particular interest in acrylic polymerisations and which often depend on a three- or four-stage addition of the redox components, gives rapid polymerisation at relatively low temperatures, and also adds to technical efficiency.

Further developments which have improved the range of available synthetic latices are based on core–shell copolymers, in which each particle varies in composition, usually by one of more changes of monomer feed in the manufacture (Chapter 3, Section 2.4.3). The major advantage is that film formation may take place at a lower temperature than from a latex based on homogeneous particles of the same overall composition. This enables tougher and harder films to be available.

As already indicated in previous chapters, properties such as freeze–thaw resistance have increased the usefulness of latices, and even such properties as corrosion resistance have been imparted by suitable additives, e.g. phosphated emulsifiers.

Perhaps the major source of improvement in properties has been the imparting of crosslinking properties. Sometimes, as explained in Chapter 5, films require thermal treatment, but in others crosslinking may take place on filming or in a relatively short period after the filming or an impregnating operation.

It must be stressed that the practical requirements of a latex are paramount in formulating a product suitable for application. These include the entire viscosity spectrum of a latex, the rate and nature of drying, especially in continuous operations such as paper coating and some textile operations, the minimum film formation temperature and the ability to be compounded, e.g. with pigments, so as to give desired application properties. On the other hand, development of improved products has often been delayed because of the insistence of technologists in matching a single viscosity specification, e.g. a measurement in Krebs units under one specific set of rotation conditions, or even the time of eflux from a Ford Cup. The various technical latices are best considered under the major class headings, depending on the monomers included. Some general properties and treatment will also be included at the risk of some overlap with earlier chapters. Tables indicate the properties of a number of latices, as produced or marketed in various countries at the time of writing. This list in no way claims to be complete, and the inclusion of the products of one manufacturer must not be interpreted as a recommendation to the exclusion of other products. It should be noted that whereas the nature of the monomers from which the polymers are formed is often disclosed fairly fully, it is rare that any details are given about the stabiliser systems, except in the most general terms such as 'anionic' or 'colloid stabiliser'. Products containing the various polyvinyl alcohols or cellulosic thickeners are, however, often described as such. Nevertheless, information as to the nature of the processes can often be acquired from patents and a few theoretical papers.

An effort will be made to present information so that properties are described in a comparable manner. However, to alter the text of specifications describing properties such as viscosity may increase confusion if trade literature is studied independently, and in general, therefore, they have been retained as described by the manufacturers.

2 POLYVINYL ACETATE AND RELATED POLYMERS

2.1 Polyvinyl acetate

Vinyl acetate is currently one of the cheapest monomers in Europe and, coupled with ready availability, it has become one of the most popular starting points for products such as emulsion paints and adhesives of all types. Unlike vinyl chloride or butadiene, it is liquid at ambient temperatures, thus facilitating handling. The solid polymer has a second-order transition point (T_g) of 25–28 °C, making plasticisation necessary for some purposes. However, as has already been indicated (Chapter 4, Section 5), this is easy to achieve in emulsion, unlike polymers and copolymers of vinyl chloride. In addition, the weathering properties of films containing polyvinyl acetate and most copolymers are very good, especially in regard to ultraviolet light resistance and oxidative degradation. With regard to these properties, they are generally superior to copolymers of styrene and butadiene, which have been of comparable price in the United Kingdom and rather cheaper in the United States.

There is an appreciable difference in the properties of polyvinyl acetate depending on the overall molecular weight. The nature of stabilisers causes slight variations and are most marked where the various polyvinyl alcohols are used as stabilisers; in the latter case this also depends on the process involved, which may cause appreciable variations in properties. These may become a major factor if polymerisation takes place under conditions of very high chain branching and grafting (Chapter 3, Section 3.2.2). If prepared under these conditions, the polymer is no longer soluble in ethyl alcohol with a small water content.

Polyvinyl acetate and its copolymers have a marked tendency to 'creep' or to have 'cold flow'. This causes a certain amount of difficulty in some adhesive applications and is overcome as far as possible by the use of high molecular weight polymers. Films are generally tough and abrasion resistant, but have poor resilience and elasticity. The polymer is stable at all normal temperatures, unlike polymers of vinyl chloride.

Chemically, polyvinyl acetate is hydrolysed comparatively easily, the acetate units being replaced by hydroxyls under both acid and alkaline conditions. The unit equation is

$$-CH_2CH(OOCCH_3)- \xrightarrow{H_2O} -CH_2CHOH- + CH_3COOH$$

Similarly, the polymer is easily submitted to alcoholysis using methanol or another alcohol [3, 4]. Certain copolymers, such as those including vinyl esters of branched long-chain acid esters and ethylene are hydrolysed with much more difficulty [5–7].

It is therefore obvious that polyvinyl acetate does not have an exceptionally strong water resistance, using the latter term in its most general sense, and moisture can be absorbed comparatively readily, especially when films have a high quantity of stabiliser present. This in itself is not intrinsically objectionable as it enables films to breathe and to transmit moisture when desirable, as in a paint used over brickwork or wood. A typical example occurs under winter conditions in parts of the United States, where an emulsion paint is used externally, but there is a humid atmosphere in a kitchen on the internal side.

The polymer is soluble in a wide range of solvents, including most ketones and esters with short hydrocarbon chains, methanol, 95 % ethyl alcohol, isopropanol, glycol ethers and their esters, and most chlorinated hydrocarbons. It is soluble in benzene, and toluene when hot, but xylene is a tolerated diluent rather than a solvent. If these solvents are water immiscible, they can be added directly to the emulsions with stirring, but water-soluble solvents require aqueous dilution before adding to the latices and additions should be limited.

The plasticisation of polyvinyl acetate has already been described (Chapter 4, Section 5). One of the principal applications in which plasticised polyvinyl acetate latices have been utilised is that of the cheaper emulsion paints, as described in Chapter 10 in Volume 2. The addition of plasticisers to latices often has the effect of increasing the viscosity, mainly because of the higher solids content. It is seldom necessary to prepare an independent plasticiser emulsion. It should be noted that some plasticisers such as tritolyl phosphate enter the polymer particles much slower than, for example, dibutyl phthalate. Various proprietary polyester plasticisers may also be included in polyvinyl acetate latex compositions, especially where a particular property is required, e.g. the absence of migration into a substrate.

In order to include sufficient plasticisation for a continuous film at ambient temperature, about 5 % of plasticiser on total weight is desirable. Thus polyvinyl acetate latex adhesives are often plasticised to this degree to increase film and hence bond strength, although this may cause a slight increase in cold flow.

Most technical latices with a solid content 50 % or greater tend to have a comparatively large particle size, of the order of 0.5–1 µm, although it is rather less with most copolymers. This relatively coarse particle size has a certain advantage in that it discourages excessive migration into porous substrates.

Polyvinyl acetate and related polymers 355

Whilst the original polyvinyl acetate latices were based on various polyvinyl alcohol stabilisers, a current major class is based on latex stabilization with hydroxyethyl cellulose combined with an anionic surfactant of the sulfate or sulfonate type, occasionally a phosphate, and optionally a non-ionic surfactant based on a polyoxyethylene condensate with long-chain alkyl alcohol or with an alkaryl phenol[†], as described in Chapter 2. A few latices which are free from water-soluble polymers are marketed. They tend to be of fine particle size (<0.5 μm) and are often prepared with an emulsifier system including ethylene oxide–propylene oxide block copolymers.

Other stabilisers of low molecular weight (phosphate ester surfactants) and water-soluble polymers such as sodium polyacrylates or dextrins have been available at various times. Many of these technical latices are prepared for 'captive' use only, and the range of products manufactured is probably considerably greater than those which are marketed. Some commercial formulations are included in Tables 6.1 and 6.2, together with copolymers of vinyl acetate.

Table 6.1 Homopolymer latices of Rhodopas ®(Rhodia, France)

	A010	A012P	A013P	A015P	A018
Type	All homopolymers				
Plasticiser DBP on solids	Nil	13 %	20 %	Nil	Nil
Stabiliser	Polyvinyl alcohol				
Solids content (%)	56 ± 1	54 ± 1	54 ± 1	62 ± 1	50 ± 1
pH	4.5 ± 0.5	4.5 ± 0.5	4.5 ± 0.5	4.5 ± 0.5	4.5 ± 0.5
Viscosity[a]	27 000 ±3000	6500 ±1500	5500 ±1500	29 000 ±3000	35 000 ±5000
Particle diameter (nm)	1–2	1–2	1–2	1–3 (0.1–0.3)	0.5–2
Surface tension (mNm^{-1})[b]	—	—	—	—	—
MFT (°C)[c]	+15	0	0	+10	+20
Freeze–thaw stability at −10 °C[d]	rev	nonrev	nonrev	rev	rev
Characteristics	Adhesive applications	Good covering power		Rapid bond	Adhesive applications

[a]Brookfield RVT (mPa s).
[b]No data provided.
[c]Minimum film-forming temperature.
[d]rev = reversible, nonrev = non-reversible on freezing and thawing.

[†] The use of alkaryl phenol derivatives is being phased out.

Table 6.2 Copolymer latices of Rhodopas ®(Rhodia, France)

	AM 021	AM023 S	AM054	AD 310	AO 485	AVD 900
Type	Maleic ester Copolymers	Copolymers	Acrylic ester Copolymer	'Versatic' and acrylic Copolymer	'Versatic' and acrylic Copolymer	Olefin Copolymer
Plasticiser DBP on solids	Nil	Nil	Nil	Nil	Nil	Nil
Stabiliser	Non-ionic	Non-ionic	Non-ionic Cellulosic	Anionic Cellulosic	Non-ionic Cellulosic	Anionic
Solids content (%)	55 ± 1	55 ± 1	53 ± 1	55 ± 1	50 ± 1	50 ± 1
pH	5.0 ± 0.5	5.0 ± 0.5	5 ± 0.5	5 ± 0.5	4.5 ± 0.5	4.5 ± 0.5
Viscosity[a]	1500 ±500	10000 ±5000	2500 ±500	1600 ±300	2500 ±500	6500 ±1500
Particle diameter (nm)	0.5–1	0.5–2	0.3–0.6	0.4–0.8	0.3	0.25
Surface tension (mNm^{-1})[b]	—	—	—	—	—	—
MFT (°C)[c]	+15	+1	+4	−10	+9	+6
Freeze-thaw stability at −10 °C[d]	—	—	rev	—	—	—
Characteristics and applications	Packaging Heat seal	Adhesive PVC, textiles paints	Exterior Semigloss and coats	Exterior Semigloss on plastics	Most paints Including coats on plastics	Thick facade Paints and

[a]Brookfield RVT (mPa s).
[b]No data provided.
[c]Minimum film-forming temperature.
[d]rev = reversible, nonrev = non-reversible on freezing and thawing.

Table 6.3 Homopolymer latices of Mowilith® (Clariant[a])

All latices in this series are fundamentally polyvinyl alcohol stabilised.

Mowilith type	Solids (%)	Plasticiser type (%)	Approximate particle diameter (μm)	Viscosity at 20 °C, Brookfield RVT tn TVT 20 rpm	pH	MFT (°C)	Freeze–thaw[c] stability
D	50	Nil	1–3	160–350P sp6	3–4	15	Yes
DN	50	Nil	1–3	25–55P sp4	3–4	15	Yes
D	60	Nil	0.5–4	180–360P sp6	3–4	15	No
DN	60	Nil	0.5–4	75–135P sp5	3–4	15	No
DH	50	Nil	0.5–3	250–450P sp6	5.5–6.5	14	Yes
DH	55	Nil	0.5–5	250–500P sp6	4–5	16–18	Yes
D025	54	10.8DBP[b]	1–3	170–300P sp6	3–4	0	Yes
D05	58	19.3DBP	1–3	170–300P sp6	3–4	0	Yes
D222	54	4.3/6.5 DBP/TTP[c]	1–3	170–300P sp6	3–4	0	Yes
DLHR	[d]	[d]	Not indicated	12–220P sp6	6–7	0	Yes

[a] Marketed by Harlow Chemical Company in the United Kingdom.
[b] DBP = dibutyl phthalate.
[c] TTP = tritolyl phosphate, also known as tricresyl phosphate
[d] This latex contains a coalescing solvent.

Table 6.4 Copolymer latices of Mowilith® (Clariant[a])

Mowilith type	Solids (%)	Comonomer (%)	Approximate particle diameter (μm)	Viscosity at 20 °C, Brookfield RVT fn 20 rpm sp6	pH	MFT (°C)
The following latices are stabilised with polyvinyl alcohol:						
DM1	55	n-BM	0.3–2	25 ± 10	3–5	12
DM 1H	51	n-BM	0.3–2	50 ± 30	3–5	−4[b]
The following latices are stabilised with a protective colloid:						
DM 2H	51	n-BM[c]	0.3–2	20 ± 7P sp3	4–5	−10[b]
DM 2HB	51	n-BM[c]	0.3–2	2–20P sp2	3.5–4.5	−14[b]
The following latex is stabilised with an emulsifier + protective colloid:						
LDM 2110	50	Veova	≤1	120 ± 30P sp5	3.5–5	6
The following are copolymers of vinyl acetate with ethylene and are stabilised with polyvinyl alcohol and a surfactant:						
DM 104	55	Ethylene	0.1–1	80 ± 30 P sp5	4–5	6
DM 130	50	Ethylene	0.1–1	80 ± 30 P sp5	4–5	−5[b]
The following is a terpolymer of vinyl acetate, vinyl chloride and ethylene and is stabilised by emulsifiers only:						
DM 128	50	As above	0.1–0.5	20 ± 13P sp3	6.8	−6[b]

[a] Marketed in the UK by Harlow Chemical Company.
[b] This column indicates the T_g, not the MFT, which cannot be below zero.
[c] Copolymer will dibutyl maleate.

Table 6.5 Copolymers of vinyl acetate and ethylene, possibly with a third monomer produced by Air Products (USA). (See also Table 6.12 for additional latices containing vinyl acetate and ethylene). Reproduced by permission of Air Products and Chemicals, Inc

Product (Airflex®)	124	125	140	100HS	199
Latex solids(%)	52	52	55	55	50
pH	5.5	5.5	4.5	5.5	6.0
T_g (°C)	−16	0	0	7	24
Viscosity (cp)	225^a	700^b	2250^c	425^b	175^a
Particle size (nm)	0.29	0.15	0.8	0.17	0.16
Particle charge	Non-ionic	Non-ionic	Non-ionic	Slightly anionic	Slightly anionic
Principal applications	Paper coating sizing	As 124	Heat seal paper size	All paper applications	Paper coat
Special properties	Self crosslink	As 124	PV alcohol stabilised	High gloss	Good high shear rheology

[a] Brookfield LVF No. 2 spindle.
[b] Brookfield LVF 60 rpm No. 3 spindle.
[c] Brookfield RVF 20 rpm No. 3 spindle.

Table 6.6 Copolymer latices produced by Enichem (Italy)

Ravemul Vinavil	T33	T40	M19	C26	023	HC(4)
Copolymer type	VV^a	VV	VV	BM^b	VV	Acrylic
Solids(%)	55	55	50	55	50	45
Viscosity (mPa s)	3000	3000	20 000	9000	1500	<400
pH	5	7.5	3	4	4	3.5
Particle size (μm)	0.5–1	0.5–1	0.3–2	0.5–1	0.2–0.6	0.1–0.6
MFT^c (°C)	12	30	3	3	13	14
Emulsifiers	A/N^d	A/N	PV alcohol	A/N	A/N	A/N
Applications	Paints	Metal coats	Adhesives	Paints	Heat seal adhesives	Building adhesives

[a] Copolymer with vinyl ester versatic acid® (see Chapter 2).
[b] Copolymer with dibutyl maleate.
[c] Minimum film formation temperature (see Chapter 4).
[d] Anionic–non-ionic.

2.2 Copolymers of vinyl acetate

2.2.1 Types of copolymer

Copolymers of vinyl acetate were a natural development in the search for latices which would overcome the disadvantages of 'external' plasticisation and are often referred to, rather erroneously, as 'internally plasticised' products. Amongst the disadvantages of external plasticisation are losses of plasticiser by migration into substrates or into surfaces with which films are in contact, losses by volatility, hydrolysis or degradation, and possibly the effects on organic lake pigments that have a tendency to solubilisation and subsequent recrystallisation, as for instance Hansa Yellow. Resistance to acid or alkaline hydrolysis is generally increased considerably by the inclusion of a comonomer.

The type of comonomer selected in various cases depends on availability and cost, but apart from commercial considerations, chemical factors are the ease of copolymerisation and the plasticising efficiency on a weight-to-weight basis. The principal monomers which have been included in copolymerisation formulations are the vinyl esters of other acids, as elucidated in the next paragraph, and esters of acrylic, maleic and fumaric acid, and occasionally of itaconic acid.

Vinyl esters that have been included in copolymerisation formulations are vinyl caprate $CH_2{:}CHOOCC_9H_{19}$, vinyl 2-ethylhexanoate (Vynate® 2-EH, Union Carbide) $CH_2{:}CHOOCC_4H_8CH(CH_2H_5)CH_3$, vinyl stearate $CH_2{:}CHOOCC_{17}H_{35}$ and the mixed vinyl esters of alkyl acids with branching on the α- atom, as described in Chapter 1, Section 6.2. The simplest monomer in this series, vinyl pivalate $CH_2{:}CHOOCC(CH_3)_2CH_3$ is currently available as Vynate® Neo-5. The vinyl esters of acids with chains of higher carbon numbers are available, *inter alia*, as Veova® (Shell Chemical Company), with a number indicating the average number of carbon atoms in the acid chain. One typical monomer with ten carbon atoms is Veova®10, also known as Vynate® Neo-10 (Union Carbide). Exxon Chemical also provide similar vinyl esters of Neo acid, a similar branched chain type. Note that during polymerisation it is desirous that about 3–5 % of the total vinyl acetate monomer should be added towards the end of polymerisation to avoid any surplus of the pivalate type of monomer remaining unpolymerised, especially as there may be an appreciable after-odour.

The r_1/r_2 ratios of most vinyl esters are sufficiently close for copolymerisation to proceed without undue difficulty. In addition, polyvinyl acetate in dilute solution may behave like a soluble polyelectrolyte [8]. There is little difficulty in copolymerisation with any of these esters except for vinyl stearate. Unless this is finely emulsified so as to present a large surface, it tends to produce a heterogeneous product on copolymerisation, which is probably a physical blend of polyvinyl acetate and copolymers very low in comonomer, together with polyvinyl stearate or copolymers of the latter with small quantities of vinyl acetate. The very finely emulsified vinyl stearate may have its water solubility, normally very low, increased appreciably by some emulsifiers. Vinyl stearate was claimed to be a valuable comonomer about 1958–9, but technical difficulties in the manufacture of the monomer have led to its virtual abandonment.

Vinyl propionate $CH_2:CHOOCCH_2CH_3$ is an alternative monomer, readily available, which has been used as an alternative to, or in conjunction with, vinyl acetate. Typical grades are produced by BASF (Germany) and as Vynate® L-3 (Union Carbide). There are technical latices available with this monomer.

In addition, copolymers of vinyl acetate with ethylene have been known in emulsion form for about three decades. Ethylene is an exceptionally efficient internal plasticiser as a copolymer with vinyl acetate, but its efficiency may actually be a disadvantage as relatively little is needed to plasticise a vinyl acetate copolymer to a desirable level for film formation. This does not justify the cost of the pressure equipment required. In consequence, terpolymers also including vinyl chloride have been marketed, increasing the useful quantity of ethylene, which can be included in the copolymerisation.

A number of copolymers have been described which include higher olefines such as 1-butene $C_2H_5CH:CH_2$ and a number of other olefines with an even number of carbon atoms, with unsaturation in the 1 : 2 position. However, all these higher olefines have allylic unsaturation, and the actual amount that can enter a copolymer is small.

A number of terpolymers and multipolymers with several different monomer units have been claimed at various times [9–11]. These copolymers, as indeed do most copolymers, tend to have a balance of properties of those of the individual monomers used in respect of the polymers that they produce. This is a general but not an invariable rule. In the case of maleate and fumarate esters, it is one of the notional polymer properties, since they are difficult to homopolymerise.

The copolymerisation of vinyl acetate with esters of maleic and fumaric acid, mainly those of butanol and 2-ethylhexanol, usually proceeds without difficulty in the emulsion form. It is possible that there may be some isomerisation between fumaric and maleic acids during the reaction, as happens in polyester production, but this has not been proved conclusively, and kinetic experiments have quoted different reactivity ratios for the various esters, fumarates in general being the more reactive:

Vinyl acetate (r_1) diethyl fumarate (r_2) = 0.011 : 0.444

Vinyl acetate (r_1) diethyl maleate (r_2) = 0.17 : 0.043

Copolymerisation of vinyl acetate with acrylic esters, although apparently proceeding very well experimentally, needs to be carefully controlled since the acrylic esters are very much more reactive than vinyl acetate [12]:

Methyl acrylate (r_1) vinyl acetate (r_2) = 6.7 ± 2.2 : 0.29 ± 0.119

2-Ethylhexyl acrylate (r_1) vinyl acetate (r_2) = 6.7 : 0.0111

In order for a reasonably homogeneous copolymer to form, it is necessary that a 'delayed addition' procedure should be adopted in which the rate of addition of monomers is restricted to the rate of overall polymerisation so that there is no build-up of vinyl acetate, the slower monomer to polymerise.

In the past many manufacturers did not take into account the differences in the relative rates of polymerisation of acrylic esters and vinyl acetate. As a result many reputed 'acrylic copolymers' were mixtures, with compositions ranging from almost pure acrylic polymer to almost pure polyvinyl acetate. These copolymer blends may produce apparently clear films when dried, as the various polymer components tend to homogenise each other. The physical properties of derived films may be considerably different from those of 'true' copolymers, the flexibility of films from the latter being generally much greater. This is most marked with copolymers of 2-ethylhexyl acrylate and vinyl acetate since the two homopolymers are not mutually compatible.

An extensive investigation of copolymers of alkyl maleate and alkyl fumarate esters with vinyl acetate has stressed the importance of even copolymerisation. A 20 % dibutyl fumarate–vinyl acetate copolymer film demonstrated a difference in elongation of 100 times, depending on the method of preparation. The best elongation, 403 %, was obtained when monomers were continuously added at the rate at which they polymerise [13].

2.2.2 Equivalence of 'internal' and 'external' plasticisation

As almost all monomers that are copolymerised with vinyl acetate are added for the purpose of giving additional flexibility, it is convenient to assess their efficiency for this purpose by comparison with dibutyl phthalate, which may be regarded as a standard. The efficiency of all the common ester monomers increases with the length of the alkyl chain, whether of the alcohol in the case of acrylates, fumarates, maleates and itaconates or of the acid in the case of the higher vinyl esters. The upper limit has not been precisely defined, but is in general alkyl chains of between eight and twelve carbon atoms. Branched chains, as would be expected, are less efficient in producing flexibility than straight chains, although esters of the slightly branched 2-ethylhexyl alcohol are popular because of the ready availability of the synthetic alcohol.

With the most efficient plasticisers, of which n-heptyl acrylate, 2-ethylhexyl acrylate, 2-ethylhexyl fumarate and 2-ethylhexyl maleate may be included, the efficiency is about 60–70 % of that of dibutyl phthalate. The vinyl esters of synthetic mixed C_7–C_9 straight chain alcohols may also be available with similar plasticising efficiency. The Veova® 10 type of monomer is less efficient than the above, its equivalent efficiency being about 40 %, approximately 25 % of comonomer on total monomers being required as equivalent to 10 % of dibutyl phthalate (on total polymer plus plasticiser weight). This, however, is advantageous as a 25 % copolymerisation of Veova® 10 is required to upgrade the water and alkali resistance of the copolymer to approximately its maximum level, which is not far short of the average 100 % acrylic copolymer.

As already indicated, ethylene is an efficient internal plasticiser for vinyl acetate copolymers. Some surveys have been made of its efficiency as a comonomer [14, 15]. Whilst solid resins of this type may contain about 55 % of ethylene, 10 % would be a more appropriate figure for a copolymer in a paint or paper coating, which indicates that the plasticising efficiency of ethylene is higher than that of dibutyl phthalate, possibly as high as 150 %. This is probably because the molecular weight of each ethylene unit is very low, only 28, and theoretically a molar percentage rather than a weight percentage should be compared. The alkali resistance of films from commercial products seems to be of the same order as those from the Veova® type already mentioned, which is probably due to a high degree of grafting, causing steric hindrance to ready hydrolysis.

Because of this very high plasticising efficiency, the special interest of ethylene copolymers with vinyl acetate is in paper coating, which demands a very high pigment loading and hence requires a more flexible film that is required with emulsion paints. Paper coating is described in Chapter 13 in volume 3. The relatively high content of ethylene justifies the expensive pressure plant needed for this type of copolymerisation to make it an economically viable proposition.

Interesting monomers which were introduced in about 1968 are diethylene glycol itaconate

$$CH_2.COO(CH_2CH_2O)_xH$$
$$CH_2=C.COO(CH_2CH_2O)_yH$$

where $x + y = 2.8$, and di(propylene glycol) itaconate

$$CH_2COO(CH_2CH(CH_3)O)_xH$$
$$CH_2=C.COO(CH_2CH(CH_3)O_yH$$

where $x + y = 2.7$. These monomers were reported to polymerise well with vinyl acetate using a conventional delayed addition formula with a hydroxyethyl cellulose/sodium lauryl sulfate stabiliser system, but as far as is known they are not currently available commercially [16]. Almost all comonomers reduce to some extent the ease of hydrolysis of vinyl acetate copolymers, the maleates and fumarates rather less so because of their relatively high polar character, the acids having two carboxyl groups on adjacent carbon atoms.

In copolymerisation formulations, very similar systems to vinyl acetate homopolymerisations can be adopted. A general rule, however, applicable to the higher acrylates, is that above 15–20 % of comonomer there should be a gradual reduction of water-soluble polymer, especially hydroxyethyl cellulose, and a corresponding increase in non-ionic surfactant. There is no reason why the ratio of comonomer should be limited in the case of monomers that homopolymerise, and for special purposes the comonomer with vinyl acetate may be raised to 40 % or greater. However, the formulation may need modification to approximate closer to that of an acrylic ester, and care should be taken to allow for the different rates of polymerisation, unless an uneven copolymer is deliberately required. The latter may be desirable if polymers are required for damping of vibration or sound, as will be described in Chapter 17 in Volume 3.

Particle sizes of most copolymer latices tend to be rather lower than those of the corresponding homopolymer, probably because of the smaller hydration shell. Further details of copolymers will be included in later chapters on specific applications, with special reference to emulsion paints.

2.2.3 Alkali-soluble copolymers

Vinyl acetate copolymerises well with crotonic acid, the r_1 (vinyl acetate)/r_2 ratios being 0.33 : 1. Emulsion polymerisation occurs quite readily (see Chapter 3, Section 3.4 and Table 3.12). About 5 % of crotonic acid with 95 % of vinyl acetate produces a copolymer that is quite stable in a latex at a pH of about 4. A standard hydroxyethyl cellulose/surfactant stabiliser system (see Chapter 3, Table 3.5) operates quite well. The crotonic acid should be dissolved in the monomer to ensure even copolymerisation. Technical crotonic acid seems to slow the rate of copolymerisation somewhat, although it is uncertain whether this is due to traces of impurities in the monomer or to charge transfer characteristics of the crotonic acid. It is possible to increase the crotonic acid content to 15–20 %, but at the upper level complete polymerisation becomes difficult.

Plasticisation is possible as with polyvinyl acetate, the major difference being, of course, the ready attack by dilute alkali, which is desirable in many applications, e.g. the repulping of paper coatings, as will be shown later. The latices are directly soluble in caustic alkalis, ammonia solution, borax, sodium carbonate and sodium bicarbonate. Amines are not recommended, as by-products and secondary reactions give rise to coloured products. Solubility starts between a pH of 6.5 and 7 and is complete at about 8.5, the viscosity rising considerably. In order to obtain a solution that is easy to handle, it is desirable that the final solids solution is reduced to 20–25 %. The solubilised films are not completely clear unless a water-soluble plasticiser is included.

Excess of alkali should be avoided as it leads to further hydrolysis, this time of the vinyl acetate units, causing a further increase in viscosity. Excess or borax may also cause thickening, probably because of a reaction with polyvinyl alcohol caused by hydrolysis. As a rule it is desirable to use the alkali solubilised products as soon as possible after they are formed. Molecular weight information of commercial products is not available, but these are probably relatively low, with a viscosity average molecular weight of about 50 000.

The lower limit of addition of crotonic acid to ensure good solubility is almost 5 % on total monomer. In this technical series of crotonic acid copolymers, the whole of the crotonic acid groupings must be neutralised in order to give a clear solution.

Terpolymers of vinyl acetate, vinyl esters of higher fatty acids or the Versatic® acids (branched chain fatty acids in which the carbon atom adjacent to the carboxyl is tertiary; see Section 2.2.1) and crotonic acid are formed readily since the reactivities are comparable, and the products have somewhat similar alkali solubilities. The main distinction is the increased flexibility due to a plasticising comonomer. Whilst the last named have the advantage that the dried films, such as those from ammonia solution, have considerable water resistance, and even alkali resistance, a somewhat greater excess of alkali is required to effect solution. The maximum quantity of the vinyl esters of branched acids that may be included is about 20 % [17].

Methacrylic acid has proved rather unsatisfactory because of its very high relative reactivity, but acrylic acid copolymerises in a more reasonable manner in a delayed addition process. However, copolymers of vinyl acetate with acrylic acid tend to increase in viscosity from a pH of 3.5–4 upwards, which causes handling and manufacturing difficulties. If a terpolymer is formed with 10–20 % of a maleate or fumarate ester, solution only occurs at a higher pH, and the product becomes more useful [18].

The major technical advantage of the alkali-soluble products is the increase in pigment binding power, which makes them useful for applications requiring high loadings of pigments and fillers, as in binders for china clay in paper coating. The property of thickening rather than solution on addition of alkali may be useful on some occasions. Certain products, e.g. Acronal 500D (see Table 6.7 and Figure 6.1 later), have this property, the increase in viscosity being of a high order, from a free-flowing latex almost to a gel by the addition of 2 % of 25 % ammonia solution on emulsion weight. Whilst the method of preparation of this product has not been revealed, similar products can be prepared as terpolymers containing approximately 48 % of an acrylic ester, 48 % of vinyl acetate and 4 % of acrylic acid. The acid is copolymerised with part of the acrylic ester and the balance of vinyl acetate and the acrylic ester by continuous addition immediately afterwards. By this means vinyl acetate is not copolymerised directly with the acrylic acid.

The property of alkali solubilisation is not reversible. If the alkaline solutions are acidified, they coagulate to a solid mass, which is difficult, if not impossible, to re-solubilise.

2.3 Typical commercial products

The latices listed here are typical of the various ranges available and must not be construed as advertising matter for these products. Note that European technical literature may refer to them as 'dispersions'. The amount of information available varies between different manufacturers.

A number of vinyl acetate copolymers available technically are self-crosslinking, although exact details are not often given. Details of the various possibilities are given in Chapter 5. It is probable that most crosslinking latices are based on derivatives of acrylamide.

3 ACRYLIC POLYMERS AND COPOLYMERS

3.1 Properties of acrylic polymers

The ready availability of a range of both acrylate and methacrylate esters enables polymers with a wide range of properties to be prepared. These include the wide range of film flexibilities possible, some indication of which has already been indicated in Chapter 1, Section 6.4. In addition to the alkyl esters, some ethoxyethyl esters of both acrylic and methacrylic acid are commercially available; for the purpose of this section, derivatives of both acrylic and methacrylic acid will also be referred to as 'acrylics'.

Of ever increasing importance are the various crosslinking reactions possible with acrylic polymers containing reactive monomers, as has already been indicated in Chapter 5. Many derivatives of acrylamide are now available technically, enabling a wide and versatile range of products to be prepared. A further noteworthy property is the wide variations possible even in polymers of fixed monomer composition due to differences in overall molecular weight and molecular weight scatter. Changes in the stabiliser system may also cause major differences in application properties, e.g. due to viscosity profiles.

Almost all acrylic monomers may be polymerised very readily in emulsion form, and in general they copolymerise readily with each other; this applies to derivatives of methacrylic acid as well as acrylic acid. It is possible to make homopolymers of the various acrylic esters, but in practice there are only a limited number in production, most of the products of commerce being copolymers. In many cases these are highly specialised products, often based on heterogeneous particles (see Chapter 4, Section 7). Polybutyl methacrylate has been available in latex form, optionally slightly plasticised, for specialised adhesives or coatings. Likewise polyethoxyethyl methacrylate has been available as an article of commerce, giving a T_g of 0 °C, and has been applied in leather finishing.

The formulation of acrylic latices depends on the properties required in specific applications, one of the major ones being film flexibility, which is governed by the T_g. In many cases the nature of compounding operations, especially the properties of a pigmented product, controls the latex formulation. As a general rule, the properties of an acrylic polymer are colligative of the monomers used, although this is not necessarily true of the emulsifier systems. Because of this, the balance of monomers is very often a function of their cost measured against the degree of flexibility or other properties that they impart. Thus methyl methacrylate is often used as the hardening agent with butyl acrylate or 2-ethylhexyl acrylate.

Styrene and less frequently acrylonitrile are used as hardening monomers. Each, however, has some disadvantages. Styrene, unless present in relatively small quantities, gives rise to yellowing of films that are exposed to ultraviolet light. The inclusion of even small quantities of acrylonitrile, which has an efficiency as a hardener about double that of styrene, tends to cause complications with the wetting and adhesion of films. Acrylonitrile, because of its high water solubility (7.9 % at 40 °C), may also give rise to some complications or irregularities during emulsion polymerisation. Its toxic character also mitigates against its use.

Methyl acrylate behaves rather exceptionally in some conditions. The physical properties of films show very high gradation with molecular weight change, the tensile strength of a latex film of molecular weight 2700 being 10.4 kg cm^{-2} with an elongation of film at break of 1600 %, whilst a polymer of molecular weight 81 700 shows a tensile strength of 79.8 kg cm^{-2} and an elongation of 970 % [19]. Methyl acrylate often has the apparent property of increasing elasticity and toughness to a degree disproportionate to its relative flexibility when included in copolymers. An early account of emulsion polymerisation shows that the emulsifier system has a profound effect on the properties of the final polymers [20]. A change to a cationic emulsifier system may also have major effects on properties.

Formulations that produce a satisfactory acrylic latex are different from those required for vinyl acetate. Colloids such as hydroxyethyl cellulose are generally omitted as they tend to produce latices of large particle sizes, 1 μm or above, or else to cause destabilization. The reason usually atributed to this is 'bridging' of particles, caused by grafting on to the cellulose derivative. Polyelectrolytes such as the salts of polymethacrylic acid or copolymers including methacrylic acid are preferred, often with other emulsifiers. Many technical latices are prepared using only anionic or cationic emulsifiers, although nonionic surfactants are often included. Even at 50 % of solids or above, most latices have a low viscosity, with particle sizes of the order of 0.2 μm or even less, giving a characteristic bluish appearance to the latex. Most technical latices contain between 40 and 50 % non-volatile solids.

In order to maintain the stability of the latices, which have a very fine particle size, a relatively high proportion of surfactant, 5–7 %, is normally required. It is obvious that the very high surface area of these fine particle size latices, which increases inversely with the square of the average diameter, will require a greater concentration of surfactant to cover the particle surfaces than latices of coarser particle size. This high concentration of emulsifier does not necessarily affect the water resistance of films from these latices with fine particle size of the order of 0.1 μm.

One important modification of most latex formulations must be mentioned. It is highly desirable to include a small quantity of a polymerisable acid, of the order of 1–2 % in most formulations. This is conveniently acrylic or methacrylic acid, preferably the latter, since it is more compatible with most acrylic esters and its partition coefficient into water is less favourable. The addition of the polymerisable acid is preferably included with the monomer, and not in the water phase during delayed addition reactions.

Other polymerisable acids such as itaconic acid, or even a half-ester of this acid or other dibasic acids, are alternatives. Thus, half-esters of dibasic acids, e.g. maleic acid, have occasionally been quoted in the literature, but there may be stability problems since half-esters often have a strong tendency to disproportionate into the diester and the original acid by a reaction such as

$$2CH_3OOCCH{:}CHCOOH = CH_3OOCCH{:}CHCOOCH_3$$
$$+ HOOCCH{:}CHCOOH$$

The presence of the acid helps stability since the polar groups act to a limited extent as internal emulsifiers. Acids up to 2 % by weight of monomers in copolymers have little or no effect on the water or alkali resistance of resultant films. The alkali resistance of acrylic polymers is of a higher order than with most vinyl ester copolymers. Methacrylic units in polymers are virtually unhydrolysable.

Acid groups in the polymers assist in giving improved freeze–thaw stability to the latices (Chapter 4, Section 3.1). In addition they assist in wetting out pigments [20]. Excess of acid groups, between about 10 and 15 %, depending on molecular weight and the balance of monomers, produces latices that are usually stable on manufacture, but of low pH, <3. These latices tend to be completely soluble in alkalis, including amines, and may be used for special purposes, e.g. when pigmented for high gloss paints. The amine-solubilised products are also useful in obtaining crosslinked products and stoving finishes (Chapter 5 and Chapter 11 in Volume 2).

Other monomers, e.g. glycidyl methacrylate, which impart potential crosslinking properties, are often included in formulations. At a neutral pH, only slight hydrolysis of the epoxide group, of the order of 7 % as determined analytically, occurs. It should be noted that most commercial latices claimed to have crosslinking properties do not have their formulations fully disclosed.

During a polymerisation process with mixed monomers, such as one including methacrylic acid and acrylic ester and styrene, it is wise to check polymerisation throughout the reaction to ensure that it is reasonably even; otherwise the quality of final products may be affected and there will also tend to be irreproducibility of products [21]. This does not apply to special products designed to be of the core–shell type.

3.2 Properties of acrylic resin films

Acrylic polymers in the form of films tend, in general, to have greater elasticity and the resilience than corresponding films of copolymers of vinyl acetate with similar second-order transition points. The films have much greater elongation at break, which tends to make them useful under conditions when considerable strain may occur, such as may be caused by temperature changes and corresponding unequal expansions of the substrate relative to the film. Acrylic polymers also have considerably better pigment binding power than vinyl acetate polymers and copolymers. This is partly because of the greater elasticity of the acrylic polymers and also because of the much smaller particle sizes. The very high elasticity makes acrylic polymers suitable for applications where high flexing is normal, as in leather coatings.

General tests on films from acrylic latices have shown that resistance to blanching is normally very good, especially having regard to the level of stabilisers present. The rate of water extraction seems exceptionally low, compared with films from a styrene/butadiene latex with similar general characteristics [22]. The resistance to ultraviolet light of films from acrylic and copolymer latices is generally good, with the proviso that the styrene content is not excessive. Certain frictional properties of the films may have special application in textile or leather finishes.

The much greater resistance of acrylic ester polymers to hydrolysis compared with the corresponding vinyl esters, with which they may be isomeric, should be noted, e.g. methyl acrylate and vinyl acetate. This is probably because the carbonyl group C:O, through which attack by alkali occurs, is attached to the chain in an acrylic polymer, and thus enjoys some steric protection. This is augmented in polymers including methyl methacrylate where the angular methyl group gives further steric protection.

Many of the technical latices containing groups capable of potential crosslinking include the methylolamide group. $CONHCH_2OH$ (Chapter 5, Section 8). This group is unique in that polymers prepared with it tend to crosslink spontaneously at slightly acid pH levels, but at neutral pH they are stable in latex form. Details of other groups capable of crosslinking, which are given in Chapter 5, may also be present, but there are few to which the term 'spontaneous' applies on filming, and there are a limited number in commercial production. Carboxyl and hydroxyl groups require an additional group with which they can react, although it is possible to include both in the same formulation without precipitation, e.g. acrylic acid and hydroxyethyl acrylate. In this case there does not seem to be any spontaneous crosslinking of dried films, but they are strengthened by 'hydrogen bonding', which may also reduce solubility. Other reactive groupings, e.g. the epoxide group in glycidyl methacrylate, which can be copolymerised in emulsion form with care (Chapter 5, Section 6), require the addition of a characteristic catalyst. This might have to be included in a 'two-pack' system, combined immediately before application, and may be used in conjunction with a conventional epoxide.

The solvent resistance, with special reference to the more polar solvents, of most acrylic polymers is of a better order than vinyl acetate copolymers in that they are more resistant to the lower alcohols. Polymers based on acrylic esters with an alkyl chain of 4–5 carbon atoms and above tend to be soluble in petroleum-type solvents such as technical 'white spirit'. The addition of polymerisable acids tends to reduce this type of solubility, as does the inclusion of methyl acrylate in a monomer mixture [23, 24].

In almost all countries acrylic monomers are considerably more expensive than vinyl acetate, on the one hand, and butadiene and styrene, on the other. This difference is high and will probably remain so, and whilst it varies between the different countries, is of the order of 150 % to double that of vinyl acetate, and rather higher in comparison with butadiene and styrene, although the last named has increased in price considerably over recent years. In consequence, acrylic latices find application in speciality uses, where their high cost is a secondary consideration, or else in high-quality products, as in some emulsion paints, where the additional advantages that they confer, e.g. high pigment loading, outweigh the additional cost. It should be noted that as polymerisation costs of the various types of monomers are approximately the same, the proportionate difference in cost of the finished latices is not as great.

Many acrylic-type latices are copolymers, with high acid contents, as already mentioned. They may find application as such in, for example, temporary coatings removable by alkali, in the solubilised state as latex stabilisers during polymerisation or as dispersing agents for pigments.

Table 6.7 illustrates a range of Acronal latices recommended for different aspects of the coating industry. This table also gives a further selection of acrylic latices produced by the BASF Company, indicating the wide range of uses of acrylic latices in adhesives. Other latices will be indicated in the successive chapters on specific applications. The range which is suitable for specialised adhesives of all types is indicated in the extreme right-hand column, and many will be quoted in later chapters. It should be noted that Acronal 14D is of a somewhat different type to the others in that it is of very high viscosity and of relatively coarse particle size, being especially suitable for a range of bonding with textiles, paper and leather, and also to laminate 'plastic' films to wood, paper and textiles. Acronal 500D, which has the property of thickening on making it alkaline (see Figure 6.1), has been considered with copolymers of vinyl acetate.

Note the following on Acronal nomenclature. The monomers are indicated by a letter in front of the numeric code:

A = acrylate or methacrylate homopolymer
S = styrene copolymer
V = vinyl ester copolymer

An S *after* the numeric code indicates that the product is self-crosslinking. The first digit of the numeric code refers to the glass transition temperature

Table 6.7 Acronal latices (BASF)

Product	Solids (%)	pH	Viscosity at 23 °C (mPa s)	Shear rate (s^{-1})	Density (g cm^{-2})	Particle size (μm)	Minimum filming temperature (°C)	Elongation at break (%)	Glass transition temperature (°C)	Properties and applications
Acrylic latices for coatings										
Acronal 18D	50	7.5–9	500–1300	100	1.04	0.1	13	300		All purpose, weather resistant
Acronal A509	50	7.5–9	200–700	250	1.04	0.1	5	700		Weather resistant
Acronal A603	50	7.5–8.5	30–200	100	1.05	0.2	16	150		Self-crosslinking, wet adhesion
Acronal A200	70	6.5–8	150–500	100	1.04	0.3	<1	1,200	−43	low odour, flooring adhesive
Acrylic–styrene latices										
Acronal S456	57	7.0–8.5	140–200	250	1.04	0.2	<1	>2,500	−6	very flexible, mortar modification
Acronal 290D	50	7.5–9	700–1500	100	1.04	0.1	20	500		All purpose
Acronal S361	50	8.5–10.5	400–1000	100	1.08	0.2	1	700	−25	Self-crosslinking
Acronal S563	50	6.0–7.5	100–450	250						
Acronal 567D	50	7.5–9	700–1700	100	1.03	0.1	1	2000	−5	Flexible at low temperature
Acrylic latices for adhesives										
Acronal A310S	55	4.5–6.5	30–130	250					−20	Laminating adhesives
Acronal S400	57	7–8.5	140–200	250					−3	Building adhesives
Acronal S300	50	8.0–9.0	50–300	25	1.03		<1	1,000	−22	Special coatings
Acronal V271	65	4–5	100–400	25					−40	Building adhesives, sealants
Acronal 4D	50	6–7.5	15–40	250					−40	Pressure-sensitive adhesives
Acronal 14D	55	6.5–8.5	4300–6000	25					+68	Laminating, packaging adhesives
Acronal 50D	50	3–5	20–40	250				53	−45	Pressure-sensitive adhesives
Acronal 81D	60	4.5–5.5	1000–1800	25					−53	Building adhesives, sealants
Acronal 500D	50	3.5–4.7	20–40	250					−45	Pressure-sensitive adhesives
Acronal A310S	55	4.5–6.5	30–130	250					−18	Paper-converting adhesives
Acronal V205	69	3.5–5	800–1600	250					−21	Laminating adhesives
									−45	Pressure-sensitive adhesives

T_g, which also gives an indication of the film hardness:

1 = below − 45 °C   6 = +6 to + 15 °C
2 = −45 to − 25 °C   7 = +16 to + 25 °C
3 = −25 to − 16 °C   8 = +26 to + 45 °C
4 = −15 to − 6 °C   9 = higher than + 45 °C

4 VINYL HALIDE POLYMERS

4.1 Polyvinyl chloride

Vinyl chloride is a low-cost monomer, and the excellent toughness, solvent resistance, water resistance, electrical insulation properties and in most conditions fire resistance make it a major article of the 'plastics' industry as distinct from the auxiliary application uses of most of the other polymers that have been considered. Polyvinyl chloride in sheet form is a tough horny mass of specific gravity 1.4. It cannot be dispersed in plasticisers in the cold. This property has been used to technical advantage, since the paste which is formed in plasticisers thickens very slowly, but can be gelatinized on strong heating.

A stabiliser must be included with polyvinyl chloride, as with most other halogenated addition polymers, to prevent decomposition by heat and light. The reaction under these conditions may be simply expressed as the loss of hydrogen chloride to produce a unit formulation of the type —CH$_2$CHClCH:CHCH$_2$CHCl—. The double bonds have an activating effect on the adjacent bonds and thus cause a 'zipper mechanism' to operate. The polymer then takes the form of a number of conjugated bonds in random distribution amongst vinyl chloride units, depending on the points of attack. The conjugated system induces colour formation and other undesirable features such as brittleness.

The principal stabilisers include metal salts such as lead stearate, cadmium and barium salts, various compounds including the epoxide $\overset{O}{\underset{CH-CH}{/\backslash}}$ group, such as, for example, epoxidized soya bean oil, and some relatively low molecular weight epoxide resins, as well as organic tin compounds, e.g. tin dibutyl maleate. A number of references consider stabilisers in further detail, and many of them used in conjunction have a synergistic effect. The stabilisers may be added to the compounded polyvinyl chloride latices, and details may be found in the various applications.

Many technical products are available in the form of copolymers, which make plasticisation easier, with a lower plasticiser requirement. Most of these copolymers contain about 30 % of vinyl acetate, and have been known as basic materials for the fabrication of long-playing (LP) record, although this, now obsolescent, application is not based on latex application.

A number of copolymers of vinyl chloride with other ester monomers, such as the maleates and fumarates, have been described, and earlier products prepared in Germany contained acrylic esters in which the acrylic esters were added in stages to ensure more even copolymerisation. Copolymers of vinyl chloride with a minor proportion of vinylidene chloride are also known. The problems of ensuring even copolymerisation are considerable as the reactivity ratios vary markedly. Thus styrene copolymerises with only minor proportions of vinyl chloride, whilst with vinyl acetate r_1(vinyl chloride) = 1.68 and r_2(vinyl acetate) = 0.23, thus indicating that vinyl chloride enters the polymer at a faster rate. However, with methyl acrylate the reactivities are r_1(vinyl chloride) = 0.083 and r_2(methyl acrylate) = 9.0 [25].

4.1.1 Compounding of polyvinyl chloride and copolymer latices

Latices, if supplied unplasticised, often require postplasticisation. Phthalate esters are amongst the most popular, and the cheapest for this purpose. Possibly 2-ethylhexyl phthalate is the most used, but allied esters are used as alternatives, including di-iso-octyl phthalate, butyl benzyl phthalate and several technical mixtures. Small quantities of tritolyl phosphate may be included if special fire resistance is required. If low-temperature flexing properties are required, the adipic acid esters are recommended, including 2-ethylhexyl adipate, di-isodecyl adipate and the corresponding n-alkyl esters. On the other hand, if absence of migration is the essential property desired, together with good electrical insulation properties and complete non-volatility, as with cables, one of the polyester types of plasticiser should be included. Preplasticised latices have some advantages.

A typical recommendation for adding plasticiser is to mix it with oleic acid, add an appropriate amount of ammonia to the latex and mix the two slowly with vigorous stirring. It should be noted that the pH of a polyvinyl chloride latex is usually about 9.5 or upwards, as this aids the stability of the polymer to dehydrochlorination. However, excessive shear may tend to cause mechanical instability. Most latices of this type are prepared in the absence of a water-soluble polymer as stabiliser, and they have in consequence low viscosity and fine particle size, of the order of 0.2 µm or less.

As a modification it is sometimes advisable to prepare an independent plasticiser emulsion, which may be with ammonium oleate or with a non-ionic polyoxyethylene surfactant. About 2–5 % of either oleic acid or a non-ionic surfactant is suggested as a suitable quantity for preparing the plasticised emulsion by either of the above methods. It is sometimes possible to stir a plasticiser directly into the latex if the surplus surfactant is sufficient to give adequate emulsification, The problems of film formation have already been discussed (Chapter 4, Section 5.2).

A number of thickeners may be used with the latex. They will of course vary with the method of preparation of the latex, and the manufacturer's literature

Vinyl halide polymers 373

should be consulted in each specific case. Sodium alginate and various soluble polyacrylate salts have been suggested, including the ammonium salt of polymethacrylic acid. With the latter, it is suggested that part of the plasticised latex is stirred into the polymethacrylate solution until the mass is homogeneous, and the rest of the latex is added gradually. The thickener is 0.5–3 % by weight of the plasticised polymer. It is possible to increase the viscosity of a low-viscosity technical latex to as high as 270 P at 1 % of thickening agents per 100 part of polyvinyl chloride, the viscosity increasing over about 7 days. The viscosity then falls slightly.

Pigmentation may be achieved by the conventional method of grinding the pigments with water and a wetting agent in a ball mill or triple roll mill, and then blending with the plasticised latex. Whilst a range of pigments and fillers may be added, depending on the end use, special mention may be made of titanium dioxide, which is added to give density to coloured film, with talc, mica or china clay as suitable fillers. Talc decreases tack, whilst china clay gives opacity. Colloidal carbon blacks, pure Middle Chrome GNS and iron oxides are also mentioned as suitable pigments if coloured compositions are required. The pigment paste is added after plasticisation and thickening. Figures 6.1 and 6.2 show the effect of filler loading on tensile strength and elongation and also the effect of passing the critical pigment volume concentration when elongation falls abruptly to almost zero. Silica or diatomaceous earth may be included with pigments for flatting purposes.

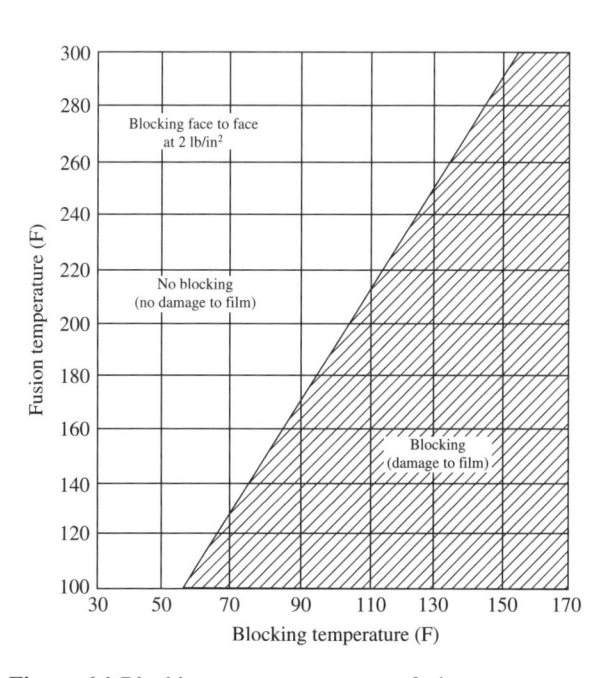

Figure 6.1 Blocking temperature versus fusion temperature.

Most technical emulsions are already stabilised sufficiently for normal application. Under severe conditions of processing the addition of extra stabiliser is advised. Some stabilisers, such as calcium stearate, may be dispersed with the pigment paste. A 1 % addition on polymer is suggested. Other stabilisers such as the epoxide type, which are often liquid, may be blended with the plasticiser, as may a miscellaneous selection of stabilisers including organotin derivatives, e.g. tin dibutyl maleate, and some organic phosphites. The epoxide stabilisers have some plasticising action in their own right.

The surface tack on films derived from plasticised polyvinyl chloride and allied latices may be troublesome in practice, causing undesirable blocking or sticking. Whilst increasing the filler may partially overcome the problem, alternative methods are sometimes considered. One method of removing tack is to add a carnauba or candellila wax in the proportion of three to ten parts per hundred parts of polymer. A suggested formulation for the emulsification of the wax is

Wax (carnauba or candellila)	20 parts by weight
Oleic acid	5 parts by weight
Ammonia solution (28 %)	1.4 parts by weight
Water	73.6 parts by weight

Ammonium oleate is often used in these auxiliary mixes since on drying only oleic acid remains. Thus 'wetting back' problems, caused by the addition of more surfactant to the latex, are avoided. On drying, the wax forms a protective coating on the surface which reduces tack. Surface tack may also be reduced by using the highest possible gelation temperature. (Figure 6.1). A further method is to spray an unplasticised latex of polymethyl methacrylate or polystyrene on to the surface of the film. These dry to powders and act by absorbing surface plasticiser.

4.2 Polymers and copolymers of vinylidene chloride

Vinylidene chloride $CH_2:CCl_2$ is an alternative halogenated vinyl monomer available commercially. Although the monomer is liquid at ambient temperatures (b.p. 31.7 °C), which may ease handling problems, it tends to polymerise spontaneously at ambient temperature, and is stabilised with phenol or methyl hydroquinone. With the former, the monomer should be distilled before use. Peroxides are readily formed by the monomer in contact with oxygen, and it is advised that the monomer be washed with a 5 % sodium bisulfite solution before polymerisation is commenced.

The monomer is rarely homopolymerised, since polyvinylidene chloride is insoluble and rather intractable, requiring the addition of considerable amounts of stabiliser to avoid decomposition. Most of the stabilisers for vinyl chloride are suitable. Most emulsion polymerisation processes describe the preparation of copolymers, which include copolymers with vinyl chloride. Vinyl chloride and vinylidene chloride when copolymerised seem to have a mutual plasticising effect on each other, probably be reduction of crystallinity. The two monomers copolymerise more slowly than either separately and vinylidene chloride enters the copolymer more rapidly than vinyl chloride [26, 27]. Most well-known copolymers are combined with acrylic esters. Emulsion polymerisation is carried out preferably using hydrogen peroxide as initiator or a low-temperature redox system. The use of a trace of ferrous salt, often included in these systems, should be avoided, since discoloration of the polymer is catalysed by iron salts. Many latices are prepared solely to obtain the dry polymer by subsequent precipitation.

A formulation for the polymerisation of a vinylidene chloride copolymer is given in Chapter 3, Table 3.24. A number of other references are available [28–32]. Copolymers are often referred to as PVDC.

Technical latices have fairly good chemical and mechanical stability, and it is advisable that exposure to light should be avoided. Likewise, contact with most metals should be avoided to remove the risk of discoloration by reaction with chlorine in the polymer. Conventional thickeners such as cellulose derivatives and polyacrylate or methacrylate salts may be employed. Polyvinyl alcohol has also been recommended, and a relatively low viscosity polymer, about 88 % hydrolysed polyvinyl acetate, is advised for some vinylidene chloride copolymer latices.

Because they are fundamentally of low viscosity, it is advised that most latices should not be subject to excessive mechanical shear. For pumping a diaphragm pump is preferred to a piston type. Although these latices have some resistance to freezing and thawing, it is inadvisable to submit them to these conditions. There is a tendency for electrolytes to cause coagulation where only anionic emulsifiers have been included, but this may be circumvented by the addition of non-ionic emulsifiers. Antifoaming agents may be added as required.

Most of the technical latices are film forming at about 10–20 °C, but if low-temperature flexibility is desired, an independent plasticiser emulsion may be prepared. A suggested formulation is a 50 % dispersion of dibutyl sebacate, using 2.5 % of sodium lauryl sulfate as emulsifier. The plasticiser emulsion and the latex are blended.

A copolymer of an acrylic ester and vinylidene chloride will enable films to be formed even at 0 °C, although the characteristic properties of vinylidene chloride copolymers are retained. It is suggested that whilst this polymer is amorphous in the emulsion, on drying it develops a high degree of crystallinity [28]. The film remains flexible, but has a hardness adequate to promote a good resistance to blocking. X-ray diffraction tests have verified the development of crystallinity.

Between the second-order transition point T_g and a point designated as T_i, the point of inflection on the curve of torsional modulus versus temperature,

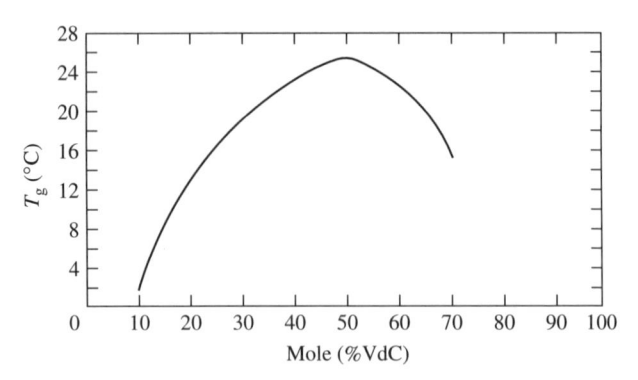

Figure 6.2 Glass transition temperature in relation to copolymer composition.

the polymer passes through a state in which it is tough, yet flexible [29]. This point T_i is usually at about 300 kg cm^{-2} for amorphous polymers. The torsion modulus of films from vinylidene chloride–acrylic ester copolymers does not have a sharp drop in modulus below the brittle point, but tends to level off at about 500 kg cm^{-2}. This degree of hardness promotes good blocking resistance, but retains adequate flexibility. Films of these copolymers remain amorphous until 115 °C, when they melt as an elastic amorphous mass, thus permitting ready heat sealing.

Amongst other specific advantages claimed for latices of vinylidene chloride with acrylic esters is better light stability than unmodified polyvinylidene chloride. Wet and dry abrasion tests have given excellent results on the Gardner linear scrub tester. In the case of wet abrasion, it is immaterial whether the film is water wet, wetted with a detergent solution or wetted with ethanol.

Under certain conditions, copolymers of acrylic esters and vinylidene chloride show a higher T_g than a corresponding polymer of either would show individually. Polyvinylidene chloride has a very low T_g in the pure state, −18 °C, although its normal physical properties do not reflect this because of the presence of a crystalline structure which may be up to 25 %. A corresponding value for polyethyl acrylate is −22 °C. If a molecular model of a 1 : 1 alternating copolymer is prepared, it will be found the chlorine and ester side groups, because of their size, prevent free rotation about the C—C single bond, although it is possible in either homopolymer separately. As a result, a copolymer of this composition has a higher T_g than either homopolymer, this being about 28 °C. The corresponding minimum film-forming temperature (MFT) also reaches a maximum, although at a considerably higher vinylidene chloride content, because of the nature of the monomer, which tends to promote crystallinity (Figures 6.2 and 6.3).

The degradation of vinylidene chloride copolymers by light follows a similar dehydrochlorination to that of vinyl chloride polymers, producing a diene conjugated structure which is a chromophore, giving a yellow-brown

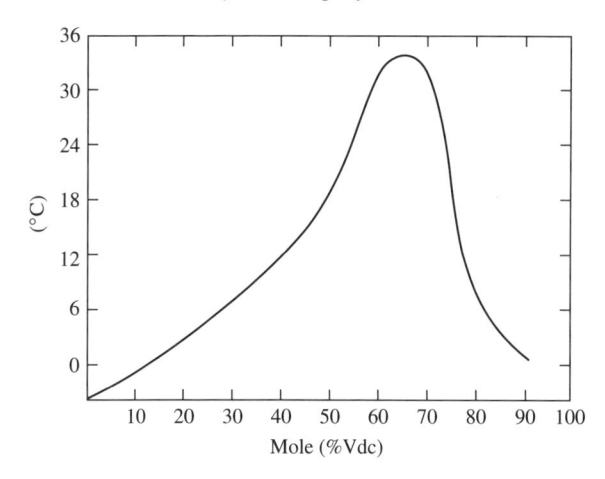

Figure 6.3 Minimum film-forming temperature in relation to copolymer composition. (From Scott, Bader & Company).

coloration. Even though a 50 % copolymer is not in practice an ideal structure, there are rarely more than four units of vinylidene chloride in a block, this being insufficient to cause coloration on dehydrochlorination.

Specific properties which render copolymers of vinylidene chloride in latex form attractive to users include resistance to flame spread, low water vapour transmission (MVT) rates and low permeability to gases. These rates increase slightly when the polymer is plasticised. If high temperatures must be withstood for fairly prolonged periods, a copolymer is best employed. This characteristic property of polyvinylidene chloride copolymer films, the high resistance to both water vapour and oxygen, can be controlled by the actual content of vinylidene chloride in the copolymers. One commercial latex has a water vapour transmission rate of g per 100 in^2 per 24 h at 100 F and at ambient temperature of 2.00. If this product, characterized as a 'medium' content vinylidene chloride copolymer, is replaced by a high (>90 %) copolymer, the oxygen permeability can drop to as low as 0.20, whilst the heat seal temperature rises to 135 °C. A very fine particle size latex will tend to have a high surface tension, indicating that emulsifier has been completely absorbed on to the particles.

In general the use of elevated temperatures tends to improve the physical properties such as the tear strength and tensile strength of films. Drying is usually performed by preheating with infrared heaters before high-velocity hot air drying. This technique avoids blister formation.

The chemical resistance of films from these latices is generally good, except when modified by plasticisers. Only concentrated nitric acid, glacial acetic acid or aniline seem to have the property of complete attack or solution, whilst sodium hydroxide causes some straining and embrittlement, depending on the concentration. In general, the coatings are resistant to all alcohols and polyols. Inorganic acids, standard solutions of metal salts including permanganates, aliphatic hydrocarbons, saturated ring hydrocarbons such as cyclohexane, and solutions of detergents. There will be swelling in aqueous amines, ketones, chlorinated and nitrated aliphatic hydrocarbons. There may be complete solution in pyridine, cyclohexanone, dimethylformamide and tetrahydrofuran.

The low gas transmission rate from films from vinylidene chloride copolymers is of the order of only 1 % of that of an uncoated polyethylene film, and these films are exceptionally useful in food packaging to avoid loss of flavours, and to prevent odours. This high impermeability applies to oxygen, carbon dioxide, liquid water and water vapour. It is caused by natural crystallisation within the finished film, and full barrier properties are normally achieved within 10–15 days at ambient temperatures. The impermeability to oxygen renders polyvinylidene chloride copolymers useful in packaging foodstuffs, preventing rancidity. There is also very good barrier properties against aromas and smells which can alter the taste and quality of packed foodstuffs. Films are also useful in providing a greaseproof layer for fatty foods.

Polyvinylidene chloride films form wafer thin transparent coatings on materials such as cellophane, polyethylene, aluminium, cardboard and plastics giving the desired protective properties without effecting the mechanical properties of the carrier material. The coating of aluminium should be avoided if the coated material is retained in a damp atmosphere as there are traces of hydrochloric acid release at elevated temperatures. Film forming latices may be applied directly to the base packaging material by dip, spray or roller application. The latices may be used as a seal to printed flexible packaging. They are heat sealable, the exact temperature depending on the formulation, and is usually in the range of 130–150°, but may be lower in some formulations.

A series of articles discusses aspects of PVDC coatings (33–6).

Formerly external plasticisation was used for polyvinylidene chloride latices, but this is now obsolete, and in general 'internal' plasticisation is preferred. If it is required to thicken a latex, a salt of a di-isobutylene–maleic anhydride copolymer is one possibility. The additive raises the blocking temperature. Polyvinyl alcohol may also be used as a thickener where the latex is applied at an acid pH, but this results in some cases in a decrease in block resistance of the finished coating.

These latices are useful and may be suitable for coating paper and paperboard (see also Chapter 13 in Volume 3). It should be noted that there may be traces of monomers in commercial products, and although manufacturers keep this at a very low level, it is recommended that latices be used in well-ventilated conditions.

Tables 6.8 and 6.9 indicate the typical properties of some vinylidene chloride copolymer latices which are available commercially. These are the Polidene series of Scott, Bader & Company, and the Diofan® series of Solvin.

Table 6.8 Properties of Polidene® latices

Reference	33-048	33-004	33-065	33-075	33-038
Solids content (%)	55	50	60	55	45
pH	4.0–5.0	3.0–4.5	2–3	3.0–5.0	3.0–4.5
Viscosity at 25 °C (P)	0.5–2.5	0.05–0.3	0.5–1.0	0.2–1.0	0.05–0.30
Particle size (μm)	0.25	0.25	0.20	0.30	0.25
Emulsifiers	Anionic	Non-anionic	Anionic	Non-anionic	Anionic
Specific gravity at 25 °C	1.12	1.22	1.28(20 °C)	1.17	1.16
Minimum film-forming temperature	<2	4	10–15	18	20
Freeze–thaw stable?	Unstable	Unstable	Unstable	Unstable	Unstable
Special type	No	No	Carboxylated	No	Curable
Applications	Textiles	Flexibility	Metal primers rust conversion coats	Gloss paint	Non-wovens

Table 6.9 Properties of Diofan® latices (Solvin, Belgium)

Diofan number	193D	232	A716	A736
Solids content (%)	55	45	59	58.65
pH	2–4	1–2	2–3	2–3
Specific gravity (g cm^{-3})	1.29	1.22	1.32	1.32
Oxygen permeability (cc.μm/d.m^2.b)	54	12	36	44
Water vapour permeability (g.μm/d.m^2)	25	11	21	24
Surface tension (mNm^{-1})	28	45	53	45
Minimum film-forming temperature (°C)	16	14	15	17
Film density	1.67	1.67	1.67	1.67
Applications	Barrier films for paper, polymers and metal foil. The A series feature transparency and low foam.			

It is noted that in general the pH of these latices falls slowly, and adjustment should be made with ammonia solution, but the pH should not be raised above 4.5 in most cases (or occasionally 5.0), otherwise irreversible discoloration may occur. Care should be taken that water used on any dilution or addition is completely free from iron, in order to avoid discoloration. It is noted that some of the latices are indicated as self-cure and have excellent water resistance. All give degrees of fire resistance.

The Diofan series shown in Table 6.9 have typical crystallisation rates and oxygen permeability. Also shown is the effect of temperature on oxygen permeability on the derived coatings of these copolymers.

Under special conditions it is possible to obtain a film directly from a composition containing a vinyl chloride–vinylidene chloride copolymer and emulsified alkyds, which have been chosen to give compatibility.

4.3 Fluorocarbon latices

Polymers of fluorinated unsaturated hydrocarbons are well known; e.g. polytetrafluoroethylene, popularly referred to as PTFE, is derived from C_2F_4, the analogue of ethylene. A number of specifications describe the direct emulsion polymerisation of this monomer [37–39]. An unusual emulsification is performed by using about 0.5 % on water weight of 1,4,4,6,7,7-hexachlorobicyclo(2,2,1)-5-heptene 2,3-dicarboxylic acid or a salt thereof [40]. This acid will be best recognised as the condensation product of hexachlorocyclopentadiene and maleic anhydride and may be written as

$$\begin{array}{c} \text{Cl} \\ \diagup \text{C} \diagdown \\ \text{ClC} \quad | \quad \text{CHCOOH} \\ \| \text{Cl-C-Cl} | \\ \text{ClC} \diagdown \quad | \quad \diagup \text{CHCOOH} \\ \text{C} \\ \text{Cl} \end{array}$$

For polymerisation about 0.1 % of disuccinic acid peroxide is used as initiator on water weight, and the addition of 1–5 % of a light hydrocarbon oil is advised for extra stability. The autoclave must be filled under pressure and about 5 ppm by weight of water of iron powder is an advised addition.

The polymerisation, which takes place at about 85 °C for 1–20 h, is remarkable in that the mineral oil does not emulsify and may be separated directly from the final latex. The solid content is lower than with most other latices, from 10 to 35 %. These products are mainly prepared for coagulation and isolation of the polymer rather than for direct use as latices.

The general physical properties of PTFE cannot be given in detail here, except to quote the high degree of crystallinity of the polymer, its high melting point, 324 °C, which increases with pressure, and its great chemical inertness. A useful review is available [41].

A number of technical latices of PTFE manufactured in Great Britain by ICI are available. One of these, 'Fluon' Dispersion GP1A, contains 60 % by weight of polymer and 4 % by weight of an anionic surfactant, the particle size average diameter being 0.15–0.2 μm and the viscosity 5 cP. The polymer is similar to moulding and extrusion grades, with a low coefficient of friction and outstanding electrical properties. Its applications include the impregnation of glass cloth, graphite and porous metals. Impregnated glass cloth is useful in electrical insulation, as in printed circuit boards and in chemical plant. Decorative glass fibre curtains may be treated to improve their drape. An unusual application is as an alternative to epoxies in impregnating graphite heat exchangers, thereby raising both their efficiency and working temperature. In processing, the wetting agent should be removed by heating to 380–400 °C, and if necessary sintering at these temperatures.

An acid-stable aqueous latex of PTFE is obtainable as 'Fluon GP2', and may be applied under acid conditions to give a primer coat on metal, to be followed by GP1 as a finishing coat. A further PTFE latex, 'Fluon MM1', contains an unspecified film-forming ingredient which reduces the sintering temperature to 90 °C, enabling paper, wood and carbon steel to be coated with the polymer. No preliminary roughing of a substrate is needed, but the maximum service temperature is much lower.

A few typical patents are included herewith. It is uncertain whether the quoted latices are produced in commercial quantities.

Copolymers of ethylene and perfluoroethylene have been prepared in emulsion in the presence of the fluorinated hydrocarbon $CF_3C_3F_6CH_2CH_3$ with ammonium perfluoro-octanoate as surfactant. The polymer is isolated as a powder [42]. A latex based on a fluorinated (meth)acrylate with formula $CF_3(CF_2)_nC_2H_4OCOCR=CH_2$ (R = H or CH_3, $n = 2$–13), prepared with a fluorinated surfactant, is suitable for coating porous surfaces, rendering them oil repellant [43]. Latices of copolymers of hexafluoropropylene $CF_3CF:CF_2$ and vinylidene fluoride $CH_2:CF_2$ have been disclosed, and also latices containing perfluorinated(methyl vinyl ether) $CF_3OCF:CF_2$ [44]. Tetrafluoropolymers have been prepared in emulsion with unsaturated perfluorinated monomers including carboxylic end groups [45].

5 STYRENE POLYMERS AND COPOLYMERS

5.1 Polystyrene

Styrene has been one of the cheapest monomers, although subject to considerable price increases in the 1980s and early 1990s. Details of the monomer are given in (Chapter 1, Section 6.1). The relatively high boiling point (145 °C at 760 mm) makes handling and emulsion polymerisation easy, but may cause some difficulty in removing the last traces of unpolymerised monomer from a latex. Vacuum or other types of strippers may have to be employed, although difficulties are sometimes caused by foaming. Polystyrene in bulk is a hard and almost colourless polymer, with a T_g about 100 °C, varying to some extent with molecular weight, which for emulsion polymers is usually very high. The principal use, however, is in bulk for the manufacture by various types of moulding of a large variety of 'plastic' goods, such as cups and ornaments.

As expected, emulsion polymers when dried give friable films, even at elevated temperatures. Polystyrene latices have been use in the unplasticised state for a number of purposes, such as spraying on to coated polyvinyl chloride to avoid surface tack. These latices have been included in compositions for the impregnation of leather (Chapter 14 in Volume 3).

If continuous films are required, plasticisers are added to the latex, either before or after the polymerisation process. Polystyrene, however, is not an easy polymer to plasticise. Dibutyl phthalate may be used, but is effective only within comparatively narrow limits, the optimum being about 28–32 % on total weight. Insufficient plasticiser tends to produce films with a 'cheesy' appearance, whilst excess produces a soft product of little film strength.

The speed at which different plasticisers enter polystyrene latex particles varies markedly. Some hydrocarbons derived from various petroleum extracts have a coalescing time at ambient temperature of over 6 h, and do not give a continuous compatible film on drying the latex at ambient temperatures. This disadvantage may be overcome by blending this type of plasticiser with a phthalate ester, a 50 % dibutyl phthalate–50 % aromatic petroleum extract mixture being quite suitable. A highly compounded product, intended as an emulsion polish, contains small quantities of tributoxyethyl phosphate as plasticiser (see Chapter 15 in Volume 3).

If continuous films are required, styrene copolymers are normally necessary. In the case of acrylic ester copolymers, the styrene content is usually below 50 % by weight, and these copolymers are described in conjunction with polyacrylates (Section 3).

A highly compounded latex, intended as an emulsion polish, contains small quantities of tributoxyethyl phosphate as plasticiser (see Chapter 15 in Volume 3, Section 2.5). An unusual type of latex in which dibutyl phthalate is the plasticiser has been described in Chapter 3, Section 5.1.2. The key emulsifier is the salt of a vinyl acetate–maleic anhydride copolymer, and this type of latex plasticises with apparent ease to give a flexible film that is less sensitive to plasticiser variation than most other types.

There has been more theoretical work on the polymerisation of styrene than with any other monomer. Detailed descriptions of the chemistry of emulsion polymerisation of styrene and general properties of the polymer have been made in earlier volumes, which are still very useful [46, 47].

Styrene polymerises comparatively easily, a special feature of the monomer being that it forms radicals which are relatively stable due to resonance stabilization. Thus, during polymerisation, it is desirable to guard against oxygen inhibition, which can be ignored with vinyl acetate, the radicals of which are highly unstable. Styrene does not form copolymers with vinyl acetate and virtually inhibits polymerisation of the latter.

Conventional methods of emulsion polymerisation may be operated with styrene, and a very wide range of temperature has been the subject of much theoretical study. This also applies to emulsion polymerisation, especially with regard to the micellar theory of Harkins and its expansion by Smith and Ewart (Chapter 3, Section 8.2). Various delayed addition techniques are advised in order to obtain latices of fine particle size.

One of the earliest techniques for emulsion polymerisation was the use of a soap in the form of sodium oleate. An improved type of formulation, producing a product of submicroscopic size, is shown in Table 6.7 (Chapter 3, Table 3.18). Table 3.19 in Chapter 3 gives formulations using a vinyl acetate–maleic anhydride copolymer salt as emulsifier [48]. This copolymer is worthy of further development as an emulsifier. Products with other formulations are available technically, the major application, as already indicated, being in emulsion polishes. An interesting modification of this type is the inclusion of an alkaline shellac in the formulation, the resultant polymer being almost certainly a polystyrene grafted on to a shellac base. A number of formulations will be given in Chapter 15 in Volume 3, as well as descriptions of some technical styrene latices.

Examples of the use of sulfonate emulsifiers have been given in Chapter 3, and a wide variety of surfactants probably give latices of reasonable stability at solids content from 33 to 50 % with stabiliser concentrations from 4 % to about 8 %, the particle size varying with both the amount of stabiliser, the method of polymerisation, whether all the monomer is added initially, gradually or by a continuous process, and the temperature. In most cases only the products of larger particle size are claimed to be freeze–thaw stable, although general methods of achieving this are available (Chapter 4, Section 3.1). Sodium dodecyldiphenyl ether sulfonate is a suitable emulsifier.

Most styrene latices may be compounded as previously described (Chapter 4, Section 6), and a variety of possible applications have been suggested, especially for the plasticised product. These include coatings for wood, plaster, cement, hardboard and other non-metallic surfaces. As a general rule polyelectrolyte types of thickener are advised in preference to cellulosics. Suitable types include polyacrylate and polymethacrylate salts, and, as already mentioned, salts of styrene–maleic anhydride copolymers.

Other possible applications are as a size for fibreglass prior to coating with polyester (Chapter 17 in Volume 3), as an additive to cement (Chapter 15 in Volume 3) and for the impregnation of leather (Chapter 14 in Volume 3). It is problematic as to whether latices for any of the suggested applications are manufactured currently in united kingdom, with the probable exception of the manufacture of emulsion polishes and possibly some 'in-house' products.

5.2 Styrene–butadiene copolymers

The major part of the production of styrene–butadiene copolymers concerns, of course, the production of synthetic rubbers, with which we are primarily not concerned. The types of interest here are those in which butadiene has been reduced to between 60 and 30 % of total monomers, examples of which have already been given in (Chapter 3, Table 3.20). These latices are fully polymerised with a solids content of about 40–55 %. The films are generally extensible and may be slightly tacky with higher levels of butadiene content. The products are almost invariably of low viscosity, of the order of 20–100 cP, and of fine particle size, about 0.2 μm in diameter.

Butadiene enters the copolymers faster than styrene, the r_1 (butadiene):r_2 ratios being 1.4 : 0.78. In consequence, there is a tendency for the copolymers to become richer in styrene towards the end of the reaction. There may also be a tendency for impurities to arise due to side reactions, such as traces of 1-vinyl 2-cyclohexene $CH_2{:}CHCHCH_2CH_2CHCH{:}CH_2$, formed by dimerisation of butadiene. Because of the reactive nature of butadiene, which has two double bonds, films from the copolymers always have a 'gel content' expressed as the amount of insoluble matter in acetone, and which is caused by a limited amount of crosslinking. Other monomers, in particular acrylonitrile, may be included or substituted for styrene.

Another variation of the monomer composition is the inclusion of a small amount of an acid, e.g. itaconic acid or methacrylic acid. Apart from questions of stability, this inclusion of acid produces 'self-cure' characteristics when the film is dried. The latices containing acid are of interest in backings for carpets (Chapter 12 in Volume 3). It is also possible to produce latices containing 2-vinylpyridine or other vinylpyridines as a third monomer. These have application for uses in the adhesion of tyre cord to rubber (Chapter 8, Section 6.4).

Whilst films of this type generally have high water resistance and high chemical resistance, their behaviour in the presence of strong light is relatively poor, both due to the cumulative effect of ultraviolet radiation, which tends to attack both moieties in the copolymer, and also to a slow oxidation caused by the vulnerability of the double bond which remains in polymerised butadiene. There has been a tendency, therefore, to replace the butadiene–styrene latices in such applications as emulsion paints, especially for exterior use. In other

Table 6.10 Styene–butadiene carboxylated copolymer (Baystal® P7105)

		Test method
Solids content	50 %	DIN 53 189
pH	6.5	DIN 53 785
Viscosity (Brookfield RVT-100, spindle 2,100 rpm)	100 mPa.s	ISO 1652
Surface tension	45 mN m^{-1}	DIN 53 593
Glass transition temperature	16 °C	

Application:
This latex gives a medium stiffness polymer, suitable for finishing coated paper. It can be used with all commercially available pigments. It should be added to a previously dispersed pigment with gentle agitation followed by viscosity and pH adjustment as required.

Table 6.11 Styrene–butadiene copolymer (Bunatex® F2420)

Technical information:
  This latex is an aqueous latex of a reinforced styrene–butadiene (S–B) copolymer, with a high solids content, an emulsifier potassium salt of a fatty acid and a particle distribution similar to natural rubber latex.

Characteristic data:

Property	Value	Unit	Method
Solids content	67.0	%	DIN 53 563
pH	11		DIN 53 606
Surface tension	37	mN m^{-1}	DIN 53 593
Viscosity (Brookfield LV No.2, spindle 3,30 rpm) Maximum	2200	mPa s	ISO 1652

Properties:
  The latex is film-forming. Vulcanisation is as in natural rubber. Addition of an anti-oxidant is desirable to prevent autoxidation of films in air.
  Add this before processing.
  The latex is suitable for manufacture of latex foam articles. The vulcanised foam exhibits a favourable hardness-to-density ratio and excellent resilience. The tear strength and elongational break of the vulcanisate of the latex allow great flexibility in mould design.

fields of application such as paper coating, where this consideration does not apply, there is a wide use of these latices.

Two examples of technical latices are included in Tables 6.10 and 6.11. They describe latices available from Polymer Latex GmbH & Co. KG, a joint subsidiary of Degussa AG and Bayer AG (Germany).

6 POLYMERS BASED ON ALIPHATIC HYDROCARBONS

The principal monomer in the class of simple aliphatic hydrocarbons is, of course, ethylene. Polyethylene, or polythene as it is popularly known, has been in production since 1939. The direct production of polyethylene latices has been claimed as early as 1945, and has now been an article of commerce, along with copolymer latices, for about three decades. There is a large number of patents covering the polymerisation and copolymerisation of ethylene, mostly, but not invariably, by the use of high pressure and possibly a solvent to increase the solubility of ethylene.

Investigations on the direct polymerisation of ethylene in emulsion have shown that there is an appreciable density difference in the polymer as the temperature of polymerisation is raised, the densities being about 0.955 at the lowest practicable working temperature and 0.910 at the highest. At the same time, the polymers, which are soft and flexible after low-temperature polymerisation, become tougher and less flexible as the temperature of polymerisation is raised. This seems to be due to an increase in chain branching as the temperature of polymerisation increases. It is to be noted that some of the earlier polyethylene latices were formed by a secondary emulsification of polyethylene, preferably after an oxidation treatment to improve emulsification properties.

One of the principal applications of emulsion polyethylenes is as a component of emulsion polishes. Some details of available polymers will be found in Chapter 15 in Volume 3.

Most latices including ethylene available commercially are copolymers. As in most cases vinyl acetate is the major comonomer, they have been included with vinyl acetate copolymers. In recent years terpolymers including both vinyl acetate and vinyl chloride have been marketed, enabling the relative amount of ethylene in these terpolymers to be increased. Ethylene is possibly the cheapest monomer, but because of its volatility, losses in transit and handling are considerable, and relatively expensive pressure plant is required. To justify the latter, a relatively high proportion of ethylene is desirable in a copolymer. The high efficiency of ethylene as a softening comonomer is due to its low molecular weight, which gives effective internal plasticisation at low proportions in the molecular chain. This may actually be a disadvantage in applications such as emulsion paints where the inclusion of only about 5 % in a vinyl acetate copolymer will give the normal desired flexibility at ambient temperatures. Hence the inclusion of vinyl chloride is desirable to enable the ethylene proportion to be increased. Suggested formulations are given in a specification [49].

Emulsion resins of acid-modified ethylene–vinyl acetate resins, which are ionically crosslinked, have been available in the United States (Elvatex® D Dispersions Du Pont de Nemours). The fine particle size claimed seems to be an indication that they are prepared by direct polymerisation. The polymers have relatively high film-forming temperatures, and still higher temperature at which they attain maximum film strength. Other non-crosslinked latices are available in which ethylene is the principal monomer. They are shown in Table 6.12. Modified products are available containing 25–60 % of wax on solids, thus providing excellent water vapour barriers in film form.

As a whole they have good heat seal properties, excellent high hot tack and can be formulated, especially with the ionomers, into a range of non-blocking, high-performance coatings. They are especially suitable where lightweight coatings are required. The ionomer latices have good adhesion properties to

Table 6.12 Vinyl Acetate–Ethylene copolymers and terpolymers (Wacker, Germany)*

Product[a]	EP1	EP14	EP17	EV2	EZ26	CEF10	CEF19
Solids (%)	50	55	60	50	56	50	47
Emulsifiers[b]	PVAI	PVAI	PVAI	HPV	CD + OS	OS	OS
Viscosity (Brookfield)[c]	9000	5500	3800	11 000	3100	2700	900
Viscosity (mPa s)	±3000	±1500	±1000	±4000	±700	±800	±400
Spindle	5	4	3	5	3	3	3
MFT(°C)	0	+1	+3	−3	+6	+17	+17
pH	4.5	4.5	4.5	6	4.5	6	6
Particle size (μm)	0.5–2	1	0.9	0.5–1	0.6	0.1	0.1
Tensile strength (N mm^{-2})	3	5	6	0.3	6	4	15
Elongation (at break %)	800	700	700	3,000	3,000	550	500
Adhesive, plastic	+	+	+	−	−	+	−
Adhesive, floors, foam, tiles	+	+	+	+	−	−	−
Emulsion paints	−	−	+	−	+	−	−
Paper coating	−	−	−	−	−	−	+
Textile coating	−	−	−	+	−	−	+

[a] Latices commencing with C contain vinyl chloride in addition.
[b] PVA-polyvinyl alcohol, CD-cellulose derivative, OS-surface active substance, HPV-high polymer.
[c] There is a very wide difference in viscosity measurements between the Epprecht rheometer (not shown above) and the Brookfield viscometer, the former giving viscosities that are approximately double that of the Brookfield.

metal foil and paper substrates. The series, in general, finds interest as heat-sealable overprint coatings, adhesion primers, non-skid coatings, varied texile applications and as binders for pigmented coatings.

It has been claimed that polypropylene may be converted directly to a latex. A suitable polypropylene consists of a linear, essentially unbranched, head-to-head polypropylene having a crystalline content, as calculated from density measurements of 50–100 % by weight, the particle size varying from 0.1 to 20 μm. The normal simple procedure for post-emulsification may be applied, the addition of a solvent being optional. Thus 30 parts of water, 10 parts of a 12 % aqueous solution of ammonium polymethacrylate and 100 parts of 70 % crystalline polypropylene are converted directly to a paste which may be diluted with water to form an emulsion. A similar process, using a steam-heated roller mill, is operated by adding 10 % of oleic acid, a non-ionic surfactant, and finally ammonium polymethacrylate [50].

The products may be applied by standard methods, including knife, roller, immersion, spraying and brushing. They are recommended for the production of

---

* Now Air Products Europe.

textile laminates, followed by pressing. They may also be used for impregnating porous materials such as cork, which afterwards may be converted to shaped bodies; the possibility of producing filaments by extrusion is suggested.

Emulsion copolymers, mainly with vinyl acetate, of 1-ethylenic monomers containing longer carbon chains have been claimed in patents. However, only small proportions of these monomers can be included in copolymers because of their allylic character, and it is doubtful if any are included currently in commercially available latices. See also Table 6.4 for additional latices containing ethylene.

7 VINYLPYRROLIDONE COPOLYMERS

Polymers of vinylpyrrolidone are water soluble, but copolymers in emulsion form are available with various other monomers. A series has been available including acrylic esters, styrene and vinyl acetate as comonomers. The exact compositions have not been disclosed, although it is a simple matter to determine them analytically, e.g. using the nitrogen content or by physical methods such as infrared spectroscopy. All are potentially film forming, although in the case of styrene and vinyl acetate comonomers, films are brittle at ambient temperature.

These latices are reported to have good mechanical, freeze–thaw and chemical stability. They have wide compatibility with other latices and good tolerance for pigment and filler loading. The compatibility extends to starches and dextrins. Films are in general colourless, tough and have high cohesive strength. Because of the built-in polarity they tend to give excellent adhesion to a variety of substrates. Inherent hydrophilic characteristics enable polymer films containing a polyvinylpyrrolidone to have good oil and grease resistance and dye receptivity. Films with the lowest vinylpyrrolidone content have the least water sensitivity. Not unnaturally, films from vinyl acetate copolymers are more sensitive to water than those from styrene or acrylic esters.

The films may be crosslinked by heating to 150 °C, preferably in the presence of 1 % of a dibasic acid such as succinic acid or a free radical source such as ammonium persulfate. The latter, however, causes the entire latex to gel after standing for a few days at ambient temperature. The reaction of curing is probably that of the opening of the pyrrolidone ring.

The vinylpyrrolidone copolymers have good adhesion to a wide variety of substrates. Claims have been made that various grades of polymer formerly manufactured by GAF (USA) are suitable as coatings for aluminium, steel, paper, particularly sensitized paper, cotton, fabric laminates, wood veneer, polyfluorohalocarbon, polyamide and polyvinyl chloride.

Products in this series may be used for pressure-sensitive adhesives to low-energy surfaces; a copolymer with 2-ethylhexyl acrylate is a good adhesive for polyethylene, polypropylene and fluorocarbon polymers. This copolymer, blended with rosin esters, produces a corrosion-resistant adhesive for bonding aluminium foil to kraft paper.

Because of their properties, polyvinylpyrrolidone latices find application in remoistenable adhesives (Chapter 8). A product with a high pyrrolidone content, although developing strong and quick tack in contact with moisture, has the best resistance to blocking at higher temperatures, and may be plasticised with triacetin $CH_2(OOCCH_3)CH(OOCCH_3)CH_2(OOCCH_3)$ to speed up remoistenability and quick grab.

A wide variety of other applications claimed includes the unusual one of a pearling agent for detergents and shampoos. However, as far as is known, there is no current Western manufacturer of these copolymers, probably because of price disadvantage.

8 THE TESTING OF LATICES

This section gives some general methods for the testing of latices, the films derived from them and in some cases the properties of the polymers on which they are based. In general, considerable latitude can be allowed in the method of test, e.g. in determining solids content.

8.1 Total solids content

A sample of the latex is weighed into a tared aluminium dish about 4 cm in diameter, 2 g being a convenient weight. Unless the latex is at ambient temperature, the dish should be covered with an aluminium lid during weighings. As a variant, a covered Petri dish may be used to weigh and dry the latex. The dish is heated in an oven, preferably, but not essentially, of the circulating air type. The temperature should be about 110 °C, although for some polymers that 'skin' badly about 120 °C is preferable. The recommended time is, 1 h although in some cases a recheck after 2 h is desirable. An alternative is to use a vacuum oven, preferably giving a pressure of <10 mbar at 105 °C, in which case the time may be reduced.

After drying in a desiccator for 20 min the dish is weighed. Any new type of polymer should be rechecked to constant weight. It is possible that some polymers may be decomposed in air, e.g. by oxidation, such as ethoxyethyl methacrylate, and the solids content may have to be determined by a special method, e.g. vacuum drying. If a plasticiser is present a 'blank' check should be made by the addition of a known quantity to an unplasticised latex. If a transient plasticiser is present it may be desirable to prolong the heating to remove the plasticiser completely.

The above method may be unsatisfactory with some plasticised latices because of their volatility. Alternatively, a 'blank' check may be made on a polymer latex with known plasticiser content to determine the best empirical method. It may be necessary to make a correction, depending on the surface area of the dish and the temperature, particularly if dimethyl phthalate is present. Duplicate results should not vary be more than 1 part in 500.

8.2 Monomer content

Estimation of unpolymerised monomer may be made by physical or chemical methods. The former can only be used when the monomer is highly insoluble, as in the case of styrene, butyl acrylate or higher ester acrylates.

About 25–50 g of latex to which a further 100 g of water is added is distilled in a standard Dean and Stark type tube, adapted for solvents lighter than water. In this method the volume of unchanged monomer may be read directly. The method is, of course, uncertain with mixed monomers unless it is possible to perform a further test on the distillate, which must be recovered quantitatively. An infrared spectroscopic test may be used to confirm the character of the distillate, which in some cases contains decomposition products formed by side reactions, as well as, or rather than, monomer.

8.2.1 Chemical methods—vinyl acetate

For most monomers a simple bromination is sufficient, vinyl acetate being typical. This is achieved most practically by the bromate–bromide technique as follows.

A standard solution is prepared by dissolving 125 g of sodium bromide NaBr (or 145 g of potassium bromide KBr) in 20 ml of water, which has been dissolved in 2-methoxyethanol, made up to 1 litre and 1.3 ml of bromine added till the solution is complete. About 2 g of latex is weighed into an iodine flask, followed by 25 ml of oxyethanol, which is also added to a second flask acting as a 'blank' control. Add 10 ml of bromine solution from a burette to each flask, with 1 ml of potassium iodide KI solution (approximately 20 % w/v) in the neck of the flask as a seal. After standing for 15 min in the dark, the seal is allowed to drain into the flask.

The stoppered flasks are shaken so that excess bromine reacts with the potassium iodide. The contents of the flask are next titrated with N/10 sodium thiosulfate solution which has been standardised with potassium iodate, to the first end point seen, when the yellow colour of the iodine is just discharged.

One molecule of vinyl acetate absorbs two atoms of bromine, the excess liberating its equivalent of iodine. The reaction with thiosulfate is

$$2\ Na_2S_2O_3 + I_2 = Na_2S_4O_6 + 2\ NaI$$

Thus 1 ml of N sodium thiosulfate = 0.043 g of vinyl acetate, as the equivalent of the latter is half of its molecular weight in this reaction. Hence, vinyl acetate monomer as a percentage of the total latex weight is given by

$$\text{Monomer} = \frac{(x - y) \times 0.043 \times n \times 100\ \%}{w}$$

(where)

x = ml of sodium thiosulfate in the blank titration
y = ml of sodium thiosulfate required in the test
n = normality of the sodium thiosulfate solution
w = weight of latex

Minor variations may be necessary; e.g. a very viscous latex may be slightly diluted with water and the 2-methoxyethanol increased if necessary. The bromine added should be at least 50 % in excess of that required, and preferably double.

8.2.2 Other monomers

Variations of the bromine method are the most satisfactory for most purposes. If a monomer is readily distilled, it is best to test the distillate. This procedure can often be improved by adding a known quantity of carbon tetrachloride or chloroform before distillation. This assists in the formation of an azeotrope with vinyl acetate and can even be used with vinyl acetate as an alternative method.

Although some monomers such as styrene brominate fairly readily, it is not particularly easy to brominate or to titrate in emulsion form since the precipitated latices absorb both bromine and iodine irregularly. For some monomers, although not for vinyl esters, it may be best to hydrolyse by warming the latex with about twofold the excess of alkali required to hydrolyse all monomer present before attempting any analysis. This is suitable for maleates and fumarates, and possibly acrylate monomers, although this will depend on the entire latex being dispersible in a solvent suitable for bromination.

Other methods not involving bromination may be used. The most interesting is that of Das [51]. It is based on the fact that the following reactions occur:

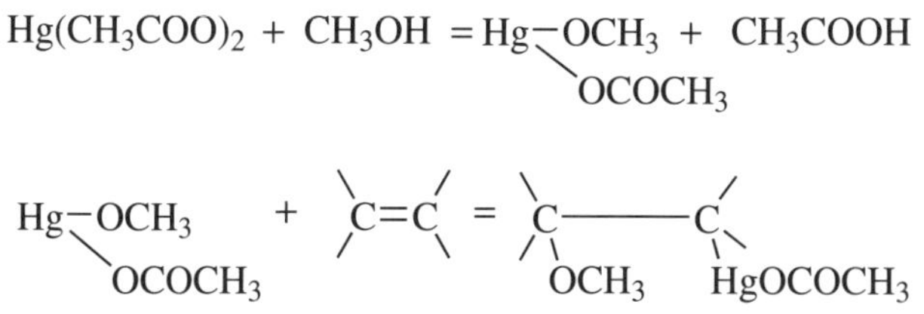

$$Hg(CH_3COO)_2 + CH_3OH = Hg\begin{matrix}-OCH_3\\ \diagdown OCOCH_3\end{matrix} + CH_3COOH$$

$$Hg\begin{matrix}-OCH_3\\ \diagdown OCOCH_3\end{matrix} + \begin{matrix}\diagdown\\ /\end{matrix}C=C\begin{matrix}\diagup\\ \diagdown\end{matrix} = \begin{matrix}\diagdown\\ /\end{matrix}C\!-\!\!-\!\!-\!C\begin{matrix}\diagup\\ \diagdown\end{matrix}$$
$$\quad\quad\quad\quad\quad\quad\quad\quad\quad\quad\quad\quad OCH_3\quad HgOCOCH_3$$

Thus one equivalent of acetic acid is liberated for each double bond. The sample is diluted with glycol–chloroform (1 : 1) and the sample titrated with 0.1N HCl in the same solvent mixture in which mercuric acetate acts as a base, and is titrated with thymol blue. The mercury addition product takes up an extra equivalent of acid by the equation:

$$\underset{OCH_3}{>C}\!\!-\!\!\underset{HgOCOCH_3}{C<} + HCl = \underset{OCH_3}{>C}\!\!-\!\!\underset{HgCl}{C<} + CH_3COOH$$

The difference between the milliequivalents of mercuric acetate and that of acid in the titration gives unsaturation in millimoles.

On the other hand, the most modern method for the estimation of volatile monomers in a nominally finished latex is by gas–liquid chromatography (GLC). It is desirable to use a method, such as with a hydrogen ionization detector, that is insensitive to water. This is by far the quickest way of estimating monomers with reasonable accuracy as long as proper blanks are performed with the instrument.

8.3 Specific gravity

With a sample of viscosity less than 0.5 Pa s, which is not aerated, a standard Westphal balance or a typical equivalent may be used where a standard weight is examined in the air and in the latex.

For viscous and aerated latices the problem is difficult. De-aeration in a vacuum may involve some water loss. The use of a tall measuring cylinder accurately calibrated in which a known volume of the sample is weighed will give an approximate result. Standard dilution may also be used if it assumed that the change in specific gravity with dilution is linear.

8.4 pH stability

A standard pH indicator is used, with a saturated calomel reference electrode and a glass electrode. Test samples of 5 ml are placed in tubes and 1 ml of N mineral acid or standard alkali is added, the tubes shaken and observed, and the pH measured. If there is no obvious instability, the stoppered samples are examined after 24 h. The limits of stability are recorded.

8.5 Freeze–thaw stability

This test is to some extent arbitrary in that the chosen temperatures and size and nature of the containers may vary. A convenient method is to use 250 ml or 500 ml bottles which are filled with latex and subjected to 18 h of freezing at $-20\ °C \pm 1\ °C$, followed by 6 h of thawing at 20 °C. The cycle is repeated at least 5 times, but is examined after each cycle for gelation or precipitation. Viscosity variation may be examined visually, but for a formal test see Section 8.6. Microscopic examination may be used to test the extent of agglomeration, but the latex is not considered unstable if these agglomerates readily disperse.

Pigmented and other compounded compositions are treated in the same way.

8.6 Emulsion viscosity

In some cases the pretreatment of a latex affects the viscosity characteristics, as with some polyvinyl acetate latices stabilised with polyvinyl alcohol. In these cases pretreatment must be carefully specified.

It is desirable to use a comprehensive instrument giving the full history of the sample over a shear range up to $10\,000$ s^{-1}, preferably with a water-cooled system to avoid temperature rise. See Chapter 4, Section 2 for modern instruments that are used. The apparatus is best stored in a constant-temperature humidity room. Operation is most convenient with a direct plotter giving stress against strain, and conversion tables supplied with the instrument enable a viscosity conversion to be made.

For more routine purposes a simple cylinder revolving in a wide outer case may be used, as in the Brookfield series of viscometers. To obtain truly meaningful viscometers, it is more desirable to measure the shear between two adjacent cylinders. Viscosities with the Brookfield series tend to be double those based on concentric cylinders with a narrow gap between them. In any test method or specification the instrument used should be specified, and any variation in testing, e.g. rotation speeds or cylinder types, should be specified—likewise the diameter of any cylinders. Some current instruments give a direct display.

8.7 Particle size

Originally the only method of measuring particle size was by direct microscopic observation, which could give magnification by about 400 at maximum. The phase contrast microscope, with oil immersion, increases magnification to about 600 times. The emulsion is diluted at least 10 times before a drop is placed on a slide, and a cover glass is pressed out. This can only measure particle diameters, but can give an approximate order of size from 0.4 µm upwards.

An electron microscope will give a photograph of the actual particles, enabling a count to be made, but it is too elaborate for routine purposes. In the past two decades there have been major advances in instrumentation.

The disc centrifuge became a standard instrument about 30 years ago and a typical instrument is shown in Figure 6.4. The principle of the instrument is that of a two-layer technique, derived from the Stokes equation

$$V = \frac{2r^2(d - d^i)g}{9\pi - 1}$$

where

V = velocity of sedimentation
r = radius of globules

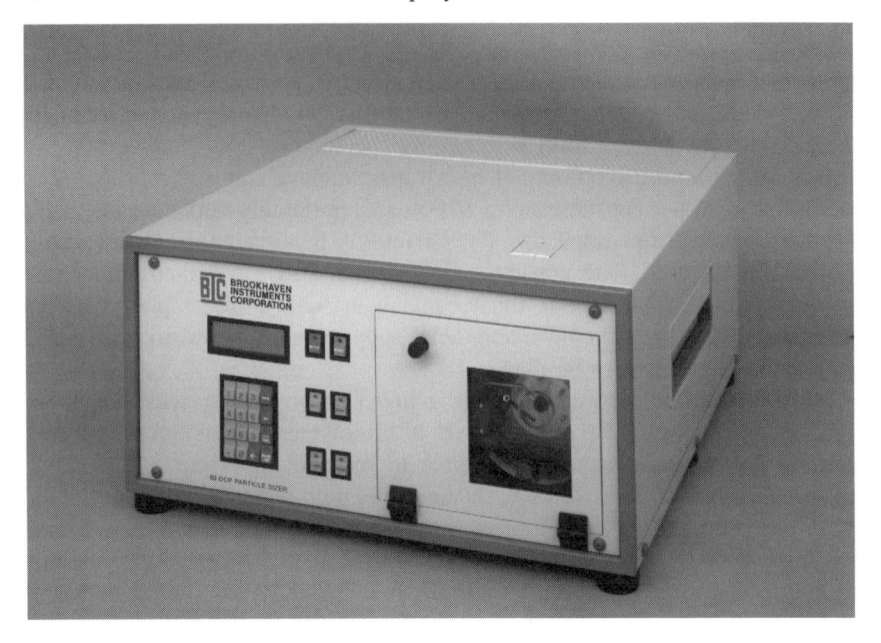

Figure 6.4 Brookhaven BI-DCP disc centrifuge particle size analyser. (Courtesy Brookhaven Instruments Limited.)

d = density of dispersed phase
d^i = density of medium
π = viscosity of medium
g = gravitational constant

A small sample is injected into the apparatus, in which the spin fluid has a higher density than the product under test. Glycerol–water, or sucrose solution, is most satisfactory with latices that are diluted to about 4 % before testing. The apparatus is operated by collecting the various samples and either estimating them chemically or more simply estimating by weighing. The disc centrifuge sedimentometer is capable of operating at 8000 rpm. The time of settling is deduced from the equation

$$T = \frac{6.299 \times 10^9 \Omega \log_{10} R_2}{N^2 d^2 \delta \rho R_1}$$

where R_2 is the radius in cm reached by a particle of diameter d μm with a density difference (g cm^{-3}) of $\delta \rho$ between the particle and spin fluid of viscosity Ω poises for a time T minutes from a radius R_1 when the disc is spinning at N rpm. Practically, some computerised results for the equation have been calculated and tabulated with variables of speed and time, these being used in obtaining actual results.

It is necessary to guard against a number of physical difficulties, such as interface 'streaming' (undesired mixing of the liquids). The instrument is designed to reduce this to a minimum, but if necessary either the density or the viscosity of the suspending liquid is increased to eliminate the problem [52].

The disc centrifuge was originally designed for measurement of pigment size, but has been found to be well adapted to latices of a wide range of particle size. The particle size of a polyvinyl acetate latex is shown in Figure 6.5. In this case a 4 % latex was used, with distilled water as the spin fluid and the dilution fluid being distilled water–methanol (60 : 40) so as not to swell the particles. Addition of a buffer layer, in this case methanol–distilled water as above, is desirable between the spin fluid and the injected sample. Figure 6.6

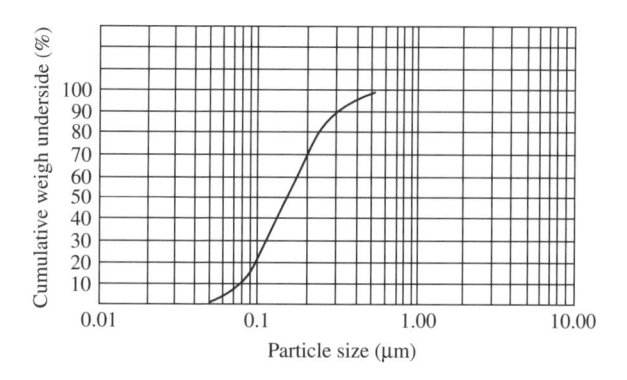

Figure 6.5 Particle size distribution of polyvinyl acetate latex versus percentage of cumulative weight undersize.

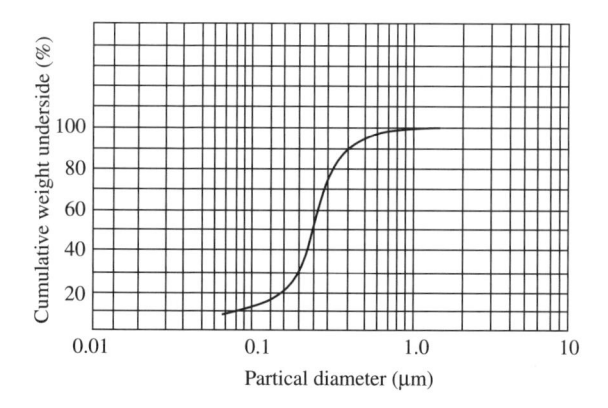

Figure 6.6 Particle size distribution of polyvinyl chloride latex versus percentage of cumulative weight undersize.

shows a similar sized distribution graph for a polyvinyl chloride latex in which the spin fluid is 20 % of glycerol in water, the buffer layer being distilled water, and the latex layer diluted with distilled water and one drop of non-ionic surfactant.

Another method by which particle size may be estimated in a relative manner is by turbimetry, using a spectrophotometer. There is a large number of suitable commercial instruments. The principle of the method is that light transmission will increase when passed through a dilute emulsion and that the finer the particle size, the greater will be the transmission. The theory has been examined to some extent [53]. Experiments confirm that for monodisperse latices the value of the slope of the log turbidity versus log wavelength plot appears to yield an average strongly weighted by the diameter (D^4 to D^5), but particles above 1.5 µm diameter are not counted by this method.

Practically, the latex is diluted to 0.00333 % solids, which is a useful experimental level. The solution is placed in the cell of the spectrophotometer; then either the appropriate filter is inserted or the appropriate wavelength switch is operated, and the deflection is read on the indicator, absorption being the normal reading. The method may be used as a 'pass or fail' test, i.e. the absorption must be below a stipulated figure, or, alternatively, the absorption at a number of wavelengths can be graphed against standards and any discrepancies noted. Thus there may be a tendency to agglomeration at certain particle sizes, causing irregularities in the curves. An instrument of this type normally covers a wavelength greater than that of visible light, with the Unicam SP 600 series being a typical commercial model covering 335–1000 nm. The ultraviolet end of the spectrum produces interesting results since absorption by the water phase affects the result.

A more sophisticated method of the use of light scattering is by an instrument which will give the particle size and scatter, provided that the instrument has been calibrated with a latex or dispersion of known particle size. The particle diameter can be measured down to 0.02 µm. Particle diameters are calculated with the aid of Mie theory according to the pattern of scattering observed. Tungsten lamps are preferable for creating good scatter in smaller particles. The scattering pattern does not depend directly on the particle diameter, but rather on the relationship between the wavelength used and the particle diameter. A suitable instrument is provided by Horiba of Kyoto, Japan, in the United States at Irvine, California and Ann Arbor, Michigan and in the United Kingdom at Moulton Park, Northampton.

8.8 Soap titration; surface tension measurements

'Soap titration' is a method of measuring particle sizes as surface areas, and as such is best considered separately. It has been used mainly for acrylic latices and seems to operate where the particles are not completely covered with surfactant. It is doubtful whether it could be readily operated with a latex stabilised with a partially hydrolysed polyvinyl acetate and a large surplus of surfactant. Although the word 'soap titration' has been used in the literature to describe the method [54], the word 'soap' is best avoided and kept to its literal meaning of the alkali salt of a fatty acid; the word 'surfactant' or emulsifier should be substituted.

Surface tension is measured with a standard Du-Nuoy ring tensiometer. It has been found experimentally that absorption of surfactant is very rapid, and it is best to titrate with the same surfactant that has stabilised the latex, if only one has been used. A typical curve of surface tension versus ml of surfactant is shown in Figure 6.7, but it is best to plot the apparent concentration at the critical micelle concentration at the break in the curve against concentration. Only the slope of the line is required, which is equal to S_n, the grams of absorbed soap per gram of emulsion solid, and the intercept is the critical micelle concentration for the surfactant in the water phase of the latex (Figure 6.8).

The total surfactant per gram of polymer (initially on the particles and part of the emulsifier system + the added surfactant) multiplied by the effective surface area of 1 gram of surfactant is determined by the Gibbs adsorption isotherm (Chapter 2, Section 1.2.1). If T (the interfacial excess of the adsorbed component) can be calculated, the surface area is derived by the equation

$$A = \frac{10^{16}}{TN}$$

where N = Avogadro number = 6.06×10^{23}.

The isotherm is measured by differences in surface tension between an aqueous solution and a hydrocarbon such as hexane, which gives an

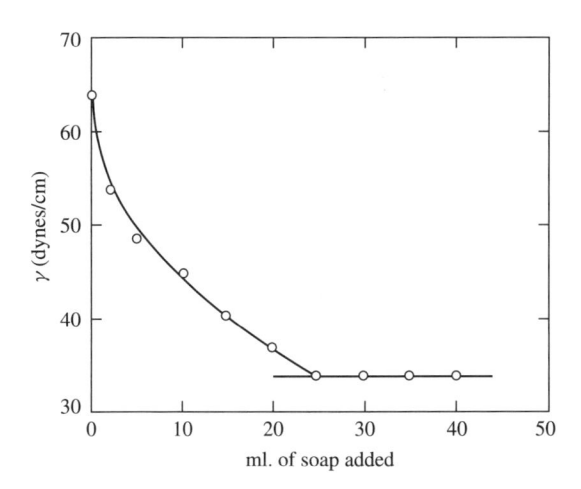

Figure 6.7 Typical curve of surface tension (γ) versus ml of soap.

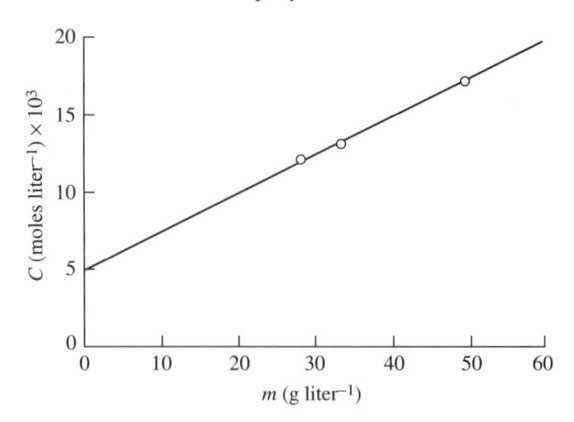

Figure 6.8 Typical curve of c versus m.

environment similar to a polymer rather than to an aqueous solution–air interface. There are already a number of known values to be applied. These include:

Sodium lauryl sulfate	61 Å^2
Triton X-202 (an alkyl aryl polyether sulfate)	62 Å^2
Triton X-405(a polyethylene oxide non-ionic alkylaryl type)	88.5 Å^2

A general account of measurement of particle sizes of latices including surfactant titrations is available [55].

8.9 Settling and sedimentation

A simple dilution test is adequate. Thus, a latex may be diluted to 20 %, placed in a stoppered measuring cylinder at ambient temperature and the cylinder stood for 3 days without being disturbed. After this time the meniscus is read and the degree of settling noted. In a few cases where the polymer is lighter than water it may be necessary to estimate the clear layer at the base. In this case a careful blank is necessary to ensure that graduations are correct. This method can also be adapted for examining sedimentation on standing.

8.10 Mechanical stability

A standard high-speed stirrer may be used after filtering a latex through fine wire or fine nylon, 100 mesh or 250 mesh. Suitable instruments include the 'Waring Blendor', 'Atomix' and the Hamilton–Beech apparatus, which is designed as a test for stability. After stirring for a standard time, any precipitate is filtered and weighed. It is expressed as a percentage of the original latex weight or alternatively of the original solids, based on 100 % polymerisation of volatile monomers.

8.11 Minimum film temperature (MFT)

The MFT is essentially a latex property, as against the second-order transition point (T_g), which is a polymer property (Chapter 4, Section 4.3). A suitable apparatus has been described already (Chapter 4, Section 4.3).

8.12 Relative molecular weight (viscosity)

Whilst the average molecular weight of a latex polymer may be determined by any of the well-known standard methods [56–59], the normal practical interest in emulsion polymers is one of viscosity measurement only, since it is the simplest method of obtaining a relative figure for the molecular weight (Chapter 1, Section 4.2). The distribution of molecular weight in a latex polymer is not normally of interest in routine testing, but if it is required, then various fractionation procedures are used. In the case of some polymers, fractionation by chemical composition, rather than by molecular weight, is desirable. Sometimes this is best performed from an entirely chemical standpoint, as, for example, by the controlled hydrolysis of reputed copolymers of vinyl acetate and acrylic copolymers.

There is another complication with emulsion polymers caused by the presence of stabiliser. No fixed rules can be given to allow for this. In some cases, solution and reprecipitation of the polymer is possible, but there is a risk of losing light 'tails', which remain soluble in spite of the presence of precipitating solvent. If surfactants of low molecular weight are the only ones present, it is usually possible to ignore them and to take only the actual amount of polymer in making a solution in an appropriate solvent. All solutions should be filtered, at least through a 250 mesh gauze, and preferably through a coarse sintered glass or equivalent filter. If there is a colloid of water or alkali-soluble polymer present, some difficulty is encountered in assessing the correct weight to be taken for viscometric molecular weights. If the colloid is known to be completely insoluble in the solvent chosen, it may be ignored provided that the solution is filtered. In some cases, however, especially with systems stabilised with polyvinyl alcohol, some grafting is known to take place, and it may be better to treat the colloid as part of the polymer.

If it is difficult to form a solution, it may be possible to disperse the latex as a whole in a suitable solvent or solvent mixture, particularly if the solvent is water soluble; an interesting case is that of vinyl acetate latices stabilised with gum acacia. Whilst films from latices of this type are difficult to disperse completely in ethyl alcohol, it is found in general that these latices will disperse as a whole in ethyl alcohol. It is suggested that with a 50 % vinyl acetate homopolymer latex, 37.7 g of water should be added for each 20 g of latex, followed by 143 g of industrial methylated spirits (UK technical ethyl alcohol). In some cases a small quantity of benzene or toluene may replace part of the ethyl alcohol if it is found that polymer solubility is improved.

There is often exceptional difficulty in dissolving latex polymers due to crosslinking or grafting to stabilisers. Plasticisers or solvents, when present at the start or during a polymerisation, can be ignored completely for viscosity purposes.

8.12.1 Viscosity measurements

The type of viscosity measurements that have significance in estimating relative molecular weights are dilute solutions, 0.4 % being generally very suitable. The same solvent must be chosen in comparing samples of similar resins, since the relative viscosity depends to some extent on the efficiency of the solvent. Thus with polyvinyl acetate, benzene solutions tend to give higher relative viscosities than solutions in 95 % ethyl alcohol at the same concentrations. These viscosities should be measured in a thermostat, usually at 25 °C. Results are expressed to give a relative viscosity (η_{rel}) of between 1.1 and 2.5.

A number of suitable viscometers are available. These are of the type which depends on the polymer solution falling under its own weight between two marked levels. These include the Ostwald and the Ubbelohde viscometers.

A full treatment of the connections between viscosity and molecular weight cannot be given here. If the viscosity of a polymer solution is η and that of the pure solvent is η_0, the ratio $(\eta - \eta_0)/\eta_0$ is the specific viscosity η_{sp} and the ratio η_{sp}/c, where c is the concentration, is called the *reduced viscosity*. The reduced viscosity is found to fall gradually as concentration falls and graphically it approximates to a straight line. The limiting value of η_{sp} as c tends to zero is the *intrinsic viscosity*, a quantity of considerable theoretical importance. Extrapolation for dilute solutions is extremely simple graphically. The relationship between viscosity molecular weight, which is between the number average and weight average, but closer to the latter, is given by the modified Staudinger equation

$$[\eta] = KM\alpha$$

where K and α are constants for any solvent/polymer system, but must be predetermined. For polymers with flexible chains, α is between 0.5 and 0.8. For determination of the constants, a fractionated polymer is used in which $M_n \cong M_v \cong M_w$.

8.13 General film measurements

General properties of films obtained from latices include film extensibility, film clarity and, if desired, properties such as blanching time when wet or blocking properties, the latter being of interest in both adhesive applications and any in which coated objects are stacked. There are many tests including ASTM, BS and ISO specifications covering these various properties, but in most cases comparative tests can be devised simply. Many are included in later chapters.

8.14 Determination of charge

All latex particles have a negative or positive charge, even if they are described as 'non-ionic'. In order to determine the nature of the charge, a small electrophoretic cell can be set up and the direction of movement of the particles determined, preferably after diluting the latex with distilled water to about 10 % concentration.

A microcell can be set up from two microscope covers that have been cemented together, with a small area cut out of the upper one in which the diluted latex is placed and a micro-platinum electrode fixed at either end of the cell. If a third cover glass, slotted to allow for the electrodes, is placed on top of the latex and the current from a small battery is switched on, the particles can be observed directly under a microscope, negatively charged particles moving to the anode and positively charged ones to the cathode. A method based on the electrophoretic migration rate and an apparatus for measuring the electric charge in the vicinity of a colloid particle have been described [60].

Other ways of determining the charge are to use a larger cell and to observe the visual accumulation of discharged latex particles on one of the electrodes. Alternatively, if an oil-soluble dyestuff in a suitable solvent is used to colour the particles, the direction of intensification of the colour can be observed. The difficulty with the latter method is that most dyestuffs themselves are charged ions and in some cases would precipitate a latex of opposite charge.

Mixing with known latices of positive or negative charges, preferably colloid free and free from an appreciable quantity of non-ionic surfactants, is another method for determining charges since oppositely charged latices tend to precipitate mutually. The method is not infallible, especially if there are large quantities of non-ionic emulsifiers in the latex under test.

REFERENCES

1. I.G. Farbenindustrie, BP 358,534, 1931
2. H. Mark and H. Fikentscher, USP 2,068,424, 1937
3. H. Warson, in *Ethylene and Its Industrial Derivatives*, S.A. Miller (ed.), Ernest Benn, 1969, Ch. 12, Sec. 8.1, pp. 1019–26
4. F.L. Marten and C.W. Zvanut, in *Polyvinyl Alcohol Developments*, C.A. Finch (ed.), Wiley, 1992, Ch. 3, pp. 57–76
5. H. Warson, in *Properties and Applications of Polyvinyl Alcohol*, C.A. Finch (ed.), Society of Chemical Industry Monograph No. 30, 1968, pp. 46–76
6. R.K. Tubbs, H.K. Inskip and P.M. Subramanian, in *Properties and Applications of Polyvinyl Alcohol*, C.A. Finch (ed.), Society of Chemical Industry Monograph No. 30, 1968, pp. 88–103

7. T. Okaya and K. Ikari, in *Polyvinyl Alcohol Developments*, C.A. Finch (ed.), Wiley, 1991, Ch. 8, pp 195–267
8. V.T. Stannett, R.H. Marchessault and E. Vanso, *Dissertation Abstracts*, 1964, Ch. 2, Sec. 3.3.v
9. A.A.W. Pateman and S.A. Miller (British Oxygen), BP 828,957, 1960
10. J.D. Nolan (Permutit), BP 1,125,612, 1968
11. H. Warson and D. Kinsler (Dunlop), BP 1,285,055, 1972
12. I. Mintzner, *Plaster und Kaut.*, **10**, 250 *et seq.* (1963)
13. R.A. Case and L.O. Raetherger, *Off. Dig.*, **36**, 947 *et seq.* (1964)
14. G. Loehr, *Plast. Rubber, Mater. Applic.*, **4**, 141 *et seq.* (1979)
15. G.E.J. Reynolds, *J. Oil Col. Chem. Ass.*, **53**, 399–410 (1970)
16. Pfizer Limited, Chemical Division, *Paint Manuf.*, **38**(11), 37 (1968)
•17. H. Warson and G.A. Reed (Dunlop), BP 1,144,316, 1969
18. H. Warson and D.S.W. Dargan (Dunlop), BP 1,088,634, 1967
19. I.S. Avetisyana, K.A. Pospelova and P.I. Zubov, *Kolloidny Zhurnal*, **25**, 278 *et seq.* (1963)
20. W.C. Mast and C.H. Fisher, *Ind. Engng. Chem.*, **41**, 790 *et seq.* (1949)
21. H. Warson, *Peintures, Pigments, Vernis*, **43**, 438–46 (1967)
22. N.J. Timmons, *Off. Dig.*, **25**, 922 *et seq.* (1953)
23. H. Warson, *Polym. Paint. Col. J.*, **180**(4265), 507–8,510 (1990)
24. B.B. Kine and R.W. Novak, in *Encyclopedia of Polymer Science and Engineering*, 2nd edn., Vol. **1**, Wiley, New York, p. 211 *et seq.*
25. F.R. Mayo and C. Walling, *Chem. Rev.*, **45**, 191 *et seq.* (1950)
26. R.C. Reinhardt, *Ind. Engng. Chem.*, **35**, 422 *et seq.* (1943)
27. J.J.P. Staudinger, *Br. Plast.*, **19**, 381 *et seq.* (1947)
28. D.M. Gibbs and R.A. Wessling, in *Kirk–Othmer Concise Encyclopedia of Technology*, Wiley, New York, 1985, p. 1224 *et seq.*
29. J.W. Vanderhoff, M.S. El-Aasser and K.C. Lee, *J. Appl. Polym. Sci.*, **45**, 221 *et seq.* (1994)
30. M. Suzuki *et al.* (Kureha Chemical Ind.), Eur. P 242,234, 1987
31. P. Akers *et al.* (Courtaulds), Eur. P 254,418, 1988
32. M.A. Tamela *et al.* (B.F. Goodrich), USP 5,344,867, 1994
•33. H. Warson, *Polym. Paint. Col. J.*, **161** (3817), 462–4 (1972)
•34. I. Williamson, *Br. Plast.*, **23**, 87 *et seq.* (1950)
•35. G.H. Elschnig, A.F. Schmidt, K. Goetz and F. Witt, Paper, Film and Foil Converter, Parts 1 to 12, October 1968–September 1969
•36. F. Witt, Paper, Film and Foil Converter, Part 1, August 1977; Part 2, September 1977
37. Du Pont de Nemours, BP 631,570, 1949
38. P. Kappler, (Elf Atochem), Eur. P 708,118, 1996
39. Du Pont de Nemours, BP 689,400, 1953
40. Du Pont de Nemours, BP 783,742, 1957
41. W. Grot, *Encyclop. Polym. Sci. Engng.*, **16**, 577–648 (1989)
42. A. Funaki and T. Takakura, (Asahi Glass), JP 340,716, 1994, *Chem. Abstr.* **123**, 33897 (1995)
43. H.S. Wu and E.W. Kaler (W.L. Gore), PCT 22,928, 1994; *Chem. Abstr.*, **123**, 144917 (1995)
44. H. Aihara *et al.* (Daikin Kogyo), PCT 17,876, 1996; *Chem. Abstr.*, **125**, 117082 (1996)
45. P. Colaianna *et al.* (Ausimont), Eur. P 648,787, 1995; *Chem. Abstr.*, **123**, 170599 (1995)
46. C.E. Schildnecht, *Vinyl and Related Polymers*, Wiley, New York and Chapman & Hall, London, 1952
47. R.H. Boundy and R.F. Boyer, *Styrene, Its Polymers, Copolymers and Derivatives*, Reinhold Publishing, New York, 1952
48. H. Warson *et al.* (Vinyl Products), BP 887,356, 1962
49. N.L. Clark (Unilever), Eur. P 255,363, 1988
•50. Montecatani, BP 874,173, 1961
51. M.N. Das, *Anal. Chem.*, **26**, 1084 *et seq.* (1954)
52. A.E. Loebel, *Off. Dig.*, **31**, 200 *et seq.* (1959)

References

53. S.H. Maron, P.E. Pierce and I.N. Ulevich, *J. Coll. Sci.*, **18**, 470 *et seq.* (1963)
54. J.G. Brodnyan and G.L. Brown, *J. Coll. Chem.*, **15**, 76 (1960)
55. J.G. Brodnyan and R.E. Zdanowski, Proc. 47th Annual Meeting, Chemical Specialities Manufacturing Association, 1960, pp. 177–81
56. G. Odian, *Principles of Polymerisation*, 3rd edn, Wiley, 1991, pp. 19–24, 350–1
57. H.G. Elias, in *Structure and Properties in Macromolecules*, Plenum, New York, 1984, pp. 301–71
58. *Encyclop. Polymer Sci. Eng.*, Vol. 10, Wiley, 1987, pp. 1–19
59. A.L. German, A.M. Hark, and H.A.S. Schoonbrood, in *Emulsion Polymerisation & Emulsion Polymers*, P.E. Lovell, and M.S. El-Aasser (eds), Wiley, (1997)
60. H. Schuller, *Kolloid-Z, Z. Polym.*, **211**, 116 *et seq.* (1966)

7

Unsaturated Polyester and Non-vinyl Emulsions

H. Warson

1 INTRODUCTION

The main object of this volume is to consider applications of latices formed by addition polymerisation of ethylenic compounds, the process of polymerisation taking place in emulsion form. However, some other classes of resins may be prepared in emulsion form, the term 'latex' not being generally used for condensation resins such as polyesters and alkyds, aminoplasts and phenoplasts. Epoxide resins, siloxanes and polyurethanes are in general formed by stepwise addition reactions, and the term 'latex' is sometimes applied to them. Polyurethanes may also be prepared by a condensation process. Unsaturated polyesters are crosslinked by a vinyl-type addition, but for the purposes of this chapter will be included here.

Each class will be considered individually, but briefly, together with the elementary chemistry of the processes involved. Most condensation processes can be available in a number of stages of the reaction and may be cured with the aid of a catalyst and heat, the term 'thermoset' often being applied. The products normally available are of low molecular weight, usually between 1000 and 10 000, although some of the simplest condensates such as dimethylolurea and the simplest condensate of epichlorhydrin and 'Bisphenol A' (Section 5) are below 1000 in molecular weight. It is for this reason that a distinction is sometimes made between a 'resin' and a 'polymer', the latter term being normally applied to a vinyl addition polymer of higher molecular weight.

The principles of emulsification vary. The case of the emulsification of a resin solution will not be considered further in most cases. Where epoxide resins are part of a crosslinking composition including a vinyl-type polymer, usually an acrylic polymer, information is already included in Chapter 5, Section 6. Some methods of special interest depend on a specific chemical structure enabling emulsification to take place readily. Thus, a resin containing carboxyl groups on neutralization may provide for emulsification rather than true solution. Excess of hydroxyl, and sometimes epoxy groups, may also give rise to favourable conditions for emulsification. Equally, under certain conditions the presence of amino groups may make the formation of a cationic product possible. In some cases a further condensation occurs, sometimes with an additional catalyst, often an acid or an alkali, on drying or stoving. A secondary reaction or polymerisation which may involve crosslinking may also occur, as with alkyds or polyesters.

The various classes, most of which are thermosetting resins, will be surveyed individually.

2 ALKYDS AND POLYESTERS

2.1 Polyesters

A polyester is, in its simple form, prepared from an acid and a polyol, the functionality of each being two or above. Thus, ethylene glycol and adipic acid, to quote a simple case, would condense as follows:

$$HOC_2H_4OH + HOOCC_4H_8COOH \xrightarrow{-H_2O} HOC_2H_4OOCC_4H_8COOH$$

A reaction with a further molecule of ethylene glycol may occur to give $HOC_2H_4OOCC_4H_8COOC_2H_4OH$.

It will be seen that products of this type may be extended in chain length indefinitely and that they are capable of containing unreacted carboxylic groups or hydroxyl groups, or possibly both at the ends, depending on the ratios of the reactants. These products are known as polyesters. Polyesters have many interesting applications, including coatings and impregnations, but possibly the most striking one is a synthetic fibre which is essentially a polyterephthalate of ethylene glycol:

$$-(C_2H_4OOCC_6H_4COOC_2H_4)_n-$$

The simple equation shown produces a thermoplastic polyester. (The polyester fibre is exceptional as it is highly crystalline.) If either the polyol or the acid has its functionality increased, it is possible to build up a three-dimensional network which will ultimately gel. This is the case if glycerol $C_3H_5(OH)_3$ or pentaerythritol $C(CH_2\text{-}OH)_4$ is used as the polyol. Full theoretical treatments of problems involved in gel formation are available [1].

A polyester prepared in emulsion contains 80 mol % of 3-hydroxybutyrate and 20 % of 3-hydroxyvalerate units. It has 67 % of particles with density <1.18 and is blended 90 : 10 with an acrylic latex and titanium dioxide to give a semi-gloss paint formulation [2].

A number of technical polyesters contains unsaturated acids capable of addition polymerisation as components. The principal acid is maleic acid, which is included in the reaction in the form of the readily available maleic anhydride:

$$\begin{array}{c} CHCO \\ \parallel \quad\quad\;\;\diagdown \\ \quad\quad\quad\;\; O \\ \parallel \quad\quad\;\;\diagup \\ CHCO \end{array}$$

This enables the polyester when formed to polymerise further by addition polymerisation of the unsaturation bond in maleic acid.

$$-OC_2H_4OOCH{:}CHCOOC_2H_4OOCCH{:}CHCO-$$

is a unit of this type of polyester molecule. Other glycols, acids or anhydrides and products such as hydroxyacids may be included. Acids or alcohols with only single functionality are sometimes included as chain stoppers.

A practical polyester is prepared with a mixture of phthalic anhydride or a similar saturated acid or anhydride and maleic anhydride. It is reasonable to assume that there is a random structure of the two acids. In order to crosslink it effectively, since each unit contains several double bonds, polymerisation is initiated by means of an ambient redox system, preferably after mixing with a readily polymerisable monomer such as styrene. Special precautions have to be taken, firstly to avoid premature gelation, which is often the result of an irregular reaction, and secondly to prevent air inhibition, for which the references 1 and 2 should be consulted.

Emulsions may be formed from low molecular weight prepolymers, e.g. from adipic acid, diethylene glycol and trimethylolpropane, emulsifying with sodium lauryl sulfate [3]. In this case a polyester is not thermoset by addition polymerisation of a double bond, but by further condensation on stoving, since the trifunctional trimethylolpropane is a potential crosslinking polyol. A suggested catalyst is $N,N(1)$-bis-1,2-ethylene isosebacimide.

A possible emulsification may be envisaged by the inclusion of a polyoxyethylene glycol $-HO(C_2H_4O)_n-H$ as a polyol component. This should make it possible for a polyester to be self-emulsifying, although it might affect the water resistance of films. This is largely overcome by ambient polymerisation of unsaturated acid groupings, alternatively on stoving, if there is a component present of high functionality capable of further addition.

In some cases polyesters, and alkyds, are referred to as 'water reducible'. This in general refers to a polyester or alkyd in an organic solvent which is water miscible, e.g. a glycol ether, and which is capable of dilution with water without precipitation. The term is used fairly loosely and may refer to a resin containing acid groups solubilised by a volatile alkali, e.g. ammonia or an amine, but which becomes insoluble after filming, especially when crosslinking takes place, as with an alkyd containing drying oil fatty acids. 'Water reducible' sometimes refers to a resin, which is converted directly to an emulsion on dilution with water.

A technical branched water dispersible polyester (Eastman AQ) is described, which is suitable for hot melt adhesives. It has good adhesion to polyolefin films and may be recycled [4].

A Japanese specification describes a plant in which there is high vibration and baffle plates to ensure emulsification. A typical polyester including dimethylpropionic acid is formed in a continuous process and is neutralised first with triethylamine to form a prepolymer solution. It is then dispersed in dilute aqueous ethylenediamine with distillation of the MEK to obtain an emulsion with 55 % of solids, viscosity of 510 mPa s and average particle diameter of 3.52 μm [5].

Polyester emulsions have been prepared in which excess carboxyl groups are neutralised with sodium acetate added during the condensation, in which one of the glycols is bisphenol A-4-ethylene oxide adduct. Hexamethylmethoxymelamine may be added in some paint formulations [6] (see also Chapter 11 in Volume 2).

In a similar process including bisphenol A–ethylene oxide (4 mol) adduct, in which tetrabutoxytitanium is the catalyst, the sodium salt of the carboxy-terminated polyester is dissolved in tetrahydrofuran and then dispersed with water, after which the tetrahydrofuran is distilled. The latex is suitable as a PET coating [7].

In tests for hydrolytic stability, polyesters derived from 2-butyl-2-ethyl-1,3-propanediol $HOCH_2C(C_4H_9)(C_2H_5)CH_2OH$ and 1,4-cyclohexane-dicarboxylic acid:

$$\begin{array}{c} \diagup CH.COOH \\ CH_2 \quad CH_2 \\ | \quad\quad | \\ CH_2 \quad CH_2 \\ \diagdown CH.COOH \end{array}$$

gave the most hydrolytic-resistant water-borne polyesters [8].

The stability, particle size and size distribution of polyester emulsions can influence the properties of the coating film [9]. Hydrolytic stability is also a function of the glycol and diacid components [10].

2.1.1 Polyester emulsions with vinyl monomers

An unsaturated polyester of low molecular weight is dissolved in a vinyl or vinylidene monomer with a vinyl-type monomer, e.g. styrene. To prevent excessive functionality a blend of maleic anhydride and phthalic anhydride or tetrahydrophthalic anhydride is used with the polyols, one example quoting a blend of propylene glycol and diethylene glycol. Emulsification is with the acetate salt of an alkylamine, e.g. $C_{12}H_{23}NH_2$, and a nonylphenol condensed with a high ratio of ethylene oxide, a trace of hydroquinone being added. The emulsion, which after polymerisation may contain 80 % of solids, in initiated with a redox system. Applications recommended for good adhesion include plasticised polyvinyl chloride sheet, leather, wool, textiles, rubber, paper and ceramics. The polymerisation of these unsaturated polyesters is air inhibited or retarded [11].

Another specification quotes 3 % of acid phosphate esters on reactive organic components to form stable emulsions of polyesters with styrene (7 : 3).

A polyethylene glycol also forms 10 % of the composition, which is stable for at least one month even if dibenzoyl peroxide is added [12].

A water-soluble polyester of molecular weight 11 000 is formed from dimethyl terephthalate, dimethyl-5-sodium sulfoisophthalate and diethylene glycol (92 : 8 : 100); 160 parts of this polyester are mixed with methyl methacrylate, butyl acrylate and dibenzoyl peroxide (160 : 22 : 20 : 2), with water to give a final solids content of 40 %, after heating for 3 h at 75–85 °C. The latex formed gives a coating on a polyethylene tetraphthalate film to give a coating with good adhesion, and is unaffected by water at 70 °C after 1 h.

The polymerisation of acrylic monomers in the presence of these preformed polyesters is illustrated in Table 7.1. Applied to a PET film and heated to form a coating, there is good adhesion and no change in water at 70 °C for 1 h [13].

2.1.2 Water-in-oil emulsions

There has been some interest in polyesters that have been emulsified to form water-in-oil (W/O) emulsions [14]. Their use is normally limited to films and impregnations in which the water can escape (but see further text). In addition, as a volatile monomer such as styrene is usually present, the curing conditions must be very carefully regulated to avoid the loss of styrene. Diallyl phthalate $C_6H_4(OOCCH_2CH:CH_2)_2$, which is often included in polyester formulations, is much less volatile than styrene. W/O polyester emulsions can be formed as a mass, e.g. in moulds, and either all the water can be driven off, producing a porous product, or else under ambient temperatures much of the water is only slowly released, if at all. This has been found to have little effect on the overall physical properties, but effects a considerable cost reduction.

An unsaturated ester prepolymer may be emulsified with a base of pK_a value above 6, which should be above a critical value of 0.3–0.5 mole of base per gram of polyester [14]. These polyesters may be polymerised directly in a W/O emulsion using standard initiators such as benzoyl peroxide to form a

Table 7.1 Acrylic copolymer formed in self-emulsified alkyd [13]

Alkyd MW 11 1000	
Dimethyl terephthalate	92
Dimethyl 5-sulfoisophthalate	8
Diethylene glycol	11 000
Monomers	
Methyl methacrylate	220
Butyl acrylate	20
Dibenzoyl peroxide	2
Polyester (as above, net weight)	160
Water—polymerisation is at 75–85 °C for 3 h	600

product, apparently touch dry, which loses water very slowly. The presence of water reduces the normal characteristic isotherm.

A conventional polyester in styrene is blended with another (70–90 : 10 : 30) which has been prepared with about 3 % of a polyethylene glycol, MW circa 1500, the balance being ethylene glycol or propylene glycol. The hydrophilic properties imparted by the polyglycol are sufficient to enable the whole composition, which contains about 50 % of styrene, to be emulsified. Another emulsifier may also be present. For the best results, 2,4,6-tris[(dimethylamino)-methyl]-phenol or allied compound is added as a promotor with MEK peroxide, the initiators being added before curing at 100 °C. Talc may be an optional addition. Mouldings from the hardened products do not lose water on ageing [15].

Water-in-oil emulsions may also be formed from a copolymerisable ester and a monomer together with a non-ionic surfactant including a short polyoxyethylene chain (HLB 2–8) and an alkali, preferably an amine of low viscosity, e.g. triethanolamine, which neutralises the end acid group. A thixotropic thickening agent, e.g. magnesium or aluminium silicates or fumed silica, may be added. The usual initiators are present [16].

The stability of W/O emulsions, stabilised by neutralization with excess acid, may be improved if the molecular weight of the polyester is increased by prolonged esterification to 1800–100 000. The molecular weight may also be increased by reaction with an epoxide or a di-isocyanate. Normal initiators are required, but no further emulsifiers. On curing there is a contraction of about 21.5 %. Eighteen examples are quoted [17].

Another claim from the same source describes a diethylene glycol–maleic anhydride polyester which is directly converted to a W/O emulsion by the addition of saline water. An interesting application is the addition to a corroded water storage tank. Even if water is present, the W/O emulsion, due to its density being >1, sinks through the water and cures on the base in 45 min to give a coating which protects the metal from further corrosion [18].

A monomeric metal or amine salt of acrylic acid in aqueous solution is emulsified in a typical styrene/polyester mixture, also forming a W/O emulsion. During this process the acrylic salt is also polymerised independently in the water droplets, forming a stable emulsion, possibly because of the formation of a stable skin around the water particles. There may be a small amount of grafting on to the polyester [19].

Water-in-oil polyesters have been exploited commercially, the advertised products containing 50–80 % of added water and with a particle diameter of about 2.5 μm [20]. The products are described as water-extended polyesters and can be moulded readily to produce resilient products. A typical material with 50 % of water has a tensile strength of about 2000 psi, compressive strength of 500 psi and flexural strength. Products of very high water content (75–90 %) can be made to lose water in an oven, producing a foam that can replace balsa wood.

2.2 Alkyds and allied products

2.2.1 Theoretical and miscellaneous preparations

An alkyd may be considered as a special case of a polyester in which a multifunctional polyol such as glycerol or pentaerythritol is partially esterified with a fatty acid, or mixed fatty acids which are derived from natural oils. These fatty acids range principally from C_{12} to C_{18}, and may be saturated, as lauric acid $C_{11}H_{23}COOH$ or stearic acid $C_{17}H_{35}COOH$. Unsaturated acids include the nondrying oleic acid $C_{17}H_{33}COOH$ and the drying oil fatty acids. Amongst the latter are included the non-conjugated *cis*-linoleic acid $CH_3(CH_2)_4CH{:}CH.CH_2CH{:}CH(CH_2)_7COOH$ and the conjugated *cis*-dienoic acid $CH_3(CH_2)_4CH{:}CHCH{:}CHCH_2(CH_2)_7COOH$, whilst the trienoic conjugated eleostearic acid $CH_3(CH_2)_3CH{:}CHCH{:}CHCH{:}CH(CH_2)_7COOH$ occurs as the triglyeride in tung oil.

The semi-synthetic dehydrocastor oil fatty acid with two CH:CH units, containing both the 9 : 11 and 9 : 12 isomers, is also a constituent of alkyds. Alkyds are made in practice in a process by which the natural oils containing the fatty acids are reacted with glycerol or another polyol. Details are given in standard volumes [21].

The alkyds have long been used as paint media. Non-drying alkyds are often combined with aminoplasts (Chapter 11 in Volume 2). Alkyds based on drying oil fatty acids may either dry at ambient temperature, with the appropriate addition of 'driers' such as lead, manganese and cobalt salts, or they may be used as stoving finishes.

Mathematical expressions are given for predicting the direction of emulsification and the cmulsion type for highly viscous alkyd emulsions, taking into account the surface activity, the concentration of oligomer in the organic phase and the concentration of surfactant in the aqueous phase [22]. A Russian study of the stability of alkyd emulsions with mono-, di- and triethanolamine concludes that monoethanolamine gives the best results, but they all have identical effects on stability and all repress foam formation [23]. An Indian study by size frequency analysis shows that, in general, the peak in the number distribution curve is about 1.5 nm and increases with time to a larger diameter range. The stability of emulsions prepared with mixed anionic emulsions is better than those prepared with anionics alone [24].

Studies of the drying of alkyds show that there is little difference whether they have been prepared in a solvent, e.g. a white spirit (petroleum fraction, b.p. circa 150 °C) solution of a linseed oil modified alkyd, or an emulsion that has been stabilised with ethoxylated nonylphenol and neutralised with ammonia. Cobalt naphthenate is include as a drier, preferably after emulsification [25]. However, one recent paper comments on the long drying time of alkyd emulsions and limited colloidal stability [26].

A number of methods of preparing alkyd emulsions are available. An early specification describes the emulsification of a glycerophthalic anhydride,

Alkyds and polyesters 411

which functions by reacting with surplus carboxyl groups to form a self-emulsifying composition from which the solvent is not removed [27].

One method combines the use of ammonia with the addition of blended non-ionic surfactants, the optimum conditions being to blend about 1 % of these surfactants with the alkyd and to add the balance of surfactants to the water phase. The emulsion is formed by an inversion method, i.e. adding the water phase to the alkyd until inversion occurs and maintaining the temperature at the early stage close to 100 °C. Driers, usually as the naphthenates of lead, cobalt and manganese, are not added until immediately before phase inversion to avoid premature gelation [28]. Typical alkyds suitable for emulsification are given in Table 7.2 and the emulsification is given in Table 7.3. The final emulsions have a solids content of about 50 %. They are of a thick creamy consistency and can be readily homogenised.

Tests on the film properties of these emulsified alkyds show that alkyds 1 and 3 were tough dry in 15 min, very slightly longer than a comparable 60 % alkyd in xylene, with a full dry time of 30–32 min, ascertained by the time an aluminium foil left no mark on the film with a 20 g weight above for 10 s. For Sward hardness, 10 is comparable, but is increased to 18 if the alkyd emulsion has 1 % of non-volatiles of ammonium polyacrylate added. This latter may take some part in crosslinking, possibly by a graft process. On baking for 1 h at 150 °C, hardness increased to 16 for alkyd 1, whilst alkyd 2 gave a very good rating of 30. Water resistance on immersion of the stoved films for various times or temperatures showed that there was only slight whitening compared to a solvent-based alkyd.

Table 7.2 Alkyds for emulsification [28]

Type	Weight (g) (moles in brackets)		
	Short oil	*Medium oil*	*Medium oil*
Trimethylolethane	843.5 (7)	676 (5.6)	—
Ethylene glycol	—	—	186 (3)
Pentaerythritol	—	—	435 (3)
Phthalic anhydride	1036 (7)	867 (5.8)	888 (6)
Tall oil fatty acids[a]	1211 (4.2)	—	1386 (4.8)
Soya bean fatty acids	—	778 (2.8)	—
Benzoic acid	—	179 (1.5)	—
Xylene	—	115	136

Process:
  For alkyds 1 and 2 heat at 245 °C until the acid number is <10; distil xylene. For alkyd 3, heat half of phthalic anhydride and ethylene glycol for 30 min, add balance, heat at 190 ° for 1h and then at 245 °C until the acid number is <10.

[a]Tall oil fatty acids are semi-drying, and a residue from paper making.

Table 7.3 Emulsification of alkyd of Table 7.2 [28]

	Weight (g)
Alkyd resin (any example in Table 7.1)	400
Nonylphenoxy polyoxyethanol, water soluble (Igepal CO-880)	1.33
Nonyl phenoxy polyoxyethanol, oil soluble (Igepal CO-430)	2.67
Water phase	
Water	400
Ammonium hydroxide, 28 %	4
Igepal CO-880	1.33
Igepal CO-430	2.67

Process:
Add emulsifier to alkyd, stir at 120 °C until dissolved, cool to 90–100 °C, increase stirring to high speed, add the water phase at 3–5 ml min^{-1} until inversion occurs (sudden drop in viscosity) and then increase the rate to 10–15 ml min^{-1}.

The following may be added optionally after the first 50 ml of the water phase has been added:

Emulsified lead naphthenate	0.5 % (Pb on alkyd weight)
Emulsified cobalt naphthenate	0.05 % (Co on alkyd weight)
Emulsified manganese naphthenate	0.025 % (Mn on alkyd weight)

The general principles of forming alkyd emulsions apply in other quoted cases where the variant may be in the nature of the polyol or the acid used in the alkyd composition, or the non-ionic surfactant or amine neutralisant in the stabiliser balance. Cellulose ethers are sometimes included in the formulations [29–33].

An alkyd is obtained by first performing the alcoholysis of safflower or other drying oil with ethylene oxide. This is followed by reacting the excess hydroxyl groups with maleic anhydride and again reacting with glycol. The temperature should be a little below 350 °F to avoid a Diels–Alder addition. This compound is unusual in that it can copolymerise with vinyl acetate. Emulsification is effected with Igepal CO 990, a non-ionic surfactant of high ethylene oxide content, the emulsifier being optionally divided between the alkyd and water. Alternatively, a phosphate-type surfactant (Gafac RE 610) may be employed.

Polymerisation is with the aid of a redox system, rigidly excluding air. The inclusion of alternative monomers, such as acrylic esters, styrene or vinylidene chloride is possible. The copolymerisation with vinyl acetate is worthy of further examination, especially as in an example the lead oxide catalyst for the alcoholysis has not been removed [34].

Technical alkyd emulsions are prepared from linseed oil with phthalic anhydride and pentaerythritol. They have a viscosity of 10–30 mPa s, pH of 6 and particle diameter of 0.6 μm, but they are not freeze–thaw stable, Non-ionic surfactants increase stability, optimally a nonylphenol polyoxyethylene with 16–20 ethylene oxide units. Alkyd emulsions tend to give rather softer films than those prepared in solvents, due to a plasticising effect of the emulsifiers. Tests made on chalky substrates confirm that alkyd emulsions have a penetration similar to those based on white spirit [35].

See also reference [55] on environment studies, Section 2.2.5. Reviews describe water-thinnable alkyds, *inter alia* [36–7].

2.2.2 Pigmented alkyd emulsions

An alkyd is prepared with a drying oil or semi-drying oil fatty acid so that the fatty acid is in the range 50–70 % by weight and the acid value is preferably <20. It is converted to an emulsion by the addition of a nitrogeneous base of pK^a 8.7–10, e.g. 2-amino-2-methyl-1-propanol or diethanolamine, an anionic or non-ionic surfactant at 0.2–0.4 %, preferably the latter, and water. The emulsion may be pigmented, optimally at 20–35 % pigment volume concentration, to form a paint with good adhesion to glass. Titanium dioxide should be dispersed directly into the emulsion for the maximum gloss. Driers are added as required. These compositions may be blended with polyvinyl acetate or acrylic esters latices [38].

Typical alkyd emulsions may be used in various paints. A technical product based on isophthalic acid is stated to be useful for flat oil paints, all sealers, exterior house paints and metal coatings. Some trouble may occur if the driers cause poor wetting and 'fish-eyes' may form, but this is overcome by satisfactory emulsification.

A pigmented alkyd emulsion has found application in a thixotropic latex paint [39]. An emulsion based on a chlorinated alkyd from which solvent has been removed is suitable for coating submarine interiors [40].

A coating formulation containing modified alkyd resins, pigment, calcium and cobalt salt driers, turpentine, water, ammonia, triethanolamine oleate and alkylaryl sulfonate forms a stable emulsion in water. The coating has high gloss and adhesion, and is fast drying [41].

2.2.3 Alkyd emulsion blends

There has been some reported investigations into blends of alkyd emulsions with latices, and other resins. Details are given in Chapter 5 of reactions where there is crosslinking.

A copolymer with vinyl caprate gives considerably better compatibility with alkyds than a copolymer with 2-ethylhexyl acrylate. (A copolymer with

Veova® would probably be equal to that with vinyl caprate.) Long oil alkyds probably have better compatibility and easier emulsification [42].

It has been claimed that only small amounts of alkyds can be added to acrylic latices, otherwise there may be storage problems, and also that the alkyd emulsion particles tend to cluster on filming, causing uneven film strength. However, this probably depends on the alkyd formulation, the relative balance of monomers in the acrylic latex and the nature of emulsification. It is known that both linseed oil and some alkyds can be coemulsified with a polyvinyl acetate copolymer latex, and this assists adhesion in some cases.

A Czech patent quotes a butyl acrylate–methyl methacrylate latex in which an alkyd of oil length 65, and other additives are emulsified to form a wood coating [43]. Other blends of alkyd emulsions with acrylic latices are claimed [44].

Alkyd/acrylic hybrid systems are prepared by polymerising acrylic monomers in the presence of colloidal alkyd droplets. There is some retardation of polymerisation due to the unsaturated groups in the alkyd, but polymerisation could be raised to nearly 100 % by appropriate conditions of polymerisation. These hybrids do not behave in the same way as mixtures, as shown by the minimum film formation temperature, and film formation is good. In some cases a synergistic behaviour is observed where the hybrid had properties superior to a simple mixture [45].

There is thus some contradiction in the literature on the subject of mixed alkyd and vinyl-type emulsions.

An alkyd–nitrocellulose emulsion is formed from the non-oxidizing Resyl®99, which is emulsified with a pentaerythritol modified rosin (Cellyl®104), whilst the nitrocellulose, plasticised with dibutyl phthalate and diluted with diethylene glycol monoethyl ether, is emulsified independently, a phosphate ester surfactant (Gafac®RE6109) being suggested. A high shear blender should be used for emulsification, which gives particle sizes of 0.1–0.8 μm. The volatile solvents are distilled and the two emulsions are blended so that there are equal quantities of nitrocellulose and alkyd, with a water content of 44.3 % [46].

Alkyd resins may be blended with melamine derivatives (see Chapter 5, Section 8.8 and this chapter, Section 3) to form a mixed condensate, in which the N-methylol groups—$NHCH_2OH$ or the methoxymethyl groups —$NHCH_2OC_2H_3$ react with carboxyl or hydroxyl groups of the alkyd. In this case a non-drying fatty acid is adequate as the long-chain component, since these products are invariably thermoset. The low molecular weight melamine deivatives are water soluble or require a small quantity of a cosolvent such as butanol, and blend readily with the alkyd emulsion. In Chapter 5 further details of the reactions involved are given, which are similar to those of carboxy-containing acrylate latices.

A typical preparation quotes an alkyd based on a polyethylene glycol which reacts together with pentaerythritol and with a drying oil by alcoholysis, and subsequently reacts with both phthalic anhydride and isophthalic acid to an acid value of 15, followed by neutralization with aqueous triethylamine to obtain an emulsion of 42.5 % of solids with a viscosity of 3000 cP. This is blended with a hydroxylated melamine resin in the proportion of melamine to alkyd of 1 : 4. On stoving a 70 μm film at 121 °C, a flexible film of Sward hardness 44 is obtained, with high adhesion and resistance to solvents [47, 48].

Blends of maleic–anhydride-modified drying oils with an etherified hexamethoxymelamine, thinned with methylethyl ketone and diethylene glycol monoethyl ether, are emulsified with aqueous diethylamine to form a 43.4 % emulsion, which may be pigmented. On dilution with 300 % of its weight of water, the emulsion is suitable for use in an electrocoating bath [49].

2.2.4 Styrenated alkyd emulsions

The process by which styrene, and occasionally vinyl toluene or acrylic esters, are reacted with drying oils is well known. Whilst there may be a certain amount of homopolymerisation, investigation has shown that the principal reaction is one by which chains of addition polymer are formed across several molecules. This process, however, only works well with conjugated double bonds, e.g. eleostearic acid. The reaction of forming styrenated oils may be illustrated:

In general, non-conjugated oils do not function in this way, but a graft polymer may be formed in some cases; e.g. when *tert*-butyl hydroperoxide is used as an initiator, it may remove a hydrogen atom from the carbon α- to the double bond, producing a new radical from the unsaturated acid, which in turn may start a polymerisation chain for the monomer. Under conditions of lower temperature, below 140 °C, the normal temperature of reaction, a Diels–Alder-type addition may occur, which limits the possibilities for polymerisation. A suitable reference should be consulted for further details of possible reactions [50].

Unfortunately, it is not possible to form a styrenated oil in emulsion form directly since the reaction temperature is far above that which is obtainable in normal reactors, even under some pressure. Several methods have been suggested to circumvent this difficulty. In one case, an oil-modified alkyd is prepared, and this is used as the solvent for the direct polymerisation of a vinyl-type monomer until about 40–70 % of the latter has polymerised. Water and an emulsifying agent together with a base, in particular lithium hydroxide, is added, and the reaction continued as a w/o polymerisation until 60–85 % of the monomer has reacted. The emulsion is then inverted by lowering the temperature until all monomer has reacted. This process avoids undue hydrolysis of the alkyd [51].

Examples quote a blended linseed oil–tung oil alkyd, also containing pentaerythritol and isophthalic acid, of acid number 16–18, or alternatively, the oils are replaced by a soya oil–dehydrated castor oil blend, with a final acid number of about 12.3. Styrenation is performed by heating 50–45 parts of the alkyd with about 2.5 % of dibenzoyl peroxide at 80 °C, adding about 4–6 % of blended non-ionic surfactants and about 1 % of lithium hydroxide after about 2 h at 80 °C. The product is heated for a further 2 and cooled to 60 °C, when inversion occurs. Heating is maintained for several hours until the bulk of styrene has reacted.

The final emulsions are reasonably stable, having particle diameters under 3 µm, and often considerably lower. Films 'dry' very rapidly, especially if methyl methacrylate replaces styrene. Another type of reaction is possible in emulsion. Very stable emulsions are formed which can be coated on to wood, metal or stone, drying to a hard durable state [52].

A direct emulsification has been described in which a styrenated dehydrocastor oil (DCO) from which surplus monomer has been removed is thinned with xylene, followed by further additions of DCO and the butyl ether of ethylene glycol. An emulsifier solution contains the sodium salt of an alkylbenzene sulfonic acid, borax, and colloidal aluminium silicate. The pH is maintained at 8–10 by the addition of ammonia solution after emulsification in a homogenizer. The particle size of the emulsion, with about 60 % of non-aqueous components is about 5 µm. It may be used as the basis of an emulsion paint [53].

2.2.5 Ecological and environmental studies

Self-emulsifying alkyds, whether air-drying or stoving, are stable to high shear, are suitable for glossy top coats in all colours and have good freeze–thaw resistance. They contain <3 % of organic solvents, may be recycled and are biodegradable [54–56].

Reviews on low volatile organic content (VOC) paint formulation focus specifically on the environmental impact of water-thin alkyd resins and give formulations that meet European Community standards [57–59].

3 AMINOPLASTS

The aminoplasts are related to the acrylamide derivatives, already described in Chapter 5, Section 8. Lightly condensed aminoplasts lend themselves more to application in aqueous solution rather than in emulsion form. However, the nature of the condensation gives a product which tends to be soluble in the first stage. This alkaline condensation produces methylol products, which are often modified to produce methylol products, which are often modified to produce the ethers, such as the dibutyl ether of dimethylolurea:

$$\begin{array}{c} NHCH_2OC_4H_9 \\ | \\ CO \\ | \\ NHCH_2OC_4H_9 \end{array}$$

Doubts have been expressed as to the symmetrical nature of this product.

The nature of the condensation is very similar to condensations of methylolacrylamide polymers and their ethers (Chapter 5). Aminoplasts are also used, is aqueous form, as an addition to an acrylic polymer with an active group capable of crosslinking (Chapter 5, Section 8).

Some direct emulsions, can, however be manufactured. Thus a specification claims the preparation of emulsions formed from partially etherified butyl ethers of methylolmelamine or methylolurea. The butanol is replaced by propylene glycol by heating with the glycol at 100 °C under vacuum. The product can be dissolved in water to give a 33 % solution. To 150 parts of this solution are added 100 parts of the original 75 % butanol solution of the butoxy ether originally used are added. The resultant product is a stable emulsion, which after addition of an ammonium salt can be used as a binding agent for paper or wood [60].

The most interesting modifications of aminoplasts are cationic products formed by coreaction with an amine of multifunctional characteristics. Amines, like amides, react with formaldehyde, the reaction being exactly analagous;

$$R.CH_2NH_2 + HCHO \longrightarrow RCH_2NHCH_2OH$$

If a multifunctional amine is added during an aminoplast condensation, the second stage of the reaction can have the effect of combining the two types of condensation. This reaction, in its simplest form, can be regarded as an ether condensation between two methyol groups. In this reaction, a melamine formaldehyde condensate is not a basic product, since the melamine amino groups are only very weakly basic:

$$\overline{M}.NHCH_2OH + HOCH_2NXR = \overline{M}NHCH_2OCH_2NXR$$

where M is a melamine residue, X is hydrogen or an alkyl group and R is an alkyl or cyclic residue.

The addition of the amines has the effect of imparting a positive charge to the composition as a whole, and under some conditions a salt may be completely water soluble. Under other specified conditions, an emulsion may be formed directly [61]. A suitable amine is 3,3′-iminobispropylamine $HN(CH_2CH_2CH_2NH_2)_2$. The exact final state of the product, which is reacted with a polyol to reduce sulfate tolerance, is not entirely clear from the specification. Materials of this type are of interest as beater additions for paper.

An alternative method that has been suggested for the emulsification of aminoplasts is the use of aqueous alkaline casein and thiourea after solution in a higher alcohol such as butanol. Aqueous solutions of gum tragacanth, starch or methyl cellulose prevent migration of the resin after absorption on a fibrous material [62].

An interesting patent quotes an aminoplast emulsifier, which may be prepared from hexamethylolmelamine tributyl ether with a polyethylene glycol and triethanolamine, presumably by transetherification of the alkylol ether groups with the alkylolamine and also with the glycol to provide an overall cationic charge. This emulsifier may be used to prepare an emulsion from a mixture of the butyl ether and a long chain, e.g. arachyl ether, of hexamethoxymethylmelamine, together with paraffin. The product, suitably diluted, may be utilised as an impregnating bath for rendering cotton goods water repellant [63].

A blend of urea, a formaldehyde–methanol mixture (Formcel®) and a dialkylethanolamine, with a trace of water and at a slightly alkaline pH, is reacted to form an intermediate. A fatty acid, e.g. a high rosin tall oil fatty acid, and melamine are added, the reaction continuing at a lower pH, for which the addition of maleic anhydride is used as a control, the final pH being 2.0–2.2. The end product may be dispersed in water on addition of alkali, and has utility as a binder for glass fibres [64].

Hexahydroxymethylmelamine is emulsified after dissolving in a solvent, e.g. monochlorobenzene, to which is added a water-soluble solvent miscible with the first, e.g. an ethoxyethanol type. The emulsifiers may be soaps prepared *in situ* or non-ionic surfactants, e.g. $C_9H_{19}C_6H_4(OC_2H_4)_x OH$, where

$x = 30-50$. It is recommended that the emulsifier is added partly to the water and partly to the organic phase. The derived films are strongly water repellant [64, 65].

An emulsion prepared at 3-20 % of solids from a polyhexahydroxymethylmelamine, a long-chain amine salt, the butyl ether of diethylene glycol and monochloracetic acid is emulsified by continuous addition of water, followed by distillation of the solvent present. The emulsion film is cured at 150-155 °C for 3 min. The emulsion may be used to impregnate fibre [66].

A benzguanamine resin has been treated with a polyvinyl alcohol solution to give a stable emulsion useful as an adhesive for plywood [67].

An emulsion may be formed from an aminoplast prepolymer and further condensed in emulsion in a composition which contains kerosene. On curing, a porous polymer is formed by evaporation of the kerosene [68].

A wool fabric is tested with an emulsion containing an etherified melamine together with a polyamide–epichlorhydrin resin, followed by overcoating with a latex of a soft acrylic polymer. This assists in obtaining bright pastel shades by pigment dyeing [69].

Several blends are claimed. An emulsion is formed from a hydrogenated polybutadienediol, MW 400, with a technical aminoplast (Cymel 1141) and butoxyethylene glycol (80 : 20 : 18). These are condensed with a catalyst, heated and triethylamine in water (2 : 5) added to maintain the pH > 9, giving an emulsion with no coagulation or phase separation. This is cast on aluminium and heated at 175 °C to cure, giving a clear, glossy, peelable elastomeric film [70].

A textile impregnant is formed by blending a melamine–formaldehyde resin, modified with a carboxylate containing carboxyl groups with an aqueous solution of mixed esters of tetraphosphoric acid esters of mixed ethylene glycol and pentaerythritol to which is added a polysiloxane and a fluorinated surfactant, the result, presumably in emulsion form, being an impregnant for wood, plastics, cellulosics and rubber, giving intumescent properties [71].

4 POLYAMIDE EMULSIONS

Polyamides are formed by the reaction of a diamine and a dibasic acid, or mixtures thereof. A lactam, such as caprolactam,

$$\begin{array}{c} CH_2CH_2NH \\ CH_2 | \\ CH_2CH_2CO \end{array}$$

containing the two reactive groups in a single molecule may also be condensed to form an amide (by itself it forms Nylon 6) or a copolymer. Polyamides can be used in conjunction with epoxides in emulsions as coatings (Section 2.4).

One of the standard methods of emulsifying polyamides is to utilise salts of relatively low molecular weight polyamides with surplus amine functionality [72].

In a modification polyamides with low but appreciable carboxyl and amine values are dissolved in about their own weight of isopropanol with 1 % (on polyamides) of polyoxymethylene (MW 60000) and 6 % of a 4 % sodium polyacrylate solution, a total of 200 % of water being added whilst distillation takes place of an azeotrope of isopropanol and water. The resultant emulsion, about 34 % of solids, 58 % of water and 8 % of isopropanol, may have apparently coagulated, but it can be re-emulsified by the addition of 10 % on polyamide of concentrated ammonia solution, giving a stable emulsion. An alternative is the use of a technical polyamide. The emulsions may be blended with various water-soluble polymers or with many other emulsions or latices. In some cases resin precursors, e.g. epoxidized oils, may be directly emulsified into this latex. The emulsions find application in coatings and adhesives, and application is greatly facilitated by the use of emulsions rather than polymer melts [72, 73].

Amongst methods of emulsification, copolymers of 15–25 % of acrylic and/or methacrylic acid with acrylamide are suggested as direct dispersants [74]. Another formulation, giving a cationic product, is based on using salts of low molecular weight polyamides with surplus amine functionality.

It is claimed that improved stability of typical polyamide emulsions will be obtained if a modifier, which may be a phenol, an aromatic glycol or a low molecular weight epoxide, is included in the composition, the resin of which should be soluble in hot n-amyl alcohol.

Preferred modifiers are resorcinol diglycidyl ether

$$m\text{-}C_6H_4(OCH_2\overset{O}{\overset{/\backslash}{CHCH_2}})_2$$

and the condensate of epichlorhydrin and bisphenol A, which is one of the simple epoxide resins (Section 5). Optimum conditions are a solution of 15–25 % by weight of the polyamide and modifier in n-amyl alcohol, together with up to 2 % of sodium lauryl sulfate in the water phase and 0.1–2 % of alkali or alkali carbonate to neutralise the residual acid groupings in the polyamide. A solution of partially hydrolysed polyvinyl acetate, with percentage hydrolysis of 78–90 % and with a 4 % solution viscosity of 20–45 cP by the Hoepler falling ball method, must be added to ensure that the resultant emulsion is stable. Amyl alcohol should be removed as an azeotrope in 3–8 h.

Dried films, e.g. those with a mixed polyamide of hexamethylene diamine adipate and hexamthylenediamine sebacate, give clear films, but tend to whiten

Epoxide resins and blends

Table 7.4 Polyamide emulsion [75]

	Weight (g)
Caprolactam–hexamethylenediamine adipate–hexamethylene diamine terpolymer	240
n-Amyl alcohol (water saturated)	1140
Distilled water	60
The above is refluxed for 2 h; then add:	
Epichlorhydrin–bisphenol A condensate ('Epon 828')	60
Emulsify the above in a 'Kady' mill with:	
Distilled water	1200
Sodium lauryl sulfate, 30 %	4.5
Sodium carbonate	0.5
Finally add:	
Polyvinyl alcohol, 10 % solution ('Elvanol' 50–42)	300

Finally distil a 90 % amyl alcohol azeotrope to give a 29.6 % of solids emulsion of high mechanical stability.

with water unless they are heated to the melting point, 150–160 °C [75] A typical example is shown in Table 7.4.

The emulsion formed is useful as a minor additive to a wash-and-wear (drip dry) textile formulation containing 4–12 % by weight of an N-methylol condensate of melamine, urea, ethyleneurea or dihydroxyethyleneurea, 0.5–2 % of an acid metal salt such as zinc nitrate, 0.5 % of the quoted polyamide emulsion, 20–30 % of a modified phenolic or aromatic glycol and 5–15 % of polyvinyl alcohol on the total weight of polyamide plus modifier. Examples in the patent also include a non-ionic polyethylene emulsion at about 10 % of the N-methylol compound. This type of emulsion is also used in waterproofing.

In one case a vinyl type polymer, solubilised through carboxyl groups, is used as a stabiliser for polyamide emulsions [76]. Many polyamide emulsions are used as cures with epoxides. They are included in Section 5.

5 EPOXIDE RESINS AND BLENDS

The low molecular weight epoxides used for the production of cured resins are sometimes referred to as 'resins', although this is not correct. The simplest aromatic epoxide in general use is a condensate of bisphenol A, the formula

of which is

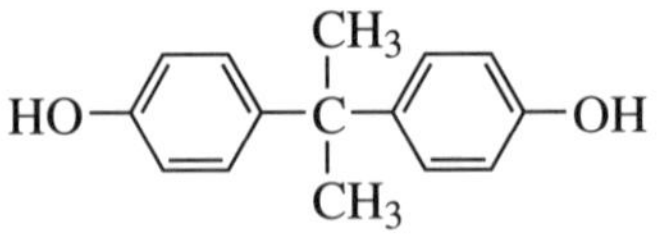

$$HO-\underset{}{\bigcirc}-\underset{\underset{CH_3}{|}}{\overset{\overset{CH_3}{|}}{C}}-\underset{}{\bigcirc}-OH$$

with epichlorhydrin

$$CH_2-CH-CH_2Cl$$
$$\diagdown O \diagup$$

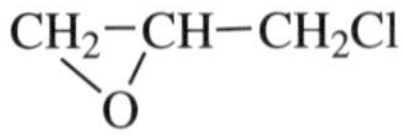

in the presence of alkali, producing bisphenol A–diglycidyl ether. This is capable of further condensation via the epoxide groups with further bispenol A and epichlorohydrin, providing a series of epoxide condensates:

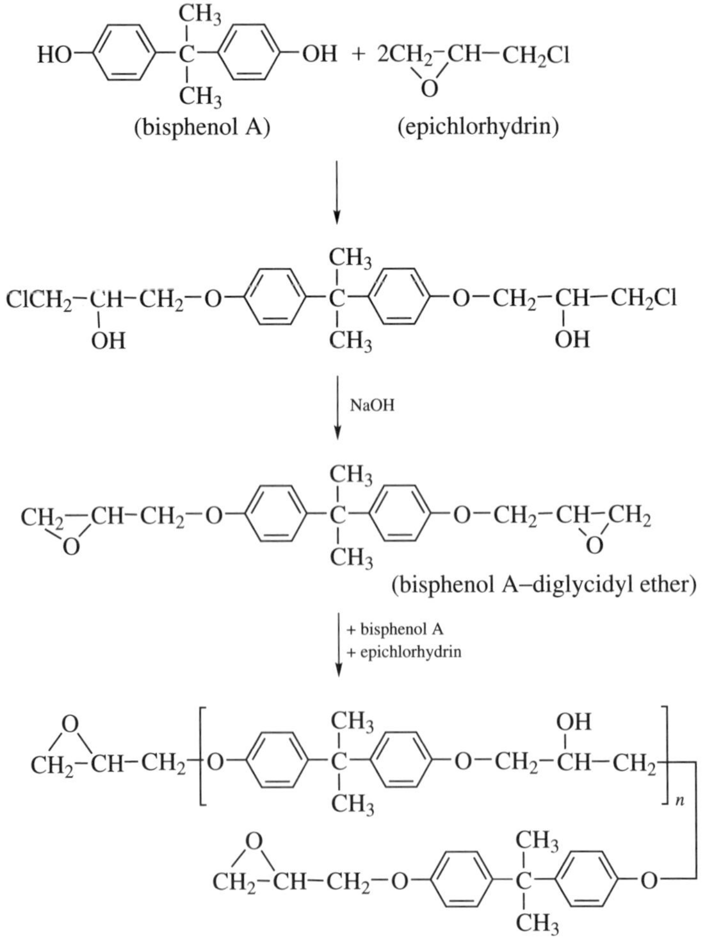

$$HO-\bigcirc-\underset{\underset{CH_3}{|}}{\overset{\overset{CH_3}{|}}{C}}-\bigcirc-OH \; + \; 2CH_2-CH-CH_2Cl$$

(bisphenol A) (epichlorhydrin)

↓

$$ClCH_2-\underset{\underset{OH}{|}}{CH}-CH_2-O-\bigcirc-\underset{\underset{CH_3}{|}}{\overset{\overset{CH_3}{|}}{C}}-\bigcirc-O-CH_2-\underset{\underset{OH}{|}}{CH}-CH_2Cl$$

↓ NaOH

$$CH_2-CH-CH_2-O-\bigcirc-\underset{\underset{CH_3}{|}}{\overset{\overset{CH_3}{|}}{C}}-\bigcirc-O-CH_2-CH-CH_2$$

(bisphenol A–diglycidyl ether)

↓ + bisphenol A
+ epichlorhydrin

$$CH_2-CH-CH_2-\Bigg[O-\bigcirc-\underset{\underset{CH_3}{|}}{\overset{\overset{CH_3}{|}}{C}}-\bigcirc-O-CH_2-\underset{\underset{OH}{|}}{CH}-CH_2-\Bigg]_n$$

$$CH_2-CH-CH_2-O-\bigcirc-\underset{\underset{CH_3}{|}}{\overset{\overset{CH_3}{|}}{C}}-\bigcirc-O-$$

Epoxide resins and blends

These condensates are marketed under various trade names. Before cure, they may be referred to as 'polyepoxides'. Many will be referred to in this text by their trade names. Textbooks on technical literature as supplied by manufacturers may be consulted [77].

5.1 Initial condensation in emulsion

References to the reaction of a bisphenol with epichlorhydrin in emulsion are scarce. A method is described whereby this occurs in the presence of a suitable emulsifier or colloid and a water-miscible solvent. The emulsifiers may include conventional anionic sulfonates and also non-ionic types such as condensates of epoxides with glycols or cationic stabilisers such as condensates of epoxides with diamines. The product is emulsified with excess water [78].

5.2 Amine salt emulsification

The original methods for emulsifying epoxides has been either by direct emulsification, often with a solvent, or with the aid of an amine, sometimes in the form of a salt, the amine often combining the emulsification function with that of a hardener. Hardening (curing) agents are generally of three types, all based on polyethylene amines, the standard curing agent for epoxide resin precursors:

(a) amines prepared by the reaction of the amine with a fatty acid, referred to as amidoamines;
(b) similar condensates formed condensates from dimer fatty acids, known as polyamides;
(c) amine adducts prepared by a controlled reaction of the amine with an epoxide resin.

There is a tendency for β-aminoamides to lose a second molecule of water to form an imidazoline [79]:

$$\begin{array}{c} \overset{C}{|} \quad \overset{C}{|} \\ N-N(CH_2CH_2)_nCH_2CH_2NH \\ \diagdown \quad \diagup \\ C_{17}H_{29-35}C \end{array}$$

In one example, 5.1 parts of a polyepoxide of molecular weight 900 are dissolved in 9.4 parts of a 1 : 1 butanol–toluene mixture; 1 part of ethylenediamine (85 %) is added at 50–60 °C and stirred for 1 h. Next, 0.7 parts of dimethylaminomethylphenol, the catalyst, is added with a mixture of ethylene glycol monoethyl ether and butanol (4.7 : 1). This is followed by 2 parts of 80 % formic acid, or lactic acid. The mixture is homogenised into an emulsion by the addition of 75.5 parts of water [80].

5.3 Non-ionic emulsifiers

The alkylphenyl polyglycol emulsifiers, blending types with high and low ethylene oxide content, are suitable as emulsifiers both for the bisphenol A type of epoxides and also aliphatic glycidyl ethers. An example quotes an emulsion in which a silicone is included. It may be pigmented to give a semi-gloss enamel [81].

A technical epoxide (Eponite® 100) to which 15 % of its weight of a technical polyamine is added is emulsified with a solution of an alkylphenol polyglycol (9.5 moles of ethylene oxide). With the addition of zinc formaldehyde sulfoxylate, it is useful as an antistatic treatment for polyester fibres and also reduces discoloration [82].

Epikote® 1004 may be emulsified with a fatty alcohol, e.g. 2-ethylhexyl alcohol, and either an aliphatic alcohol adduct with 20 ethylene oxide groups or a polyglycol ester of an acid, e.g. oxalic acid, and then neutralised with a diamine, e.g. isophoronediamine, which also acts as a hardener. Application is with cement [83].

The ethylene oxide–propylene oxide condensates (see Chapter 2, Section 3.6) may also be used as stabilisers, preferably by dissolving these emulsifiers in the epoxide phase, especially a low molecular weight type, e.g. Epon® 828, and gradually adding water until the emulsion phases are reversed. An epoxide emulsion of this type may be cured with a condensate of a phenol and a polyamine, e.g. one prepared from *p*-chlorophenol and tetraethylenepentamine. They may be pigmented [84].

A formulation is shown in Table 7.5 for a blend that also includes styrene–acrylic latices. Ionic, cationic or amphoteric emulsifiers may be used as alternatives [85].

Other emulsifiers useful for Epikote® 828, alone or in mixtures, are polyoxyethylene sorbitan mono-oleate and monostearate, and polyoxyethylene glycerol borate mono-oleate and dioleate [86]. A Russian specification quotes the use of long-chain aliphatic esters (as distinct from ethers) of polyglycol ethers at 10–20 °C in forming epoxide emulsions. Oleic acid salts at 5–10 %

Table 7.5 Epoxide latex [85]

	Weight (g)
Epon 828	100
Butoxytriethylene glycol	35
Nonylphenyl condensate + 25 ethylene oxide groups	4
Ethylene oxide–propylene oxide condensate (Pluronic F68)	2
Water	50
Aromatic amine (Ciba H850)	50
Water	150

of resin precursor or a corresponding percentage of bentonite are also included as additional stabilisers [87].

Polyoxyethylene nonylphenol diglycidyl ether is a suitable emulsifier for bisphenol A–diglycidyl ether. It is formed by reacting polyoxyethylene nonylphenyl ether and toluene (940 : 360), which are heated to reflux to remove water and then mixed with stannic chloride for 1 h, followed by addition of epichlorhydrin (16 : 202) at 55–60 °C for 6 h, and finally adding 50 % of aqueous sodium hydroxide after cooling [88].

A very high resin content (70–80 % of total weight) emulsion is formed when a polyepoxide is emulsified with an alkoxylated colophony, 25–35 ethylene oxide units per molecule of the latter being the optimum; 5.25 % of emulsifier (on precursor) is the optimum for an 80 % solids emulsion, the particle size being <0.5 μm. The major part of the emulsifier is added to the epoxide phase, the balance being water. The emulsion should be stable for at least two years [89].

5.4 Miscellaneous emulsifiers

An alternative method of emulsification is the use of a partial phosphate of a long-chain alcohol or alkylphenol which may be partially ethoxylated. Curing agents, e.g. amines, are added as required. The phosphates are considered to be more satisfactory than most sulfates and sulfonates in forming stable emulsions of particle size 0.5 μm. Certain curing agents, e.g. some water-stable boron trifluoride BF_3 complexes, enable a cure to take place on drying without the application of heat. These emulsions are suitable for electrodeposition [90].

The reaction of 3–10 parts of a boric acid ester with 100 parts of liquid epoxide produces a partially cured self-emulsifying resin [91].

A water-soluble polyester is able to emulsify an epoxide resin precursor. An example is a dimethyl sulfoisophthalate–dimethyl terephthalate–dodecanedioic acid–ethylene glycol–isophthalic acid–trimellitic anhydride copolymer (12 : 168 : 52 : 174 : 72 : 88), acid value 45, softening point 100, MW 350 : 400 Acid value 45 was mixed with 182 parts of epoxide resin, mw 350–400 followed by neutralization with aqueous ammonia to obtain an emulsion. It is stable for ⩾6 months and cures in 2 min at 170 °C [92]. Another complex carboxylated polyester, also partially extended with a diisocyanate, and with a polyoxyethylene surfactant, is also a suitable emulsifier for epoxide prepolymers [93].

Another interesting example of direct production of emulsions is obtained by the addition of an imidazoline type of cationic surfactant and also a 'Pluronic®' ethylene oxide–propylene oxide condensate. Preferred epoxies have molecular weights from 300 to 900 and an epoxide equivalent (the molecular weight required to give one equivalent of epoxide) of 140–550.

This emulsion is useful for the treatment of bundles of glass fibres suitable for laminates, together with a coupling agent, e.g. a siloxane of formula $H_2NCH_2CH_2CH_2Si(OC_2H_5)_3$ and a fibre lubricant [94].

An allied emulsion is prepared from bisphenol A–diglycidyl ether, a dimer acid, formed from a drying oil, triethylenetetramine, ammonia, a silicone release agent and the siloxane $H_2NC_2H_4NHC_3H_6Si(OCH_3)_3$, the latter two being in catalytic quantities. This emulsion is used as a binder for glass fibres which have had a previous coating of the same silicon derivative, together with a wetting agent and an acetylenic alcohol. Cure is at 210 °C [95].

An ammonium salt of an unsaturated acid, e.g. maleic acid, is included with the polymerisation of vinyl or acrylic esters in emulsion in the presence of an epoxide prepolymer. Either N-vinylpyrrolidine or a polyvinyl alcohol is present. The latex may be pigmented with titanium dioxide. The cure is with a diamine neutralised with oxalic acid and/or formic acid. A phenoplast is an alternative to the epoxide. Application is also as a concrete additive [96].

5.5 Cycloaliphatic epoxides

The cycloaliphatic epoxide 3,4-epoxycyclohexylmethyl-3,4-epoxycyclohexane carboxylate (Epoxide ERL 4221)

is able to crosslink acrylic latices including acrylic acid. It is emulsified with Triton®-200, Triton®-305 and water (75 : 1.61 : 1.5 : 75) and blended with an acrylic latex based on methyl methacrylate, butyl acrylate and acrylic acid (45.9 : 38.7 : 4.5), where the emulsifiers including the polymerisable surfactant 2-sulfoethyl methacrylate (see Chapter 2, Section 3.9) are in a ratio of 50 : 3.6. Heating films gave crosslinking by the epoxide groups reacting with the carboxyl groups in the polymer. The crosslinked films show improvements in film hardness, solvent and heat resistance, but water resistance is still poor [97].

Aliphatic diglycidyl ethers, e.g. propylene glycol diglycidyl ether or butanediol diglycidyl ether, may be condensed with diamines on heating, examples quoted being ethylenediamine, hexamethylenediamine or tetraethylenepentamine. The resultant condensate doubles the function of an emulsifier and also of a hardener for a standard bisphenol A–epichlorhydrin epoxide resin precursor. However, the pot life of this material is limited to about 2 h. Films cast from this emulsion are touch dry in 3 h and tack free in 6–7 h [98].

5.6 Water-in-oil emulsions

A water-in-oil includes as emulsifier a carboxyl-terminated adipic acid–diethylene glycol polyester which is neutralised with diamines. Water is held up to 50 %. The film on hardening retains water drops [99] (see also reference [100]).

5.7 Epoxides containing halogens

Table 7.6 illustrates the modification of the polyepoxide by retaining 1 % of chlorohydrin units and also the inclusion of a polyvinyl alcohol. Emulsification takes place at 60 °C, storage stability being improved compared with an emulsion of the diepoxide without residual chlorohydrin units [101].

The effect of fluorine-containing compounds is considered in terms of free energy changes and the lowering of surface tension. Thus fluorination of the two central methyl groups of bisphenol A–diglycidyl ether results in the lowering of surface tension by 10 dynes cm^{-1}, whilst fluorinated surfactants have an allied effect in assisting the displacement of water from the surface of the polymer. Hydrogen bonding caused by the hydroxyl in the epoxide bonding with surface metal oxide may also assist in promoting adhesion [102].

5.8 Epoxide–polyamide emulsions

The original epoxide–polyamide finishes were oil based and have amine end groups which enable the crosslinking process to take place. They may be utilised to provide corrosion-resistant finishes (see Section 5.7.1). These mixed emulsions are suitable as binders for cork and cellulosic materials generally, whilst at low concentrations (3–10 % on total resins) they may be applied to fabrics, providing a good hand (Chapter 12), stiffness and a degree of water repellancy.

They are also of interest as glass fibre sizes, giving very high flexural strength to laminates, and are also of value as sizing agents for glass 'preforms' and for glass mats that are to be subsequently laminated into various shapes and objects. They may also be used as coatings for paper and as wet seal and heat seal adhesives. Equally, baked pigmented compositions, at 150 °C for 30 min, give high gloss and the film resists impact when coated on to tinplate. Other uses which may be mentioned are as a concrete block filler and as a corrosion

Table 7.6 Epoxide emulsion, modified resin precursor [101]

Polyepoxide, modified as text	50
Polyvinyl alcohol, unspecified type	2
Polyethylene glycol nonyl phenyl ether	2
Water (as required)	

inhibitor for metal surfaces, to which the polyamide–epoxide combination has particularly good adhesion.

Suitable polyamides are derived from polymeric fatty acids, formed by dimerisation and trimerisation of linseed oil fatty acids and diethylenetriamine ($H_2NC_2H_4)_2NH$) so as to have a residual amine value of 80–90, to which is added 0.3–0.4 equivalents of formic acid. A standard bisphenol A–epichlorhydrin epoxide is used at approximately the same weight, both being dissolved in suitable solvents, e.g. the polyamide in toluene–isopropanol and the epoxide in toluene–methylethyl ketone. Emulsification is effected by adding water until phase reversal takes place, and the organic content, including solvents, is about 33 %. The emulsions formed by this procedure are rather viscous, of a mayonaise-like structure [103].

It is possible to emulsify the solutions by adding acetic acid to a polyamide which is amine terminated, emulsifying the solvent solution as a whole to form a cationic emulsion in which a diepoxide could be emulsified. Whilst with a solids content of 35 % the products could be sprayed easily, the mixing and handling procedures are lengthy, the solvent level is too high and pigmentation levels limited [104].

Di- or trimerised linoleic acid is reacted with a glycidyl-type epoxide, a glycol or glycol ether being the reaction medium. The addition of acetic acid produces an emulsion which may have a pH of 6–8. The emulsion may be utilised as a felt-resisting finish for wool [105].

It is desirable to eliminate solvents as much as possible, and also with a two-pack system. An aliphatic epoxide type monomer has been added to a bisphenol A type, giving a reduction in viscosity. A polyamide with 70 % of solids and an amine value of 255 has a strong cationic effect, causing self-dispersion, part of the polyamide acting as an emulsifier for the rest and also emulsifying the epoxides. The polyamide is also an efficient pigment deflocculating agent [106].

Compositions containing esterified epoxide resins, together with polyamides or other condensates such as phenoplasts or aminoplasts with pigments and xylol, may be emulsified simply with a fatty acid salt [107]. The products are useful in electrophoretic coatings (Chapter 11 in Volume 2).

5.8.1 Epoxide–polyamide blends in coatings

Gloss coatings may be obtained by grinding titanium dioxide in the water-reduced epoxide blend, and then adding the polyamide at a pigment–binder ratio of about 0.7–1.0. Semi-gloss coatings are obtained by the addition of talc. These coatings should not be force dried above 100 °C. The level of gloss may be increased by the addition of ethylene glycol monoethyl ether as coalescing agent.

Table 7.7 gives some formulations for corrosion-resistant primers based on the technical epoxides Genepoxy® M195 with an epoxide equivalent of 195, which is self-emulsifiable, and the polyamide Versamid 265-WR70, with an

Epoxide resins and blends

Table 7.7 Corrosion-resistant primers [108]

	Weight (lb)		Weight as solids	
	1	2	1	2
Genepoxy M195, 51.5 %	306	205	157.5	205
Water	221	194	—	—
Lead silicochromate	290	—	290	—
Strontium chromate	—	120	—	120
Talc	75	—	75	—
Red iron oxide	140	—	140	—
Mineral fibre	—	120	—	120
Versamid	193	252	135	176
		Formulation 1 (%)	Formulation 2 (%)	
Pigment volume concentration		31	17.2	
Non-volatiles by weight		65	49.2	

amine value of 225. In this table, formulation 1 gives the maximum possible packing of pigments and fillers that can be dispersed in the epoxide component. It yields a flat finish, and a dry film thickness up to 125 μm is possible without trapping water in the drying film. Formulation 2 provides a very good balance of flow and anti-ageing properties and is particularly good in avoiding pinholing and air entrapment. Further formulations for corrosion-resistant coatings will be included in Chapter 8 and Chapter 11 in Volume 2 [108].

The anticorrosive properties of polyamide–epoxide blends are discussed in some detail [109]. A mixture of Epikote® 834 and ethyl formate (47.78 : 9 g) is mixed with a pigment paste (109.6 g) and a hardener solution prepared from Versamid® 140 (a polyamide), butoxyethyl alcohol and the hydrocarbon solvent Solvesso 100 (47.78 : 18.58 : 40.42). Water is added until an O/W emulsion is formed. This may be used to give a coating on aluminium for aircraft [109].

5.9 Epoxide emulsions—miscellaneous blends and reactions

5.9.1 Reaction with vinyl-type monomers

An epoxide resin precursor is dissolved in styrene and acrylic ester monomers (7 : 3) and irradiated with 1.5 MeV to form a graft copolymer. This is dispersed in butanol–cyclohexanone with heat and converted to an emulsion after addition of 2-dimethylaminoethanol and water [110].

An epoxidized polybutadiene is reacted with dimethylamine and the resultant polymer, including amino groups quaternised with epichlorohydrin. Styrene and 2-ethylhexyl acrylate are copolymerised in emulsion using this quaternised compound as stabiliser. The resultant latex is added to paper stock as a size [111].

An ammonium salt of an unsaturated acid, e.g. maleic acid, is included with the polymerisation of vinyl or acrylic esters in emulsion in the presence of an epoxide prepolymer. Either N-vinylpyrrolidine is present or a polyvinyl alcohol. The latex may be pigmented with titanium dioxide. The cure is with a diamine neutralised with oxalic acid and/or formic acid. A phenoplast is an alternative to the epoxide. Application is also as a concrete additive [112].

A styrene–ethyl acrylate–methacrylic acid copolymer in ethylene glycol monobutyl ether (300 : 210 : 90 : 288) with Epikote® 1007 in the same solvent with 2-dimethylaminoethanol (500 : 333.3 : 4.8) is heated to 80 °C. This is diluted with water to a 20 % dispersion and then blended with a 30 % solution of a phenoplast (360 : 60) to form a 22 % emulsion on dilution with ammonia solution. This is coated on to a tinplated can, dried at 200 °C for 5 min, giving a corrosion-resistant coating [113].

Table 7.8 describes the production of a vinyl-type polymer in the presence of phosphated epoxide esters. The final modified latex is suitable for coating beverage can interiors [114].

An epoxide resin precursor is dissolved in styrene and acrylic monomers (7 : 3) and irradiated with 1.5 MeV electrons to form a graft copolymer. This is dispersed in butanol–cyclohexanone with heat and converted to an emulsion after addition of 2-dimethylaminoethanol and water [115].

5.9.2 Miscellaneous

A joint emulsion is prepared with a polyester as in Table 7.9. Note that a water-in-oil emulsion, which is first formed, is inverted [116]. Epoxide resin emulsions, prepared with non-ionic surfactants and sodium polymethacrylate, may be blended into emulsions with adducts of the maleic anhydride-drying oil type [117].

An emulsion of complex character has been described containing both an epoxide and an aminoplast. The former is a 60 % emulsion of bisphenol A–diglycidyl ether emulsified with 1.8 % on epoxide of a polyethylene oxide–cetyl alcohol ether and 4 % of polyvinyl alcohol. The aminoplast emulsion was formed by adding a solution of 48 parts of a partially crosslinked non-ionic surfactant to 48 parts of water, emulsifying in conjunction with 320 parts of methylolmelamine dibutyl ether in 75 % butanol and obtaining a 50 % O/W emulsion by further addition of 64 parts of water. The blend consists of 8.4 parts of epoxide to 90 of aminoplast emulsions, which is adjusted to pH 7. Zinc borofluoride $Zn(BF_4)$ is added as a catalyst before stoving [118]. An emulsion of an epoxide and a phenol-modified aromatic hydrocarbon, with lactic acid and polyamidoazoline, is disclosed [119].

A mixture of a dicyclopentadiene–vinyl acetate copolymer, formed under pressure in xylene at 260 °C for 3 h, butyl glycidyl ether and a block polyoxyethylene–polyoxypropylene copolymer (70 : 30 : 3) are mixed after heating and then cooled to 40 °C with water (80) to give a stable emulsion.

Table 7.8 Blend including phosphated epoxide esters [114]

A. Epon® 828, 95 % in xylene	1149 g
Bisphenol A	614 g
Butoxyethanol	310 g
Sodium acetate	0.52 g
The above is heated at 90–175 °C for 5 h and then phase B is added at 125 °C	
B. Butoxyethanol	310 g
Polyphosphoric acid in 20 g butoxyethanol	7.32 g
Then add over 2 h.	
C. Water	34 g
Butanol	888 g
Methacrylic acid	283 g
Styrene	148 g
Ethyl acrylate	4 g
Dibenzoyl peroxide	38.5 g
Butoxyethanol	32.4 g
The polymer is mixed as below to form an opalescent emulsion.	
D. Water	3687 g
Dimethylaminoethanol	162 g
Then blend with nominal solids weight.	
E. Cymel 303 (aminoplast)	5

Table 7.9 Epoxide–polyester emulsion [116]

Polyester (Villion® GV-230, acid 59, softening point 93 °C)	1000 g
A. Epoxide (Epikote 1004, softening point 98 °C)	930 g
Coumarone resin (Coumarone® G-90, softening point 98 °C)	70 g
Benzoin	4 g
Modaflow®	2 g
Ethoxyethyl alcohol	300 g
Titanium dioxide	500 g
Stir the above at 95 °C and add the following.	
B. Polyoxyethylated alkylphenol (HLB 15.9)	40 g
Triethylamine	32 g
Water	1000 g
The water is added dropwise at 85–90 °C, cooled to 43 °C and then added as follows to invert the emulsion from W/O to O/W.	
C. Water	1622 g
The inversion on cooling is the major feature.	

This is blended with a Japanese diepoxide (Epiclon FM 80) to give an emulsion, and gives a stability greater than 1 month [120].

Of a rather different character is the use of an epoxide resin precursor with various polyamines to give water-soluble products, which, with the aid of a volatile acid, produce cationic emulsifiers. These will emulsify 5–15 times their own weight of polymers such as polysiloxanes, chlorinated rubber and polyethylene, pigments being optionally included. The emulsifiers will crosslink on heating, possibly with the disperse phase if it is reactive [121].

Water-borne two-pack coatings based on a low particle size epoxide–novolak emulsion have a high crosslink density and a rapid through drying time, good film formation and high hardness development at an early stage of the cure. However, the pot life is fairly short and there is some brittleness of coatings, which tend to peel when exposed to salt spray and humid conditions [122].

5.10 Applications

A technical epoxide, DER330, is emulsified with an alkylphenoxy-poly(ethyleneoxy) ethanol and methyl cellulose, with the addition of 11 % of diacetone alcohol on the epoxide and water to give a 51 % emulsion. This is diluted to a size for glass fibres with 12.4 % of this emulsion, silanes and polyvinypyrrolidone [123].

Czech researchers recommend that epoxide ester emulsions are suitable for impregnations and coatings for concrete, plaster, wood and cardboard. If the ester is based on a fatty acid with multiple unsaturation, a cobalt drier should be added [124].

A bisphenol A-derived diepoxide is emulsified with lauryl alcohol and lauric acid together with oxalic acid and is blended into cement [125].

Water-emulsifiable epoxy resin systems suitable for hydraulic materials, e.g. for improved resistance to frosts, salts and carbonation, are prepared by the addition of selective hardening agents based on a polyamine, a polyamidoimidazoline and/or a polyaminoamide. The procedure is as in Table 7.10; A is the hardener preparation.

A mortar is formed which is easily flowable and with a water–cement ratio of 0.36 and density of 2.270 kg m^{-3}. The mortar has 7-, 28- and 360-day compressive strengths of 45, 67 and 103 N mm^2 with bending strengths of 8,14 and 17 N mm^2 respectively. The mortar resists carbonation and salts [126].

An emulsion is formed from Epon 828 and a polyoxyethylenenonylphenyl sulfate ammonium salt and a polyoxyethylene emulsifier (190 : 10–140). This is added to a cement mix so that the resin–cement ratio is 0.11 : 1 [127].

A number of other specifications give formulations suitable for addition to cement as in Table 7.11 [128]. An epoxide emulsion (unspecified formulation) may be combined with an EVA latex in a comparable ratio, together with hardeners, dispersants and trichlorethane, to form compositions suitable for coating

Table 7.10 Water emulsifiable epoxide with polyamino hardener [126]

A + B form the hardener.	
A. Polyamino–imidazoline including polyaminoamide (amine equivalent 95)	700 g
Bisphenol A diglycidyl ether	100 g
Heat for 2 h at 90 °C and then add B.	
B. Lactic acid	10 g
Phenol-modified aromatic hydrocarbon	190 g
Cool, to give viscosity of 8600 mPa s and then add C.	
C. Bisphenol A type epoxide resin precursor; viscosity of 12 000 mPa s at 25 °C, epoxide equivalent 190	190 g
Hardener A + B above ⎫ premix	195 g
Water ⎭	330 g
D. Application. To the above add E.	
E. Water	1.45 kg
Portland cement	5 g
Sand	15.5 g

Table 7.11 Epoxide additive for cement [128]

A. Epoxide emulsion	
Water	80 g
Polyvinyl alcohol, 10 % aqueous	100 g
Silicone defoamer	5 g
Triethylenetetramine, oxalic acid salt	73 g
Rutanox VF 2913	200 g
Mix the above for 1 h at <70 °C and then blend as B.	
B. Portland cement	100 g
Water	44 g
Sand (<1 mm)	230 g
Sand (1–2 mm)	130 g
Epoxide emulsion as A	29 g

walls, floors and for water proofing [129]. Epoxide emulsions may be employed for high-performance coatings, printed circuit board laminate fabrication and flexible packaging materials. A pigmented lacquer emulsion with low solvent content, suitable for treating litho plates, is formed as in Table 7.12 [130]. Printed wiring circuit boards, of which the first stage of manufacture is by impregnating glass cloth with a laminating resin, originally solvent based, may now be impregnated by a technical epoxide emulsion [131].

434 Unsaturated polyester and non-vinyl emulsions

Table 7.12 Pigmented epoxide laquer emulsion [130]

	Parts by weight
Water	650
Poly(N-vinyl-N-methylacetamide)	50
Polyepoxide, equivalent 450–625; softening point 65–75 °C	75
2-Phenylethanol	70
Phosphoric acid, 85 %	2.5
Talc	220
Carbon black dispersion (in 1 : 3 ethylene glycol–water) with circa 1 % non-ionic dispersants	30

6 PHENOPLASTS

The reaction of phenols and formaldehyde results in the formation of methylol (CH_2OH) groups in positions *ortho* and *para* to the hydroxyl group. Products which are formed in the first stage are

[structures: 2-hydroxybenzyl alcohol (ortho-methylolphenol) and 2,6-bis(hydroxymethyl)phenol] and

The product formed depends on the ratio of phenol and formaldehyde respectively. These methylol compounds tend to be water soluble. If phenol is in excess and acid catalysts are included, the methylol groups tend to condense further with loss of water, giving methylene bridges. If there is only one methyl group per molecule crosslinking cannot occur, and a linear polymer is formed, the molecule formula of which in an idealised form is

[structure: linear novolak chain with three phenol units linked by $-CH_2-$ bridges, each phenol bearing an OH group]

These compounds are known as novolaks. Under acid conditions, the intermediate phenolic alcohols are not isolated. In alkaline conditions, which are the best for condensing with excess of formaldehyde, branched chain structures are formed as there are at least two methylol groups on some molecules. These condense further to form crosslinked insoluble and infusible structures containing both methylene $:CH_2$ and ether. CH_2OCH_2. crosslinks, and possibly other structures.

The above is a highly simplified account, but more details of the chemistry of phenolic resins of this type are available [132]. A number of modifications are possible. Thus the use of a partially substituted phenol such as *ortho-* or *para*-cresol restricts the activity of the phenols by blocking potential positions of substitution. Under some conditions, drying oils or simple esters of their fatty acids can react with intermediates, probably by a reaction which involves direct addition across the double bond. The oil-modified phenolics have better compatibility with natural oils and produce the equivalent of an internal plasticisation.

It is possible to prepare emulsions at an intermediate reaction stage in the preparation of novolaks by careful control of conditions and the inclusion of a suitable surfactant system. The literature in the past has been somewhat obscure on this subject and many preparations do not make it clear whether the product precipitates on formation or remains emulsified.

Some of the earliest preparations of phenolic resins were prepared by dissolving a phenolic resin, optionally modified with tung oil, in benzene, adding to water containing sodium oleate and bentonite and distilling surplus solvent. Alternatively, triethanolamine was added directly to the resin and water was added with stirring until inversion occurred giving an O/W emulsion [133].

Another process includes a small quantity of glycerine in a condensate, afterwards emulsifying with ammonia and soya protein, using a gear pump for emulsification [134]. These phenoplast emulsions may be used in beater additions in paper treatment (Chapter 13 in Volume 3).

A further procedure is to react the sodium salt of a phenol with an alkylene oxide to produce a phenolic ethylene oxide condensate $C_6H_5(OCH_2CH_2)_nOH$, which will also react to form a novolak provided that $n = 1$, higher condensates being weakly reactive. It is also possible to react a novolak with ethylene oxide to form an ethoxylated product, which may be esterified in the normal manner with an unsaturated acid, such as soya acid, which further reacts with maleic anhydride, adding across the double bond to form a maleinised product. By forming the ammonium salt this will either become soluble or will be self-dispersible in emulsion form if the ethylene oxide addition is low or zero [135].

In an example a phenol–formaldehyde novolak resin (90 g), prepared with phenol–formaldehyde in a ratio of 1 : 0.65 with an oxalic acid catalyst is reacted with 100 g of tall oil fatty acid at 240 °C, entraining water with xylol and then reacting with 30 g of maleic anhydride under nitrogen at 200–220 °C for several hours.

Water (116 ml) is added after partially cooling to give a viscous emulsion, which can be neutralised with ammonia (SG 0.91), followed by adjusting to 40 % of solids with 15 ml of isopropanol and 43 ml of water. This product is described as a 'dispersion', and stoved films, after adding water-dispersible cobalt drier, gives hard resistant films with good adhesion to most substrates, and also good flexibility.

In an alternative process, an aqueous dispersion of a phenol resol is obtained by reacting the alkali-condensed product with 0.2–0.4 moles of ethylene oxide for each 100 g of resol in water in the presence of polyvinyl alcohol [136].

Emulsions may be produced by reacting phenol and formaldehyde in the presence of oleamide and a base. The oleamide acts as a plasticiser, making the condensate suitable for impregnating paper used in printing circuits, and must be capable of being punched without shattering [137]. Condensates of phenol and formaldehyde may be dispersed in vinyl acetate–dibutyl fumarate latices, prepared by conventional methods and used as sizes for wood fibres [138].

Other developments in the preparation of phenoplast emulsions have been mainly based on specific emulsifiers. Thus sodium β-naphthalene sulfonate–formaldehyde condensate at 3–15 % of dry resin may be used as a dispersant for a preformed p-chlorophenol phenoplast, an optional additive being sodium stearylpolyoxyethylene sulfate or an allied surfactant. The resin is first prepared as a damp filter cake to which about one-third of its weight of water is added with the dispersant. This is converted to an emulsion with a high-speed agitator or similar device. The 40 % emulsion has a viscosity of 30 cP and a particle size of about 1 µm. These emulsions give uniform coatings on paper with special reference to copying paper [139].

Many specifications claim the inclusion of a water-soluble polymer, natural or synthetic. In one case identical alkaline solutions of phenol and formaldehyde are condensed to different degrees, giving viscosities of 200–250 and 100 cP respectively. These are blended in a 1 : 4 ratio to give a condensate of viscosity of 600–800 cP at 21 °C. This is emulsified, typically with sodium 2-ethylhexyl sulfate and carboxymethyl cellulose, forming an emulsion with a viscosity of 390 cP at 21.5 °C, suitable for use as a plywood adhesive when applied as a curtain coater [140].

Mixed phenols, e.g. p- and m-cresol, with about 10 % by weight of p-toluenesulfonamide and formaldehyde are dispersed in polyvinyl alcohol (d.p. 2000) solution. The dispersion formed is adjusted to pH 8.4 with aluminium hydroxide and triethylamine and condensed after being given extra stabilization both with a non-ionic surfactant and with the cationic surfactant diethylaminoethyloleylamide. The inclusion of polyvinyl alcohol gives the emulsion formed stability for 7 months at 20 °C [141].

A simple formulation claims the formation of an emulsion from a terpene-modified phenolic resin, polyoxyethylene nonylphenyl ether, polyvinyl alcohol and water (500 : 30 : 20 : 550) to give an emulsion of good storage stability at 20 °C. It is miscible with many vinyl-type latices, including a chloroprene latex with which it forms an adhesive suitable for veneers [142].

An alkaline mixture of >1.5 moles of formaldehyde to 1 mole of a cresol–resorcinol mixture is condensed until a slight turbidity is developed. It is added to a methanol–water (0–4 : 6–14) mixture and reacted at pH 3–3.5, after which organic acid is added until the dark brown mixture is turbid. At this stage the addition of polyvinyl alcohol at 10 % of the original reactant weight, together with about 4.5 % of gum acacia, produces, on homogenization, an emulsion of 40–45 % of solids with >6 months ambient stability. A uniform film is obtained on drying [143].

Another emulsion including phenol, formaldehyde, polyvinyl alcohol, hexamethylenetetramine and tripolyoxyethylene phosphate linoleate is condensed in aqueous solution to give an emulsion. It is intended as a binder for wood fibre [144].

Solvent solutions of an acrylic polymer, an epoxide (Epikote® 1007) dispersed in water containing 2-dimethylaminoethanol, are blended with an A stage phenolic resol solution, based on bisphenol A, to which further ammonia is added. It is suitable for coating tinplate on which it is dried at 200 °C [145].

An emulsion is formed by heating phenol, formaldehyde, a polyurethane ionomer and 48 % sodium hydroxide (329 : 379 : 490 : 13.2) at 80 °C for 5 h to give a polymer which dispersed in 300 parts of water to give a 40 % product, exhibiting no sedimentation for $\geqslant 8$ months. On heating at 150 °C for 30 min, a clear film, resistant to water and acetone, was formed [146].

7 SILICONES

7.1 Elementary chemistry

Silicones are organosilicon polymers, the structures of which depend on the —Si—O—Si skeleton in which an organic group, often methyl CH_3, is attached to each silicon. The simplest silicones are fluids of formula

$$CH_3 \mathrm{-\!\![\!-Si.O-\!\!]\!\!-} \begin{matrix} [CH_3] \\ | \\ [CH_3] \end{matrix} \begin{matrix} CH_3 \\ | \\ Si-CH_3 \\ | \\ CH_3 \end{matrix}$$

They are formed by hydrolysis of the substituted chlorosilanes, e.g. trimethylchlorosilane $Si(CH_3)_3Cl$.

The degree of condensation depends on the number of hydroxyl groups in the molecule. Silanes containing only one hydroxyl group can only condense to form dimers as

$$CH_3 - \begin{matrix} CH_3 \\ | \\ Si \\ | \\ CH_3 \end{matrix} - O - \begin{matrix} CH_3 \\ | \\ Si \\ | \\ CH_3 \end{matrix} - CH_3$$

Mixtures of di- and trichlorosilanes, when hydrolysed, enable branched and crosslinked structures to be built up. However, small quantities of monohydroxysilanes act as chain terminators. Difunctional units polymerise in chains and sometimes in rings to form products of the fundamental structure

$$-\underset{\underset{CH_3}{|}}{\overset{\overset{CH_3}{|}}{Si}}-O-$$

Trifunctional units give rise to branched and also to crosslinked structures, which may be represented as

$$\begin{array}{c} \quad\quad CH_3 \quad CH_3 \\ \quad\quad | \quad\quad | \\ -O-Si-O-Si-O- \\ \quad\quad | \quad\quad | \\ \quad\quad O \quad\quad O \\ \quad\quad | \quad\quad | \\ -O-Si-O-Si-O- \\ \quad\quad | \quad\quad | \\ \quad\quad CH_3 \quad CH_3 \end{array}$$

It will be observed from this idealised formula that the larger the molecule, the closer it becomes in formula to silica SiO_2, and highly hydrolysed and crosslinked silanes ultimately resemble silica in properties, as the latter has a multilattice structure.

It is possible to hydrolyse chlorosilanes with alcohols so as to leave residual alkoxide groups rather than hydroxyl ones. Also, by starting with a silane which has not had all of its hydrogen atoms replaced by chlorine, it is possible for the condensates to contain some residual hydrogen atoms, making for enhanced reactivity. This will also tend to liberate a certain amount of hydrogen gas on base hydrolysis.

The reactivity of silicones, whether by chain extension or by crosslinking, is enhanced by a catalyst, typically acids, bases and metal salts. Zinc salts are included in a one-pack system if low reactivity is required. Other catalysts are indicated in Section 7.2. The hydroxyl group of silicone resins is capable of reacting with the hydroxyl groups of alkyds, epoxides and phenolic resins. As the ether condensation of a silicone[†] may take place in aqueous medium, it produces interesting possibilities in emulsion form. Because of the wide variation possible in the formulation, silicones may vary from oils to hard resinous solids.

A major interest of silicones is in their heat resistance, which makes them invaluable for conditions of high-temperature exposure. There are many other valuable properties, such as in most cases a tendency to water repellancy. However, this depends entirely on the formulation of the silicone, and under some conditions, as indicated later, silicones may have emulsifying properties.

[†] The term 'ether condensation' is not strictly correct as —Si—O—H is acidic.

Silicones

The term 'siloxane' is normally used for silicones containing oxygen bridges. A recent general account of the chemistry is available [147].

7.2 Some methods of emulsification

7.2.1 General principles

High molecular weight siloxanes may be formed from hydroxyl-ended siloxanes of lower molecular weight under strongly acid or alkaline conditions by a process analogous to emulsion polymerisation, although it is actually a condensation [148]. A wide temperature range is possible, optimally at 25–80 °C.

The general formula of the starting product is $RSiO_{(4-n)/2}$, where R is a hydrocarbon radical and $n = 1-3$. The starting viscosity is not critical as long as it is not too high, and R may contain hydrocarbon units as well as additional siloxane units. The most satisfactory emulsifying agents are cationics of the quaternary type, but non-ionic and anionic types may be employed, although the latter should not be included under conditions of acid catalysis.

Recent chinese experiments in the emulsion polymerisation of γ-aminopropyl-substituted siloxanes have been performed with anionic and cationic emulsifiers. The effects of catalyst ratio, molar ratios and concentrations of monomers and temperatures are considered [149].

Continuous mass production of an organopolysiloxane latex is achieved in a two-stage mixing plant with stator-and turbine-type rotors. Polysiloxane fluid or gum, emulsifying agent and water are injected into the first stage. A wide range of concentrations of polysiloxane is possible [150].

7.2.2 Cationic emulsifiers

In an example 8 g of a quaternary surfactant (Arquad® 2HT, Armour Chemicals) is dissolved in 283 g of octamethylcyclotetrasiloxane. This is added to 243 g of water, made alkaline with ammonia, and heated to 70 °C for 10.5 h. The polymeric product has a viscosity of 3000 cSt.

Further examples quote quaternary ammonium salts as emulsifiers, with ammonia as catalyst, at from 1 mole per 100 atoms of silicone to 1 mole of alkali per 50 000 atoms of silicon. Acid catalysts are required in much higher proportion, sometimes as high as 80 % of siloxane weight, HCl and some sulfonic acids being preferred. Emulsions may be above 50 % in concentration, the reaction taking place over several hours at 25–80 °C. Examples quote the polymerisation of octamethylcyclotetrasiloxane and cyclic ethylmethylsiloxane. In some cases the polymer viscosity rises on standing for some days or weeks by further condensation [151].

Other emulsifiers of polysiloxane oils are formed by the acid neutralised addition product of an epoxide with a polyamine. This type of cationic emulsification, which is fairly general, may be performed by first reacting 100 g of the glycidyl ether of 4 : 4 dihydroxydiphenylmethane (epoxide number 0.43) with 12.5 g of diethylenetriamine and 24 ml of methanol for 10 min at 45 °C, and then by adding 8 ml of glacial acetic acid. This gives by itself a fine particle size emulsion in appearance [152].

A reaction of 150 g of a copolymer of hydrogen siloxane and ethyl siloxane dissolved in 100 g of methylene chloride is mixed with 160 g of the above emulsion, 500 g of water and 8 mL of glacial acetic acid to obtain a stable latex of pH 4.2. This latex, filled with silica, is useful as a waterproofing agent for textiles (Section 7.3 and Chapter 12 in Volume 3).

A 42 % methanolic solution of $(CH_3O)_3SiC_3H_6N^+(CH_3)_2C_{18}H_{37}Cl^-$ is mixed with water (48 : 852 mL) and then homogenised with $(CH_3)_2O$-terminated poly-dimethylsiloxane(20 cSt), giving a creamy white emulsion, not separating during 24 h [153].

More complex emulsifiers for polyorganosiloxanes are formed, *inter alia,* by reacting cyanamide or an allied compound, e.g. dicyandiamide, ethyleneurea or benzoguanidine, with bisphenol A–diglycidyl ether or other epoxide, followed by reaction with an ethanolamine. An example quotes a reaction in isobutanol, followed by excess dilute acetic acid. This product is capable of emulsifying a solution of poly [(hydrogen methyl)siloxane] in perchlorethylene, about 10 % of emulsifier on polymer being included [154].

It is possible to prepare self-emulsifying silicones by reacting a silicone with terminal hydroxyls with an isocyanate (Section 8). Reaction of the product with an amine, such as bisaminopropylmethylamine, followed by dimethyl sulfate will function as a quaternising agent, enabling a direct emulsion to be prepared. It is also possible, by the choice of compounds with an active hydrogen instead of amines, to obtain an anionic emulsion [155].

7.2.3 Anionic emulsifiers—mainly sulfates and sulfonates

A range of sulfates and sulfonates are described as emulsifiers and the final emulsion is passed through a cationic exchange resin to obtain the surfactant in acid form, preferably below pH 2.5. An example quotes the emulsification of octamethyltetrasiloxane, the sodium of salt of dioctylsulfosuccinic acid being the preferred emulsifier, the reaction continuing for 7 h at 70 °C, followed by 4 h at 25 °C. The viscosity of the product, although dilute, may exceed 10 000 cP. Applications include that of a release agent and a water repellant [156].

Anionic surfactants, e.g. ethoxylated sulfate esters of general formula RO $(C_2H_4O)_nSO_3H$ (R = alkyl or aryl), are included in a formulation with 7 % of n-octamethylcyclotetrasiloxane. Only 7 % of added water is included in the first instance. The homogenised emulsion is polymerised for 7 h at 70 °C and is claimed to give an 87.1 % emulsion of poly(octamethylcyclotetrasiloxane) [157].

A polyorganopolysiloxane, formed with the aid of an anionic surfactant, is contacted with a cationic ion exchange resin in its acidic, i.e. salt, form. The ion exchange resin reacts with the original anionic surfactant producing a simple salt, and simultaneously a lowering of the pH, since the cationic surfactant base is much weaker than the anionic surfactant acid, usually a sulfate or a sulfonate. This lowering of the pH causes the siloxane to polymerise, this effect being observed from 15 to 90 °C. The resultant polysiloxane emulsions may be crosslinked with the aid of a catalyst, e.g. dibutyltin dilaurate. An example quotes triethoxyphenyl silane with sodium lauryl polyethoxysulfate as emulsifier, which on heating with Amberlyst® 15 gives a high polysiloxane yield [158].

Hydroxyl-terminated dimethylsiloxane, tetraethyl silicate, water, dodecylbenzenesulfonic acid and acidic silica gel (850 : 38.3 : 768 : 19.5 : 32.5) are mixed to form a latex. The films have 272 % elongation [159].

A homogeneous mixture of an alkylbenzene sufonate, an alkylbenzene polyoxyethylene sulfonate, ethoxylated castor oil, a hydroxyl-terminated dimethyl siloxane (120 cSt) and water (0.5 : 1 : 0.25 : 35 : 2) is aged for 18 h, diluted with water (61.25) and neutralised with triethanolamine to give a stable latex of viscosity 560 000 cSt [160].

7.2.4 Non-ionic surfactants and phosphate surfactants

Stable aqueous latices of dimethylpolysiloxane are prepared with an alkyl polyoxyethylene ether and poly(oxyethylene sorbitol) oleate mix as non-ionic surfactants, together with anionic phosphates of an ethylene oxide–nonylphenol adduct. The method suggested is to form a polymer paste of 360 cSt viscosity with the emulsifiers and a small quantity of water. This is followed by mixing with excess water and sodium benzoate to give a stable latex, circa 35 % of solids, unchanged after several months [161].

Another useful emulsifier proved to be a 2.5 % solution of polyvinyl alcohol, including 8.5 % of residual acetate groups. This produced an emulsion somewhat disperse in character. Possible applications are indicated in Section 7.3.

7.2.5 Hydrophilic silicones as emulsifiers

A hydrophilic silicone can be formed by hydrolysing a siloxane containing —Si—H with unsaturated glycols. The resultant product is of general value as an emulsifier, being useful for such products as linseed oil and waxes, but it will also emulsify other siloxanes, which may be dissolved in xylene if necessary [162].

Other modified siloxane products are formed by reacting polyethers containing olefinic bonds with the H—Si group in an organohydrogen polysiloxane or by reacting an epoxide-containing organohydrogen polysiloxane with a polyether containing at least one alcoholic hydroxyl group [163]. They are useful in preparing emulsions of other organo-polysiloxanes.

These silicone-based emulsifiers are sometimes decomposed by water and heat. In general, re-emulsification of the polymer is not possible.

A polymerisable emulsifier based on siloxanes has the formula shown

$$X_aR^1_{3-a}SiO-(SiO)_b-(SiO)_c-(SiO)_d-SiR^1_{3-a}X_a$$

with substituents R^1, R^1, R^1 on top and R^1, X, Y on bottom respectively.

where $R^1 = C^1$–6 alkyl and/or phenyl, $X = (CH_2)_3O(C_2H_4O)_nR^2$, where R^2 is a carboxyl, phosphono or sulfo group, Y = 3-(meth)acryloyloxypropyl, a = 0, 1, b = 1–20, c = 0–10, d = 1–10, $1 \leqslant a+c \leqslant 10$ and n = 5–40.

A typical compound is formed from a dimethylsiloxy-terminated siloxane including dimethylsiloxy, diphenylsiloxy and methylmethacryloyloxypropylsiloxy units. It is reacted consecutively with polyethylene glycol monoallyl ether, hexahydrophthalic anhydride and 2-dimethylaminoethanol to neutralise. This gives a reactive emulsifier, which is polymerised in emulsion with butyl methacrylate to give a film-forming polymer [164].

Further highly efficient emulsifiers are formed by reacting a carboxy silicone and an alkyl aromatic alcohol alkoxylate containing 1–20 carbon atoms in the alkoxylate group. The polyoxyalkylene groups present enable the compounds formed to have the desired emulsification properties [165].

7.2.6 Problems of hydrogen release

Cationic emulsifiers including standard pyridinium salts and morpholinium salts, e.g.

$$CH_2CH_2OCH_2CH_2N^+Cl^-$$

with $C_{12}H_{25}$ substituent on N

inhibit hydrogen cleavage in poly(methyl)hydrosiloxane emulsions. This emulsification occurs at a pH between 2.5 and 5.0. There is a slight hydrogen release, but it is less than 1 % of that with other types of cationic stabiliser, which differ in that the positive N atom is not part of the ring [166].

Another type of cationic surfactant recommended for polysiloxanes with residual hydrogen is represented by the class of ethoxylated N-alkyltrimethylenediamines or N-alkylamines (alkyl > 12 atoms). An example quotes the hydrochloride of an ethoxylated hexadecylamine and dimethylpolysiloxane to which small quantities of perchlorethylene and toluene are added. The surfactant is added at about 8 % on polymer weight. The latex formed is approaching 50 % solids and has pH 2.2 [167].

If an alkyl polysiloxane still contains active hydrogen, stabilization is sometimes necessary to prevent hydrolysis liberating hydrogen. Solutions in lower alcohols or acetone, which include partial fatty acids of polyhydroxy compounds or acrylamide alkyl sulfates, form emulsions spontaneously when added to water including an aminocarboxylic acid such as glycine or a buffer such as potassium citrate. Alternatively, a 40 % solution of a methyl hydrogen polysiloxane with up to 2 % of an amino acid is emulsified with an alkylamine salt or a quaternary ammonium salt, the emulsified pH being adjusted to between 2.5 and 5.0. The optimum viscosity of the silicone oils is 500–1000 cst [168].

7.2.7 Theoretical developments

Russian workers have investigated the emulsification of a resin of formula

$$-\underset{\underset{|}{\overset{|}{OR'}}}{\overset{\overset{|}{R}}{Si}}-O-\underset{\underset{|}{\overset{|}{O}}}{\overset{\overset{|}{R}}{Si}}-O-\underset{\underset{|}{\overset{|}{OR''}}}{\overset{\overset{|}{R}}{Si}}-O-\underset{\underset{|}{\overset{|}{O}}}{\overset{\overset{|}{R}}{Si}}-O-\underset{\underset{|}{\overset{|}{O}}}{\overset{\overset{|}{R}}{Si}}-O-$$

where R is phenyl, R' ethyl and R" n-butyl. This (known as F-9) was emulsified to a 60 % solution in toluene, with varied conditions of stirring and temperature. The best emulsifier was casein as an ammoniacal solution (the percentage is not indicated), which showed a peak particle size in the distribution curve of 0.3 μm diameter, only increasing to about 0.5 μm after 19 months, the emulsion remaining perfectly stable [169]. Examination of some high-viscosity polydimethylsiloxane latices of particle diameter 0.03–0.3 μm has indicated that some creaming occurs, although this is not visible [170].

A summary of the reactions with cyclosiloxane stresses the role of water as a component in the equilibrium determining molecular weight:

$$—CH_3SiOSi(CH_3)_2- + H_2O \rightleftharpoons 2(CH_3)_2SiOH$$

There is an equilibrium constant, as in bulk polymerisation, of about 15 % of cyclosiloxanes to total siloxanes. It is considered that during the reaction the cyclic monomer is converted to an intermediate species, presumably a silanol, at or near the surface of the monomer droplet. This intermediate undergoes a series of condensations and hydrations, including reconversion to mixed cyclic products which diffuse back to the monomer droplets. Polymer formation results from condensation of the siloxanols in the aqueous phase. These precipitate and are stabilised by adsorption of surfactant on the surface and orientation of the surfactant. The process continues with further formation of siloxanols and their conversion to polymer.

There is a general reduction of particle size during the polymerisation, as has been already noted in some patents. Diffusion of the hexamethyl cyclosiloxane, unlike the tetramethyl cyclic compound, is much slower than the potential polymerisation rate, and hence in the former case R_p diminishes with particle size [171].

Precrosslinked poly(organosiloxane) particles are synthesized by emulsion polycondensation/polymerisation of alkoxysilanes and cyclic siloxanes. The crosslink density and organic functionality can be varied widely. Graft polymers are obtained by subsequent emulsion polymerisation [172].

The condensation of methyltrimethoxysilane in the presence of the surfactant benzethionium chloride occurs in micro-emulsions. The condensate can be functionalised with azo groups which are capable of grafting reactions with vinyl monomers, producing a core–shell structure [173].

Based on the phase behaviour of an ethanol/water/non-ionic surfactant/methylphenyl polysiloxane quaternary system, the hydrophile/lipophile balance of non-ionic surfactants is greatly influenced by the addition of ethanol. In general, stability against creaming is controlled by Brownian movement of fine emulsion droplets of submicrometre order [174].

7.3 Textile and paper applications of silicone organopolysiloxane) latices

7.3.1 General properties required

There is in general a wide range of potential applications to textiles. The general function of the silicones is that of imparting water repellancy, but in some cases silicone latices may be used as softening agents.

Silicone emulsions require a catalyst, which reduces both the time and temperature required for the silicone fluid to harden or crosslink on the fibre surface [175–177]. Lead, zinc and tin compounds function satisfactorily as catalysts, provided that there are residual hydrogen atoms on the silicone. Titanium salts are useful with groups of intermediate activity, such as Si-OH and Si—OH and Si—alkoxy groups as distinct from the non-reactive Si—CH_3 groups. Many proprietary catalysts are available. The surface action is in some cases chemical combination of the active groups in the fabrics with the silicone. Almost all standard fabrics, including cotton, rayon, wool, nylon and polyester, may be treated with suitable silicone latices, together with a catalyst as required. The proportion of the catalyst may vary from 10 % upwards of the silicone weight. Zirconium oxychloride may be used under some conditions. The addition of a melamine aminoplast, which also promotes crease resistance, is often recommended, and a typical composition would give a pick-up of 2 % of the silicone and 5 % of the crease-resistant resin at 10 % mangle expression (Chapter 12 in Volume 3).

The Russian work referred to previously [169] indicates that the characteristics of the emulsifier may seriously interfere with one of the characteristic properties for which silicones are used, namely the imparting of hydrophilic and water-repelling properties. One of the principal applications is in textile impregnation. Some very stable non-ionic silicone latices enable water repellancy to be developed after application and drying. In general latices may include either an alkali catalyst or dodecylbenzenesulfonic acid (1.5–4 % on permethylcyclosiloxane). The molecular weight may vary from 49 000 to

172 000 as the reaction temperature falls from 90 to 25 °C, the corresponding polymer viscosities varying from 5200 to 1 260 000 cP; the particle diameter varies from 500–5000 Å and is lower than in the initial emulsion. The polymerisation is terminated by neutralization.

Suitable textile latices may be obtained from the rapidly hydrolysing alkoxysilanes, which in acid conditions are rapidly converted to emulsions of particle diameter 70–100 Å. Advantages claimed, especially with polyester knit fabrics, include improved resistance to wrinkling, wet or dry, improved and variable hand and spot and stain resistance. A special application is improvement in the durable press performance of polyester/cellulosic blends [178].

Methyl-H-polysiloxanes of general formula

$$\begin{array}{c} R' \\ | \\ --SiO-- \\ | \\ R' \end{array} \begin{array}{c} H \\ | \\ (SiO)_{\overline{a}} \\ | \\ CH_3 \end{array} \begin{array}{c} CH_3 \\ | \\ -(SiO)_{\overline{b}} \\ | \\ CH_3 \end{array} \begin{array}{c} R'' \\ | \\ -Si--CH_3 \\ | \\ R'' \end{array}$$

where $a = 3-600$ and $b = 0-200$ (substitution of the CH groups is possible, as are enol end groups), may be emulsified with polyoxyethylene-type non-ionic surfactants, typical cationic surfactants and polyvinyl alcohol. For extra stability 0.1–10 % of polysiloxane may be added. Emulsification takes place by first mixing the emulsifiers with the polymer and then adding water. Toluene may be added to the polysiloxane if necessary to reduce viscosity.

The specific interest of this preparation is that it is coemulsified with a polyamide which has been reacted with epichlorhydrin, about 5–15 parts of this resin being included per 100 parts of polysiloxane. The joint emulsion is applied to textile fabrics with a catalyst, e.g. a zinc or a dioctyl or dibutyl tin salt, and heated for a few minutes at 100–80 °C, giving a water-repellent finish [179]. This latex gives a finish on cotton cloth with low water absorption. Likewise, cationic products are also useful in imparting water resistance to cotton or cotton/polyester textiles [180].

Latices prepared with ethoxylated amines and having some residual active hydrogen directly attached to silicon impart excellent water resistance to cotton poplin after mixing with a zirconium compound as catalyst [167].

A composition containing a technical latex, a 40 % solids dimethylpolysiloxane (SM-2013), with Aircoflex 500, an ethylene–vinyl acetate copolymer latex and Aircoflex 46–3, a similar carboxylated latex, and trichlorethylene, which swells the polymers, together with dibutyltin dilaurate catalyst, is applied to polyester/cotton fabric and dried at 120 °C to 9 % solids pick-up, followed by a standard-type impregnation and cure with dimethylolurea. The cured polymers possess breathability and durability, and are suitable for rainwear [181].

An aqueous emulsion containing 35 % hydroxyl-terminated polydimethylsiloxane is mixed with an aqueous emulsion containing the reaction product of this polydimethylsiloxane with $H_2N(C_2H_4)NHC_3H_6Si(OCH_3)_3$, an emulsion containing 40 % of $CH_3Si(OC_2H_4OCH_3)_3$ and 25 % of bis(2-ethylhexyl dilaurate(sic)[1] emulsion (90 : 4.5 : 5 : 0.5). This is suitable for polyester treatment, the final cure being at 150 °C. There is good fastness to dry cleaning solvents [182].

A latex suitable for providing a release coating and which has rapid thermosetting properties is based on emulsifying mixed cyclic pentamers and tetramers of dimethylsiloxane with the aid of dodecylbenzenesulfonic acid and heating at 85 °C for 1–3 h. This is followed by the addition of methyltrimethoxysilane, followed by heating at 47.5 °C for 15–18 h and subsequent neutralization. Even after diluting to 8 % prior to application, and adding dibutyltin dilaurate, the catalysed latex is stable for 6 months. The latter may be applied to parchment, and hardened for only 30 s at 93 °C. Secondary adhesion, as measured by adhesion to a steel plate, indicates that the parting force is reduced to a minimum at a content of 12.6 mol % of monomethylsiloxane in the polymer [183].

Stable aqueous emulsions of methylsilsesquioxane have been used to confer upon fabrics the properties of dullness, slide resistance and perspiration resistance [184]. Antistatic and lubricating finishes on synthetic fibres are obtained by coating with latices containing 100 parts of dimethyl polysiloxane, 1–20 parts of calcium chloride or lithium chloride and 20–40 parts of a polyol such as glycerol [185]. Amongst specific applications, silicone emulsions may be used with foam-back fabrics.

The chemistry and application of siloxane latices, particularly in waterproofing, is given in a referenced review [186].

7.3.2 Glass fibre treatment

Mixed methylphenylsiloxane emulsions, only one of which is crosslinked, together with a fluoropolymer latex and a graphite dispersion, form an impregnating bath for fibreglass fabrics, the impregnation being fixed at high temperatures. Flexibility varies according to the ratio of the components [187].

An organosilane, which is used in a latex together with polyvinyl acetate, has the object of reducing static electrical charges when used as a size for glass fibres [188]. A cationic silicone latex, a cationic lubricant and urea, suitably diluted, are used to size chopped glass fibers [189]. A cationic latex, curable at 177 °C, gives an abrasion finish suitable for glass cloths [190].

7.3.3 Paper treatment

Treatment for paper depends on characteristic non-adhesive properties and water-repellant properties [191]. Silicone emulsions may be applied to both

---

[1] Copied from Chem. Abstr. Error uncertain, possibly dibutyltin dilaurate.

Silicones

absorbent and non-absorbent papers. Kraft and parchmentised kraft, either with or without previous resin treatment, wallpapers with a resin film, and sulfite-based papers for cups and for cartons may be treated successfully. The concentration of the silicone latex may vary between 2 and 20 %. For general methods of application see Chapter 13 in Volume 3.

Silicone latices may be applied in conjunction with other types of paper finishes, such as vinyl acetate copolymer latices for wallpapers, and also as part of the composition for backing papers for adhesive-coated materials. There is also a general application for release papers.

7.4 Silicone latices in surface coating and building

The properties of silicones make them obvious candidates for a surface application where water resistance is desirable. The use of silicones on porous brickwork is well known [192]. Various polysiloxanes may be used on building boards containing free lime provided that an acrylic, an acrylic–styrene or a butadiene–styrene latex is blended with them over a wide range of ratios. Typical silicone latices are shown in Table 7.13 [193].

Cyclic dimethylsiloxane oligomer and γ-methacryloyloxypropyltrimethoxysilane (95 : 5) are reacted in aqueous dodecylbenzenesulfonic acid and its sodium salt at 85 °C for 4 h to give a silicone emulsion. This is used as a seed for the emulsion polymerisation of methyl methacrylate, butyl methacrylate and methacrylic acid (342 : 540 : 18) with a potassium persulfate initiator at 70–80 °C for 2 h. The latex, of particle diameter 55 nm, gives a transparent film [194].

Emulsion blends containing 13–20 g of dry resin mixture, containing about 20 % by weight of silicones, were applied to standard building boards, which absorbed about 410 g of water $m^{-2} h^{-1}$. After rapid air drying followed by heating at 50 °C for 2 min, a test after 24 h showed that water absorption dropped to 40–50 g $m^{-2} h^{-1}$. These emulsion blends may be pigmented [195].

Table 7.13 Silicone latices [193]

	Parts by weight	
	1	2
Methylphenyl polysiloxane (Me + Ph)–Si = 1.5 : 1		
Methyl polysiloxane (Me–Si = 1.6 : 1)	—	15
Stearic acid	2.5	—
Morpholine	1.1	—
Octylphenyl polyglycol ether	—	3
Toluene	15	—
White spirit	0.7	—
Water	65.7	67

Organosiloxane latices in which methyl or phenyl are directly bonded to silicon, and at least 50 % of a silicon-bonded organic group should be monovalent hydrocarbon radicals, are suitable for application to stone surfaces, e.g. chippings for road surfaces. The adhesion of bituminous materials is improved. These latices may also be used for buildings [196].

Silicon-containing polyesters with silicon–carbon links are prepared by treating unsaturated polyesters with hexamethyldisilazane to block the hydroxyl groups, grafting silyldichloride groups to pendant polymer chains via hydrosilation and hydrolysing the SiCl bond to the stable silanediol. The latices formed give hardened films spontaneously [197].

In general, the addition of sesquisiloxane latices to other types of latex such as polyacrylates, polyvinyl chloride and styrene–butadiene has the effect of reinforcing the vinyl-type polymer, in some cases nearly doubling the tensile strength [198]. A siloxane latex that has been treated with a cationic ion exchange resin is suitable in the neutralised form for inclusion in emulsion paints. It may be applied to fresh plaster under difficult weather conditions [199]. Coatings for the protection of buildings consist of an emulsified silicone binder and a styrene–acrylic ester latex. The silicone latex reduces water absorption without affecting the water vapour permeability [200].

Other applications of silicones will be found in Chapter 10 in Volume 2, with emulsion paints, and Chapter 16 in Volume 3, on building applications. A number of rather complex siloxane blends are emulsified with and blended with dioctyltin dilaurate and further emulsifier. The modified latex shows good storage stability, paintability and mould-release properties and is thus useful for release coatings [201].

7.5 Miscellaneous applications

One of the major characteristics of silicone latices, imparted by fluid silicones, is as a good release coating, preventing the adhesion of most types of tacky materials. A typical commercial latex, Dow-Corning HV-490, an anionic latex prepared from 100 000 cst dimethyl polysiloxane, of 35 % solids content, has excellent release characteristics for many rubber, plastic and metal surfaces. It has good wetting ability in almost all cases. It may, however, cause difficulty with subsequent painting, in which case the surfaces may be washed with solvent and then with detergent, followed by immersion in alcoholic potassium hydroxide and rinsing in clear water before painting. The non-stick properties make this latex useful to add to aerosol starches, since the slip characteristics make for excellent lubricity and ease of ironing.

Some silicone latices contain curable polymers such as linear siloxanes with at least 10 units per molecule and a sesquisiloxane, and are suitable for depositing insulating coatings on wires, coatings on medical devices and decorative coatings on wood, metal and concrete [202]. Other polysiloxanes modified by di-isocyanates, as already described [156], dry to a tacky mass useful in cosmetics, hydraulic fluids, cleaning compositions and dispersants.

8 POLYURETHANES

8.1 Chemistry of isocyanate adducts

The chemistry of the polyurethanes, one of the most versatile series of polymers, is bound up with the chemistry of the very reactive isocyanate —NCO group. This group reacts with almost every compound containing an active hydrogen, including water. If both compounds are bifunctional, chain extension becomes possible. One of the isocyanates most regularly used in commerce is tolylene diisocyanate $CH_3C_6H_3(NCO)_2$, which is a mixture of the 2 : 4 and 2 : 6 isomers in most cases. Some typical chain extensions are:

(a) With a glycol:

$$n\,R(NCO)_2 + n\,HOC_2H_4OH = [-RNH\overset{O}{\overset{\|}{C}}OC_2H_4O\overset{O}{\overset{\|}{C}}NHR-]_n$$

(b) With water:

$$R(NCO)_2 + 2H_2O = R(NHCOOH)_2 = R(NH_2)_2 + 2CO_2$$

Water is chain terminating. However, since the amines themselves, which are formed by decomposition of the carbamic acid, can take part in chain extension to form a substituted urea as in (c) below, water is for this purpose bifunctional.

(c) With diamines:

$$nR(NCO)_2 + nR'(NH_2)_2 = [-RNHCO\,NHR'OCNHR'NH-]_n$$
poly-substituted urea

(d) With carboxylic acids:

$$n\,R(NCO)_2 + n\,R'COOH = [RNH\overset{O}{\overset{\|}{C}}R'NH\overset{O}{\overset{\|}{C}}R-] + 2\,CO_2$$

The reactivity of di-isocyanates varies widely. Thus, *para*-tolylene di-isocyanate is 7–8 times as reactive as the *ortho* modification, and electron-withdrawing substituents in a benzene ring have a very powerful positive influence on the reactivity of di-isocyanates.

There is wide variation in the compounds reacting with di-isocyanates. Primary amines are the most reactive, followed by primary alcohols, water, secondary and tertiary alcohols, other urethanes, carboxylic acids and

450   *Unsaturated polyester and non-vinyl emulsions*

carboxylic acid amides—in that order. Catalysts such as tertiary amines, tin salts and especially organotin salts such as dibutyltin dilaurate have a strong accelerating effect on isocyanate/hydroxyl reactivity.

Another essential valuable concept is that of a blocking agent. This is, in general, a simple monomeric reagent capable of reacting with the isocyanate group without chain extension, and includes alcohols, phenols, thiols, e.g. C_6H_5SH, and tertiary amines. These react normally with the active NCO group, but the adducts are decomposed at high temperature, preferably in the presence of a polymeric and non-volatile reactant, and are thus 'unblocked' at higher temperatures. The volatile moiety is removed at the reaction temperature, whilst the polymeric compound reacts to form the final polyurethane.

Useful technical products are based on a blocked di-isocyanate, usually masked by a phenol. A prepolymer can also be masked in the same way. When used with a suitable polyol, temperatures of about 150 °C with a 30-min cure are required. This type is of considerable interest in the preparation of latices.

Further useful products include the so-called 'urethane oil', in which the di-isocyanate is reacted with hydroxyl groups in a modified natural oil or an alkyd. A third popular class of isocyanate is the 'moisture-cured' type, in which the condensate of excess di-isocyanate with a polyol is allowed to cure in film form by the moisture of the air. This, as indicated above, is via an unstable intermediary which decomposes to an amine, with the slow loss of carbon dioxide. If the polyol contains more than two reactive groups, there is obviously a strong latent crosslinking propensity.

An alternative method of handling polyurethanes is by a two-pack system, which may consist of an independent isocyanate or prepolymer and polyol. In this case mixing is immediately before application, the polyols being either polyesters which are hydroxyl ended, or sometimes polyethers derived from polyethylene oxide. Another method is to rely on a catalyst, which may contain a hydrolylic component, to accelerate the curing by atmospheric moisture.

This short account must suffice for the chemistry of polyurethane formation. More detailed surveys are available [203, 204].

8.2 Polyurethane latex

It might be considered impossible to form a latex from a polyurethane other than by post-emulsification of a completely reacted product. However, at least three routes have been found to be practical. One of these is the use of a blocked isocyanate, or prepolymer, with which the polyol can be mixed if it is water soluble, or coemulsified if it is not. It is possible to react an isocyanate end group with an amine so that the product as a whole will be self-emulsifying; a simple emulsification with anionic or cationic reagents is also possible. In the latter two cases there may be a slow, but controllable, chain extension by water, or even a crosslinking, but this can be controlled so that the product as a whole does not precipitate.

The term 'emulsion polymerisation' can apply to the polyisocyanate/polyurethane systems, although, of course, it is a different phenomenon to vinyl addition polymerisation. The term 'latex' is generally used in reference to the emulsions of polyurethanes. A polyurethane has in general both urethane, -NH-CO-O and urea NHCONH groups in the polymer chain. A preponderance of the urea groups is desirable to obtain the most desirable characteristics of the polymers.

8.2.1 Emulsions via blocked copolymers

This is a simple system for preparing a water-based polyurethane. Prepolymers are normally prepared before the blocking reaction. Note that in many cases some solvent is present, and it may be distilled at a later stage.

In a typical preparation, a hydroxyl-ended polyester is prepared from a condensate of 4.8 moles of ethylene glycol, 1.2 moles of propylene glycol and 5.5 moles of adipic acid, removing 97 % of theoretical water until the acid number is <3. Of this product 0.25 equivalents are reacted with 0.54 of mixed 2,4-and 2,6-tolylene di-isocyanate at 50–80 °C under nitrogen for 4 h, The equivalent weight of the final product is about 1400. Then 1000 parts of this prepolymer by weight having an active. NCO content of 3.17 % are reacted at 55 °C in 450 parts of toluene including 57 parts by weight of methyl ethyl ketoxime, the blocking agent:

$$\begin{array}{c} CH_3 \\ \diagdown \\ C{:}NOH \\ \diagup \\ C_2H_5 \end{array}$$

Finally, after the reaction is complete, 48 parts of $N,N,N'N'$-tetra(2-hydroxypropyl)ethylenediamine are added to form a prepolymer of about 70 % concentration, giving a ratio of active. NCO–blocking agent–curing agent (active hydrogen) of 1.15 : 1 : 1.

Then 100 parts of the prepolymer solution are emulsified in 99 parts of water with the aid of 10 parts of a 1.9 mixture of sorbitan monolaurate and polyoxyethylene monolaurate, and 5 parts of hydroxyethyl cellulose; 7 parts of a lead octoate catalyst are then added. The latex is prepared by adding the water phase to the solution until phase reversal to an o/w latex forms, producing a stable product at 35 % of solids. Other latices also include colloidal silica [205]. These latices are used to treat textiles (Chapter 12 in Volume 3).

In another example, a prepolymer formed from toluene di-isocyanate(TDI) and propylene oxide–trimethylolpropane condensates is blocked with ε-caprolactam or phenol. A latex is formed with a polyalkylene glycol ether surfactant or a propylene oxide–ethylene oxide block copolymer, including also the polyols used for chain extension, the NCO–OH ratio varying from 1 : 1 to 10 : 1. The latex, with the addition of catalysts, may be used for paper impregnation, providing good wet tensile strength and light stability after curing at 105 °C for 90 min [206]. A polymethylenepolyphenylene di-isocyanate, blocked with caprolactam in a solvent which may be tetrahydrofuran, is dispersed in a large excess of 1 % aqueous lauryl pyridinium bromide [207].

A specification quotes a range of blocking agents including n-butanol, a polyoxyethylene monolaurate, diethanolamine or caprolactam. These are claimed to form emulsions directly and are suitable as pressure-sensitive adhesives. The blocking agents are removed at various temperatures, as already indicated, enabling further chain extension and crosslinking to take place [208].

8.2.2 Emulsification by salt formation—cationic

A spontaneous type of emulsification by salt formation has been investigated by research workers in the laboratories of Farbenfabric Bayer in Germany and Wyandotte Chemical Corporation and W.R. Grace in the United States, *inter alia*. In most cases they are cationic. A theoretical study has been made including the use of various macroglycols and of hydroxyamines, e.g. N-methyldiethanolamine for the formation of prepolymers which are still .NCO ended. Chain extension may occur slowly due to the water itself, or other active chain extenders may be added. Emulsions are formed by salt formation, e.g. as the hydrochloride. When film formation takes place, it is considered that the strong ionic forces involved cause the films to resemble ionically crosslinked vinyl compounds. In this, polyurethane latices differ from vinyl and acylic latices in that even polyurethane latices which give very hard films nevertheless form continuous films at ambient temperatures due to strong electrostatic forces [209, 210].

In a typical procedure, a prepolymer prepared from a di-isocyanate and a hydroxy-terminated polyester or polyether is reacted with an alkyl diethanolamine as follows:

$$3\ OCN \sim NCO + 2\ RN(CH_2CH_2OH)_2 \longrightarrow OCN \sim \underset{H}{\overset{R}{\underset{|}{N}}} \sim \underset{H}{\overset{R}{\underset{|}{N}}} \sim NCO$$

The partly extended urethane is added with high-speed mixing to aqueous 3 % acetic acid, when an emulsion forms spontaneously. Curing of the emulsion occurs as water diffuses into the prepolymer globules and reacts with the isocyanate end groups as follows:

$$\text{OCN} \sim \underset{\underset{H}{|}}{\overset{\overset{R}{|}}{N^+}} \sim \underset{\underset{H}{|}}{\overset{\overset{R}{|}}{N^+}} \sim \text{NCO} \xrightarrow{H_2O} \underset{\underset{H}{|}}{\overset{\overset{R}{|}}{N^+}} \sim \text{NHCONH} \sim \underset{\underset{H}{|}}{\overset{\overset{R}{|}}{N^+}} \sim + CO_2$$

An alternative method is to add a water-soluble diamine. If limited quantities of triethanolamine are included in the original prepolymer a crosslinked system becomes possible since this is trifunctional with respect to hydroxyl [211].

The addition of 0.96 parts of triethanolamine and 2.2 parts of methyl diethanolamine per 100 parts of prepolymer having about 3.6 % of NCO content has the effect of increasing tensile strength whilst reducing the tendency for tetrahydrofuran to dissolve the film as a whole.

The stability of a series of latices was fairly good, a sediment of larger particles which formed after a few days being readily dispersed. Most particles were relatively large, 2–3 μm. Although further additives were not essential, polyvinylpyrrolidone at 1.5 % of additive improved the mechanical stability, as did hydroxyethyl cellulose at 0.5–1 % of additive, the latter also having a thickening effect. Pigmentation with titanium dioxide or a non-ionic dispersed carbon black is possible. Alkalis cause precipitation, but mixing with a non-ionic polyvinyl acetate latex is possible. These latices retained a certain amount of toluene, used initially as a solvent diluent. A number of variants have been disclosed; e.g. the urethane prepolymer may be directly quaternised in a solvent which is afterwards removed [212–224].

A series of polyurethanes containing about 1 % of quaternary ammonium groups may be prepared by direct quaternisation of polyurethanes containing tertiary amino groups. The use of alcohols containing halogens and tert-aminoalcohols enables a combined di-isocyanate polyaddition and quaternisation to take place simultaneously. The polymers are soluble in polar solvents in the salt form, and addition of water followed by evaporation of the solvent enables solvent-free cationically charged latices to be formed. There is no need to add any further emulsifiers, since many forms of secondary reactions can be used to induce salt formation.

Crosslinking is possible in the particles themselves due to chain extension by water where there is surplus isocyanate. Formaldehyde may be used, preferably directly on the film, whilst bifunctional quaternisation, e.g. by 1,4-bischloromethylbenzene, also induces quaternisation.

Extremely fast drying and self-curing are features of many of these systems, although this can be a disadvantage in some cases. The films do not become brittle and high gloss is a feature, as is adhesion to a wide range of substrates. Cationic types, however, are difficult to pigment. General disadvantages of these latices are very low critical pigment/volume concentrations and intercoat adhesion [225].

A prepolymer with excess NCO groups, formed from a di-isocyanate and a polymeric diol, is emulsified with an ethylene oxide–propylene oxide block copolymer, hydrogenated 4,4-diphenylmethane di-isocyanate being suggested as suitable. An NCO–OH ratio of 1.5 : 2.0 is advised. If necessary a solvent such as a ketone or a *tert*-alcohol may be present. Other emulsifiers may be included, and it should be noted that if an alkali soap is used, the carbon dioxide liberated tends to neutralise it.

This prepolymer emulsion is chain-extended, with a diamine, polyoxypropylenediamine $H_2N[CH(CH_3).CH_2O]_n$-$CH_2CH(CH_3)NH_2$ or dimethylpiperazine, piperazine, being quoted. Hexamethylenediamine is not suitable. The films have elastomeric properties [226].

It is possible to use an acid containing hydroxyl groups, e.g. citric, tartaric or 2,2-bis(hydroxymethyl)propionic acid for chain extension, followed by neutralization with an amine. Thus the last-named acid may be reacted with TDI to form a prepolymer with 5.7 % of free NCO groups. The polymer is esterified and simultaneously emulsified with dilute triethylamine to give a latex of 35 % of solids. This may be blended with an alkali-soluble polyacrylate as a thickener to give a product suitable as an adhesive in laminated or flocked textiles [227].

A polyurethane solution is emulsified with the aid of a water-soluble crosslinking reagent, preferably a polyol containing *tert*-nitrogen, and a nonionic surfactant, preferably at 4–6 % of the organic solution. It is desirable to age the latex for 5–10 days. The particle size is <2 μm. This latex may be applied to textiles with the addition of N,N'-dimethylethylene, its use being for textiles with wash–wear and durable press characteristics [228].

A salt-forming polyurethane may be utilised in the emulsification of another polyurethane not containing salt groups, which may be present at from 40–99.5 % of total polymers. The presence of a solvent is optional [229, 230].

Stable latices are formed by reacting polytetramethylene glycol (MW 2000, hydroxyl number 56.7) with 2,2-dimethylpropionic acid and bis(4-isocyanatocyclohexyl)methane (615 : 148 : 485) to form a prepolymer. This is reacted with triethylamine, water and isophoronediamine (1000 : 27 : 773 : 26) to form a latex [231].

8.2.3 Emulsification by salt formation—anionic

It is possible to prepare an anionic latex by using compounds containing active hydrogen, which also contain carboxyl or sulfonate groups [232–234]. Many specifications describe anionically charged latices, usually with sulfonate groups built into the polymer chain, which are thus self-emulsifiable. An aromatic polyurethane prepolymer is reacted with a sulfonating agent, e.g. chlorosulfonic acid. The sulfonated polymer is suitably neutralised, followed by the addition of water to give simultaneous chain lengthening and emulsification. The final latices are considered to be suitable for emulsion paints [235].

Propane sultone $CH_2CH_2CH_2OSO_2$ is reacted in benzene with diethylenetriamine to give a sulfonic acid of the triamine, readily converted to the sodium salt. This is reacted with a typical prepolymer based on TDI to produce a latex with simultaneous chain extension [236, 237].

Aminosulfonates such as sodium 2-[β-(cyclohexylamino)propionamide]-2-methylpropane sulfonate as a 50 % solution may be added to prepolymers formed from hexamethylene di-isocyanate (HMDI) and adipic acid–glycol polyesters at 80 °C, the temperature being held at 80–90 °C for 150 min. It is followed by the addition of further water and formaldehyde. The resultant latices are viscous, producing elastic films at pH 3–4 [238].

Sulfamic acid has been used to emulsify a polymer in which there are excess NCO groups [239].

Further sulfonate-type self-dispersing polyurethanes may be prepared by heating sodium isothionate with ethylenediamine to form sodium 5-amino-3-azapentane-1-sulfonate. These derivatives, on addition to prepolymer, form emulsifier-free latices from which colourless, tough, light-resisting films are derived [240].

A prepolymer derived from hexamethylene di-isocyanate (HMDI) and a polyester in acetone has been reacted with mixed potassium salts of a 1 : 1 ethylenediamine sultone addition product, together with potassium lysinate $NH_2(CH_2)_4CH(NH_2)COOK$. On distilling the acetone a 44 % stable latex is obtained from which films with excellent water resistance are derived [241].

Anionically charged stable latices where no further stabilisers are necessary are formed from the metal, ammonia or amine salts of a diaminocarboxylic acid which is reacted with an NCO-ended prepolymer and another compound containing active hydrogen atoms [242].

Another rather unusual method of providing self-emulsification is by the use of the 1 : 1 adduct of ethylenediamine and potassium acrylate, i.e. potassium N-β-(aminoethyl)-β-alaninate, which may be reacted with the prepolymer formed from HMDI and hydroxyl-ended polyethylene glycol adipate. The addition takes place in acetone, but the resultant polymer is self-emulsified when added to water. The dried films are featured by low water uptake [243].

Other methods may be adapted to ensure self-emulsification. Thus the reaction product of a polyethylene glycol (MW 5000–10 000) with an aromatic epoxide, e.g. bisphenol A–diglycidyl ether, is disclosed. The molecular weight of the emulsifiers varies from 12 000 to 24 000. Standard chain extenders such as diamines, e.g. dimethylpiperazine, are used. Thus the emulsifier also plays the part of a chain extender. A very wide range of applications is also claimed, including coatings, impregnations for latices to reduced abrasion and rub resistance, non-woven fabrics, industrial adhesives and metal and wood industrial coatings [244].

In a mixed system for emulsification, a prepolymer is mixed with a solution containing diaminobiuret, a sulfosuccinic acid salt surfactant, a wax, presumably predissolved, and the reaction product of a caprolactone glycol

456   *Unsaturated polyester and non-vinyl emulsions*

and propanesultone. This forms a latex, stable for at least 6 months under ambient conditions, which produces a film of good tensile strength and high elongation [245].

The preparation and properties of anionic polyurethane latices prepared from isophorone di-isocyanate and dimethylpropionic acid and containing polybutadiene polyols are discussed [246].

8.2.4 *Direct (chain extension) emulsification*

It is practicable to emulsify a prepolymer directly provided that the reactivity is not excessive. In general, the latices should be applied within 24 hours of preparation. A suitable latex has not more than one crosslinking hydroxyl group for each 1200 units of molecular weight and a ratio of free isocyanate groups to total hydroxyl groups of between 1 : 2 and 2 : 1. A typical prepolymer is formed by heating 288 parts of toluene 2,4-di-isocyanate with 1000 parts of polytetramethylene ether glycol of molecular weight 1000 at 80 °C for 4 h. The latex is formed by diluting the prepolymer formed with one-third of its weight of toluene and emulsifying with 4–6 parts of a non-ionic or an anionic surfactant in a homogenizer to give about 30 % of solids. The direct emulsification of a prepolymer containing 5.77 % of NCO and with MW 500–10 000 has been claimed. Sodium dodecylbenzene sulfonate is the emulsifier, the operation being purely mechanical. The latex has a particle size of 0.5 μm and pH of 5.3. It is stable to shear without creaming for at least 3 months [247].

Isocyanate prepolymers may be extended by a polyethylene oxide compound containing active hydrogens, usually as hydroxyl. These prepolymers become self-emulsifying [248–254]. They are of particular use for the impregnation of leather and paper [255].

Non-ionic urethane latices having improved low temperatures have been described [256]. Non-ionic emulsifiers may optionally be mixed with polyvinyl alcohol, producing a viscous latex, with a solids content of 17.5 % [257].

Fatty acid salts of alkali metals at 0.3–3 % of polymer, together with 0.2–1 % of an alkyl or alkylaryl sulfonate or sulfate have been used in a direct emulsification of a polyaryl isocyanate. This latex is suitable as a coating for cellulose [258].

Adipic acid–1,6-hexanediol–neopentyl glycol copolymer, methoxypolyoxyethylene, MDI and piperazine hexahydrate (550 : 42 : 250 : 95) are used to prepare a 40.7 % latex which forms a tough elastic coating on glass. No additional emulsifiers are required [259].

The same inventors also claim the formation of a 55 % latex from a 1,6-hexamethylene di-isocyanate, which is chain-extended to form a latex free from emulsifiers by first reacting a blend of the polyester above (OH value 58, MW 1935) and polyethylene glycol monomethyl ether with HMDI (967 : 35 : 103) and then with 2,4-butanediol and trimethylolpropane, in methylethyl

Polyurethanes

ketone and isopropanol (45 : 0.4 : 807 : 403). After aqueous dilution and removal of the volatile solvents in vacuum, the resultant latex has a solids content of 45.2 %. It shows good dispersibility for aluminium hydroxide [260].

A sulfonated polyether, an adipic acid–1,6-hexanediol–neopentyl glycol polyester (OH Value 56.3), isophorone diisocyanate and ethylenediamine (11.2 : 118.5 : 22.2 : 1.5) with water form a 55 % latex [261].

Other prepolymers are formed with natural oils, e.g. diglycerides or alkyds containing residual hydroxyl groups. Solvents that may be later distilled are optional. In this case all isocyanate groups are substantially reacted with hydroxyl before emulsification, and driers such as oil-soluble salts of lead, manganese or calcium should be added to assist drying; i.e. the unsaturated fatty acid units derived from the oil dry by oxidation and polymerisation.

The procedure is to grind the pigments as for an oil composition and then to emulsify this in non-ionic or anionic surfactant solutions, or a mixture thereof, most oxyethylene ethers and many sulfates or sulfonates being suitable. In an example, a solution of 2 % of sodium lauryl sulfate together with 2 % of methyl cellulose has been found to be suitable. Optimum pigment volume concentrations are between 30 and 42 % [262].

8.2.5 Vinyl–urethane block copolymers

A simple, but ingenious method includes urethane and vinyl monomer on the same chain, in some cases with crosslinking. If a prepolymer prepared with an excess of .NCO units is reacted with a vinyl monomer containing hydroxyl, e.g. propylene glycol monoacrylate, a vinyl-terminated prepolymer is formed, and the remaining isocyanate groups can be destroyed with further chain extension by the action of excess water, forming a very viscous product.

This viscous product can be directly emulsified by high-speed agitation with various ratios of other monomers, e.g. vinyl acetate, methyl acrylate, styrene or vinylidene chloride, using optimally about 2 % by weight of polyvinyl alcohol with an anionic sulfate or sulfonate-type surfactant, although a varied system can be used. Polymerisation follows the normal routine, as shown in Chapter 3.

Typical products are remarkable for very high tensile strength, the tensile modulus of the resultant films still showing reasonable elongation at break. A feature of these copolymers is a tendency to resist blocking [263].

In some cases chain addition can be made to isocyanates of molecular weight 300–20 000 in emulsion form by first using an active hydrogen giving emulsifying properties and chain-extending with a polysiloxane containing active hydrogen (see Section 7) [264, 265].

8.2.6 Vinyl polymerisation in a polyurethane latex

A polyurethane latex (PU) as in A in Table 7.14 is the water phase for an acrylic ester polymerisation as in B. The final mixed latex is suitable as a coating [266].

Table 7.14 Acrylic latex formed in a PU latex [266]

	Parts by weight
A. Polypropylene glycol	49
Dicyclohexylmethane diisocyanate	176
Dimethylpropionic acid	70
N-Methylpyrrolidone	106
Heat the above at 80 °C and then add:	
Triethylamine	48
Hexanediamine	5
Water	456
Then add B, using the normal emulsion polymerisation process.	
B. Aqueous latex above	400
Water	456
Methyl methacrylate	215
Butyl acrylate	65
tert-Dodecyl mercaptan	0.8
tert-Butyl hydroperoxide, 10 % aqueous	10
Sodium formaldehyde sulfoxylate	10
Additions are during 3 h at 70 °C maintaining this for 1 h.	

An allied product is disclosed in which 2-ethylhexyl acrylate–styrene (7 : 3) are polymerised in the presence of a latex of particle diameter 0.006 μm, formed from TDI, polyoxypropylated bisphenol A polyamines, dibutylamine and succinic anhydride. The latices have good gloss and water resistance [267].

Ethyl acrylate may be polymerised conventionally in a latex formed by chain extension of a prepolymer formed from methylenebis(isocyanatocyclohexane) with ethylenediamine, the monomer being in a proportion comparable with the polyurethane. The resultant films have high tensile strength and elongation [268].

8.3 Applications of polyurethane latices

8.3.1 General properties, including adhesives

Polyurethane latices are of interest in applications to textiles, paper, leather and in some cases as special purpose paints. In general, they may be pigmented in the normal manner. Adhesion is excellent to almost all surfaces, including glass, metals, wood, paper, masonry and fabrics [269].

A range of pigments has been found compatible with anionic latices including some of the phthalocyanines and some carbon blacks, especially when predispersed. Certain technical pigments or their pastes are also compatible wih non-ionic poyurethane latices and tests have shown good colour retention. Various titanium dioxide pigments may be used where white colour is desired, using 1 % of tetrasodium pyrophosphate as a dispersant.

A latex formed from polypropylene glycol, TDI, diethylenetriamine, triethylenetetramine, succinic anhydride and ammonia, of particle diameter 0.006 µm, is mixed with glycerol diglycidyl ether at 2 % by weight, giving an adhesive with good water resistance and peel strength [270]. A 60 % solution of TDI in trimethylpropane–ethyl acetate is blended (3 : 100) with a 71.5 % solid vinyl acetate–butyl acrylate–2-ethylhexyl acrylate–methacrylic acid–methacrylamide (8 : 30 : 25 : 3 : 0.6) latex. This forms a pressure-sensitive adhesive giving excellent adhesion of corona discharge-treated polyethylene film to stainless steel and also giving good anchoring to untreated polyethylene film [271, 271A].

8.3.2 Textiles and allied applications

General textile and allied applications of polyurethane latices include coatings and saturants, binders for non-woven fabrics and for dipped goods such as gloves and footwear, and as finishes for synthetic and natural leather. The properties imparted are clarity, high abrasion resistance, high tensile and tear strengths, good colour stability and good adhesion to substrates. Coatings may be applied by any of the conventional methods, including trailing blade, air knife and reverse roll coating (Chapters 12 and 13 in Volume 3).

Cotton fabric specimens can be coated with a thickened latex and fused for 10 min at 150 or 200 °C for a few seconds under pressure. Most latices, except those with the softest texture, show a tendency to cloth failure on a peel test.

The latices of blocked copolymers which have already been described (Section 8.2.1) are considered to be exceptionally useful for imparting crease and abrasion resistance to fabrics. On application, a typical latex, including the active hydrogen compound, is diluted to about one-third with water, added on to the cloth, i.e. scoured and bleached cotton print cloth, to a wet pick-up of 100 %. A cure of 5 min at 150 °C is adequate to provide good tensile strength, tear strength and crease recovery, abrasion resistance and initial crease recovery being still further improved by extending the cure to 10–15 min.

Isocyanate derivatives of some fatty acids may be used in emulsion form to improve the water resistance of cellulosic and fibrous materials and leather [272]. Latices containing vinyl polymers and extended polyurethanes are suitable for the production of felted textile materials [273]. A foamed polyurethane emulsion, prepared directly from a solution, is used to impregnate a staple fibre web in the manufacture of a non-woven fabric [274]. Latices based on technical non-ionic and cationic aliphatic polyurethanes are used in textile finishes, improving, *inter alia*, the bond, appearance and hydrophilicity [275].

A water-dispersible polyamide is formed from ethylenediamine, isophthalic acid, adipic acid and sodium dimethyl 5-sulfoisophthalate (600 : 332 : 365 : 148). This is then dispersed in water (3000) to give a solution of, pH 11.8. Hexamethylene di-isocyanate is dispersed therein to give a stable 42.1 % latex suitable for finishing polyester fabric [276].

A latex of a phenol-blocked prepolymer of polyethylene glycol and 2.4-TDI with a crosslinking agent such as triethanolamine provides a finish for polyester fabric, the cure temperature being 170 °C. Some formulations also contain vinyl-type polymers. Thus a prepolymer of polybutylene glycol–HMDI in toluene/petroleum ether solution may be converted directly to a latex with a paraffin sulfonate, after blending with an emulsion of a copolymer containing 2-hydroxypropyl acrylate as a comonomer. It is utilised to give cotton fabric a crease-and abrasion-resistant finish of good dimensional stability [277].

Freshly emulsified polyisocyanates of relatively low molecular weight may be used in conjunction with other synthetic latices, especially polyacrylates, for finishes to textiles, whether they are of natural or synthetic fibre. Better resistance is given to creasing and abrasion, with improved tensile strength and dimensional stability [278].

8.3.3 Paper applications

Paper applications are of considerable interest, including beater addition saturation and coatings (Chapter 13 in Volume 3). The superior abrasion resistance makes them valuable in packaging. In saturation operations, the increase in fold strength is noteworthy, whilst burst strength is superior to untreated papers. For methods of treatment for both beater addition and saturation see Chapter 13 in Volume 3. The properties of the saturated papers not unnaturally vary with the physical properties of the polyurethane latex selected for use and the degree of addition. Thus, fold factor increases with softness at the expense of the burst factor.

A mixture of a hexadecylketene dimer, TDI extended with polytetramethylene glycol and diethylenetriamine, and water is emulsified, being stable for 10 days, and is suitable for paper sizing [279]. Specific suggestions for treated papers include gasketing, abrasion-resistant cartons, backing for masking and pressure-sensitive tapes, map paper, book covers, luggage labels, waterproof adhesive papers, release papers and imitation leather binders [214, 280].

A combination of a suitable vinyl-type latex with a chain-extended and blocked di-isocyanate produces compositions suitable for a wide range of applications under the broad heading of 'adhesives'. These include various laminates, especially non-woven and flocked fabrics. Typical latices suitable for this blend contain both a hydroxyl and a carboxyl group, e.g. a monomer balance of 10 parts of butyl acrylate with 5 parts of methacrylic acid and 3 parts of ethylene glycol monoacrylate, prepared with mixed anionic–anionic surfactants with redox initiation.

Polyurethanes

The di-isocyanate compound is a blocked polymer prepared from tolylene di-isocyanate and excess of an adipic acid polyester of high hydroxyl number, which is blocked by m-cresol. The compounded latices used for cotton cloth deposited with rayon piles contain 100 parts of 45 % vinyl latex and 20–27.5 parts of 20 % polyurethane latex. The final product, after drying and curing, is highly abrasion resistant [281].

8.3.4 Coatings

Pigmented latices previously quoted can be used for surface coatings, and are especially useful as undercoats for wood, even giving a fair degree of protection without top coats, provided that they are not badly weathered [262]. These polyurethane latices can be applied to wet wood and the top coat presents no problems. Recoatability on any substrate is good. These latices are formulated with driers, preferably oil soluble, which are added before emulsification, but water-soluble driers may also be used. The term 'driers' as used here applies to the traditional paint driers. Otherwise, the principles of formulation are not unlike those of 'classical' vinyl-type emulsion paints (Chapters 9 and 10 in Volume 2). A small quantity of solvent, however, is often required.

In tests on coatings, the hard type of polyurethane has a very low, almost negligible, loss on the Tabor abrasion test, although this rises to the same level as acrylics with soft coatings.

A saturated polyester which has been reacted with a di-isocyanate is emulsified by the addition of triethylamine to give a latex. This is ball-milled with calcium carbonate, talc and carbon black to give an anti-corrosive coating [282].

Polyesters of hydroxyalkane monocarboxylates of dibasic acids, the latter with at least four carbon atoms, or lactones, e.g. caprolactones, can be converted to latices with polyurethane prepolymers, with salt formation via the carboxyl groups taking place on neutralization. The use of diaminocarboxylic acids is an alternative. These latices are of special interest in the coating of polyvinyl chloride sheet, with the advantage that they avoid the 'quick grab' that is a defect with products based on solvents. Thus, if necessary, repositioning is possible after application [283].

The use of radiocurable polyurethane latices as coatings has been claimed. Thus a latex prepared from a 3 : 2 maleic acid based polyester acrylate–bisphenol A diglycidyl ether acrylate, dimethylpropionic acid, hexamethylene di-isocyanate, isophorone di-isocyanate, ethylenediamine and water (150 : 17 : 30.8 : 40.7 : 3.6416) with a theoretical NCO content of 1.39 % gave films with pendulum hardness at 174 s, which after curing had a pencil hardness of 2H and an Erichsen indentation of 6.5 mm [284].

Solvent two-pack polyurethane coatings (2K-PUR), which came into the market in about 1970, have become the dominant class of resins for automotive refininishing paints and commercial vehicle finishes. Aqueous coatings are now available. Blends of low molecular weight and hydrophilically modified di-isocyanates based on hexamethylene di-isocyanates are preferred. These are added to an aqueous dispersed polyol phase before application. Mixtures of hydrophobic and hydrophilic polyisocyanates are used, giving a pot-life greater than 3 hs. Some organic solvent is required for the polyisocyanate [285].

8.3.5 Leather treatment

A primer for leather is formed from an acetone solution of a prepolymer from hexamethylene di-isocyanate and polypropylene glycol ammonia neutralised and dispersed in an alkyl ether phosphate solution [286]. Polyurethane (PU) latices prepared from 2,2-bis(hydroxymethyl)propionic acid, TDI and polyoxytetramethylene are considered to be superior to acrylic latices as leather coatings [287]. A number of other PU latex adhesives for bonding both natural and synthetic leather have been described [288]. Directly prepared latices may be used to impregnate leather, and so provide a material with high scuff resistance [289].

A prepolymer from HMDI, an adipic acid/mixed glycols polyester in acetone, is chain-extended in water with ethylenediamine and propane sultone, together with potassium hydroxide to neutralise, forming an emulsion from which acetone is distilled. After pigmentation with the addition of ammonium chloride, this latex provides a glossy, microporous finish for a polyurethane leather substitute [290].

8.3.6 Sizes for glass fibres and miscellaneous

4,4'-Dicyclohexylmethane di-isocyanate dissolved in toluene is dispersed in an aqueous solution of an ethylene oxide–propylene oxide copolymer emulsifier, and reacted, rather unusually, with hydrazine hydrate solution to form a 40 % polyurethane latex. This is compounded, *inter alia*, with a siloxane to form a size for glass fibres [291].

A composite of Plaster of Paris and a polyurethane–polyurea latex is useful for dental and surgical purposes [292].

REFERENCES

1. G. Odian, *Principles of Polymerisation*, 3rd edn, Wiley, 1991; D.H. Solomon, *The Chemistry of Organic Film Formers*, 2nd edn, Robert E. Krieger Publishing, 1977
2. L.P. Taylor (ICI), BP 2,291,648, 1996
3. J.F. Abere (Minnesota Mining & Manufacturing), USP 3,266,921, 1966; equivalent to BP 1,014,302, 1966
4. R.A. Miller, S.E. Geiorge and J. Adhs. *Sealant Counc.* **26**(1), 209–18 (1995)
5. T. Toda *et al.* (Sekisui Chemical), JP 120091, 1996; *Chem. Abstr.*, **125**, 117557 (1996)
6. K. Kano and Y. Sugito (Dainichiseika Color), JP 238,254, 1995; *Chem. Abstr.*, **124**, 59338 (1996)
7. T. Nakamura *et al.* (Teijin), JP 268,189, 1995; *Chem. Abstr.*, **124**, 89694 (1996)

References

8. T. Jones and J. McCarthy, *Coat. Community Care, 14th Int. Conf.*, 1994, pp. 1–19 (PRA Teddington); *Chem. Abstr.*, **124**, 10997 (1996)
9. M.K. Sharma, Proc. Engng Found. Confon. Dispersion, Aggregation, 1992 (published 1994), pp. 313–26; *Chem. Abstr.*, **124**, 346077 (1996)
10. J.M. McCarthy, *Paint India Ann.*, **93**, 95–6, 98–9, 104–5, 107–8 (1995); *Chem. Abstr.*, **124**, 178882 (1986)
11. Soc. Chimiques des Charbonnages, FP 1,591,600, 1970
12. R.R. Rabenold (PPG), USP 3,539,441, 1970
13. T. Inaba (Takamatsu Yushi), JP 268,001, 1995; *Chem. Abstr.*, **124**, 120276 (1996)
14. H. Kazuyuki, M. Itaru and K.J. Hirotaro, *J. Appl. Polym. Sci.*, **12**(1), 13 *et seq.* (1968)
15. S. Vargiu *et al.* (SIR), GP 2,063,247, 1971
16. L.B. Pedighian (Vistron), USP 3,529,169, 1971
17. R.H. Leitheiser (Ashland Oil), FP 2,018,217, 1970
18. R.A. Coderre and R.H. Leitheiser (Ashland Oil), USP 3,608,773, 1971
19. T. Tsubakimoto *et al.* (Nippon Shokubai), GP 2,047,305, 1971
20. Ashland Chemical Company, Kentucky, prior to 1970
21. Z.W. Wicks, F.N. Jones and S.P. Pappas, *Organic Coatings, Science and Technology*, Vol. 1, Wiley, 1992, Ch. 10; W.T. Elliott, in *Surface Coatings, Raw Materials and Their Uses*, 3rd edn, Surface Coatings Association of Australia, Chapman & Hall, 1993, Ch. 5 pp. 76–109
22. V.V. Verkholantsev and I.V. Shvaikolovskaya, *Kolloidn. Zh.*, **49**(6), 1178–82 (1987); *Chem. Abstr.*, **108**, 113344 (1988)
23. N.A. Smirnova, *Tr. Vses. Nauch-Issled Inst.*, **1965**(25), 421 *et seq.* (1965)
24. M. Yaseen *et al.*, *Indian J. Technol.*, **2**(7), 227 *et seq.* (1964)
25. J. Hires and E. Krjar, *Chem. Prum.*, **21**(12), 603–6 (1971)
26. G. Oestberg, *Surfactant Sci. Ser.*, **61**, 327–41 (1996)
27. L.G. Little (Hercules), USP 2,378,320, 1945
28. W.M. Kraft and J. Wiesfield (Tennecco Chemicals), USP 3,223,658, 1965
29. H.C. Cheetham and R.J. Myers (Resinous Products & Chemicals), USP 2,308,474, 1943
30. R.L. Broadhead (Standard Oil, Chicago), USP 3,269,967, 1966
31. H.R. Gamrath and R.A. Cass (Monsanto), FP 1,394,008 1965
32. R. Rutkowski (Glasurit-Werke, M. Winkmann), GP 1,273,730, 1968
33. American Cyanamid, BP 562,573, 1944
34. L.O. Cummings (Pacific Vegetable Oil), USP 3,620,989, 1971
35. T. Fjeldberg, *JOCCA*, **70**, 278–9, 281–3,285 (1987)
36. H. Blum, P. Hoehlein and J. Meixner, *Farbeu. Lack*, **9**(5), 342–5 (1988); J. Beetsma and A.D. Hofland, **46**(2), 53–7 (1996)
37. S.-N. Gan and K.-T. Teo, *Surf. Coat. Intt.*, **82**(1), 31–6 (1999)
38. Celanese, BP 1,223,033, 1971
39. E. Arntson (Archer-Midland-Daniels), BP 1,013,367, 1965
40. A. Weinberg, *Chem. Abstr.*, **67**, 3762 (1967)
41. M. Espinosa Flores and L.M. Cespedes Paez, *Ing. Cienc. Quim.*, **11**(2–3) 61–5 (1987) (in Spanish); *Chem. Abstr.*, **108**, 77184 (1988)
42. H. Warson and R.M. Levine (Vinyl Products), BP 906,117, 1962
43. D. Vodova *et al.*, Cz. P 242,286, 1987; *Chem. Abstr.*, **124**, 263528 (1996)
44. M. Yamamoto (Kansai Paint), JP 331,286, 1995; *Chem. Abstr.*, **124**, 263528 (1996)
45. T. Nabours, R.A. Baijards and A.L. German, *Prog. Org. Coat.*, **27**(1–4), 163–72 (1966)
46. F.J. Keene (Du Pont), USP 3,615,792, 1971
47. Archer-Midland-Daniels, Neth. P Applic., 66–11, 659, 1967
48. California Research Corp., USP 3,133,032, 1965
49. O.W. Huggard (Mobil Oil), USP 3,519,583, 1970
50. D.H. Solomon, *The Chemistry of Organic Film Formers*, 2nd edn, Robert E. Krieger Publishing, 1977, p. 119 *et seq.*

51. D.F. Percival *et al.* (Chevron Research), USP 3,306,866, 1967
52. J.F. McKenna (Pittsburg Plate Glass), USP 2,941,968, 1960
53. P. Narashinham *et al.* (Council of Scientific Research), India P 112,301, 1969
54. W. Weger, *Tech. Dev. Coat.: Their Applic. and Uses*, 2nd Middle East Conf., 1995, 16 pp., paper 1; *Chem. Abstr.*, **124**, 59219 (1996)
55. W. Weger, Adv. Coat. Technol. Conf., 1995, paper 8; *Chem. Abstr.*, **125**, 61020 (1996)
56. V. Digernes and E.M. Ophus, Adv. Coat. Technol. Conf., 1995, paper 24; *Chem. Abstr.*, **125**, 61021 (1996)
57. A. Hofland, *Pint. Acobados Ind.*, **37**(219), 14–23, 1995; *Chem. Abstr.*, **124**, 59106 (1996)
58. E. Ophus and V. Digernes, Coat. Community Care, 14th Int. Conf., 1994, pp. 1–13; *Chem. Abstr.*, **124**, 10995 (1996)
59. G. Hardeman and J. Beetsma, Coat. Community Care, 14th Int. Conf., 1994, pp. 1–10; *Chem. Abstr.*, **124**, 10996 (1996)
60. L. Schibler (Ciba), Swiss P 435,317, 1967
61. American Cyanamid, BP 894,833, 1962
62. Ciba, Swiss P 255, 102 and 262,963, 1949
63. A. Hiestand (Ciba), GP 2,015,197, 1970
64. H.J. Deuzeman and H. Lumley (Fiberglass Canada), USP 3,624,426, 1971
65. L. Wyokoff and D.W. Wurmser (Ciba), USP 3,624,426, 1971
66. D.C. Fielding and J. Massey (ICI), BP 1,253,213, 1971
67. T. Tsubakimoto *et al.* (Nippon Shokubai), JP 09,420, 1971; *Chem. Abstr.*
68. S. Nishida and H. Noda (Toyo Rubber), JP 21,446, 1971
69. J.B. Angliss and M. Lipson (Australian Commonwealth Scientific and Industrial Research), GP 2,107,189, 1971
70. D.J. St Clair and J.R. Erickson (Shell International Research), Eur. P 698,639, 1996
71. A. Rudolf (Oetker Chem. Fabric Budenheim), Israel P. 100,165, 1995; *Chem. Abstr.*, **124**, 292554 (1996)
72. P.T. Judd (Grace), BP 1,901,295, 1969
73. P.T. Judd (Grace), GP 2,057,691, 1970
74. BASF, FP 1,475,381, 1967; H. Bille *et al.* (BASF), FP 1,496,263, 1967; BASF, Neth. P Applic. 07,321, 1966
75. D.E. Tuites (Du Pont de Nemours), USP 3,386,940, 1968
76. D.E. Peeman (General Mills), USP 3,582,507, 1971
77. D.H. Solomon, *The Chemistry of Organic film Formers*, 2nd edn, Robert E. Krieger Publishing, 1977, Ch. 7, pp. 187–210; Z.W. Wicks, F.N. Jones and S.P. Pappas, *Organic Coatings, Science and Technology*, Vol. 1, Wiley, 1992, Ch. 11, pp. 162–83
78. Dynamit Nobel, BP 1,088,496, 1967
79. F.H. Walker and M.I. Cook, *Polym. Mater. Sci. Engng* (ACS), **77**, 379–80 (1997)
80. Nippon Kasei, JP 26,092, 1963
81. C.D. Grieco and A.R. Sacriston (R.T. Vanderbilt), GP 2,037,523, 1971
82. J.J. Hirschfield (Monsanto Chemicals), USP 3,371,052, 1968
83. H. Remer *et al.* (C.W. Huels), USP 4,514,467, 1985; J.-V. Weiss *et al.* (C.W. Huls), USP 4,442,245, 1985
84. R.R. Pettit (American Pipe & Construction), FP 1,472,061, 1967
85. R. Fromique and E. Lopez (Soc. Chimique de Gerland), FP 2,464,977, 1981
86. Y. Suzuki (Fujiko), JP 15,956, 1979; *Chem. Abstr.*, **91**, 5931 (1979)
87. E.M. Blyakhman and Sh.M. Rizentuler USSRP 293,028, 1971
88. Manufacture de Produits Chimique Protex, JP 19,188, 1977; *Chem. Abstr.*, **88**, 154769 (1978)
89. A. Giller and O. Jacobi (Ch. Weke Albert), BP 1,244,424, 1971
90. M.K. Carter (Shell), USP 3,634,348, 1972
91. J.G. Zora and D.V. Todd (Koppers), USP 3,301,804, 1967

92. T. Wakabayashi *et al.* (Gooh Chemical Industry), JP 37,8121, 1986; *Chem. Abstr.*, **105**, 79933 (1986)
93. Zakrocki *et al.* (Bayer), Eur. P 25,139, 1981; *Chem. Abstr.*, **95**, 44216 (1981)
94. R.L. Kolek and E. Eilerman (Pitsburgh Plate Glass), Belg. P 645,489, 1964
95. J.P. Stalego (Owens Corning), USP 3,562,081, 1971
96. J.V. Weiss (C.W. Huels), GP 3,345,399,19
97. S. Wu and M.D. Soucek, *Polym. Prepr. Am. Chem. Soc.*, **38**(1), 492–3 (1997)
98. O.L. Nikles (Resyn Corp.), GP 2,053,359, 1972
99. Wm Goebel and W. Von Bonin (Bayer), GP 1,495,843, 1971
100. C.A.M. Hoefs and P. Oosterhoff (Akzo), Eur. P 244,905, 1987; *Chem. Abstr.*, **108**, 133504 (1988)
101. A. Saito *et al.* (Nippon Synthetic Chemical), JP 107,349, 1976; *Chem. Abstr.*, **86**, 17664 (1977)
102. A. Herczeg T. Gatwood and X. Callis, *SAMPE*, **6**(1), 25–30 (1970)
103. D. Aelony and H. Witcoff (General Mills), Can. P 578,315, 1959; equivalent to USP 2,899,397, 1959
104. W.J. Fullen, *Paint Varnish Prod.*, **58**(7), 23 *et seq.* (1968)
105. H. Abel (Ciba-Geigy), GP 2,054,173, 1971
106. W.R. Grace, FP 11,530,051, 1966
107. Celanese Coatings, BP 127,706, 1968
108. W.J. Fullen, *Paint Varnish Prod.*, **58**(7), 23 *et seq.* (1968)
109. H. Wittcoff, *J. Oil Co. Chemists Ass.*, **47**, 273–88 (1974)
110. S. Egusa *et al.*, *J. Appl. Polym. Sci.*, **34**(6), 2177–86, (1987); *Chem. Abstr.*, **108**, 6897 (1988)
111. Nippon Oil, Hoshimitsu Chemical, JP 205,596, 1982; *Chem. Abstr.*, **99**, 97072 (1983)
112. J.V. Weiss (C.W. Huels), GP 3,345,399, 1995; equivalent to USP 4,622,353, 1986
113. K. Shizawa and M. Ueno (Toyo Ink), JP 250,034, 1986; *Chem. Abstr.*, **106**, 178216 (1987)
114. P.P. Winner (SCM), USP 4,600,754, 1986
115. S. Egusa *et al.*, *J. Appl. Polym. Sci.*, **34**(6), 2177–86 (1987)
116. A. Kubo *et al.* (Shinto Paint), GP 3,518,486, 1985
117. F.S. Shahade and R.M. Cristenson (Pittsburgh Plate Glass), USP 3,293,201, 1966
118. A. Maeder and R. Aenishaenslin (Ciba), Swiss P 464,514, 1968
119. H. Remer *et al.* (Ciba-Geigy), GP 3,331,730, 1983; equivalent to USP 4,514,467, 1983 and Eur. P 96,736, 1983; Ruetgerswerke, GP 3,319,675, 1983
120. A. Fujii and A. Tamaki (Mitsubishi Electric), JP 11,946, 1979; *Chem. Abstr.*, **90**, 169720 (1979)
121. H. Enders and H. Deiner (Chem. Fabric Pfersee), USP 3,320,197, 1967; equivalent to BP 1,071,162, 1967
122. D.H. Klein and K.C. Jork, *Surf. Coat. Int.*, **81**(2), 72–6 (1998)
123. R.M. Haines and R. Wong (Owens-Corning Fiberglass), USP 4,448,911, 1984
124. J. Hires, and J. Kinco, *Konf. Vyuziti Dispersi (Latexu)*, **1971**, 46–64 (1971) (in Czech); *Chem. Abstr.*, **77**, 7392 (1972)
125. H. Reimer *et al.* (Huels), GP 3,222,529, 1984; equivalent to Eur. P 96,736, 1984 and USP 4,514,467, 1984
126. K.H. Hermann *et al.* (Ruetgerswerke), Eur. P 103,908, 1984; equivalent to GP 3,331,730 and GP 3,319,675, 1984
127. R. Miller and J.M. Rizer (Research One), USP 4,501,830, 1985
128. J.V. Weiss (Huels), Eur. P 147,553, 1985
129. T. Akao, JP 270,6690, 1987; *Chem. Abstr.*, **108**, 133507 (1988)
130. G. Sprinschik (Hoechst), GP 3,006,964; Eur. P 34,788, 1981
131. J.C. Hedrick, C. Sensenich, Viehbeck and K. Papathomas, *Polym. Mater. Sci. Engng*, **76**, 174–5 (1977)

132. D.H. Solomon, *The Chemistry of Organic Film Formers*, 2nd edn, Robert E. Krieger Publishing, 1977, Ch. 9, pp. 253–62; P. Swaaraj (ed.) *Surface Coatings, Science and Technology*, 2nd edn, Wiley, 1996, pp. 161–89
133. Bakelite, BP 461,649–50, 1937
134. P.K. Porter (Westinghouse Electric), USP 2,436,328, 1948
135. R.W. Hall and D.J.R. Massey (Distillers), BP 1,045,715, 1966
136. Beck, Koller, BP 1,056,073, 1967
137. R. Dijkstra (Phillips), USP 3,355,407, 1967; equivalent to BP 1,071,360, 1967
138. Reichhold, FP 1,440,952, 1966
139. K. Koguchi *et al.* (Mitsui Toatsu), GP 2,064,155, 1971
140. R.A. Jarvi (Simpson Timber), USP 3,591,535, 1971
141. K. Kobayashi *et al.* (Dainippon Ink), JP 27,256, 1971
142. T. Sasaki and K. Noguchi (Sumitomo Durez), JP 100,104, 1996; *Chem. Abstr.*, **1255**, 88757 (1996)
143. N. Konoishi *et al.* (Oshika Shinko), JP 20,526, 1970
144. T. Watanabe *et al.* (Dainippon Ink & Chemicals), JP 67,799, 1996; *Chem. Abstr.*, **125**, 60112 (1996)
145. K. Shozawa and M. Ueno (Toyo Ink), JP 250,024, 1988; *Chem. Abstr.*, **106**, 178216 (1987)
146. N. Shibahara and T. Kawasaki (Dainippon Ink & Chemicals), JP 27,356, 1996; *Chem. Abstr.*, **124**, 262625 (1996)
147. H. Mayer, *Surf. Coat. Int.*, **82**(2), 77–83 (1999)
148. Midland Silicones, BP 785,174, 1957; K. Ariga and I. Kodama (Shin-Etsu Chemical), GP 1,802,424, 1969
149. Z. Du and C. Zhou, *Gaofenzi Tongxun*, **1984**(4), 277–82; *Chem. Abstr.*, **102**, 167272 (1985)
150. M. Hosokawa *et al.* (Dow Corning-Tordy Silicone), Eur. P 761,724 1997
151. J.F. Hyde and J.R. Wehrly (Dow-Corning), USP 2,891,920, 1959
152. Farbenfabric Bayer, BP 1,036,083, 1966; Chemische Fabric, BP 1,011,027, 1966
153. L.M. Blehm *et al.* (Dow Corning), Eur. P 181,182, 1986
154. W. Bernhelm and H. Deiner (Chem. Fabric Pfersee), GP 1,965,068, 1972
155. W. Keberle and H. Niederpruem (Farbenfabric Bayer), BP 1,128,642, 1968; equivalent to FP 1,481,512, 1967 and Neth. P Applic. 66–07,150, 1966
156. Shinetsu Chemical, BP 1,228,527, 1971
157. M. Ikoma (Shin-Etsu), JP 41,038, 1971; *Chem. Abstr.*, **77**, 6061 (1972)
158. Shin-Etsu, BP 1,228,527, 1971
159. D.J. Huebner and J.C. Samm (Dow Corning), Eur. P 166,397, 1966
160. G. Koerner *et al.* (Th. Goldschmidt), GP 3,216,585, 1983
161. H.L. Brooks (Stauffer-Wacker Silicone), GP 2,014,174, 1970
162. Midland Silicones, BP 793,501, 1958
163. Farbenfabric Bayer, BP 1,036,083, 1966
164. H. Oda *et al.* (Nippon Paint), JP 120,0856, 1996; *Chem. Abstr.*, **125**, 117510 (1997)
165. A.F. O'lenick (Siltech), USP 5,523,445, 1995; *Chem. Abstr.*, **125**, 145627 (1997)
166. W. Bernheim and H. Deiner (Chem. Fabric Pfersee), GP 1,917,701, 1971
167. Chem. Fabric Pfersee, GP 2,032,381, 1972
168. Farbenfabric Bayer, BP 977,822, 1964; H.-H. Steinbach (Bayer), USP 3,306,759, 1967
169. K.P. Grinevich *et al.*, *Sov. Plast., Eng. Ed.*, August, 65 (1963)
170. M.R. Rosen, *J. Coll. Interface Sci.*, **36**(1), 155–6 (1971)
171. A.E. Bey, D.H. Weyenberg and L. Seibles, *Polym. Prepr. Am. Chem. Soc.*, **11**(2), 995–6 (1970)
172. M. Geck *et al.*, *Organosilicon Chem. ii(Muench Silicontage)*, 2nd edn, 1996; *Chem. Abstr.*, **125**, 143463 (1996)

References

173. F. Baumann et al., *Organosilicon Chem. ii(Muench Silicontage)*, 1996, pp. 665–71; *Chem. Abstr.*, **125**, 143462 (1996)
174. T. Suzuki, M. Kai and A. Ishida, *Yagagaku*, **34**(11), 938–45 (1985); *Chem. Abstr.*, **104**, 70708 (1986)
175. Dow-Corning, BP 1,113,543, 1968
176. G.W. Madaras, *J. Soc. Dyers & Colourists*, **74**, 835 et seq. (1958)
177. Farbenfabric Bayer, GP 1,155,593, 1963
178. R.J. Rooks, *Text. Chem. Col.*, **4**(1), 47–8 (1972)
179. M.L. Camp (Rhone-Poulenc), GP 2,047,919, 1971
180. Chemische Fabric Pfersee, GP 1,917,701, 1971; BP 1,300,250, 1972
181. C.R. Crabtree and M.A. Thomas (Deering-Milliken), USP 3,649,344, 1972
182. Wacker-Chemie, JP 134,072; *Chem. Abstr.*, **104**, 52041 (1986)
183. M.E. Sorkin (Dow Corning), USP 3,624,017, 1971
184. Shinetsu Chemical, BP 1,228,517, 1971
185. J.R. Cekado (Dow Corning), FP 1,560,728, 1969
186. V.A. Shenai, S. Sanjanwala and X. Murthy *Text. Dyer Printer*, **15**(17), 25–8 (1982)
187. G. Wiedemann and H. Frenzel BP 1,252,070, 1972
188. Owens-Corning, BP 1,250,194, 1971
189. D.G. Brown and D.L. Motsinger (PPG), USP 4,393,414, 1983
190. Dow Corning, BP 979,640, 1965
191. J.W. Weil (Dow Corning), GP 1,241,790, 1967
192. A.K. Simcox and P.A.J. Gate, *Master Builders J.*, September 1955
193. Union Chimique de Belge, BP 946,776, 1967
194. A. Yanagese et al. (Mitsubishi Rayon), JP 40,912–3, 1997 *Chem. Abstr.*, **126**, 252502–3 (1997)
195. Union Chimique Belge, GP 1,235,790, 1967
196. P.M. Burill (Midland Silicones), BP 1,165,813, 1972
197. S.F. Thames and J.M. Evans, *J. Paint. Technol.*, **43**(558), 49–53 (1971)
198. J. Cokada (Dow Corning), USP 3,355,399, 1967
199. Shin-Etsu, BP 1,228,527, 1971
200. R. Krebbs and H. Kober, *Paint Ink Int.*, **10**(1), 24–6, 1997; *Chem. Abstr.*, **126**, 270919 (1997)
201. T. Osanawa et al. (Dow Corning–Toray Silicone), JP 40,866, 1997; *Chem. Abstr.*, **126**, 251976 (1997)
202. Dow Corning, BP 1,127,152, 1968
203. E.N. Doyle, *Development and Use of Polyurethane Products*, McGraw-Hill, 1971; D.H. Solomon, *The Chemistry of Organic Film Formers*, 2nd edn, Robert E. Krieger Publishing, 1977, pp. 211–31; J.W. Rosthauser and J.G. Williams, *Polym. Mater. Sci. Engng*, **50**, 344–52 (1984)
204. Z.W. Wicks, F.N. Jones and S.P. Pappas, *Organic Coatings, Science and Technology*, Vol. 1, Wiley–Interscience, pp. 186–211; P. Swaraj (ed.), *Surface Coatings, Science and Technology*, 2nd edn, Wiley, 1996, Sec. 2.6, pp. 284–311
205. Thiokol Chemical, BP 996,208, 1965
206. C.H. Howell (Diamond Shamrock), USP 3,519,478, 1970
207. L.W. Georges (Firestone), USP 3,642,553, 1970
208. D.C. Bartizal (3M), GP 2,141,805, 1970
209. O. Lorenz, G. Poppel and V. Uerlinge, *Kaut. u. Gummi Kunst.*, **24**(12), 641–6 (1971)
210. O. Lorenz (Bayer), GP 2,436,017, 1976
211. S.P. Suskind, *J. Appl. Polym. Sci.*, **9**, 2451 et seq. (1965)
212. O. Bayer and D. Dieterich (Bayer), BP 1,078,202, 1967
213. W. Keberle, A. Reisch and D. Dieterich (Bayer), BP 1,092,028, 1967; equivalent to FP 1,491,744, 1967
214. Wyandotte Chemical, BP 1,122,077, 1968

215. G.M. Wagner (Hooker Chemical), BP 1,207,727, 1970
216. H. Witt and D. Dieterich (Bayer), BP 1,143,309, 1969
217. R.J. Vill and S.P. Suskind (W.R. Grace), USP 3,264,134, 1966
218. D. Dieterich, O. Bayer and J. Peter (Bayer), USP 3,388,087, 1968
219. O.M. Grace and J.M. McClellan (Wyandotte), USP 3,401,133, 1968
220. Du Pont de Nemours, Can. P 583,572, 1964
221. Farbenfabric Bayer, FP 1,521,170, 1968
221A. C.M. Hansen, *Ind. Eng. Chem. Prod. Res. Dev.*, **13**(2), 150–2 (1974)
222. Farbenfabric Bayer, FP 1,528,227, 1968
223. Farbenfabric Bayer, FP 1,548,467, 1967
224. Farbenfabric Bayer, GP 1,237,306, 1967
225. D. Dieterich, W. Keberle and R. Wuest, *J. Oil. Col. Chem. Ass.*, **53**(5), 363–79 (1970)
226. R.L. Rowton (Jefferson), BP 1,243,604, 1971
227. D.T. Hermann and K.H. Remley (American Cyanamid), BP 1,250,266, 1971
228. G.M. Wagner (Hooker), BP 1,207,727, 1970
229. A. Reischl and D. Dieterich (Bayer), BP 1,176,252, 1970
230. D. Dieterich, GP 2,035,729, 1972
231. D. Garry (Textron), GP 2,744,544, 1978
232. W. Keberle, H. Wieden and D. Dieterich (Bayer), BP 1,128,568, 1968
233. W. Keberle and E. Muller (Bayer), BP 1,146,890, 1969
234. W. Keberle and W. Thoma (Bayer), BP 1,942,992, 1971
235. R.C. Carlson (3M), FP 2,014,990, 1970
236. K. Matsuda *et al.* (Kao Soap), GP 2,536,678, 1976
237. W. Keberle and D. Dieterich GP 2,035,729, 1972
238. D. Dieterich *et al.* (Bayer), GP 2,035,729, 1972
239. Farbenfabric Bayer, FP 2,005,413, 1972
240. D. Lesch and W. Keberle (Bayer), GP 2,035,732, 1972
241. W. Keberle and D. Dieterich (Bayer), GP 2,035,729, 1972
242. W. Keberle and G. Oertel (Bayer), USP 3,539,483, 1970
243. D. Dieterich and W. Keberle (Bayer), GP 2,034,479, 1972
244. P. Davis and O.M. Grace (Wyandotte), FP 2,001,362, 1969
245. I. Suzuki *et al.* (Asahi Chemical), JP 41,307, 1971
246. R.H. Boutier, *Proc. 24th Water-borne and High Solids, Powder Coat. Symp.*, 1997, pp. 216–24
247. Y. Kuwashima and S. Gazama (Takeda Chemical), JP 01,234, 1972
248. E. Mueller (Bayer), BP 1,044,267, 1966
249. Wyandotte, BP 1,077,257; equivalent to FP 1,477,193, 1967
250. W. Keberle and D. Dieterich (Bayer), BP 1,125,277, 1968; equivalent to FP 1,483,587, 1967
251. Wyandotte, BP 1,111,043, 1968
252. S.L. Axelrod (Wyandotte), USP 3,294,724, 1966
253. O.M. Grace and P. Davis (Wyandotte), USP 3,563,943, 1971
254. P. Shirota *et al.* (Dainippon Ink), JP 03,594, 1971
255. R.I. Berger and M.I. Youker (Du Pont de Nemours), USP 3,178,310, 1965
256. P. Davis and O.M. Grace (Wyandotte), USP 3,563,943, 1971
257. A. Hanson (UGB), GP 2,146,888, 1972
258. G.M. Wagner and W.J. Vullo (Hooker), USP 3,617,189, 1971
259. T. Masuda and H. Ozawa (Dainippon Ink & Chemicals), JP 270,614, 1987; *Chem. Abstr.*, **108**, 133536 (1988)
260. T. Masuda and H. Osawa (Dainippon Ink & Chemical), JP 260,811, 1987; *Chem. Abstr.*, **108**, 114359 (1988)
261. J. Fock and D. Schedlizki (Th. Goldschmidt), GP 3,633,421, 1987
262. S.T. Bowell and H.J. Kiefer (Glidden), USP 34,210,302, 1965

263. W.R. Grace BP 1,132,887, 1968
264. Farbenfabric Bayer, FP 1,505,790, 1967
265. K. Damm, H. Steinbach and W. Noll (Bayer), USP 3,398,172, 1968
266. T. Gomi et al. (Mitsui Toatsu), JP 230,863, 1987; *Chem. Abstr.*, **108**, 16928 (1988)
267. T. Sakai (Kao), JP 241,902, 1987; *Chem. Abstr.*, **108**, 15223 (1988)
268. P. Loeweigheit and K.A. Van Dyk (Wico Chemical), Eur. P 189,945, 1986
269. S.P. Suskud, *J. Appl. Polymer Sci.*, **9**(2), 451 et. seq. (1968)
270. T. Sakai and M. Dobashi (Kao), JP 297,375, 1987; *Chem. Abstr.* **108**, 37338 (1998)
271. J.W. Rosthauser and K. Nachtkamp, in *Advances in Urethane Science and Technology*, K.C. Frisch and D. Klempner (eds.), Vol. 10, 1987, pp. 122–162
271A. M. Furomoto and Y. Harano (Daicel Chemical), JP 199,642, 1987; *Chem. Abstr.*, **108**, 132988 (1988)
272. L.F. Elmquist and R.K. Kamal (General Mills), FP 1,487,786, 1967
273. W. Klebert, K. Schafer and G. Becker (Bayer), BP 1,083,625, 1967
274. C.W. Leupold et al. (Vereignite Papierwerke Shicckedanz), GP 1,619,191, 1971
275. G. Prelini and A. Trovali, *Tinctioria*, **84**(11), 25–9 (1987)
276. Dainippon Ink & Chemicals, JP 149,417, 1981; *Chem. Abstr.*, **96**, 105752 (1982)
277. W. Wunder et al. (Bayer), USP 3,639,157, 1972
278. F. Reich, *Text. Veredlung*, **2**(7), 441 et. seq. (1967)
279. Kao Soap, JP 04,897, 1983; *Chem. Abstr.*, **99**, 55297 (1989)
280. J.M. McLellan and L.C. MacGugan, *Rubber Age (NY)*, **100**(3), 66 et seq. (1968)
281. Y. Sato (Takedo Chemical), USP 3,401,135, 1968
282. H. Maki et al. (Daiichi Kogyo Seiyaku), JP 246,972, 1997; *Chem. Abstr.*, **108**, 188554 (1988)
283. Bayer, FP 2,008,761, 1970
284. W. Paulus et al. (BASF), Eur. P 942,022, 1999
285. M. Sonntag, *Surf. Coat. Int.* (9), 456–9 (1999)
286. H. Antiger et al. (Henkel), GP 3,625,442, 1987
287. B. Pang et al., *Huaxue Shiji*, **9**(4), 226–30 (1987)
288. S.J. Ayuso and J. Barges, Sp. P 513,373
289. R.I. Berger and M.I. Youker (DuPont de Nemours), USP 3,178,310, 1965
290. G. Balle et al. (Bayer), GP 2,041,550, 1972
291. Y. Tamaki and H. Takegawa (Dainippon Ink & Chemicals), JP 292,658, 1987; *Chem. Abstr.*, **108**, 168648 (1988)
292. P. Mueller et al. (Bayer), GP 3,320,217, 1984

8

Polymer Latices in the Formulation of Adhesives

C.A. Finch

1 INTRODUCTION

Adhesives from natural products, such as animal glue (protein based) starches and natural gums (both carbohydrate based), have been used for many years: the tombs of Egyptian mummies are bonded with animal glues and the same type of adhesive is still used for violins and similar musical instruments, where highly stressed wood-to-wood joints are required. The systematic study of adhesives and the mechanisms of adhesion processes became important with the development of modern aircraft frames during the First World War. The studies were closely followed by the production of polymer latices (also known as latexes, emulsions or dispersions—the terms are more-or-less interchangeable) in the succeeding decades, with the large-scale manufacture of the long-known vinyl monomers (mainly vinyl acetate) and their conversion to polymer latices, as alternatives to natural rubber latices, which are based on the photosynthesis of isoprene (2-methylbutadiene) within the rubber tree *Hevea brasiliensis*. Much of the early, and important, work on the production of latex polymers has been described previously [1–3]. Many types of adhesives have also been described [4, 5]. A more recent summary of adhesives in general has been published [6]. There are several scientific journals concerned with adhesion and adhesives [7–10] and some trade magazines [11, 12], although very little of the discussion and information presented is related to latex-based systems.

A brief discussion of the theory of adhesion is presented in Section 2. Many practical applications of polymer latices depend upon the control of the property of adhesion-it is also essential in the formation of a satisfactory paint film, which can be considered as an adhesive film bonded to only one substrate. In this chapter, however, the preparation, properties and performance of latex-based systems used to bond two substrates are considered. In most cases, the bond is sufficiently strong as the bonding forces are greater than the forces that hold the materials of the substrates together, so separation may result in failure of the substrate rather than that of the adhesive. Such failure can occur readily in, for example, wood-to-wood bonds, where one of the materials will break whilst the adhesive bond is still intact.

Adhesives have many functions, as bonding agents for surfaces, such as wood, paper or metal, or as binders for fibres or powders. Some of these applications are discussed later in chapter 12 on textile applications, in chapter 13 on paper coating and in chapter 16 on building products, all in volume 3. The principal applications discussed in this chapter are those in which porous substrates, notably paper and wood, form at least one of the adherents. Such applications represent a major part of the industrial applications of water-based adhesives prepared from synthetic polymer latices. Latices of polyvinyl acetate, sometimes plasticized, either internally or externally, acrylate polymers and copolymers (including acidic and crosslinkable comonomers), polyvinyl chloride, polyvinylidene chloride and their copolymers, polychlorobutadiene ('chloroprene') and other butadiene copolymers, notably styrene–butadiene copolymers, have all been investigated for possible use in adhesives, and are discussed in this chapter. Adhesion is low or zero below the second-order transition temperature, T_g (see Section 4.1.2), irrespective of type, on wood or ceramic substrates, but, if polyvinyl alcohol is present in the aqueous phase, this will provide some adhesion. A suitable surfactant, normally part of the latex system, improves compatibility with each component and with the substrate.

Adhesive applications involving natural rubber latices are not considered specifically, apart from those mentioned in Section 7.

1.1 Terminology of adhesives

The terminology of adhesives is complex and, to some extent, confusing. The summary in this section provides a general introduction to the terms employed in this chapter. Section 10 presents a Glossary, in which the terms are defined in more detail.

Adhesion is the state where two substrates are held together by interfacial forces. These may be chemical forces, for specific adhesion, or interlocking action, for mechanical adhesion. In practice, both types of adhesion occur, in differing proportions, for most bonds. *Cohesive* forces are the internal forces of the adhesive bond, which depend on the chemical bonding forces within the adhesive. An *adherent* is a body which is held to another body by an adhesive. The *substrate* is any material on to which an adhesive has been coated. A *laminate* is a product, either rigid or flexible, made by bonding together two or more layers of materials.

The *adhesive* is a mixture of compounds capable of holding the materials together, by surface forces. Alternative terms for adhesive include glue, paste, gum, mucilage, adhesive cement and bonding agent. Adhesives may be described:

(a) by their *physical* form, such as liquid or paste adhesive;
(b) by *chemical type*, such as epoxy-resin adhesive;
(c) by the *material* to be bonded, such as wood or paper adhesives;
(d) by the *method of application*, such as cold-setting, contact, pressure-sensitive, heat-activated, solvent-activated or delayed-tack adhesives.

The *bond strength* is the unit load required to break an adhesive assembly, with failure occurring at or near the plane of the bond. The load may be applied in tension, compression, shear, impact, cleavage or peel, depending on the geometry of the assembly. An adhesive is *set* by solidification, by chemical action or by physical action, such as evaporation.

The *setting time* is the time between joining of the adherents and the setting of the adhesives. The *setting rate* is the increase in strength of the adhesive bond with time. This rate is not constant and often decreases towards the end of the setting time. *Curing* is the process of setting of an adhesive by chemical reaction (usually polymerisation, polycondensation or polyaddition). The *minimum film forming temperature* (MFFT) is the temperature at which a latex or a latex-based adhesive no longer dries to form a clear homogeneous film. A *lap joint* is made by placing one adherent partly over another and bonding the overlapping parts.

Viscosity behaviour is an important characteristic of adhesives, affected by formulation and by temperature, which is a partial description of their flow properties and rheological behaviour. Different viscosities and flow properties are required for different end uses and methods of application (e.g., blade-coating, roller-coating, nozzle application). Adhesives generally show non-Newtonian viscosity behaviour. *Thixotropy* of viscous fluids, including adhesives, is the property of reduction of viscosity with increasing shear. This shear-thinning affects the process of application of adhesives to substrates.

Tack or *initial tack* is the ability of an adhesive to form a partial bond immediately after the adhesive and the surface have been brought into contact under low pressure.

Blocking is the (usually undesired) property of adhesion between touching layers of materials under pressure during storage or use.

The *critical temperature* and *critical humidity* are, respectively, the points at which a given degree of blocking occurs. The *dry strength* of an adhesive is determined after drying (or conditioning under standard conditions for a known time), by loading the formed joint under standard conditions.

Creep of an adhesive is the dimensional change under load. Creep at ambient temperatures may be known as 'cold flow'.

Sizing or *priming* is the process of applying a material to a surface, usually porous, to allow absorption of an adhesive, to improve later adhesion (the term 'sizing' also has other meanings in textile and paper technology).

2 SURVEY OF THEORIES OF ADHESION

The science of adhesion serves a far wider field of human knowledge than the technology of adhesives, though it is impossible to follow the latter rationally without knowing something of the former [13].

Several different theories of adhesion have been developed over the years, describing the mode of action of adhesives and the mechanisms by which adhesives operate under conditions of use. On the whole, the different theories are not in conflict, but represent a series of refinements of the basic mechanism. They extend from the simple theory of mechanical entanglement, followed by electrostatic and absorption theories, to the diffusion theory. Common to all of these theories is the pattern that they provide reasonably satisfactory explanations of some basic phenomena, but largely fail to explain other aspects, and have only limited predictive value. The adhesion of an adhesive, of whatever chemical nature, is not a consistent process and several mechanisms are involved. Because of this, it is not possible to confirm or reject the significant factors of particular theories [14]. Detailed accounts of theories of adhesion (with limited discussion of latex-based adhesives) have been presented [15–17].

2.1 Basic theory

The adhesion of a polymer to a substrate depends greatly on the physical state of the polymer. It is difficult to generalise about an adhesion without referring to the physical state of the adhesive, since the viscosity of a polymer latex and the resulting dried polymer film vary considerably as the latex changes from a 'rubbery' to a 'glassy' state. Several historical reviews of the mechanism of adhesion have appeared [18–20].

The oldest theory of adhesion is the mechanical theory, which considers that adhesion is based on mechanical anchorage in the pores and surface irregularities of the adherent. The nature of the adherent surface is, consequently, of major significance: several characteristics must be considered, including the freedom from (or the presence of) non-adherent materials (such as surface oils and grease, or oxide layers), i.e. the cleanliness of the adherent surfaces. The importance of this factor is relevant to many porous and semi-porous substrates, such as wood, paper and paper products, leather and many textile fibres [21]. In the present context, this suggests that the viscosity must be the main variable in the performance of the adhesive. With solution adhesives, viscosity is usually a function of the concentration and molecular weight of polymers dissolved in a solvent mixture, which may be lost by evaporation and/or absorption into porous substrates. In latex-based systems, the situation is more complex: in general, the viscosity of a latex depends on the stabiliser system, including the protective colloids, and any self-stabilising monomers present, and only to a limited extent on the polymer concentration, except at very high solids contents. The mechanical forces of adhesion may also be affected by the latex particle size and particle size distribution, in relation to the porosity and imperfections of the adherent surfaces. It has been suggested that mechanical interaction is only a technical means of obtaining strong adhesive bonding, notably in structural adhesives. Here, mechanical adhesion may be a method of creating a larger surface contact area for physical absorption or a large polymer volume (of adhesive in the bond) to sustain deformation during the bond-breaking process. Although interlocking may, in some cases, affect the strength of the adhesive bond, this contribution towards joint strength from surface roughness may be explained by other mechanisms of bond formation.

Sometimes, the forces of adhesion are primary valence forces. Electrostatic (polar) bonds may form when copolymers containing acid functions are compounded into adhesives for metallic surfaces. Covalent bonds may occur rarely, but coordination functional bonds may be of more significance. According to the *electrostatic theory* [22], adhesion forces between adherent and substrate depend on contact or transfer potentials, which may cause buildup of an electrical double layer at the boundary and corresponding Coulomb attraction forces between the two components. Examination of these forces, however, suggests that they contribute only slightly to the considerations concerned with the design of adhesives of practical value.

Physical absorption is common to all types of adhesion systems. *Adsorption theory* regards adhesion as a special property of phase interfaces [23]. The secondary valence forces, known as *Lifshitz–van der Waals interactions*, have three components. They include orientation forces of molecules that have permanent dipoles and induction forces from induced dipoles, which arise when one molecule only has a permanent dipole moment. The dispersion (London) forces are the most important, since these arise from the interaction of electron forces. These bonds consist mainly of bonds, such as —OH, —O—, and $=NH-N=$, with bond energies of 3–7 kcal mol^{-1}. It has been estimated that $\sim$80 % of the cohesion forces in most organic systems arise from dispersion forces.

For these forces to be effective, distances between the adhesive and the adherent must converge towards molecular intervals, requiring spreading of the adhesive over the whole surface of the substrate. This suggests that good wetting of the adherent is necessary for good absorption and diffusion. Adhesion is therefore affected by the ratio between the surface energies of the adhesive and of the adherent: the specific surface energy of the adhesive must be lower than that of the adherent. For this reason, materials with high surface energy, such as metals, and those with medium surface energy (wood and paper) may be bonded relatively easily. As the surface energy decreases, as in polymer films, especially polyolefins, adhesion becomes more difficult, and almost impossible (poly(tetrafluoroethylene)). Surface energies may, however, be increased by surface treatment (e.g. corona discharge treatment of polyolefins). For an adhesive to be effective (i.e. it should approach the adherent at such close range that molecular forces operate) it must wet the surface of the adherent. Complete wetting of surfaces is desirable; i.e. there should be a zero contact angle. When two surfaces are separated by a thin layer of liquid with zero contact angle, strong adhesion occurs. However, almost all liquids, even those with a high contact angle, wet surfaces to some extent, so there is some limited adhesion. The contact angle, θ, is a good inverse measure of wetting and, hence, of adhesion. This is also shown by

$$W_{SL} = \lambda_L(1 + \cos \Theta)$$

where W_{SL} is the work of separation of the solid and liquid surface and λ_L is the surface tension of the liquid.

The additional effects of surface roughness of the substrate and of microscopic variations in smoothness mean that adhesion using a wetting liquid becomes a complex phenomenon. In addition, occluded air bubbles can create a point of weakness which can be significant in terms of the reduction in strength of the overall bond. Adsorption theory suggests that the adhesion forces between adhesive and adherent are not reciprocal, and so seem not to show full correlation between surface energies and adhesive bond strength.

Diffusion theories of adhesion are of particular interest in relation to the properties of pressure-sensitive adhesives [24]. It is suggested that adhesion depends on the mutual penetration of adhesive and substrate. The 'reptation' (derived from the term 'reptile') model can be applied successfully to explain tack, 'green strength', healing and welding of polymers. This model, proposed by de Gennes [25], assumes that a flexible polymer chain diffuses in a fixed three-dimensional mesh of obstacles which the chain does not cross. The depth of penetration of the adhesive into the adherent need only be a few Å. It has been shown that a highly branched polymer will have a greater adhesion than a linear polymer. The effect of molecular weight is more complicated: diffusion increases as molecular weight decreases, as does the probability of diffusion. However, since cohesive strength (and adhesive bond strength) increases with molecular weight, there is the possibility of an optimum value in many cases. The degree of polydispersity of molecular weight also caused complications with high molecular weight polymers. Small variations in low molecular weight polymers in compounded adhesives can have a marked effect on adhesion properties. However, limits of the diffusion theory appear when considering adhesion of polymers to glass, where this theory does not appear to have much relevance. Besides the Lifshitz–van der Waals interactions, there are short-range (<1 Å) forces due to donor–acceptor interaction or acid–base interactions. Molecular bonding is between the van der Waals and chemical bonding. Physical absorption is based on van der Waals interactions, whilst chemisorption is related to molecular bonding. The acid–base interaction, which is the predominant form of molecular interaction, can be represented as

$$A \text{ (acid)} + :B\text{(base)} \longrightarrow A : B \text{ (acid-based complex)}$$

which actually involves both covalent (homopolar) and ionic (heteropolar) factors. From this interaction, the work of adhesion can be deduced in thermodynamic terms.

The role of chemisorption in adhesion is significant. Since adsorption is one of the more important mechanisms in achieving adhesion, diffusion and wetting are only kinetic means of obtaining good polymer adsorption at the interface. By this means, the polymer molecules can reach intimate contact so that either (or both) Lifshitz–van der Waals (long-range) interaction or short-range (acid–base) interaction may take place. For short-range interaction, the molecules should be within ~ 4 Å, so that they may achieve both chemical and physical adsorption.

Chemical bonding, based on the primary covalent formed at the interface, is the strongest form of polymer adhesion, and can be enhanced further by the introduction of coupling agents [26]. The major theories of adhesion by adsorption and diffusion have been correlated to a useful extent [27].

In summary, polymer adhesion is divided into three types:

(a) rubbery polymer–rubber polymer (R-R) adhesion;
(b) rubbery polymer–glassy polymer (R-G) adhesion;
(c) rubbery polymer–non-polymer (R-S) adhesion.

R-R adhesion can be explained by diffusion; R-G contact probably involves both diffusion and adsorption, although diffusion adhesion is considerably less significant with a glassy polymer. It is doubtful whether diffusion adhesion plays a part in R-S systems, of which glass and metal are the most common non-polymer components. More recently, the literature on current theories of adhesion have been summarized [28]: the two main mechanisms are considered to be mechanical interlocking and adsorption, although, as indicated above, electrostatic forces and diffusion may be significant in some limited instances. The mechanism of mechanical interaction leads to suggestions of alteration of the roughness on the adhesive surface in order to increase adhesion. Since latex adhesives, as discussed in this work, are frequently used to bond porous surfaces, such as wood, paper and leather, this is largely true, although all of these are fibrous materials, so successful bonding involves penetration of the adhesive between the fibres, thus partially embedding the dried (or crosslinked) adhesive film. This explanation is reasonably satisfactory for the types of substrates where latex-based adhesives are employed, but requires refinement when load-bearing adhesive bonds with metallic substrates are considered. Adsorption interactions in solids are of different types, depending on the nature of the material, but, in general, include forces of chemical origin–covalent, ionic and metallic bonding, and hydrogen bonding—as well as those of physical origin—van der Waals' forces, dipole interactions and dispersion forces. The quantitative size of these forces can be calculated, giving ideal values of the strength of particular materials. In practice, much lower values (often by several orders of magnitude) are measured, depending on the material and the irregularities, flaws and defects of adhesive films, especially with latex-based adhesives, which normally contain several components. An important feature is that the forces are only significant over very short distances (of a few Å only). This means that, to be effective, adhesives must be in very close contact. All normal adhesives should, therefore, be mobile liquids which can wet, and then flow, on a surface, penetrating the irregularities and surface roughness so that intimate contact can be achieved. For fibrous materials, the interpenetration on a macroscale is fairly clear, but metal/metal oxide surfaces require complex surface examination techniques to reveal this on a microscale. From this it is clear that the effects of wetting and the thermodynamics of liquid/solid interfaces are important. Detailed examination of these phenomena [22] suggests that there are two particular factors of particular importance: the interfacial free energy [29], and the nature of the non-dispersion polar forces, including acid–base interactions and, especially, hydrogen bonding [30]. Spectroscopic examination of adherent surfaces also suggests that primary covalent bonding may also be involved, to a more limited extent, especially where surface pretreatments are used [31].

2.2 The function of adhesives: latex adhesives

In mechanical terms, the function of an adhesive is to displace weakly held molecules from the two surfaces of the joint, fill the gaps between them and attach itself firmly to both. To achieve this function successfully, it is necessary to control the rheological and physical properties—notably the flow properties, rate and ease of application and speed of drying. An adhesive should also develop sufficient mechanical strength for the intended application, which often means that the ideal strength is at least that of the substrates. Amongst load-bearing substrates, this is usually attainable with wood and most plastics, but is not ordinarily achieved with a bond to glass. Chemical aspects of the drying process may involve a curing reaction resulting in crosslinking. Physical aspects, in many cases, involve evaporation of solvent or suspending medium (usually water), with shrinkage of the adhesive polymer layer, possibly resulting in cracking and separation of the adhesive system. During the application of a compounded latex adhesive, water is removed either by absorption into the substrate, followed by slow evaporation after diffusion, or by direct evaporation, which can be at the edges of the latex adhesive film. In practice, both mechanisms of water removal take place, depending on the particular conditions. Alternatively, the water may be partly evaporated from a single open surface—a method adopted with pressure-sensitive adhesives.

Evaporative shrinkage is notably important with latex-based products, since the film formation is a secondary process. It is usual to employ polymeric components in latex adhesives which have a minimum film forming temperature (MFFT or MFT), T_g, below that of film formation, either by a suitable choice of polymer, or by addition of a solvent or plasticizer which will effectively lower the T_g. However, when this method is employed the creep or cold flow of the adhesive film is increased, and may lead to joint failure with time. Alternatively, an adhesive that will form a bond that does not 'creep' significantly can be made by including a crosslinking system into the composition. This may crosslink either at ambient temperature (when storage—the 'shelf-life'—of the composition can be a problem) or by a thermosetting mechanism. In some cases, e.g. where polyacrylamide and its derivatives are part of the polymer system, a crosslinking catalyst may be added immediately before application of the adhesive to the substrate. This use of a two-part polymer system introduces further complications, since the working life (when the mixture is applied to the surfaces to be bonded) must also be controlled at a convenient time span for the assembly of components, which will be affected by the ambient temperature and by the temperature of the substrate. *Creep* may also be avoided, in some cases, by increasing internal dispersion forces, usually by the introduction of hydrogen-bonding systems, without necessarily invoking crosslinking systems. Some examples of formulations designed to meet these requirements are described later in this chapter.

The action of the wetting agent is of major importance in a latex-based adhesive, especially with high-porosity substrates such as cellulosic materials (wood and paper). Conditions of drying can also affect the formation of the adhesive bond and its performance, whether measured as shear strength or by peel properties, since this may depend on the tendency of the latex stabilizer to concentrate on one of the other surfaces during drying. However, the presence of a high molecular weight protective colloid (which will give high viscosity in the latex adhesive compound) may prevent excessive absorption into a porous adherent by more readily blocking the pores.

3 LATEX-BASED ADHESIVES: SOME PRACTICAL ASPECTS

This section, and the succeeding sections, are intended to provide critical guidance on the formulation of latex-based adhesives for specific applications and the considerations that affect such design, with examples illustrating the factors involved and their relative importance.

3.1 Introduction

Polymer latexes are used in adhesives for several reasons, since they:

(a) are relatively low in cost;
(b) have many different applications;

(c) are water-based, with relatively low hazard in use;
(d) allow great flexibility in compounding;
(e) can be formulated for many different applications;
(f) can have a wide range of viscosities;
(g) can have a wide range of rheological properties, including variations in thixotropy.

Drying of the adhesive film is affected by the nature of the substrates being bonded. If these are impervious to water, then slow drying may be expected, since the moisture can only escape from the edges of the adhesive area. The rate of drying increases with the porosity of the substrate, so that improved drying rates can be achieved with most cellulosic substrates and with many porous building materials. Added solvent may also increase the rate of drying in some cases, forming a comparatively weak bond to the substrate (the reverse may occur if the substrate is a solvent-sensitive thermoplastic film), but this will tend to increase with time.

3.2 Functions of adhesives

At low temperatures (down to -10 to -20 °C), there should be some flexibility of the dry adhesive film; otherwise, at temperatures below the T_g, the adhesive film may crack, and so form a weak bond. However, if drying takes place below the T_g, a strong bond may not be formed. The T_g of a typical polyvinyl acetate homopolymer is $\sim +35$ °C. This T_g can be altered to a considerable extent by the degree of plasticisation: external plasticisation depends on the addition of a further plasticizer to the adhesive composition, whilst internal plasticisation involves including a copolymer in the system (either in addition to the homopolymer or replacing it). In general, the lowest cost option is the use of external plasticisation, but the long-term performance of the bond tends to be poor and, for many purposes, plasticisation is essential. In some cases, it may be desirable to use an adhesive which, after application, will dry to form discrete particles, giving a bond with properties that have something of a 'spot-welding' effect, with bonding occurring by single particles of adhesive between the adherents. This principle can be used with adhesive compositions based on a polyvinyl acetate homopolymer, where a bond without stiffening effects is required, such as that of textile bonding for fabric interlining.

Most adhesives are applied by machine, including roller, blade or air knife coatings and extrusion systems. Spray gun methods may also be employed. In most roller coating systems, a medium viscosity adhesive is required. It is usual practice to keep the adhesive reservoir permanently stirred, to avoid the effects of drying and surface skinning. A rapid 'quick dry' time is desirable, to ensure an even film thickness; this is obtained by formulation of the adhesive to give a quick thixotropic set, with a rapid increase in viscosity during drying.

These properties are significantly affected by the type of substrate. Whilst rapid absorption or 'strike-in' of the adhesive into a porous substrate, such as paper, may be desirable, with other substrates, such as brick or stone, this effect should be avoided. The rate of absorption into substrate, especially of adhesives based on polyvinyl acetate homopolymers, may be reduced by the presence of a high molecular weight, high-viscosity water-soluble polymer (e.g. a cellulose ether) in the formulated adhesive or, more generally, by the use of a polymer latex with a relatively high particle size (over 1 μm).

Although all latex adhesives have water as the suspending medium, the drying time, which affects the rate of development of the strength of the adhesive bond, can be varied significantly by alterations in formulation, since the rate diffusion of water through the latex film and the rate of evaporation from the surface may vary. The latter depends on the rate of diffusion of water through the polymer film as it is formed. Some protective colloids tend to absorb water, and so reduce the rate of evaporation [32].

Solvents are often added to compounded latex adhesives, as they have a marked effect on important factors, such as film formation and setting time. Such additions tend to be empirical, but are widely used. Water-immiscible solvents (such as many hydrocarbons) do not greatly affect the setting time, but may evaporate slowly from the adhesive film, because of strong polymer–solvent interactions. Water-soluble solvents, which are not usually polymer solvents and are also hygroscopic, tend to retard evaporation, in relation to their own evaporation rates. Water-miscible solvents, which also solubilise the polymer, such as ethanediol monoethyl ether, $C_2H_5.O.C_2H_4.OH$ and similar oxygenated ethers, retard the evaporation of water and, hence, the setting rate, but also tend to remain in the adhesive film as transient plasticisers, evaporating away slowly to leave a hardened polymer film. This type of solvent is added to the compounded adhesive (usually with additional water to prevent precipitation). A permanent plasticiser, used to provide some long-term flexibility to the adhesive film, may also be added to the latex system. Some examples of this type of formulation are described later (see, for example, Section 4.1.4).

Tolerance of a compounded adhesive to ethanol (or methylated spirits) is sometimes required. Most latex systems are precipitated by addition of ethanol, unless this solvent is diluted with water. If the polymer is ethanol soluble, a gel may be formed, but some polyvinyl acetate latexes have a tolerance to ethanol.

For most purposes, an adhesive composition should be resistant to freezing during storage. Many polymer latices are not freeze–thaw stable, but compounding, as above, with solvents and plasticizers introduces sufficient freeze–thaw stability for most applications.

Difficulties of the surface wetting of adherents are unlikely to occur with latex-based systems, since the original latex used contains surface-active agents in significant quantities. Difficulties of adhesion are therefore reduced to those of specific adhesion between adherents or difficulties of application (generally due to unsuitable rheological behaviour). Some protective colloids and surfactants used in latex production tend to reduce tack [33]. Adhesive compositions normally also include minor, but essential, additives as biocides and antifoam agents; many of these are based on medium-chain fatty alcohols.

3.3 Practical requirements of adhesives

Most adhesives are required to be stable to chemical agents and to be resistant to water and extremes of temperature. Apart from these, most requirements depend on the mechanical strength to which the final joint is subjected. In addition, the performance requirements also make the distinction between wet stick and dry stick and instant and delayed tack.

A wet stick adhesive is required to stick the two adherent surfaces whilst it is still in the wet state, or at least in the sticky (high-viscosity) state. In dry stick adhesives, the adhesive is applied to the principal adherent and becomes adhesive to another adherent when suitably activated by, for example, pressure, heat or a solvent (including water). Adhesive compositions of this type are widely used in the formation of laminates. Delayed tack adhesives must be heated to obtain the desired properties, but those properties may be retained for a significant time after cooling. These adhesives are usually formulated to include solid plasticisers, such as dicyclohexyl phthalate $(C_6H_{11}OOC)_2C_6H_4$, which are not usually effective until heated with a polymer and retain equilibrium only after cooling (see Section 5.5).

There are various types of pressure-sensitive adhesives, including those that allow the two adherents to be separated after application, with the adhesive remaining cleanly on one surface without substrate damage, e.g. paper tear, taking place. Such adhesives should have a film cohesion strength greater than that of the adhesion (see Section 5.2). Quick grab pressure-sensitive adhesives are those which, after the minimum drying time, provide an effective bond with minimum pressure. This property may be obtained by suitable plasticisation or by the addition of a suitable fugitive solvent, soluble in the aqueous phase. Other types are described in Section 5.

Adhesives with a high bond strength are formulated using high molecular weight resins, usually with some limited crosslinking. It is preferable that the crosslinking system should retain some flexibility by the use of resins including long-chain esters, e.g. the divinyl ester of a long-chain dibasic acid, such as sebacic acid, or a polyglycol dimethacrylate, such as tetraethylene glycol dimethacrylate. Adhesive film properties can vary markedly with molecular weight, high tack at low molecular weights (typically, 10^4) being replaced by a greater bond strength at higher molecular weights (typically, 10^6).

Many latex-based adhesive bonds are required to be insensitive to water, although most have, under normal conditions of use, adequate resistance to unfavourable atmospheric and weather conditions. Some water sensitivity is due to the presence of the emulsifiers used in the preparation of the latex component, and polyvinyl acetate homopolymer systems are sensitive to both water and to alkalis. Adhesives based on polyvinyl acetate latexes alone, although relatively low in cost, are, therefore, unsuitable for use in bonds likely to be exposed to high humidity or to alkalis. Some copolymers (either polyvinyl acetate–acrylate or acrylate alone) alleviate this problem. When both adherents are cellulosic (e.g. paper-to-paper or wood-to-wood), most latex-based adhesives are suitable, and polyvinyl acetate homopolymer or copolymer based compositions may be used. The protective colloid systems of the latexes and the nature of any thickeners present in the adhesive formulations also affect the moisture sensitivity of the bond: broadly speaking, cellulose ether-type colloids and thickener-based formulations are less water sensitive than other types, such as those using partly hydrolysed polyvinyl alcohols as colloid or thickener. In addition, the mechanical and rate-of-setting properties are significant in relation to the properties of the adhesive films obtained. The rheology and adhesive properties of a latex are directly dependent on the type and level of protective colloid and/or surfactants used as stabilisers. These factors, and the stirring conditions during manufacture, can influence the size development and size distribution of the particles in the latices.

An apparently minor, but sometimes important, requirement is that the compounded latex-based adhesive should not foam during manufacture or during application, e.g. when being pumped or being applied to the substrate using a mechanically operated coating system. Many proprietary anti-foam agents are available, usually hydrocarbon oil or silicone based. These can be effective, but should be added sparingly, since they are expensive and, also, their use may reduce the strength of the adhesive bond. A low-cost emulsion of linseed oil fatty acids (25 g in 2.2 L of water) has been employed to reduce foam in a 45 % latex of acidic acrylate copolymers used as a removable protective coating (which may be considered as a stable film-forming adhesive with unusually low bond strength) [34].

A useful review of polymer latices for use in adhesives has been published [35]. Many manufacturers also provide information on formulations in their technical literature; some is mentioned in the following pages.

4 SPECIFIC ADHESIVE TYPES

Simple latex adhesives have several preferred applications, which depend on their known performance and cost-effectiveness. In general, they are most widely used with porous substrates, notably wood and cellulose products, such as paper and board, but there are many exceptions. Compared with solvent-based systems, they are usually lower in cost, on a weight-for-weight basis, but this advantage may not apply when 'hot-melt' formulations, which contain no solvent, are considered. In addition, many formulated adhesives contain a significant proportion of filler, which reduces cost but may also reduce

Specific adhesive types

performance. In designing adhesives for specific purposes, many factors, apart from the latex composition and method of manufacture, are significant, and depend principally on the method and conditions of application and the final performance required by the adhesive bond between the substrates. In production processes using adhesives, there is frequently a conflict between the lower-cost water-based systems, where the rate of adhesive bonding may depend on the evaporation rate of water controlling the speed of bonding, and the more rapid bond formation, which may be possible with solvent-based systems, coupled with toxicity and fire hazards and the costs of solvent recovery, or with 'hot-melt' systems, where the rate of heat removal from the molten bond is important. In practice, such considerations mean that adhesive formulations are designed with the method of application very much in mind.

Most adhesives are applied by machine, using rollers, doctor blades, air knife coaters, extruders or spray guns. The viscosity characteristics required for each method are different. For most roller-coating applications, moderate viscosity is required, with a rapid increase in viscosity of the coated film on evaporation. Added solvents may affect the setting rate and are chosen to suit particular substrates.

The performance behaviour required from an adhesive bond may include many factors:

(a) short-term adhesion, for bonding disposable articles;
(b) long-term adhesion, with minimal 'creep' for bonding furniture or constructional materials;
(c) high bond strength, for bonding high-performance substrates (usually obtained by formulation from high molecular weight resins, sometimes with crosslinking);
(d) degree of resistance to environmental exposure, short term or long term;
(e) moisture resistance (for constructional bonds);
(f) stability to chemical reagents (including water);
(g) stability to extremes of temperature;
(h) low cost (for bonds in disposable articles);
(i) ability to bond to 'difficult' substrates, including many 'plastic' substrates, such as PVC, and additives (such as plasticisers);
(j) surface treatments of substrates (e.g. metallization, or gloss treatments).

In addition, with pressure-sensitive adhesives, there may be requirements for:

(a) 'wet stick' (where the bonded surfaces need only stay together whilst the adhesive is in the wet state);
(b) 'dry stick' (where the bond is formed by pressures);
(c) 'quick grab' (where an adhesive, after drying, is required to form an adhesive bond with minimal pressure) or

(d) 'delayed heat seal' (where the adhesive retains the seal properties after cooling, usually by incorporation of solid plasticisers).

The manufacturing cost of latex-based adhesives is frequently critical. In most cases, the lowest cost of raw materials can be obtained by using polyvinyl acetate homopolymer-based latices, with added fillers, at the lowest solids concentration suitable for the bonding application required.

4.1 Polyvinyl acetate homopolymer-based adhesives

Polyvinyl acetate (which has several abbreviations, PVAc (preferred form), PVA or PV-OAc) has the working formula

$$-(CH_2.CH(O.CO.CH_3)_n)-$$

It is a major component of latex-based adhesives, which have the advantage of relatively low cost and are of major interest for use with cellulosic substrates, such as wood and paper [36], where it has been used for many years. Such adhesives have some disadvantages: they have a tendency to 'creep'. If the adhesive is applied to a vertical or inclined surface, there will be some downward migration of the adhesive, resulting, in time, in bond failure. In a test of construction adhesives under load, polyvinyl acetate-based adhesives (with added crosslinking) has fair creep resistance [37]. Most latices in current use are made using a polyvinyl alcohol stabiliser and many formulations have changed little since their original publication in the BIOS and FIAT reports (reported German practice) in the late 1940s [38]. Most adhesive-grade polyvinyl acetate latices have a solids content of 55 % w/w, which includes several minor components, apart from the major proportion of polyvinyl acetate. The principal minor component is up to 5 % of cellulosic water-soluble polymer (dextrin, starch or gum arabic) or, more often, water-soluble polyvinyl alcohol, of which a wide range of grades is available [39, 40].

This is produced in quantity by solution polymerisation of vinyl acetate monomer to polyvinyl acetate, which is then hydrolysed, in a methanol–water mixed solvent to polyvinyl alcohol. The type normally used in latex polymerisation is a partially hydrolysed grade, with 87–89 % hydrolysis (saponification value = 140) in a range of viscosities.

In production of polyvinyl acetate latices, there is a preference for using a mixture of grades of polyvinyl alcohol of different viscosities (and, therefore, of different polymer chain lengths). Low to medium viscosities are preferred, since use of high-viscosity polyvinyl alcohols tends to lead to very high viscosity latexes, which are difficult to handle. The viscosity of aqueous polyvinyl alcohol solutions is measured, by convention [41], using a Hoeppler falling ball method, as the viscosity (in centipoise, cP) of a 4 % aqueous solution at 20 °C. Typical commercial grades are low viscosity (at 4–6 cP), medium viscosity (at 21–25 cP) and high viscosity (at 35–45 cP). These are, however, nominal figures and there can be significant variation in performance between commercial grades from different manufacturers, with apparently similar specifications. Such variations are due to differences in manufacturing methods for the polyvinyl acetate from which the polyvinyl alcohol is produced. The resulting polyvinyl alcohols may have different degrees of chain branching or grafting, differing end groups on the polymer chain and variations in molecular weight distribution. During hydrolysis of polyvinyl acetate to polyvinyl alcohol, there are further possibilities of the introduction of other variations, such as random or sequential removal of acetate groups, and of the introduction of unsaturated groups by removal of water during the hydrolysis. Because of these factors, there is likely to be some variation in the properties of latices produced with commercial polyvinyl alcohols from different sources, even if the viscosity and degree of hydrolysis are nominally similar. The latex viscosity and the rheology of the latex are also likely to show major variations due to differences in the polymerisation process. This is especially true with the various 'delayed addition of catalyst' methods employed. In general, the lower the amount of free monomer present at the commencement of polymerisation and the higher the temperature of gradual addition, the higher will be the viscosity at constant solids content. It is probable that this viscosity variation is due to variations in the degree of vinyl acetate monomer grafting on to polyvinyl alcohol. Increased grafting leads to a reduction in the available amount of stabiliser, which results in destabilization of the latex. The viscosity of suitable polyvinyl alcohol-based latices for compounding into adhesives may be changed by high shear during stirring or in coating out the adhesive film, which can cause a permanent reduction in viscosity, due either to the permanent reduction of hydrogen bonds or to removal of the hydration layer. This behaviour can occur in many latices; e.g. the long-established Mowilith D, a polyvinyl acetate homopolymer made by Hoechst AG, drops in viscosity during shearing from about 110 P to about 10 P.

Most latices used in adhesives include some particles with rather large diameters and have a wide range of sizes—typically, from ~0.5 to 2.5 μm. Such a range allows some penetration of the particles into porous substrates, but excess surfactant should be avoided, since this may reduce surface tension and excessive penetration into the substrate. The solids content of most latices for adhesives is 50–55 % before addition of plasticizer, although for economic reasons (to reduce transport costs, for example) some adhesives-grade latices with a solids content up to 60 % are available. A high molecular weight may be preferred to obtain maximum strength and minimum 'creep' of the adhesive bond. The tendency to 'creep' is a principal disadvantage of polyvinyl acetate homopolymer latex-based adhesives, but may be overcome by the introduction of some crosslinking agents, also at the expense of long-term stability.

Adhesives employing latices stabilised with polyvinyl alcohol give much greater tensile strength on adhesion than is obtained with other types of stabiliser, especially with wood-to-wood adhesives. It is believed that this is due to the graft structure of the polymer, which provides a suitable balance of hydroxyl groups for forming the adhesive bond. Most latices stabilised with polyvinyl alcohol, when suitably formulated, show good adhesion to glass and ceramics.

An example of one typical range of commercially available polyvinyl acetate homopolymer latices, with the recommendations of the manufacturers for suitable applications, is shown in Table 8.1 (see also Table 8.89). Many others are shown in the tables in Section 9.

4.1.1 Adhesive latices for compounding

Manufacturing processes are available for preparing very highly grafted or branched polyvinyl alcohol homopolymers in latex form. These processes add both monomer and the aqueous phase (which includes surfactants and stabilisers) gradually to a 'heel' of preformed latex. Typically, ~5 % of polyvinyl alcohol, based on added monomer, is included in the aqueous phase. Such latices have particle sizes in the range of 1–3 µm: the bulk viscosity varies according to the details of the methods of preparation. Polymerization normally depends on a redox-type initiator system, with hydrogen peroxide and sodium formaldehyde sulphoxylate $HO.CH_2.SO_2.Na$ added to the latex during preparation, in separate streams. An example of the detailed production technique is shown in Table 8.2. With this type of latex, there is good tolerance to ethanol, and polymers prepared by this process appear to be insoluble in this solvent. The setting speed of the latices is good, with good performance in wood-to-wood compression adhesion.

Many other alterations in formulation are possible, and the molecular weight and its distribution may vary widely. This variation is conveniently indicated by a viscosity index measured at a standard concentration, often 0.4 % of polymer (w/v) in ethanol or in benzene, or in a solvent mixture which will dissolve the latex as a whole. Viscosity variations, which will depend on the molecular weight, affect the heat seal characteristics of the coated polymer film: temperatures from 85–115 °C have been quoted for various technical products. The second-order transition temperature, T_g, also varies with molecular weight, and is a measure of the lowest temperature at which a continuous adhesive film can be formed. Latex polymers for adhesives have been prepared by methods involving variation of the mixed monomer feed composition during the polymerisation: this technique is claimed to produce particles with T_g values that vary with particle diameter. Typical latices, such as terpolymers of acrylic acid, butyl acrylate and methyl methacrylate, are useful for bonding to metallic surfaces [42].

Latices containing polyvinyl alcohol, either as stabiliser or thickener, are unstable to borax, except under carefully controlled conditions of a narrow

Table 8.1 Some typical commercially available polyvinyl acetate latices and their recommended applications (Vinavil SpA)

Name	Siolids content (%)	Viscosity (Pa s)	pH	Particle size (μm)	MFT (°C)	Emulsifying system	Adhesive applications			
							Wood	Paper	Textile lamination	Heat seal
Ravemul O 12	60 ± 1	30 ± 5	4	0.5–2	14	Polyvinyl alcohol	*	*	*	
Ravemul O 13	60 ± 1	3 ± 0.5	4	0.3–3	14	Polyvinyl alcohol	*	*	*	
Ravemul O 16	50 ± 2	40 ± 7	4	0.3–2	14	Polyvinyl alcohol	*	*	*	
Vinavil RP	55 ± 1	7 ± 2	4.5	0.2–0.6	15	Anionic–non-ionic				*
Vinavil SA	52 ± 1	2.5 ± 1	4	0.2–0.6	15	Anionic–non-ionic			*	*
Vinavil KA/SR	50 ± 1	50 ± 1	4.5	0.6–2	14	Polyvinyl alcohol	*	*	*	*

range of pH (~4.0–5.0), so they cannot be compounded with borax-modified dextrins. There is no difficulty with other types of dextrins, which may be used as stabilisers in the latex polymerisation of vinyl acetate to form types of latices with good compatibility with starches and other polysaccharides. Some methods [43, 44] have described polyvinyl acetate latices prepared with hydroxyethyl starch, hydroxypropyl starch or hydroxyethyl cellulose as high molecular weight stabilisers suitable for adhesives. Such latices are borax stable. Vinyl acetate (and other monomers) have been polymerised in starch solution (10–200 % on monomer) to give latices from which the dry films had good hardness and heat resistance [45]. Typically, vinyl acetate monomer (175 g) has been polymerised in a solution of starch gum (50 g) in water (250 g) to form a latex (43 % of solids) with a viscosity of 37 P at 0 °C and 20 P at 30 °C, with an MFFT of 0 °C.

Coated films made from polyvinyl acetate compositions have some water sensitivity, which may cause difficulty under conditions of high humidity, but use of copolymers with vinyl esters of branched chain fatty acids (e.g. vinyl Versatates(® Shell Chemical Company)) tends to overcome this difficulty.

4.1.2 Plasticisers

The flexibility of polyvinyl acetate adhesive films is limited, but can be increased by copolymerisation (internal plasticisation) or by the direct addition of plasticisers (external plasticisation). A wide variety of plasticisers can be employed: they are normally added to the latex, with stirring, as it is cooled after manufacture. The plasticisation of a latex increases the solids content and tends to increase the viscosity of the system and to lower the T_g of the formulation. With wood-to-wood glues, addition of small amounts of plasticiser (e.g. <5 %) tends to increase the strength of the adhesive bond, but also increases the tendency of the joint to 'creep'. For this reason, wood-to-wood glues are prepared, where possible, from polyvinyl acetate latices with molecular weights as high as possible, or with crosslinking additives.

Dibutyl phthalate and di-isobutyl phthalate, (which is lower in cost but is rather less effective) are widely used as plasticisers for adhesive systems, imparting adequate flexibility and film softening with relatively low volatility. Diethyl phthalate may be used when fast tack is required. Benzyl butyl phthalate $(C_6H_5O.CO).C_6H_4.(O.CO.C_4H_9)$ is considered to be a good alternative to dibutyl phthalate in most applications. It is normally slightly more expensive, has lower volatility than dibutyl phthalate, but gives improved moisture resistance and toughness to the adhesive film.

The performance of plasticisers may be compared, relative to a composition using a standard latex and dibutyl phthalate, by the extent to which they allow 200 % elongation at 25 °C of a test strip 500 mm long and 6.2 mm wide in

Table 8.2 Preparation of highly branched polyvinyl acetate latex (final solids content: 5.45 %)

	Weight (g)
Phase 1	
Base latex (Vinac XX-210) (Air Products & Chemicals Inc.)	400
Distilled water	37
Na formaldehyde sulphoxyalate	0.5
The above are placed in a 3 L flask with stirrer, thermometer, two dropping funnels, burette and condenser. The flask is immersed in a water bath and heated so that the modified 'heel' is at 70 °C	
Phase 2	
Polyvinyl alcohol (88 % hydrolysed) (viscosity of 4 % aq solution at 20 °C = 25 cP) as 13.8 % aq solution	145
Polyvinyl alcohol (88 % hydrolysed) (viscosity of 4 % aq solution at 20 °C = 45 cP) as 10 % aq solution	55
Hydrogen peroxide (3 %)	25
Distilled water	87
Na bicarbonate	0.9
Phase 3	
Vinyl acetate	425
Surfynol P (a non-ionic surfactant including acetylenic units)	2
Phase 4	
Na formaldehyde sulfoxylate	1.5
Distilled water	35
Phases 2, 3 and 4 are added in separated streams to the polymerizing mixture which is allowed to increase to 89 °C over 3–3.5 h, then held for 1 h at 80 °C, by which time the free monomer content should be <0.5 %.	

a tensile test apparatus in which the moving jaw travels at 33 cm s^{-1}. Under these conditions (which are arbitrary, but convenient) latices plasticised with 80 % of dimethyl phthalate, 55 % of dimethyl sebacate or 110 % of ethyl phthalyl ethyl glycollate are equivalent in plasticising power.

Depending on the characteristics of the particular latex, addition of dibutyl phthalate can also lead to an increase in the setting time. Conversely, an increased setting rate means a decreased 'open' time, which may make application of the adhesive to the substrate more difficult.

A further increase in plasticiser content causes reduction in the setting rate (probably due to increased flow (or creep) of the plasticised polymer). Other variables that can affect the setting rate include the addition of fillers and the content of polyvinyl alcohol [46, 47]. The effect of the protective colloid is significant: the setting rate of a polyvinyl acetate increases with increasing polyvinyl alcohol content and may increase with increasing plasticisation to an optimum of 10 % dibutyl phthalate (on polymer) and then decrease with increasing plasticisation. These effects may be expected to occur, to different extents, with other additives. The optimum is less marked with a lower polyvinyl alcohol content and is barely noticeable when a cellulose ether stabiliser is used instead of polyvinyl alcohol: in this case, the setting rate is very low in the absence of plasticiser. This effect is not general to all polymer latices. With a 10 % plasticised polyvinyl acetate latex, the solids content affects the setting rate: for raid setting the optimum is about 56 %. Other factors apart from the plasticiser level and the protective colloid can affect the setting rate: these include the filler content and the temperature and humidity of application.

Benzyl butyl phthalate (see above) may be used as a satisfactory, if slightly more expensive, plasticiser rather than dibutyl phthalate in most applications, since it has lower volatility and provides the adhesive film with improved moisture resistance and greater toughness. They are reported by Monsanto (USA) to have fairly good plasticising efficiency and are non-toxic, so are acceptable for use in adhesive compositions for food packaging. Several compounds of this type have been proposed, based on the general formula:

$$C_nH_{2n+1}.O.CO.C_6H_4.CO.O.CH_2.CO.O.C_nH_{2n+1}$$

where

Methyl phthalyl methyl glycollate $n = 1$
Ethyl phthalyl ethyl glycollate $n = 2$
Butyl phthalyl butyl glycollate $n = 4$

For bonding overlays to paper, the methyl ester has been found to be useful, because of its superior light stability, ability to dissolve cellulose acetate and ability to provide quick tack when added to adhesive formulations. Other benzoate esters are also of interest as plasticisers for polyvinyl acetate latex-based adhesives. These include the 'Benzoflex' range (Velsicol Chemical Corp., USA) [48], which are:

Diethyleneglycol dibenzoate 'Benzoflex 2-45':

$$C_6H_5.CO.O.C_2H_4.O.CO.C_6H_5$$

Dipropyleneglycol dibenzoate 'Benzoflex 9-88':

$$C_6H_5.CO.O.CH(CH_3).CH_2.CH(CH_3).CH_2.O.CO.C_6H_5$$

These compounds have a plasticising efficiency similar to that of dibutyl phthalate and superior to that of butyl benzyl phthalate, especially at low levels of addition. Data on the tack, heat-seal and blocking characteristics obtained [49] with several different plasticisers compounded with polyvinyl acetate are shown in Figures 8.1 to 8.6. It is notable that the heat-seal behaviour of both benzyl butyl phthalate and Benzoflex 2-45 are similar (see Figure 8.5), although Benzoflex 2-45 and dibutyl phthalate are more sensitive at low plasticizer levels below 20 phr (parts per hundred of polymer). The blocking temperatures indicated show similar behaviour to that of the heat-seal behaviour. Benzoflex 9-88 has the highest blocking temperature under comparable conditions of plasticisation (see Figure 8.6).

Polyethylene glycol ethers of monohydric phenols with short-chain alkyl groups attached to the phenol function are also used as plasticisers for polyvinyl acetate [50]. Such compounds (which may be mixed esters of mixed chain length alkyls) are water soluble and surface active: the resulting plasticised films show permanent tack. A typical example is triethylene glycol monophenyl ether: the technical product Pycal 94 [51] is probably of this type. Ethoxyethanol, ethylene glycol monoethyl ether $C_2H_5.O.CH_2.CH_2.OH$ (trade names Cellosolve, Oxitol and others), the ethyl and butyl esters and the acetate [52] may be added to polyvinyl acetate latices in small amounts (1–2 %w/w on polymer) as coalescing agents to aid film formation and improve stability of the formulation and flexibility of the resulting adhesive film. There are

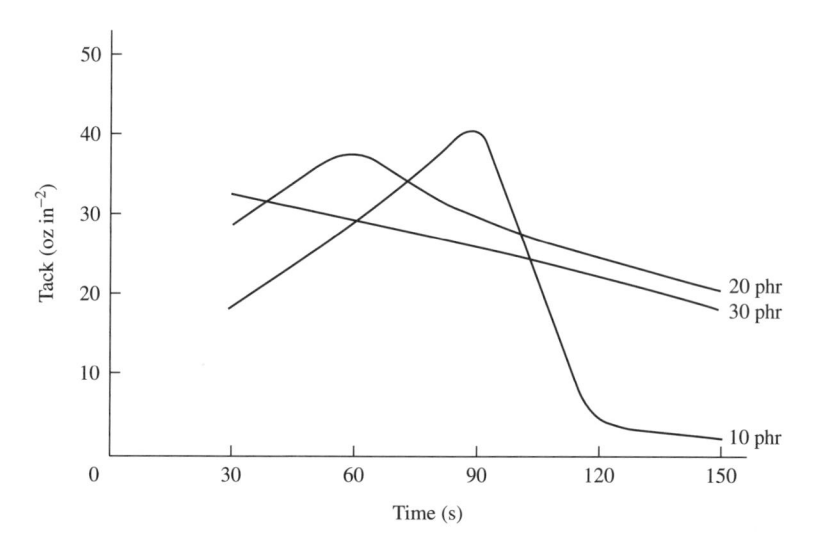

Figure 8.1 Tackifying effect of butyl benzyl phthalate.

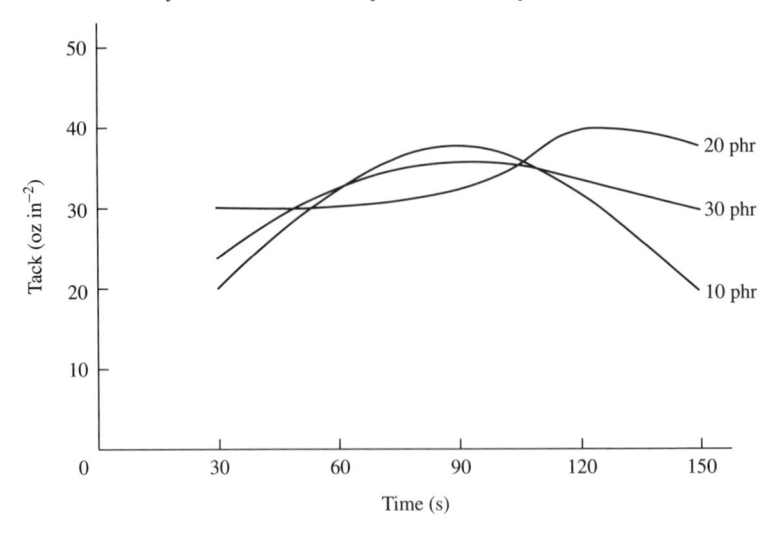

Figure 8.2 Tackifying effect of dibutyl phthalate.

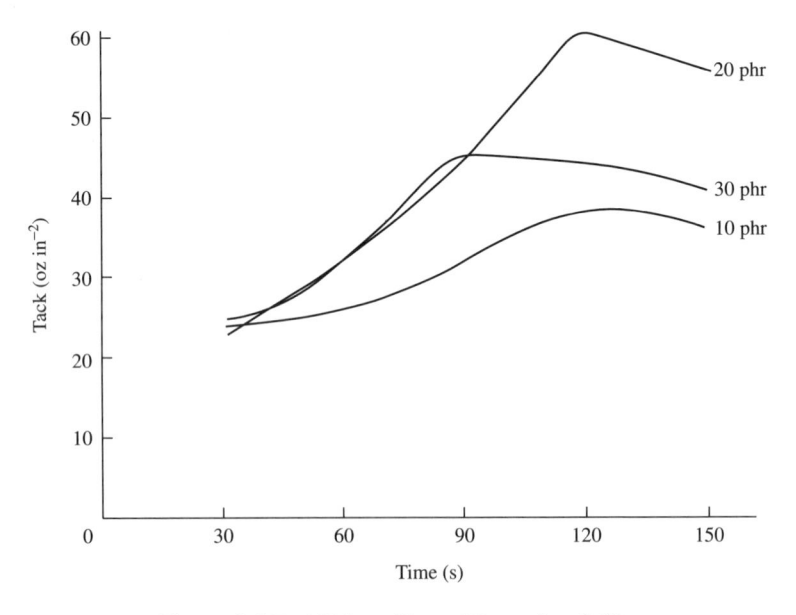

Figure 8.3 Tackifying effect of Benzoflex 9-88.

also several proprietary products, which appear to be mixed esters, sold under trade names, with broadly similar behaviour in latex adhesive formulations. A typical product is Polysolvan O [53], reported to be butyl hydroxyacetate $HO.CH_2.CO.O.C_4H_9$. 1-Phenoxy-2-propanol (Dowanol PPH) is effective with acrylic latices. It has low water solubility, does not retard evaporation and

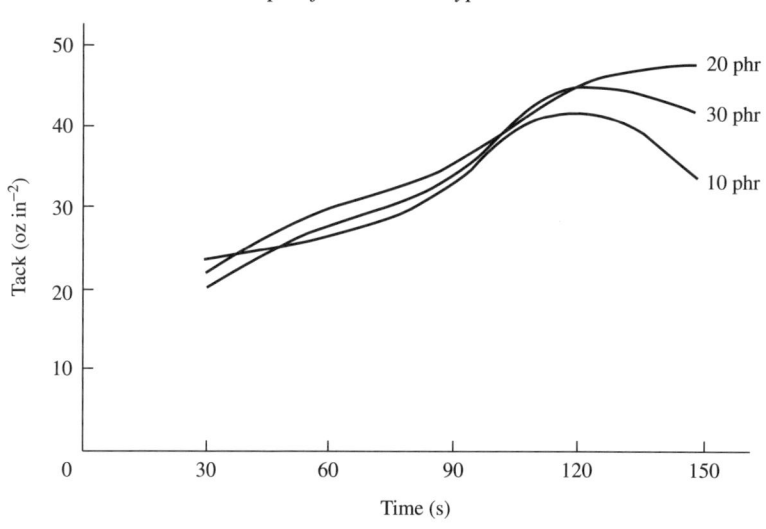

Figure 8.4 Tackifying effect of Benzoflex 2-45.

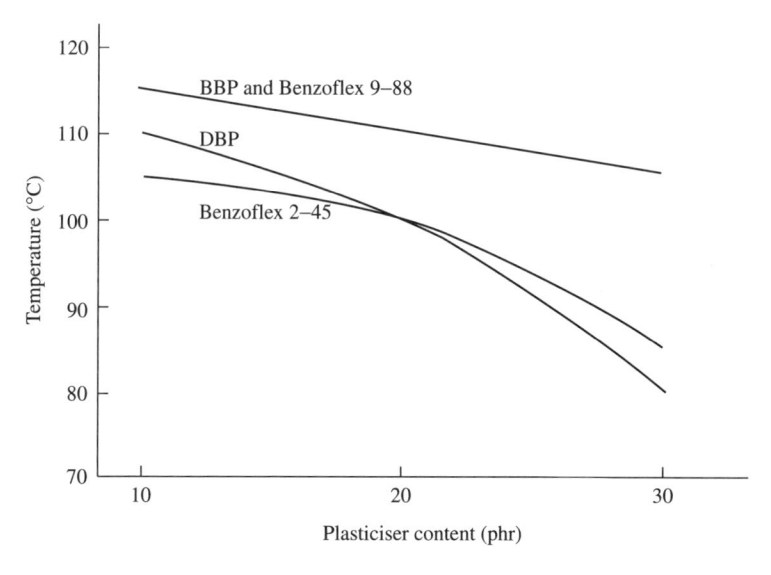

Figure 8.5 Minimum heat-seal temperatures of plasticised polyvinyl acetate.

minimises loss into a porous substrate [54]. Some compounds of this type may be more cost effective in use than lower cost 'conventional' plasticisers, as they are used in smaller amounts.

Phosphate esters are used when flame resistance, oil resistance and low volatility are required in addition to normal adhesive properties. Typical

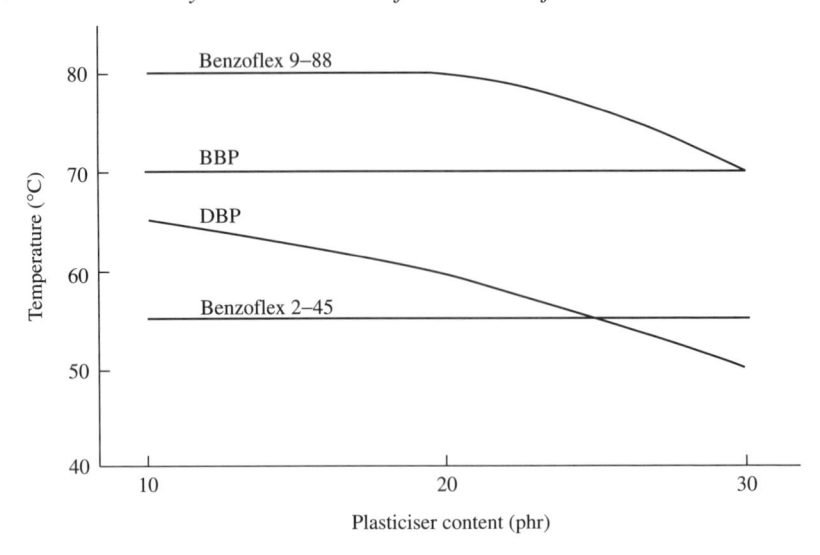

Figure 8.6 Blocking temperatures of plasticised polyvinyl acetete.

compounds employed include 2-ethylhexyl diphenyl phosphate (which is non-toxic), tolyl diphenyl phosphate and triphenyl phosphate [55]. The 2-ethylhexyl compound is reported to have slightly better plasticising efficiency than the other compounds. Solid plasticisers, including diphenyl phthalate, dicyclohexyl phthalate and p-toluenesulfonamide, have special interest as components of delayed tack adhesives, and are considered in more detail in Section 5.5 below.

Mixtures of o- and p-toluenesulfonamides, o- or p-CH$_3$.C$_6$H$_4$.SO$_2$.NH$_2$, have been used to impart good grease resistance and quick tack: the low-temperature flexibility of the film is less good than that obtained with dibutyl phthalate, but bonding of rubber to metal is improved. Whilst external plasticisation (with the additives mentioned in this section) is relatively simple and inexpensive, the resulting adhesive film may embrittle and fail, due to migration of plasticiser into the bonded substrate and, to a lesser extent, into the atmosphere. The former occurs especially when a polyvinyl acetate adhesive composition containing plasticiser is used to bond a 'plastic' film. In addition, when a synthetic resin film is used as a laminating component, migration of plasticiser will soften the film, with loss of cohesive strength, and a weakening of the bond. To reduce this effect, internal plasticisation of the polymer is employed, in which part of the polyvinyl acetate chain is replaced (during the manufacture of the latex) with another monomer with a flexible side chain. Typically, polyvinyl acetate copolymers containing a proportion of maleate or acrylate ester are more suitable for compounding into adhesives for bonding to films of polyvinyl chloride or polyvinylidene chloride [56]. Such latices, suitable for adhesives, have been made from a 1 : 1 : 1 mixture of vinyl acetate, dibutyl maleate and isopropenyl methyl ether, stabilized with a buffered nonionic surfactant (Tergitol NPX) and hydroxyethyl cellulose [57]. However, this type of latex may be considered obsolete in view of the improved performance obtainable from latices based on vinyl acetate–ethylene copolymers [28] (see, for example, Section 4.2.2), due to the high efficiency of ethylene as a plasticising monomer when incorporated directly into the backbone of the polymer chain, giving an exceptionally low T_g. Copolymers of vinyl acetate with 24 % (w/w) of ethylene, vinyl laurate or vinyl Versatate® have a T_g of -9, $+11$ and $+23$ °C respectively. The ethylene–vinyl acetate copolymer-type latices also contribute significantly to improved adhesion to polyvinyl chloride-type substrates and provide the basis for adhesive compositions with much faster setting rates than those obtained using polyvinyl acetate–maleate or acrylate copolymer latices.

T_g can be determined approximately using a well-known formula, the use of which depends on the availability of suitable data. The determination is

$$\frac{1}{T_g} = \frac{w_1}{T_{g1}} + \frac{w_2}{T_{g2}} + \cdots + \frac{w_n}{T_{gn}}$$

where $w_1, w_2, \ldots, w_n$ are the weight fractions of the respective monomers and $T_{g1}, \ldots, T_{gn}$ are the respective T_g of the homopolymers corresponding to these monomers (presumed to be of the same M_n or M_w average molecular weight [58].

4.1.3 Tackifiers, adhesion promoters and other additives

'Tack' is a generic term which described one or more of a number of phenomena relating to the speed of bond formation of adhesives. It is the ability of two materials to resist separation after bringing their surfaces into contact for a short time under a light pressure.

Two types of tack may be considered:

(a) *autoadhesive* tack, from a pair of similar materials;
(b) *adhesive* tack, from a pair of dissimilar materials.

The property of tack suitable for the intended application should, in general, be formulated into the adhesive compound by the addition of tackifier resins [59]. Many materials have been used as additives, to alter properties of latex-based adhesives. Three principal types of tackifier resin are employed:

(a) rosins,
(b) terpenes and
(c) hydrocarbons.

Within this broad scope, tackifiers are often based on petroleum or wood-based rosins (the latter are called *colophony* in some countries) or their derivatives, such as 'terpene gums' or 'ester gums' (of varying composition and chemical structure), with little solubility in water but useful solubility in hydrocarbons or water-compatible organic solvents. Available hydrocarbon resins (usually 'of proprietary composition', which means 'of less than specific composition', since they are largely prepared from a mixture of low-cost polymerisable hydrocarbon monomers) include, amongst many others:

(a) The Escorez range [60], in a range of grades:
    1300 Series of mixed aliphatic, aromatic, branched chain resins, mainly intended for pressure-sensitive adhesives;
    5300 Series of water-white hydrogenated resins, for premium purposes;
    Escorez 5000 prepared from hydrogenated dicyclopentadiene.
(b) Hercoprime [61] resins, intended for polyolefin-to-metal adhesives.
(c) Aromatic resins, mainly based on hydrogenated polystyrene, α-methylstyrene and indene in different proportions, including Regalez, Petrorez and Nevex [62].
(d) Mixed aliphatic–aromatic resins are available, including Sta-Tac resins [63]. These are mainly intended as additives to rubber- or alkyd-based compositions and to hot-melt adhesives based on solid ethylene–vinyl acetate resins.
(e) Wingtack resins [64], both solid and liquid types, mainly synthetic polyterpenes, derived from polymerised piperylene and isoprene, with $M_n < 500$, liquid at 5 °C, with good compatibility, mainly used in pressure-sensitive adhesive systems. Wingtack 95 is prepared [65] by cationic polymerisation of

Piperylene	60 %
Isoprene	10 %
Cyclopentadiene	5 %
2-Methylbut-2-ene	15 %
Dimer	10 %

and has a softening point of 80–115 °C.
(f) Wood rosins are produced by the thermal polymerization of natural or synthetic terpene hydrocarbons (mainly β-pinene and dipentene), which include the following.
(g) Zonarez polyterpene resins [66], soluble in aromatic solvents (notably toluene), and available in a wide range of melting points. The Zonarez 7000 series are based on dipentene, whilst the Zonarez B series are based on β-pinene. They are widely solvent compatible.
(h) Nirez polyterpene resins (the same trade name is used for synthetic hydrocarbon resins). Pale, colour-stable resins, available in a range of melting points.

Specific adhesive types

It is common practice to include 1–2 % (w/w on polyvinyl acetate resin solids) of tackifying hydrocarbon resin in hydrocarbon solution (either mixed aliphatic or mixed aromatic solvent, depending on the nature of the tackifying resin chosen) in a latex-based adhesive formulation. One example, typical of many, describes the formation of a tackifier resin by the reaction of a petroleum resin in butoxyethylene glycol as solvent, with 3 % w/w of maleic anhydride, followed by neutralization with ammonia. Tackifiers may also be used in emulsion form.

Typically a tackifier emulsion, with 53 % solids, can be formed by [67]

Ester gum, 70 % in toluene	200 g
Polyvinyl alcohol (87–89 % hydrolysed, low viscosity) (Poval 200E)	30 g
Water	100 g

This emulsion can be blended with an adhesive latex, including a film former, and a thickener, to give a composition with good tack and a high bond strength. In another example, mixed aliphatic–aromatic petroleum resin, m.p. 55 °C, has been blended with aqueous sodium crotonate and a non-ionic surfactant (30 % aq.), and polymerized using *tert*-butylperoxy-2-ethylhexanoate in dioctyl phthalate for 1 h at 90 °C, then cooled to 55 °C, and mixed with water to yield an oil-in-water (O/W) emulsion that can be added to acrylic latices, as a tackifier for pressure-sensitive adhesives [68]. Improved tackifiers have been used to enhance the properties of adhesives for bonding rockwool insulating boards to epoxy-coated iron sheet: these tackifiers are based on a styrene–butadiene latex, with added surfactants (as wetting agents and stabilisers) [69].

A relatively simple system for improvement of tack and adhesion of polyvinyl acetate latex-based adhesives involves addition (60 : 40) of an emulsion formed by melting 50 % of hydrogenated wood rosin and dimethylamine as emulsifier. The product has high peel strength when used to bond PVC [70].

Another system suggested (mainly for acrylic latex adhesives) includes a proprietary tackifier, N,N-diglycidylaniline (which may be an adhesion promoter), a surfactant, toluene and water (50 : 95 : 5 : 2 : 100), which is blended with an acrylic latex (50 : 1). The resulting adhesive may be coated on to polyester film and coated with release paper [71]. Tackiness in carboxylated styrene–butadiene latices is reduced by addition of hydroxypropyl starch (and probably several other starches) [72].

Several adhesion promoters for polyvinyl acetate latex-based adhesives are known: their use depends mainly on economic factors, since these adhesives have adequate bonding performance for most purposes (see also Section 6.1.6). Many wood adhesives for laminating purposes include urea–formaldehyde resins in the formulation. This gives a degree of crosslinking (giving good bonding and moisture resistance) at the expense of the storage stability properties of the system [73]. Alternatively, phenol– or resorcinol–formaldehyde resins may be added [74]. These are also useful in rubber-to-fabric adhesives (see Section 7).

Aziridine derivatives have been proposed. A copolymer containing 4 % of methacrylic acid reacted with 2-methylaziridine in vinyl acetate adhesives (ensuring that the aziridine does not react with the relatively easily hydrolysed ester group) is reported to improve adhesion to wood [75] (see also Section 6.1.6).

Diallyldimethylammonium chloride has been added as an adhesion promoter to a conventional pressure-sensitive adhesive, based on a latex of 2-ethylhexyl acrylate, vinyl acetate and acrylic acid (74.5 : 22 : 3.5), giving significant adhesion to stainless steel [76]. Allyl carbamate $H_2N.COO.CH_2.CH=CH_2$ has been included (0.6 % on vinyl acetate) in polyvinyl acetate copolymer latex adhesives, including ethylene–vinyl acetate latices (stabilized with a polyvinyl alcohol—a mixture of Gelvatol 20-30 and 20-60), to reduce solubility of the polymer in organic solvents (presumably by grafting and crosslinking) to give improved adhesion to surfaces such as polypropylene [77]. Low-toxicity solvents (including alkyl butyrates, ethylene glycol butyrate, d-limonene and C_5–C_{20} petroleum distillates) have been used with 1,1,1-trichloroethane in latex-based packaging adhesives [78].

Sodium polyphosphate (five parts) has been added to an acrylic latex adhesive used for bonding a substrate containing $CaCO_3$, apparently improving peel strength on adhesion to polyethylene at higher temperatures and humidity [79]. The presence of sequestering agents is also believed to improve adhesion. Chromium compounds have been added to polyvinyl acetate latex-based adhesives, to improve bond strength and water resistance, although there is the disadvantage of discoloration. To avoid this difficulty, the use of (more expensive) zirconium compounds has been suggested. Water resistance and bond strength of polyvinyl acetate-based wood adhesives is improved by addition of 5 % of a water-soluble zirconium complex, such as the reaction product of glycerol and chlorohydroxyoxyzirconium. Many other zirconium compounds, such as ammonium zirconium carbonate, zirconium nitrate and other zirconium chelates [80–83], have been used. A copolymer latex from vinyl acetate and acrylic acid, with a polyvinyl alcohol stabilizer, has been blended with zirconyl chloride and adjusted with acetic acid to pH 3.5 to form adhesives for wood claimed to have improved wood-to-wood bond strength, water resistance and storage stability [84]. A graft copolymer latex, prepared by redox polymerization of vinyl acetate on to a partly hydrolysed polyvinyl alcohol with 0.03 % of triallyl cyanurate as crosslinker, viscosity 120 mPa and added 3 % of dibutyl phthalate, and either chromium nitrate or zirconium oxychloride, forms an adhesive with bonds with good wet strength [85]. Some suitable compositions for use are suggested below (see Section 4.1.5). Addition of an isobutene–maleic anhydride copolymer has been claimed to improve the adhesive performance of latices [86].

The gelation of latices may be reduced by breaking the hydrogen bond formation in the composition. Typically, the addition of dicyanamide (~5 % on polymer content) to polyvinyl acetate latices for adhesives reduces the MFT from 14.5 to 0 °C, with little reduction in adhesive bond strength [87]. Similarly, addition of ammonium, sodium, potassium or lithium isocyanates tends to stabilize the viscosity of a polyvinyl alcohol-based latex, and at pH < 6 will prevent gelation even if boric acid is present. These latices show rapid setting, with good wet and dry adhesion [88]. The flow properties of polyvinyl acetate latices (with polyvinyl alcohol stabiliser, prepared using a tartaric acid–hydrogen peroxide redox initiator) are improved by addition of ~10 % of urea, ammonium thiocyanate, guanidine hydrochloride or tetrabutyl ammonium bromide [89]. Addition of ⩽5 % of thiourea or an alkali metal cyanate with ⩽5 % of polyvinyl alcohol and a small amount of boric acid to polyvinyl acetate latices makes them suitable for use as adhesives. They have good tack, are redispersible and can form films at low temperatures (⩽0 °C) [90].

Most water-based adhesives are intended to have long-term resistance to bacterial degradation, both in the liquid form and as the dry film. Although cleanliness in manufacture is important, it is also necessary to include preservative in the formulation. Many preservatives have been proposed (for a list, see, for example, reference [91]); examples include *p*-chloro-*m*-cresol, which has been used for many years, the proprietary Dowicil range [92] (the active ingredient is 1-(3-chloroallyl)-3,5,7-triaza-1-azaadamantane chloride) and the *Proxel* range [93]. 1,2- Dibromo-2,4-dicyanobutane has also been suggested; addition of 0.025–0.05 % to a polyvinyl acetate latex is reported to ensure preservation against bacterial growth for up to 18 months. Alternatively, preservation is achieved by using 0.075 % of 1,2-dibromo-2,4-dicyanobutane together with 0.075 % of 2,4-(thiazolyl)benzimidazole for a higher level of preservation of polyvinyl acetate latex mixtures. Acrylic and styrene–butadiene latices may be treated similarly. The properties and applications of other biocides in surface coatings under other conditions are mentioned in Chapters 9, 10 and 11 in Volume 2.

The biodegradability of biocides has to be considered: different types have different degrees of biodegradability while some are believed to present environmental hazards. Many biocides mentioned in the First Edition of this book, although functionally effective, are now unacceptable under current regulations.

Defoamers are normally added to adhesive compositions intended for pumping systems and machine use. A simple mixture of medium-chain fatty alcohols (e.g. octanol) can be adequate, but several proprietary systems are available, e.g. Tamol 731 (a 25 % solution of a sodium polyacrylate). Silicone defoamers, widely used in latex paints, should be used with caution in adhesive systems, since they have significant surface activity and excessive addition may affect adhesion properties.

4.1.4 Fillers

Many latex adhesive formulations include fillers, which are more or less inert, but they help to:

(a) reduce cost;
(b) increase solids content (and, therefore, rate of development of bond strength);
(c) increase viscosity;
(d) reduce penetration of adhesive into the substrate;
(e) sometimes toughen the adhesive film;
(f) improve gap-filling performance.

They also tend to raise the MFT and, especially with organic fillers in wood adhesives, increase wear on cutting machines used on the bonded articles. Organic fillers increase the possibility of microbial degradation. The level of filler addition affects adhesive performance. High loadings will reduce the strength of the adhesive bond. Organic fillers are used only up to 5–10 % (w/w on polymer solids) as they increase viscosity and decrease bond strength, but some inorganic fillers are used at loadings up to 50 %. Some fillers, such as wood flour, are dispersed in mineral spirits, to allow the formation of a free-flowing powder before addition to a latex [94]. The relevant properties of the principal fillers are shown in Table 8.3.

4.1.5 Some typical formulations

In general, the lowest-cost vinyl acetate latex-based adhesives are formulated using polyvinyl acetate homopolymers, although their performance, both in application and bonding, is likely to be inferior to those formulated from vinyl acetate copolymers. For these reasons, such formulations are used mainly in long-established applications, and only minor variations have novelty. Even these types, when examined in the context of the manufacturing cost of the total system (including production rates and yield of the required desired bonded products), may not be the most economic for every purpose). A general formula for polyvinyl acetate latex-based adhesives is shown in Table 8.4 [96]. This general formula indicates the wide range of variation in formulation of adhesives for different applications. This variation in performance with plasticiser level is indicated by the data of Table 8.5 [97].

If ungelatinised starch is included in a polyvinyl acetate latex-based adhesive composition, rapid setting is obtained by heating the wet adhesive film above the gelation point of the mixture [98]. Similarly, a polyvinyl acetate latex,

Table 8.3 Physical properties of commonly-used adhesive fillers. (Adapted from Gouding [95])

	Chemical name	Specific gravity	Oil absorption (%)[a]	Machinability[b]	Effects on product
Wood flour	—	0.8–1.2	100–300	Good	Viscosity increase
Starch	—	1.0–1.2	15–30	Good	Viscosity increase
Clay or kaolin	Hydrated aluminium silicate	2.6	35–45	Fair	
Whiting—uncoated	$CaCO_3$	2.7	25–35	Fair	Viscosity increase
Whiting—coated	$CaCO_3$	2.7	20–30	Fair	Slight viscosity increase
Talc	Magnesium silicate	2.8	35–45	Poor	
Silica	Silicon dioxide	2.65	20–30	Poor	
Barytes	$BaSO_4$	4.45	6–12	Poor	Density increase
Gypsum	Hydrated calcium sulphate	2.4	20–30		Thickens on hydration

[a] An approximate indication of the ability to disperse into an adhesive formulation.
[b] When used in wood adhesives—ease of machine cutting of adhesive bonds when used as a structural component.

Table 8.4 Polyvinyl acetate latex-based adhesives: general formulation

	Parts by weight
Polyvinyl acetate latex; 55 % solids (polyvinyl alcohol stabilised)	100
Plasticizer	10–50
Clay filler	0–30
Dextrin or starch	1–1000
Preservative	0–2
Stabiliser	0–2
Wetting agent	0–0.2
Water	0–100
Defoamer	0–2
Reodorant	0–1

with added gelatinised and ungelatinised starch, is the basis of satisfactory adhesives for laminated papers: gum karaya is used as a thickener [99].

A useful formulation for a multipurpose polyvinyl acetate latex-based adhesive, with good film-forming properties, which retains stability and application flow properties after repeated freeze–thaw cycles (to $-20\ °C$), is shown in Table 8.6 [100]. An adhesive composition for corrugated board, in which the additives improve the performance significantly, is shown in Table 8.7 [101].

A borax-stable polyvinyl acetate latex may be prepared using hydroxyethyl or hydroxypropyl starch or methyl-α-D-glucoside as the stabilizer [102]. A plasticised polyvinyl acetate latex or a vinyl acetate–butyl acrylate copolymer latex, with carboxymethyl cellulose (6 % on monomer) as stabiliser, has been mixed with a similar weight of dextrin to form a stable adhesive system [103]. Starch-based latexes, when formulated, require an increased biocide content for satisfactory long-term storage.

Latex polymerization using a redox catalyst of vinyl acetate with hydroxyethyl cellulose as high molecular weight stabiliser gives a borax-stable latex with pseudoplastic rheology, suitable as a general-purpose adhesive for tube winding and paper lamination. Borax stability is achieved with the absence of polyvinyl alcohol [104]. A basis for a polyvinyl acetate latex-based adhesive for plywood veneers is shown in Table 8.8. This basic formulation is, unusually, strongly acid (pH 1.2), which may cause veneer staining. It is coated (at 300 g m^{-2}) on to a veneer, then cold-pressed, followed by hot-pressing at 120 °C, resulting in a bond strength of <15 kg cm^{-2} with 100 % cohesive failure [105]. It should be noted that the formulation, as reported, does not specify the type of clay used (which may affect the pH) and also does not specify the defoamer or biocide that may be included.

A polyvinyl acetate latex-based wood adhesive, which may, after addition of formaldehyde, be applied to birchwood and pressed for 30 s, then develops a

Table 8.5 Plasticised adhesive compositions from polyvinyl acetate latices

Application	Performance	Polyvinyl acetate (% by weight of composition)	Dibutyl phthalate (% by weight of polyvinyl acetate)
Padding paper bonding cloth	Wet adhesive bond: high tack not required	22–50	20–40
Shoe cements	Dry adhesive bond; strong bond; long tack duration	19–45	35–60
Shoe fabric bonding agents	Wet adhesive bond; high tack; weak bond adequate	11–40	60–180
Automatic bonding of labels	Dry adhesive bond; heat sealing	22–50	20–40

Table 8.6 Multi-purpose polyvinyl acetate adhesive with good freeze–thaw performance

Polyvinyl acetate latex (55 % solids) (including dibutyl phthalate (10–15 % on polymer))	50 g
Polyvinyl butyral (69 % acetal, 27 % hydroxyl) (10 % solution in butanol–ethane diol (1 : 1))	60 g
Limestone powder	300 g
Kaolin	30 g
Fungicide[a]	3 g
Water	97 g

[a] In early formulations of this type, an alkyl tin fungicide was used, but in current practice this should be replaced by a fungicide with regulatory approval.

Table 8.7 Polyvinyl acetate latex-based adhesive for corrugated board

	Parts by weight
Polyvinyl acetate latex (50 % solids)	100
Methyl cellulose (2 % aqueous solution)	100
Dibutyl phthalate	3
Ethane diol	2.5
Corrosion inhibitor	0.5

Table 8.8 Polyvinyl acetate latex-based veneering adhesive

Polyvinyl acetate latex (55 % solids)	100 parts
Clay	60 parts
Hydrochloric acid (10 %)	6 parts

Table 8.9 Polyvinyl acetate latex-based wood bonding adhesive

Polyvinyl acetate latex (41 %)	200 parts
Polyisobutene–maleic anhydride–maleimide terpolymer (2 : 1 : 1) (31 % in aqueous ammonia)	104 parts
$CaCO_3$	60 parts

high compression shear rate (<150 kg cm^{-2}) after 72 h, as shown in Table 8.9 [106]. Alternatives to formaldehyde include dialdehyde starch or 1,4-dioxan-2,3-diol (both significantly more expensive). Another fast-curing system also uses dialdehydes [107]. Like the preceding formulation, the minor additives used, such as preservatives, are not reported.

4.2 Polyvinyl acetate copolymer-based adhesives

With advances in copolymer latex technology, the use of simple homopolymers in adhesive formulations has decreased, compared with other systems based on copolymers, which, although greater in initial materials cost, tend to be more cost effective in use. Publications that have appeared since the First Edition of this work are mainly concerned with the production and formulation of internally plasticised polymers, prepared either by introduction into vinyl acetate of flexible side chains, such as 2-ethylhexyl acrylate, or by copolymerisation with ethylene. The former type of polymerisation can be carried out at atmospheric pressure, with simple manufacturing equipment, whilst copolymerisation with ethylene requires high-pressure equipment. In practice, this difference means that copolymerization with ethylene is essentially a larger-scale operation, with more sophisticated, and expensive, manufacturing equipment. A further extension of the concept of polyvinyl acetate copolymers is the introduction of crosslinking monomers into the polymer chain, as discussed in Section 4.2.3 below.

4.2.1 Polyvinyl acetate copolymer latices with other esters

Latices of vinyl acetate copolymerized with plasticising comonomers are available from many manufacturers. An adhesive [108] suitable for bonding plasticised polyvinyl chloride sheet can be made from a mixture of a vinyl acetate and 2-ethylhexyl acrylate with a conventional anionic–non-ionic surfactant and hydroxyethyl cellulose stabiliser, with acrylic or methacrylic acid and a wetting agent. The composition, when coated, shows little shrinkage during ageing, probably due to matching adhesion and cohesion forces. With higher proportions of plasticising comonomers, these latices may be used as the major component of compounded pressure-sensitive adhesives, e.g. for self-adhesive wallpaper. When coating on to a roll form, an exterior release coat, based on a styrene or vinylidene chloride copolymer, is also required [109].

Some further examples of adhesive systems using vinyl acetate copolymer latices include the following:

(a) A copolymer latex of vinyl acetate and dibutyl maleate (77 : 23), plasticised with glycerol and bis-(methoxyethyl)phthalate, has been applied in high-speed bonding of various substrates to viscose (Cellophane) and polyethylene terephthalate labels to various substrates. The system is reported to be particularly useful for damp surfaces [110].

(b) A copolymer latex, useful as a general-purpose adhesive, has been prepared from vinyl acetate, dibutyl maleate and isopropenyl methyl ether, stabilised with a buffered non-ionic surfactant (Tergitol NPX) and hydroxyethyl cellulose [111].

(c) Wood-to-polyvinyl chloride laminating adhesives have been prepared from vinyl acetate–ethyl acrylate copolymer latices, modified by the addition of 5 % of a low molecular weight (~9000) polymethyl acrylate latex, showing improved adhesion compared to the unmodified polymer. The resulting bond shows rapid set and wood failure [112].

(d) A copolymer (100 parts) of vinyl acetate, ethyl acrylate and methacrylic acid with sodium lauryl sulphate as surfactant has been mixed with isopropanol, heated to 80 °C, and then blended with glycerol (5 parts) and a powdered soap to form a solid adhesive [113]. The satisfactory production of this type of adhesive from this formulation will depend on the detail of the method of mixing of the components.

(e) Vinyl acetate copolymers may also be prepared by grafting. Typically, a heat-resistant adhesive can be prepared from a polyvinyl acetate latex, polyvinyl alcohol stabilised, grafted with methyl methacrylate using an ammonium persulphate initiator [114].

(f) Water-resistant wood and plastics adhesives have been made from a 50 % polyvinyl acetate latex mixed (4 : 1) with a 40 % poly(diallyl)phthalate latex. A free-radical generating agent, *tert*-butyl perbenzoate, is added before application of the mixture to the substrate, presumably to cause further crosslinking of the diallyl phthalate [115]. A similar system involves copolymerisation of vinyl acetate with tetra(allyloxy)ethane (82 : 12), with 11 % of dibutyl phthalate. The product is suggested for lamination of wood panels. A further variant on this theme (probably lower in cost) includes *N*-methylolacrylamide (7 %) as the crosslinking monomer: the suggested adhesive formulation also has added wheat flour and urea [116]. (See also Section 4.2.3 below for other crosslinking adhesive formulations.)

4.2.2 Ethylene–vinyl acetate copolymer latex-based adhesives

Latex copolymers of vinyl acetate with ethylene, usually known as *E-VA latices* and, less often, as *acetoxylated polyethylene latices*, first became available during the 1960s. The major component is vinyl acetate, with stabilization systems generally similar to those used for other vinyl acetate copolymer latices. A typical latex composition has an ethylene–vinyl acetate ratio of 16 : 84, with a stabilization system of polyvinyl alcohol and an alkylphenol ethoxylate (10 parts) as surfactant [117]. Copolymerization of ethylene with vinyl acetate and an unsaturated acid, formed on a pre-prepared 'seed' latex, with polyvinyl alcohol as stabiliser in both 'seed' and prepared latex, is reported to give a good general-purpose adhesive [118]. These latices have many applications in adhesive systems [119]. Typical specifications of adhesive grades of E-VA latices are shown in Table 8.10. Table 8.11 summarises the properties of Wacker-Chemie products of similar type and application.

The general aspects of the importance of the aqueous phase in the preparation of this type of latex have been indicated [120]. In a colloid-stabilised system, the addition of a small amount of a water-soluble polymerisable

Table 8.10 Properties of Airflex adhesive grade E-VA Latices (Air Products & Chemicals Inc., Allentown Pennsylvania)

	Airflex 400	Airflex 410 (carboxylated polymer)	Airflex 300	Airflex 465	Airflex 7200	Airflex 920 DEV (carboxylated polymer)
Solids content (%)	55.0	55.0	55.0	65.0	72–74	53–55
Viscosity (Brookfield RVT; 60 rpm)	18–27	2.5–9	18–27	8–13	15–30	80–200[a]
pH	4–5	4.8–5.5	4–5	4.5–5.5	4.0–5.5	4.3–5.5
Stabilization type	PV alcohol	Cellulosic	PV alcohol	PV alcohol	PV alcohol	PV alcohol
Borax stability	Coagulates	Stable	Coagulates	Coagulates	Coagulates	Coagulates
Tolerance to:						
Ethanol	Fair	Excellent				
Aromatic solvents	Excellent	Excellent				
Chlorinated solvents	Excellent	Excellent				
Typical film properties:						
T_g (°C)[b]	22	7	17	−5	0	−20
Adhesion to glass	Good	Excellent	—	—	Good	Excellent
Film flexibility	Fair	Excellent	Good	Excellent	Excellent	Excellent
Gloss	Moderate	High	High	Good		Clear
Film clarity	Slightly hazy	Excellent	Good	High	Clear	Excellent
Heat seal temperature (°C)	125–133	105–110	—	—	—	
Kraft to kraft						
Water resistance	Excellent	Excellent	Fair	Excellent	Excellent	Excellent
Wet tack	Good	Good	High	Good	High	High
Adhesion to:						
Wood	Excellent	Good	Good	Excellent		Excellent
Kraft paper	Excellent	Good	Excellent	Excellent		Excellent
Coated paper	Excellent	Good	Excellent	Good		Excellent
Polyester	Excellent	Good	Excellent	Excellent		Excellent
PVC(vinyl)	Excellent	Excellent	Not used	Excellent		
Cellulose acetate	Excellent	Good	Excellent	Excellent		
Glass	Good	Excellent	—	Good	Excellent	
Aluminium	Good	Not used	Good	Good		
Comments:	Compatible with SB-R latices; long open time	Sprayable; long open time; cationic compatible	High thickening response	Fast setting. Medium thickening response allows formulation to high solids (<80 %)	Used for particle board and film lamination; high-speed packaging	High ethylene content; good wet tack; used for coated packaging films, bottle labels; fast setting; shear thinning

[a] At 20 rpm.
[b] Measured by differential thermal analysis.

Table 8.11 Properties of Vinnapas adhesive-grade E-VA latices (Wacker-Chemie GmbH, Munich, Germany)

	EP 1	EV 2	EAF 60
Solids content (%)	50 ± 1	50 ± 1	60 ± 1
Viscosity (mPa s)[a]	9000 ± 3000	$11\,000 \pm 4000$	$12\,000 \pm 4000$
Stabilizer type	Polyvinyl alcohol	High polymer compound	Surfactant
Particle size (μm)	0.5–2.0	0.5–1.0	0.1–0.8
pH	~4.0	~4.0	~5.5
Minimum film forming temperature (MFT) (°C)	0	0	0
Film T_g (°C) Cold break temperature[b]	0	-15	-35[c]

[a]Brookfield RVT 20 rpm, spindle 5 (ISO 2555; 23 °C)
[b]To DIN 53 372.
[c]By differential scanning calorimetry.

monomer simplifies the manufacturing process and yields lower viscosity latices at high solids, which is useful for adhesive applications. The water-soluble comonomer is CH_2: $CR'R''$, where R' is —H or —CH_3 and R'' is selected from

 —CO.NHX.X
 —COO$(CH_2)_2$.OH
 —COOH
 —SO_3Na
 or pyrrolidinyl

and X = —H; —CH_2OH, —$C(CH_2)_2SO_3H$ (or a salt).

Films from these copolymer latices are normally compatible with most other anionic or non-ionic vinyl acetate and vinyl acetate copolymer latices. The latices can be compounded like other vinyl acetate latices, and may include some standard plasticisers and, if necessary, solvents. Grades of E-VA latices are available which include small amounts of vinyl chloride and crosslinking agents, as discussed in the following Section 4.2.3.

The latices show good compatibility with various resin derivatives, including modified hydrocarbon and wood resins (mentioned in Section 4.1.3 above), which may be added in hydrocarbon (usually aromatic) solvent. The addition of oxygenated solvents, such as butan-2-one (methyl ethyl ketone, MEK), to a plasticised formulation may show a significant lowering of viscosity. Most E-VA latices will accept high loadings of fillers and pigments, so can be formulated economically into adhesives for the bonding of many substrates. Adhesives based on E-VA latices show good adhesion to substrates of polyvinyl chloride, polyvinylidene chloride (Saran and

Specific adhesive types

copolymers), polystyrene, urethane foams, coated viscose films, almost all natural and synthetic fibres, and aluminium foils. They also tend to form strong heat seal bonds between polyvinyl chloride and polystyrene laminates. Improved peel strength (for bonds including aluminium, paper, polypropylene or polyester) in E-VA latex-based formulations has also been claimed by addition of γ-mercaptopropyl trimethoxysilane [121].

Many examples of the preparation and use of E-VA latices in adhesives have been reported. These include:

(a) Copolymerization of vinyl acetate with ethylene, with polyvinyl alcohol as stabiliser, a non-ionic surfactant and a persulphate–sodium formaldehyde sulphoxylate buffered redox initiator (added at 55–60 °C), at a working pressure of 60 kg cm^{-2}. The resulting coating made from the copolymer (16 % ethylene) has high wet strength and dry peel strength, low creep and high tensile strength. The ethylene–vinyl acetate latex may be modified by addition of polyvinyl alcohol (d.p. = 1700), which has been crosslinked by previously treating with glyoxal (3 %) (2.5 % on latex weight), so increasing the viscosity from 53 to 1200 P. The product is reported to be stable over 1 month [122].

(b) Copolymerization of vinyl acetate with ethylene, with standard stabilisers, in the presence of ~0.5 % of rosin gives a copolymer latex containing 18.5 % ethylene, which provides a water-resistant bond in plywood–polyvinyl chloride lamination. (Since simple rosin, which has considerable hydrocarbon unsaturation, will show considerable retardation of polymerization, the additive is likely to be a hydrogenated or disproportionated rosin [123].)

(c) High solids (65–75 %) ethylene–vinyl acetate latices have been prepared under pressure, using vinyl acetate and 5–40 % of ethylene, with polyoxyethylene non-ionic surfactants and a sulphonated polyvinyl alcohol (Gosheran L-3266, from Nippon Goshei, Osaka) (0.5–10 %, based on the vinyl acetate monomer content); the resulting latices (particle size of 0.6–1.2 μm) can be formulated into adhesives which give good final adhesion, with high-speed application [124].

(d) An ethylene–vinyl acetate copolymer latex, prepared on a 'seed' latex with additional polyvinyl alcohol, is an adhesive for 'vinyl' plastics (especially plasticised PVC films) [125].

(e) Ethylene–vinyl acetate copolymer latices, stabilised by low molecular weight polyvinyl alcohol (d.p. 100–500) and surfactants, have been used for bonding PVC sheet to cellulosic substrates [126].

(f) An E-VA latex, mixed with soluble starch, will heat seal untreated polyethylene film or rigid polyvinyl chloride sheet; the addition of starch is claimed to avoid blocking [127].

(g) Addition of acetone to an ethylene–vinyl acetate copolymer latex (with careful addition of the acetone, with control of stirring) enables the mixture to function as an adhesive to laminate polystyrene foam to polystyrene film, without warping or bubble formation [128].

(h) An improved adhesive has been prepared by reacting vinyl acetate monomer (10–200 parts) with an ethylene–vinyl acetate latex (100 parts, as solids), with additional polyvinyl alcohol and a persulphate initiator, presumably by grafting. The resulting product had good mechanical stability, low temperature adhesion and good resistance to thermal creep [129].

(i) A paper adhesive with good low-temperature properties can be made by polymerising vinyl acetate on to an E-VA seed latex, as shown in Table 8.12 [130]. The final latex mixture has a viscosity of 15–20 P, good freeze–thaw resistance, and shows rapid adhesion to paper.

The preparation of many terpolymer latices has been claimed, including:

(a) An ethylene–vinyl acetate–methyl methacrylate (V-OAc–MMA with ratio 7 : 3) latex, prepared with polyvinyl alcohol as stabiliser and a redox initiator, by addition of part of the liquid monomers under pressure at 40 bar to form an adhesive for wood with good heat and moisture resistance [131].

(b) A compounded adhesive for lauan plywood, with good strength, has been prepared from one of the terpolymers:

> Ethylene–vinyl acetate–crotonic acid (20 : 79 : 1)
> Ethylene–vinyl acetate–N-methylolacrylamide (20 : 76 : 4)
> Ethylene–vinyl acetate–methacrylamide (70 : 74 : 6)

Table 8.12 Latex for paper adhesive prepared using 'seed' polymerisation

Polyvinyl alcohol (88 %: d.p. 1700)	5.5 g
Ethylene–vinyl acetate latex 'seed': 55 % solids	40 g
Hydrogen peroxide (aqueous solution)	0.5 g
Tartaric acid	0.2 g
Water	100 g
To the above, add gradually (over 2–3 h) and maintain at 80 °C:	
Vinyl acetate	100 g
Finally, add, with stirring:	
Dibutyl phthalate	20 g

The resulting terpolymer may be blended with a di-isocyanate adduct and a clay filler [132].

(c) Blended latices of vinyl acetate, vinyl chloride and ethylene with those of ethylene and vinyl acetate have been used as adhesives. Such latices should have <80 % of polymer insoluble in benzene. A latex of vinyl acetate, vinyl chloride and ethylene (20–70 %) ($T_g = 0$–30 °C) may be blended with an ethylene–vinyl acetate (18 : 82) latex, with added toluene, to form an adhesive suitable for laminating particle board to decorative polyvinyl chloride sheet [133].

(d) An ethylene–vinyl acetate copolymer latex, made with hydroxyethyl cellulose as stabiliser, a nonylphenol ethoxylate (10) as surfactant and a redox initiator contained 15–40 % ethylene (with <30 % of benzene-insoluble polymer). It is suitable as an adhesive for bonding polypropylene [134].

(e) An ethylene–vinyl acetate–2-ethylhexanoate copolymer latex (typically 35 : 47 : 18), with a balanced monomer ratio, can provide good adhesion to non-polar surfaces used in packaging, such as polyethylene and polypropylene films, acrylic- or UV-cured lacquers, aluminium foil and polyvinyl chloride sheet. The components employed are shown in Table 8.13. This mixture was polymerised using this initiator system, followed by a second initiation stage using *tert*-butyl hydroperoxide [135].

(f) An adhesive for lamination of wood veneers with good resistance to boiling water immersion may be prepared as shown [136] in Table 8.14. Replacement of the graft copolymer by a polyvinyl alcohol results in delamination. Water resistance is probably due to hydrogen bonding or interesterification of the hydroxyl groups (including those of the cellulosic components) with the carboxylic groups of the graft copolymer.

There are detailed differences in performance and, especially, in compatibility with other products between E-VA latices from different sources, so it should not be assumed that apparently similar products from different

Table 8.13 Adhesive used in packages

Ethylene	559 g
Vinyl acetate	839 g
2-Ethylhexanoate	466 g
Polyvinyl alcohol (Gohsenol GL-05H)	39.3 g
Perlankrol PA (ammonium alkylphenol sulphate) (surfactant)	32 g
Ferric chloride	0.01 g
Sodium formaldehyde sulphoxylate ($NaSO_2.CH_2O.2H_2O$)	0.01 g
$Na_2S_2O_4$	As required

Table 8.14 Wood veneer adhesive with boiling water resistance

Ethylene–vinyl acetate (3 : 17) latex (56 % solids)	20 parts
Aqueous acrylic acid–vinyl alcohol (3 : 17) graft copolymer	5 parts
Water	5 parts
Gypsum hemihydrate	15 parts

manufacturers may be formulated identically. However, the general properties, performance and relatively low cost of E-VA latices, for purposes where a more-or-less flexible bond is required, make their use a highly cost-effective option where they are available economically (from large-tonnage manufacturers).

4.2.3 Crosslinking latex-based adhesives

Use of crosslinkable vinyl polymers always presents a compromise between the 'shelf-life' of a formulation (which depends on the chemical stability of the polymer system in storage under recommended conditions before application) and the required hardness, reduction of 'creep', resistance to moisture (or solvent) or weathering conditions after application to a substrate (as either a coating or an adhesive bond). Crosslinking agents may be part of the principal polymer system, added reactive polymers or reagents that chemically react with components of the polymer system [137, 138].

There are two basic approaches:

(a) introduction of functional groups into the main chain, with later addition of a multifunctional additive able to react with the functional group of the main chain;
(b) introduction of separate multiple functionalities able to react with the polymer chain, providing the sites for intermolecular crosslinking.

There are many variations on these two themes, for many different applications, which are discussed elsewhere in this work. Those principally concerned with adhesive applications are considered here.

The simplest crosslinking agent for polyvinyl acetate is formaldehyde, H.CHO, available as a 38 % solution in water ('formalin'). Although inexpensive, it is much too reactive for most purposes and has a well-known, and unpleasant, odour. The related dialdehyde, glyoxal OHC.CHO, is available from several manufacturers, usually as a 30 % solution. It is stable below pH 8.0 and is preferably stored below pH 7.0. At this pH, it is stable when added to a latex below this pH and only reacts in the dried latex film, presumably by acetal formation, to increase the water resistance and reduce the 'creep' of the resulting adhesive bond. It reacts with the hydroxy groups always present in a polyvinyl acetate latex. Glyoxal is particularly useful in polyvinyl alcohol-stabilised latices, and is added at 10–20 % on the weight of polyvinyl alcohol.

Addition of excess glyoxal may weaken adhesion, probably by saturation of the hydroxy groups in the polymer molecules [139].

Other reactive compounds, such as dimethylolurea $CO(NH.CH_2OH)_2$, hexamethylolmethylamine and various urea-formaldehyde and melamine-formaldehyde resins have been used with polyvinyl acetate latices in specific formulations [140]. In general, the formaldehyde condensates employed are based on those that were originally developed for 'minimum-iron' finishes for cellulosic textiles. The rate of crosslinking is, in effect, controlled by the pH of the mixture, so this is selected to provide an adequate 'pot-life' for the system. In a typical example, the 'pot-life' of a polyvinyl acetate latex/urea-formaldehyde/oxalic acid adhesive system increases with oxalic acid concentration. The highest initial bond strength is obtained with a latex–formaldehyde ratio of 1 : 1 to 1 : 3 [141–143]. Other reactive compounds include:

(a) A cationic latex has been made from vinyl acetate with includes, unusually, a secondary or tertiary aminoacrylate ester. This was reacted with epichlorhydrin, resulting in quaternisation, with the basic formulation shown in Table 8.15. The resulting mixture is held for 3 h at 60 °C and then diluted to 40 % of solids and 5 % (w/w on latex) of dibutyl phthalate added as plasticizer. The final product is a wood adhesive claimed to have better water resistance than a technical polyvinyl acetate–urea-formaldehyde adhesive [144].

(b) A complex monomer mixture of vinyl acetate, butyl acrylate, vinyl crotonate, acrylic acid, sodium vinyl sulphonate and N-methylolacrylamide (90 : 10 : 0.1 : 1 : 0.4 : 4.8) was copolymerised with ethylene to form a copolymer containing 25 % ethylene, which could be crosslinked by heating at

Table 8.15 Formulation of a cationic latex

Water	1475 parts
Potassium persulphate	4 parts
$CH_2{:}CH.CO.O.CH_2.CH_2.NH.CH_3$	3 parts
$HS.CH_2.CO.O.CH_2.CH_2.O.CO.CH_3$	3 parts
Ferric chloride hexahydrate (0.5 % aq)	3 parts
Then add to the above, over 3 h at 80 °C, and maintain at this temperature for 1 h further; then add:	
Vinyl acetate	1000 parts
Hydrogen peroxide (5 % aq.)	80 parts
Then, at 50 °C, add 5 % NaOH to pH 7, followed (over 3 h) by:	
Epichlorhydrin	44.3 parts

150 °C for 5 min giving a solvent-resistant binder for non-woven fabrics [145].

(c) Epoxy resins, although relatively expensive, are also effective crosslinking agents in latex-based adhesive systems. The substantially monomeric Epikote 838 may be used, with diaminodiphenylmethane as the crosslinking agent, in the presence of a polyvinyl acetate latex. The reaction probably takes place on the hydroxyl groups of the polyvinyl acetate chain, formed by hydrolysis. A filler such as dolomite may be added to the mixture, to reduce cost and improve the working properties [146]. A similar system, based on an acrylic latex, caused the alkaline to increase in viscosity, which, when mixed with an epoxide using a polyamine curing agent and a clay filler, gave a coating adhesive with a viscosity of >200 P [147].

4.2.3.1 Some typical formulations

Several patents suggest the use of crosslinking agents in polyvinyl acetate latex-based adhesives, including phenolplasts (resols) in ratios of 0.3–4.0 of vinyl acetate units (as partially hydrolysed polymer) for each phenolplast unit, such as 2,4,6-trimethylolphenol $C_6H_3(CH_2OH)_3$. Such products are used for bonding porous substrates, but, if an oxidizing acid catalyst is added, the adhesive composition can bond stainless steel to aluminium, wood, porcelain or glass [148].

Features from each of several methods of crosslinking may be combined in adhesive formulations. Some examples include:

(a) Grafting vinyl acetate monomer on to polyvinyl alcohol (low molecular weight: d.p. 400–2000) during redox polymerisation (H_2O_2 and tartaric acid) produces a wood adhesive which combines bonding strength with the degree of grafting [149].

(b) The water resistance of polyvinyl acetate adhesives is improved by copolymerisation of vinyl acetate with N-methyloxymethylallyl carbamate [150] or, alternatively, by adding a substituted phenol (e.g. m-cresol or 3.5-xylenol (30 % w/w on latex)) to formaldehyde and a hardener [151].

(c) A polyvinyl acetate latex (preferably alcohol-tolerant) mixed with a resorcinol-modified phenol–formaldehyde resin and fillers can be used for bonding wooden beams [152].

(d) The pot-life of an adhesive made from polyvinyl acetate [I]; urea-formaldehyde [II] and oxalic acid [III] increases with the ratio of [I] : [II] at high concentrations (or absence) of [III]. The highest wood-to-wood initial bond strength was obtained at [I] : [II] ratios of 1 : 2 to 1 : 3 [153].

(e) Vinyl acetate copolymer latices including 2 % acrylic acid may be converted to pale coloured crosslinked adhesives by addition of ∼12 %

of a phenol-formaldehyde resol, such as hexamethoxymethylmelamine, and a water-soluble salt of zirconium (or, possibly, antimony, vanadium or bismuth). Zirconium nitrate (7 %) is a typical addition. The pot-life may be extended by addition of 12 % of methanol (w/w on latex) of glycol acid (3 %). Over 50 formulations based on these features have been suggested for wood-to-wood bonding [154]. Related formulations describe addition of both a phenoplast and hexamethoxymethylmelamine to the vinyl copolymer latex and hardeners including zirconium tetranitrate, zirconium oxychloride and chromium nitrate. The resulting wood-to-wood bonds show wood failure, rather than adhesive failure, after immersion in boiling water [155]. An example of the type of formulation employed is shown in Table 8.16 [156].

(f) Paper-to-plywood bonds may be formed using a 1 : 1 mixture of a polyvinyl acetate latex and a urea-formaldehyde resin with a paste of wheat flour mixed with ammonium chloride (as the latent crosslinking catalyst) [157] and then coated and heated at 40–70 °C.

(g) A polyvinyl acetate latex, with a smaller proportion of resol, mixed with methyl cellulose (0.5–1.0 w/w on latex) and a filler, has been used to bond porous paper to metal for oil filters. The mixture has good coating properties, probably because of the convenient rheological properties imparted by the added methyl cellulose [158].

(h) A polyvinyl acetate latex with a major proportion of urea-formaldehyde resol compounded with $CaCO_3$ snide diethylene glycol has been used as a coating adhesive for abrasive cloth, for dry polishing. The presence of the glycol improves elasticity and aids retention of the abrasive grains, improving performance life of the product [159].

(i) The use of phenol derivatives with metal salt crosslinking agents has been suggested, with a typical mixture being [160].

2,4,6-Trimethylolphenol (70 % aq solution)	329 parts w/w
$Cr_2(O.CO.CH_3)_6 \cdot 2H_2O$	160 parts w/w
Water	300 parts w/w

(j) When used to laminate birchwood a polyvinyl acetate latex (50 % of solids) gives a tensile strength of 4.7 kg cm^{-2}; when 17 parts of the above mixture are added, the tensile strength increases to 22 kg cm^{-2} [161]. Copolymer latices of vinyl acetate with glycidyl acrylate or methacrylate, mixed with trimethylolphenol and then with an acid catalyst, set rapidly at ambient temperatures when used as laminating adhesives, giving waterproof bonds. A phenolplast and a polyvinyl alcohol may also be added [162].

Other reactive monomers may be copolymerised with vinyl acetate, both to plasticise the resulting film and to insolubilise the adhesive bond:

Table 8.16 Typical formulation of polyvinyl acetate latex adhesive with acid/metal salt catalysed crosslinking agents

Polyvinyl acetate latex, 50 % solids; $M_w \sim 45\,000$ (polyvinyl alcohol stabilised)	85 g
Para-formaldehyde	2 g
Nitric acid (3 %)	9 g
Chromic nitrate	0.5 g
Cupric nitrate	0.5 g
Na carboxymethyl cellulose	2 g
Dibutyl phthalate	1 g

(a) Polyethyleneimine (added at 2 % w/w on latex) to a polyvinyl acetate latex (50 % solids) significantly improves the water resistance of the bond of a plywood laminate [163].

(b) A vinyl acetate–acrylamide copolymer latex (6–15 % of acrylamide) has been prepared using a polyoxyethylene phosphate ester surfactant and *tert*-butyl hydroperoxide–sodium bisulphite as the initiator (buffered at pH 5–7). Plasticised with 3 % dibutyl phthalate, the resulting adhesive system gives high strength maple-to-maple wood bonds [164].

Vinyl acetate can be copolymerized with *N*-methylolacrylamide, added as a continuous stream, using polyvinyl alcohol and an anionic surfactant as the stabilising system. Aluminium chloride (as an acid source) is added to form a wood-to-wood adhesive [165]. *N*-Methylolacrylamide and *N*-methoxymethyl allyl carbamate $CH_3.O.CH_2.NH.CO.O.CH_2.CH:CH_2$ are also employed, by polymerization on to a fine particle latex with a redox initiator and polyvinyl alcohol as stabiliser. Typical monomer ratios are vinyl acetate–dibutyl maleate–reactive monomer of 900 : 500 : 28 [166]. Other monomers suggested for use with vinyl acetate include diallyl phthalate [167], tetra(allyloxy)-ethane $(CH_2:CH.CH_2.O)_4C_2H_2$, crotonic acid [168] and itaconic acid.

A 'water-in-oil' emulsion process (a 'reverse emulsion') has been used to polymerize vinyl acetate with 2-ethylhexyl acrylate and glycidyl methacrylate with *tert*-butyl peroxypivalate as initiator, by gradual addition to hexane (as the continuous phase), forming a latex with a viscosity of 1.3 P at 30 °C. The latex was stable for 1 month and is suitable for adhesive applications [169].

A suggested laminating adhesive for elm plywood requires the polymerisation in the presence of an E-VA latex and a carboxylated styrene–butadiene latex (from which trace styrene should be removed, as this will inhibit vinyl acetate polymerisation). The outline of the polymerization is shown in Table 8.17 and the formulation of the adhesive in Table 8.18 [170]. After coating the wood surface and laminating, the bond is pressed

Specific adhesive types

Table 8.17 Vinyl acetate polymerisation for lamination adhesive

Polyvinyl alcohol (7 % aq)	300 ml
Vinyl acetate	15 g
Ethylene vinyl acetate latex (unspecified solids)	300 g
Carboxylated styrene–butadiene latex (unspecified solids)	420 g
Na_2HPO_4	4 g
Potassium persulphate	5.3 g
The mixture is heated at 65–80 °C for 1 h to give a 38–39 % latex.	

for 60–70 min at 120 °C. Thermally activated adhesives, based on *core–shell* latices, for use in the printing industry are also based on the use of such latices, in which the shell has a $T_g > 20$ °C and a thermally softenable core with a $T_g < 20$ °C, with a difference of at least 10 °C between the T_g values of the core and shell, typically with a core–shell ratio of 1 : 1. The core composition a with T_g of 3 °C, is prepared from methyl methacrylate ethyl acrylate and hexane diol diacrylate (20 : 80 : 3) (w/w), and the shell composition, with a T_g of 42 °C, is methyl methacrylate, ethyl acrylate and N-methylolacrylamide (56 : 39 : 5). To form an adhesive composition, this latex mixture is blended with inorganic, wax or polymeric particles, an optical brightener, UV absorber(s) and a further core–shell polymer latex [171].

4.2.3.2 Release coatings

'Blocking' of polyvinyl acetate-coated adhesive surfaces can be reduced by application of release coatings [172]. When the adhesive layer is based on mixed latices of polyvinyl acetates, ethylene–vinyl acetate copolymer, natural rubber (high ammonia; 62 % solids) and water (11 : 7 : 70 : 10), with added ammonia and defoamer, satisfactory release properties may be obtained by coating with a dispersion of micronised waxes blended with a polyamide or polyurethane-micronised dispersion [173], such as:

Micronised dispersion of polyamide or polyurethane	70–99 %
Defoamer	0.1–0.5 %

with

Waxy slip lacquers, polyethylene waxes, micronized hydrocarbon waxes, or natural waxes (e.g. carnuaba wax or beeswax)	to 100 %

4.3 Polyvinyl propionate latices in adhesives

Latices from vinyl propionate, —$(CH_2.CH(O.COCH_2.CH_3))_n$—, are made by BASF AG, the principal producers of propionic acid derivatives. These

Table 8.18 Crosslinking adhesive for elm-to-plywood lamination

Latex prepared according to Table 8.16 (38–39 % solids)	100 g
Melamine-urea aminoplast	100 g
Wood flour	100 g
Water	60 g

Table 8.19 Basic properties of polyvinyl propionate latices (measured to German Standard DIN 51 562)

Product type	Propiofan 5D	Propiofan 6D
Solids (%)	50	50
Viscosity (P)	100–600	3–17
Particle size	Coarse	Fine

long-established latices are sold as Propiofan 5D and 6D; their basic properties are shown in Table 8.19.

4.4 Acrylic copolymer latices

Many specialised adhesives are based on acrylic esters. Applications of acrylic latexes are wide. They are considered in more detail later both in this and other chapters. However, with a few exceptions, acrylic latices are more expensive than plasticized polyvinyl acetate latices (especially ethylene–vinyl acetate copolymer latices) so are only employed where the requirements for the particular application are not satisfied by use of low-cost vinyl acetate-based products. Acrylic-based latices have the advantage of high tack, so are used in laminating adhesives and pressure-sensitive, heat-seal and delayed heat-seal tapes, where the higher tack of their soft film is an advantage. They are widely used for their ability to bond to 'plastic' films (especially PVC-based foils); some types can bond metals, such as aluminium foils, to both porous and impermeable substrates.

A wide range of acrylic ester monomers, with different hydrocarbon chain lengths is available, from methyl acrylate $CH_2{:}CH.CO.O.CH_3$ to 2-ethylhexyl acrylate $CH_2{:}CH.O.CO.CH_2.CH(C_2H_5).C_4H_9$ and several monomers with reactive side chains, which can be used to produce latices with a wide range of properties. These range from the 'hard' (almost tack-free) polymethyl acrylate to the very 'soft' (and tacky) polymers made from long-chain acrylate monomers, which may be viscous liquids at low molecular weights. Methyl acrylate is rather toxic because of its volatility and is partly water soluble (5.2 %), which can cause problems of latex stability. Methacrylate monomers available economically include methyl methacrylate $CH_2{:}CH(CH_3).O.CO.CH_3$, butyl methacrylate (a component of heat-seal adhesive formulations (see Section 5.4) and hydroxyethyl methacrylate

('HEMA') $CH_2{:}CH.(CH_3).O.CO.C_2H_4.OH$. Functional acrylic esters, such as hydroxyethyl acrylate $CH_2{:}CH.O.CO.C_2H_4.OH$, hydroxypropyl acrylate and the related methacrylates are sometimes included in small amounts (2–4 %) in latex adhesive formulations to aid adhesion.

Acrylic acid $CH_2{:}CH.COOH$ or, preferably, methacrylic acid $CH_2{:}CH\text{-}(CH_3).COOH$ is included, in similar small amounts, in copolymer latices for adhesives, to improve the freeze–thaw stability, the pigment and filler binding power and the wetting ability of the polymer latices. It is suggested that acrylate copolymer latices for adhesives can be defined by the T_g, which should be $< -10\ °C$ within the concentration range 40–70 %. In this context, a typical example is based on butyl acrylate, 2-ethylhexyl acrylate, ethyl acrylate, methyl methacrylate and acrylic acid (80 : 6 : 3 : 9 : 2). This latex can bond steel with high instant peel strength [174].

Other comonomers employed with acrylic monomers include styrene $CH_2{:}CH.C_6H_5$ and acrylonitrile $CH_2{:}CH.CN$, which increase the T_g of the copolymer film. Apart from the potential economies of including a proportion of lower-cost monomers in the latices, the use of these comonomers is necessary to alter the properties of the adhesive to suit the conditions of application and use of the product. A high styrene content may result in poor adhesive properties. The use of acrylonitrile introduces some toxicity difficulties, both in manufacture and application of the adhesives. Acrylic latex copolymers with controllable siloxane functionality have been developed: some of these are useful. They are prepared by free radical and cationic-initiated polymerisation, with at least one linear polysiloxane precursor monomer and at least one polysilane or polysiloxane precursor monomer with both free radical-polymerisable and siloxane crosslinking functionalities and a composition, typically, of butyl acrylate, methyl methacrylate, methacrylic acid, octamethylcyclotetrasiloxane and methacryloxypropyltrimethoxysilane (246 : 221 : 6 : 48 : 5) in water (350 parts) in the presence of dodecylbenzenesulphonic acid as surfactant and ammonium persulphate as initiator, forming a 30 % solids latex of pH 2 with an average particle size of 44 nm [175].

Many acrylic latexes also contain a small amount of a cross-linking agent, such as N-methylolacrylamide. Typically, up to 5 % of this monomer is included in the formulation, to ensure that the latex is self-crosslinking at slightly acid pH, with heating. Acrylic latexes for UV-curable adhesives, giving good peel strength when coated on to polypropylene, polyethylene terephthalate (PET) or polyvinyl chloride, have been prepared from the following components [176], which make up the major parts of the formulation, in general terms:

$\geqslant C_1$—C_{12}-alkyl(meth)acrylate	60–98 %
Vinyl monomer	2–40 %
Disproportionated rosin with added acrylate rubber	5–50 %

A typical, more specific, example of the system is

2-Ethylhexyl acrylate	82 parts
N-Vinyl pyrrolidone	8 parts
Carboxyethyl acrylate	10 parts

with

Superester 115	20 parts
Darocure 2959	1.36 parts
Acrylic rubber	3 parts
Hexanediol diacrylate	0.07 parts

The actual compositions and detailed methods of manufacture of acrylate latices depend on the properties required from the system and on the reactivity ratios of the monomers chosen (see Chapter 6 for a more detailed discussion).

Some acrylic latices are formed in a two-stage process, usually resulting a major difference in T_g in each component of the polymer if each section is polymerised separately. These are often described as *core–shell* latices. Typically, a latex made using a sodium lauryl sulfate as stabiliser and potassium persulfate as initiator is formed with a core of polystyrene (100 parts) and a shell of a copolymer of 2-ethylhexyl acrylate, ethyl acrylate and acrylic acid (65 : 30 : 5) [177]. The adhesive from this latex shows good peel strength. For heat-sealable adhesives, a core–shell copolymer latex with soft core particles surrounded by hard shell polymers can be prepared by first polymerising a mixture of styrene, 2-ethylhexyl acrylate and acrylic acid (25 : 25 : 2), followed by a second polymerisation of the same monomers in the ratio 50 : 2 : 3 (see Section 5.4.1).

Core–shell latices have been prepared using reactive surfactants alone [178]. Typically, butyl acrylate (167 parts) in water (156 parts) was added to Elinmol JS 2 (a reactive surfactant) (3.3 parts), Adeka Reasoap NE12 (a further reactive surfactant) (5 parts), with aqueous H_2O_2 (0.2 parts)/ascorbic acid (1 %) (as redox catalyst), followed by methyl methacrylate (72 parts), to form a core–shell type latex (49.9 % solids, pH 3.3, viscosity 3.4 mPa s, average particle diameter 278 Å), which is suitable as a tack-free adhesive to paper and fibres, with good water resistance and film transparency. Stable invert core–shell latex copolymers useful as adhesives have also been prepared by a first-stage polymerization, forming a low molecular weight hydrophilic material, which becomes water soluble on a pH change, followed by a second stage, polymerising monomers in an 'invert core–shell', followed by adjusting the pH to dissolve the first monomers. Typically, ethyl acrylate, methacrylic acid, butyl mercaptopropionate and water (80 : 20 : 2 : 500) is the first stage, polymerised at 80 °C for 15 min, followed by addition of methyl methacrylate (100 parts) over 30 min, forming a latex (pH 2.5; optical density 1.4). When the pH is increased to 9.5 with ammonia, the optical density dropped to 0.37 [179]. Various vinyl ethers have been used as comonomers with acrylic esters, but their use is limited, since they tend to hydrolyse in the acid conditions used for latex polymerisation. A further example of this approach involves the preparation of a mixed latex, in which a monodisperse polystyrene latex is used as a 'seed' for a butyl acrylate–methacrylic acid copolymer 'shell'. In one study [180] a range of comonomer ratios was used to investigate the particle morphology in relation to reaction kinetics and time. It was concluded that the amount of acid in the shell copolymer was the main factor in controlling the particle shape (generally, non-spherical) and morphology. Irregularities in particle shape increased with reaction time: such variations in particle shape, even in a relatively simple 'model' system, may be expected to influence adhesive performance when latices are formulated for their coating characteristics.

Commercial acrylic latices are normally made using polyelectrolytes as the basis of the stabiliser system (rather than cellulose ethers). They can, in general, be compounded with most standard fillers (see Section 4.1.4), such as calcium carbonate and china clay. Most acrylic latices are supplied at low viscosity (to ease bulk handling) but can be thickened either with soluble polyacrylate salts or by the addition of acrylic acid, followed by raising of the pH. Plasticisers are rarely used, except for special purposes, such as heat-seal adhesives (see Section 5.4).

Latices for pressure-sensitive adhesives may be prepared using a 'seed' stage, without conventional surfactants or water-soluble monomers. An essential component is a crosslinking agent, such as diallyl phthalate (1–10 %). As an example of the method, the components given in Table 8.20 are heated together at 50 °C for 30 min, to form the seed particles. The polymerisation is taken to completion and then followed by pH and viscosity adjustment, to form a latex adhesive which may be coated on to kraft paper and dried at 110 °C, forming a paper which is suggested as a protective layer for acrylic sheet [181].

Inverted core–shell latices can be prepared in two stages. The first stage is preparing a latex from the mixture given in Table 8.21, by heating at 80 °C

Table 8.20 Pressure-sensitive adhesive

Ethyl acrylate	40 parts
$K_2S_2O_8$	0.4 parts
Na thiosulfate	0.6 parts
$CuSO_4$	
Followed by:	
2-Ethylhexyl acrylate	880 parts
Diallyl phthalate	60 parts
2-Hydroxyethyl methacrylate	12 parts
Acrylic acid	25 parts
Reactive surfactant	20 parts
$K_2S_2O_8$	1.5 parts

Table 8.21 Preparation of inverted core–shell latex

Ethyl acrylate	310 parts
Methacrylic acid	78 parts
Butyl mercaptopropionate	7.8 parts
Na lauryl sulphate	8 parts
Na dodecyl diphenyloxide disulphonate	8.5 parts
Ammonium persulphate	2 parts
Water	580.7 parts
100 parts of this latex are mixed with:	
Butyl acrylate	433 parts
Methacrylic acid	10 parts
1,4-Hexanediol diacrylate	4 parts
Polyoxyethylene nonyl phenyl ether	15 parts
Ammonium persulphate	1.3 parts
Water	366.7 parts

for 30 min at pH 5–6 (adjusted with ammonia). The resulting core–shell copolymer latex is heated at 80 °C for 60 min and the pH adjusted from 5.5 to 7.0–7.5, that raising the viscosity from 0.75 to 10 P. The product, when coated on to polyester film (1 mm thick), forms an effective pressure-sensitive adhesive [182].

Multiphase thermoplastic acrylic elastomers, useful in adhesives (and in related mastics, caulks and coatings), can also be prepared by multiphase latex polymerisation. Typically, a triple-phase system has been prepared from a terpolymer latex of butyl acrylate, acrylonitrile and methacrylic acid, followed by polymerisation of butene glycol dimethacrylate in the presence of this latex and finally, *in situ*, formation of a copolymer of methyl methacrylate and butyl acrylate. The resulting three-phase polymer has markedly improved tensile strength and elongation properties compared with those of a two-phase polymer [183].

Many points of current production practice in the manufacture of acrylic latices are shown in the following extract [184]:

Production of latex copolymer of butyl acrylate–methacrylic acid (85 : 15)
  Heat 300 g of water to 90 °C in a reaction kettle, under a nitrogen blanket (to exclude oxygen).
  Add ammonium persulphate (1 g) in water (20 g), together with 17.8 g of 45 % solids, 100 nm particle size butyl acrylate–methyl methacrylate–methacrylic acid latex copolymer (as 'core latex'), forming *mixture A*.
  Rinse *mixture A* with 20 g of water (to remove residual salts from 'core latex' particles).
  Allow *mixture A* to equilibriate at 85 °C.

Prepare a monomer *mixture B* from water (300 g), butyl acrylate (850 g), methacrylic acid (150 g) and 26 g of a solution of sodium dodecylbenzene sulphonate (23 % in water).

Add *mixture A* to *mixture B* over 3 h.

Over the same period, add to *mixture A* a solution of 11 g of ammonium persulphate in 195 g of water.

When additions are complete, rinse feed lines with 20 g of water (to ensure complete addition of catalyst).

Hold reaction at 85 °C for 30 min and then cool to 60 °C.

Add 0.03 g of ethylene diamine tetraacetic acid ('EDTA') (complexing agent) and 20 g of 0.1 % solution of ferrous sulphate ($FeSO_4.7H_2O$).

Prepare *mixture C*, of 0.33 g of 70 % *tert*-butyl hydroperoxide dissolved in 6.7 g water, with 0.2 g of sodium formaldehyde sulphoxylate ($NaHSO_2$. HCHO) in 6.7 g of water.

Add *mixture C* to the reaction.

Stir for 20 min.

Add a further equal amount of *mixture C*.

Allow the resulting latex to cool to room temperature.

A 2-ethylhexyl acrylate–styrene (9 : 1) copolymer latex (50 % solids) has been stabilised with a neutralised low molecular weight butyl acrylate–acrylic acid (4 : 1) copolymer (50 % on monomers), yielding a high-viscosity adhesive with good storage stability and high peel strength in bonding to stainless steel [185].

Many examples of commercially available latices for adhesives are shown in tables later in Section 9 below.

4.4.1 Some typical formulations

The formulation of a general-purpose acrylic latex is shown in Table 8.22 [186]. Vinyl acetate is widely used as a comonomer in acrylic latices for adhesives, as a lower-cost component. If an acidic comonomer such as acrylic

Table 8.22 Typical formulation for an acrylic latex for adhesives

2-Ethylhexyl acrylate	99 g
Butyl acrylate	30 g
Vinyl acetate	22.7 g
Acrylic acid	5.3 g
Water	150 g
Polyoxyethylene octylphenyl ether betaine ester chloride (surfactant)	4 g
2,2′-azobis-(*N*,*N*-Dimethyleneisobutyramidine) (initiator)	As required

acid or methacrylic acid is included in the composition, there may be a marked increase in latex viscosity upon addition of ammonia solution, probably due to the hydrolysis of the partly solubilised vinyl acetate units. Many latices containing acrylic acid behave similarly; e.g. Acronal 500D® (from BASF AG) increases in viscosity from ~0.2–60 P when the pH is raised from 4.0 to 9.5. The mixture is heated with stirring at 40 °C for 30 min, and then at 60 °C for 90 min. The initiator is added at a rate sufficient to maintain these temperatures.

Acrylic latex-based adhesives should have an optimum balance between cohesion and adhesion for maximum peel strength. Introduction of low levels of multifunctional monomers, such as glycol dimethacrylate, increase shear strength, but reduce tack and peel strength, and cause a sharp fall in quick tack. Crosslinking agents such as urea–formaldehyde resins require a high temperature for satisfactory reactivity. Zinc acetate (>0.5 %) also increases shear strength, but tends to reduce tack [187]. Improved flow and coating properties in acrylic latex-based adhesives can be obtained by adding 2 % (on adhesive weight) of 10 % aqueous polyoxyethylene dilauryl ether (M_w = 6000) to a latex copolymer of 2-ethylhexyl acrylate, vinyl acetate and acrylic acid (55 % solids), with added hydroxyethyl cellulose (4 % aqueous), 50 % dispersed TiO_2 and ethane diol (87 : 7.5 : 5 : 1.2) [188]. A suitable latex for the coating of pressure-sensitive tapes can be obtained using the monomers of Table 8.23, with a radical polymerisation [189].

An alternative system, from the same organisation, for preparation of alkali-soluble latices for adhesive tape coatings uses a reactive surfactant, typically by copolymerisation of 2-acryloxypropyl hexahydrophthalate and butyl acrylate in 1,5-bis-(3,6,9-trioxadecyloxy)-1,1,3,3,5,5-hexamethyltrisiloxane, followed by neutralizing with NaOH and adding ethanediol monomethyl ether. This formulation, when applied to a polyester film, shows no adhesion to metal and has good corrosion resistance [190].

Good temperature and mechanical stability are obtained in an acrylate copolymer latex prepared from butyl acrylate (150 g) and methyl methacrylate (150 g), with a non-ionic ethoxylate surfactant (Nonipol 200) (6.3 g) and mercapto-terminated polyvinyl alcohol (14.7 g) as protective colloid [191].

Table 8.23 Acrylic latex copolymer for pressure-sensitive tape coating

2-Ethylhexyl acrylate	20 parts
Butyl acrylate	74 parts
Methacrylic acid	3 parts
N-Vinyl pyrrolidone	3 parts
Hitenol N 08	Surfactant
Aqualon RN 20	Cellulose ether stabilizer

4.4.2 Crosslinking systems

Crosslinking agents, such as N-methylolacrylamide, are frequently included in acrylic latices, in proportions up to $\sim$5 %. As an example, diacetoneacrylamide CH_2:$CH.CO.NH.C(CH_3)_2CH_2.CO.CH_3$ may be included as a comonomer with methacrylic acid, with 2-ethylhexyl acrylate as the principal component [192]. Crosslinking usually increases adhesive strength, although in some cases (such as pressure-sensitive adhesives (see Section 5.1) and heat-seal adhesives (see Section 5.4)) there is some loss of tack if copolymers including crosslinking agents are included in the formulations. Other adhesives have been prepared using a polyethyl acrylate latex, with a poly-N-methylolacrylamide copolymer as protective colloid. This gives a water-resistant film after addition of formic acid, a rosin soap and aluminium sulphate solution [193]. An outline formulation of an acrylic latex adhesive includes a butadiene–styrene (60 : 40) latex, with added polyacrylamide ($M_w = 5 \times 10^6$), giving much improved flow stability [194]. The use of aziridines as components of crosslinking systems for adhesives is also mentioned in Section 6.1.5.

The solvent resistance of water-based latices using acrylic esters depends on the number of carbon atoms in the side chains of the acrylic ester monomers employed. With short-chain ester monomers, resistance to oils and gasoline is good. With longer-chain ester monomers, solvent resistance is much poorer, but can be improved by copolymerisation with acrylic or methacrylic acids, or with N-methylolacrylamide (as above). Resistance to chlorinated solvents is also improved. Polyvalent metal ions, e.g. amine–ion complexes, can function as crosslinking agents, improving the shear resistance of the bond, especially where the film of the latex has a $T_g < -15$ °C.

The performance of pressure-sensitive adhesives containing crosslinking agents can be improved (in terms of better high temperature, peel strength and shear properties) by electron beam curing, without sacrificing ambient temperature performance. The latex copolymers have a T_g below the temperature of use. The added multifunctional crosslinking agents are, typically, pentaerythritol triacrylate [195]. Similar multifunctional agents have also been suggested as a component of acrylic latices for pressure-sensitive adhesives with good elevated temperature adhesion and guillotine properties [196].

Aqueous acrylic polymer adhesives have been prepared by latex photopolymerisation, to form adhesive sheets containing dispersed water droplets, prepared by water-in-oil (W/O) emulsions of mixtures of (meth)acrylate ester monomers with functional monomers (containing amide or hydroxyl groups) with a photoinitiator, typically from a mixture of butyl acrylate, acrylic acid (100 : 3 parts by weight, or pbw), an acrylic rubber latex (3), polyoxyethylene (ethylene oxide 50), nonyl phenyl ether (5) and water (25), forming a W/O emulsion, then mixed with Irgacure 651 and irradiated with UV light to yield an irradiated sheet containing 19 % of water [197].

Other examples of acrylic latices for use in adhesives are mentioned below.

4.5 Vinylidene chloride and other halogenated polymer latices in adhesives

Vinylidene chloride latices for adhesives are prepared from this monomer together with a minor proportion of a hydrophilic monomer, such as methyl methacrylate or methacrylic acid (which are also soluble in the principal monomer). The latex is usually prepared by a two-stage process, with an ionic surfactant, such as sodium sulphoethyl methacrylate, using a 'seed' latex to control particle size. The initiation of the polymerization is usually carried out with *tert*-butyl hydroperoxide with sodium formaldehyde sulphoxylate [198]. Similarly, a copolymer latex based on vinylidene chloride, methyl acrylate and methacrylic acid (8 : 1 : 1), partly neutralised with ammonia, can be used to provide a heat-seal coating on corona-discharge treated polypropylene. Vinylidene chloride–methyl acrylate (9 : 1) copolymer latices have been suggested as a top coat (over an amorphous vinylidene chloride-rich copolymer coating), to reduce the water vapour transmission rate, to reduce blocking and to provide satisfactory heat-seal temperatures (<121 °C) [199] (see Section 5.4.1). A latex copolymer of chloroprene (2-chloro-1 : 3-butadiene) with methacrylic acid, stabilised with polyvinyl alcohol, is reported to have good mechanical and electrical properties at low viscosity and can be blended with hydrocarbon resins (see Section 4.1.3) to give pressure-sensitive adhesives with aggressive tack and cohesive strength. A formaldehyde-releasing resin may also be incorporated [200, 201]. A vinyl chloride–vinyl acetate (19 : 1) copolymer latex blended with a polyvinyl chloride emulsion polymer, plasticised with bis-(2-ethylhexyl)phthalate forms adhesives with high shear strength [202].

Other vinylidene-containing latexes may be used in heat-seal adhesives. For examples, see Table 8.28 in Section 4.5.1 and Tables 8.69, 8.70 and 8.83 in Section 9.

Only a relatively small number of polychloroprene latices for adhesives are commercially available: the majority are anionic. A few are non-ionic, prepared using a polyvinyl alcohol stabilising system. Most polychloroprenes are gel types, partially crosslinked, and so insoluble, whilst the solid types of polychloroprene have the sol form, being soluble inorganic solvents. Other components in polychloroprene latex formulations which significantly affect performance may include:

(a) Coalescence aids. The choice of coalescing agent depends on the T_g and MFT of the polymer and on the tackifying agents present. Typically, a mixture of butyl carbitol (ethoxyethanol) and propylene glycol is used.

(b) Thickeners (to improve latex stability). Suitable materials include starch, dextrins, guar and xanthan gums, ammonium caseinate, cellulose ethers, sodium carboxymethyl cellulose derivatives and polyacrylates, or inorganic thickeners, such as modified bentonites.

(c) Zinc oxide (as stabiliser).

(d) Solvent (either toluene, which is absorbed into the polymer, so aiding coagulation under pressure at bond formation, or heptane, a non-solvent, which can reduce setting times).

(e) Fillers (to increase solids content and improve water and solvent resistance; typically, colloidal silica is used).

(f) Curing agents (although polychloroprene latex adhesives are usually designed to be used without curing, thiocarbanilide (1–2 phr) may be added).

(g) Wetting agents (to improve bonding to 'difficult' surfaces and to increase penetration into porous substrates; minimum amounts should be used as they are likely to interfere with contactibility (bond formation), tack life and reduction of water resistance in contact bonds).

(h) Stabilisers (to increase mechanical stability during mixing, spraying or spreading; preferred surface active agents and stabilisers include alkali salts of rosin acids, caseins and some ammonium salts, polyvinyl pyrrolidine, and also conventional surfactants such as sodium lauryl sulphates, sulphosuccinates and non-ionic alkyl phenol ethoxylates.

(i) Since most polychloroprene latex formulations are anionic and do not suffer from microbial attack, only formulations using non-ionic surfactants at pH 7 require an antimicrobial agent.

With these constraints, polychloroprene latex polymers have some limitations in their compounding behaviour. Chloroprene latex-based copolymers have been prepared by copolymerising chloroprene monomer (100 parts) with acrylic acid (0.5–20 parts) (or another ethylenically unsaturated carboxylic acid) in the presence of a water-soluble protective colloid, such as polyvinyl alcohol (88 % hydrolysed; viscosity 4–6 cP) (3 parts) in water (100 parts). The resulting latex was compounded:

Latex as above	100 parts
Nipsil ER	5 parts
2,6-di-*tert*-Butyl-4-methylphenol (antioxidant)	2 parts
ZnO	2 parts
Rosin acid ester	50 parts
Na polyacrylate	2 parts

to form an adhesive with good water resistance, which forms a good bond to aluminium.

Polychloroprene latices with added 2,3-dichloroprene, initiated with potassium persulphate and added aqueous ethanolamine, have been converted into adhesives, with satisfactory heat and discoloration resistance, for bonding canvas by addition of ZnO dispersed in casein (1 % solution), and aqueous

Na polyacrylate (3 %) [203, 204]. Similar latices of chloroprene with other monomers, in the presence of protective colloids and a tackifier resin, have been prepared by polymerisation using the basic formulation:

Chloroprene	1000 parts
2,3-Dichlorobutadiene	250 parts
Methacrylic acid	30 parts
Octyl mercaptan	3 parts

This mixture, with added polyvinyl alcohol (88 % hydrolysed, d.p. 600) as protective colloid in water at 40 °C, can be polymerised using potassium persulphate and Na anthraquinone-α-sulphonate as initiator. The resulting agent, formulated with ZnO, a β-naphthalenesulphonic acid–formaldehyde condensate (as tackifying resin) and Na polyacrylate, forms an adhesive with greater bond strength compared with a similar latex without the 2,3-dichlorobutadiene component.

The properties of a wide variety of pressure-sensitive adhesives formulated from polychloroprene latices are mainly dependent on [205–207]

(a) the colloidal properties of the latex,
(b) surfactants,
(c) tackifiers and
(d) solvents (if present).

Low-shear mixers are preferred for producing stable adhesives: the charge and pH of the additives have to be balanced before addition to the latex.

The following selection of polychloroprene latex adhesive formulations, for different applications, is mainly based on Neoprene® (DuPont) latices:

(a) 'Quick-break' adhesives depend on the use of Neoprene® 572 and NPR 5587 latices, which coagulate readily under relatively low pressure, with rapid crystallisation, leading to the rapid development of a bond of moderate strength. A soft resin with an antioxidant is added to increase the specific adhesion and early bond strength, without reducing the final adhesion:

Neoprene® NPR 5587	100 parts
Non-staining antioxidant	2 parts
ZnO	5 parts
Rosin ester	20 parts

If a high bond strength is required, a higher-gel polymer, Neoprene® 572, can be used.

(b) In pressure-sensitive adhesives, Neoprene 115 latex, a non-ionic carboxylated copolymer containing amorphous polymer, may be preferred, when

Specific adhesive types

formulated with a good antioxidant and a suitable resin. A typical formulation is:

Neoprene 115	100 parts
NaOH	to pH 10
Picconal A400 (an aqueous emulsion of a thermoplastic aromatic hydrocarbon copolymer resin with a softening poiints at 85 °C) (Hercules Inc.)	5 parts
Cymel 301 (a Blocked melamine-formaldehyde resin) (Cytel Inc., formerly American Cyanamid Inc.)	20 parts
Antioxidant (Wingstay L) (Goodyear Inc.)	2 parts

The resulting adhesive has good, long-lasting tack and good peel adhesion (of vinyl-coated fabric to steel), which increases over 8 days. A balance of properties can be achieved by using blends of soft and hard resins. Combinations of the carboxylated non-ionic latex with other polymer latices, such as natural rubber and acrylate copolymer of carboxylated styrene and butadiene, show a significant increase in creep resistance, sometimes with partial loss of tack.

(c) The same carboxylated non-ionic latex has been suggested for film lamination adhesives, for bonding polyester to polypropylene, to high- and low-density polyethylene and to aluminium foil, and also low-density polyethylene to aluminium foil and to viscose film. Suitable formulations are based on:

Neoprene 115	100 parts
Picconal A400	95 parts
Picconal A501 (emulsion of an aromatic hydrocarbon resin with a softening point at 49 °C) (Hercules Inc.)	20 parts
Antioxidant	2 parts
ZnO	As required

(d) Foil lamination adhesives for bonding aluminium foil to paper or board are also based on Neoprene latices, since they show:

> Rapid development of strength for high-speed laminating;
> Good embossing properties in the resulting laminates;
> Good hot water resistance.

Where required by the end use, the ingredients of foil laminating adhesives should be chosen to satisfy relevant regulations, notably in relation to approval for food contact applications and absorption of odour. For use on

530   *Polymer latices in the formulation of adhesives*

high-speed laminating machinery, formulations are prepared at 35–45 % solids, with a viscosity of ~5000 mPa s. A formulation for a sprayable polychloroprene-based pressure-sensitive adhesive is shown in Table 8.24 [208]. Some typical formulations for sprayable adhesives, based on products from Bayer AG, are shown in Tables 8.25 and 8.26. The resulting adhesive at pH 9.0 (viscosity 100 poise at 50 % non-volatiles) can be thixotropic and can be sprayed on to uneven surfaces, improving the surface smoothness and allowing pressure bonding with a thickened adhesive layer.

4.5.1 Some typical formulations

Some typical formulations for laminating adhesives based on polychloroprene latices are shown in Table 8.27.

Table 8.24 General formulation for sprayable polychloroprene-based pressure-sensitive adhesive

Polychloroprene latex	100 parts
Rosin ester tackifier	20 parts
Na polyacrylate solution (thickener)	2 parts
ZnO	5 parts
Antioxidant	2 parts
Ammonia solution (25 % aq)	0.5 parts

Table 8.25 Butadiene copolymer latices in adhesives (Bayer AG)

Adhesive component	Solids content (%)	Formulation 1 (parts by weight)	Formulation 2 (parts by weight)	Function
Dispercoll® C 84	55	70	70	Polychloroprene latex
Dispercoll® C 74	58	30	30	Polychloroprene latex
Emulvin® W	20	2	2	Surfactant stabiliser
ZnO (activated)	33	6	2	Stabiliser (dispersed in 15 % w/w of Tamol (naphthalene–sulfonic acid condensate Na salt (5 % solution))
Vulkanox® DDA-EM	50	2	2	Antioxidant (non-staining)
Oulotac® 90 D 2	50	30	—	Tall oil resin
Dermulsene® 92	50	—	30	Terpene phenol resin
Fluorotenside FT 448	10	0.1	0.1	Fluoroalkyl wetting agent
Properties:				
pH		11.5	11.3	
Viscosity (mPa s)		100	150	
Solids content (%)		52	52	

Specific adhesive types

Table 8.26 Basic coating formulation using polychloroprene latex (Bayer AG)

Adhesive component	Solids content (%)	Formulation (parts by weight)
Dispercoll C 84	55	182
Emulvin W solution	20	3.0
ZnO (activated) (dispersed in 15 parts of 5 % aq Tamol (naphthalenesulfonic acid condensation product Na Salt)		7.5
Vulkanox DDA-EM 50		4.0
Solvent emulsion (containing 50 % ammonium caseinate solution, 5 %)		0–60

Table 8.27 Foil laminating adhesives using polychloroprene latices (Du Pont)

	Solids content (%)	Formulation 1 (parts by dry weight)	Formulation 2 (parts by dry weight)	Formulation 3 (parts by dry weight)
Neoprene® latex 842A	50	100	100	—
Neoprene® latex 115	47	—	—	100
Antifoam	—	—	—	0.1–0.2
ZnO	50	5	5	2–5
Antioxidant	2	2	2	2
Ammonium caseinate	10	40	30	—
Sodium silicate	10	0.25	—	—
Filler	10	—	—	0–80
Ammonium silicofluoride	10	—	1.5	—
Reactive resin[a]	100	—	—	0–5
Ammonium chloride[a]	100	—	—	0–0.2
Thickener (typically Na polyacrylate)		To 3000 mPa s	To 6000 mPa s	

[a]Addition of a small amount of trimethoxymethylmelamine enhances water resistance on ageing or heating. This effect is accelerated by addition of ammonium chloride as catalyst.

Fibreglass and rock wool panels can be laminated to aluminium foil. Fibreglass scrim ('FSK') and kraft paper composite–PVC panels are also used. Polychloroprene latices can be used for such laminations, as they do not detract from the fire retardance of the composites. Typically, FSK adhesives are based on Neoprene 115 latex (100 parts) formulated with hydrated alumina ($\geqslant 200$) and a trimethoxymethylmelamine compound (to improve water resistance) with a thickener (such as Na polyacrylate). The resulting adhesive should be pH 7–8, as Al foil is sensitive to acids and alkalis. In practice, for suitable mechanical stability on coating lines, the adhesive composition should also include defoamer and antibacterial agents. Coating adhesives for such systems can be 'froth applied' to achieve a satisfactory coating with a low application rate. Low gel latices are preferred (such as Neoprene NPR 3911), with a mixture of a non-ionic surfactant, such as a nonylphenyl ether ethoxylate and Na lauryl sulphate and Na dioctyl sulfosuccinate. Large particle size alumina may be employed, since the possibility of pinholing in a facing adhesive is small. Polychloroprene latices have also been used in adhesives for fabric–rubber bonds in the manufacture of V-belts and timing belts, and coated fabrics (see Section 7 below).

Polychloroprene latices have been formulated with terpene–phenol resins into adhesives [209]. Contact adhesives of this type employ a polychloroprene latex, with an added terpene–phenol latex (40 %) and hydroxyethyl cellulose (2 % aq) (10 : 3 : 1). They can be used for spray-gun application to steel, with good adhesion and little spray-gun clogging [210]. Low molecular weight polychloroprenes or methyl methacrylate copolymers are used, with a fatty acid ethoxylate surfactant (10 %) and/or an alkylphenyl polyether sulphate (3.5 %) [211]. An adhesive is formed by blending a technical chloroprene latex with a polymerised rosin in toluene, which is then emulsified with triethanolamine/ammonia and oleic acid (polymer–rosin ratio = 2 : 1) and 3 % (w/w on solids) of zinc oxide and an oxidant [212].

A 50 % polychloroprene latex (60 g) with a proprietary tackifier (Alresat PT) in toluene (20 g) and a stabiliser forms a low-viscosity contact adhesive suitable for spraying and bonding foams [213]. A thickened polychloroprene-based pressure-sensitive adhesive has been formulated as in Table 8.24. The resulting adhesive, at pH 9.0 (viscosity 100 P at 50 % non-volatiles), can be thixotropic and can be sprayed on to uneven surfaces, improving the surface smoothness before allowing pressure bonding with a thickened adhesive layer [214].

The properties of some commercially available polychloroprene latices suitable for adhesive formulations are shown in Table 8.28. Others are mentioned in Tables 8.69, 8.70 and 8.83 in Section 9. Carboxylated polychlorprene latices stabilised with polyvinyl alcohol have particle sizes of 0.3 mm and surface tension ~50 dyn. The latices have non-Newtonian characteristics, with relatively high viscosity, and form good contact adhesives, possibly superior to acrylic latex adhesives. Such polychloroprene latices are suitable as contact adhesives: they bond well at 13 °C, and are crosslinked by ZnO. Acrylic latices bond equally well at pressures >4.5 kg cm^{-2}, but are less effective at lower pressures and relative humidity (RH) ~50 %; at low RH (~ 10 %) both series bond equally well [215].

Polychloroprene latices can be used in heat-resistant adhesives, prepared from polychloroprene (100 pbw), 2,3-dichlorobutadiene (1–100 pbw) and, optionally, vinyl carboxylic acid monomers. The polymer should be completely soluble in CHCl$_3$ at 5 % concentration. A blend of chloroprene; 2,3-dichlorobutadiene and methacrylic acid (87 : 10 : 3) (w/w) was polymerised using anionic surfactants, with n-dodecyl mercaptan as the chain transfer agent, and the latex was then mixed with emulsified tackifier resins and antioxidants to form a good bonding agent for S-BR, with good water and creep resistance [216].

A high-temperature adhesive formed from a 20 % latex of hexafluoropropylene and tetrafluoroethylene, with 8.8 g of graphite milled into 400 g of latex, can laminate polyamide sheet, retaining an adhesive bond up to 200 °C [217]. Two component adhesives, with a long pot-life and good thermal resistance, designed to bond cellular polyolefin sheets (used in vacuum moulding) are based on a carboxyl-modified chloroprene rubber latex (with a tackifier) and an oxazoline-substituted latex-based hardener:

Component A:
   Carboxyl modified chloroprene latex LC-501   100 parts by weight
   Yukaresin KE 904 (tackifier)   80 parts
   Borchigel L 75   1 part

Component B:
   CX WS 120 (oxazoline hardener)   5 parts

After mixing, the adhesive mixture is kept for 2 h and then coated on to a wood panel and a cellular polypropylene sheet. The surfaces are press-laminated (140 °C for 15 s) to form a test piece with a 180 °C peel strength of 3.0 kg for 25 mm [218]. Table 8.28 lists the properties of a range of polychloroprene latices.

4.6 Butadiene copolymer latices

As butadiene is a relatively low-cost monomer (in bulk), with the major proportion of its use in production of solid synthetic rubbers, copolymer latices including butadiene are often economically attractive for many purposes. Carboxylated styrene–butadiene latices are widely used in applications involving concrete products (see also Chapter 16 in Volume 3). There are also many speciality applications based on butadiene latices with various monomers, some of which are discussed in this section.

Table 8.28 Properties of some commercially available polychloroprene latices suitable for adhesives formulation

Polymer latex[a]	Solids content (%)	Viscosity (mPa s)	Gel type	Crystallization
Neoprene 400	50	15	Fast	Fast
Neoprene 571	50	23	High	Medium
Neoprene 650	60	400	Medium	Very slow
Neoprene 450	41	350	Sol	Very slow
Neoprene 572	50	23	High	Very fast
Neoprene 635	58	350	Very high	Medium
Neoprene 671	57	60	Medium	Medium
Neoprene 115	46	200–500	High	—
Baypren SK	50	35	Medium	Slow
Baypren MKB	55	150	Medium	Fast
Baypren GK	32	1.0	Medium	Very slow
Baypren D	58	150	High	Slow

[a]Neoprene is from E. I. Du Pont de Nemours Inc. and Baypren from Bayer AG.

Low-viscosity (70–100 cP) carboxylated copolymer latices (48–50 % solids) of butadiene and vinylidene chloride (1 : 1) at pH 8 are used as adhesive coatings. No cure is required [219]. A ∼25 % latex of butadiene, acrylonitrile and acrylic acid (125 : 72 : 9) (obtained by polymerisation for 24 h at 30 °C, with an H_2O_2–Fe^{2+} initiator) can be blended with equal parts of a polyurethane latex to form an adhesive [220].

A polybutadiene (12.5 %) latex has been dissolved with 2-ethylhexyl acrylate, vinyl propionate and styrene and then emulsified (with an acrylic ester–acrylic acid copolymer of controlled M_w) and polymerised to form a copolymer with high adhesive strength to steel, polyethylene and polypropylene [221]. The vinyl monomers are presumably grafted on to the polybutadiene.

Carboxylated styrene–butadiene latices, in which rosin derivatives have been dispersed, can form pressure-sensitive adhesives with good long-term peel strength [222]. Different methods of incorporation of the tackifier (which can affect the conditions of polymerisation) have been suggested (see Section 5.1.2.9 below).

Blends of two polybutadiene-based latices have been used as pressure-sensitive adhesives: both are copolymers of styrene, butadiene and itaconic acid, with different monomer ratios, (33 : 65 : 2) and (78 : 20 : 2), and T_g values. The second latex (T_g of −10 to +50 °C) acts as the tackifier for the first (T_g of −70 to −5 °C). A typical adhesive, composed of equal parts of the two latices, used to bond polypropylene to paper, showed a high value in the 180 °C peel test, decreased by ∼20 % in the absence of the second latex [223].

A latex suitable for bonding paper to aluminium can be prepared from the components given in Table 8.29. The resulting mixed latex has an MFT of

Table 8.29 Preparation of a latex for bonding paper

Stage 1	
Water	1900 parts
K oleate	75 parts
$K_2S_2O_8$	4.5 parts
Butadiene	500 parts
Styrene	300 parts
Ethyl acrylate	700 parts
tert-Butyl mercaptan	7.5 parts

Polymerise at 70 °C to 70 % completion (=36 % solids). Then:

Stage 2	
Latex, as above	500 parts
$K_2S_2O_8$	3.5 parts
Butyl acrylate	165 parts
Methyl methacrylate	350 parts
Styrene	200 parts

Polymerise at 70 °C for 8 h.

~38 °C, with good antiblocking properties. The two parts of the latex blend have different T_g values [224].

Small amounts of organic solvents, preferably with low odour, may be polymerised into vinyl copolymer latices, during preparation with a similar two-stage polymerisation, to improve bond strength and blister resistance. The solvent should have some limited water solubility. An example of the components used with this approach is shown in Table 8.30 (the original does not indicate the amount of water used in the system).

The components of Stage 1 are reacted at 70 °C for 2 h. The Stage 2 reactants are then added and the reaction continued for a further 16 h. The resulting copolymer latex, with added $CaCO_3$, clay, oxidized starch and water, when

Table 8.30 Styrene–butadiene latex copolymer with added solvent

	Stage 1	Stage 2
Butadiene	5	23
Styrene	8	46
Methyl methacrylate	3	12
Itaconic acid	2	—
Acrylic acid	2	—
Cyclohexane (solvent)	2	—
tert-Butyl mercaptan (chain transfer agent)	0.15	0.5
$K_2S_2O_8$	1	—
Na dodecylbenzenesulfonate	0.15	0

coated on to paper, give good adhesive strength without blistering. Without cyclohexane, the coating shows some blistering. Low-viscosity styrene–butadiene adhesives have been claimed, notably as blends with polyphosphate or polycarboxylate salts. A copolymer latex of styrene, butadiene, methyl methacrylate, acrylonitrile, itaconic acid and acrylic acid (55 % solids 10 % methanol soluble) has a viscosity of 6 P, which is reduced to 1.5 P by addition of sodium polyacrylate (1.5 %) [225]. Water-based adhesives have been based on blends of butadiene–acrylonitrile copolymer latices (such as Perbunan latex 8239) and polyurethane–polyurea dispersions (prepared by the reaction of an adipic acid–butanediol polyester and a sulfonate-containing diamine and an amino alcohol, with (optionally) a polyisocyanate (Desmodur DA). The products, for lamination and assembly of wood, plastics and leather, have low activation temperature, good adhesion and heat resistance [226].

Other diene copolymer latices useful in adhesives (and also in carpet backing and in paper coating) have improved mechanical stability and lower coagulum levels. They are prepared from a conjugated diene, a comonomer, a chain transfer agent non-reacting hydrocarbon and a halogenated aromatic hydrocarbon or heterocyclic compound, typically from butadiene–styrene–methyl methacrylate–2-hydroxyethyl acrylate–fumaric acid, dodecyl mercaptan (as the molecular weight controlling chain transfer agent) and iso-octane. The resulting latex has a low coagulum content and there is little reactor fouling [227].

Water-resistant adhesives have been prepared from blends of a novel copolymer latex (100 parts) with polyvinyl alcohol (20–200 parts) and isocyanates (20–250 parts) [228]. Typically, the latex is prepared with a 1 : 1 styrene–butadiene 'seed' stage, using a surfactant with the general formula $R.C_6H_4.O(C_2H_4O)_n.SO_3.X$, where $R = C_{6-14}$ alkyl, $n = 2-75$ and $X = Na$, K or NH_4. The resulting adhesive has a viscosity of 100 P, increasing to 180 P in 30 days. Details of the preparation of styrene–butadiene–vinyl pyridine latices for rubber-to-fabric adhesives are given in Section 7 below.

4.6.1 Some typical formulations

This section includes compositions and comments on the design, manufacture and formulation of latices, mainly butadiene–styrene and acrylic copolymers for specific adhesive applications. A composition for the high-speed coating of printing paper has been prepared from a latex from the following:

Butadiene	40 parts
Styrene	25 parts
Methyl methacrylate	20 parts
Acrylonitrile	10 parts
Itaconic acid	1 part
Acrylic acid	2 parts
Hydroxyethyl methacrylate	2 parts

After polymerization, the latex is neutralised to pH 7 with 25 % sodium hydroxide and zirconium ammonium carbonate (1 part, as Zr) is added as an insolubilising crosslinking agent. Some latex-based adhesive systems have been proposed for the production of laminates, which include a polyamine–amide adhesion promoter. Typically, a styrene–butadiene–itaconic acid copolymer latex has been mixed with 6 % (on latex solids) of a polyamine–amide (from methyl acrylate and ethylenediamine). After adjustment to 35 % solids, the mixture was coated on to corona-discharge-treated polypropylene film, which was then laminated under pressure to polyvinylidene chloride-coated polypropylene film, with a marked improvement in peel strength, compared with that of the bond made without added adhesion promoter [229]. Latices based on butadiene, for coating polyvinyl chloride sheet for bonding to cement board, have been prepared [230] from methacrylonitrile (10–75 %) and a conjugated diene (25–65 %), with (optionally) ⩾40 % styrene or another aromatic vinyl compound, and minor components, typically:

Butadiene	6 parts
Methacrylonitrile	56 parts
Methacrylic acid	8 parts

The bond obtained is claimed to be superior in peel strength to that obtained using an acrylic latex.

Latex polymers containing conjugated dienes, with low residual odour, can form adhesives with good mechanical stability and blister resistance. Typically, these are based on a complex monomer blend of butadiene, styrene, methyl methacrylate, acrylonitrile, itaconic acid, acrylic acid, N-methylolmethacrylamide and α-methylstyrene (34 : 25 : 15 : 8 : 1 : 1 : 3 : 12) (w/w), polymerized using $K_2S_2O_8$ as initiator, Na dodecylbenzene sulfonate as surfactant and $tert$-dodecyl mercaptan as the chain transfer agent, for 11 h at 60 °C. The resulting (odourless) latex has a gel content of 60 % [231].

Addition of tall oil (1.7–2.4 %) (which can be considered, chemically, as an unsaturated long-chain plasticiser) to a styrene–butadiene copolymer latex has been reported to increase the pot life and the bond strength of the resulting adhesive composition [232].

4.7 Other latex types

Increasingly, some non-vinyl latices are found to have applications in adhesives. These include:

(a) Latex adhesives prepared from dispersions of preformed polyolefins (Cl content of 5–40 %), typically a 30 : 70 molar but-1-ene–propylene copolymer chlorinated to 21 % Cl content, dissolved in toluene (20 % solids), homogenised in aqueous isopropanol with potassium oleate as surfactant and the solvent stripped, were converted to a latex adhesive, and good adhesion to polypropylene sheet and good resistance to weathering [233].

(b) Chlorosulfonated resin latices for adhesives based on suspensions of a chlorosulfonated $C_{2-8}\alpha$-olefin polymer (Cl content of 20–70 % w/w; S content 5–10 %) (the polymer may be a chloro-olefin homopolymer with an unsaturated acid or anhydride, typically based on a chlorosulfonated graft-modified ethylene–propylene copolymer) have been suggested for adhesives and coatings. The latices are effectively stabilised by the sulphonate groups formed by *in situ* hydrolysis of the chlorosulfonic groups on the polymer chain, so no additional surfactants are required for stable latices [234].

(c) A polyurethane dispersion with tackifiers can be prepared from polycaprolactone diol (100 parts), isophorone di-isocyanate (70.5 parts) and the dimer $(HOCH_2)_2CH.CH_2.COOH$ (10.6 parts) in butan-2-one (45 parts), then stirred with a hydrogenated rosin (90 parts) and triethylamine (34 parts) in butan-2-one (22.6 parts) at ~50 °C and poured into ethylene diamine (0.5 parts) in water (500 parts), to form a stable dispersion [235].

5 LATEX-BASED ADHESIVES WITH SPECIFIC FUNCTIONS

For many adhesive formulations the function (in terms of the method of application, the adhesion properties, manufacturing cost and performance life) is the defining feature and the physicochemical methods by which this is attained is secondary to the nature of the product. This section describes a wide selection of adhesive formulations that has been suggested and evaluated in relation to their function.

5.1 Pressure-sensitive adhesives

Pressure-sensitive adhesives are permanently tacky mixtures, which bond two coated surfaces when pressure is applied. They are based on mixtures of polymers (often natural rubbers) and other substances in organic solutions, solvent-free thermoplastic mixtures or as polymer latices with additives to control their performance in specific applications. This latter type is discussed here and in Section 4.5 above, where polychloroprene latex-based adhesives have already been discussed. In general, pressure-sensitive latices are applied to a carrier, to form a tacky film, after evaporation of the water. The surface to be bonded is then pressed against the tacky layer. The adhesive film is formed on a non-adhesive paper and transferred to its final carrier.

5.1.1 Pressure-sensitive adhesion and terminology

Pressure-sensitive adhesives are used with a wide variety of substrates, and particular copolymer formulations have been designed for coating on to, for example polyethylene, polypropylene and polyvinyl chloride-based films. Such self-adhesive films and tapes are widely used in packaging, with many other applications.

Certain types of pressure-sensitive adhesives are intended to form a permanent bond: attempted separation of the two components will cause tearing of the substrate. Other types, such as those used for labels, or *decals*, for glass, are designed to be removed easily when required, without leaving residual polymer on the glass surface. Many variations between these two 'extremes' have been developed, including adhesives that can be removed easily immediately after application but develop a permanent bond with time.

Pressure-sensitive adhesives form an immediately usable adhesive bond when a dry adhesive-coated surface is brought into contact with an uncoated substrate under light contact pressure at room temperature. This ability, known as 'tack' (see Section 4.1.3 above), requires that the adhesive should make rapid contact with the substrate, as with a liquid-to-surface contact. Balancing this necessary 'flow' property with adequate cohesive strength in the adhesive film is essential to the design of the adhesive formulation [236]. Bond strength depends on the affinity of the pressure-sensitive adhesive for the substrate, the degree of initial tack in the product and, especially, on the internal cohesive strength of the polymer blend. This cohesive strength depends on several factors, including the molecular weight of the polymer component and the degree of controlled cross-linking, either during the formation of the adhesive polymer, during a controlled secondary reaction, or by hydrogen bonding (e.g. of carboxyl or amino groups) in the polymers, although not necessarily within the same molecule.

In latex-based systems, with the advantages of zero or low (non-aqueous) solvent formulations, the design of the copolymer compositions in relation to the application is of particular importance; this is discussed below. Many compositions are based on acrylic polymers, as these have good resistance to ageing and weathering, but other monomers may be included for reasons of cost or performance. Suitable acrylic monomers (with plasticizing side chains) can form polymers that do not require additional plasticisation. Monomer selection and composition ratio determines specific adhesion, low-temperature bonding performance and the retention of flexibility in the bond.

The design of the adhesive should also relate to the method of coating employed. The obvious considerations of the nature of substrates and the ability to 'wet' the surfaces rapidly, the type of bond required (permanent, removable, heat-sealable, delayed-tack, creep resistant), ageing and release properties [237] are common to most adhesives. The size of the contact area with the substrate depends mainly on:

(a) the elastic modulus and viscosity of the adhesive, which may be considered as a viscoelastic liquid of high tenacity;
(b) the ratio between the specific surface energy of the adhesive and that of the substrate;
(c) the contact pressure and contact time.

The quality of the bond, however, also depends on the internal cohesive forces of the adhesive layer and its resistance to shear deformation, possibly at elevated temperatures; which may be partly governed by crosslinking and by the molecular weight of synthetic polymers present in the adhesive. In general, all measures that improve the cohesion reduce the adhesive properties and also the tackiness. Increased shear and peel strength can only be achieved with reduced adhesion and tackiness. These competing factors must therefore be identified and considered in the design and formulation of pressure-sensitive adhesives.

Basic factors to be considered [238, 239] in pressure-sensitive adhesive coating performance include:

(a) mechanical (and shear) stability of the latices;
(b) dispersibility of polymer solids (to avoid 'grit' on rollers);
(c) over-rapid drying (to avoid 'skinning');
(d) levelling of the coated film;
(e) avoidance of striations and other marks during drying;
(f) minimisation of the use of anti-foams (to avoid 'fish-eyes' in the coated film).

Some practical consequences of these requirements are:

(a) If the carrier with the adhesive layer has to separate 'cleanly' from the substrate (e.g. with 'masking tapes'), then the adhesion to the carrier must exceed the cohesion within the adhesive layer and, in turn, the adhesion to the substrate. This is known as adhesive failure.

(b) If a permanent bond is required, the adhesion to the substrate and carrier must be greater than the cohesion, so that cohesive fracture takes place when the bond is broken. This is known as cohesive failure.

Predictive methods have been developed for the production of improved pressure-sensitive acrylic adhesives for flexible PVC films, based on the measurements of the residence rate and migration rate of plasticiser in the film, as functions of the adhesive formulation. A good shrinkage rate is also acquired [240].

5.1.2 Types of latices formulated into pressure-sensitive adhesives

5.1.2.1 General

The general nature of latices suitable for formulation into pressure-sensitive adhesives have been indicated above and many examples of detailed formulations have been published. These include:

(a) An acrylic copolymer latex, with the general formula below, has a T_g of $-40\,°C$ and is recommended for pressure-sensitive adhesives [241]. In this case, a surfactant-free polymerisation system is preferred, with variations of the product being obtained by altering the amount of chain transfer agent. Postaddition of a nonylphenyl ethoxylate surfactant (3 % w/w) is suggested, so that a good balance between the peel strength and shear strength can be obtained. Typically, a modified butyl acrylate copolymer is often the basis of improved pressure-sensitive adhesives, such as a toughened polymer based on poly-2-ethylhexyl acrylate, as indicated in the following formulation:

1. Water   390 g
2. Sodium bicarbonate   2 g
3. $K_2S_2O_8$ 3.5 g
4. Butyl acrylate   546 g
5. Methyl methacrylate   30 g
6. Acrylic acid   24 g
7. tert-Butyl perbenzoate (initiator)   2.2 g
8. Sodium formaldehyde sulfoxylate (in 20 ml of water) 2.2 g

The manufacturing technique involves dissolving components 2 and 3 in component 1, then adding a mixture of components 4, 5 and 6 at 2.5 g min^{-1} at 80 °C and maintaining the temperature for 4.5 h. The temperature is then lowered to 60–70 °C, adding redox components 7 and 8 separately, and the polymerization completed.

(b) A simple acrylic latex for formulation into pressure-sensitive adhesives can be prepared from:

2-Ethylhexyl acrylate	74.5 parts
Vinyl acetate	22.0 parts
Acrylic acid	3.5 parts

which are polymerized in emulsion with $K_2S_2O_8$ as initiator. To improve adhesion, 0.5 % (on solids) of diallyldimethylammonium chloride may be added to the latex before coating on to tape [242]. Ethylene–vinyl acetate copolymer latices are the basis of a major group of pressure-sensitive adhesives, blended with other latices containing minor proportions of acrylic, maleic or fumaric esters [243].

Formulations for pressure-sensitive adhesives which are non-tacky to the touch but form a permanent bond under pressure have been suggested [244], using the general principles described above, with a mixture of a tacky adhesive latex (such as an acrylic ester latex copolymer) and a non-tacky latex (such as a styrene–butadiene copolymer) or a non-tacky water-soluble polymer, such as polyvinyl alcohol (typically a 87–89 % hydrolysed medium-viscosity grade, or starch or cellulose derivatives) and a filler (typically colloidal silica or titanium oxide). The non-tacky component is included to provide 'anti-block' characteristics during handling and storage. The inert filler is included to absorb machine oil from the paper printing process. Too much non-tacky material and absorbent filler will reduce the sealing properties. The composition is 25–40 % solids, pH 7–10, with viscosity of 2–20 mPa s, and is coated to a thickness of ~5–10 µm.

(c) Mixed copolymer latices are used in the formulation of pressure-sensitive adhesives. In a simple example, equal parts of high and low molecular weight copolymer latices of ethyl acrylate and acrylic acid (9 : 1) were coated on to polyethylene, giving good peel strength and bleed resistance [245]. In more complex systems, polymerization of a second latex is carried out in the presence of the first latex. In an example of this approach [246], a latex was prepared from:

Acrylic acid	14 g
Butyl acrylate	35.5 g
2-Ethylhexyl acrylate	35.5 g
Polyoxyethylene methacrylate	15 g

The resulting polymer has a T_g of 32 °C and M_n or 2760.

A further latex is then prepared from this latex, with added butyl acrylate (or other C_{1-12} alkyl (meth)acrylate, 2-ethylhexyl acrylate, methyl methacrylate and tetraoxyethylene diacrylate (a crosslinking agent)), yielding a final mixed copolymer (depending on the monomer ratios employed) T_g of 48 °C, suitable for use as pressure-sensitive adhesives for laminates.

(d) Another type of acrylic latex suitable for pressure-sensitive applications may be formed in a two-stage process, from a polymer with a 20–60 % gel fraction, followed by further polymerisation [247] with components and given in Table 8.31.

Stage 1 is added at 80 °C over 5 h, and then heated under nitrogen with crosslinking to give polymer particles (with a 43 % gel fraction). Stage

Table 8.31 Acrylic latex for pressure-sensitive applications

	Stage 1 (g)	Stage 2 (g)
A. Butyl acrylate	80	20
2-Ethylhexyl acrylate	15	28
Trimethylolpropane	5	—
Methacrylic acid	—	2
B. $K_2S_2O_8$ (as 0.5 % solution)	100	—

Latex-based adhesives with specific functions

2 is added over 3 h, and the resulting latex kept at 80 °C for 1 h. The object of the method is that Stage 1, with the gel fraction, imparts strength, whilst Stage 2 ensures good flow properties of the latex. The same group have also prepared sprayable aerosol pressure-sensitive adhesives with a relatively high acid content; e.g. acrylic acid–butyl acrylate–ethyl acrylate (9 : 90 : 10) of M_w 500 000, using a water-soluble non-ionic polymer, is neutralized with 10 % ammonia, diluted to a 30 % hydrosol (average particle size of 0.05 nm) and mixed (100 parts) with ethoxyethanol (1.5 %), followed by dimethyl ether, to give an aerosol pressure-sensitive adhesive [248]. A similar system, from tacky elastomeric microspheres useful in aerosol adhesives (or, combined with a hot melt adhesive on the substrate, forming a positionable hot melt adhesive system), employs a polymer obtained from iso-octyl acrylate (100 parts) water (300 parts), Lucidol 98 (catalyst) and Sipomat DS-10 (surfactant) (2 parts) and is reacted at 25–85 °C under nitrogen to form a 25 % suspension [249]. Other modifications include preparation of a similar hydrosol (10 parts) from a 35 % blend of 2-ethylhexyl acrylate, methyl methacrylate, acrylonitrile and acrylic acid (75 : 15 : 5 : 5), which is neutralized, mixed in water (90 parts) with butyl acrylate (40 parts), methacrylic acid (3 parts) and divinylbenzene (0.5 parts) and polymerized using ammonium persulphate for 3 h at 70 °C. Coated on to a tape, this formulation has high peel strength to both stainless steel and cotton [250]. A latex (viscosity ~5 P) from butyl acrylate, acrylonitrile and acrylic acid (16 : 4 : 1), coated on to a corona-discharge-treated polyethylene film primed with 5 % butyl acrylate–dimethylaminomethyl methacrylate and dried at 90 °C, gave an adhesive tape with good peel strength and an anchoring strength for >100 repeated bondings [251].

Another mixed latex adhesive system (reported by Rohm & Haas Inc. [252]) with improved latex stability has been prepared, using the components given in Table 8.32. In manufacture, B1 and B2 are added as a pre-emulsion to A simultaneously (over 2 h at 80–83 °C). Then, after 15 min, C is added at (20 min at 50–60 °C), followed by D1 and D2, until polymerization is complete. The product can be used to coat polyester film.

(e) Latices for pressure-sensitive adhesives (and also for waterproofing purposes) with good adhesion to both paper and polyester have been prepared from long-chain alkyl methacrylates (70–99 %) and hydrophilic monomers (1–3 %), typically by the latex copolymerisation of lauryl methacrylate, ethane diol dimethacrylate and methacrylic acid (372 : 8 : 20), with Levenol WZ surfactant and *tert*-butyl hydroperoxide and ascorbic acid as initiator [253] (see also references [254]).

(f) Contact adhesives can be formulated from core–shell latices (see Chapter 4). Typically, a core is prepared from a monomer mixture of ethyl acrylate, methyl methacrylate and acrylic acid (5 : 13 : 2), with dodecyl mercaptan (0.2) as the chain transfer agent and $K_2S_2O_8$ (0.2) as initiator, giving a copolymer core with M_w of 30 000 and T_g of 4 °C. The shell is composed of a butyl acrylate–ethyl acrylate (55 : 25) copolymer, prepared using $K_2S_2O_8$ (0.1), so forming a latex with a T_g of 46 °C, with an average particle diameter of 0.8 μm. The product, applied to birch plywood at 60 g m^{-2} and pressed at 3 kg cm^{-2}, developed satisfactory strength after 7 days [255].

Some review articles of polyacrylate and related latices suitable for various types of adhesives for tapes and for contact adhesives have been published [256]. The parameters affecting water resistance, and relevant experimental methods have been discussed [257] and general test methods for determining adhesive and tack properties have been reviewed [258]. Typical examples of non-tacky pressure-sensitive adhesives are shown in Table 8.33.

5.1.2.2 Vinyl ester polymer latices

Pressure-sensitive adhesives, suitable for temporary bonding between aluminium and oily steel, have been prepared from a mixture of a polyvinyl isobutyl ether latex, an acrylic latex (T_g of 30 °C) and an alkyl acrylate copolymer [259]. Many other variants have been suggested, including a latex copolymer of approximately equal parts of vinyl acetate and 2-ethylhexyl acrylate, stabilised with hydroxyethyl cellulose and non-ionic surfactant, with an added antistatic agent (typically, stearoamidopropyldimethyl-β-hydroxyethyl ammonium nitrate) that has been used as a pressure-sensitive adhesive [260]. The following are examples:

(a) A pressure-sensitive adhesive, with a pot-life of 48 h, has been prepared from a latex copolymer of 2-ethylhexyl acrylate, vinyl acetate and itaconic

Table 8.32 Mixed latex adhesive system

	Weight (g)
A. Water	450
Na dodecylbenzene sulfonate (surfactant)	0.48
NaHCO$_3$	0.43
B1. Na dodecylbenzene sulfonate	3.7
Butyl acrylate	537
Ethyl acrylate	224
Acryloxypropionic acid	54
B2. Water	50
(NH$_4$)$_2$S$_2$O$_8$	1.3
C. FeSO$_4$·7H$_2$O (0.15 % aq)	6
EDTA (ethylenediamine tetraacetic acid)(1 % aq solution)	2
NH$_4$OH (11.2 % aq solution)	50
Trimellitic anhydride–hydroxyethyl acrylate (1 : 1 molar)	48
D1. (NH$_4$)$_2$S$_2$O$_8$ (2.6 % solution)	15.4
D2. Na formaldehyde sulfoxylate NaCH$_2$(OH)SO$_2$	5.12

Latex-based adhesives with specific functions 545

Table 8.33 Non-tacky pressure-sensitive adhesives

	Parts by weight		
Tacky latex Robond PS 60[a]	68	70	78
Non-tacky latex Rhoplex B 85[a]	12	—	—
Dow 620[b]	—	10	—
Polyvinyl alcohol: Vinol 523[c]	—	—	2
Colloidal silica	20	20	20
Properties:[d]			
Blocking (%)	0	0	0
Seal (%)	100	100	100

[a] Rohm & Haas Inc.
[b] Dow Chemical Company Inc.
[c] Air Products & Chemicals Inc.
[d] Coated on to No. 24 bond paper, with a blank cover sheet at 100 psi (7 days at 65 °C).

acid (88 : 10 : 2) (40 % solids) to which is added 1 % (on latex weight) of 50 % butoxymethylmelamine.

(b) Latex polymers suitable for pressure-sensitive adhesives for bonding paper can be based on vinyl acetate and related vinyl alkanoates polymerised in the presence of polyvinyl alcohols, with a small amount of chain transfer agents, notably propionaldehyde, which reduces the time of the resulting adhesive [261].

(c) Vinyl acetate copolymer latices have been developed into systems described as hot-melt adhesives and claimed to be non-polluting and capable of preparation in low-cost equipment. Typically, the following components are polymerised at 100 °C, to form a low-solids (28 %), low-viscosity (1 Pa s) latex:

Vinyl acetate	220 parts
Butyl acrylate	80 parts
Ethoxylated stearylamide (surfactant)	15 parts
Polyacrylic acid	4.5 parts
Water	650 parts

Coated on to paper (at 200 μm thickness), followed by drying, this latex produced [262] blocking-resistant sheets with good adhesion to porous and non-porous substrates when pressed at 160 ° for 3 s.

(d) Grafted copolymers are also employed. A grafted 40 % latex copolymer of acrylic acid, 2-ethylhexyl acrylate and vinyl acetate (2 : 70 : 28) (T_g of 30 °C) was grafted with 60 % of a styrene–divinylbenzene mixture, forming a latex copolymer with particle diameter ∼0.4 μm. This mixture, coated on to paper tape, showed good adhesion and antiblocking properties [263].

(e) A range of pressure-sensitive adhesives with good room- and low-temperature performance ($T_g < 30$ °C; gel content of 50–70 %) is based on copolymers of alkyl acrylates, vinyl esters, dicarboxylic acid diesters and unsaturated monocarboxylic acids, with a reactive surfactant, a chelating monomer and a chain transfer agent. A typical example is obtained by latex polymerisation of 2-ethylhexyl acrylate, bis-(2-ethylhexyl) maleate, acrylic acid and methacrylic acid, with Alipal CO-443 and Igepal CO-887 (nonylphenol(30)EO, 70 % solids) as surfactants [264].

(f) Mixtures of latices of butyl acrylate and 2-ethylhexyl acrylate, in which one latex has a very high M_w, have been recommended for high-tack adhesive coatings for polyester-based pressure-sensitive tapes [265].

(g) Relatively low cost copolymer latices for pressure-sensitive adhesives have been prepared using mixtures of vinyl and acrylic monomers (including maleic anhydride and vinyl laurate) with a colloidal emulsifier (a polyalkylene oxide) and protective colloids (typically, carboxymethyl cellulose (d.p. 500–600) and free radical initiators) [266].

A terpolymer latex of vinyl acetate and 2-ethylhexyl maleate formed under ethylene pressure is claimed to be a satisfactory pressure-sensitive adhesive, with good peel strength. Similarly, latices from vinyl alkanoates (notably vinyl acetate), ethylene and bis-(2-ethylhexyl)maleate (or the related dioctyl maleate or fumarate) may be used in pressure-sensitive adhesives for bonding plasticised polyvinyl chloride; the T_g of the polymer particles should be from −45 to −25 °C [267].

Ethylene–vinyl acetate–acrylic ester copolymer latices containing polyvinyl alcohol have also been suggested as components of pressure-sensitive adhesives; the acrylic ester should have a T_g of at least −50 °C when homopolymerised, and should be 50–85 % of the copolymer. A typical copolymer latex including butyl acrylate, ethylene and vinyl acetate (56 : 12 : 32) bonds well to oriented polypropylene film, to form a pressure-sensitive tape, with tack and bond strength comparable to that obtained with a commercial acrylic latex [268].

5.1.2.3 Acrylic copolymer latices with added polar groups

The applications and use of acrylic latices, in relation to their characteristics, in pressure-sensitive adhesives, including methods of bar coating and slot-die extrusion technology, have been reviewed [269]. The monomer composition for a typical acrylic copolymer latex composition for a pressure-sensitive adhesive is:

Butyl acrylate	50 parts
2-Ethylhexyl acrylate	47 parts
Acrylic latex	3 parts

When coated on to cellulose diacetate or polystyrene tape, the bond strength between stainless steel and the prepared tape is ~580 g per 20 mm. [270]. The following latex composition has been suggested for coating polyester adhesive tape [271]:

2-Ethylhexyl acrylate	80
Ethyl acrylate	15.5
Acrylonitrile	2.5
Acrylic acid	2.0
Water	102.5
Sulfonated diphenyl ether (Dowfax 2A1) (45 % aq) (surfactant)	0.8
Naphthalene-formaldehyde condensate sulfonate (Daxad 11)	0.5
$K_2S_2O_8$	0.25
Sodium bisulfite	0.03

In more detail, a suitable latex for pressure-sensitive adhesives can be prepared from the following components, including the necessary initiator and molecular weight controller [272]:

A. Butyl acrylate	784 g
Acrylic acid	16 g
n-Dodecyl mercaptan (molecular weight controller)	
B. Water	704 g
Surfactant	16 g
Azobiscyanovaleric acid, 10 % ammoniacal, pH 10	16 g

Monomers are added under N_2 for 2 h at 70 °C, then held for 2 h, and pH adjusted to 7.2, yielding a latex with ~52 % solids content. The gel fraction is ~38 % and viscosity ~120 P. The resulting latex, when coated on to both sides of viscose tape, showed high bond strength to stainless steel.

Pressure-sensitive adhesives with good adhesion to non-polar polymers have been prepared by latex polymerisation of a solution of rosin esters (see Section 4.1.3) (including >60 % of (unsaturated) dehydroabietic acid) in a mixture of alkyl (meth)acrylate esters; typically 2-ethylhexyl acrylate, butyl acrylate, rosin glyceryl ester (70 % of dehydroabietic acid), acrylic acid and N-methylolacrylamide (85 : 13 : 3 : 1.8 : 0.2) have been polymerised with Na polyoxyethylene sulfate as surfactant and ammonium persulphate as initiator, to a 53 % solids latex, viscosity 1.2 P, pH 2.1 and neutralised with ammonia. The resulting adhesive, coated on to polyester film and dried for 2 h, formed a pressure-sensitive tape with good adhesion to polyethylene and stainless steel [273].

An adhesive with good release properties can be prepared by coating successively from a mixture of latices from the same monomers, used in different ratios, so that a 'three-layer' structure is obtained [274]. In each case, a formaldehyde-based crosslinking agent is included, although 2-ethylhexyl acrylate is not present (which is unusual for pressure-sensitive compositions). The suggested compositions are given in Table 8.34. The systems are used for application to metals.

A polymer latex based on butyl methacrylate (558 parts), with added methyl methacrylate (15 parts), acrylonitrile (15 parts), and acrylic acid (12 parts), with crosslinking by ethylene glycol dimethacrylate (6 parts), and dodecyl mercaptan (0.4 parts) as molecular weight controller, has been polymerised as a pre-emulsion, with sodium lauryl sulphate and polyoxyethylene alkyl phenyl ether Na salt as surfactants and ammonium persulphate as initiator, and is claimed to be suitable for preparing a pressure-sensitive adhesive with high film elongation before break [275]. Pressure-sensitive adhesives for bonding to polyolefin films have been based on blends of alkyl(meth)acrylates with tackifier resins emulsions (tackifier resin softening point <100 °C and particle diameter ~1 nm). A coagulation-resistant adhesive was formed from acrylic acid, 2-ethylhexyl acrylate, 2-hydroxyethyl methacrylate and methyl methacrylate (0.5 : 81.5 : 3 : 15) with a tackifier and antifoam agent [276]. Other latices for pressure-sensitive adhesives have been prepared using anionic surfactants which include at least one sulfonic acid group and one carboxylic acid group, with a redox-type initiator. Typically, a latex may be formed from acrylic acid, 2-ethylhexyl acrylate and vinyl acetate (3 : 82 : 15) with polyoxyethylene-2-ethylhexyl ether sulfosuccinate disodium salt as

Table 8.34 Adhesive with good release properties

Composition	Parts by weight		
	Stage 1	Stage 2	Stage 3
Butyl acrylate	85	65	85
Acrylonitrile	15	15	15
Acrylic acid	5	5	5
Polyethylene dimethacrylate	0.1	0.1	0.1
Lauryl mercaptan	0.1	0.1	0.1
Methyl methacrylate	—	20	—
Parts of above composition in final blend (as solids)	35	30	35
To each of the above are added:			
Melamine resin	5	5	5
Release agent	0.1	0.1	0.1

surfactant [277]. Low-viscosity pressure-sensitive acrylic latices have been prepared in the presence of polymerisable emulsifiers [278], using a mixture of the components given in Table 8.35.

The resulting latex has a viscosity of 40 P, and shows good wettability on release paper. Without added ammonia, the latex viscosity is >100 P, with poor wettability. Another system, also employing polymerisable surfactant uses a mixture of 2-ethylhexyl acrylate, methyl methacrylate, acrylic acid and ethyl acrylate (60 : 20 : 5 : 15), with Latemul 120A (a polymerisable surfactant) (4 %) and polyoxyethylene nonylphenyl ether (1 %) as dispersant and potassium persulphate as initiator. The resulting polymer latex (neutralised with ammonia) is coated on to polyester film, giving good adhesion to polyester film, with cold water resistance [279].

Pressure-sensitive adhesives based on latices from (mainly) methacrylate esters have been claimed [280]. Typically, the mixture of monomers given in Table 8.36 is used to produce a monomer emulsion. 10 % of this emulsion was polymerised, using an ammonium persulfate initiator at 70 °C, to form

Table 8.35 Low-viscosity pressure-sensitive acrylic latex

Butyl acrylate	94 parts
Acrylic acid	3 parts
Acrylamide	1 part
Polyoxyethylene lauryl ether Na sulfate	1 part
Water	50 parts
Aqueous ammonia	(unspecified)
This mixture is added dropwise at 82 °C over 3 h to:	
Water	40 parts
Ammonium persulfate $(NH_4)_2S_5O_8$	0.2 parts
Antox HS 60 (proprietary polymerisable surfactant)	2 parts
Butyl acrylate	2 parts

Table 8.36 Preparation of a monomer emulsion

2-Ethylhexyl acrylate	80 parts
Methyl methacrylate	15 parts
Acrylic acid	1 part
2-Hydroxyethyl acrylate	3 parts
n-Dodecyl mercaptan	0.03 parts
This was added to a solution of:	
Water	30 parts
Neocol P	0.4 parts
Y 100	0.2 parts
EA 140 (non-ionic surfactant), HLB (hypophlic-lipophile balance) = 14	(amount not stated in original)

a 'seed' latex. Following this, the remainder of the monomer mixture was added over 5 h, to yield a latex with particle size ~0.9 nm, which has good wettability and can be defoamed readily.

Latices for pressure-sensitive adhesives have also been prepared using a 'seed' stage process, without surfactants or water-soluble monomers. Diallyl phthalate (1–10 %) is included as a crosslinking agent. Typically, the components in Table 8.37 have been heated at 50 °C for 30 min, to form 'seed' polymer particles. After pH and viscosity adjustment, the resulting latex can be used as an adhesive for coating acrylic sheet, giving a bond with high peel strength and good ageing resistance [281–283].

2-Ethylhexyl acrylate has been used as the principal monomer in several formulations of latices suitable as the basis of pressure-sensitive adhesives [284, 285]. The same monomer has also been employed in the formation of pressure-sensitive adhesives able to accept high tackifier concentrations without loss of strength, mainly by employing longer chain length alkyl acrylates with surfactants and dextrins, typically with a 2-ethylhexyl acrylate–methyl methacrylate (55 : 2.9) copolymer with 2.5 % dextrin, for bonding a polyester tape coated with this mixture to glass and polyethylene [286].

Mixed polymer latices of 2-ethylhexyl acrylate, vinyl acetate and acrylic acid (5 : 14 : 1) have been prepared by conventional latex polymerisation techniques to have low and high degrees of polymerisation, and are then neutralised with ammonia and initiated by addition of benzoyl peroxide (2 %). The strength of the coated bond increased markedly compared with that of the coated bond made using high d.p. polymer alone, presumably due to the crosslinking of the low d.p. polymer [283].

2-Ethylhexyl acrylate (75 g) has also been used in a pressure polymerisation, with ethylene (50 °C at 50 bar) and vinyl acetate (248 g) and a small amount

Table 8.37 Preparation of pressure-sensitive adhesives using a 'seed' stage process

Ethyl acrylate	40 parts
Copper sulfate	0.004 parts
Sodium thiosulfate	0.6 parts
Potassium persulfate (initiator)	0.4 parts
Water	350 parts
The following components are then added and the polymerisation taken to completion:	
2-Ethylhexyl acrylate	880 parts
Diallyl phthalate	80 parts
2-Hydroxyethyl methacrylate	12 parts
Acrylic acid	25 parts
Eliminol JS 2 (a reactive emulsifier)	20 parts
Potassium persulfate	1.5 parts

Latex-based adhesives with specific functions

of acrylamide. Mixed sulfonate–non-ionic surfactants and $(NH_4)_2S_2O_8$–Na formaldehyde sulphoxylate as a redox initiator were used to form a latex, reported to have good peel resistance with glass and PVC [287]. In a further development, an ethylene–vinyl acetate copolymer latex suitable for coating heat-resistant pressure-sensitive flexible tapes for adhering to glass was prepared in two stages [288] (see Table 8.38).

In Stage 1, the mixture is heated to 50 °C under ethylene (60 bar) pressure, with ammonium persulfate–Na formaldehyde sulfoxylate as the initiator system. After reaction, the components of Stage 2 are added, yielding a 57 % latex containing 30 % ethylene. Addition of 2-hydroxyethyl acrylate gives a marked improvement in adhesive properties.

An adhesive with good adhesion and peelability to a wide range of substrates has been prepared from 100 parts of a latex copolymer of 2-ethylhexyl acrylate, acrylic acid, vinyl acetate and methyl methacrylate (80 : 5 : 5 : 1), mixed with 60 parts of a 50 % aqueous solution of wheat starch and 1.5 parts of a polyepoxide crosslinking agent. The adhesive, coated on to paper at 25 g m^{-2}, was heated treated at 40 °C for 7 days [289].

Acrylic copolymer latices suitable for pressure-sensitive adhesives and coating kraft paper tapes have been formulated with bituminous emulsions (particle size of 0.5–3.0 nm, preferably ~0.95 nm). The T_g of the resulting adhesive system should be −60 to −30 °C. Typically, the monomer mixture:

2-Ethylhexyl acrylate	68 parts
Methyl acrylate	30 parts
Acrylic acid	2 parts
N-Methylolacrylamide	3 parts

is polymerised at 75–85 °C to form a 60 % latex, with particle size of 4–5 μm. Addition of 2 % of bitumen gives a marked improvement in adhesion to polyethylene film, compared with a latex without added bitumen [290].

Table 8.38 Copolymer latex for coating heat-resistant pressure-sensitive flexible tapes

	Stage 1	Stage 2
Water	5200	530
Ammonia	—	18
Surfactants	145	285
Acrylamide	21	—
Acrylic acid	70	120
Vinyl acetate	2160	226
Vinyl laurate	840	600
2-Hydroxyethyl acrylate	—	600
2-Ethylhexyl acrylate	400	—

A basic monomer blend for pressure-sensitive adhesives with good water resistance is [291]

Butyl acrylate	30 g
2-Ethylhexyl acrylate	800 g
Methacrylic acid	10 g
Thioglycollic acid (included to control the degree of polymerization of the resulting polymer)	5 g

Vinyl acetate–ethylene pressure-sensitive copolymer latices, prepared using a copolymerisable surfactant and similar acrylates, can be used for pressure-sensitive adhesives which bond well to polyolefin surfaces [292]. A typical copolymeric oligomeric surfactant has the formula

$$C_9H_{18}.C_6H_4.O.CH_2.CH[O.(CH_2.O)_{10}.SO_3.NH_4].CH_2.O.CH_2.CH=CH_2$$

and is used in the formulation given in Table 8.39.

Alternative mixtures suggested include:

(a) A blend of a 2-ethylhexyl acrylate, methyl acrylate and acrylic acid (85 : 13 : 2) copolymer latex with an equal weight (of solids) of a copolymer latex of 2-ethylhexyl acrylate, vinyl acetate and ethylene (15 : 65 : 20). It is claimed that including the former latex increases the strength of a paper-to-polythene bond significantly [293].

(b) Other pressure-sensitive adhesives for bonding polyolefin films have been prepared by blending alkyl (meth)acrylate latexes with emulsions of tackifier resins (softening point $<100°$ and particle diameter ~ 1 μm). Typically, a coagulation-resistant adhesive may be prepared from a copolymer latex of acrylic acid and 2-ethylhexyl acrylate, 2-hydroxyethyl methacrylate and methyl methacrylate (0.5 : 81.5 : 3 : 15) (100 parts w/w) with 20 parts of

Table 8.39 Pressure-sensitive copolymer latex prepared used a polymersable surfactant

2-Ethylhexyl acrylate	264 parts
Acrylic acid	5 parts
Vinyl acetate	113 parts
Potassium persulfate	15 parts
This is polymerised at 78 °C with a mixture of:	
Water	379 parts
Acrylamide (40 % aq solution)	10 parts
Polymerisable surfactant (typically that given above)	15 parts
Ethylene (at 50 bar pressure)	Addition over 4 h

a rosin tackifier (Aquatec 5527) (softening point 27° and particle diameter ~1.3 μm) a thickener (1 part) and a defoamer [294].

Alternative pressure-sensitive formulations may incorporate a tackifying resin into the monomer mixture before polymerisation, typically using an aromatic hydrocarbon resin (<10 % w/w), with a M_w of 500–5000, forming a latex with a T_g of −70 to +10 °C from a mixture of resin (Escorez ECR-149), acrylic acid, ethyl acrylate and 2-ethylhexyl acrylate (10 : 1.6 : 8.4 : 30) [295]. Other latices for bonding polyolefin films have been prepared by latex polymerisation of either styrene, α-methyl styrene, acrylonitrile, methyl methacrylate or vinyltoluene, in the presence of tackifier emulsions (e.g. terpene resins, rosin esters or petroleum resins, with a softening point 65–130 °C), and then adding, in a second stage, butyl (meth)acrylate [296].

(c) Tacky, pressure-sensitive adhesives, with initial repositioning ability, long-term performance and good water and aliphatic solvent resistance, have been prepared by latex polymerisation of a reactive silicone acrylate, an unsaturated carboxylic acid and at least one alkyl acrylate. Typically [297], a mixture of butyl acrylate, methyl acrylate, acrylic acid and silicone acrylate (RC 300) was copolymerised in an aqueous surfactant solution with dodecyl mercaptan (as the chain transfer agent), using potassium persulfate as initiator, at 80 °C.

(d) Good peel strength to polyethylene and similar surfaces with pressure-sensitive adhesives based on acrylic ester copolymer latices by employing a surfactant based on sodium (or ammonium) sulfosuccinate esters. Typically, a 40 % latex of butyl acrylate, methyl methacrylate, methacrylic acid and N-butoxymethacrylamide (87 : 7 : 7 : 3) (T_g of −42 °C) is used [298].

(e) Other pressure-sensitive adhesives for bonding polyolefin films have been made from blends of alkyl meth(acrylate) latices with emulsions of tackifier resins, with a softening point of <100 °C and particle diameter ~1 μm. Typically, a coagulation-resistant adhesive was prepared from a copolymer latex.

(f) A copolymer latex of 2-ethylhexyl acrylate, ethyl acrylate, acrylonitrile and acrylic acid (50–95 : 4–40 : 0.5–5 : 0.5–5) (preferably in the middle range of these ratios), stabilised with mixed sulfonated emulsifiers, is suitable as a pressure-sensitive adhesive for tapes with good adhesion to polyester and other synthetic polymer films [299].

(g) The preparation of pressure-sensitive adhesives from a copolymer latex based on 2-ethylhexyl acrylate, with acrylic or methacrylic acid, diacetoneacrylamide (~25 % of the total monomer content) and, optionally, methyl acrylate, has been claimed [300].

Apart from those mentioned earlier in this section, some pressure-sensitive systems for metal surfaces have been developed. These include:

(a) A copolymer latex of 2-ethylhexyl acrylate, acrylic acid and methacrylate (97 : 3 : 0.3) has been prepared (as the ammonium salt shows very high bond strength to stainless steel when a minor proportion of an ethylene–vinyl acetate (7 : 18) copolymer latex is included in an adhesive formulation). The glycidyl methacrylate acts as a cross-linking agent [301].

(b) Peelable adhesive papers have been prepared using latex copolymers. Such papers, which adhere to stainless steel, are prepared by coating (5–20 gm^{-2}) a base paper with an adhesive latex of acrylic acid, butyl acrylate, 2-ethylhexyl acrylate, hydroxyethyl acrylate and methyl methacrylate (1 : 42 : 45 : 2 : 10) with Na dodecylbenzene sulfonate (1 %), which is then foamed (expansion ratio 1 : 2.5), dried (1 min at 110 °C) and bonded to a release paper. The resulting foamed film adheres to stainless steel, with no residue adhering to the substrate [302].

(c) C_{4-12}-alkyl (meth)acrylic esters have been redox-polymerised (e.g. 2-ethylhexyl acrylate with acrylic acid) to form copolymer latices which can be blended with a non-ionic tackifier to yield a product that bonds PVC to paper [303].

(d) Latices for pressure-sensitive adhesives have been made by variation of the monomer feed (typically butyl acrylate, methyl methacrylate and acrylic acid) ratio with time. This has been claimed to produce multidisperse particles with T_g values that vary with particle diameter, giving adhesive systems useful for bonding films to metallic surfaces [304]. In a latex copolymer suitable for use in pressure-sensitive adhesives, a typical ratio of the principal monomers (2-ethylhexyl acrylate and butyl acrylate) is 5 : 4, with, in addition, 2–5 % of glycidyl methacrylate (2,3-epoxypropyl methacrylate); 1–5 % of acrylic or itaconic acid may also be added. The latex is neutralised and ZnO may be incorporated to form a pressure-sensitive adhesive for kraft paper [305].

(e) Latices suitable for use in pressure-sensitive adhesives have been prepared using water-soluble chain transfer agents, such as dihydropyran. These are odourless compared with mercaptan-type agents. Typically, a 57 % latex was prepared from 2-ethylhexyl acrylate, acrylic acid, methyl methacrylate, vinyl acetate, acrylonitrile and dihydropyran (80 : 2 : 8 : 8 : 2 : 0.2) with initiation by sodium persulfate at 90 °C. Paper coated with this latex showed good adhesion to polyethylene film and to stainless steel [306].

(f) A copolymer latex of 2-ethylhexyl acrylate and methyl acrylate (13 : 7) has been mixed with 5 % of n-propyl trimethacrylate monomer:

$$C_2H_5.C(CH_2O.OC(CH_3) : CH_2)_3$$

and used as a coating. Irradiation of the coating gave an adhesive tape with high shear strength [307].

(g) Block copolymer emulsifier systems (ethylene oxide–propylene oxide copolymers) grafted with acrylic acid, with polyvinyl alcohol present as viscosity controller and latex stabiliser, have been used in pressure-sensitive adhesive systems, which have the advantage of easy recycling [308].

(h) Aqueous acrylic pressure-sensitive adhesives have been prepared by a two-stage latex copolymerisation process. The first stage involves polymerisation of C_{1-4} alkyl (meth)acrylates polymerised using an initiator with a long half-life to form a 'seed' latex, followed by a second stage with pre-emulsified monomers and initiators continued to completion. Alternatively, the two stages may be carried out separately and then equal parts of the two latices mixed—typically from 2-ethylhexyl acrylate, ethyl acrylate and acrylic acid (70 : 27 : 3), with 2,2′-azobis(2-methyl-n-(phenylmethyl)-propionodiamidine) dihydrochloride as initiator at 50 °C for 8 h. A second latex, with the same monomers in the same proportions, using $(NH_4)_2S_2O_8$ as initiator was blended in equal parts and used to prepare an adhesive tape with an adhesive strength of 910 g per 20 mm when applied to stainless steel [309].

(i) Solid peelable adhesives, based on polyacrylate microspheres, have been developed. These are prepared, with diameter ~50 μm, by suspension polymerisation of alkyl (meth)acrylates using oil-soluble initiators and surfactants, typically from 2-ethylhexyl acrylate with benzoyl peroxide and Na dodecylbenzene sulfonate. The suspension (100 parts), with Na stearate (5 parts) and glycerol (20 parts) forms a solid adhesive stick with good adhesive properties [310].

(j) A redispersible powder for a water-borne pressure-sensitive adhesive (for reclaimable paper and board) has been prepared, typically, from a latex polymerisd from 2-ethylhexyl acrylate, vinyl acetate, mono-octyl maleate and acrylic acid, in the presence of polyoxyethylene 8000 ($M_w > 3000$). The acid component may be neutralised before, during or after polymerisation [311].

5.1.2.4 Copolymer latices with multiple unsaturation

Latices based on copolymers of a conjugated diene and styrene sulphonic acid can be used in pressure-sensitive adhesives, with added fatty acid metal salts and a tackifier. A typical latex can be prepared from isoprene–styrene sulfonic acid, with zinc stearate and a hydrocarbon resin [312]. Other latices containing diene monomers, acrylic esters and polymerisable acids can form pressure-sensitive adhesives without the aid of further tackifiers. Typical monomer ratios are formed from:

Butadiene	54 g
Butyl acrylate	15 g
Acrylic acid	2 g
Styrene	38 g
tert-Dodecyl mercaptan	1.8 g

The resulting latex shows high bond strength, with good water and heat resistance. The presence of the mercaptan, as a chain transfer agent, controls the degree of polymerisation of the latex copolymer [313].

Latices suitable for pressure-sensitive tapes and carpet backings have been prepared by blending two separately prepared latices with different T_g values: the 'soft' latex acts as 'tackifier' for the other 'hard' latex. The 'tackifier' latex ($M_w \sim 2000$) was prepared from a butadiene–itaconic acid–styrene (20 : 2 : 78) mixture, with CCl_4 and tert-butyl mercaptan as the chain transfer agents. Use of the tackifying latex increases the peel adhesion of the blend markedly [314]. The same 'tackifier' was mixed (1 : 1) with a similar copolymer (composition ratio 65 : 2 : 33) latex ($M_w < 5000$) to form an adhesive for polypropylene film to paper, with a 180° peel strength of 108 kg cm^{-2}; without the tackifier latex, the bond strength was 86 kg cm^{-2}. A further formulation, formed on a 'seed' latex in the presence of the same chain transfer agent, is initiated by a persulfate, followed by tert-butyl hydroperoxide; the resulting latex may be highly polydisperse. The basic formulation is:

Carboxylated styrene–butadiene latex	57.1 g
Styrene	576 g
α-Methylstyrene	24 g
Butadiene	573.6 g
tert-Dodecyl mercaptan	17.4 g

The resulting adhesive shows high peel strength [315]. An acrylic latex (49 %) (prepared from butyl acrylate, methacrylic acid and tert-butyl mercaptan (28.5 : 1.5 : 0.3) by adding $(NH_4)_2S_2O_8$ in water (0.17 : 31.7) as initiator over 5 h at 80 °C and then heating for a further 3 h) was blended (100 : 6) with a syndiotactic 49 % polybutadiene latex, with added Na polyacrylate as thickener [316], to form an adhesive with a viscosity of $\sim$20 P and good adhesion to polyethylene. A diene-type copolymer latex, T_g of 44 °C, was formed using an oligomeric surfactant, $(C_8H_{17}.S(CH_2.CHCN)_m(CH_2CH.COOK)_n$ (where $m + n = 16$; $m/(m + n) = 0.5$), butadiene, styrene, methyl methacrylate, hydroxyethyl acrylate and itaconic acid (1.5 : 51 : 36 : 10 : 10 : 1 : 2), and an initiator system. The resulting latex is mixed with a tackifier (35 phr) and thickened with Na polyacrylate before coating paper at <60 μm, forming a pressure-sensitive sheet with 180° peel, good adhesive strength and creep resistance [317].

Another method reported includes a diene copolymer based on a mixture of butadiene, isoprene, methyl methacrylate, butyl acrylate and acrylic acid (46 : 31 : 15 : 5 : 5), with 2.1 % of acid in the water phase, suggesting that there is some independent polymerisation in this phase, possibly with other monomers. It is then mixed with a tackifier. The resulting latex can be coated to give pressure-sensitive tape with good tack, peel strength and creep resistance [318].

Some blends of polymer latices for pressure-sensitive adhesives based on acrylate copolymers with acid groups with styrene–butadiene copolymers have been described. These include, typically, a 1 : 1 blend (on latex solids) of a latex of acrylic acid, 2-ethylhexyl acrylate and polyoxyethylene monoacrylate (2 : 96 : 2) with a latex of butadiene, methacrylic acid and styrene (50 : 2 : 48), which, coated on to kraft paper, displayed a good balance of tack, peel strength, holding power, elasticity and low-temperature adhesion [319].

5.1.2.5 Acrylic copolymer latices with amphoteric/basic comonomers

Pressure-sensitive adhesives with good peel and high shear strength may be prepared from acrylic esters, e.g. iso-octyl acrylate and betaine-type comonomers, such as N-(alkyldimethylammonio)methacrylamidate:

$$CH_2=C(CH_3).CO^-\ ^+N(CH_3)_2.CH.CO.R$$

where (R = alkyl or phenoxy radical) [320]. Acrylic-based pressure-sensitive latices have been made from polymerisable quaternary surfactant-type monomers, which are unsaturated in one of the alkyl groups attached to quaternary nitrogen. However, the counter-ion sulfonate group is attached to a lipophilic hydrocarbon chain. Therefore, when ethyl acrylate is copolymerised, additional anionic surfactant is included to give an overall negative charge on the polymer particles [321].

Copolymer latices for peelable adhesives (Section 5.2) can be prepared by a conventional redox polymerization, using surfactants including a polymerisable surfactant (typically Na sulfoethyl methacrylate, with 88–90 % of a mixed alkyl acrylate monomer (average C_8 alkyl) and a small amount of a zwitter-ion, 1,1-dimethyl-1-(2-hydroxyoctyl)amine methacrylamide [322].

Pressure-sensitive adhesives with inherent tack and good low-temperature properties have been prepared using a three-component anionic surfactant system. Typically, a monomer mixture of butyl acrylate, methacrylic acid and acrylic (97 : 1.8 : 1.2) has been latex polymerised with a surfactant mixture of Na dioctyl sulphosuccinate, an ethoxylated C_{10-12} alcohol acid sulphosuccinate Na_2 salt and ethoxylated Na lauryl sulfate, with an added tackifier [323]. Good peel adhesion is obtained on glass, high- and low-density polyethylene film and corrugated board.

5.1.2.6 Acrylate–acrylonitrile copolymer latices

The effects of different polymer blends and tackifiers on nitrile rubber and Hycar latices used as adhesive bases have been discussed in relation to standard test parameters [324]. A useful copolymer latex for contact adhesives can be prepared using redox initiation with conventional delayed monomer addition techniques, using a monomer mixture of 2-ethylhexyl acrylate, methacrylic acid, acrylonitrile and vinyl acetate [325]. Another two-stage latex system, suitable for bonding to corona-discharge treated polyethylene sheet, has been based on acrylonitrile, butyl acrylate, 2-ethylhexyl acrylate and methacrylic acid (4 : 5 : 10 : 1), with added glycidyl methacrylate and dimethyl aniline as the 'anchor' coat. After heating for 1 min at 90 °C, this was further treated with a latex adhesive of 2-ethylhexyl acrylate, butyl acrylate, acrylonitrile and acrylic acid (14 : 3 : 3 : 3) and then crosslinked by treatment with an electron beam [326]. In a similar process, with minor variations, after addition of benzophenone, UV radiation was used as the curing mechanism [327]. A crosslinkable latex for adhesives for bonding floor covering was prepared in two stages. In the first stage, 25 % of a mixture of 2-ethylhexyl acrylate, acrylonitrile and acrylic acid (790 : 90 : 18) was partially polymerised and then added, together with trimethylolpropane triacrylate (2.25 parts), until completion. The resulting latex retains tack for >3 h after application [328]. Acrylic copolymer latices in combination with anti-tack agents have been developed as adhesives for bonding vibration-damping, rubber-backed flooring materials, typically a latex copolymer (80 parts) of 2-ethylhexyl acrylate and acrylonitrile (90 : 10) blended with $CaCO_3$ (17.7 parts) and other additives (1.3 parts). This forms an adhesive with a wet and dry bond strength better than those of a two-pack epoxy adhesive [329].

5.1.2.7 Crosslinking acrylate copolymer latices

An acrylic–butyl acrylate copolymer latex, partially crosslinked with zinc acetate, was formulated, together with sodium ammonium polyacrylate, polyvinyl methyl ether (a water-soluble thickener) and dibutyl phthalate (as plasticiser), into a pressure-sensitive adhesive for labels: a silicone-type release paper was employed. The adhesive bonds well to the substrate, but can be removed with water [330]. Pressure-sensitive adhesives for removable labels have also been formed from a latex copolymer of butyl acrylate, 2-ethylhexyl acrylate, vinyl acetate and 1,1-dimethyl-2-hydroxymethyl-3-oxobutyl acrylamide (10 : 81 : 5.5 : 3.5), with a proprietary amine salt as accelerator. Other amides may be substituted, notably N-(butoxymethyl)acrylamide $CH_3.CH=CH.CO.NH.O.C_4H_9$ or other acrylamide derivatives [331, 332].

Adhesives for corona-discharge treated polyolefins, both polethylene and polypropylene, with good adhesion over a wide temperature range, have been prepared using a latex from [333] made of components given in Table 8.40.

Table 8.40 Preparation of adhesives for corona-discharge treated polyolefins

Water	100 parts
Disproportionated rosin (K salt)	3 parts
Then add over 8 h at 80 °C:	
Styrene	19.5 parts
Butadiene	50 parts
2-Ethylhexyl acrylate	25 parts
2-Hydroxyethyl acrylate	0.5 parts
Acrylic acid	1.1 parts
tert-Dodecyl mercaptan	1.5 parts
Also, add simultaneously:	
Na persulfate in water	As initiator
Na caseinate	0.3 parts
Na dodecylbenzene caseinate	0.3 parts
Water	20 parts

The mixture is then heated for a further 1 h, adjusted to pH 8.0 and excess monomer removed by steam stripping. The resulting latex is thickened with 0.5 parts of Na polyacrylate and then coated on to release paper.

A removable pressure-sensitive adhesive, suggested for tapes and similar substrates requiring high peel strength, has been prepared from [334]

Na styrene sulfonate	3.92
$NaHCO_3$	0.48
$K_2S_2O_8$	0.3
Na dodecylbenzene sulfonate	1.18
Iso-octyl acrylate	341
N-tert-Octylacrylamide	47
Water	47

The long-chain acrylamide (N-tert-octylacrylamide) derivative included in the copolymer improves removability of the coated polymer film from the substrate.

A polyester film has been coated with a modified latex:

Butyl acrylate–acrylic acid copolymer latex (50 %)	100
Maleic anhydride-modified latex (45 %)	30
Butoxyethanol	15
Morpholine (30 %)	12
Zn acetate (aq solution)	1

This modified latex is coated on to the polyester, dried for 2 min, cooled to 20 °C and bonded to steel plate, with improved bond strength. The maleic anhydride modification gives marked improvement in adhesion properties [335, 336].

A further formulation for a pressure-sensitive adhesive for polyester film is based on a latex of butyl acrylate, 2-ethylhexyl acrylate, styrene, acrylic acid and acrylamide (21 : 20 : 3 : 1.5 : 1.25). This latex (200) is blended with bisphenol A (25), dimethylaminophenol (2.2) and water (20), held at 90 °C for 6 h, followed, after a further 2.5 h, by additional dimethylaminophenol (4.4), resulting in a modified latex (47 % solids, pH 9.5–10.5) with particle diameter of 150–160 nm, which is coated on to polyester film, dried, pressed on to sailcloth, and their cured (130 °C for 10 min), giving a bond with good peel strength [337]. Similar adhesives with good cohesion and warp resistance have been claimed, based on copolymer latices of alkyl (meth)acrylate and unsaturated carboxylic acid (T_g < 250 °K and gel content 20–60 %) and other polymer latices, with T_g > 250 K. Typically, a 2-ethylhexyl acrylate–acrylic acid (96 : 4) copolymer latex was mixed with a 2-ethylhexyl acrylate–acrylic acid–styrene (96 : 4 : 30) latex, (T_g of 370 °K), to form an adhesive with good adhesion to polyester film and cohesion on a Bakelite resin surface [338].

A peelable pressure-sensitive adhesive has been formed from diene monomers, where methyl methacrylate and butyl acrylate may copolymerise unevenly with the dienes. It is suggested that this unevenness reduces the adhesive strength and increases the cohesive strength of the bond [339].

The use of core–shell latex copolymers, with hard shells and a softer outer shell, in pressure-sensitive adhesives has been claimed [340]. In one example, microparticles (diameter 0.1 nm) of polymethyl methacrylate were coated with butyl acrylate–2-ethylhexyl acrylate–acrylic acid shells to give microparticles (diameter 0.3 nm). When coated on to a base, the resulting films have low peel resistance. Alternatively, a reverse process has been suggested, in which a 'soft' core of iso-nonyl acrylate–acrylic acid (99 : 1) copolymer with a 'sheath' of iso-butyl acrylate–2-ethylhexyl acrylate–acrylic acid (54 : 44 : 2) copolymer is crosslinked with Epikote® 812 (epoxy-resin) and 1,8-diazacyclo[5.4.0]undec-7-ene to form a peelable adhesive. Alternatively, an OH-terminated core in the same shell as above, crosslinked by a di-isocyanate prepolymer, may be employed.

Peelable adhesives that do not leave residues include polymer latices with pendant oxazoline groups and tackifiers. Typical monomer blends employed are [341]

	A	B
2-Isopropenyl-2-oxazoline	7.5	—
Styrene	22.5	33
Butadiene	70	65
Itaconic acid	—	2

The above latices are blended in the ratio 60 (A) : 40 (B), with tackifer Foral® 85 [342], forming a product with rapid adhesion to polyester film, and giving no residue upon peeling.

Polymer mixtures of C_{4-8} alkyl acrylates, as principal monomers, with smaller ratios of acrylic acid, using diacetoneacrylamide or N-methylolacrylamide as crosslinking agents, have been proposed as pressure-sensitive adhesives [343]. Removeable pressure-sensitive adhesives are improved by the addition of small amounts of organofunctional silanes, as in the latex monomer formulation [344]:

Butyl acrylate	1230
Methyl methacrylate	150
3-Methacryloxypropylrimethoxysilane	2.4

This latex can be applied to silicone-coated release paper, dried and transferred to paper. The resulting tape has better initial and matured peel strength than that obtained with an adhesive made without the silane. Addition of a xanthan gum to an acrylic-based pressure-sensitive adhesive latex improved adhesion to polyester film and reduced the transfer of unwanted adhesive to another substrate [345]. A copolymer latex (30 % solids) of butyl acrylate, 2-ethylhexyl acrylate and N-methylolacrylamide, compounded with ammonia and a thickening agent, requires heating after coating to give maximum bond strength [346].

Pressure-sensitive adhesives for bonding to polyolefin films have been prepared from blends of alkyl(meth)acrylate latices with dispersions of tackifier resins, with a softening point <100 °C, and particle diameter ~1 μm (see Section 5.1.2.2 above). Pressure-sensitive adhesives for label coating can be prepared from a crosslinking acrylate copolymer latex made from 2-ethylhexyl acrylate, butyl acrylate, methyl methacrylate, 2-hydroxyethyl methacrylate, ethylene glycol dimethacrylate and acrylic acid (61.5 : 20 : 14.9 : 3 : 0.1), with n-dodecyl mercaptan (0.01 parts) as the molecular weight controller. The resulting polymer had a gel ratio of 50–60 %, with a <15 % ionic bond content in the gel, which had a degree of swelling of 35–45 % [347]. Labels prepared using this system show good cutting properties [348, 349].

5.1.2.8 Diene–monomer latices

The most widely used diene monomer latices in adhesives are those based on styrene and butadiene, where the applications include (because of their relatively soft dry film) bonding of PVC-based vinyl flooring to jute fabric–foam backed carpet or to carpet tiles. Latexes with ⩽68 % solids, with 20–25 % of styrene, are normally employed and carboxylated latic, which have increased adhesion, are also widely used. Pressure-sensitive systems should be prepared with non-migratory surfactants. Compatible tackifiers (see Section 4.1.3) are often based on technical rosin derivatives.

A latex copolymer of isoprene, butyl acrylate and lauryl methacrylate (typical ratio of 7 : 3 : 5) was prepared by graft polymerisation with a mixture of a rosin salt and sodium lauryl sulfate as surfactant and $K_2S_2O_8$ as the initiator and then precipitated. The resulting polymer, dissolved in butan-2-one, was used as a base for grafting of butyl acrylate and acrylic acid and gave a pressure-sensitive adhesive with good low-temperature properties [350], which appears to depend on the molecular weight of the polymers (although only within a particular series—in this case, a solvent-based system). However, with a range of ethylene–vinyl acetate copolymer latices *made by similar methods*, with similar aqueous phases, there is little relation between the MFT and monomer composition [351]. A latex copolymer of isoprene : butyl methacrylate : lauryl methacrylate (typical ratio 7 : 3 : 5) has been prepared by graft polymerization, with a mixture of a rosin salt and sodium lauryl sulphate as emulsifier, and $K_2S_2O_8$ as initiator, then precipitated. The resulting polymer, dissolved in butan-2-one, used as a base for grafting of butyl acrylate and acrylic acid, gave a *pressure sensitive adhesive with good low temperature properties* [352], which appear to depend on the molecular weight of the polymers (although only within a particular series-in this case, a solvent-based system). However, with a range of ethylene:vinyl acetate copolymer latices, with similar aqueous phases, there is little relation between MFT and monomer composition [353]. Pressure-sensitive adhesives for bonding card have been prepared from filled blends of natural and synthetic rubber latices, typically:

Natural rubber latex	100
Butyl acrylate–styrene copolymer latex (with a maleic acid–styrene copolymer as the protective colloid)	50
Silica gel (average particle diameter 1.5 μm)	20
Tapioca starch	20

A coating of this blend on paper showed negligible tack, with good anti-blocking properties, peelability, printability and receptivity to water-borne inks [354].

5.1.2.9 Heat-seal latex adhesives

Adhesives from ethylene–vinyl acetate copolymer latices, prepared by copolymerisation of vinyl acetate with ethylene under pressure, show good heat seal properties at lower temperatures, better than those from 'ionomer'-type ethylene–vinyl acetate latices, especially for bonding aluminium foil to paper. However, both types of ethylene–vinyl acetate latex-based adhesives will heat-seal more readily than those prepared from polyvinylidene chloride latices, as the former type will allow faster machine application speeds [355]. A series of copolymer latices and their suggested formulation into heat-seal adhesives for different substrates has been tabulated (see Table 8.41).

In addition to the 'conventional' types of latex mentioned above, many specialist latices have been suggested. Adhesives which can be activated by heat are often based on core–shell copolymer latices (see Section 4.4

Latex-based adhesives with specific functions

Table 8.41 Copolymer latices and their formulation (Hoechst AG)

Monomer basis	Substrates to be bonded
Vinyl acetate	Paper, cardboard
Vinyl acetate–maleate esters	Paper, textiles
Ethylene–vinyl acetate	Paper, varnished paper, polyolefin films, polyester, PVC, polystyrene, aluminium
Vinyl acetate–ethylene–vinyl chloride	Paper, textiles, leather, artificial leather
Acrylic esters–methacrylate esters and styrene–acrylate esters	Paper and aluminium foil

and earlier sections for a general description and methods of manufacture), consisting of latex particles with a soft core of polymer of with a with low T_g, surrounded by an outer shell of 'hard' polymer with a high T_g. A simple heat-seal adhesive of this type may be formulated by dispersing 40 % (w/w on monomer content) of dicyclohexyl phthalate, a solid plasticiser, into the composition. The adhesion of flocks to base fabric can employ a postpolymerisation dispersion of a reactive ethylene–vinyl acetate copolymer, with a filler, which may be $\geqslant 1$ part of a powdered polyamide or polyester with a thermally reactive crosslinking agent, such as a methylated methylolmelamine [356].

Alternatively, a heat-seal adhesive based on a core-shell latex consists of an ethyl acrylate–methyl methacrylate–vinylidene chloride–N-methylolacrylamide–acrylamide (5 : 5 : 90 : 1 : 1) copolymer, made by redox initiation. This will laminate polyethylene and polypropylene sheets (coated at 1.9 g m^{-2}), making films heat-sealable, with blocking at 110 °C, and little blocking at higher temperatures [357]. A hot-melt adhesive has been prepared from an ethylene–vinyl acetate copolymer latex, with a powdered heat-seal resin (such as a polyamide or polyester) dispersion and a crosslinking agent containing —NH.CH$_2$.OR or —N(CH$_2$.OR)$_2$ groups, such as methylated methylolmelamines [358]. This has been suggested for transferring flocks and has good washing stability. Similarly, an aqueous dispersion of a hot-melt polyester resin, blended with a crosslinked acrylic–methacrylic alkyl ester copolymer latex thickener and a linear polyacrylate latex (100 : 28 : 3.6 parts), forms a paste-like adhesive for bonding textile fabrics [359].

A core–shell system, probably with some grafting, suitable for adhesive coating of polyolefin tapes for bonding to paper or stainless steel, is based on a polymer blend of

Ethylene vinyl acetate latex, 50–55 %	40
2-Ethylhexyl acrylate latex	20
2-Ethylhexyl acrylate	87
2-Hydroxyethyl methacrylate	3

After neutralization, the latex is heated to 100 °C, suggesting that both hydroxyl–carboxyl exchange and an interesterification reaction occur in the curing process [360]. Heat-seal adhesives have also been made with a core layer and 10–30 % of an inactive outer layer with $T_g \sim 50$ °C above that of the core polymer, using a latex made by polymerising butyl acrylate (18 parts) with methyl methacrylate (4 parts) as initiator ($K_2S_2O_8$ at 70 °C for 150 min, followed by additional initiator for 3 h). Upon coating on to glass at 40 °C, the bond strength increased, probably caused by rapid increase in diffusion with temperature [361]. Adhesives based on similar principles, using 'core–shell' or 'layered' latices, form coated films that do not peel spontaneously, but will peel from substrates when required. These are formed from a blend of a copolymer of 2-ethylhexyl acrylate, vinyl acetate and acrylic acid (T_g of −80 to −50 °C), mixed with an ethyl acrylate–2-ethylhexyl acrylate latex, with a 'layered' structure [362].

Apart from the general principles indicated above, many other latex-based heat-seal adhesive formulations based on copolymer latices have been suggested. These include:

(a) Copolymerisation of propylene with methyl acrylate and vinylidene chloride (6 : 90 : 4) in a heat-seal latex. This can reduce the heat-seal temperature from 110 to 90 °C, with improvement of bond strength and blocking resistance of the adhesive film [363].

(b) Laminated polyethylene/polypropylene, when coated with a methyl acrylate–methyl acrylate–methacrylic acid (21 : 27 : 2) copolymer on one side and a vinylidene chloride–methyl acrylate–methacrylic acid (87.5 : 7 : 5.5) latex (with 5 % of carnuaba wax and 0.5 % of talc added) on the other side forms a heat-sealable packaging film with low water vapour and oxygen permeability [364].

(c) Both corona-discharge treated biaxially oriented polypropylene and polyester film can be coated with a low molecular weight copolymer latex, (T_g of 30–70 °C) of ethyl acrylate–methyl methacrylate–methacrylic acid (36.6 : 55 : 8.4), giving good adhesion to viscose film after 1 min at 110 °C. The preferred molecular weight is $<\sim 50\,000$, preferably $\sim 24\,000$. At higher $M_w (<150\,000)$ only poor adhesion is observed [365], such as ethyl acrylate–methyl methacrylate–methacrylic acid (36.6 : 55 : 8.4) terpolymer, with a heat-seal temperature of 110 °C [366].

(d) A latex based on a copolymer of $M_w \sim 30\,000$, of methyl methacrylate, ethyl acrylate and methacrylic acid (55 : 36.6 : 8.4), with sodium dodecylbenzene sulfate as stabiliser, was polymerised with potassium persulphate (1.4 %) as initiator in the presence of a mercaptan (0.6) as the molecular weight controller. The product was then neutralized with ammonia and coated on to corona-discharge treated biaxially stretched polypropylene film, to give good adhesion to viscose film [367].

Latex-based adhesives with specific functions 565

(e) A solid resin of ethylene and acrylic acid (melt flow index 50 and acid number 156) has been postemulsified to a latex (25 % solids) and then mixed with tackifying wood rosin (7 : 3) and coated at 35–40 μm on to low-density polyethylene film. This, when heated at 120 °C for 5 s, gave a bond strength of 0.7 kg m^{-2}, compared with 0.2 kg cm^{-2} for the same latex without tackifying rosin [368].

(f) A copolymer latex of methyl acrylate, methylmethacrylate and methacrylic acid (49.9 : 48.1 : 2) (T_g of 49 °C) and average particle diameter ∼0.35 μm, mixed with 40 phr of fine silica (as a non-slip agent), formed a heat-sealable, blocking-resistant heat-seal adhesive, applied to biaxially oriented polypropylene, with high optical clarity and good bond strength [369].

(g) Adhesion to steel by polyester film is achieved when the latter is coated with a latex of vinyl acetate, 2-ethylhexyl acetate and bis-(2-ethylhexyl)-maleate (51 : 68 : 17) under ethylene (60 bar pressure) with a persulfate redox initiator and an anionic surfactant, with hydroxyethyl cellulose as the protective colloid [370]. However, improved adhesion to steel (compared with a vinyl acrylic latex alone) is given by corona-discharge treatment when coated with a blend of latices [371]:

Latex A (43 % solids):		
Vinyl acetate	33 %	100 parts
2-Ethylhexyl acrylate	17 %	
Butyl acrylate	50 %	
Latex B (50 % solids):		
Vinyl acetate	19 %	30 parts
Vinyl chloride	45 %	
Ethylene	36 %	

Modified polyvinyl acetate latices, based on vinyl acetate and acrylic esters with crotonic acid (typically 10 parts of total monomer content), blended with an ethylene–vinyl acetate latex, polyethylene, a hydrocarbon resin, stearic acid and stearamide (10 : 6 : 3 : 0.2 : 0.2), form heat-seal adhesives for coating papers, which, after drying, can be bonded to textiles [372]. Alternative vinyl acetate copolymer latices (claimed to be non-polluting and to be made in conventional non-pressure (low-cost) equipment) have been made from

Vinyl acetate	220 parts by weight
Butyl acrylate	80 parts
Ethoxylated stearylamide	15 parts
Polyacrylic acid	4.5 parts
Water	650 parts

This mixture, polymerised at 100 °C, yielded a latex (28 % solids, viscosity 1 Pa s), that can be coated on to paper (200 μm), and dried, to give blocking-resistant adhesive sheets with good adhesion to porous and non-porous substrates when pressed at 160 °C for 3 s [373].

A graft copolymer latex from vinyl chloride, (g)ethyl acrylate, acrylonitrile, acrylic acid and N-methylolacrylamide (45 : 49.5 : 2.8 : 1.1 : 1.6) is suitable as a heat-seal adhesive, especially for bonding heavy-duty nylon fabric to a foam padding base, for motor vehicle seats [374]. A heat-seal adhesive has been prepared from a polyvinyl alcohol latex (50 % solids), stabilised with polyvinyl alcohol (3.5 % on monomer) and mixed with a $\sim$ 50 % wax emulsion and an ethylene–carboxylic acid copolymer. The presence of the latter is claimed to improve blocking resistance and to reduce the heat-sealing temperature from 120 to 80 °C [375]. Another polyvinyl alcohol-containing latex, combined with an ethylene–vinyl acetate copolymer (MFT of 25 °C) and a polyethyleneimine, Epomine P1500 (1 : 15 : 100), imparts heat-seal properties to polypropylene. The presence of polyvinyl alcohol (87–89 % hydrolysed) increases the heat-seal strength by up to 400 % [376]. Alternatively, an ethylene–vinyl acetate (3 : 1) copolymer, prepared from a solid copolymer by postemulsification, can be blended with rosin ester and wax emulsions. A film (10 nm thick) coated on to polystyrene may be heat-sealed to aluminium foil, polyester or polypropylene film [377]. A copolymer latex from styrene, acrylonitrile and acrylic acid (24 : 4 : 2), neutralised with ammonia to viscosity of 40 P at 64 % solids, pH 9.0, T_g of 12 °C, has been applied to parchment paper and to aluminium foil, heated at 100 °C for 1 min and then dried. After 24 h in water at 20 °C, 100 % cohesive failure of the paper bond was observed [378]. Acrylic latices with minimum film-forming temperatures above ambient temperatures have been blended with thermoplastic resins and plasticisers (typically, mixed using a screw extruder at 110 °C), with

Butyl methacrylate–N,N-dimethylaminoethyl methacrylate (as acetate)–lauryl methacrylate (71 : 62.9 : 25.4) latex (25 % solids)	20 g
Glyceryl esters of rosin acids	100 g
Water	70 g

The resulting latex mixture (MFT $\sim$ 65 °C) was further mixed with water (1220 g) and 830 g of a mixed resin of dicyclohexyl phthalate and ethylene–vinyl acetate (72 : 28) resin, yielding a 40 % latex with an average particle size of $\sim$ 0.5 μm. This was applied to paper and dried at 20 °C to form a non-tacky film (20 g m^{-2}), which becomes tacky at 100 °C in 10 s, giving a laminate with high peel strength when pressed against polyethylene film [379].

A copolymer latex activated by high-frequency heating, based on acrylic acid, acrylonitrile, vinyl chloride and vinylidene chloride (0.5–1 : 2.8 : 5–17 : 75–88), can be plasticised with latices from tributyl acetyl citrate or bis-(2-ethylhexyl)-adipate. Alternatively, 25–100 parts of butadiene–styrene–methyl methacrylate (48–70 : 22–32 : 4–6) latex, with 2–6 parts of a second latex of methyl methacrylate, 2-hydroxyethyl acrylate or itaconic acid, give a weather-resistant bond for flexible polyurethane foams [380].

Postemulsification of a copolymer by a latex has also been employed as a method of preparing heat-seal adhesives. An acidic copolymer latex of acrylic acid, butyl methacrylate, methacrylic acid and stearyl methacrylate (1 : 5 : 3 : 2), neutralized with ammonia, has been used to emulsify a solid ethylene–vinyl acetate copolymer to form a 53 % solid dispersion, which was then blended with a polyvinyl chloride latex (1 : 9 on solids) to form a peelable heat-seal adhesive, used for adhering lids to containers [381].

Urethane-primed polypropylene coated with a polyvinylidene chloride latex (50 % solids), with added silica (particle diameter 3 nm) and a 30 % dispersion of carnuaba–paraffin wax (3 : 2) and water (100 : 0.15 : 5 : 11), give a film with a heat-seal strength of 75 g per 15 mm at 100 °C [382]. A heat-seal adhesive for a porous base, such as 'Vinylon' paper (polyvinyl alcohol-based), can be used as an impregnant and dried by heat (100 °C for 3 min, followed by 150 °C for 1 min under pressure). The impregnating mixture is [383]

Chloroprene latex (50 % solids)	100 g
ZnO dispersion (50 % solids)	5 g
Diallyl-p-phenyleneamine	2 g
Terpene-phenol resin	80 g
Vulcanisation accelerator	5 g

Heat-seal adhesives for high-temperature-resistant bonds have been prepared using fluorinated copolymer latices (see also Section 4.5).

5.2 Removable pressure-sensitive adhesives

Removable pressure-sensitive adhesives are not well defined. Peel force may depend on adhesion, but also on other factors: the viscoelastic properties of the coating adhesive, the stiffness of the adherent, the rate of bond separation and the temperature. The general principles of peel adhesion and the transition from adhesive to cohesive failure, which results in removable pressure-sensitive adhesion, are discussed above in Section 5.1.1. Standard tests have been devised to measure the performance of pressure-sensitive adhesives, including peel adhesion. Such tests are carried out at a constant peel rate: it is expected that a removable pressure-sensitive tape will strip off cleanly from the adherent, leaving no visually noticeable residue. This 'adhesive failure' occurs at, or near, the adhesive–adherent interface. Many adhesives of this type show a transition from cohesive to adhesive failure at elevated temperature, or at low rates of peel, but such transitions may not be observed with adhesives formulated from high molecular weight components, including crosslinking systems. This section describes formulations for adhesives of this type.

5.2.1 Acrylic and other copolymer latices

Peelable adhesive tapes have been prepared from a blend of a latex copolymer (including a polyepoxide crosslinking agent) and a starch solution. The copolymer latex may be based on 2-ethylhexyl acrylate, acrylic acid, vinyl acetate and methyl methacrylate (80 : 5 : 5 : 10). The resulting adhesive is based on [384]:

Latex copolymer (as above)	100 parts
Denecol EX 830 (crosslinking agent)	1.5 parts
Wheat flour starch	60 parts

The mixture, applied to base paper (64 g m^{-2}) and then dried and heat-treated at 40 °C for 7 days, gave an adhesive tape with good adhesion and peelability in relation to a range of substrates. Polypropylene coated with a polyurethane has been further coated with a 50 % latex of polyvinylidene chloride (Saran L-502) with silica (particle diameter ∼3 nm) and a 30 % wax dispersion (carnuaba wax–paraffin wax 3 : 2) and water (100 : 0.15 : 6 : 11). The film showed a bonded heat-seal strength of 75 g per 15 mm at 100 °C [385].

Peelable water-based pressure-sensitive adhesives with good adhesion to rough surfaces, with elongation-to-break > 3000 %, have been prepared [386], based mainly on latex copolymers of C_{4-12} alkyl(meth)acrylates, typically by copolymerisation of 2-ethylhexyl acrylate–ethyl acrylate–acrylic acid (86 : 12 : 2) (w/w) at 50 °C, using 2,2azobis(2-amidinopropane) dihydrochloride (0.03 parts by weight) as initiator. A polyethylene terephthalate was coated with the adhesive, which showed an elongation to break of 5600 % and adhesive strengths of 220 and 300 gm^2, initially and after 1 h at 80 °C.

5.3 Contact adhesives

A contact adhesive is a type of pressure-sensitive adhesive in which two coated surfaces adhere on contact, with minimal pressure. Earlier water-based contact adhesives were based on latices of polychloroprene of relatively low molecular weight. Tackifiers (usually based on rosin or rosin derivatives (see Section 4.1.3 above)) were included to provide initial tack. Polychloroprene resin latex types may also include a *p-tert*-alkylphenol–formaldehyde resin (in a similar amount to that of the polychloroprene), forming contact adhesives with high peel strength [387]. Surfactant may also be included; the rosin or terpene–phenolic resin may also be included with 10 % of a fatty acid ethoxylate and/or a 3.5 % alkylphenyl polyether sulfate solution [388].

For contact adhesives, it appears that an acrylic polymer with $T_g \sim 0$ °C provides the optimum 'green strength' when used with phenolic resin dispersions as the tackifier system. Partial compatibility of the phenoplast and the acrylic polymer is desirable [389]. Carboxylated polychloroprene latices (typically Neoprene® or Baypren® latices; see Table 8.28) are suitable, bonding well at 13 °C when crosslinked with ZnO. Acrylic adhesives bond equally well at pressures >4.5 kg cm^{-2}, but are less effective at lower pressures, especially at relative humidities ∼50 %. At lower humidities (<10 %), both types bond equally well [390].

Aqueous latices for contact adhesives have been prepared from a complex mixture of monomers, based on a variable blend of vinyl acetate, vinyl laurate and 2-ethylhexyl acrylate (40–70 %), with acrylic acid (10–40 %), methyl methacrylate (10–15 %), acrylamide (0.1–2 %), hydroxyethyl acrylate (0.5–3 %), sodium vinyl sulfonate (0.5–3 %), and N-methylolacrylamide (0.5–3 %), by latex polymerisation, with stepwise addition of the mixture, using free radical initiators [391]. Other latices for contact adhesives have also been prepared by stepwise polymerisation using radical initiators from a mixture of

C_{2-12} carboxylic acid vinyl ester   40–70 %
C_{4-12} acrylic acid ester   10–40 %

with other monomers, including crosslinking agents such as N-methylolacrylamide and hydroxyethyl methacrylate, and also 2-ethylhexyl acrylate, acrylamide, acrylic acid and vinyl acetate.

Chlorosulfonated olefin resin latices (typically from polypropylene), together with toluene, have been post-emulsified in water with triethylamine, and a coalescing agent can also form stable systems suitable for use as adhesives [392]. Another latex, which may also be considered as a pressure-sensitive adhesive, has been obtained from 2-ethylhexyl acrylate, with added crosslinking monomer and unsaturated acid. The latex is formed by addition of a pre-emulsion to the water phase (see Table 8.42). A conventional persulfate initiator is employed, whilst the monomer mixture is added to the water phase at 85 °C for 3.5 h, with a 10 % aqueous solution of ammonia added at the same time. This polymerisation is maintained for 1.5 h, to yield a 52 % latex, pH 2.1, stable for >3 months, which can be coated on to stainless steel at 150 g m^{-2} and then bonded to a canvas fabric by roller pressing [393].

5.3.1 Some commercial formulations

Adhesive tapes may be prepared with a water-based formulation by using a precoat as a barrier coat, of a mixture of equal parts of an acrylic latex, a styrene–butadiene latex and a natural rubber latex, before applying a 'conventional' rubber–resin adhesive [394]. A relatively simple formulation for a

Table 8.42 Latex used as a pressure-sensitive adhesive

	Pre-emulsion (g)	*Water phase (g)*
2-Ethylhexyl acrylate	375	—
Styrene	125	—
Divinyl benzene	0.4	—
Na dodecylbenzene sulfonate	4	1
Water	250	180
Na pyrophosphate	—	2
Itaconic acid	—	2

pressure-sensitive adhesive latex employs a copolymer of butyl acrylate (80 parts) with styrene (18.5 parts) and itaconic acid (1.5 parts), using Na dodecylbenzene sulfonate as emulsifier and $Na_2S_2O_8$ as initiator (3 hr/85°) to give a latex copolymer with a T_g of -34 °C which is thickened with Na polyacrylate [395].

A general formulation suggested for an economical chloroprene-based contact adhesive includes a polychloroprene latex–polyvinyl acetate latex (M_w of 500–20 000)–terpene or petroleum resin (as tackifier) (100 : 10–100 : 20–200) [396]. An acrylic polymer, $T_g \sim 0°$, provides the maximum 'green strength' for a contact adhesive. A modification of a standard composition, based on a polychloroprene latex, includes a metal hydroxide. The method proposed involves formation of a latex mixture [397] from the components given in Table 8.43.

Another modification of the polychloroprene type of adhesive depends on preparing an emulsion of a rosin phenol in toluene, with water and a surfactant of a polyoxyethylene alkylphenol ether (50 : 25 : 30 : 5), which is mixed with a carboxylated polychlorprene latex (Neoprene® 115), aqueous ammonia (28 %), ZnO and water (200 : 3 : 3 : 4), yielding a water-dispersed contact adhesive (see reference [255]). One wooden face has been coated with anionic polychloroprene latex (with added ZnO and a rosin ester dispersion as tackifier) and the other face with a cationically charged polychloroprene. Bonding occurs by coagulation on contact, with development of bond strength over 1 h [398]. Adhesive bond strength increases rapidly over 10 days when the wood surfaces are bonded with a blend of a polychloroprene latex (62 % solids) with addition of 5 % of a 85 % starch phosphate [399].

Phenolic resins are most effective as cotackifiers in latex form. Even water can be considered to act as a cotackifier. Some compatibility between the phenoplast and the acrylic polymer is desirable [400]. As an example of this approach, a contact adhesive with good tack, peel and shear strength can be prepared from a copolymer latex of 2-ethylhexyl acrylate, (meth)acrylic acid, acrylonitrile and vinyl acetate (70–90 : 0.5–3 : 1–5 : 4–12), by the normal method of delayed monomer addition with redox initiation [401]. Alternatively, a latex from ethyl acrylate, methyl methacrylate and acrylic acid, with

Table 8.43 Polychloroprene-based latex

Dimerized rosin	50 g
Toluene	25 g
Polyoxyethylene alkylphenyl ether	5 g
This emulsion is added to:	
Carboxylated polychloroprene latex	200 g
Ammonia (with an NH_3–COOH ratio $> 1 : 5$)	3 g
ZnO (dispersed in water)	3 g

polyvinyl alcohol as the protective colloid, has been blended as

Copolymer latex, as above	188.7
Hydroxyalkylated alkylphenol phenoplast (70 % in xylene)	71.5
Octyl phenoxyphenol	25
Fungicide	0.3
Water	25

The mixture is adjusted to pH 5.9–6.0, with dilute ammonia solution [402].

Vinylidene chloride copolymers have been claimed as components of several contact adhesives. A blend (76 : 24) of a copolymer of vinylidene chloride, acrylic acid: 2-ethylhexyl acrylate (2 : 70 : 28) (copolymer A) with another crystalline copolymer, typically of vinylidene chloride, 2-ethylhexyl acrylate and vinyl chloride with ~84 parts of vinylidene chloride and 2-acrylamidopropane sulfonic acid as the internal latex stabiliser, (T_g of 0–30 °C), form a contact adhesive. A related blend consists of equal parts of copolymer A (T_g of 50 °C), with another copolymer of vinylidene chloride, butyl acrylate and 2-ethylhexyl acrylate [403]. Chloroprene has been copolymerised with an unsaturated carboxylic acid, typically methacrylic acid, in the present of a glycol ether, such as 3-methyl-3-methoxy-1-butanol, with polyvinyl alcohol as the protective colloid, forming an adhesive for bonding canvas to canvas [404].

Another blend consists of an amorphous copolymer of vinylidene chloride (with T_g of 50 to <0 °C and M_w 1–7 × 10^5) and acrylic ester copolymer, (T_g of 0–80 °C) and a proprietary tackifier with a plasticizing monomer [405]. A typical adhesive, (T_g of 12 °C) is based on a latex, (T_g of 20 °C) from butyl acrylate, methyl methacrylate and acrylic acid, with the T_g reduced to −12 °C by addition of a proprietary tackifier [406].

A typical pressure-sensitive contact adhesive employing a diene latex is prepared from a mixture of butadiene, styrene and itaconic acid (50 : 48 : 2) with Picconal A800E, a modified wood rosin-based (abietic acid type) technical tackifier [407]. Another diene-based formulation employs an isoprene–methacrylic acid (97 : 3) copolymer, prepared using a polyvinyl alcohol stabiliser, with pH adjusted to 8, mixed with 5 % of a 55 % latex of an α-methylstyrene–vinyltoluene copolymer. The product, coated on to polyester, acts as a contact adhesive [408, 409].

Dresinol 155 (a proprietary rosin ester (40 % emulsion) used as a tackifier) has been mixed with a butadiene–styrene–vinylpyridine latex (Nipol 2518FS) (see Section 7.2) and an acrylic latex (Primal® LC-40) (55 % solids, T_g of 9 °C) to give an adhesive that can bond ethylene–propylene copolymer to slate, canvas to iron sheet and polypropylene to itself [410]. The use of a wood-based abietic-type resin, or a derivative, soluble in the monomers, avoids the need for a solvent for the resin. A high-strength coating for polyester film has been based on a butadiene–styrene–acrylic acid–Balsam Resin WVX (6 : 3.8 : 2 : 1) blend [411].

572   *Polymer latices in the formulation of adhesives*

A polychloroprene latex with a small amount of methacrylic acid comonomer has been blended with 10–40 parts (on solids) of a low molecular weight epoxide resin and then with a polyamide (50–150 % of the amount needed to cure the epoxide), a rosin-type tackifier and ZnO, to form a composition that will bond wood to aluminium. The addition of the epoxide–polyamide improves the tensile shear strength by 2–3 times at ambient temperature and at 82 °C, and also improves the strength after 16 h immersion in water [412]. An ethylene–vinyl acetate copolymer latex has been claimed as a contact adhesive for two polyvinyl chloride sheets, but pressure is required to improve the bond strength [413].

5.4 Heat-seal adhesives

Heat-sealing using latex adhesives occurs when the dried adhesive film forms a rapid bond between coated surfaces when minimal pressure is applied. The heat-seal temperature depends on several factors, including the nature of the polymer latex, the plasticisation (internal or external), additives (including stabilisers) and the molecular weight of the polymers employed. Use of an external plasticiser usually means a lowering of the heat-seal temperature. Addition of polyvinyl alcohol as a stabiliser tends to increase the heat-seal temperature, but, since some grades are water-sensitive, also increases the tendency for the coated film to block under conditions of high humidity. This effect is minimal when cellulosic-based latices are employed.

5.4.1 General

Latex copolymers have many applications as heat-seal adhesives, including the precoating of papers, labels, films, textiles, etc., for subsequent bonding by heat-sealing to another material. Latex-based heat-seal adhesives are often based on polyvinyl acetate homo- and copolymer latices, with subsequent formulation for particular applications. With some substrates, storage conditions of coated products can present some difficulties [414]. In these cases, non-latex systems based on filled hot-melt adhesives are alternatives.

5.4.2 Ethylene–vinyl acetate latex copolymers

Ethylene–vinyl acetate copolymer latices, formed under high pressure with additional monomers, can also be formulated into pressure-sensitive adhesives. Such latices may contain minor proportions of acrylic, maleic or other unsaturated esters. The resulting adhesives can combine the principal desirable features of those obtained from both acrylic copolymer and tackified rubber latices [415]:

(a) high initial grab,
(b) good resistance to oxidation and discoloration,

Latex-based adhesives with specific functions 573

(c) clarity,
(d) adhesion to low-energy surfaces,
(e) compatibility with plasticised polyvinyl chloride surfaces.

A strong heat-seal latex for biaxially stretched polypropylene can be prepared by the emulsification *in situ* of a polyvinyl alcohol solution with a mixture of a solid ethylene–vinyl acetate resin with chlorinated polypropylene, paraffin wax and stearamide, followed by addition of polyethyleneimine [416].

Other applications of related copolymer systems are discussed in Sections 5.1.2.8 and 5.1.2.9.

5.4.3 Acrylate latex copolymers

A series of four-component latices for pressure-sensitive adhesives have been formed from vinyl acetate–2-ethylhexyl acrylate and bis-(2-ethylhexyl)maleate (51 : 69 : 17) under ethylene pressure (60 bar), using hydroxyethyl cellulose as protective colloid, with a persulfate/Na formaldehyde sulfoxylate system as initiator. The resulting products can be formulated into useful adhesives for steel [417].

5.5 Delayed tack adhesives

Delayed tack adhesives coated on to substrates are tack free at ambient temperatures but can be activated by heat to a tacky state, which can last from a few minutes to several days. The basic principle is that the adhesive formulation includes a solids crystalline plasticiser, with only limited plasticising action at ambient temperatures. Upon thermal activation, the plasticiser dissolves in the polymer component, exerting a normal softening and tackifying effect on the adhesive, which may be 'supercooled' and then last for a significant time. The principal solid plasticisers used include:

(a) diphenyl phthalate,
(b) dicyclohexyl phthalate,
(c) dimethyl cyclohexyl phthalate,
(d) aromatic sulfonamide derivatives.

These adhesives are widely used in heat-sealable label systems. Several examples are given in the First Edition of this work.

5.5.1 Some commercial formulations

Some more recently suggested formulations for delayed tack adhesives include a range reported [418] to give good adhesion to paper, good retention and antiblocking properties, based on an ethylene–vinyl acetate latex (details not stated in the original, but see Section 4.4.2):

Ethylene–vinyl acetate latex	184 parts
Dicyclohexyl phthalate	316 parts
Maleic anhydride resin solution	53 parts
Colloidl silica (antiblocking agent)	160 parts

Other heat-sensitive delayed tack adhesives with good adhesion, bond strength and antiblocking properties have been prepared using the formulation [419]:

Styrene–butadiene latex	184 parts
Dicyclohexyl phthalate (silica coated)	348 parts
Maleic acid rosin solution	50 parts
Rosin ester dispersion	158 parts

5.6 Remoistenable adhesives

Remoistenable adhesives are used on envelopes, postage stamps and some labels. They represent a significant part of the use of water-sensitive coating systems, although their range of application is probably being eroded by the growth in pressure-sensitive systems. It is necessary that the dry composition, as coated on to the substrate (usually paper), is completely free from tack, but will regain adhesive properties on moistening.

5.6.1 General

Originally most remoistenable adhesives were based on carbohydrate polymers, notably natural gums such as gum arabic (also known as gum acacia) and some dextrins. However, these gums can have difficulties with supply and specification, and synthetic latices and water-soluble polymers are now important components of remoistenable adhesives, especially as latex-based remoistenable adhesives provide improved paper-to-paper bond strength, so that it is difficult to separate the bond without paper tearing taking place.

5.6.2 Some commercial formulations

A simple removable adhesive has been prepared from a stable polyvinyl alcohol-based polyvinyl acetate latex, to which is added 5 % of $CaCl_2$ solution, which allows the resulting adhesive films to be removed readily by addition of water. Without the added salts, removal requires extensive rubbing [420].

Copolymers of acrylamide with acrylic esters (especially ethyl acrylate and/or vinyl acetate) including a non-ionic surfactant can form remoistenable adhesives for wallpaper. A dry coating of 10 g m^{-2}, remoistened with

30 g m^{-2} of water, gave a high peel strength and had a satisfactory ageing performance. Use of a copolymer latex prevents the paper curling which occurs when a polyacrylamide alone is used [421]. Other *wallpaper adhesives* with low water permeability can be based on 1 : 1 blends of ethylene–vinyl acetate–vinyl chloride terpolymer latices with ethylene–vinyl acetate latices. These may be coated at 200 g m^{-2} on to the paper, forming a film with water permeability reduced by 10 times [422]. Ethylene–vinyl acetate (1 : 4) latices containing polyvinyl alcohol can form remoistenable adhesives for bonding PVC to wood, with satisfactory peel strength. Copolymer latices of vinyl acetate with Veova® (vinyl versatate) comonomers are also suitable [423].

5.7 Quick-tack and contact-grab adhesives

Many different types of 'soft' copolymer latices (see above, especially acrylic latices in Section 4.4) and polychloroprenes (see Section 4.5) have been used (with suitable additives) as the basis of quick-tack (or quick grab) adhesives. The green strength of these adhesives, when used, for example, to bond wood in the form of plywood, depends on the contact area of the bond, the cohesive strength of the polymer and the method of application (bonding pressure, bonding temperature and time) [424].

5.7.1 Some commercial formulations

Variations on the general principles of adhesive formulation of latex adhesives have been suggested, including:

(a) An ethylene–vinyl acetate copolymer latex (49–50 % solids, with polyvinyl alcohol stabiliser) has been mixed with a *p-tert*-butylphenol-formaldehyde (20 : 3) to form a cloth laminating adhesive, with a high initial grab [425].

(b) A mixture of a self-crosslinking acrylic dispersion, a phenolic resin dispersion and a self-crosslinking polyurethane dispersion forms a water-based contact adhesive with extended tack retention time [426].

(c) A copolymer latex, prepared by latex polymerisation of isoprene and $CH_2=CH.CO.NH.C(CH_3)_2.CH_2.SO_3Na$ ('AMPS') (32.5 : 1) at 45 °C, mixed in equal proportions with petroleum resin, has been cast on to polyester film and then fused for 2 min at 100 °C, forming a contact adhesive with good 'quick stick' at 180 °C peel strength [427].

(d) A contact adhesive has been prepared from a 90 : 10 mixture of an acrylic copolymer, with a T_g from -60 to -30 °C (typically obtained by latex polymerisation of 2-ethylhexyl acrylate, butyl acrylate, styrene, methacrylic acid and glycidyl methacrylate (50 : 25 : 25 : 3 : 4) to form a 57 % solids latex, of viscosity 2.8 P, T_g of -46 °C) and a rosin ester with a softening point of 70–130 °C [428].

(e) Pressure-sensitive adhesives, which are tack-free to the touch but form a permanent bond under moderate pressure, can be formed from a blend of a tack latex (which provides the adhesive/sealing bond), a non-tacky latex (which gives non-blocking properties) and a filler (which absorbs oil, which can impair bonding). A typical formulation employs a blend of commercially available latices:

Robond PS-60 (acrylic latex)	56.6
Cab-O-Sil 5M (silica filler)	8
Latex 620 (Dow) (styrene–butadiene)	24

Applied to copying paper, at 10–12 nm (dry thickness), this blend form a tack-free coating, which bonds well under pressure to uncoated paper [429].

5.8 Redispersible components for pressure-sensitive adhesives

Redispersible water-based pressure-sensitive adhesives, which can be used in the production of recyclable paper and paperboard, can be obtained by latex copolymerisation of unsaturated acids with alkyl (meth)acrylates and other monomers. The acidic monomer is neutralised either at the start of or during polymerisation, which is carried out in the presence of a polyalkylene oxide plasticiser ($M_w > 3000$). A typical latex has been prepared from 2-ethylhexyl acrylate, vinyl acetate, mono-octyl maleate and acrylic acid in the presence of polyoxyethylene (PEG 8000) ($M_w \sim 3000$) [430, 431].

Inherently tacky water- or alkali-dispersible pressure-sensitive adhesives employ an alkyl acrylate copolymer with <20 % of acrylic and/or methacrylic acid, with a T_g of −15 to +150 °C, prepared with sufficient chain transfer agent to allow the polymer to be repulpable, with the recovered paper fibre substantially free from adhesive, suggesting that a low polymer M_w is necessary. Preferred latex compositions include 85.5 % of a latex of 2-ethylhexyl acrylate, vinyl acetate, ethyl acrylate, acrylic acid and methacrylic acid (43 : 12.5 : 12.5 : 13.5 : 4.5), blended with 4.5 % of a copolymer of 2-ethylhexyl acrylate, butyl acrylate, methacrylic acid and itaconic acid and 10 % of rosin ester as the tackifying agent, with n-butyl mercaptan as the chain transfer agent [432].

Redispersible powders for water-borne pressure-sensitive adhesives (mainly for reclaimable paper and board) have been prepared similarly, typically from a latex obtained by polymerisation of 2-ethylhexyl acrylate, vinyl acetate, mono-octyl maleate and acrylic acid in the presence of polyoxyethylene 8000 (molecular weight > 3000). The acidic component may be neutralised before, during or after polymerisation.

Other repulpable adhesive compositions (for pressure-sensitive systems and for laminating and ceramic tile fixing) also comprise two copolymer latices, with differing acidic contents. Typically, one copolymer latex (butyl acrylate and methacrylic acid), with a low acid content, has been blended with a second copolymer (butyl acrylate, methyl methacrylate and methacrylic acid) of a higher acidic content. The second latex has been used as a 'seed' for the polymerisation of the monomers of the first latex at 85 °C, using a *tert*-butyl hydroperoxide/Na formaldehyde sulfoxylate initiator system, yielding a mixture with good peel strength and good repulpability [433].

6 ADHESIVES FOR SPECIFIC APPLICATIONS

Previous sections of this chapter have mainly described latices in terms of their polymer composition. This section discusses latex adhesives in relation to their required function in bonding specific substrates and types of adhesion.

6.1 Wood adhesives

Many formulated latices, of many types, have been considered and used as wood adhesives [434]. The chemical nature and morphology of wood is outside the scope of this work, but it is convenient to note that wood is a blend of cellulose and lignin (a mixture of condensed phenolic compounds). Wood is slightly hydrophilic and is readily swollen by water, which is, sometimes, the cause of warping of timber articles. The adhesive strength depends on specific adhesion, the result of intermolecular action between van der Waals' forces of the wood and adhesives, and on the polarity. Blocking polar groups in cellulose of wood surfaces by acetylation of chlorosilanes reduces adhesion of the resin (by ~70–80 %). In bonding wood, the strength of the adhesive bond should be comparable with or greater than that of the timber. Adhesion of single-component adhesives and two-component systems, including chromium nitrate $Cr(NO_3)_3$, causes a slight increase in the MFT. Fillers such as $CaCO_3$ and $CaSO_4$ also influence the time and temperature of adhesion [435]. The most widely used water-based systems are those based on vinyl acetate polymer and copolymer latices, stabilised with polyvinyl alcohol and with a fairly large particle size (~1 µm). Small particle size latices, with a high surfactant content, tend to be absorbed into the porous wood substrate (especially softwoods) with loss of adhesion. High M_w is desirable to reduce the tendency to 'creep', which becomes more significant if the products are plasticised. Adhesion is low or negligible below the T_g, irrespective of type. If partly hydrolysed polyvinyl alcohol is present, this gives adhesion in its own right. Latices made with polyvinyl alcohol as stabilization are much more effective in wood adhesives than those made with other stabilisers. Since the polyvinyl alcohol of the colloid base is subjected to some degree of grafting by the monomer, the resulting grafted product may have the optimum hydrophilic–hydrophobic balance to give the maximum surface interaction, increasing both bond strength and film strength. Variations in the method of manufacture of polyvinyl alcohol (which will affect the degree and type of grafting) can significantly affect the performance of polyvinyl acetate latex adhesives into which they are incorporated. It has been claimed that a polyvinyl acetate latex, made with polyvinyl alcohol with 90 % of $M_w < 10\,000$, and the remainder of 3000–4000, gives optimum stability and performance in wood adhesives [436]. In practice, polyvinyl alcohols of nominally identical viscosity in aqueous solution, degree of polymerisation and degree of hydrolysis, from different manufacturers, can affect the properties of the resulting polyvinyl acetate latices. The detailed preparation of a wood adhesive based on a graft copolymer system is described in Section 6.1.2.

The correct level of plasticiser addition is also significant. If the dibutyl phthalate content of a polyvinyl acetate latex-based wood adhesive is increased from 0 to 15 %, the strain values of the adhesive layer (at a constant rate of increase of tangential stress) increases; this also increases with temperature from 20 to 60 °C. A typical plasticiser content is 3–5 % of dibutyl phthalate [437]. Compatibility is improved by the presence of a good surfactant system, since, where the surfaces of the latex particle are saturated, the emulsifier/stabiliser system must make the first contact with the substrate. If the operating temperature is $< T_g$ of the polymer, the latter will be absorbed on to the substrate [438]. 'Wetting' between the adhesive and the substrate is important; there is some evidence that a monolayer of water is desirable for good bonding between the wood and a polymeric adhesive.

Fillers are added to wood adhesive compositions to improve creep resistance and to reduce cost. Simple wood adhesives based on polyvinyl alcohol are not suitable for bonding wood joints that are in constant contact with water [439].

The viscosity of technical wood adhesives is usually ~5–20 mPa s, but such adhesives may be diluted, or a higher viscosity used, to suit particular application machinery. A good wood adhesive can form a bond with greater tensile strength that of the wood itself. Some wood adhesive systems for laminating adhesives are designed for bonding wood substrates to printed or embossed papers, for decorative laminates, which do not involve a wood-to-wood bond but a wood-to-paper or plastic film (usually plasticised polyvinyl chloride). Such systems are discussed in several of the following sections, and one dependent on the chemical systems involved.

Water-resistant wood adhesives have been prepared using crosslinking vinyl acetate copolymer-based latices. Typically, a vinyl acetate copolymer latex is prepared with 7.7 % of acrylamide(on vinyl acetate), by redox polymerisation at ~40 °C with *tert*-butyl hydroperoxide and sodium bisulfite, with a polyoxyethylene phosphate surfactant. N-Methylolacrylamide may be used instead of acrylamide, thus allowing additional crosslinking [440]. Polyvinyl alcohol may be used as the stabiliser, with acidic crosslinking with $AlCl_3$ [441]. The system may be post-thickened by addition of methyl cellulose [442]. Typically,

Adhesives for specific applications 579

a wood adhesive can be prepared by reacting [443]

Vinyl acetate	100
Isobam 304 (isobutene–maleic anhydride copolymer)	16
Polyvinyl alcohol (3–6 cP, 87–89 % hydrolysed)	4
H_2O_2	0.45
Tartaric acid	0.3
Water	120.7

to yield a latex with a viscosity of 160 P at 30 °C, which is useful as a wood adhesive and also as an adhesive in the production of paper tubes.

A range of polyvinyl acetate latex-based wood adhesives has been developed which incorporate several of the above features [444]. In these adhesives, the principal additives, apart from partly hydrolysed polyvinyl alcohol (which provides wet tack) and plasticisers, are wood flour (as filler) and glyoxal (as the crosslinking and insolubilising agent, claimed to have a synergistic effect on each other and to improve the moisture resistance of the adhesive bond) and xanthan gum (a semi-synthetic water-soluble microbial (polysaccharide) gum used as thickener and viscosity-controlling agent, which controls the 'wet tack' and rheological properties, notably 'thixotropy', 'pseudoplasticity' or 'shear-thinning' of the adhesive). The presence of wood flour is claimed also to increase water resistance and to make the adhesive bond sandable and stainable when dry. A typical adhesive composition of this type is shown in Table 8.44.

Other crosslinkable compositions involving latex copolymers containing N-methylolacrylamide have been reported, including:

(a) Blends of 17 parts of a 50 % composition with 100 parts of a 50 % solids latex, copolymers of vinyl acetate with either N-methylolacrylamide

Table 8.44 Thixotropic polyvinyl acetate latex-based wood adhesive

	Parts by weight
Deionised water	54.6
Polyvinyl acetate	33.5
Polyvinyl alcohol (partly hydrolysed)	4.0
Wood flour	6.2
Kelzan-S[a]	0.5
Glyoxal[b]	0.04
Ethyl *para*-hydroxybenzoate (preservative)	0.1
Benzoflex 9–88[c]	0.2
Propylene glycol	0.6

[a] A dispersible grade of xanthan gum (Merck & Co).
[b] As 40 % aqueous solution.
[c] Dipropylene glycol benzoate (coalescing agent) (Velsicol Chemical Corp.) (see Section 4.1.2 above).

(96 : 4) or monocyanoethyl maleate (96.5 : 3.5) used to bond yellow birchwood veneers, showed marked improvement in retention of bond strength after immersion in water. The maleic half-ester copolymer gave tensile results better than the N-methylolacrylamide copolymer [445].

(b) A copolymer latex of vinyl acetate and N-methylolacrylamide, with $SnCl_4$ at 8 % w/w and a filler, showed improvement in adhesion of beech wood compared to polyvinyl acetate alone [446].

Other examples of the preparation of crosslinking vinyl acetate copolymer latices are mentioned in Section 4.2.2.

Many latex-based wood adhesives are based on mixtures or modifications of latices, including:

(a) A technical acrylic latex blended with an ethylene and vinyl acetate (∼2 : 1), with ∼7 % of dibutyl phthalate, forms a high-strength adhesive [447, 448]. A latex made from vinyl acetate and crotonic acid (360 : 19) has been saponified with sodium hydroxide to form a modified polyvinyl alcohol (presumably with carboxyl groups attached to the main polymer chain). A small amount (∼2 %) has been added to a urea-formaldehyde resin-based adhesive with $Al_2(SO_4)_3$ and $AlCl_3$ as acid catalysts, to form a laminating adhesive for plywood [449]. Details of the methods for the preparation of polyvinyl acetate–urea-formaldehyde resins have been described. Typically, 75 parts of a vinyl acetate copolymer latex (40 % solids) have been mixed with 25 parts of a urea-formaldehyde resin (50 % solids) [450].

(b) In another system, similar in concept but different in detail, a polyvinyl acetate homopolymer latex was mixed with a dispersion of an ethylene–acrylic acid copolymer and then grafted with methacrylic acid to yield a suitable adhesive for wood veneers [451].

(c) Carboxyl groups to improve adhesion may be introduced by coating plywood with an α-olefin–maleic anhydride copolymer latex, overlaid with a grid of polystyrene foam and then covered with a further plywood sheet coated with a polyvinyl acetate latex and heated to 100 °C. The bond is, presumably, also a heat-seal bond [452].

(d) Carboxylated latices based on diene monomers have also been used in wood adhesives. Typically, a 55 % latex of butadiene, styrene and methacrylic acid (35 : 63 : 2) (particle size ∼0.2 nm) mixed with a urea-formaldehyde aminoplast, wheat flour and water mixture (65 : 35 : 20 : 20) can bond lauan wood to walnut, with pressing at 110 °C [453].

(e) A similar latex, with filler, thickener and a modified polyamide (an epichlorhydrin: adipic acid: diethylene triamine condensate) is claimed to form a wood adhesive, also suitable for textile applications [454].

(f) A carboxylated butadiene–styrene latex (48 % solids), thickened with carboxymethyl cellulose and mixed with Al(OH)$_3$, an isocyanate (Desmodur® 44) and an epoxy resin, Epikote® 812 (100 : 0.3 : 40 : 3.5 : 1.5) gave an adhesive for wood veneers with good resistance to immersion in boiling water [455].

(g) A vinyl acetate–allyl carbamate (99.5 : 0.5) copolymer latex (53 % solids), prepared on a 5 % fine particle size 'seed' latex with redox initiation, and a partly hydrolysed polyvinyl alcohol stabiliser, with dibutyl phthalate (3 %) phenol (1.5 %) and 10 % (on latex weight) of 15 % aqueous polyvinyl alcohol, gave a wood adhesive suitable for maple-to-maple wood bonding [456].

(h) A wood-textured caulking compound can be made from a mixture (3 : 1) of an acrylic latex and wood flour (particle size of 140–600 μm), dispersed in 70 % of its own weight of mineral spirits (an aliphatic hydrocarbon mixture) to 'wet' the flour surface and form a free-flowing powder [457].

(i) Some 'two-part' latex-based wood adhesive systems have been suggested, especially where good water resistance of the adhesive is required. A polyvinyl acetate homopolymer latex, filled with talc or CaCO$_3$, may be coated on to plywood, followed by a coat of a urea aminoplast mixed with wood flour and ammonium chloride. When applied to substrates, this will bond a veneer, with pressing at 5 kg cm^{-2} at 120 °C for 1 min [458]. Similarly, wood adhesion is improved by mixing a polyvinyl acetate latex with an epoxy resin (Epikote 828) and a curing agent (based on the reaction product of Epikote 827 with an imidazoline derivative of linseed oil fatty acids) [459].

(j) Polyurethane systems have also been suggested. One surface of a plywood can be coated with a polyvinyl acetate latex, stabilised with a non-ionic surfactant, with dibutyl tin laurate as the preservative (which may now be considered to be unacceptable) and the other surface coated with a 4,4′-diphenylmethane di-isocyanate copolymer. After pressing and curing for 24 h, the bond will maintain high strength in water [460].

(k) Plywood sheet can be bonded to form a laminate with decorative paper by coating with polyvinyl acetate latex at 80–85 g m^{-2}, followed by drying at 130–140 °C for 20 s, with pressing. A urea–aminoplast and a pigment can also be added [461].

(l) An ethylene–vinyl acetate copolymer latex has been blended with polyvinyl alcohol and a water-soluble resorcinol–formaldehyde resin (see also Section 7 below) and neutralized to pH 7.0 to form an adhesive for plywood. Crosslinking can be increased by addition of glyoxal [462]. Good water resistance, possibly with some bond flexibility, may be expected from this system.

(m) Expandable thermoplastic microspheres (Dow XD 2378) (5 % w/w) have been included in an adhesive for plywood consisting of equal parts of a polyvinyl acetate latex (40 % solids) and a melamine-formaldehyde resin solution (60 % solids), presumably to control the rheological behaviour. The bond is resistant to water at 70 °C [463].

(n) An adhesive with good peel strength at low temperature can be prepared from [464]

Ethylene–vinyl acetate latex (51 % solids)	100 g
Dibutyl phthalate	14 g
Nitrile latex (46 % solids)	4.6 g

6.1.1 Polymer latices for wood adhesives

The preparation of a polyvinyl acetate latex with a polyvinyl acetate–acrylic acid grafted stabiliser is described in Table 8.45. The presence of a graft copolymer improves the latex cohesion; a crosslinking glycidyl ether significantly improves adhesion after immersion in water [465].

Adhesives for bonding wooden floor structures have been based on 'core–shell' latices, with core $T_g < -10$ °C and shell $T_g > 50$ °C, prepared typically from a core formulation of 2-ethylhexyl acrylate, vinyl acetate and

Table 8.45 Adhesive formulation based on block graft copolymer latex

A. Latex stabilizer:	
Saponified polyvinyl acetate (—SH terminated)	10 g
Acrylic acid	10 g
Potassium bromate	0.1–0.2 % on monomer
Water	~150 g
The above mixture is reacted with a 14.4 % graft copolymer solution.	
B. Latex polymerization:	
Vinyl acetate	100 g
Graft copolymer solution (as above)	62.5 g
Water	45 g
Ammonium persulfate	0.1–0.2 % on monomer
The above is polymerised to 50–52 % latex.	
C. Final adhesive:	
Latex from B	100 g
Talc	50 g
Ethylene glycol diglycidyl ether	3 g

Adhesives for specific applications

acrylic acid (75 : 23 : 2) and a shell formulation of styrene and divinylbenzene (95 : 5), with a core–shell ratio of 12 : 8. The formulated adhesive, filled with $CaCO_3$, clay and talc, will bond wooden floors that are claimed not to squeak in use [466].

Adhesives for bonding wood have been based on vinyl acetate homo- and copolymers mixed with a variety of partly and fully hydrolysed polyvinyl alcohols as protective colloids. The polyvinyl alcohols employed are:

(a) partly hydrolysed (87–89 % saponified), degree of polymerisation 2000–2400 (also, notably, with a blocking rate of 0.46–0.50);
(b) fully hydrolysed (>96 % saponified), degree of polymerisation 1600–1800.

6.1.1.1 Modifications and additives

A minor problem with polyvinyl acetate latices used in wood adhesives is discoloration caused by traces of iron salts (used in the initiator systems) with tannic acid in the wood. This can be avoided by including salts whose anions form colourless complexes with ferric ions and with an unsaturated carboxylic acid. Such modified adhesives may be used for veneers and in plywood manufacture. Typically, a 60 % solids, polyvinyl alcohol-stabilized, polyvinyl acetate latex was blended with 0.5 % of sodium fluoride, crotonic acid (2 %) and $CaCO_3$ [467]. Discoloration and corrosion of processing equipment by polyvinyl acetate latex-based wood adhesives may be reduced by adding reductive discoloration inhibitors (e.g. sodium bisulfite or ethanolamine) into conventional adhesives [468].

Several modifications can be made to the basic latex in order to improve water resistance and to eliminate 'creep' as far as possible. Some of these involve 'building in' a small amount of crosslinking in the formulation, sufficient to improve the adhesive performance without significantly affecting the storage life of the compound. Amongst additives suggested have been the addition of 1.5 % of butyl glycollate $C_4H_9.O.CO.CH_2OH$ with 7 % of polyvinyl alcohol, so giving a suitable adhesive for beechwood laminate, which is resistant to boiling water [469, 470].

Polyvinyl acetate latices suitable for adhesives can be prepared by latex polymerisation of vinyl acetate in the presence of a complex salt of a Group IVB metal (such as Zn) with an α-hydroxy and/or an α-oxo acid (e.g. glyoxylic acid $H.CO.CO.OH$). The resulting low viscosity latex can be mixed with a hardener ($AlCl_3$) to give an adhesive with a long pot-life, which forms water-resistant bonds between wood and other porous substrates [471].

6.1.2 Aminoplasts and phenoplasts in wood adhesives

Wood adhesives have been prepared from mixtures of acrylic latices and thermoplastic or thermosetting resins, especially urea-formaldehyde condensates and phenoplasts [472, 473]. As an example, typical of many, polyvinyl acetate latex (30–40 parts) may be mixed with 60–70 parts of a thermally convertible phenol- or urea-formaldehyde resin, to form an adhesive for a fibrous backing for holding the sheets forming individual plies of plywood together before lamination. Because of its inherent resistance, the backing tape does not need to be removed [474]. An aminoplast has been blended with polyvinyl acetate latex to form an adhesive for laminating lauan wood at 170 kg m^{-2} pressure to paper with a simulated design [475]. A butadiene–styrene–α-methylstyrene dimer copolymer latex blended with a 9 : 1 urea-formaldehyde–melamine-formaldehyde resin mixture gave a bond strength with plywood of 15.4 kg cm^{-2}. Using an acrylic copolymer latex and a thermoplastic resin (Isobam 06), mixed (55 : 45) with the copolymer latex as before, an adhesive was obtained [476] with a bond strength of 19.5 kg cm^{-2}.

6.1.3 Glycidyl and epoxy additives

A wide range of glycidyl- and epoxy-containing monomers has been suggested as potential crosslinking agents for wood adhesives, with the general objective of improving the mechanical strength of the adhesive bond and its resistance to creep and to external moisture. Latex adhesives have been prepared from epoxy-containing monomers (0.3–20 %), aldehyde- or ketone-containing monomers (0.3–20 %) and other monomers, in the presence of water-soluble polyhydrazines and/or amines with active hydrogen atoms; e.g. styrene–methyl methacrylate–butyl acrylate–glycidyl methacrylate–diacetone acrylamide (110 : 100 : 255 : 20 : 15) latex with a protective colloid were polymerised to a 50 % solids latex, 100 parts of which, mixed with adipic dihydrazide (0.75), formed a wood adhesive which formed bonds with good compressive shear strength under dry test conditions. The bond strength was reduced by $\sim$50 % by immersion in water at 100 °C [477].

Use of hydrazine derivatives in similar systems for wood adhesives is discussed in Section 6.1.5 below. Glycidyl-type additives are also included in more complex systems. These are mentioned in later sections, especially Sections 6.1.5, 6.1.7 and 6.1.8 below.

6.1.4 Addition of metal complexes

Complexes or salts of some metals, notably of aluminium, chromium or zirconium, can be added to latex-based wood adhesives to form crosslinking bonds so that they improve adhesion to substrates and resistance to moisture. A general problem with the addition of chromium compounds is the development of colour, due to interactions with minor components of wood, possibly due to the presence of trace amounts of phenolics. This is not, normally, of major importance, except where the adhesive bond is displayed (as in some furniture bonds). There may be problems of stability with trivalent chromium components, which can be avoided in most cases by using non-ionic stabilisers, such as polyvinyl alcohols. Vinyl ester copolymer latices for wood laminate bonding have been developed, which include polyvalent metal salts (notably those based on zirconium), mixed with mercapto-group-containing vinyl acetate copolymers [478]. Typically, vinyl acetate and mercaptoacetic acid ($HS.CH_2.COOH$) are polymerised in methanol with azobisiso butyronitrile as initiator to produce a mercapto-terminated polyvinyl alcohol (d.p. 130) and then converted with acrylic acid to a graft copolymer. This copolymer is used as stabiliser for a persulfate-initiated latex polymerisation of vinyl acetate which was blended with talc and a Zr compound to give a wood adhesive with good adhesion and moisture resistance. Ethylene–vinyl acetate copolymers containing protective colloids of water-soluble polymers with added $ZrCl_2$ solution have been claimed to give satisfactory wood-to-wood bonds which can pass an immersion test [479].

The use of ammonium zirconium carbonate (as a colourless crosslinking agent) has been described [480]. Two separate adhesive phases are prepared, each of which is coated on to a single face of separate pieces of wood, which are then pressed together (10 min at 7 kg m^{-2}) and stored for 24 h, to form an adhesive bond that has twice the bond strength of a bond prepared from a polyvinyl acetate latex alone. The components are shown in Table 8.46.

In a more complex system, a vinyl acetate–acrylic acid (98 : 2) copolymer latex may be crosslinked by addition of a water-soluble aminoplast, with either a *p*-toluenesulfonic acid salt or a zirconium salt (immediately before application to the substrate), as given in Table 8.47.

Adhesion to Douglas fir is achieved using a coating thickness of 3.2 mm on each face, followed by high-pressure clamping and ageing for 7 days. As an alternative, a soluble zirconium complex (the reaction product of chlorohydroxyoxyzirconium as a 5 % additive) has been suggested [481]. Other alternatives include titanium or zirconium compounds, added with lactic acid and ammonia solution, for use as a plywood adhesive [482].

Table 8.46 Adhesive bond formed in two phases

Adhesive liquid A:	
Acetoacetylated polyvinyl alcohol	100 g
Polyvinyl acetate latex (50 % solids)	20 g
Na_2CO_3 (45 %) in water	2 g
Coating thickness	50 g m^{-2}
Adhesive liquid B:	
Acetaldehyde (20 % solution in water)	100 g
Coating thickness	5 g m^{-2}
Ammonium zirconium carbonate	2 g

Table 8.47 Crosslinking of a complex system

Vinyl acetate–acrylic acid (98 : 2) copolymer latex (50 % solids)	70 g
Water	3.1 g
Hexahydromethacrylamide	8.35 g
Methanol	9 g
Phenol–formaldehyde (1.13 : 2.24) resin (25 % solids)	9.55 g
Before application to the substrate, add:	
Zirconium nitrate (48 % aqueous)	0.022 g

The adhesive properties of acid-catalysed modified polyvinyl acetate latices containing phenolplasts are improved by microwave treatment at 2450 MHz for 200–500 W. To satisfy a standard adhesive test, 10 % of phenoplast is required compared with 20 % of phenoplast without it. For teak and maple wood, this irradiation improves the bond strength. $ZrOCl_2$ is recommended as a hardener, requiring longer irradiation than with HCl, but causing less wood tissue damage [483].

6.1.5 Addition of aziridines, oxiranes and related compounds

Other nitrogen-containing crosslinking functions have been introduced into the vinyl chains of latex polymers, including the aziridine group, which behaves rather like the oxirane group:

$$\begin{array}{cc} -\underset{\diagdown\;\diagup}{C-C}- & -\underset{\diagdown\;\diagup}{C-C}- \\ N & O \\ \text{Aziridine} & \text{Oxirane} \end{array}$$

A carboxyl-containing latex based on a terpolymer of ethyl acrylate, methyl methacrylate and methacrylic acid was reacted with an aziridine to form an adhesive useful for plywood. When aziridines such as ethyleneimine or propyleneimine are employed, the resulting product contains amino ester groups, typically

$$-(C=O).O.CH(R_4).CR_2R_3.NHR_1$$

where $R_1, R_2, R_3, R_4 = $ H or alkyl. In formation of these groups, the azidirine ring is broken, but the carboxyl group reacts with the C atom, which then ruptures the H bond [484]. Typically, a terpolymer latex (42 % solids) of ethyl acrylate, methyl methacrylate and methacrylic acid (57 : 39 : 4) ($M_w > 10^6$), prepared with Na lauryl sulfate as surfactant, is heated slowly to 50 °C for 30 min, held at this temperature for a further 30 min and then pH adjusted to 9.5, giving a viscosity of 2.4 cP at 24 °C. The amino N content was 0.8 %, indicating a 53 % reaction. Used in plywood lamination, the bond strength of the adhesive is about twice that of a similar adhesive in the absence of the imine reaction. However, short chain-hydrocarbon aziridines are toxic and should be treated with caution. Polyethyleneimine is believed to be less toxic, and the addition of 2 % on latex weight (50 % solids) improves adhesion of a plywood laminate.

Adhesive films for laminating silicon wafers, without corrosion, have been prepared using crosslinking latex polymers containing aziridine compounds. Typically, a latex copolymer from 2-ethylhexyl acrylate, methyl methacrylate, methacrylic acid and 2-hydroxyethyl acrylate, adjusted to pH 8.5, was treated with diphenylmethane-4,4'-bis(1-aziridinecarboxamide) and applied to a corona-discharge treated ethylene–vinyl acetate copolymer film. The adhesive coating was then bonded to a silicon wafer [485].

Polyfunctional azidirines are reported to be good temperature crosslinking agents for carboxylic group-containing latices. Such latices have been prepared by copolymerisation of 2-ethylhexyl acrylate with acrylic acid (110 : 3), with azobisiso butyronitrile as catalyst [486] and a crosslinking agent including aziridine groups. In another system, acrylic–polyurethane latex adhesives have been prepared from a mixture of an acrylic acid–N-butoxymethacrylamide-ethylene–vinyl acetate latex, a polyurethane latex (12.5), aziridine (1.4) and ZnO (3.6) [487]. Acrylic copolymer latices containing aldehyde or ketone groups and hydrazine derivatives have been developed as wood adhesives. A typical basic composition is

Methyl methacrylate	220
2-Ethylhexyl acrylate	168
Acrylic acid	8
Acrylamide	4
Diacetone acrylamide	13

which is polymerised in an aqueous solution of a terpolymer of acrylamide, acrylic acid and diacetone acrylamide (at pH 7–8). The resulting latex is blended with a (95 %) saponified copolymer of vinyl acetate and diacetone acrylamide (80 parts) and the hydrazide of an acrolein–acrylamide–acrylic acid–butyl acrylate–styrene copolymer (70). Wood bonds made with this adhesive showed good compressive shear strength initially, reduced after immersion in water and further reduced after immersion in boiling water [488]. Similarly, use of a mixture with toluene di-isocyanate (TDI) with 20 % of trimethylol trimethacrylate and a curing accelerator is also satisfactory. Use of the trimethylol compound gives a major improvement in wood-to-wood bond strength, especially when immersed in water [489].

6.1.6 Addition of isocyanates

Di-isocyanates have been used as crosslinking agents for several types of latex-based adhesives with other additives, as mentioned above. A simple system is based on the addition of a di-isocyanate to polyvinyl acetate latex mixed with wood powder, clay and flour to give a water-resistant adhesive for plywood [490]. A group of several patents describes the addition of polyisocyanates to different latices, including ethylene–vinyl acetate, styrene–butadiene, butadiene–methyl methacrylate and isoprene–styrene–hydroxyethyl acrylate–itaconic acid copolymer latices. The polyisocyanate (~15 % on polymer) is usually added in dispersion in a plasticiser. Fillers may also be incorporated [491].

A plywood adhesive can be prepared from a diene latex (typically a commercial styrene–butadiene or an isoprene–styrene (7 : 3) latex) or a polyvinyl acetate latex, with polyvinyl alcohol, $CaCO_3$ or wheat flour as filler, and added polyisocyanate dispersed in dibutyl phthalate. The resulting composition may be stabilised against gelling by addition of citric acid or another dibasic organic acid [492]. Plywood can also be coated with a moisture-sensitive polyurethane as a sealant, then with a polyvinyl acetate latex and, finally, with decorative paper. It is dried at 70 °C [493].

A wood adhesive has been prepared from a polyvinyl acetate latex with benzoyl peroxide (50 % paste in dibutyl phthalate) and an unsaturated isocyanate from 2,4-toluene di-isocyanate and 2-hydroxyethyl methacrylate. This mixture forms a water-resistant adhesive bond with birchwood [494]. Similarly, use of a toluene di-isocyanate and 20 % of trimethylol trimethacrylate with a cure accelerator has been suggested. Use of the trimethylol compound gives a major improvement in wood-to-wood bond strength, especially after water adhesion [495].

Crosslinkable adhesives for bonding wood laminates have been based on hydroxyl-containing copolymer latices (polyvinyl alcohol stabilised), crosslinked with a polyisocyanate, giving water-resistant bonds, typically from vinyl acetate, butyl acrylate and 2-hydroxyethyl acrylate (160 : 40 : 10), which are pigmented with $CaCO_3$ and then mixed with diphenylmethane di-isocyanate, the curing agent [496]. One-pack adhesives for bonding plywood to polyvinyl chloride sheet have been prepared from an ethylene–vinyl acetate copolymer latex (20–80 %) and an acrylic latex with a T_g from −20 to +20 °C (typically Desmocoll VPKA 8481) (5–50 %). The product has good bond strength at low temperature and good resistance to water and thermal creep [497].

Aqueous adhesives with good heat and water resistance can be prepared from mixtures of copolymer latices (I) with anionic (aqueous) polyurethane resins (II) and aliphatic polyisocyanates (III), with the solids ratio of the components being (I)/(II) = 0 : 10 to 30 : 70; (III)/[(I) + (II)] = 1 : 100 to 30 : 100 :

Sumikaflex 480 (copolymer latex)	64 parts by weight
Hydran HW 333 (anionic polyurethane)	36 parts
Water-soluble polyurethane thickener	2 parts
Toluene	5 parts
Coronate HX	3 parts

Adhesives for specific applications 589

This composition bonds polyvinyl chloride sheet to a plywood substrate. Water and heat resistances at 60 and 80 °C are good and adhesion at 5° is also good [498].

Latex copolymers, obtained by polymerisation of monomers in the presence of modified polvinyl alcohol (containing $CH_2{:}CH(CH_2)_nOH$ units, where $n = 2-20$), have been used as wood adhesives, blended with an isocyanate [499]. A typical latex has been prepared by latex polymerisation of vinyl acetate (100 parts by weight) in the presence of a modified polyvinyl alcohol (as above) containing 2 mole % of oct-7-en-1-ol, with a tartaric acid redox initiator, which is then blended with a partially saponified polyvinyl alcohol (35 parts), $CaCO_3$ (65 parts) and a polymethylene polyphenyl isocyanate (Millionate MR) (100 parts). The resulting wood adhesive, used to bond two wooden panels by pressing for 24 h at 24 °C, had an initial bond strength of 32 kg cm^{-2} and 210 and 105 kg cm^{-2} after ageing (7 days at 20 °C) and after immersion in boiling water respectively.

Water-borne adhesives with good open times have been obtained by blending a vinyl acetate–vinyl chloride–ethylene copolymer latex (46 % solids) with polyvinyl alcohol (Poval 217; 15 % aqueous) and $CaCO_3$ (40 : 40 : 20)(w/w) with diphenylmethane di-isocyanate (15 %) as the curing agent [500].

6.1.7 Diene copolymers with crosslinking

Most water-based adhesives consist of vinyl or acrylic polymer latices, but many patents have claimed the use of diene polymers, usually with additives assisting crosslinking, to improve bond strength and water resistance and to reduce creep. Typical of many examples is a blend of carboxylated diene copolymer latices, filled with $CaCO_3$ and potassium pyrophosphate (so adding a small proportion of oxygenated groups), giving an adhesive with good water resistance [501]. One wood face was coated with anionic polychloroprene latex (with added ZnO and a rosin ester dispersion) and the other face with a cationically charged polychloroprene latex. Bonding occurs by coagulation on contact, with the strength increasing over 1 h [502]. Adhesive strength also increases with longer times; it grows rapidly over 10 days, when wood surfaces are bonded with a blend of 62 % polychloroprene latex with 5 % addition of starch phosphate (85 %) [503]. An adhesive for lauan wood has been prepared from an isoprene–methacrylic acid–styrene (9 : 1 : 10) terpolymer latex, blended with a non-ionic surfactant, glycerol diglycidyl ether, $Ca(OH)_2$ and the Na salt of an isobutene–maleic acid copolymer [504].

A two-stage latex was first prepared by formation of a copolymer of vinyl acetate, ethyl acrylate and acrylic acid (49 : 49 : 2), with tert-dodecyl mercaptan (0.75) as the chain transfer agent (probably yielding a vinyl acetate-rich polymer with variable viscosity). A second stage follows, with a styrene–butadiene–acrylic acid (56 : 36 : 2) latex, also using *tert*-dodecyl mercaptan (0.72), to yield a latex of 33 % solids. This latex was blended with a 15 % solution of partly hydrolysed polyvinyl alcohol, giving an adhesive which bonded plywood to sailcloth with little reduction of strength on aqueous immersion [505].

A butadiene–styrene latex (including 3 % of acrylic acid) can be compounded with a 1,3-trimethylolpropane-2,4-toluenedi-isocyanate adduct, polyethyleneimine and flour. Alternatives include addition of a salt of an isobutene–maleic acid copolymer, $Ca(OH)_2$, urea and glycerol diglycidyl ether. The products are reported to have a high wet strength [506]. A butadiene–methyl methacrylate copolymer latex, blended with an alkaline resorcinol–formaldehyde-releasing crosslinking agent, has been suggested as an adhesive for birchwood, and also for cement [507]. Glycerol diglycidyl ethers have been found to give better results in crosslinking than bisphenol A. A mixture of a bis-(diglycidyl ether) of 2,2-p-phenylenepropane, with a polyamide and a carboxylated styrene–butadiene copolymer latex, forms a suitable adhesive for a wood veneer. Alternatively, an acrylic copolymer or an ethylene–maleic acid copolymer, with a filler, may be used [508].

A wood adhesive may be prepared from the reaction product of a polybutadiene ($M_w \sim 1650$) first reacted with maleic anhydride and then with an octylphenol resin and mixed with a styrene–butadiene latex and ZnO. Alternatively, a resorcinol–formaldehyde resin may be added [509].

6.1.8 Some commercial formulations

For both economic and technical reasons, ethylene–vinyl acetate copolymer latices are widely used in many adhesive formulations. Typical of many is a mixture of an ethylene–vinyl acetate latex (56 %) (4 parts) within aqueous acrylic acid–vinyl alcohol (3 : 17) graft copolymer (1 part), with water (1 part) and gypsum $CaSO_4.0.5H_2O$ (3 parts) for use as a wood laminating adhesive, which does not delaminate upon immersion in boiling water. If the graft copolymer is replaced by polyvinyl alcohol, delamination occurs. Hydrogen bonding and weak interaction between hydroxyl groups (including cellulosic hydroxyl groups and carboxyl (.COOH) groups in the acrylic graft copolymer) both improve water resistance [510]. Other ethylene–vinyl acetate copolymers, including a polyvinyl alcohol stabiliser prepared by redox initiation and plasticised with dipropylene glycol dibenzoate, can be formulated into adhesives suitable for laminating wood with plasticised PVC sheet. A related terpolymer including 1 per cent of crotonic acid can be formulated into a suitable adhesive for lauan wood [511].

The same wood can be bonded to rigid PVC with a blend of ethylene–vinyl acetate with 20 % (by weight) of either PVC or natural rubber latex [512]. Some ethylene–vinyl acetate copolymer latices, which contain 5–40 % of ethylene, with a mixture of 88 % hydrolysed polyvinyl alcohols and conventional surfactants, are suitable for lamination of wood to plastic products, with improved peel strength and reduced creep compared with similar latices [513]. Adhesives with good adhesion to hydrophobic surfaces have been prepared from mixtures of ethylene–vinyl acetate latex with a water-dispersible polyester–polyurethane, typically

>Ethylene–vinyl acetate (18 : 90) latex (56 % solids)   100 parts
>Ethylene–vinyl acetate–vinyl chloride (10 : 15–30 : 30–55) latex (50 % solids)   30 parts
>Polyester–polyurethane (40 % solids)

This system bonds flexible polyvinyl chloride to plywood, with good peel adhesion [514]. A highly branched, high molecular weight polyvinyl acetate latex (see Section 4.1), 3 % plasticised, was mixed with 15 % of a (commercial) vinyl methyl ether–maleic anhydride copolymer (which is water soluble, hydrolysing slowly to the acid form in solution). This addition is reported to improve the cold flow (creep) of wood adhesives, without reducing shear strength [515].

A mixture (200 : 60 : 104) of a commercial polyvinyl acetate latex with $CaCO_3$ and a 31 % ammoniacal solution of a terpolymer (2 : 1 : 1) of isobutene, maleic anhydride and maleimide, crosslinked with formaldehyde solution, can bond birchwood (after pressing for 30 s) to form bonds with compressive shear strengths of 12 and 153 kg cm^{-2} after 0 and 72 h respectively, compared with 2 and 25 kg cm^{-2} if the formaldehyde is omitted [516]. Copolymers of vinyl acetate and acrylic acid have been mixed with a range of amido acids, $CH_3.C_nH_{2n}.CO.NH.CH_2.COOH$ (where $n = 0$ to 17), and TiO_2 paste to form a wood adhesive [517].

A polyvinyl acetate latex (50 %) with 40 % poly(diallyl) phthalate latex, with added *tert*-butyl perbenzoate before application, gives a water-resistant adhesive, suitable for wood and plastics lamination. The two components may be grafted during application [518]. A related system employs a copolymer latex of vinyl acetate and tetra(allyloxy)ethane $(CH_2=CH.CH_2O)_4CH_2CH_2$ (88 : 12), plasticised with dibutyl phthalate (11 %) to form an adhesive which is suitable for bonding lauan wood panels, with high laminating strength. Alternatively, the comonomer may be *N*-methylolacrylamide [519].

Incorporation of silane-containing monomers, typically *m*- or *p*-$NH_2C_6H_4.O.CH_2CH_2CH_2.Si(OR)_3$ or $C_6H_5.NH.CH_2CH_2CH_2.Si(OR)_3$, with up to 80 % of plasticiser (on polymer weight) in polyvinyl acetate latices gave wood adhesives with improved moisture resistance of the adhesive bond after immersion, compared with using polyvinyl acetate alone [520].

A vinyl acetate–acrylic ester copolymer latex, with 12–25 % of a dibasic acid such as tartaric acid and 2.5–5 % of urea or, alternatively, an all-acrylic latex, can bond plywood to decorative paper [521]. Alternatively, lauan plywood is given a first coat of a filler latex, followed by a polyvinyl acetate latex pigmented with TiO_2, followed by pressing paper on to wood [522].

Addition of a silane-treated clay to a vinyl acetate–crotonic acid copolymer latex used for adhering decorative paper to plywood improves, as expected, the overall water resistance [523]. Similarly, plywood may be laminated to paper embossed with a tile pattern, using a polyvinyl acetate latex followed by application of a polyester cover [524].

Fillers may be added to polyvinyl acetate wood adhesives without a significant loss of bond strength, even with 20 % addition to a composition. Many fillers have been suggested, including a mineral silicate [$MgO.Al_2O_3.3SiO_2$]$_2 \cdot H_2O$ ('Plastoritz'), which is believed to react with carboxyl groups to form [525]

$$[MgO.Al_2O_3.3SiO_2][R.COO^-] + H$$

A relatively economical wood adhesive formulation for furniture has been prepared, using a non-ionic surfactant and a persulfate initiator from a vinyl acetate–di-isobutyl maleate copolymer (7 : 3) latex loaded with baryta [526]. This system may be subject to difficulties with creep behaviour of the joint.

Acrylic latices have also been suggested for wood adhesives. A dispersible acrylic latex including acrylic or itaconic acids (3–5 %) was spray-dried with added $CaCl_2$ or $Ca(OH)_2$ (27.5 %) and redispersed by addition of Na N-(hydroxyethyl)ethylene diamine triacetate (5.3 %) to form a 10 % dispersion. The product can be used as a wood adhesive and has good bond strength [527].

Heat- and water-resistant adhesives for bonding polyvinyl chloride sheet to wood have been based on homo- or copolymer latices (T_g of −50 to +100 °C), prepared in the presence of a mercapto-terminated polyvinyl alcohol (which will provide both protective colloid and chain-terminating activity) in a two-stage process, typically as given in Table 8.48. The pH of the Stage 1 components is adjusted to pH 3.5. On completion of this reaction stage, the Stage 2 reactants are added and the polymerisation carried out to completion; it is then adjusted to pH 6.8 with ammonia [528].

Fast curing two-component latex systems, with good storage stability, useful for bonding plywood, paper and construction board have been prepared using an isobutene–styrene–methyl methacrylate–acrylonitrile–acrylic acid–itaconic acid, with $CaCO_3$, a viscosity modifier and a polyepoxide. The second component is an aqueous solution of glyoxal and hexamethylenetetramine [529].

Table 8.48 Preparing heat- and water-resistant adhesives

	Stage 1	Stage 2
Modified polyvinyl alcohol	15	—
Water	280	—
Methyl methacrylate	50	100
Butyl acrylate	50	100
$(NH_4)_2S_2O_4$	(not stated)	—

6.2 Latex-based adhesives for paper bonding

Latex adhesives for paper-to-paper bonding are well established and can be designed for many different performance requirements, whilst related adhesives for packaging, a major industrial sector, are mainly designed for more demanding applications, since requirements include the ability to apply adhesive coatings by high-speed machinery and to form reliable bonds between paper, both coated and uncoated, and plastic substrates. They are, however, being partially replaced by solvent-free hot-melt adhesives for higher-speed applications.

6.2.1 General

For many purposes, simple adhesive formulation for paper-to-paper adhesion, with application properties superior to those of (lower cost) starch-based systems, can be based on polyvinyl acetate homopolymer latices, with added plasticiser and thickeners. For machine application, more complex formulations may be required. Latices for paper coating applications, with added fillers, represent a further major use of polymer latices. These are discussed in Chapter 13 in Volume 3.

Many configurations have been suggested. As a simple example, it is possible for an adhesive system for bonding paper to substrates to be arranged in two layers: an adhesive (e.g. a formulation based on a polyacrylate or a polyvinyl acetate latex) may be applied to the substrate and, as the upper layer, a coating of a polyvinyl alcohol solution in water, containing a wetting agent. This system is activated by water and is suitable as a remoistenable adhesive for wallpaper, labels, posters and similar substrates [530]. Alternatively, a polyvinyl acetate homopolymer latex (47 % solids) with a polyoxypropylene nonyl phenyl ether (4 %) as surfactant and partly hydrolysed polyvinyl alcohol as stabiliser, prepared using a hydrogen peroxide/tartrate redox initiator system, is suitable for impregnation of paper to give improved adhesion of the paper in contact with water [531]. Acrylic or methacrylic esters, sometimes copolymerised with other esters, with a high content of an unsaturated acid as an additional monomer, may be used combined with an aminoplast, to make a water-resistant adhesive for paper [532]. Adhesives for labels and paper tapes, with $T_g < -41$ °C have been claimed, based on formulations using a copolymer latex from 2-ethylhexyl acrylate, acrylic acid and α-methylstyrene (590 : 12 : 2) [533].

Vinyl polymer and copolymer latices have been used in some bookbinding applications, although there has been significant increase in the use of hot-melt adhesives (outside the scope of this work) in this field. Latex-based adhesives are not suitable for all bookbinding applications, especially where high-speed operation with rapid setting is required, but they are used in padding and general bookbinding operations where adhesion of paper to paper, paper to leather and paper to synthetic leather is required. The particles of polyvinyl acetate latex-type adhesives enter the pores of the paper structure through pores <5 μm in diameter, so they do not enter the paper fibrils but will bond the paper itself.

A highly stable plasticised polyvinyl acetate latex, preferably ethanol resistant, is useful as a bookbinding adhesive for 'traditional' methods. Alternatively, a polyvinyl acetate latex with excess polyvinyl alcohol and with carboxymethyl cellulose as thickener can be applied as a narrow strip along the paper edges before sewing [534]. Polyvinyl acetate latices have been used as the sole binder for short-life volumes, such as telephone directories, and for 'tear-off' stationery pads, but the former application, at least, has now largely been replaced by hot-melt adhesives.

6.2.2 Some commercial formulations

Several wallpaper adhesives have been based on mixtures of polyvinyl acetate converted to powder form (e.g. by spray drying), which can be reconstituted at the point of use before application [535, 536]. A wallpaper adhesive was obtained from an alkali-soluble vinyl acetate–crotonic acid copolymer latex which was spray-dried and blended with methyl cellulose [537]. Similarly, a re-emulsifiable powder including polyvinyl acetate, methyl cellulose modified with glyoxal and a small amount of Na carboxymethyl cellulose serves as a wallpaper adhesive with better adhesion than methyl cellulose alone [538]. Alternatively, in a liquid system, an alkali-soluble copolymer latex (57 % solids and viscosity at 13.8 Pa s at 25 °C) based on a vinyl acetate–dibutyl maleate–acrylic acid–crotonic acid latex (490 : 450 : 175 : 18.5), stabilised with a non-ionic surfactant and a maleic acid–methyl vinyl ether copolymer and used as a paper-to-paper adhesive, gave a high strength paper tear bond to gloss paper-to-uncoated paper bond [539]. Plasticised latices of a polyvinyl acetate or ethylene–vinyl acetate copolymer, prepared with partly hydrolysed polyvinyl alcohol and having a working viscosity of 1.5–5.5 P (Brookfield viscometer, 6–60 rpm), form paper adhesives suitable for high-speed operation. The operating speed of such adhesives can be markedly increased by formulation at increased solids content of the system, with the penalty of reduced latex stability under conditions of high shear [540]. Wallpaper adhesives have also been prepared from water-swellable ionic copolymers prepared by inverse emulsion polymerisation of acrylic acid and N,N-methylenebisacrylamide, initiated by α-cumyl peroxyneodecanoate with a surfactant and an organic solvent. The resulting adhesives have a good bond strength, paste-body and slip [541].

Several formulations which include ethylene–vinyl acetate latices have been suggested as paper adhesives. These include a terpolymer including butoxymethylacrylamide for laminating polyethylene to paper [542] and an ethylene–vinyl acetate latex, stabilised with polyvinyl alcohol and added starch and alum, forming an adhesive for corrugated paperboard, with good wet adhesion [543].

Adhesives for specific applications 595

Other latices may also be used in water-borne adhesives for paper [544]. Acrylic latices have occasionally been suggested, apart from their major application in pressure-sensitive systems (see Section 5.1 above). Such latices are suitable for paper sheets impregnated with a melamine aminoplast [545]. Acrylic latices have sometimes been blended with styrene–butadiene latices to form adhesives with good grab but poor long-term bonding performance [546]. Styrene–butadiene latices with stabilising additives have been used as a dispersant for bonding the edges of paper sheets on sets of paper with microencapsulated ink components in multilayer carbonless copying sets [547]. Polyvinylidene chloride copolymer latices, sometimes plasticised with additional acrylic latex and natural rubber latex and with added alkali-soluble protein (casein, soya protein or zein) (5–20 % dry weight on latex), have been suggested for the adhesion of paper to polyethylene [548, 549]. An ethylene–vinyl chloride–acrylamide copolymer latex has been suggested for bonding polyvinylidene chloride to paper [550].

Styrene–butadiene latices have also been mixed at 60 °C with tackifiers, surfactant, plasticiser and mixed solvents. A typical formulation with good adhesive properties and storage stability for lamination of paper tape to plastics is [551]

Styrene–butadiene latex (48 % solids)	600
Ester Gum S 100	100
Dialkyl sulfosuccinate	5
Toluene	10
n-Hexane	10
Ethanol	7
Dioctyl phthalate	3

6.3 Adhesives for packaging

The performance requirements for water-based packaging adhesives are diverse. Apart from the obvious requirements of the surfaces to be bonded (which, for most latex-based packaging adhesives, involves at least one cellulosic substrate), there are often other stringent requirements, dependent on the method of application, in terms of viscosity, mechanical stability open time and rate of development and final strength of the adhesive bond. In general, water-based adhesives are adequate (and usually economical) for cellulosic substrates, but have performance limitations in their bonding ability to synthetic surfaces, including films such as regenerated cellulose, polyolefins, polyvinyl and polyvinylidene chloride and their copolymers, and polyamide and polyester films, which together represent a major fraction of the use of films in packaging applications. Many gloss lacquers on paper and board are also suitable substrates for synthetic latex adhesives. The adhesive layer in a packaging application may also be required to withstand extremes of cold without cracking (e.g. in storage freezer applications) and heat without softening (e.g. under summer display conditions). Many adhesives are used in food packaging, and it is essential that otherwise technically suitable formulations also satisfy the appropriate food packaging regulations, especially those of the US Food & Drugs Administration (FDA), which are also often accepted as equivalent to International Standards. Latex-based adhesives are often slower in performance in the rate of development of bond strength after application than water-free 'hot-melt' adhesives, although usually lower in cost.

Adhesives based on polymers of natural origin are still widely used for paper-to-paper bonding, mainly for economic reasons; such adhesives are often compounded with some synthetic components. These include vegetable products such as starch and dextrin, several natural carbohydrate gums and animal products such as casein (the protein of milk) and animal glue and gelatine (from waste bone and skin). All these materials are water sensitive and are subject to microbiological attack, unless suitable preservative agents are included in the formulations. In addition, because of viscosity effects, formulated adhesives of natural origin have a relatively low solids content, so operating speeds tend to be relatively low, and the rate of development of strength of adhesive bonds is also low. (Although this can be advantageous, since it allows the positioning of the bond to be adjusted whilst it is being formed, flexibility of operation can be important in packaging assembly operations.)

Polymer and copolymer latices based on vinyl acetate are the basis of many principal types of packaging adhesives, replacing solvent adhesives in many applications. However, these have, in turn, been replaced in some important areas by hot-melt adhesives.

Externally plasticised polyvinyl acetate latex adhesives are useful in bonding synthetic coatings on to paper, since the plasticiser or solvent assist in giving specific adhesive behaviour. Thus, low-cost homopolymer latices, including significant amounts of chlorinated or aromatic solvent, are able to bond many hydrocarbon waxed barrier coatings, but not those with coating based on polyolefins or copolymers of olefins with vinyl acetate [552]. In some cases, there are migration troubles with plasticisers. However, for bonding regenerated cellulose (viscose), aluminium foil and paper, a suitable adhesive [553] can be prepared from a polyvinyl acetate latex, 55 % solids (typically, Gelva S-55) (100 parts), butyl benzyl phthalate (9 parts), Santolite MS-80 (13.5 parts) and water (13.4 parts) [554]. A carboxylic acid is sometimes a component of the latex formulation; this addition tends to improve the initial 'grab' for 'difficult' surfaces of the resulting formulated adhesive. Polyvinyl alcohol may be a major component, sometimes with the addition of small amounts of borax, for increasing 'tack' by controlled boration (see Section 4.1.3). In these cases, the polymer latices are a significant part of the composition. The many applications of latices in adhesives for packaging have been discussed [555–560]. Acrylic latices are widely used in adhesives for bonding synthetic polymers used in packaging, notably to aluminium foil, polyethylene terephthalate and polyethylene (including corona-discharge-treated polyethylene). Examples from Rohm & Haas Inc. are shown in Table 8.49 [561].

These data indicate, as expected, that polyvinyl acetate latices have comparatively poor adhesive performance, especially with hydrophobic forces, such as polyethylene. However, terpolymers including vinyl acetate are reported to be suitable adhesives for polyethylene and polypropylene, giving satisfactory tack, with high viscosity, sufficient to avoid excessive absorption into porous substrates and to reduce the edge extrusion of adhesive by laminating pressure. A formula for preparing a typical latex is described in Table 8.50 [562].

The water phase is heated to 58 °C, then 5 % of the monomer mixture is added over 5 min, followed by initiator A and the rest of the monomer over 3 h, finally adding the remainder of the monomer at 87–89 °C and heating for 30 min. The resulting latex (54–56 % solids) has a viscosity of 390–500 P (Brookfield No. 4, 60 rpm).

Other systems suggested include:

(a) Polyethylene film, for food packaging, may be given a polyvinylidene chloride layer (to improve barrier properties for organic vapours and oxygen) by coating first with a latex mixture of a styrene–butadiene copolymer latex with a minor amount of a latex copolymer with an acrylic ester and a major proportion of plasticising monomers [563].
(b) Latex copolymers of styrene, methacrylate esters and methacrylamide have been mixed with 50–800 parts of sodium silicate to prepare adhesives suitable for paper tube manufacture [564].

Table 8.49 Peel adhesion of laminated specimens for packaging

Latex type	Aluminium foil	Polyethylene terephthalate	Untreated polyethylene	Treated polyethylene
Rhoplex AC-22	950	73	14	44
Rhoplex AC-33	580	36	10	15
Rhoplex AC-34	1130	95	31	140
Rhoplex B-10[a]	1860	1810	100	350
Rhoplex B-15[a]	1450	1160	94	360
Rhoplex E-32[b]	1460	330	59	252
Rhoplex HA-8[b]	1810	290	62	290
Rhoplex K-3[b]	190[c]	270[c]	34	400
Polyvinyl acetate	100	36	8	16
Vinyl acetate–2-ethylhexyl acrylate copolymer	760	300	11	35

[a] Crosslinkable latex copolymer.
[b] Self-crosslinkable latex copolymer.
[c] Cohesive (not adhesive) failure.
All data measured by ASTM test method D 1876. Results in lbf (pounds force).

598   *Polymer latices in the formulation of adhesives*

Table 8.50 Latex copolymer formulation for use in adhesive for bonding to polyolefins

	Parts by weight	
Monomer phase		
Vinyl acetate	33.3	
di-2-Ethylhexyl maleate	33.3	
2-Ethylhexyl acrylate	33.3	
Aqueous phase		
Nonylphenol–ethylene oxide condensate (86 % EO)	1.0	
Alkylphenyl-ethylene oxide condensate, acid phosphate ester	2.0	
Alkyl-substituted starch ether	1.5	
Polyvinyl alcohol (75–80 % hydrolysed)	1.5	
Ammonium hydroxide	0.14	
Water	71.5	
Initiator	A	B
$K_2S_2O_8$	0.48	0.04
Water	8.40	0.4
Neutralising slurry		
$NaHCO_3$		0.25
Water		0.25

Table 8.51 Adhesive composition for corrugated board and cartons

	Parts by weight
Polyvinyl alcohol (partly hydrolysed)	3.0
Urea-formaldehyde resin (65 % solution)	0.65
Polyvinyl acetate latex (55 % solids)	27.5
2-Amino butanol hydrochloride	3.0
Zinc nitrate (as aqueous solution)	
Clay filler	4.0
Water	59.3

(c) Some crosslinking systems, notably aminoplasts, may be employed in packaging. Zinc nitrate may also be included in a formulation, stable for ∼3 months, suggested for bonding paperboard for corrugated cartons (Table 8.51) [565]. This composition is relatively expensive, compared with others used as adhesives for corrugated board, and could probably be diluted with additional clay, or with starch, to reduce the cost, at the expense of some loss of performance (in terms

Adhesives for specific applications

of bond strength and reduced moisture resistance). The mixture is heated to 90–93 °C and stirred to mix the ingredients. An aqueous adhesive with improved water resistance, used in the manufacture of corrugated paper board, employs an ammonia-based styrene–butadiene latex or a carboxylated styrene–butadiene latex and starch. Typically, a styrene–butadiene–itaconic acid latex (prepared at 90 °C in the presence of ammonia) was mixed with gelatinised starch and aqueous borax solution. Wet adhesion was improved significantly, compared with that obtained using NaOH [566]. Laminating adhesives for packaging, with good adhesion to plastic materials and wax coatings, have been prepared from copolymer latices, with T_g of −35 to −15 °C, of butyl acrylate, 2-ethylhexyl acrylate and/or vinyl acetate, in the presence of hydroxyethyl cellulose or polyvinyl alcohol as stabiliser and a chain transfer agent. They are free from organic cosolvents and are formulated with clay and an aqueous tackifier [567].

(d) Packaging adhesives with controlled gas barrier properties can be prepared from blends of a conventional acrylic adhesive with low barrier properties and a latex from chlorinated vinyl–acrylic copolymers with higher barrier properties. Control of the gas barrier properties between defined limits may be achieved by altering the proportions of the two latices. For laminating two polyethylene films a blend of latices may be employed. A laminate with an oxygen transmission rate of 113 ml per 100 in^2 per day can be prepared using a blend (90 : 10) of a copolymer latex of 2-ethylhexyl acrylate, hydroxyethyl acrylate and vinylidene chloride (30 : 10 : 60)(w/w) with a copolymer latex of acrylic acid, butyl acrylate, hydroxyethyl acrylate and vinyl acetate (6 : 71 : 6 : 17) which is used to bond two polyethylene films together [568].

6.4 Miscellaneous latex-based adhesives

Many adhesive formulations have been suggested for bonding different substrates, sometimes with multiple applications. In general, these suggestions have some common factors:

(a) The latices are designed to form 'soft' films (those with a low T_g), which are flexible and tend to bond well with hydrophobic substrates, but can flow, so have poor creep resistance.

(b) May include a proportion of acidic groups (which improve adhesion to so-called 'difficult' surfaces).

(c) Have some method of crosslinking (either chemically bound, by copolymerisation of suitable monomers, or by addition of crosslinkable components to the adhesive composition, at the expense of reducing the stability and storage life of the system).

In addition to these characteristics, which determine the final performance of the coated system, other 'intermediate' properties, such as latex viscosity and rheological properties, which affect the application of the adhesive as a film, are also considered. The adhesive systems mentioned in the following sections are, generally, designed to have a balance of the above properties (and also to take account of manufacturing costs and possible hazards in use), chosen so that they can achieve the performance required.

Some secondary aspects of formulation, affecting practical performance, should also be considered:

(a) *Corrosion* at adhesive surfaces may be minimised by addition of small amounts of sodium nitrite or sodium dichromate.

(b) *Foaming* in use may be controlled by standard defoamers (mainly proprietary mixtures) used in minimum amounts for satisfactory machine performance, so that 'fish-eyes' in perfect films can be avoided, since their occurrence may cause uneven adhesion.

(c) *Low setting speeds* (the rate of development of bond strength) tend to cause isolated bulges and wrinkles in a paper substrate, which may be overcome by reducing film thickness and using an adhesive with a higher solids content. In some cases, the type of resin in the latex used may be changed, since drying performance is mainly affected by the properties of the water phase.

6.4.1 Adhesives for organic substrates

An 'anchor' coat improving the adhesion of vinylidene coatings to polyamide films (previously corona-discharge treated) consists of a latex of ethyl acrylate, butyl acrylate, acrylic acid and N-methylolacrylamide (64 : 12 : 7 : 5), prepared conventionally with a non-ionic–anionic surfactant with added maleic acid (4 % on polymer) and $MgCl_2$ (2 %) [569]. Terpolymer latices of ethyl acrylate, vinyl acetate and sodium methacrylate (50 : 50 : 7) provide good bonds between wood and polyvinyl chloride [570]. Similar copolymers, with varied amounts of acrylic salt, allow acrylic and vinyl resins to bond to themselves and to wood and leather. Other adhesives for bonding leather have been developed using polyurethanes, prepared by the reaction of polymer polyols, di-isocyanates, chain extenders and dimer diols. A typical product has been prepared by reacting a dimer diol (100 parts) with adipic acid (180 parts) at 200 °C for 24 h, then heating 100 parts of the product with isophorone di-isocyanate (135 parts) and dimethylolpropionic acid (42 parts) in butan-2-one (850) (80–85 °C for 5 h), before adding triethylamine (2 parts) and water (removed by distillation) to form a 30 % solids polyurethane emulsion (viscosity of 1 P). This adhesive will also bond ABS resin to foamed polyurethane fabric [571] (see also Chapter 15 in Volume 3).

Heat- and moisture-resistant adhesives, for bonding polyolefin foams to construction boards employ ethylene–vinyl acetate copolymer latices (ethylene content 10–40 %), with 2–6 % of acetoacetylated polyvinyl alcohol, inorganic fillers and/or phenolic resin additives [572]. 2-Ethylhexyl acylate is the main component of a copolymer latex, which, with additional finely powdered polymethyl methacrylate and a solvent, forms a suitable adhesive for polyvinyl chloride-based wallpaper [573]. A vinyl acetate–methyl methacrylate copolymer latex, prepared using a redox initiator, bonds polyvinyl chloride to backing cloth, with a typical tensile strength of $\sim$140 kg m^{-2} [574]. Another system for bonding polyvinylidine chloride films is based on a partly crosslinked styrene–butadiene–itaconic acid terpolymer latex, mixed with 6 % (on latex solids) of a polyamine–amide obtained from ethylenediamine and methyl acrylate. This modified latex (35 % solids) can be used to improve the adhesion coat of a corona-discharge-treated polypropylene film, which is then laminated under pressure with a polyvinylidene chloride-coated polypropylene film. Use of the adhesion promoter increases the peel strength by more that 300 % [575]. Corona-discharge-treated polypropylene can also be treated with an adhesive mixture (90 : 10) of a methyl methacrylate–cyclohexyl methacrylate–acrylic acid copolymer latex with a polyvinyl chloride latex, which can then be heat-sealed [576].

Polyester film may be laminated to polyethylene with an adhesive composed of a vinyl acetate–ethyl acrylate–butyl acrylate–acrylic acid copolymer latex, with an Epon® 828 I (epoxy resin) latex (33.3 % solids), polyvinyl alcohol, a surfactant and water (80 : 60 : 35 : 4 : 1). Before application, aqueous tetraethylene pentamine (16.5 %) is added as a crosslinking agent for the epoxy resin. The acrylic acid in the copolymer improves bond strength, probably by a carboxyl–epoxide reaction [577]. Other laminating adhesives have been prepared from a styrene–butadiene acrylic acid latex (prepared with ammonium dodecylbenzenesulfonate as surfactant and ammonium persulfate as initiator), neutralised with NH$_4$OH at 34 % solids and then used to laminate metallized polyester film with an ethylene–vinyl acetate–Surlyn® coextruded film, to form a water-resistant laminate. A marked reduction in adhesive strength occurred when Na dodecyl sulfonated phenyl ether was used as the surfactant [578].

Adhesives for bonding fabrics, polyvinyl chloride sheet and aluminium have been prepared from ethylene–vinyl acetate–water-soluble comonomers, using a protective colloid system of polyvinyl alcohol (d.p. 100–600, M_w 11 000–31 000) and an ethoxylated alkyl phenol surfactant, as given in Table 8.52. The initiator system is based on H$_2$O$_2$/Na formaldehyde sulfoxylate (HO.CH$_2$.SO$_2$Na·2H$_2$O) [579]. An ethylene–vinyl acetate terpolymer, also including a polymerisable comonomer, formed with a 'seed' latex and stabilised with polyvinyl alcohol, can be thickened with alkali, giving

Table 8.52 Preparation of bonding adhesive

Initial stage	
Polyvinyl alcohol (partly hydrolysed) (10 % aq)	351
Polyvinyl alcohol (fully hydrolysed) (10 % aq)	234
Water	400
Ferrous ammonium sulfate (1 % aq)	4.8
Igepal CO 887 (non-ionic surfactant)	55.7
Vinyl acetate	1387.5
Acetic acid	1.3
Second stage	
Vinyl acetate	484.5
Acrylic acid	78
Ethylene	To 550 psi

good adhesion to flexible polyvinyl chloride with little creep in the resulting bonds [580].

Another adhesive system for bonding crystalline polymers (typically, injection moulded polypropylene test pieces to nylon-based non-woven fabric) is based on mixtures of copolymer latices with graft copolymers and organic solvents. A solution of a butyl acrylate-grafted chlorinated polypropylene was mixed with a styrene–acrylic copolymer latex to form a good bond with high resistance to ultraviolet radiation [581].

A rapid-setting latex system suggested consists of a styrene–butadiene–vinylidene chloride (55 %) latex (92) with hydrated alumina (100) and 10 % Na_2CO_3 solution, which can be coated on to a vinyl (PVC) film, pressed to a fibreglass board coated with an acrylic copolymer latex, diluted to 16 % solids and thickened (to 30 P) with Na_2CO_3 [582].

Copolymer latexes have been developed for adhering photosensitive (usually silver halide based) gelatin-based 'emulsions' (or other photosensitive water-sensitive polymers) to a substrate of a biaxially oriented polyester film. Typical of many published patents in this field, the suggested copolymers consist of vinyl acetate with minor amounts of methyl or ethyl acrylate comonomer and 3–30 % of a polymerisable acid, notably fumaric or itaconic acid. Alternatively, the ester mixture may include 7–20 % of a crosslinking diallyl ester, such as a malonate, sebacate or oxalate [583]. Compositions of this type, coated at $\sim$0.3–0.4 g^{-2}, may also be used to adhere to other resin and cellulosic surfaces (typically cellulose acetate and similar esters) [584].

A 'barrier'-type adhesive with good solvent and water resistance (e.g. for motor vehicle interiors) has been prepared from a mixture of a latex of acrylic acid, acrylonitrile, butyl acrylate, ethyl acrylate and N-methylolacrylamide, with thickeners and other additives. This can be applied to polyester non-woven textile trim fabrics, to avoid penetration of hot-melt adhesives used to apply trim to motor vehicle interiors [585]. The bonding of a polyvinylidene chloride film to a polyethylene substrate can be improved by applying a paste composition of 70–98 % of a styrene–butadiene (20–80 to 80–20 %) latex, with the balance of the composition as a copolymer of vinyl acetate (30–50), 2-ethylhexyl acrylate or dibutyl maleate (70–50), giving a barrier coating impervious to water, organic vapours and oxygen [586].

An adhesive system for bonding plastics to canvas has been prepared [587] from a copolymer latex of methacrylic acid, acrylonitrile, phenylenedivinyl acetate and ethyl acrylate (3 : 12 : 10 : 75), mixed with phosphohydroxyammonium silicate (2.6 mmole), Collacryl VL (a proprietary polyacrylate thickener) (0.3 parts by weight) and NH_4OH. The surfaces are coated with the adhesive and kept at 20 °C, 65 % RH for 1 h, forming a bond with strengths of 2.5 kg in^{-1} and 2.0 kg in^{-1} after 7 days at 80 °C.

6.4.2 Adhesives for inorganic substrates

In formulating adhesives for metal-to-metal bonds (e.g. for metal-based packaging boards and metal foil-coated insulating boards), several complications must be considered. The surface of some metals (notably, the transition metals, such as iron) are polar under all conditions, whilst others, such as aluminium, are non-polar. However, some metals, including aluminium, are normally coated with a thin layer of (polar) oxide, and most metals have a surface layer of grease, which is difficult to remove and presents hazards in the process [588]. Allowance must also be made for the roughness of the metal surface, as it is desirable for the adhesive to 'wet' the surface of the metal. However, there is no problem similar to that caused by the capillary absorption of fine particles of resin into an organic material, such as paper, which, apart from altering the characteristics of the paper, creates an excessive requirement for adhesive. Use of an adhesive composition with maximum solids content, with a protective colloid, is desirable, as such adhesives tend to 'dry' by 'wicking' of the water phase into the paper-type substrate.

Adhesive requirements vary greatly with the application properties of each surface, so bonding time, temperature, oil and water resistance are important. Polyvinyl acetate homo- and copolymers have, in general, good adhesion to metal foils: usually, some plasticisation is necessary to obtain optimum adhesion, and non-migratory plasticisers are particularly useful with homopolymers, when tackifiers may be added. Such plasticised latices are more stable mechanically than butadiene-type latices, since the former include water-soluble polymeric protective colloids and are more convenient for machine operation. A general formula for an adhesive based on polyvinyl acetate for paper-to-foil lamination is indicated [589] in Table 8.53. This general type of formulation can be employed in some food packaging applications, such as laminated sachets for powdered soup and similar products.

An adhesive for bonding paper and corrugated steel foil has been based on a 40–60 % polychloroprene latex, with an added carboxylated

Table 8.53 Latex adhesive for metal foil to paper

	Parts by weight
Polyvinyl acetate latex (55 % solids)	4–75
Dibutyl phthalate	2–14
Casein[a]	0.2–10
Preservative[b]	0.1–1.0
Formaldehyde (40 % solution)	0.025–0.6
Ammonia (0.880 %)	0.25–1.9
Water	11–43

[a]Casein improves adhesion to aluminium and other metals and gives higher viscosity at low concentration. It is insolubilised by ammonia before adding to the mixture. Ammonia also neutralises the formaldehyde. The solution viscosity varies with the grade of casein used.
[b]Used to preserve the casein from bacterial degradation. o-Phenylphenol or other preservatives may be used.

styrene–butadiene (up to 30 : 70) or a carboxylated acrylic latex (each including 2–10 % of acrylic or methacrylic acid), with 5–40 % of terpene–phenolic resin, an aminoplast and polyvinyl acetate or an epoxide resin [590]. An ethylene–acrylic acid latex blended with an A-stage phenoplast (100 : 67.5) can improve the adhesion of other metals to steel, and can also be used for bonding metals to wood [591]. A copolymer latex of ethyl acrylate, vinyl acetate and sodium methacrylate (50 : 50 : 4) can be used for bonding vinyl resins to iron [592].

Paper has been bonded to copper foil using a styrene–butadiene–acrylic acid (50 : 48.5 : 1.5) latex with added polyvinyl pyrrolidone and a carboxymethyl cellulose as a high-viscosity adhesive [593]. Use of the (relatively expensive) polyvinyl pyrrolidone could be avoided by using another viscosity-modifying additive (possibly a different grade of carboxymethyl cellulose). Decorative paper can be bonded to inorganic boards with an adhesive from polyvinyl acetate latex, together with the reaction product of styrene, trimethylolpropane triacrylate, methylethylketone peroxide (butan-2-one peroxide), copper octanoate, an epoxide, a polyamide (100 : 30 : 20.5 : 130 : 20 : 25) and water. Adhesion takes place on pressing (100 °C at 10 kg cm^{-2}), forming a water-resistant bond, which appears to depend on both crosslinking and polymer grafting, together with some secondary products involving the epoxide group [594].

A formulation for bonding paper to perlite (an expanded vermiculite, typical of many volume-expanded inorganic thermal insulating materials) is based on a Portland cement-rich mixture with polyvinyl acetate latex (Table 8.54) [595].

A polyvinyl acetate copolymer latex has been used to bond an inorganic fibre mat (e.g. a rockwool-phenol resin) to a glass fibre mat, with heating at 200 °C for 3 min. A smooth surface was obtained when the resulting product

Table 8.54 Adhesive system for bonding expanded vermiculite to paper

	Parts by weight (%)
Polyvinyl acetate latex (50 %)	2–12
Na alkylsilicate (30 %)	1–5
Polyacrylonitrile (partly hydrolysed)[a]	4–15
Portland cement	33–50
Water	To required viscosity

[a] 'Partly hydrolysed polyacrylonitrile' is effectively a polyacrylonitrile copolymer, which acts as a viscosity increasing and bonding agent.

is dried [596]. A non-woven glass mat was bonded to a PVC film with an acrylic latex prior to coating with PVC to form a cushioned floor tile [597]. A polybutadiene latex (mainly *cis*-polymer) (viscosity <30 000 mPa s) can be used as a bonding primer for degreased sand-blasted sheet and stoved before coating with a hot-melt polyamide in a fluidised bed [598]. Ethylene–vinyl acetate copolymer latices have been widely used in formulations for adhering to porous surfaces. With an added miscible solvent, such as methanol, and plaster of Paris, they can be used to bond concrete to wood, with satisfactory adhesive strength [599]. Alternatively, the same polymer latex (100 parts) with 2–10 parts of tartaric or another polyacid form an adhesive with high peel strength for pulp cement board to polyvinyl chloride sheet [600].

Polymer latices of acrylic esters may be used to bond both aluminium and cellulosic foils when heat resistance is not required, but there are some technical difficulties, which can be overcome by careful formulation. Corrosion may be minimised by addition of small amounts of sodium nitrate and sodium dichromate; forming may be controlled by addition of standard defoamers (used carefully, see Section 4.1.3 above, so that 'fish-eyes' may be avoided, since their occurrence may cause uneven adhesion). Low 'setting' speeds (the rate of development of bond strength) tend to cause isolated bulges and wrinkles in the adhesive bond, which can be overcome by reducing film thickness and use of an adhesive formulation with high solids content. In some cases, a different type of resin in the latex may be chosen, since these imperfections tend to arise during drying, which is affected mainly by the properties of the water phase.

Sailcloth (heavy weight woven fabric) can be bonded to metals such as aluminium and steel (and also to itself, forming a flexible laminate) using cationic polychlorprene latices; similar anionic latices give lower peel strengths [601]. A copolymer latex prepared from 2-ethylhexyl acrylate, butyl acrylate, styrene and chloromethylstyrene (50 : 30 : 15 : 5) (with dodecyl mercaptan as the polymer chain-length controller), adjusted to pH 8.5 with triethylene tetramine, can be coated on to degreased iron and then bonded to a polyurethane foam with a polyvinyl chloride skin, for automobile interior trim [602].

A fabric laminating system for hemp fabrics used to seal natural gas pipeline joints has been suggested, which employs impregnation by styrene–butadiene–methacrolein latices [603]. However, the established tyre-cord adhesives, based on styrene–butadiene–vinyl pyridine latices appear to be more effective (see Section 7 below). An adhesive for carpets to floors uses an ethylene–vinyl acetate (1 : 3) copolymer latex with emulsified carnuaba wax and paraffin wax, and polyoxyethylene monocetyl ether. On peeling the backing of the carpet when required, the adhesive is claimed to stay with the carpet [604]. See also references [605] to [607] for related applications.

Latex-based tile adhesives with good tack retention before solidifying have been developed. They include thickening agents to prevent excessive flow on vertical surfaces, notably cellulose ethers, together with a formaldehyde–(alkyl-substituted) arylsulfonic acid salt condensate. A typical base adhesive formulation [608], useful for fixing ceramic tiles to concrete, includes styrene–butadiene latex, $CaCO_3$ (filler) and an 80 : 20 mixture of methyl cellulose and hydroxypropylmethylcellulose with a formaldehyde and Na naphthalenesulfonate (0.1 %).

6.4.3 Other applications

Disposable scouring pads can be made from high wet-strength paper with an abrasive coating on one surface (prepared from a polyvinyl acetate latex, mixed with a hard-ground melamine-type resin, such as a polymer of bis-(methoxymethyl melamine) or triallyl cyanurate) and a polyurethane pad on the other surface, bonded to the paper with a plasticised acrylic latex [609]. Porous materials, such as foundry moulds with improved moisture resistance and strength, have been prepared from powdered quartz mixed with polyvinyl acetate latices. A typical formulation employs a latex (40 % solids) (55 parts), with quartz (90 parts), a cellulose thickener (5 parts) and a wetting agent [610].

An adhesive for highly porous surfaces has been suggested, which includes a vinyl chloride–vinylidene chloride copolymer latex (41 % solids) (70 parts), a polyacrylate latex (40 % solids) (19 parts), casein (10 parts), ammonia solution (25 %) (0.5 parts) and chlorobenzene (15 parts). Such a coating may be expected to have good fire resistance, but may have some problems of toxicity [611].

7 RUBBER-TO-FABRIC ADHESIVES

The major applications of rubber-to-fabric adhesives include the industrial manufacture of vehicle tyres and related products, such as transmission and conveyor belts, hose, rubber-reinforced fabrics and inflatable rubber products. Most of these are fabric reinforced, with reinforcing cords based on viscose rayon, steel, nylon, polyester or glass fibres, bonded to the bulk rubber. The strength and reliability of the finished rubber article is greatly dependent on the rubber-to-fabric adhesive bond—indeed, failure of this bond is probably the most important mechanism of failure in service of heavy duty rubber articles. This section describes the latex products, the additives and the principal formulations employed.

In general, the simple adhesion of textile fibres to solid rubbers without coating of the fibres is inadequate, so special 'dips' are employed to coat the fibres and contribute towards providing strong adhesion between the fibres and the elastomer. The most generally satisfactory type of 'dip' used to treat textile fibres is a resorcinol–formaldehyde resin, or *resol* system, usually made *in situ* (as an 'R-F' resin), which is bonded with a natural resin or synthetic latex (or sometimes a blend of both types). The synthetic resin latex is usually based on a terpolymer of butadiene, styrene and vinylpyridine [612–615].

Adhesion of the R-F resin is probably due to the methylol groups of the resin reacting with the hydroxyl groups of regenerated cellulose (when viscose rayon is used), although other mechanisms are possible, such as hydrogen bonding and mechanical entanglement. Studies using dynamic mechanical and thermal analysis for the determination of the behaviour of resorcinol–formaldehyde resols with vinylpyridine-modified styrene–butadiene latices indicate that the films have interpenetrating network morphology, with T_g values for both rubbery and glassy phases. Higher relaxation is over a broad band from 20 to 150 °C, varying, to some extent, with composition and thermal history [616].

With polyamide (nylon 66 or nylon 6) fibres, adhesion is usually more difficult, due to increased crystallinity in the polymer fibres arising from stretching. The combination of these properties improves static and dynamic adhesion, chemical and mechanical stability, so minimising the flexing fatigue and improving the working life of the fabric-to-rubber bonded product.

Adhesion of rubber to concrete (e.g. concrete floors) can be achieved with a blend of styrene-butadiene rubber S-BR (50 % latex) (15–40 parts) with casein (3–10 parts), polymethoxymethylurea (50–55 % solution) and polyhydroxymethylurea (10–20 parts), starch, ethoxylated fatty acid (10–20 parts), a mixture of fillers, carboxymethyl cellulose (as thickener) and water [617].

7.1 Introduction

Rubber-to-fabric adhesives are mainly composed of vinyl copolymer latices (usually a copolymer of butadiene, with minor amounts of styrene and vinylpyridine) and resorcinol–formaldehyde or other phenol–formaldehyde resol-type resins. The nature of the latex and that of the resin in the 'dip' mixture employed significantly affects the performance of the adhesive bond between the fibre and the rubber and, hence, the performance of the resulting fabricated tyres or belting. The preferred formaldehyde–resorcinol ratio and the operating pH are important variables, as they significantly affect the strength of the adhesive bond (the optimum fabric-to-rubber ratio of ~2 : 1, at pH 9.3–9.9), and the rate of methylol reaction is accelerated with increasing amounts of formaldehyde. The viscosity of a matured R-F solution increases with increasing amounts of formaldehyde [618]. At a formaldehyde–resorcinol ratio of 3 : 1, the viscosity increase is sharp, with rapid onset of gelation. The optimum formaldehyde–resorcinol ratio is 2 : 1 for best adhesion. There is also a marked effect of the pH of the dip mixture on adhesion of rubber to fibre. If the pH is too high or too low, there is a loss of adhesion, which can also occur with the use of different alkalis (sodium hydroxide or ammonium hydroxide). With sodium hydroxide, optimum adhesion is obtained at pH 8–9, but with ammonium hydroxide no significant difference occurs in the wider range of pH 7–12. The pH of the latex dip affects the rate of condensation of the resorcinol–formaldehyde resin. Ammonia-stabilised latex dips appear to be more stable and show an increase in adhesion with dip pick-up weights up to ∼7 % w/w on polyester, nylon and viscose rayon fibres. The resulting adhesion to rubber is greater with polyester cord, compared with the other types.

The adhesion of the (powerfully hydrogen-bonding) resorcinol–formaldehyde resin is probably due to the methylol groups of the resin reacting with hydrogen groups from regenerated cellulose (rayon cords). However, it has been suggested that, when vinylpyridine-type latices are present, reaction occurs such that methylol groups are formed on the imino-nitrogen and these groups then combine with the resorcinol by methylene bridges [619]. A typical recipe for a resorcinol–formaldehyde condensate is shown in Table 8.55. The resulting resin solution (dry solids ∼5.4 %) has pH > 7.

Typical recommended 'dip' formulations for viscose rayon and for nylon tyre cord fabrics are shown in Tables 8.56 and 8.57. These formulations are based on information from GenCorp Polymer Products Inc. (USA), suppliers of the Gen-tac® vinylpyridine copolymer latex (41 % solids, pH 10.5, average viscosity 30 cP (Brookfield)) and a styrene–butadiene latex, S-BR 2000 (42 % solids).

Chemical bonding between the cord and the latex particles affects the structure of the adhesive film. Simple styrene–butadiene latex films on rayon have a globular structure. Carboxylated styrene–butadiene and vinylpyridine copolymer latexes are reported to produce large elongated globules, but films adhering to nylon have an elongated non-globular structure, with strong chemical bonding [620].

Table 8.55 Basic resorcinol–formaldehyde condensate

	Wet weight (kg)	Dry weight (kg)
Resorcinol	16	16
Formaldehyde (37 %)	23	8.7
Water (deionised)	348	—

Rubber-to-fabric adhesives

Table 8.56 Rayon cord adhesive 'dip' formulation (80/20)^a

	Wet weight (kg)	Dry weight (kg)
Gen-tac latex (41 %)	122	50
S-BR 2000 latex (42 %)^b	191	80
R-F resin (see Table 8.55)	266	17.3
Water	446	—
Final total solids	12 %	
pH	8.0–8.5	

^{a}An alternative is a 50 : 50 Gen-tac rayon dip.
^{b}This should be replaced by a nitrile latex if the cord is to be bonded to a nitrile rubber stock.

Table 8.57 Nylon cord adhesive 'dip' formulation

	Wet weight (kg)	Dry weight (kg)
Gen-tac latex (41 %)	244	100
R-F resin (see Table 8.55)	266	17.3
Water	60	—
Ammonia (28 %) (if required)	11.3	—
Final total solids	20 %	
pH	7.0 (minimum) (or 10.3 for ammoniacal dip mixture)	

A modified 40 % latex with antioxidant properties has been prepared [621] from

Butadiene	700 g
Styrene	150 g
4-$(C_6H_5NH).(C_6H_4).NH.CO.C(CH_3):CH_2$	15 g

This latex, blended 3 : 5 with an R-F resin (resorcinol–formaldehyde (37 %)) (11 : 16.2) has been suggested for coating nylon cord. The type of polymer latex used in a resorcinol–formaldehyde dip should be chemically compatible with the type of rubber being bonded, so that it is miscible and covulcanisable. Some suggested latex–rubber combinations are shown in Table 8.58.

7.2 Vinylpyridine copolymer adhesives

The standard composition of vinylpyridine used in resorcinol–formaldehyde type rubber latex dips is butadiene–vinylpyridine–styrene (70 : 15 : 15) (see

Table 8.58 Basic latex and rubber combinations (adapted from reference [622])

Rubber type	Latex type
Natural rubber	Butadiene–vinylpyridine–styrene latex
S-BR	S-BR latex
Polybutadiene	Butyl latex
	Halogenated butyl latex
Nitrile	Nitrile rubber latex
Polychloroprene	Polychloroprene latex
EPDM (ethylene–propylene–dicyclopentadiene)	EPDM latex
	Ethylene–vinyl acetate latex
	Chlorosulfonated polyethylene latex

reference [1], pp. 304 et seq.). The vinylpyridine monomer is relatively expensive, relative to the other components of the latex, so use of lower levels of vinylpyridine in adhesive formulations has been evaluated [622] by mixing different amounts of a 30 % vinylpyridine latex with non-carboxylated styrene–butadiene latex (2000 series). Latex compositions down to 11 % vinylpyridine show little reduction in adhesion (compared with those with 15 % vinylpyridine), with both polyester and nylon fibres.

7.3 Dip additives and composition

The performance of adhesive systems used in fibre–rubber bonding depends on the composition of the 'dips' used for coating. The performance of pneumatic bias tyres, without fabric separation, may be improved by coating the nylon fibres with a blend of natural rubber (or polyisoprene) (26–60 %), vinylpyridine polymer (10–80) and styrene–butadiene (0–70 %), followed by vulcanising [623].

A copolymer of butadiene and vinylpyridine has been prepared by latex polymerisation in the presence of disproportionated resin acids as the surfactant system [624]. This latex (100 parts) (85 % solids content, in relation to an added mixture of sodium, potassium and ammonium salts) has been mixed with a formaldehyde–resorcinol resin (18 parts) to form an adhesive suitable for styrene–butadiene rubbers.

The nature of the textile cord greatly affects adhesion to the rubber. The most difficult adhesion problems arise with polyester fibres. Other specific reinforcing fibres, including glass [625] (largely replaced by steel for rubber reinforcement), steel wires (which are coated with brass (a Cu–Zn alloy) for improved adhesion, so the functional adhesive bond is actually a brass-to-rubber bond) and aramid fibres, also require consideration.

A modified procedure for forming the R-F resin consists of reacting an aldehyde–phenol (1 : 0.6–1.5) in acid for 3–10 h at 11–32 °C and then for 0.75–10 h at pH 7–7.5. Trimers and oligomers are mainly produced, which are combined with a vinylpyridine latex and a butadiene latex to form an adhesive, used to coat glass fibre previously coated with a polypropylene dispersion containing methacryloxytrimethoxysilane (as the 'keying' agent for polymer to glass). The coated glass fibres are then applied to the rubber for bonding [626].

Typical latex dip composition recipes for different fibres are shown in Tables 8.59, 8.60 [627] and 8.61 [628]. A further composition is suggested for aramid fibres based on a phenol–formaldehyde resin (which is lower in cost than an R-F resin) and a vinyl pyridine latex [629].

Table 8.59 Manufacture *in situ* of rubber–latex dip compositions

Fibre type	Nylon/ polyester	Viscose rayon
Resorcinol	11.0	11.0
Formaldehyde	16.2	16.2
NaOH	0.3	0.3
The above resin mixture is aged (at 23 °C for 6 h) and then the following are added:		
Butadiene–vinylpyridine–styrene latex (81 : 11 : 8) (41 % solids)	244.0	122.0
Butadiene–styrene latex	—	119.0
Water	326.0	683.5
The total mixture is aged for a further 18 h at ambient temperature before use.		

Table 8.60 Single dip adhesive for polyester cord

	Parts by weight
Penacolite R 2200 (70 %)	3.4
NaOH	0.17
Water	8.8
Pexul (20 %)[a]	30.0
Vinylpyridine copolymer latex (38 %)	31.8
Water	12.6
Formaldehyde (36 %)	1.0
Water	1.0
Total	98.8

[a] A commercial polymeric resin of resorcinol and chlorophenol.

612   *Polymer latices in the formulation of adhesives*

Table 8.61 Two-stage dip for rubber-to-aramid rubber bonding

Stage 1		
Epoxide resin	2.22	
NaOH (10 %)	0.28	
Na dioctylsulfosuccinate (surfactant)	0.56	
Water	96.44	
Total	103.66	
Stage 2	Wet	Dry
a. Water	141	
Ammonia (28 %)	6.1	
R-F resin (75 %)	22	16.2
b. Vinylpyridine (41 %)	24.4	100
Water	58	
c. Formaldehyde (37 %)	11	
Water	58	
d. HAF carbon black dispersion (25 %)	60.3	
Total	600.4	135.7

An example of the preparation of a typical system for vehicle tyre bonding is shown in Table 8.62. This can include 0.4 : 1 to 9 : 1 mixtures of sodium and potassium rosin acid salts as latex stabilisers [630]. This latex (100 parts), mixed with 18 parts of a resorcinol–formaldehyde resin, coated on to the tyre fabric, gave good adhesion of nylon to natural and styrene–butadiene rubbers. Related work reports similar latices prepared by polymerisation in the presence of disproportionated rosin acid–alkali metal or ammonium salts (with <60 % dehydroabietic acid). The resulting latices are blended with 7–30 % of an R-F resin and are used for nylon-to-rubber bonding.

A multistage process to form a suitable latex at 50° employs the formulation shown in Table 8.63, where the first stage of polymerisation is substantially

Table 8.62 Manufacture of a typical 'dip' system for vehicle tyre bonding

2-Vinylpyridine	15 %
Butadiene	70 %
Water	15 %
Polymerise in water, with a naphthol–formaldehyde condensate, with:	
NaOH	0.5
Na rosin acid soap	1.6
K rosin acid soap	2.4
Dodecyl mercaptan (chain-length controller)	0.5
$K_2S_2O_8$ (initiator)	0.5

Table 8.63 Multistage process for 'dip' for coating nylon cord for bonding to natural rubber/styrene–butadiene blends

Stage 1	
K rosinate soap	4.0
Na naphthalene–formaldehyde condensate sulfonate	1.0
Butadiene–styrene–dodecyl mercaptan (75 : 25 : 0.5)	70.0
$K_2S_2O_8$ (initiator)	0.5
Stage 2	
Butadiene–styrene–vinylpyridine–dodecyl mercaptan (50 : 20 : 30 : 0.1)	
Then add:	
Latex blend, as above	100
Resorcinol–formaldehyde resin solution	18
Based on:	
Resorcinol	11
Formaldehyde (37 % aq)	16.2
Water	239
NaOH	0.3

completed before addition of the components of the second stage. This adhesive (absorbed on to nylon cord at 4 % dry weight), bonded to a 7 : 3 natural rubber/styrene–butadiene blend shows 20 % improved bond strength, compared with that obtained by a single-stage method [631]. Reversing the order of the stages of polymerisation, with intermediate removal of surplus monomer before adding the R-F resin, gives similar improved adhesion performance [632]. In a further variation, the first-stage polymerisation was 75 : 20 : 5 and the second-stage polymerisation was styrene–2-vinylpyridine–dodecylmercaptan (60 : 40 : 0.1). The resulting compounded latex showed no improved bond strength, unlike the other formulations [633]. Grafted copolymers have also been studied by preparing a copolymer latex from butadiene–styrene–acrylic acid–fumaric acid (70 : 28 : 1.5 : 0.5) and grafted with styrene (15 parts) and 2-vinylpyridine (10 parts). The resulting grafted latex (75 parts) was then treated with resorcinol (16.6 parts) and formaldehyde (14.6 parts) reacted together and converted to a latex (31 parts). The resulting 'dip' system, used with polyester cord to natural rubber, showed good adhesion, reduced significantly after heating for 30 min at 130 or 170 °C. The ratio of grafted monomers is critical for adhesive performance [634]. A styrene–butadiene (25 : 75) latex, grafted with acrylamide (6.5 %) (or an N-substituted derivative) by redox initiation with R-F resin (5.75 : 1.065, based on solids), forms an adhesive for nylon cord to mixed (natural/synthetic) rubber. After dipping and drying, the cord is heated for 1 min at 218 °C [635]. Acrylonitrile–butadiene–vinyl pyridine copolymer latices have been prepared by redox polymerisation at 10 °C using conventional surfactants, $HO.CH_2.SO_2Na$, $Na_2S_2O_8$ and a hydroperoxide as initiators. These latices, with added R-F latex, form mixtures with good adhesion and tack [636].

Many other pretreatments have been suggested, which may involve alteration of the R-F resin. Most of the additives suggested are based on isocyanates or on epoxides and glycidyl groups. These are discussed in the following sections.

Claims have been made for the use of polyepoxides and blocked isocyanates, with either a rubber latex and an azidirine derivative or an R-F resin and a standard vinylpyridine latex. The former system is preferred for adhesion of polyamide fibres and the latter method for polyester-to-rubber bonding [637, 638].

7.4 Other rubber adhesive systems

With some adherent surfaces, difficulties of surface bonding can arise, which significantly affect the performance of the bonded articles when 'conventional' low-cost treatments are employed, especially with polyester fibres and other hydrophobic surfaces. Other possible treatments are discussed in this section. They are usually more expensive.

7.4.1 Blocked isocyanates

A typical general formula of a 'blocked isocyanate' is $R(NH.COOC_6H_5)_n$, obtained by reaction of phenol with a conventional di-isocyanate, such as tolyl di-isocyanate ('TDI'), where R may denote an aliphatic or mixed radical, but other blocking agents such as hexamethyleneimine may be used. An R-F resin has been claimed using blocking agents of this type [639]. Upon heating to >200 °C, the masking groups are removed and the isocyanate, with the reactive —N:C:O groups, recovers full activity *in situ*. These groups react with active hydrogen sites, acting as crosslinking agents for rubbers [640]. Isocyanates react with both the polymer latex of the adhesive and the rubber of the coating, so producing a bond at the interface. They are also believed to react with the nitrogen of the vinylpyridine groupings and/or with the carboxyl functions of carboxylated latices. Recommended protective colloid and surfactant systems for these latices include alginate salts with dioctylsulfosuccinic acid, the sodium salts of naphthalene sulfonic acid–formaldehyde condensates and sodium carboxymethyl cellulose. Technical products suggested for these purposes include diphenylmethane-4,4'-di-isocyanate blocked with phenol and other di-isocyanates blocked with acetamide [641]. Studies of the effects of using different blocked isocyanates (including hexamethylene(ethyleneurea), hexamethylene di-isocyanate, phenol-blocked triphenylmethanetri-isocyanate and tolyleneisocyanate) and a proprietary blocked isocyanate (Hylene MP®) (an insoluble di-isocyanate powder) with different bonding systems have indicated that the diethyleneurea system is the most effective (in terms of resistance to ply separation), but also gives off highly toxic by-products on heating [642]. Amongst other variations, *p*-chlorophenol–resorcinol–diphenylmethane–4,4′-tri-isocyanate, followed by a butadiene–styrene latex and a vinyl pyridine copolymer latex, was applied directly to a tyre carcass [643], and phenol-blocked diphenylmethane-4,4′-di-isocyanate was applied to a polyester cord with *p*-chlorophenol–resorcinol–formaldehyde condensate [644, 645].

A blend of an isocyanate resin precursor and an epoxide (see Section 7.4.2 below), dispersed in water with Na dioctylsulfosuccinate, including vanadium tris(acetylacetonate), has been suggested as a tyre cord treatment, followed by curing at 145 °C and a normal treatment with R-F and vinylpyridine resins. Trimethylene trisulfone may also be included [646, 647]. The use of di-isocyanates and blocked di-isocyanates for textile-to-rubber adhesives for polyester fibres has also been proposed, e.g. use of a polybutadiene terminal glycol as a blocking agent [648] and a 36-carbon aliphatic di-isocyanate (and, optionally, tris(2-methyl-1-aziridinyl)phosphine oxide) as a pretreatment for tyre cord [649–652]. For a discussion of fabric-to-rubber adhesion with cyanurates, see reference [653].

7.4.2 Epoxides and glycidyl groups

Many specifications have claimed the use of various epoxides in the pretreatment of tyre cord, notably those made from polyamides [654–657]. Various polyepoxides of relatively low molecular weight, in conjunction with R-F resin or including an ethylenediamine curing agent, have been suggested [658–661].

Epoxy resin precursors, such as Epikote 812® and polymethylene polyphenylene isocyanate, have been used to treat the tyre cord, before treatment with R-F resin and vinylpyridine latex [662]. This epoxide precursor may also be used (∼1 : 2 parts) as a pretreatment with a vinylpyridine latex [663]. An epoxide can be blended with a phenoplast [664]. The epoxy resin precursor Epikote 282® with bisphenol A and (*N*,*N*-dimethylamino)methylphenol and a *tert*-amine in butan-2-one form a pretreatment which can be applied to poly(phenylene terephthalimide) fibres [665]. The use of reactive polyamides (e.g. Versamid 125) with conventional epoxide resins (such as Araldite® AY-105) as pretreatments has been suggested [666]. Many of these systems are difficult and unpleasant to handle. To overcome this, polyester, polyamide and viscose rayon tyre cord has been first treated with a polyurethane–epoxy resin mixture containing 5 phr of *N*,*N*′-ethylene-bisstearamide, providing protection against atmospheric pollutants [667]. They may be blended with thiourea [668]. Carbon fibres can be surface-treated for bonding to rubbers by immersing in equal parts of a vinylpyridine latex and a polymer of bis-(oxiranylmethoxy)propanol, followed by R-F treatment and heating at 210 °C for 1 min. The resulting coated fibres contain 4.5 % of modified vinylpyridine polymer and 21 % of oxirane-type polymer. They can be embedded in a natural rubber–styrene–butadiene blend, giving a peel strength of 10 kg cm^{-2} [669]. The electrical deposition of adhesive on to carbon fibres has been carried out using a blend of resorcinol–formaldehyde solution (40 % aq) (60 parts), butadiene–styrene–vinylpyridine copolymer latex (40 %) (100 parts), water (250 parts) and formaldehyde (37 %) (20 parts). An acrylic-based carbon fibre was treated for 10 s with current passed at 10 V m^2 and then dried and heat-treated for 2 min at 230 °C, giving cord with 45 % of finish. After vulcanizing (30 min at 148 °C) the cord in a rubber composition, an adhesive bond with a strength double that of a bond made without the use of electric current was obtained [670]. Aramid (poly-*p*-phenylene terephthalate) fibres have been heat-treated (1 min at 280 °C) in the relaxed state before dipping in an R-F resin–vinylpyridine latex. The product shows greatly improved fatigue resistance compared with dipped cords which have not been heat-treated [671]. Alternatively, aramid fibre cord can be first dipped in a polyamide, $[N(CH_2.O.CONH_4)CO(CH_2)_5]_{5-10}$ (10 % solution in water), then heat-treated at 230 °C for 6 min, followed by an aqueous solution of resorcinol–formaldehyde (35 %), with a latex (40 %) of butadiene–styrene–vinylpyridine, and similarly heated. Use of the polyamide is claimed to improve peel and tensile strength, compared with an epoxide base pretreatment [672].

7.4.3 Dip additives

Many other additives, apart from those mentioned above, have been claimed as adhesion improvers for fabric-to-rubber bonds. A latex copolymer of 2-acryloxyterephthalate and hydroxyethyl acrylate (2 : 1) or a copolymer of diaminoethyl- and bis(hydroxylethyl)itaconic acid have been used (at 2 % concentration) as cord pre-impregnants, followed by drying [673]. The addition of proline (2-pyrrolidinecarboxylic acid) (2 %) to a standard vinylpyridine–R-F latex dip is believed to improve heat resistance by inhibiting gassing [674].

Other additives suggested include:

For polyester fibres:

(a) During the spinning phase, ~5 % of a polyamide, with a trace quantity of copper dimethylthiocarbamate [675].
(b) Dipping with 1.5 % polyethyleneimine solution assists adhesion with both natural and synthetic rubbers, and may be used together with an epoxide [676]. Other nitrogen derivatives have been used, including *N*-containing

silanes [677]. An epoxide emulsion or polyurethane latex may also be present.

(c) Grafting on to the surface of the fibre, e.g. by butadiene with γ-ray irradiation, gives improved adhesion of cork to rubber. Acrylamide or N-methylolacrylamide may also be grafted, before final treatment [678].

(d) Azidoformates have been claimed as cord treatments. The hydroxyethyl methylacrylate derivative, formed by a notably hazardous reaction with phosgene, $COCl_2$, at 10 °C, followed by reaction with (explosive) sodium azide, yielding H_2:$C(CH_3)$.$CO.OC_2H_4.OCO.N_3$, which may be latex polymerised (0.5–5 %) with blends of acrylate or methacrylate monomers or with styrene, to yield polymer latex mixtures. These can be diluted for use as a first dipping, followed by curing at 70–350 °C for 0.5–60 min and then a further direct application of the 'dip' mixture [679].

(e) Improved high-temperature adhesion of polyester to rubber has been obtained by treatment with an Epon 812® epoxide resin precursor followed by a vinylpyridine–R-F mixture, including a blocked polyisocyanate [680].

For polyamide cords:

(f) Nylon 66 cord, impregnated with an R-F adhesive and a vinylpyridine latex, gives adhesion twice as much to natural rubber as an unimpregnated cord. Heat treatment at 160–220 °C shows maximum adhesion when first applied, but this adhesion decreases with time. The adhesion may be improved by a preliminary treatment with a sulfur chloride (S_2Cl_2)–resorcinol condensate [681].

(g) Treatment with polydimethylsiloxane ($M_w \sim 1000$) (0.5 % emulsion), followed by heating for 3 min before conventional R-F–vinylpyridine latex treatment, enhances fatigue endurance of the cord without loss of adhesion [682]. Use of an allyl chloride–formaldehyde resorcinol condensate treatment has also been claimed [683]. Improved heat resistance of the adhesive bond is obtained by addition of 0.5 % (on vinylpyridine latex solids) of N-isopropyl-N'-phenyl-p-phenylenediamine and 5 % (on the latter) of Na naphthalene formaldehyde sulfonate condensate [684].

For steel cords:

(h) Rubber-to-cord adhesion is improved by a pretreatment with a copolymer of 4-vinylpyridine, 2-hydroxy-N,N'-dimethylpropylamine methacrylimide and diacetone acrylamide (105 : 270 : 8.45) [685, 686] or, alternatively, with a polymer from (3-glycidyl oxypropyl)-trimethoxysilane, which improves adhesion [687, 688].

7.5 Modifications of resorcinol–formaldehyde resins

Many variations of the resorcinol–formaldehyde resins used in latex-based dip systems have been mentioned. They include use of butyraldehyde in the condensation [689] to prepare variations of the phenolic component and of the condensation process by using an acid catalyst [690], and reaction of the R–F resin with *p*-chloroaniline [691] or with tetrahydroxydiphenylmethane (in the presence of oxalic acid) [692], followed by blending with a vinylpyridine copolymer latex. Benzaldehyde or butyraldehyde have been used with formaldehyde [693], and a dip coadditive of a resorcinol–furfuryl alcohol–formaldehyde copolymer has also been suggested [694]. *p*-Chlorophenols have also been used in partial substitution of resorcinol [695]; with this variation, polychloroprene rubbers may be bonded to polyvinyl alcohol ('Poval') fibre [696]. The use of a blended mixture of resorcinol resole resin (A) (resorcinol–formaldehyde 1.4 : 1) and Novalak resin (B) (resorcinol–formaldehyde 1.0 : 0.7–1.0), with an A : B ratio optimally at 4.5 : 1, has also been claimed [697]. Similar blended acid- and base-catalysed phenol–formaldehyde resins are reported to improve the adhesion of tyre cord to blended natural and synthetic rubber [698–700]. A similar resin, with a phenol–formaldehyde ratio of 1 : 3, prepared using a small amount of alkaline catalyst and then blended with a vinylpyridine latex (40 % solids) and resorcinol (16.7 : 100 : 3) is claimed to give good rubber-to-cord adhesion under test conditions [701]. A complex system claimed for bonding polyester cord to rubber involves the use of resorcinol-blocked TDI, with 7 % of nonylphenol, which is then further reacted with formaldehyde and, possibly, with an epoxy-novolak resin, blended with a vinylpyridine copolymer latex [702]. A phenoplast derived from 2,6-bis-(2,4-dihydroxybenzyl)-4-chlorophenol is used in comparable ratios to that in the 'conventional' R-F resol. After treating polyester cord with this resin, the cord is immersed in a polyisocyanate solution and baked before bonding with mixed natural and S-BR rubbers [703].

Other unsaturated monomers have been suggested. These include allyl compounds, such as triallylcyanurate condensed with formaldehyde and a resorcinol derivative with <3 % Cl in the aromatic ring [704]. Alternatively, resorcinol (6 mol), condensed with 2.4 mol of allyl chloride or bromide and 6 mol of phenol–formaldehyde resin, gives a product which can be blended with natural and styrene–butadiene latices [705, 706]. A condensate of resorcinol and allyl cyanurate was further condensed with formaldehyde and then blended with a vinylpyridine latex [707]. A triallyl cyanaurate–tris-(2-hydroxyethyl)-isocyanurate–resorcinol–formaldehyde adduct has been applied directly, with a rubber latex containing formaldehyde [708].

Resorcinol reacted with S_2Cl_2 in toluene yields a polysulfide, which has been blended (4 : 1) with an R-F condensate and then compounded with a vinylpyridine copolymer latex (41 %), aqueous formaldehyde (37 %) and aqueous ammonia (18 : 244 : 11 : 10) with excess water, and aged for 24 h before use, forming an adhesive for polyester fibres, which can be vulcanised with rubber and cured (240 °C for 2 h) [709]. Other dip additives have also been suggested [710–713].

7.6 Latex modifications

Styrene–butadiene–vinylpyridine latices represent the most satisfactory system in general use for fabric-to-rubber adhesives. However, some modifications have been proposed, mainly devoted to reducing overall cost by reducing the proportion of the relatively expensive vinylpyridine monomer, with minimal loss in adhesive performance and, on the other hand, introduction of other monomers to improve performance. Typically, a 3 : 1 blend of Gentac® with Gen-Flo® 2000 (a hot-polymerised styrene–butadiene copolymer, stabilised with rosin soap) has been suggested [714] as suitable for many major applications. A more complex blend, of a conventional styrene–butadiene–vinylpyridine (15 : 70 : 15) copolymer latex with three other styrene–butadiene latices (small particle size to larger particle size of the copolymer with hydroxyethyl methacrylate) has also been suggested [715]. Copolymer latices of 2-chloro-2-hydroxypropyl acrylate with butadiene, possibly also with a natural rubber latex, have been suggested as a secondary dip after the original first dip, with a significant increase in the peel strength [716, 717]. For steel tyre cord, a latex of styrene, butadiene and methacrylic acid with added tetramethylenepentamine has been proposed [718] instead of a vinylpyridine latex. 2-Vinylpyridine copolymer latices, normally used with resorcinol–formaldehyde resin condensates for polyester-to-rubber adhesion, may be modified by addition of methacrylic acid or methyl methacrylate in typical ratios (see Table 8.64).

Adhesion of polyester tyre cord in radial tyres is improved by incorporation of a carboxylated butadiene–styrene–vinylpyridine copolymer latex into the R-F mixture. The carboxylic acid groups react with amines, so preventing them from penetrating into the cord and reacting with the ester groups. A more highly crosslinked network results from hydrogen bonding with the vinylpyridine copolymer, thus reducing the rate of migration of amines into the polyester [721]. A modification of the vinylpyridine latex is to copolymerise butadiene with 2-methyl-5-vinylpyridine and the polymerisable

Table 8.64 Modification of two latices

	Reference [719]	Reference [720]
Butadiene	70	70
Styrene	9	—
2-Vinylpyridine	15	15
Methacrylic acid	6	—
Methyl methacrylate	6	15

peroxide, dimethylvinylethynylmethyl *tert*-butyl peroxide (86 : 10 : 4), with *tert*-dodecyl mercaptan as the chain transfer agent. This is reported to improve adhesion to a Soviet synthetic rubber, probably because of the presence of substantially unchanged peroxy monomer in the terpolymer [722].

An adhesive containing the proprietary Vulcabond E (a proprietary polyurethane precursor) gave an improvement of ∼40 % on peel strength, compared with products replacing the methacrylic acid or ester with styrene [723]. In producing vinylpyridine copolymer latices, it is desirable to reduce the concentration of unreacted vinylpyridine monomer to <0.2 % by heating the latex to 60 °C at 160 mm pressure. A typical 'dip' has been prepared from an R-F resin, NaOH catalysed, mixed with latex and Vulcabond E, and used for dipping polyester cords, which are heated for 1 min at 240 °C between natural rubber sheets and then press-vulcanised for 30 min at 150 °C, giving significantly improved cord retention [724].

Polychloroprene copolymer latices have been used in dip formulations when the coating elastomer is also polychloroprene-based, as in drive belts. Resorcinol–formaldehyde prepolymers have been used with vinylpyridine copolymer latices and polychloroprene latices for many years. A typical formulation, based on Neoprene® 750 (see Table 8.69), is shown in Table 8.65.

Addition of 2 % of methacrylic acid to the polymerisation of chloroprene to a 40 % latex (with added carbon black (1 phr)) increases the adhesion of rubber to steel on vulcanisation, with an added 40 % dispersion of dinitrosobenzene (5 phr) [725]. The use of a chloroprene (or dimethylvinylpyridine) latex (solids 4.5–15.7 %) with R-F resin (0.08–4.14 %), with formaldehyde (0.01–1.1 %), NaOH (0.07–0.16 %) and a polyacrylamide-modified polyethylene–polyamine (ethoxylation of primary amino groups 23–100 %) (0.037–0.22 %) and water (78–95 %) has also been suggested [726].

Table 8.65 Polychloroprene latex—resorcinol–formaldehyde 'dip' formulation

		Parts by weight	
		Dry	Wet
Part A:	Neoprene® 750 latex (Note 1)	25–75	50–150
	Styrene: butadiene latex	0–50	1–125
	Vinyl pyridine latex (Note 2)	25	61
	ZnO	5	10
	Antioxidant	1	2
Part B:	Water	—	238.4
	NaOH	1.5	1.5
	Resorcinol	11	11
	Formaldehyde (37 %)	6	16.2

Note 1: See Table 8.69.
Note 2: From GenCorp Inc.

7.6.1 Latex dip additions

Many different additives have been used in formulated adhesives for fabric to rubber to improve the bonding of particular fibres. Addition of the Na salt of a formaldehyde-naphthalene/sulfonic acid improves the stability of the 'dip' [727]. Polyester-to-rubber adhesion is improved by addition of a di-isocyanate to the 'dip' (see Section 7.3 above), preferably in the form of masked isocyanates. A polyurethane is formed from hexamethylene di-isocyanate $OCN.(CH_2)_2.NCO$ and a mixture of polyoxyethylene and trimethylolpropane, blocked with phenol and solubilised by reaction with the Na salt of the amino acid, taurine, $H_2N.(CH_2)_2.SO_3Na$. This modified polymer, added 80 % (on solids) to a vinylpyridine latex (possibly also with an epoxide), has been suggested for coating polyester tyre cord [728]. A multiple 'dip' formulation for polyester cord includes a first-stage dip of Decanol EX-611 (a polyepoxide), Na dioctylsulfosuccinate, a butadiene–styrene–vinylpyridine copolymer latex and Hylene MP (a polyisocyanate), with a second-stage aqueous dip of R-F initial condensate, vinylpyridine latex, butadiene–styrene latex, aqueous formaldehyde, diphenylmethane-diethyleneurea, Na dioctylsulfosuccinate and pyromelliltic anhydride. The polyester cord (dried for 1 min at 230 °C after each dip) has good, thermally stable adhesion to rubber [729]. A similar dip includes a first-stage dip of Decanol EX 311 (0.5 %) in water, mixed with aqueous sodium lignosulfonate dispersion, carbon black and NaOH in water, and a second-stage dip of a resorcinol–formaldehyde, vinylpyridine latex, ammonia and Hylene MP (25 % aqueous dispersion). Tyre cords are passed through a bath of the first dip, dried (200 °C for 30 s), then passed through the second dip and dried again (245 °C for 60 s). The resulting products show good interply bonding strength [730]. Another modification, suitable for the adhesion of polyester, contains, in addition to the standard vinyl pyridine latex, a resol-type resorcinol–formaldehyde latex and another resin prepared from mixed resorcinol and 4-chlororesorcinol treated with S_2Cl_2 and then with further formaldehyde [731]. A polyester tyre cord, primed with isocyanate–epoxide composition, also containing trimethoxymethylamine and 4,4-dihydroxy-1,3-dimethylol-2-imidazolidone, was coated with a modified vinylpyridine latex to form a finish with very good hot peel adhesion [732]. The preparation of a typical 'dip' is shown in Table 8.66 [733].

An alternative method for the formulation of a vinylpyridine latex for nylon cord has been reported [734]. The components are

Water	130
Potassium rosinate	4
Na naphthalene–formaldehyde condensate	1
$K_2S_2O_8$ (initiator)	0.5
Dodecyl mercaptan	0.5
Butadiene–styrene–vinylpyridine (13 : 3 : 3) mixture	70

Table 8.66 Tyre cord 'dip' composition for polyester fibre

	Parts by weight
Resorcinol	2.7
Formaldehyde (37 % aqueous)	5.9
NaOH (10 % aqueous)	4.5
Water (first addition)	86.9
Add above to:	
Styrene–butadiene–vinylpyridine latex (41 %)	60.4
Water (second addition)	60.4
Add above to:	
Masked isocyanate dispersion (37 %)	40
Water (third addition)	160

Polymerisation is taken to 90–95 % conversion, when hydroquinone (0.1 g) is added as inhibitor. The adhesive formed from the latex (100 parts), with alkali-condensed resorcinol–formaldehyde resin (18 parts), shows much improved adhesion, compared with the use of a styrene–butadiene (30 : 70) latex alone [735].

A polyurethane latex (formed from adipic acid, ethanediol and trimethylolpropane with excess 2,4-tolylene di-isocyanate in toluene (>2 mol —NCO:1 mol —OH), followed by addition of cresol) emulsified with a non-ionic surfactant was added to a standard 'dip'. Blended vinylpyridine and styrene–butadiene latexes with a dibutyl phthalate solution (to reduce hydrolytic reaction) of diphenylmethane di-isocyanate in aqueous polyvinyl alcohol has been used as a 'dip' to treat viscose, polyamide and polyvinyl alcohol fibres. This modified 'dip' mixture, used with viscose, is claimed to increase adhesion to rubber by ~2.5 times, compared to the bonding obtained with a conventional R-F 'dip' [736]. Addition of trimethylolphenol to an R-F–vinylpyridine 'dip' approximately doubles adhesion of nylon cord to rubber [737], whilst preliminary treatment of a similar cord with a S_2Cl_2–resorcinol condensate before the main 'dip' also improves adhesion [738]. Addition of glycerol diglycidyl ether with a vinylpyridine latex, followed by a second treatment with an R-F–vinylpyridine 'dip' has also been suggested [739]. A TiO_2 aqueous dispersion (15 %), prepared using a maleic acid–maleic acid copolymer salt as stabiliser (at 1 % of pigment concentration), can be added to a 'dip' claimed to be particularly suitable for Nylon 6 cord [740]. Adhesion of polyamide cord was also improved by using a mixed novalak with 2,4,5-tris(hydroxymethyl)phenol [741]. However, use of N-hydroxymethylated nylon 66 reduced the rate of change of adhesive strength with time [742]. Other suggested additives for improving the adhesion include cuprous iodide CuI_2 and potassium iodide (each ~1 % on 'dip' solids) or ~2 % of sodium phenyl phosphate $Na_2(C_6H_5)PO_4$, which double the strength of the adhesive bond [743].

Addition of a peroxide, such as 3,3′-di-(*tert*-butylperoxy)butyrate, aids crosslinking and so improves adhesion [744]. Silica can be included in a vinylpyridine latex (in the ratio of silica to polymer solids of ~12 : 25), presumably with loss of some bonding strength [745].

Adhesion of steel cord to rubber is aided by immersion in a coupling agent of a polyphosphoric ester and a hydroxy-terminated polybutadiene in perchloroethane, followed by further treatment with a vinyl pyridine latex [746]. The formation of glass-to-rubber bonds from resorcinol–formaldehyde with a butadiene–styrene–vinylpyridine copolymer latex with vinylidene chloride ($\geqslant$60 %) and wax has been suggested [747]. Addition of a (toxic) nitroso compound, such as $C_2H_4(NO)_2$ (1 %), to a 2,2-dichlorobutadiene latex (with additional sulfonate and non-ionic surfactants) is reported to increase significantly the adhesion of natural rubber to steel [748]. Adhesion of rubber to steel can also be promoted with a blend of a vinylpyridine latex with a copolymer latex (24 % solids) of α-chloroacrylonitrile and dichlorobutadiene, possibly with glycidyl methacrylate as a comonomer [749]. In a further system, steel is first coated with a mixture of a novalak resin and chlorinated natural rubber and then with the following blend [750]:

Vinylidene chloride–methyl acrylate (55 %) (or acrylonitrile–butadiene latex)	843
p-Dinitrosobenzene	253
Phenol-formaldehyde resol resin	42
Poly-(*N*-vinyl-2-pyrrolidone)	28
Butadiene oxime-blocked 1 : 3-$(HOCH_2)_3C.C_2H_5$-TDI adduct	168
Nonylphenol-(20 EO)-ethoxylate	12
ZnO	8
Carbon black	97
Stearyl behenate	8
Water	1541

Adhesion of glass fibre to rubber may be improved by addition of small amounts of γ-aminopropyl(triethoxy)silane and furfuryl alcohol or, alternatively, addition of zinc sulfate [751]. Other suggested additives include a thiourea or thioamide, $CH_2.C:S.NH_2$ (at ~1 % of overall latex weight) in a vinylpyridine–R-F 'dip'. Addition of a microcrystalline wax and ZnO has also been suggested [752]. Addition of a standard polybutadiene latex and a standard carboxylated styrene–butadiene latex to a vinylpyridine latex improves the flexibility and fatigue resistance of the glass-to-rubber bond [753]. A modified composition, also for glass fibre, for radial tyre constructions employs a mixture of polybutadiene (non-self-curing) and vinylpyridine copolymer latices (diene–vinylpyridine ratio 7 : 3) and resorcinol–formaldehyde resin (2400 : 1153 : 102) (based on resorcinol weight) with a wax emulsion, antioxidant and ammonia [754]. Another system, for similar purposes, involves dipping glass fibre in a latex of butadiene, styrene and vinylpyridine (14 : 3 : 3), blended with a butadiene rubber and an R-F resin (40 : 60 : 7.7), followed by vulcanisation in a natural rubber–polybutadiene–S-BR blend [755]. Other glass fibre-to-rubber adhesion promoting agents include, apart from the water-soluble R-F resin, acrylonitrile-containing copolymer latices with low residual unsaturation. These are prepared in the presence of C_9–C_{20} alkyl soaps as surfactants, which have been hydrogenated [756] (using a Pd acetate catalyst) to reduce residual unsaturation to an iodine value <120 [757].

8 TEST METHODS FOR ADHESIVES

There are many test methods for adhesives and their components. Most of these relate to performance testing, and not specifically to the physical and chemical properties of latex adhesives. National and European Standards are being replaced by International Standards, published by the International Standards Organisation (ISO) in Geneva and are listed in the (well-indexed) annual ISO Handbook, which is available from National Standards authorities.

9 GLOSSARY OF TERMS RELATING TO ADHESIVES

(This is based on International Standards Committee TC61/SC 11/WG 5, especially IS 6534 *Adhesives—Vocabulary*, publications of European Committee for Standardization CEN/TC 193, British Standard 6138 : 1989 and other sources.)

This glossary is mainly concerned with terms used in connection with the formulation and application of latex-based adhesives and others with similar formulations and applications.

Adhesion
The state in which two surfaces are held together by surface bonds (*specific adhesion*) or by interlocking action (mechanical adhesion) (see also *Cohesion*)

Adhesive
A non-metallic substance able to join materials by surface bonding (*adhesion*) with a bond possessing internal strength. Specific types include:

> *Water-borne* (aqueous) adhesive, in which the solvent (or the continuous phase, in the case of an emulsion or latex) is water
> *Dispersion* adhesive, a stable dispersion of a solid polymer in a liquid continuous phase (usually water)
> *Emulsion* adhesive, a stable dispersion of a liquid hydrophobic resin in water
> *Encapsulated* adhesive, in which particles or droplets of one of the components are enclosed in a protective film (as microcapsules) to

prevent crosslinking or other chemical bonding until the protective film is destroyed

Gap-filling adhesive, an adhesive designed for filling wide gaps between surfaces, such as uneven surfaces

Hot-melt adhesive, usually solid or tacky at room temperature, which is applied as a heated viscous fluid mixture to the surface being bonded and forms an adhesive bond during the cooling process

Latex adhesive, similar to an *emulsion* adhesive or a *dispersion* adhesive, the terms being used (more-or-less) interchangeably

Solvent adhesive, *solution* adhesive, *solvent-based* adhesive, an adhesive in which the polymeric component (and any additives) is dissolved in an organic solvent (usually a mixture of solvents)

Spray adhesive, an adhesive which is applied as small particles projected under pressure on to the substrates being bonded

Sealant, a gap-filling adhesive, used to fill gaps, with some flexibility between surfaces where some movement may occur in service

Adhesive tape

A pressure-sensitive, remoistenable or heat-activated adhesive (sometimes latex based) coated on to a flexible backing or carrier. In *double-sided* (double-coated) tape, adhesive is applied to both faces of the carrier substrate

Application equipment

Roller coater, a device for the application of adhesive to a substrate by a mechanically driven roller (the doctor roller revolves from, or in the opposite direction to, the *application* roller, resulting in a 'wiping' action)

A *pick-up roller* runs in the bath or reservoir of adhesive

An *applicator roller* transfers a controlled amount of adhesive to a substrate, the applicator roller of a forward roller being called a *lick roller*

Spreader, a device for controlled application of adhesive to a substrate

Curtain coater, a machine for rapid application of liquid adhesive, in which an adherent is passed through a continuous descending stream of adhesive (the curtain); where accurate adhesive film thickness depends on control of machine speed and curtain thickness

Application methods

Single-spread adhesive, applied to only one of the adherents

Separate spread adhesive, with two parts. One part is applied to one adherent and the other part to the second adherent. The two parts react together when brought together in the bonding process

Two-way application adhesives *double-spread* adhesive is applied to both adherents

One-part adhesive, an adhesive packed in a single container, ready for use (but may require the addition of water or the presence of ambient water for setting by reaction)

Two-part adhesive, packed as two separate reactive components, which are mixed before applications

Multipart adhesive, packed as separate reactive components, which are mixed before application

Binder
The component of an adhesive composition which is primarily responsible for the adhesive forces that bond two surfaces together

Blocking
Undesired adhesion between touching layers of a material which may occur under moderate pressure during storage of stacked sheets of, for example, coated paper. The nature and degree of blocking is affected by the moisture content and humidity of the coated paper.

First-degree blocking is the adherence of two surfaces, such that, when the upper component is separated from the lower component, the lower component will cling to it, but may be parted without damage to either surface

Second-degree blocking is similar: the adherence is such that, when the surfaces are separated, at least one surface is damaged to some degree

Critical humidity of blocking is the lowest humidity at which blocking of a given degree occurs

Critical temperature of blocking is the lowest temperature at which blocking of a given degree occurs

Bloom ('blooming')
A white haze which can occur on the surface of a coated adhesive film

Bonding conditions
Application time, the period of time required for spreading an adhesive on to the surfaces specified to be coated

Open time open assembly time, the interval between the completion of application of the adhesive to the adherents and the assembly of the adhesive joint

Closed assembly time, the interval between assembly of the adhesive joint and the initiation by heat and/or pressure of the setting process in the assembled joint

Pressing time, the time for which an adhesive joint is pressed after assembly

Clamping time, the time for which an adhesive joint is clamped under mechanical pressure

Setting time, the time required for an adhesive to set under specified conditions

Bond strength
The unit load, applied in tension, compression, flexing, peel, impact, cleavage or sheer, required to break an adhesive assembly, with failure occurring at, or in the plane of, the adhesive bond

> *Green strength*, the strength of a bond determined immediately after assembling

Breaking force
The force necessary to bring an adhesive joint to the point of failure, with failure occurring in or near the plane of the bond line

Carrier
A flexible material to which an adhesive is applied, for subsequent transfer to a substrate

Cleavage
A mode of application of a force to an adhesive joint between rigid adherents which is not uniform over the whole area, but resulting in a stress concentrated on one edge

Cleavage strength
The force required to bring an adhesive assembly to the point of failure by the application of force in the cleavage mode

Coalescence
The fusion of surfaces in a system. In adhesives, the term is used to described the fusion of dispersed particle of a latex and, also, the fusion of two adhesive films when brought into contact

Cohesion
The state in which particles of a single substance are held together by primary or secondary forces

Conditioning time
The time between ceasing application of heat and/or pressure to an adhesive bond and attaining the desired bond properties

Creep
Dimensional change with time of a material, notably an adhesive joint, under load, e.g. of gravity. *Creep* at ambient temperature is also known as *cold flow*.

Delamination
Separation of layers in a laminate, due to failure of the adhesive bond, either in the adhesive itself or because of cohesive failure of the adherend

Dry strength
The strength of an adhesive bond dried under specified conditions

Elongation
Increase in length of a specimen (usually expressed as a percentage of the original length)

Equilibrium moisture content
The moisture content at which an adherent neither gains nor loses moisture when subjected to a given constant condition of humidity and moisture

Extender
An inert material, with modest adhesive properties, added to an adhesive to reduce overall cost of the product in which it is included

Filler
An inert material (usually with minimal adhesive properties) added to an adhesive to improve flow and working properties and to reduce overall cost

Fish-eyes
Random-shaped, but usually round or oval, deformations (and imperfections) in an adhesive layer

Flow
Movement of an adhesive during the bonding process before the adhesive reaches a solid state

Gelation
The process of flowing a gel. Adhesives may be classified by mode of gelation:

(a) by loss of volatile components, by evaporation of diffusion.
(b) by loss of organic solvents, e.g. from elastomer solutions;
(c) loss of water from solution (starch, animal glue, sodium silicate, etc.) or from latex (vinyl or acrylic copolymers, elastomeric polymers);
(d) by cooling (e.g. animal glue solution);
(e) by condensation polymerization, induced by heat, catalyst and or/heat, e.g. polyesters and polyurethanes.

Gel temperature
The temperature at which a system no longer flows under shearing stress

Gel time, gelation time
The measured time from the moment the adhesive is ready for use until that when the adhesive has sufficient consistency to resist cold flow under the conditions of test

Gel strength, bloom strength
A measure (in arbitrary units) of the rigidity of a gel prepared under standard conditions (see IS 9665)

Glue
A protein-based material of animal origin (from waste bone or waste hides), usually based on collagen. The term is also used as a synonym for other types of adhesive

Grab
A subjective estimation of *tack* (see below)

Handling strength
An (indeterminate) level of bond strength in a bond which allows removal of a joint from a clamping or pressing device without damage to the bond

Heat strength
Bond strength at an elevated temperature under specified conditions of temperature, load and time

Impact strength
The force necessary to bring an adhesive bond to the point of failure by means of a very high rate of shear stress development

Joint, lap
A joint made by placing one adherent partly over another adherent, with adhesive between the two surfaces, and bonding the overlapping portions

Laminate
A product made by bonding together two or more layers of materials

Modulus, shear
Modulus of elasticity in shear

Modulus, Young's
Modulus of elasticity in tension

Peel
A mode of application of a force in which one or both of the adherents are flexible, and in which the stress if concentrated at a boundary line

Pot life
The period during which a two- or mullet-part adhesive can be used, after mixing the components. The time varies with the volume and temperature of the mixed adhesives, and with the ambient temperature

Release paper
A sheet which acts as a protective layer or as a carrier for an adhesive film, which can be readily used before the film is applied to a substrate

Setting
Conversion of an adhesive by development of cohesive strength into a fixed or hardened state by chemical action (polymerisation, crosslinking, oxidation, curing, etc.) or by physical action (gelation, hydration, cooling, evaporation of volatile components)

> *Setting temperature*, the specified temperature for setting the adhesive (which may differ from that of the temperature surrounding the assembly)

Shear strength
The force necessary to bring an adhesive joint to the point of failure, by means of forces applied in shear mode

> *Lap shear strength, longtitudinal shear strength*, the force necessary to bring an adhesive joint to the point of failure, by stress applied longtitudinally, parallel to the plane of the bond line

Shelf life, storage life
The period of storage under stated conditions for which an adhesive may be expected to retain its working properties

Sizing
The process of applying a liquid phase (usually a low-solids, water-based solution of mixed polymers or a latex adhesive) to a substrate, to fill pores or other voids, and so improve absorption of a subsequently applied adhesive or otherwise improve adhesion. The term *sizing* has several meanings, which may be different in related technologies, e.g. in textile finishing, paper coating or applying surface coating to constructional materials

Solids content
The percentage by mass of non-volatile matter in an adhesive, determined under standard defined conditions

Strain
The unit change due to force in the size of a body, related to the original dimensions

Stress
The intensity at a point in a body of the internal forces (or components of force) which act on a given plane through the point

> *Yield stress*, the stress applied to a joint at which permanent deformation occurs

Stringiness
The property of an adhesive which results in 'stringiness' of filaments or threads when adhesive transfer surfaces are separated (before final setting). The phenomenon is also known as *legging*

Substrate
A material upon the surface of which an adhesive is coated for any purpose of bonding

> *Adherent*, a body which is held, or intended to be held, to another body by an adhesive. (Note that *adherent* is a narrower term than *adhesive*)

Tack
The property of an adhesive formulation which allows it to bond rapidly on contact with another surface (which may be an adherent or another layer of adhesive). *Tack* describes the ability of an adhesive substance to deform and flow, wetting the second surface on contact, so forming a bond

> *Tack force*, the force required to separate an adhesive coat from a second surface. This force may increase with time
> *Quick stick* may be used to describe the tack force immediately after the bond is made
> *Dry tack*, the property of some adhesive films to adhere to themselves, when apparently dry to the touch
> *Tack time*, the period of time for which the adhesive film is in the 'dry-to-touch' condition after application at specified conditions of temperature and humidity. This time (also known as *tack range*) depends on the pressure applied when the adhesive coats are brought together

Tensile strength
The force necessary to bring an adhesive joint to the point of failure by application of a uniform stress at right angles to the bond line.

Tension
The mode of application of a tensile force normal to the plane of a joint between adherents, and uniformly distributed over the whole area of the bond line

Water-resistant
An adhesive bond is said to be water-resistant if it will withstand prolonged contact with water whilst retaining adequate bond strength and other properties required for its intended purpose. (Note that the term *waterproof* is normally avoided, since adhesives impervious to the passage of water during a normal service life are rare)

Table 8.67 'Propiofan' polyvinyl propionate latices for adhesives (BASF AG, Ludwigshafen, Germany)

Name	Solids (± %)	Viscosity (Pa s)[a]	Shear rate (s⁻¹)	pH	T_g (°C)	Particle size (nm)	Dispersion type	Film properties	Applications
Propiofan 5D	50	6–13	25	5–7	10	~2	Non-ionic	Tacky; very flexible	Building, packaging adhesives
Propiofan 5D Extra	50	9–13	25	5–7	10	~2	Non-ionic	Slightly tacky; very flexible	Packaging adhesives
Propiofan 6D	50	0.8–2	250	5–7	<1	~0.6	Anionic	Slightly tacky; highly flexible	Packaging, laminating and flocking adhesives
Propiofan 800D	55	4.5–6.3	100	5–7	<1	~1–1.5	Non-ionic	Slightly tacky; highly flexible	High-speed packaging adhesives

[a]Measured by IS 3219.

Wet strength
The strength of an adhesive bond, determined immediately after removal from a liquid in which it has been immersed under specific conditions. (Note that with some water-borne and latex adhesives, the term is also used to describe the bond strength when the adherents are brought together with the adhesive still in the wet state)

10 APPENDIX: SOME TECHNICAL PRODUCTS FOR WATER-BASED ADHESIVES

BASF AG,	Ludwigshafen, Germany
Bayer AG,	Leverkusen, Germany
E I Du Pont de Nemours Inc.	Wilmington, Delaware, USA
Harlow Chemical Company Ltd	Harlow, Essex, UK
Malland Creek Polymers Inc.	Akron, Ohio, USA
Reichhold Inc.	Durham, North Carolina, USA
Resad Polymers Ltd	Andover, Hampshine, UK
Rhodia Ltd (Rhône-Poulenc Group)	Manchester, UK
Rohm & Haas Inc.	Philadelphia, Pennsyloania, USA
Scott Bader & Company Ltd	Wellingborough, Northamptonshire, UK
UCB SA	Vilvoorde, Brussels, Belgium
Union Carbide, Inc.	Danbury, Connecticut, USA
Vinamul plc	Carshalton, Surrey, UK
Vinavil SpA	Milan, Italy
Wacker-Chemie GmbH	Munich, Germany

Other products for particular applications, from other manufacturers, are mentioned earlier in this chapter, notably in Sections 4.1, 4.2.2 and 4.5. The products are listed in alphabetical order of the names of the individual producers and the date of compilation. The assistance of these producers in providing the data from which the information in Tables 8.67 to 8.95 is taken is gratefully acknowledged.

Table 8.68 Polychloroprene latices for contact adhesives (Bayer AG, Leverkusen, Germany)

Name	Emulsifier	TS (%)[a]	Latex density	pH	Mean particle size (mm)	Viscosity (Pa s)[b]	Minimum filming temperature (°C)
Dispercoll C 74	Anionic/non-ionic	58 ± 0.5	1.13	12.5–13.5	~160	~0.15	~5
Dispercoll C 84	Anionic/non-ionic	55 ± 0.5	1.12	~13	~160	~0.12	~5

[a] To DIN 53563.
[b] Brookfield LVF viscometer spindle 3; 30 rpm at 23 °C.

Table 8.69 Polychloroprene 'Neoprene' latices for adhesives (E I Du-Pont de Nemours Inc., Wilmington, Delaware, USA)

Type	Comonomer	Emulsifier	Emulsifier type	Chlorine content (%)	pH[a]	TS (%)	Brookfield viscosity[b]	Features	Applications
115	Methacrylic acid	Ethanol	Non-ionic	36	7	47	500 : 2 : 6 350 : 2 : 30	Carboxylated polymer; high stability	Adhesives, asphalt modification
400	2,3-dichloro-–1,3-butadiene	K salts of disproport-ionated resin acids	Anionic	38	12	50	9 : 1 : 6 9 : 1 : 30	Max Cl content; rapid crystallising; good ozone and weather resistance	Coatings
622	—	K salts of disproport-ionated resin acids	Anionic	38	12.5	60	65 : 1 : 6 65 : 1 : 30 100 max	High solids, low-viscosity, high gel polymer	Foam
671	—	K salts of disproport-ionated resin acids	Anionic	38	12.5	56	60 (max): 1 : 30	High solids, low-viscosity, medium-high gel polymer; wet gel strength	Adhesives, binders, coatings, dipped goods, elasticised asphalt and concrete, foam

Grade	Comonomer	Emulsifier	Type	Solids (%)	pH	Viscosity[b]	Spindle:speed	Characteristics	Applications
750	2,3-dichloro-1,3-butadiene	K salts of disproportionated resin acids	Anionic	40	12.5	50	10 : 1 : 6 10 : 1 : 30	High wet gel strength; low modulus, crystallisation rate	Adhesives, dipped goods
842A	—	Sodium salts of resin acids	Anionic	37.5	12.0	50	15 : 1 : 6 15 : 1 : 30	Medium strength cured films with slow crystallisation rate	Adhesives, binders, coatings, dipped goods
NPR-3911	—	K salts of disproportionated resin acids	Anionic	38	12.5	56	50 : 1 : 60	Moderate crystallisation rate; medium-low gel; high wet gel strength	Adhesives, binders, paper modification, dipped goods
NPR-5587	—	Na salts of disproportionated resin acids	Anionic	37.5	12.0	58	200 : 2 : 6	Fast crystallising low gel polymer	Coatings, adhesives

[a]Typical value at 25 °C.
[b]Viscosity in mPa s (cp): spindle number: speed (rpm) at 25 °C.

Table 8.70 Polyvinyl acetate homopolymer latices for adhesives (Harlow Chemical Company Ltd, Harlow, Essex, UK)

Grade	Stabiliser	Solids (%)[a]	Viscosity (mPa s)[b]	pH	MF-T (°C)	Applications
Emultex F500	Natural gum	56.5	2000	5	16	Borax stable; rapid setting packaging adhesives
Emultex 553	Dextrin	66	1700	4.5	7	Low-cost base for remoistenable adhesives
Emultex 580	Polyvinyl alcohol	55	12 000	4.5	16	Basic latex for packaging, wood and building adhesives
Emultex 581	Polyvinyl alcohol	46	13 000–19 000	5	16	Low-cost base for packaging, wood and building adhesives
Emultex 581R	Polyvinyl alcohol	46	10 000–18 000	6.5	4	Ready-to-use wood adhesives
Emultex 582	Polyvinyl alcohol	46	11 500	3	4	Water-resistant wood adhesives (to EN 204 (Part D2))
Emultex 585	Polyvinyl alcohol	56	4500–6500	4.5	16	Low-viscosity base latex for packaging, building and wood adhesives; near-Newtonian rheology
Emultex 586	Polyvinyl alcohol	60	19 000–24 000	4.5	16	High-viscosity base latex for packaging, wood and building adhesives
Emultex DP 3849	Polyvinyl alcohol	55	3000–5000	4	17	Base latex for fast setting packaging adhesives with good cleanability; slightly pseudoplastic; also for heat-seal adhesives and woodworking
Emultex 593	Polyvinyl alcohol	55	14 000	5	17	Pseudoplastic base dispersion for fast setting (low solids) packaging adhesives with good cleanability
Mowilith DHL[c]	Polyvinyl alcohol	50	21 500	6.5	16	Latex for high-quality wood adhesives (to BS 4071), building and packaging adhesives

[a] Determined at 105 °C for 1 h in a vacuum oven.
[b] Determined using Brookfield RVT viscometer speed 10 at 23 °C.
[c] Manufactured under licence from Hoechst AG.

Table 8.71 Vinyl acetate copolymer latexes for adhesives (Harlow Chemical Company Ltd, Harlow, Essex, UK)

Grade	Comonomer	Stabiliser	Solids (%)[a]	Viscosity (mPa s)[b]	pH	MF-T (°C)	Applications
Emultex 592	Butyl acrylate	Polyvinyl alcohol	50	11000	5.5	0	Heat-seal adhesives; adhesives for low-energy ('difficult') surfaces
Mowilith DS5[c]	Dibutyl maleate	Colloid	56	9000	4	5	Adhesives for PVC, low-energy 'difficult' surfaces and wallcoverings
Mowilith DM104[c]	Ethylene	Polyvinyl alcohol	55	2500	4	0	Packaging adhesives; plastic foil lamination; flooring and wallcovering
Mowilith DM105[c]	Ethylene	Polyvinyl alcohol	55	7500	4	0	Packaging adhesives; PVC adhesives with good heat resistance
Mowilith DM111[c]	Ethylene	Polyvinyl alcohol	55	3000	4.5	0	Fast-setting packaging adhesives with rapid development of fibre tear
Mowilith DM114[c]	Ethylene	Polyvinyl alcohol	59	3500	5	0	Fast-setting packaging and PVC adhesives, with good heat resistance
Mowilith DM132[c]	Ethylene	Polyvinyl alcohol	60	7000	4	0	Packaging adhesives; coated and printed board, film and wallcovering adhesives
Mowilith DM137[c]	Ethylene	Polyvinyl alcohol	56	7500	5	0	Pressure-sensitive adhesives for coating paper and foils; additive to other latexes to improve adhesion

Table 8.71 (*continued.*)

Grade	Comonomer	Stabiliser	Solids (%)[a]	Viscosity (mPa s)[b]	pH	MF-T (°C)	Applications
Mowilith DM155[c]	Ethylene	Colloid	55	7000	5	0	Single-pack remoistenable adhesives
Mowilith DM427[c]	Dibutyl maleate	Polyvinyl alcohol	55	22 000	4.5	4	Packaging and building adhesives
Mowilith VDM1330[c]	Ethylene	Polyvinyl alcohol	55	15 000	4	0	Packaging adhesives; lamination of foils to paper and board; rapid setting
Mowilith DM1340[c]	Ethylene /acrylic ester	Surfactant	65	2000	4	0	Flooring adhesives; lamination of foils to paper and printed board
Emultex 518	External plasticiser	Polyvinyl alcohol	40	16 000	5	2	Universal bonding and building adhesives; general-purpose adhesives
Emultex 4026	External plasticiser	Colloid	50	1500	4	5	Efficient binder for filled adhesives; film has high resistance to water
Emultex DO4440	External plasticiser	Polyvinyl alcohol	39.5	10 500	5	—	Ready-to-use washable school glue

[a] Determined at 105 °C for 1 h in a vacuum oven.
[b] Determined using Brookfield viscometer speed 10 at 23 °C.
[c] Manufactured under licence from Hoechst AG.

Table 8.72 Styrene–acrylic ester copolymer latexes for adhesives (Harlow Chemical Company Ltd, Harlow, Essex, UK)

Grade	Comonomer	Stabiliser	Solids (%)[a]	Viscosity (mPa s)[b]	pH	MF-T (°C)	Applications
Revacryl 143	Butyl acrylate	Surfactant	50	200	8.5	12	Packaging; heat-seal and cold-seal adhesives
Revacryl 309	2-Ethylhexyl acrylate	Surfactant/ polyvinyl alcohol	55	400	7.2	35	Heat-seal adhesives; anti-block additive
Revacryl 344	2-Ethylhexyl acrylate	Surfactant	56	200	8.5	10	General-purpose tile adhesives; non-slip tile adhesives; aluminium foil adhesives
Revacryl 380	Butyl acrylate	Surfactant	50	1000	7.5	15	Concrete and cement additives; self-adhesive flooring compounds
Revacryl 381	2-Ethylhexyl acrylate	Surfactant	55	2000	8	0	General-purpose binder for acrylic sealants/caulks; flexible roofing compounds; concrete admixtures
Revacryl 389	Butyl acrylate	Surfactant	47	50	7.5	23	Cement additive
Revacryl DP 4334	Butyl acrylate	Surfactant	57	1200	8	0	Cement additive with good flexibility and water resistance; joint fillers/caulks with good adhesion to many substrates
Viking 5420[d]	2-Ethylhexyl acrylate	Surfactant	55	—	8	0	Frame sealants and caulks
Viking 5488[d]	Butyl acrylate	Surfactant	50	11 500	8	16	Ceramic tile adhesives to BS 5980 (Class B)
Mowilith 9716	2-Ethylhexyl acrylate	Surfactant	50	300	7.5	9	Water-resistant ceramic tile adhesives to BS-5980 (Class AA)

[a] Determined at 105 °C for 1 h in a vacuum oven.
[b] Determined using Brookfield viscometer speed 10 at 23 °C.
[c] Manufactured under licence from Hoechst AG.
[d] Manufactured by Viking Polymers.

Table 8.73 Pure acrylate copolymer latices for adhesives (Harlow Chemical Company Ltd, Harlow, Essex, UK)

Grade	Primary monomers	Stabiliser	Solids (%)[a]	Viscosity (mPa s)[b]	T_g[c] (°C)	pH	Particle size (μm)	Applications
Revacryl 396	2-Ethylhexyl acrylate/acrylonitrile	Surfactant	58	1000	−55	4.5	0.3	Pressure-sensitive adhesives for permanent tapes and labels, medium peel and high shear strength
Revacryl 398	2-Ethylhexyl acrylate/acrylonitrile	Surfactant	60	800	−58	8	0.4	Pressure-sensitive adhesives for permanent tapes and labels, medium peel and high shear strength
Revacryl 491	2-Ethylhexyl/butyl acrylate; vinyl acetate	Colloid	58	10	−34	5	0.8	Pressure-sensitive adhesives with high peel and low shear strength for permanent adhesion; good low-temperature adhesion
Revacryl 620	Butyl/2-ethylhexyl acrylates	Surfactant	59	1200	−48	7.5	0.2	High-performance deep freeze pressure-sensitive adhesives (for transfer coating processes)
Revacryl 623	Butyl/2-ethylhexyl acrylates	Surfactant	55	250	−60	7	0.3	Pressure-sensitive adhesives (for transfer coating processes)
Revacryl 630	Butyl/2-ethylhexyl acrylate/vinyl acetate	Surfactant	68	1300	−52	3.5	0.4	Pressure-sensitive adhesives, high shear and tack; good compatibility with resin tackifiers
Revacryl DP4064	Butyl acrylate	Surfactant	57	500	−58	6.5	0.25	Pressure-sensitive, with low adhesion; used for removable/peelable labels and tapes
Revacryl 480	Butyl acrylate/methyl methacrylate	Colloid	50	2000	−4	8.5	0.5	Cold-seal and packaging adhesives

[a] Determined at 105 °C for 1 h in a vacuum oven.
[b] Determined using Brookfield viscometer speed 10 at 23 °C.
[c] Determined by differential scanning calorimetry (DSC).

Table 8.74 *Rovene* carboxylated styrene–butadiene copolymer latices for adhesives (Ameripol Synpol Corporation–Mallard Creek Polymers Inc., Akron, Ohio, USA)

Grade number	TS (%)	Styrene–butadiene ratio	Viscosity (mPa s)	pH	Particle size (nm)	Particle charge	T_g (°C)	Crosslinking[a]	Applications
9410	50	25 : 75	500	8.7	0.15	Anionic	−61	Y	Very soft latex; high MW for low-temperature adhesives
9423	50	35 : 65	300	8.7	0.15	Anionic	−45	Y	Very soft latex; general-purpose permanent label adhesives; high adhesion
6105	50	45 : 55	150	8.7	0.25	Anionic	−32	Y	PS compounding base for high tack adhesives
6130	50	45 : 55	300	8.7	0.2	Anionic	−32	Y	Compounding base for high tack PS labels
4151	50	45 : 55	300	9.0	0.15	Anionic	−30	S	Textile and non-woven fabric binder
4176	50	50 : 50	200	9.0	0.15	Anionic	−22	Y	Textile and non-woven fabric binder
4150	50	55 : 45	300	9.0	0.15	Anionic	−16	Y	Carpet and upholstery backcoating
4817	53	56 : 42	300	8.7	0.2	Anionic	−14	Y	Carpet backing
4437	53	62 : 38	300	8.5	0.2	Anionic	−9	Y	Medium handle carpet backing latex
4040	50	65 : 35	150	6.8	0.2	Non-ionic	−8	N	Adhesives, caulks, mastics, cements
4179	50	65 : 35	200	9.0	0.15	Anionic	−7	S	Polymer coatings; non-woven binder
4106	50	90 : 10	200	6.5	0.20	Anionic	+76	N	Hard latex film, with high gloss

[a]Crosslinking: Y = crosslinkable with external crosslinking agent; N = non-crosslinking; S = self-crosslinking.

Table 8.75 'Teksyn' styrene-butadiene latices for pressure-sensitive adhesives (American Synpol Corporation–Mallard Creek Polymers Inc., Akron, Ohio, USA)

Grade number	TS (%)	Styrene–butadiene ratio	Emulsifier type[a]	Viscosity (mPa s)	pH	Particle size (mm)	T_g (°C)	Applications
8330	55	0 : 100	R	200	11.2	0.17	−80	Soft polybutadiene films; high peel strength; rapid tack; rapid break; good water and low temperature resistance
8331	55	0 : 100	S	150	8.2	0.17	−80	Soft polybutadiene films; good shear stability and low temperature film flexibility; laminating natural latex compatible
4810	55	10 : 90	S	300	10.7	0.17	−62	Soft films; high peel strength; rapid tack; good water resistance
4800	55	24 : 76	R	300	8.0	0.17	−52	PS adhesives; contact adhesives; sprayable adhesives; laminating
4822	55	30 : 70	R	250	10.7	0.17	−42	Rapid adhering PS adhesives; contact adhesives; laminating; good water resistance
4823	55	30 : 70	S	300	8.2	0.17	−42	Wide application: PS and flooring adhesives; high peel strength
4824M	55	30 : 70	S	250	9.0	0.17	−42	Lower viscosity version of 4823 for PS and contact adhesives
4823L	55	30 : 70	S	250	8.2	0.17	−42	Very soft, water-resistant films; contact adhesives; PS adhesives
4825	55	40 : 60	S	250	8.2	0.17	−29	High shear/peel strength mastics; PS and contact adhesives; laminating
4826	55	40 : 60	R	250	11.0	0.17	−29	High shear strength; natural rubber compatible; rapid break
8305	55	50 : 50	S	150	8.5	0.17	−24	High shear/peel strength mastics and flooring adhesives; oil-extendable water-resistant films
8300	200	65 : 35	S	200	8.5	0.17	−15	High shear strength mastics; ceramic tile adhesives

[a] S = rosin acid; S = synthetic anionic surfactant.

Table 8.76 Polyvinyl acetate homo- and copolymer latices for adhesives (Reichhold Inc., Durham, north Carolina, USA)

Trade name and number	Type	Application	Solids (%)	pH	Viscosity[a]	Particle size (nm)	T_g (°C)[b]	Protective system
Plyamul 40351-00	Vinyl acetate homopolymer	Multipurpose adhesive base for wood and paper	54–56	4–5.5	1400–2000	0.2–3.0	30	PV-OH
Plyamul 40354-00	Vinyl acetate homopolymer	Higher viscosity version of 40351-00	54–56	4–5.5	2000–2600	0.2–3.0	30	PV-OH
Plyamul 40359-00	Vinyl acetate homopolymer	Higher viscosity version of 40354-00	54–56	4–5.5	3000–3600	0.2–3	30	PV-OH
Pace 381, 382, 383	Vinyl acetate homopolymers	Low VOC versions of above (<1000 ppm VAM)	54–56	4–5.5	As above	0.2–3	30	PV-OH
Plyamul 40305-00	Speciality vinyl acetate polymer	Adhesive base for high-speed packaging of coated paper/film	54–56	4.5–5.5	1000–2000	0.2–2.5	34	PV-OH
Plyamul 40315-00	Speciality vinyl acetate polymer	Multipurpose adhesive base for uncoated papers and film	54–56	4.5–5.5	800–1200	0.2–2.5	36	PV-OH
Plyamul 97825-00	Speciality vinyl acetate polymer	Crosslinkable base for water-resistant wood adhesives; RF curable; freeze–thaw stable	49–51	4.5–5.5	3000–4500	0.2–3	35	PV-OH
Plyamul 97897-00	Speciality vinyl acetate polymer	Borax-stable adhesive base for uncoated papers; water-resistant film	57–59	4.5–5.5	1000–2200	0.2–1.5	33	Hydroxyethyl cellulose
Plyamul 97945-00	Speciality vinyl acetate polymer	Crosslinkable base for water-resistant wood and paper adhesives; fast setting; RF curing; borax stable	50–52	5–5.5	4000–5000	0.2–2	34	Hydroxyethyl cellulose

[a] Measured by Brookfield RV at 20 rpm.
[b] Measured by differential scanning calorimetry.

Table 8.77 Elvace ethylene–vinyl acetate copolymer latices for adhesives (Reichhold Inc., Durham, North Carolina, USA)

Grade	Ethylene content	Application	Solids (%)	pH	Viscosity[a]	Particle size (mm)	T_g (°C)	Protective system
40701-00	Medium	Multipurpose adhesive base for uncoated paper and board, cotton cloth	>54	4–5.5	2000–3000	0.2–2	11	PV-OH
40705-00	High	Multipurpose adhesive base; PVC and polyester film lamination	>54.5	4–5.5	1900–2800	0.2–2	0	PV-OH
40706-00	High	High-viscosity version of 40705-00	>54	4–5.5	3000–4200	0.2–2	0	PV-OH
40709-00	Medium	Multipurpose adhesive base; PVC and polyester film lamination; good water resistance; compatible with fully hydrolysed PV-OH	54–56	4–5.5	2700–3700	0.2–2	1	PV-OH
40713-00	—	Multipurpose adhesive base; good adhesion to difficult surfaces, including PVC and metallized surfaces	>54	4–5.5	700–1300	0.2–2	−4	PV-OH
40722-00	—	Multipurpose adhesive base; fast-setting; crosslinkable; self-thickening; good adhesion to metals including Al foil; wide formulation latitude	55–57	4–5.5	700–1300	0.2–2	11	Hydroxy-ethyl cellulose
40724-00	Low	Multipurpose adhesive base for uncoated paper and PVC; fast setting; good water resistance	>54.5	4.5–5.5	1500–2500	0.2–2	19	PV-OH
97955-00	High	Multipurpose adhesive base for uncoated paper, PVC and polyester; forms remoistenable films; easy-clean version of 40705-00	56–60	4–6	1500–2500	0.2–2	0	PV-OH

[a] Measured by Brookfield RV at 20 rpm.
[b] Measured by differential scanning calorimetry.

Table 8.78 Styrene- and acrylic-based latices for adhesives (Reichhold Inc., Durham North Carolina, USA)

Grade	Type	Application	Solids (%)	pH	Viscosity[a]	Particle size (mm)	T_g (°C)	Protective system
Synthemul 97967-00	Carboxylated acrylic; high peel, low cohesive strength	PS adhesive base; good tack, UV stability; borax stable; bonds polyester films	55–57	4.8–5.3	<300	0.25–0.35	−50	Anionic
Synthemul 97982-00	Carboxylated acrylic; medium peel strength and tack	PS adhesive base; good clarity and UV stability; good coating rheology; for polyester films	55–57	4.8–5.2	500–1200	0.3–0.4	−45	Anionic
Synthemul 97982-01	Carboxylated acrylic	Lower viscosity version of 97982-00	55–57	4.8–5.2	<500	0.3–0.4	−45	Anionic
Synthemul 40402-00	Carboxylated styrene acrylic	Multipurpose adhesive base; good adhesion to metals, mastics	48–50	7.5–8.3	200–500	0.2–0.3	10	Anionic
Tylac 68202-00	Carboxylated styrene butadiene	Multipurpose adhesive base to laminating; carpet backing	43–45	9–9.2	<250	0.05–0.15	−5	Anionic
Tylac 68217-00	Non-carboxylated rosin acid emulsifier	GR-S 2002 type	51–53	9.8–10.8	<1000	0.1–0.2	−27	Anionic
Tylac 68219-00	Carboxylated styrene butadiene	Multipurpose adhesive base; for foil laminating	51–53	7.5–8.5	<500	0.15–0.25	−11	Anionic
Tylac 68200-00	Carboxylated styrene butadiene	Pressure-sensitive adhesive base; high shear and cohesive strength	50–52	8.5–9.5	<500	0.15–0.25	−9	Anionic
Tylac 68221-00	Carboxylated styrene butadiene	Pressure-sensitive adhesive base; good stability	52–54	8–9	<500	0.15–0.25	−40	Anionic
Tylac 97924-00	Polybutadiene	Adhesive formulations	44–46	8.8–9.2	100–400	0.15–0.25	−75	Anionic

[a] Measured by Brookfield RV at 20 rpm.
[b] Measured by differential scanning calorimetry.

Table 8.79 *Resad* polyvinyl acetate homo- and copolymer latices for adhesives (Resad Polymers Ltd, Andover, Hampshire, UK)

Product	Solids (%)	Viscosity (Pa s)[a]	pH	T_g (°C)	Particle size (mm)	Protection system	Applications
Homopolymers:							
A 1350	50	120	5	16	1–2	PV-OH	Base for building and packaging adhesives
A 1470	50	130	5	17	1–3	PV-OH	Base for packaging and wood adhesives
A 311	55	100	5	16	1–2	PV-OH	Medium viscosity; used for packaging and wood adhesives
A 128	55	30	5	15	1–2	PV-OH	Fast-setting medium viscosity packaging and converting paper adhesive
A 130	55	50	5	17	1–2	PV-OH	Fast-setting medium to high viscosity packaging adhesive
A 237	55	140	5	16	1–2	PV-OH	General-purpose adhesive base with good stability
A 206	55	10	5	15	1–2	PV-OH	Low-viscosity fast setting base for packaging adhesives
A 640	55	60	5	17	1–4	PV-OH	Base with Newtonian viscosity for building, wood and packaging adhesives
A 730	60	180	5	17	1–5	PV-OH	High solids base for packaging and wood adhesives
J 105	61	12	5	15	0.5–2	Dextrine	Base for remoistenable adhesives
Vinyl acetate/maleic and acrylic ester copolymers:							
PS 330	51	21	5	<0	1	Cellulose ether and surfactant	Low-viscosity base for pressure-sensitive and speciality adhesives
PS 370	54	80	5	<0	0.5–2.5	PV-OH and surfactant	Base for pressure-sensitive and speciality adhesives
EA 232	55	25	5	5	0.5–3.5	PV-OH and surfactant	Base for speciality adhesives

[a]Measured by Brookfield RVT at 25 °C and 20 rpm.

Table 8.80 'Rhodotack™'base copolymer latices for pressure-sensitive adhesives (Rhodia Ltd, Manchester, UK)

Grade	Applications	Latex characteristics		Tack	Film characteristics of final peel	Shear	Main properties	Main applications
		TS (%)	Viscosity (mPa s)					
R 294	Permanent adhesive	50.0	500	Low	High	Very high	Very high cohesive film strength; high final adhesion; high clarity; UV resistant; plasticiser tolerant	Suitable for bonding permanent and repositionable labels, PVC and other synthetic films and double-sided tapes
R 315	Permanent adhesive	50.0	700	Medium	High	High	Balanced cohesive/ adhesive bonding; high clarity: UV resistant; plasticiser tolerant	Base for permanent label adhesives; single-and double-sided tapes and decorative PVC
R 343	Permanent adhesive	49.5	1400	High	High	Medium	High tack; bonds well to polyolefins; good clarity/ UV resistance; excellent water resistance	Film labels; transfer tapes; single-and double-sided tapes

Table 8.80 (continued.)

Grade	Applications	Latex characteristics		Tack	Film characteristics of final peel	Shear	Main properties	Main applications
		TS (%)	Viscosity (mPa s)					
R 346	Low-temperature adhesive	58.0	1400	Medium	Medium	Low	Medium tack; low cohesion; high adhesion at low temperature; good adhesion to moist surfaces	Chill temperature labels; repositionable labels; abrasive sheets
R 393	Low-temperature adhesive	53.0	500	Low	Low	Very low	Low tack; low cohesion; high adhesion at very low temperatures; good adhesion to moist surfaces	Formulation base for deep-freeze labels; base for removable labels
R 300	Removable adhesive	58.0	1100	Low	Low	Low	Low final adhesion; low tack; good ageing behaviour	Removable labels
R 361	Removable adhesive	48.5	800	Medium	Medium	Very high	Very high cohesion; moderate adhesion	Repositionable labels; decorative PVC

Table 8.81 Primal latices for pressure-sensitive adhesives (Rohm & Haas Inc., Philadelphia, Pennsylvania, USA)

Product	T_g (°C)	Solids (±0.5) (%)	pH	Typical viscosity (mPa s)	Description	Principal application				
						Packaging tape	Speciality tape	Removable	Film label	Paper label
PS-83D	−48	53	9.1–9.8	200–600	High peel/tack adhesive for direct application to substrates by Meyer bar	x	x			
EP-5471	−48	54	9.0–9.5	<300	Low-viscosity version of PS-83D, for reverse gravure application and clear overlaminates	x	x			
EP-5550	−48	52	9.0–9.8	200–600	High tack and peel, with high-temperature shear resistance; for speciality double-sided Al tapes		x			
N-580	−48	55	7.8–9.0	125 <	Moderate peel/shear/tack; good overall balance of PS properties; good filler compatibility		x	x		
N-582	−48	55	7.5–8.7	200 <	Good peel/tack with high shear; wide applications		x			
PS-67	−41	50.5	9.3–9.8	1000–3000	High tack/moderate shear; for film labels and decals		x		x	x
EP-5560F	−52	53	7.5–8.5	100–400	Acrylic latex for permanent paper labels; highly tackifier compatible; good adhesion to polyolefins				x	x

Table 8.81 (continued.)

Product	T_g (°C)	Solids (±0.5) (%)	pH	Typical viscosity (mPa s)	Description	Principal application				
						Packaging tape	Speciality tape	Removable	Film label	Paper label
PS-20	−45	53.5	9.0–9.5	1000–3000	Ready-to-use adhesive for removable applications		x	x	x	x
E-1961D	−39	52.5	8.5–9.5	1000–2500	High shear resistance: used for repositionable applications		x	x		
EP-6010	−41	54.5	7.5–8.0	100–400	Ready-to-use adhesive; high tack and shear, for bonding to treated polyolefins; for file labels, polyolefins and decals		x		x	
EP-5510	−48	53	9.1–9.8	300–600	Ready-to-use adhesive, with high tack and shear strength for low-temperature applications	x	x			
E-2875	−55	52.5	9.0–10.0	750–2500	Ready-to-use adhesive; for low-temperature applications				x	x
P-91	−37	53	9.0–10.0	150–600	Ready-to-use adhesive; for high-speed tape production		x	x		
PS-90	−48	53	9.1–9.8	125–300	Ready-to-use adhesive; for high-speed tape production		x			

Table 8.82 *Primal* latices for non-pressure-sensitive adhesives (Rohm & Haas Inc., Philadelphia, Pennsylvania)

Product	T_g (°C)	Solids (±0.5 %) (%)	pH	Typical viscosity (mPa s)	Description
R-253	32	40	8.9–9.5	100<	Acrylic release coat for paper tapes; masking and packaging; good solvent resistance and release properties on ageing
P-376 LO	18	50	8.0–8.6	200<	Hydrophobic; good water resistance; for wet and dry laminating adhesives
CEC-1050	−2	59	6.5–9.5	1600<	Thermally crosslinkable; for high-bond adhesives for wet laminations
N-495	−3	58	3.5–5.5	1500–3000	Good adhesion, wet grab, rapid set to aluminium, vinyl films, wood, hardboard, paper and steel
LC-40	−5	55	4.0–5.0	150<	Soft flexible film, with rapid water release; good adhesion to viscose, vinyl and aluminium films, and other metal surfaces
L-90D	−5	41	6–9	30<	Ready-to-use acrylic latex, for flexible film laminates for food packaging
LE-1126-M	−22	60	3.5–4.5	100–500	Low-energy cure latex; crosslinkable at room temperature after addition of catalyst; good adhesion and water release; for general wet and dry lamination; also used as leather finish

(continued overleaf)

Table 8.82 (continued)

Product	T_g (°C)	Solids (±0.5 %) (%)	pH	Typical viscosity (mPa s)	Description
Primal 3362 (= CA 162)	−45	61.5–62.5	7–8	100–400	Anionic; acrylic latex for high-performance low-modulus sealants; high adhesion with low internal strength, even with long outdoor exposure
Primal CA-187	−27	60–61	6.5–7.5	130<	High solids latex for solvent and plasticiser-free adhesives for floor coverings
Primal CM-219	+19	53–54	6.8–7.8	200<	Styrene–acrylic latex with good penetration sealing and adhesion, with high wet adhesion in bonding coat formulations, notably ceramic tile adhesives

Table 8.83 *Texicote* and *Texicryl* latices for construction adhesive applications (Scott Bader Company Ltd., Wellingborough, Northamptonshire, UK)

Polymer type	Solids (%)	Viscosity (P at 25 °C)	pH	Particle size (μm)	MF-T (°C)	Applications
Polyvinyl acetate	50	90–140	4–6	1–2	15	General-purpose adhesive base; ceramic tile adhesives to BS 5980 (Class B)
Polyvinyl acetate	53	110–160	4–6	1–2	7	Universal bonding agent to BS 5270
Polyvinyl acetate	40	60–100	4–6	0.5–1.5	7	Ceramic tile adhesives to BS 5980 (Class B) (Economy version)
Modified acrylic	60	7–12	8–9	0.25	<0	Binder for water-based caulks and sealants

Table 8.83 (continued)

Polymer type	Solids (%)	Viscosity (P at 25 °C)	pH	Particle size (μm)	MF-T (°C)	Applications
Acrylic	60	1–4	5–6	0.25	<2	Binder for water-based caulks and sealants—medium elongation
Styrene–acrylic	45	0.3–1.2	9.5–10.2	0.20	15	Water-resistant ceramic tile adhesives and grouts to BS 5980 Class AA
Styrene–acrylic	50	0.5–1.5	7.5–8.5	0.20	27	Water-resistant ceramic tile adhesives and grouts to BS 5980 Class AA; no ammoniacal odour
Vinylidene chloride copolymer	55	0.5–2.0	5–7	0.20	<2	Binder for fire-retardant water-based caulks and sealants

Table 8.84 'Ucar' latices for pressure-sensitive adhesives (Union Carbide Corporation, Danbury, Connecticut, USA)

Grade number	TS (%)	Viscosity (mPa s)^a	pH	T_g (°C)	Surfactant	Applications
171	50	<100	7.0	−40		Self-crosslinking acrylic latex; high shear strength; low tack; low peel; blends with other latexes to improve shear resistance.
173	50	500	5.5	−35	Anionic	Crosslinkable acrylic latex with moderate peel and tack, and high shear; used for tapes and permanent labels on paper, PVC and other flexible laminates, polypropylene, carpet adhesives
174	50	600	6.0	−45	Anionic	Crosslinkable acrylic latex with high peel and tack, and moderate shear strength; used for polypropylene box sealing tape, permanent labels and vinyl foam composites

(continued overleaf)

Table 8.84 (*continued*)

Grade number	TS (%)	Viscosity (mPa s)[a]	pH	T_g (°C)	Surfactant	Applications
175	49	1500	5.0	−60	Anionic	Crosslinkable acrylic latex, with very high peel and tack, for low-temperature labels; blends with other labels of this series to improve peel adhesion and tack
177	50	400	5.5	−50		Crosslinking acrylic latex with balanced peel, tack and shear strength for polypropylene sealing tape, permanent labels and flexible laminates; low foaming under high-speed coating conditions
178	50	100	6.0	−37		Self-crosslinking acrylic latex with lower peel and tack (compared with 173); used for flexible laminates and automotive applications (at elevated temperatures) and other high-temperature tapes

[a]Brookfield LVT at 60 rpm and at 25 °C (No. 3 spindle).

Table 8.85 Base polymer *Ucecryl* latices for pressure-sensitive adhesives (UCB Chemicals Ltd, Vilvoorde, Brussels, Belgium)

Product	Solids (±1 %)	Viscosity[a]	pH	Characteristics and end uses
BA 90	67	750	4.5	Base polymer for permanent label formulations
DS	59	850	4.0	Base polymer for vinyl flooring sealants and adhesives; medium-high cohesion, moderate tack and adhesion
FC 88	64	900	2.0	Base for vinyl labels and decals
PC 80	64	950	6.0	Base formulation for permanent paper labels and foam tapes; good ageing characteristics, aggressive tack, self-thickenable

Appendix: some technical products for water-based adhesives 655

Table 8.85 (*continued.*)

Product	Solids (±1 %)	Viscosity[a]	pH	Characteristics and end uses
RO	60	150	2.0	General-purpose base polymers (PVC and paper labels, vinyl tapes)
913	62	950	4.5	Base polymer for non-permanent end uses, paper labels, vinyl films and labels, book covering, etc.
WB980	64	500	5.0	Base polymer for PC 80
WB 1117	64	1000	4.5	Base polymer for medical adhesives for plasters and tapes
WB 1220	59.5	850	4.5	Base adhesive for permanent polyethylene/oriented polypropylene labels
WB 1227	60	800	5.0	Higher cohesive version of WB 1220
WB 1265	60	400	7.5	Base adhesive for permanent label paper formulation and speciality tapes

[a] Brookfield viscosity in mPa s at 25 °C.

Table 8.86 Polyvinyl acetate homopolymer latices for adhesives (Vinamul plc, Carshalton, Surrey, UK)

Homopolymer	Polymer type	Stabiliser	TS (%)	Viscosity (mPa s)[a]	pH	MFT (°C)[b]	Particle size (nm)	Applications
Vinamul 8330	PV-OAc	Dextrin	65	1500	4.5	9	0.2–0.8	Envelope adhesives
Vinamul 84125	PV-OAc	PV-OH	60	2500	4.5	16	1–2	Paper, wood, ceramic tiles
Vinamul 84148	PV-OAc	PV-OH	45	12 500	4.5	16	1–2	Paper, wood, ceramic tiles
Vinamul 8481	PV-OAc	PV-OH	55	300	4.5	18	1–2	Paper, envelopes, ceramic tiles
Vinamul 99071	PV-OAc +15 % plasticiser	PV-OH	40	12 500	5.0	5	1.5–4	Ceramic tiles, cement additives, bonding aids building adhesives

[a] Average viscosity on Brookfield RVT at 20 rpm and 25 °C.
[b] Minimum filming temperature (°C).

Table 8.87 Polyvinyl acetate copolymer latices for adhesives (Vinamul plc, Carshalton, Surrey, UK)

Product	Copolymer type	Stabiliser	TS (%)	Viscosity $(mPa\,s)^a$	pH	MFT $(°C)^b$	Particle size (nm)	Applications
Vinamul 31161	VOAc-ethylene	PV-OH	60	5000	5.5	0	0.2–3	Paper, wood foil lamination, coated surfaces
Vinamul 3253	VOAc-ethylene	PV-OH	55	3250	4.5	0	0.5–2	Parer, packaging, films and foils, envelope adhesives
Vinamul 3254	VOAc-ethylene	PV-OH	50	5000	4.5	0	0.5–2	Paper, coated surfaces, films and foils
Vinamul 3265	VOAc- ethylene	PV-OH	55	3000	4.5	0	1–3	Paper, coated surfaces, envelopes
Vinamul 3267	VOAc-ethylene	PV- OH	55	5500	4.5	—	—	Paper and board, coated surfaces
Vinamul 3281	VOAc-ethylene	PV-OH	55	2500	4.5	0	1–3	Cement additive, building adhesives
Vinamul 3485	VOAc-VCl-ethylene	Surfactant	50	75	5.5	22	0.1–0.5	Ceramic tiles
Vinamul 3490	VOAc-VCl-ethylene	Surfactant	50	75	7.0	—	—	Ceramic tiles
Crosslinking systems:								
Vinamul 31102	VOAc-ethylene-crosslinker	Surfactant	59	600	5.0	—	—	Reactivatable adhesives; flooring adhesives
Vinamul 3479	VOAc-VCl ethylene-crosslinker	Surfactant	50	200	4.7	22	0.1–1	Reactivatable adhesives
Vinamul 3305	VOAc-ethylene-crosslinker	Surfactant	50	400	4.5	0	0.2–1.5	Reactivatable adhesives; flooring adhesives

^{a} Average viscosity on Brookfield RVT at 20 rpm and 25 °C.
^{b} Minimum filming temperature (°C).

Table 8.88 Acrylate-based homopolymer latices for adhesives (Vinamul plc, Carshalton, Surrey, UK)

Product	Copolymer type	Stabiliser	Non-volatiles (%)	Viscosity (mPa s)[a]	pH	MFT (°C)[b]	Particle size (μm)	Applications
Vinacryl 4344	Acrylic	Surfactant	45	50	5.0	—	—	Reactivatable adhesives
Vinacryl 4512	Acrylic	Surfactant	45	100	5.6	—	—	Removable pressure-sensitive adhesives
Vinacryl 7172	Styrene-acrylic	Surfactant	50	100	6.5	11	0.2–1	Reactivatable adhesives
Vinacryl 71264	Styrene-acrylic	Surfactant	50	1000	6.5	18	0.1–0.2	Ceramic tile adhesives

[a] Average viscosity on Brookfield RVT at 20 rpm and 25 °C.
[b] Minimum filming temperature (°C).

Table 8.89 *Vinavil* group polyvinyl acetate homopolymer latices for adhesives (Vinavil SpA, Milan, Italy)

Trade name	Solids (±1 %) (%)[a]	Viscosity (Pa s)[b]	pH	Particle size (nm)	T_g (°C)[c]	Protective system	Main adhesive applications
Ravemul O 12	60	25–35	4	0.5–2	14	PV-OH	Wood; paper; building products
Vinavil C/AR	60	35–45	5	0.6–2	15	PV-OH	Wood; paper; building products
Ravemul O 13	60	2.5–3.5	4	0.3–3	14	PV-OH	Wood; paper; textile lamination
Ravemul O 112	55	25–35	5	0.3–2	14	PV-OH	Wood; paper; building products
Vinavil RP	55	5–9	4.5	0.2–0.6	15	Anionic/non-ionic surfactant	Heat-seal adhesives; textile lamination
Vinavil SA	52	1.5–3.5	4	0.2–0.6	15	Anionic/non-ionic surfactant	Heat-seal adhesives
Vinavil PA	51	20–30	3.5	0.4–3	14	PV-OH	Paper adhesives
Ravemul O 16	50	33–47	4	0.3–2	14	PV-OH	Wood; paper; building products
Ravemul O 17	50	7–9	4	0.6–2	14	PV-OH	Wood; paper; building products
Ravemul O 40	40	30–40	4	0.6–2	14	PV-OH	Wood; paper
Vinavil MV	50	6–10	3.5	1–2	14	PV-OH	Wood; textile finishing; compounding
Vinavil KA/R	50	40–60	4.5	0.6–2	14	PV-OH	Wood; paper; building products; textile lamination

[a] By ISO 1625.
[b] Measured by Brookfield RVT at 20 °C and 20 rpm.
[c] By ISO 2115.

Table 8.90 Polyvinyl acetate homopolymer latices (Wacker-Chemie GmbH, Munich, Germany)

Grade	Polymer basis	Solids (±1 %) (%)	Viscosity (Pa s)[a]	Spindle	MFT (°C)	pH	Particle size (nm)	Plasticiser	Protective colloid system	Food acceptability (to BGA XIV)
D 50	V-OAc	50	26 ± 8	6	+14	4	1–3	—	PV-OH	+
D 50 G	V-OAc	50	30 ± 10	6	0	5	1–3	—	PV-OH	—
M 50	V-OAc	50	29 ± 10	6	+14	4	0.5–2	—	PV-OH	+
M 50/300	V-OAc	50	1 ± 0.3	3	+14	4	0.5–2	—	PV-OH	+
Z 50	V-OAc	50	4.5 ± 2	5	+16	4.5	0.5–2	—	Cellulose derivative	+
A 50	V-OAc	50	0.1 ± 0.03	1	+18	5	0.15	—	Surfactant	+
H 60	V-OAc	60	35 ± 10	6	+14	4	1–3	—	PV-OH	+
H 65	V-OAc	65	30 ± 10	6	+14	4	1–3	—	PV-OH	+
H 54/15 C	V-OAc	54	4 ± 1	3	0	4	1–3	15 % DBP[b]	PV-OH	—
H 65 C	V-OAc	65	18 ± 6	6	+9	5	1–3	7 % DBP	PV-OH	—

[a]Measured by Brookfield RTV at 23 °C.
[b]DBP = dibutyl phthalate.

Table 8.91 Vinnapas polyvinyl acetate homopolymer latices applications (Wacker-Chemie GmbH, Munich, Germany)

Application	D 50	D 50 G	M 50	M 50/300	Z 50	A 50	H 60	H 65	H 54/15 C	H 65C
Wood adhesives	**	**	*	—	—	—	**	**	—	*
'Do-it-yourself' adhesives	—	**	—	—	—	—	*	—	—	**
Untreated paper and board	**	—	**	*	—	—	**	*	—	—
Bookbinding	**	—	**	—	—	—	**	**	—	*
Wall covering adhesives	*	—	*	—	—	—	*	*	—	—
Flocking adhesives for leather and textiles	*	—	*	**	—	*	**	*	*	—
High-frequency welding adhesives	*	*	**	—	*	**	*	*	—	**

Key: * = suitable, ** = very suitable.

Table 8.92 Vinnapas ethylene-vinyl acetate copolymer latexes properties (Wacker-Chemie GmbH, Munich, Germany)

Grade	Polymer basis[a]	Solids (±1 %) (%)	Viscosity (Pa s)[b]	Spindle	MFT (°C)	pH	Particle size (μm)	Plasticiser	Protective colloid[c]	Food acceptability (to BGA XIV)
EP 1	E/V-OAc	50	9 ± 3	5	0	4.5	0.5–2	—	PV-OH	+
EP 11	E/V-OAc	50	5 ± 1	4	0	4.5	1	—	PV-OH	+
EV 2	E/V-OAc	50	10 ± 4	5	0	4	0.5–1	—	HPC	+
EP 14	E/V-OAc	55	5.5 ± 1.5	4	0	4.5	1	—	PV-OH	+
EP 400	E/V-OAc	55	2.4 ± 0.4	3	0	4.5	0.8	—	PV-OH	+
EP 24	E/V-OAc	57	12 ± 3	5	0	4.5	1	—	PV-OH	+
EP 17	E/V-OAc	60	3.5 ± 1	3	0	4.5	0.9	—	PV-OH	+
EF 41	E/V-OAc	65	3.5 ± 1	3	0	4.5	0.3	—	S	+
EAF 59	E/A/V-OAc	60	1 ± 0.3	2	0	4.5	0.3	—	S	+
EAF 60	E/A/V-OAc	60	12 ± 0.3	5	0	5.5	0.1–0.8	—	S	+
EAF 61	E/A/V-OAc	60	0.9 ± 0.3	2	0	4.5	0.3	—	S	+

[a]Monomers: V-OAc = vinyl acetate, E = ethylene, A = acrylic ester.
[b]Measured by Brookfield RTV at 23 °C.
[c]Protective colloid/emulsifiers: PV-OH = polyvinyl alcohol, HPC = high polymer compound, S = surfactant mixture.

Table 8.93 *Vinnapas* ethylene–vinyl acetate copolymer latices applications (Wacker-Chemie GmbH, Munich, Germany)

Application	EP 1	EP 11	EV 2	EP 14	EP 400	EP 24	EP 17	EF 41	EAF 59	EAF 60	EAF 61
'Do-it-yourself' adhesives	*	**	—	—	—	**	**	—	—	—	—
Untreated paper and board	*	**	—	*	**	**	*	*	—	—	—
Coated paper	**	*	**	**	*	**	**	**	*	*	*
Adhesive raw materials	—	—	—	—	—	—	—	—	*	**	**
Bookbinding	—	*	—	*	*	**	**	**	—	—	—
PVC foil for furniture	**	*	—	**	—	**	*	*	—	—	—
Plastic films	**	*	**	**	*	**	**	**	*	*	*
Wall covering adhesives	*	**	*	*	**	**	*	**	*	*	*
Flooring adhesives	—	—	*	*	*	**	**	**	**	**	*
Flocking adhesives for leather and textiles	*	*	—	*	**	**	*	*	—	—	—
High-frequency welding adhesives	—	*	—	—	*	*	—	—	—	—	—

Key: * = suitable, ** = very suitable.

Table 8.94 *Vinnapas* co- and terpolymer latices (Wacker-Chemie GmbH, Munich, Germany)

Grade	Polymer basis[a]	Solids (±1) (%)	Viscosity (Pa s)[b]	Spindle	MFT (°C)	pH	Particle size (μm)	Protective colloid[c]	Food acceptability (to BGA XIV)
50/5 VL	VOAc/VL	50	7 ± 2	6	+14	4	0.5–3	PV-OH	+
A50/25VL	VOAc/VL	50	0.075 ± 0.05	1	0	5	0.1	S	+
MV 70 H	VOAc/M	50	11 ± 5	5	0	6.5	0.5–2	S	+
MV 70 N	VOAc/M	50	0.5 ± 0.2	3	0	5.5	0.5–2	S	+
MLP 70	V-OAc/M/VL	50	10 ± 3	4	0	1–3	0.5–2	PV-OH + S	+
T53/22VL	V-OAc/VL/VC	53	9 ± 3	5	+15	4.5	0.5–2	PV-OH	+

[a]Monomers: V-OAc = vinyl acetate, E = ethylene, A = acrylic ester, VL = vinyl laurate, VC = vinyl chloride, M = maleic ester.
[b]Measured by Brookfield RTV at 23 °C.
[c]Protective colloid/emulsifiers: PV-OH = polyvinyl alcohol, S = surfactant mixture.

Table 8.95 'Vinnapas' co- and ter-polymer latices: applications (Wacker-Chemie GmbH, Munich, Germany)

Applications	50/5 VL	A50/25 VL	MV70 H	MV70 N	MLP 70	T53/22VL
Bonding to untreated paper and board	*	—	—	—	**	**
Bonding to coated paper	—	—	*	*	*	**
Adhesive raw materials	—	—	**	**	—	—
Plastics film adhesives	—	—	*	*	*	—
Flocking adhesives for leather and textiles	—	**	—	—	*	—
High-frequency welding adhesives	*	**	—	—	—	**

Key: * = suitable, ** = very suitable.

REFERENCES

1. H. Warson, *Applications of Synthetic Resin Emulsions* Ernest Benn, London, 1972
2. D.C. Blackley, *High Polymer Latices*, Vols. 1 and 2, Maclaren & Sons, London, 1966; *Polymer Latices: Science and Technology*, 2nd edn, 3 Vols., Chapman & Hall, London, 1997; *Emulsion Polymerisation*, Applied Science Publishers, London, 1975
3. P.A. Lovell and M.S. El-Aasser (eds.), *Emulsion Polymerization and Emulsion Polymers*, John Wiley & Sons, Chichester, 1997
4. I. Skeist (ed.), *Handbook of Adhesives*, 3rd edn, Van Nostrand-Reinhold, New York, 1990 (especially H.L. Jaffe, F.M. Rosenblum and W. Daniels, Ch. 21, pp. 381–400); see also D. Distler (ed.), *Wässriger Polymerdispersionen: Synthese, Eigenschafen, Anwendungen*, Wiley–VCH, Weinheim, 1999 (especially D. Urban and E. Wistuba, *Anwendungen in der Klebstoffindustrie*, Ch. 7, pp. 125–70); I. Bendek and L.J. Heymans, *Pressure Sensitive Adhesive Technology*, Marcel Dekker, New York, 1997; I. Bendek, *Development and Manufacture of Pressure-Sensitive Products*, Marcel Dekker, New York, 1999.
5. A. Pizzi and K.L. Mittal (eds.), *Handbook of Adhesive Technology*, Marcel Dekker, New York, 1994
6. D.E. Packham (ed.), *Handbook of Adhesives*, Longmans Scientific, Harlow, 1992
7. *J Adhesion*, Gordon & Breach, New York and London, from 1969
8. *J Adhesion Sci. Technol.*, VSP BV Publishers, Zeist, Netherlands, from 1987
9. *Int. J. Adhesion Adhe.*, Butterworth-Heinemann, Oxford, from 1980
10. *Adhäsion*, Heinrich Vogel Fachzeitschriften, Munich, from 1954
11. *Adhes. Age*, Argus Business Magazines, Alberta, Georgia, from 1955, Chemical Week Associates, New York, from 1999.
12. *European Adhes. Sealants*, FMJ Publications, Redhill, from 1983
13. W.C. Wake, in *Synthetic Adhesives and Sealants*, Critical Reports on Applied Chemistry Vol. 16, Wiley for Society of Chemical Industry, Chichester, 1987, p. vii
14. See, for example, N.J. DeLollis, *Adhesives, Adherents, Adhesion* Ch. 2, 1989: Malabar, Florida Robert E Krieger Publishing Co. Inc. Malabar, Florida, 1989, Ch. 2 (Previously published in *Handbook of Adhesive Bonding*, (C.V. Cagle (ed.)), McGraw-Hill, New York, 1973, pp. 2/1–2/16
15. L.-H. Lee (ed.), *Fundamentals of Adhesion*, Plenum Press, New York, 1991
16. L.-H. Lee (ed.), *Adhesive Bonding*, Plenum Press, New York, 1991
17. V.L. Vakuta and L.M. Pritykin, *Polymer Adhesion: Physico-chemical Principles*
18. J.R. Hunstberger, *J. Paint Technol.*, **39**(507), 199 (1967)
19. A.J. Kinloch, *J. Mater. Sci.*, **15**, 2141 (1980); **17**, 617 (1982)
20. L.-H. Lee, *SPIE J.*, **1999**, 6 (1993); *Chem. Abstr.*, **121**, 181 252 (1994)
21. W.C. Wake, *Polymer*, **19**, 291 (1978)
22. W.A. Zisman, *Ind. Engng Chem.*, **55**, 18 (1963)
23. L.H. Sharpe and H. Schornhorn, *Adv. Chem. Ser.*, **43**, 189 (1964)
24. S.S. Voyutskii, *Autohesion and Adhesion of High Polymers*, Interscience, New York, 1963
25. P.G. de Gennes, *J. Chem. Phys.*, **55**(2), 572 (1971); see also *Soft Interfaces*, Cambridge University Press, Cambridge, 1997, pp. 43–78
26. See, for example, L.-H. Lee (ed.), *Fundamentals of adhesion*, Plenum Press, New York, 1991 (especially E.P. Plueddemann, pp. 279–90 and J.D. Miller and H. Ishida, pp. 291–324)
27. L.H. Lee, *J. Polym. Sci., Part A-2*, **5**, 751, et seq. (1967)
28. K.W. Allen, *Int. J. Adhesion Adhes.*, **13**(2), 67 (1993)
29. L.A. Girifalco and R.J. Good, *J. Phys. Chem.*, **61**, 904 (1957)
30. F.M. Fowles, in *Physico-Chemical Aspects of Polymer Surfaces*, K.M. Mittal (ed.), Plenum Press, New York, 1983, pp. 583–603
31. See, for example, K.W. Allen, J.E.D. Spencer and B.O. Field, in *Adhesion 13*, K.W. Allen (ed.), 1988: Elsevier Applied Science, London, 1988, pp. 278–93

32. E.A. Theiling, *Farbe ü Lacke*, **12**(11), 1183 (1971)
33. S.A. Parry and P.F. Ritchie, *Adhesion*, **11**(5), 201 et seq. (1967)
34. P.G. Eisel and H. Voss (Herberts GmbH), GP 4 236 672 (1994); *Chem. Abstr*, **121**, 258 008 (1994)
35. K.-H. Michl, *Eur. Adhes. Sealants*, **6**(1), 34–6 (1989)
36. P.J. Weiss, *J. Polym. Sci., Part C*, **12**, 169 (1966)
37. B.H. River and R.H. Gillespie, Report FRSP-FPL-400, 1981; Order Ad-D 106565, 1982; *Chem. Abstr*, **96**, 200 635 (1982)
38. BIOS and FIAT Reports were published in 1946–7, describing production processes of German factories at that time, with many details of operations. The principal report on polymer latex technology is : FIAT Final Report 1102, by S.J. Baum and R.D. Dunlop, London, HM Stationery Office, London, 1947
39. For details see C.A. Finch (ed.), *Polyvinyl alcohol: Properties and Applications* Wiley, Chichester, 1973 (especially K. Toyoshima, Polyvinyl alcohol in adhesives, Ch. 16 and E.V. Gulbekian and G.E.J. Reynolds, Polyvinyl alcohol in emulsion polymerization, Ch. 17)
40. See also C.A. Finch (ed.), *Polyvinyl Alcohol: Developments*, Wiley, Chichester, 1992 (especially D.B. Farmer, Applications of polyvinyl alcohol in emulsion polymerization, Ch. 13 and C.A. Finch, *Applications of polyvinyl alcohol in adhesives*, Ch. 15)
41. C.A. Finch (ed.), *Polyvinyl Alcohol: Developments*, Wiley, Chichester, 1992 Appendix 1, p. 759
42. P. Anton and A. Mayer (Novocel SA), FP 2 664 280 (1992); *Chem. Abstr*, **117**, 132 177 (1992)
43. L.P. Kovacs (Anheuser-Busch Inc.), USP 3 769 248 (1973)
44. J.S. Van Arkel, in *Cellulose–Its Derivatives* J.F. Kennedy (ed.), Ellis Horwood, Chichester, 1985, pp. 301–10
45. D. Asai and T. Mitsuda, *JP Applic.* 93 007 (1994); *Chem. Abstr*, **121**, 135 024 (1994)
46. R.M. Mohsen, A.M. Ramadan, K.A. Shaffei and A.S. Badran, *Pigment Resin Technol.*, **22**(4), 4–7, 17 (1993); *Chem. Abstr*, **120**, 300 541 (1994)
47. E.A. Theiling, *Farbe ü Lacke*, **77**(12), 1183–92 (1971); *Adhäsion*, **12**, 428 (1972)
48. Other suppliers include: Unitex Chemical Corporation, Greensboro, North Carolina, USA; Kalama Chemicals Inc., Kalama, Washington, USA; Croda Chemicals Ltd, Leek, UK; Pentagon Chemicals Ltd, Workington, UK
49. Velsicol Chemical Corporation Technical Bulletin 35-A-2
50. J.D. Grandmer and R.H. Hunter, USP 2 889 207 (1959)
51. Supplier: Eastman Chemical Corporation, USA
52. Suppliers include: BASF AG, Ludwigshafen, Germany; BP Chemicals, UK; ICI, UK; Occidental Petroleum Inc., USA; Union Carbide, USA
53. Supplier: Hoechst AG, Frankfurt, Germany
54. J.S. Kelyman, D.H. Guthie and H.F. Hussey, *Mod. Paint Coat.*, **76**(10), 155 (1986)
55. Suppliers include: Albright & Wilson plc, Oldbury, West Midlands, UK; Bayer AG, Leverkusen, Germany
56. See, for example, H. Stecher, *Adhäsion*, **7**, 311 (1963)
57. K. Moser *et al.* (Lonza AG), BP 1 240 812 (1971)
58. S. Kordzinski and M.B. Horn (Ashland Oil Inc.), USP 3 579 490 (1971)
59. For a general account of tackifier resins in all adhesives, including latex-based adhesives, see J.A. Schademan, Tackifier resins, in *Handbook of Pressure-Sensitive Adhesives*, D. Satas (ed.), Van Nostrand–Reinhold, New York, 1982, pp. 353–369
60. Supplier: Exxon Chemicals Inc., Akron Ohio, USA and subsidiaries
61. Supplier: Hercules Inc., Wilmington, Delaware, USA and subsidiaries
62. Suppliers: Lawter Chemical Inc, Northbrook, Illinois, USA; Neville Chemical Inc., Pittsburgh, Pennsylvania, USA
63. Supplier: Reichhold Chemicals Inc., White Plains, New York, USA

64. Supplier: Goodyear Chemicals Inc., Akron, Ohio, USA
65. H.A. Pace and V.J. Anhorn (Goodyear Tire and Rubber Corp.), USP 3 577 398 (1968) (see FP 1 525 781, (1968) (BP 1 161 007), *Chem. Abstr*, **71**, 13 742(1969)
66. Arizona Chemical Co Inc., Wayne, New Jersey, USA
67. K.K. Sunstar, JP applic. 225 173 (1983); *Chem. Abstr*, **101**, 56 197 (1984)
68. G. Wouters, J. Callibaut and A. Lepert (Exxon Inc.), USP 5 268 399; *Chem. Abstr*, **120**, 136 100 (1994)
69. M. Kishi, JP applic. 150 359 (1988); *Chem. Abstr*, **109**, 212 202 (1988)
70. Nitto Electric KK, JP applic. 65 179 (1985); *Chem. Abstr*, **103**, 38 417 (1985)
71. Y. Matsumoto, (Soken Chemical Engineering KK), *JP applic.* 92 372 (1985); *Chem. Abstr*, **104**, 51 744 (1986)
72. D. Roberts *et al.* (Uniroyal Inc.), USP 3 538 025 (1970)
73. N.C. Jain and R.S. Shah, *IPIRI*, **5**(1), 31 (1975)
74. K. Kaito and H. Ishizuka (Aica KK), JP 97 639 (1976); *Chem. Abstr*, **85**, 193 821 (1976)
75. J.A. Simms (Du Pont de Nemours Inc.), GP 520 590 (1971)
76. T. Morikawa, T. Midorikawa, K. Iwasaki and H. Yamazaki (Toyo Ink Manufacturing Co.), *JP applic.* 234 781; *Chem. Abstr*, **116**, 61 268 (1992)
77. M.K. Lindemann (Chas. S Tanner Co.), USP 4 115 306 (1978)
78. E.M. Jones and P.P. Puletti (National Starch & Chemical), Eur. P Applic. 643 115 (1995); *Chem. Abstr*, **123**, 34 822 (1995)
79. T. Shimada *et al.* (Sankyo-Kokusaku Pulp KK), *JP applic.* 22 884 (1988); *Chem. Abstr*, **109**, 94 439 (1988)
80. K. Ito, Y. Kawada, and K. Kunibe (Dai-ichi Kigenso KK), JP 99 237 (1973); *Chem. Abstr*, **80**, 121 920 (1974)
81. Borden Inc., USP 3 677 883 (1972); GP 2 011 725 (1971); FP 2 053 768–9 (1971)
82. W. Shimokawa (Hoechst Gosei KK), JP 202 176 (1985); *Chem. Abstr*, **104**, 110 995 (1985)
83. H. Isaki and M. Suzuki, JP 93 449 (1974); *Chem. Abstr*, **82**, 16 266 (1975)
84. P.J. Moles, *J. Oil Coll. Chem. Ass.*, **72**(8), 301 (1989)
85. M. Yamamoto and M. Furadate (Kanebo KK), JP 53 932 (1974); *Chem. Abstr*, **81**, 121 784 (1974)
86. Wacker-Chemie GmbH, BP 1 539 664 (1979)
87. K. Nishinaka and S. Haba (Sekisui Chemical Co. KK), JP Applic. 227 481 (1990); *Chem. Abstr*, **114**, 44 608 (1991)
88. A.E. Pierce (Borden Inc.), USP 3 213 051 (1965)
89. Y. Doi and T. Kato (Showa HighPolymer Industries KK), JP Applic. 96 078 (1974); *Chem. Abstr*, **82** 99 402 (1975)
90. J.C. Baatz (Shawinigan Resins Ltd), USP 3 197 429 (1965)
91. F.H. Sharpell, *Adhes. Age*, **25**(4) 23 (1982)
92. From Dow Chemical Co., Midland, Michigan, USA.
93. From Zeneca plc (formerly ICI), UK
94. See, for example, R.E. Porzel and C.D. Bator (De Soto Inc.) USP 4 273 688 (1981)
95. T.M. Gouding, Polyvinylautate Wood Adhesives in *Wood Adhesives: Chemistry and Technology*, A. Pizzi (ed.), Marcel Dekker Inc., New York, 1983, Ch. 7, pp. 319–50; see also K.R. Geddes, Chemistry of PVA, in *Polymerization Using Continuous Reactors*, Vol. 2, 1989, Ch. 2, pp. 31–73.
96. From Monsanto Inc., Technical Bulletin PL-307
97. T.P. Gladstone Shaw, J.D.P.E. Mercier and H.M. Collins (Shawinigan Resins Ltd), BP 649 505 (1961)
98. W.J. Opie (National Adhesives Ltd), BP 906 652 (1962)
99. A.R. Robbins (Stein-Hall & Co. Inc.), USP 2 996 462 (1961)
100. Cemedine KK, JP Applic. 133 141 (1982); *Chem. Abstr*, **97**, 199 041 (1982)
101. M. Kubota and K. Highashidani (Hoechst-Gosei KK), JP 44 976 (1972); *Chem. Abstr*, **80**, 121 983 (1974)

References

102. See reference [42]
103. H. Wiest and W. Lechner (Wacker-Chemie), GP 2 349 925 (1974) (Belg. P 820 700); *Chem. Abstr*, **83**, 60 705 (1975)
104. T. Hayazaki, J. Nakajima, M. Oishi, K. Nishioka *et al.* (Sekisui Chemical KK), JP Applic. 330 009 (1994); *Chem. Abstr*, **122**, 216 168 (1995); T. Hayazaki and M. Ooishi, JP Applic. 336 582 (1994); *Chem. Abstr*, **122**, 216 173 (1995)
105. J.S. Van Arkel, in *Cellulose and Its Derivatives*, J.F. Kennedy (ed.), pp. 301–10; see also reference [44]
106. Oshika Shinko KK, JP 139 20 (1980); *Chem. Abstr*, **95**, 116 613 (1981)
107. Konishi KK, JP 53 575 (1985); *Chem. Abstr*, **103**, 23 539 (1985)
108. R.L. Burke (United Merchants Inc.), USP 3 547 852 (1970)
109. Scott, Bader & Co. Ltd., FP 2 008 156 (1970)
110. C.T. Fazioli, J. Sirota and R.A. Weidener (National Starch & Chemical), GP 2008 011 (1970)
111. See reference [57]
112. T. Tamura and S. Kondo (Matsushita Electrical KK), JP 40 879 (1970); *Chem. Abstr*, **77**, 102 677 (1972)
113. K. Kimura and W. Shimokawa, JP 43 431 (1973); *Chem. Abstr*, **73**, 105 933 (1973)
114. M. Fujii, T. Yamashita, H. Seno and K. Kuwahara (Kanebo KK,) JP 6 936 (1973); *Chem. Abstr*, **80**, 4 258 (1974)
115. E. Matsumoto and M. Yamamoto, JP Applic. 121 854 (1974); *Chem. Abstr*, **82**, 141 241 (1975)
116. N. Sakato and H. Nakamura (Shin-Etsu Chemical Co. KK), JP Applic. 130 933 (1974); *Chem. Abstr*, **83**, 11 370 (1975)
117. See, for example, W.T. Tseng, F.S. Lin and Y. Kyo, JP Applic. 286 333 (1988); *Chem. Abstr*, **110**, 174 388 (1989); JP Applic. 310 634 (1988); *Chem. Abstr*, **113**, 212 844 (1990); and JP Applic. 312 829 (1988); *Chem. Abstr*, **110**, 194 381 (1989)
118. G.M. Knutson and J. Biale (Union Oil of California Inc.), USP 3 769 151 (1973)
119. W.E. Lenney (Air Products & Chemicals Inc.), Eur. P Applic. 389 893 (1991); see also M. Nakamae, N. Fujiwara, K. Terada, D. Miyake, K. Yuki, T. Sato and H. Maruyama (Kuraray Co. KK), Eur. P Applic. 692 494 (1994); *Chem. Abstr*, **124**, 203 314 (1996)
120. Y.J. Shih (National Starch & Chemical Inc.), Eur. P Applic. 373 313 (1990); *Chem. Abstr*, **113**, 192 639 (1990)
121. Y. Fujita (Nippon Synthetic Chemical Industry KK), JP 118 467 (1979); *Chem. Abstr*, **92**, 59 666 (1980)
122. T. Ishibashi, K. Akasaka and A. Oshita (Nippon Synthetic Chemical Industry KK), JP 154 438 (1979); *Chem. Abstr*, **92**, 154 052 (1980)
123. S. Masuda, T. Ohkubo and Y. Furakawa (Electrochemical Industry KK), GP 2 019 233 (1970) (FP 2 046 302 (1970)); *Chem. Abstr*, **74**, 13 745 (1971)
124. A. Harada, T. Okamoto and F. Tsukyama (Showa HighPolymer Industries KK), JP Applic. 100 751 (1994) *Chem. Abstr*, **121**, 135 535 (1994)
125. G.M. Knutson (Union Oil of California Inc.), USP 3 661 696 (1976)
126. W.E. Lenney and J.G. Iacoviello (Air Products & Chemicals Inc.), Eur. Pat Applic. 279 384 (1988); *Chem. Abstr*, **109**, 232 097 (1988)
127. E. Inaba (Sumitomo Chemical Co. KK), JP 18 134 (1974); *Chem. Abstr*, **81**, 50 728 (1974)
128. S. Misono (Takeno KK), JP Applic. 130 691 (1979); *Chem. Abstr*, **92**, 95 308 (1980)
129. K. Tone, N. Tsuji, T. Adaki and T. Ogawa *et al.* (Shoei KK), JP Applic. 235 875 (1985); *Chem. Abstr*, **104**, 169 748 (1986)
130. K. Sato (Chuo Rika KK), JP Applic. 252 280 (1986); *Chem. Abstr*, **106**, 197 587 (1987)
131. F. Ishikawa, N. Araki and K. Okazaki (Konishe Gisuke KK), JP 13 206 (1974); *Chem. Abstr*, **82**, 18 175 (1975)
132. O. Ohhara *et al.* (Kuraray Co. KK), JP 26 345 (1974)

133. C. Tomisawa, T. Oyamada, K. Nitta and H. Tanaka (Sumitomo Chemical KK), JP Applic. 161 647 (1979); *Chem. Abstr*, **92**, 182 224 (1980)
134. T. Oyamada, C. Tomizawa and M. Domoto (Sumitomo Chemical KK), GP 2 718 716 (1977); *Chem. Abstr*, **88**, 23 906 (1978)
135. P.J. Blincow and M.T. Sarkis (Vinamul Ltd/Unilever plc), Eur. P Applic. 530 013 (1994); *Chem. Abstr*, **120**, 23 438 (1994)
136. Showa Highpolymer Co. KK, JP 12 078 (1982); *Chem. Abstr*, **96**, 163 707 (1982); T. Okamoto, T. Kondo and H. Hirano (Showa Highpolymer Co. KK), JP Applic. 170 060 (1996); *Chem. Abstr*, **125**, 198 130 (1996)
137. M. Yako, T. Kondo and H. Hirano (Denki KK), JP Applic. 93 243 (1994); *Chem. Abstr*, **121**, 11 391 (1994)
138. E.C. Hornaman (Air Products & Chemicals Inc.), Can. P 2 145 938 (1995); *Chem. Abstr*, **124**, 178 542 (1996); G.S. Casebolt (American Cyanamid Inc.), USP 2 502 510 (1952); *Chem. Abstr*, **46**, 6 435 (1952); see also USP 2 818 397 (1952); *Chem. Abstr*, **52**, 5 888 (1957)
139. AG Hoechst, BP 1 440 337 (1956)
140. R. Tomioka, Z. Ninomiya Y. Nishiyama and F. Yoshino (Dai-Nippon Ink KK), JP Applic. 91 547 (1991); *Chem. Abstr*, **91** 212 156 (1979)
141. M.C. Bricker and P.R. van Rheenen (Rohm & Haas Inc.), Eur. P Applic. 640 675 (1995); *Chem. Abstr*, **122**, 216 225 (1995)
142. T. Hayazaki, M. Ooishi and M. Hanashita (Sekisui Chemical Co. KK), JP Applic. 240 219 (1994); *Chem. Abstr*, **122**, 189 217 (1995)
143. N. Okinaga and T. Hayakaki (Sekisui Chemical Co. KK), JP Applic. 292 333 (1995); *Chem. Abstr*, **124**, 178 441 (1996)
144. H. Wiest, E. Lieb and H. Schaefer (Wacker-Chemie GmbH), GP 2 512 589 (1976); *Chem. Abstr*, **85**, 178 436 (1975)
145. M.I. Bjurvald (Casco AB), GP 3 402 134 (1974)
146. M. Nishida (Konishe Gisuke KK), JP 16 751 (1974)
147. See J. Teppema (Borden Inc), USP 2 902 458–9 (1959); W.B. Armour (National Starch & Chemical Inc.), USP 3 041 302 (1962); W.B. Armour and W.C. Kania (National Starch & Chemical Inc.), USP 3 284 280 (1966)
148. K. Motohashi and B. Tomita, *Mokazai Gakkaishi*, **25**(3), 225 (1979); *Chem. Abstr*, **90**, 204 926 (1979); K. Motohashi, B. Tomita and H. Mizumachi, *Holzforschung*, **36**(4), 183 (1982); See also H. Warson, *Applications of Synthetic Resin Emulsions*, Ernest Benn, London, 1972, pp. 289–90, 390
149. M.K. Lindemann (Ciba-Geigy Inc.), Can. P 1 074 480 (1980); *Chem. Abstr*, **93**, 96 378 (1980)
150. T. Araki, S. Hagazawa and N. Okizu (Konishi Yoshisuke Shoten KK), JP 39 358 (1970); *Chem. Abstr*, **75**, 110 825 (1971)
151. E.B. Baker (Borden Inc.), USP 3 600 353 (1971)
152. P.V. Petrov and S.N. Zigelboim *Derevobrab Prom.*, **10**, 3 (1974)
153. Borden Inc., BP 1 204 818–9 (1970); *Chem. Abstr*, **73**, 110 520 (1971)
154. S.R. Sandler (to Borden Inc.), GP 2 006 805 and 2 005 806 (1971); *Chem. Abstr*, **76**, 4 506/4 507 (1972); GP 2 011 725 (1971); *Chem. Abstr*, **76**, 46 900 (1972); GO 2 010 332 (1971); *Chem. Abstr*, **76**, 60 584 (1972)
155. J. Dufek, V. Ulbricht, L. Formaner, Z. Ditryck and S. Klicka, Cz. P 182 570 (1980); *Chem. Abstr*, **93**, 205 838 (1980); see also H. Warson, *Applications of Synthetic Resin Emulsions*, Ernest Benn, London, 1972, pp. 299–300
156. S. Naito (Marutama Mokuzai KK), JP 17 314 (1970); *Chem. Abstr*, **74**, 32 309 (1971)
157. Borden Inc., USP 3 542 706 (1970)
158. Y.I. Zetler *et al.*, USSRP 293 019 (1971)
159. National Starch & Chemical Inc., BP 1 333 000 (1968); USP 3 444 037 (1969) (BP 1 034 497); *Chem. Abstr*, **65**, 9 210 (1966)

160. W.B. Armour and R.A. Fass (National Starch & Chemical Inc.), USP 3 300 430 (1967) (BP 1 034 497); *Chem. Abstr*, **65**, 9 210 (1966)
161. W.B. Armour (National Starch & Chemical Inc.), USP 3 274 048 (1966)
162. M. Yamao and K. Maeda (Fudow Chemical Co. KK), JP 23 809 (1974)
163. N. Sakato and H. Nakamura (Shin-Etsu Chemical Co. KK), JP 24 636 (1976)
164. R. Lanthier (Shawinigan Chemicals Ltd), BP 1 092 030 (1967); *Chem. Abstr*, **68**, 22 368 (1968)
165. Borden Inc., FP 2 154 837 (1975)
166. L.K. Lindemann (Chas. S Tanner Co.), USP 3 942 735 (1976); *Chem. Abstr*, **84**, 181 320 (1976); USP 4 001 160 (1976); *Chem. Abstr*, **86** 104 401 (1977); FP 2 302 330; *Chem. Abstr*, **86**, 156 520 (1977)
167. E. Matsumoto and M. Matsumoto, JP 111 956 (1974); *Chem. Abstr*, **83**, 29 110 (1975)
168. K. Kobayashi, JP 102 646 (1975); *Chem. Abstr*, **84**, 6 145 (1976)
169. M. Sakai, H. Hano and M. Yoshino (Nippon Oil KK), JP 31 596 (1975); *Chem. Abstr*, **84**, 106 988 (1976)
170. Shoei Chemical KK, JP 99 270 (1981); *Chem. Abstr*, **96**, 20 839 (1982)
171. Y.-Y. Lu *et al.* (Minnesota Mining & Manufacturing Co. Inc.), Eur. P Applic. 788 029 (1996) (PCT P WO 96/05871); *Chem. Abstr*, **125**, 41 878 (1996)
172. See, for example, M. Bornack and K.R. NcNally (NL Industries Inc.), USP 4 810 747 (1989); *Chem. Abstr*, **111**, 117 000 (1989); B.A. McCarthy *et al.*, Belg. P 905 477 (1987); USP 4 804 573; *Chem. Abstr*, **107**, 135 436 (1987)
173. T.E. Bublitz and L.S. Timm (Findlay Adhesives Inc.), Eur. P Applic. 555 830 (1994); *Chem. Abstr*, **120**, 109 595 (1994)
174. P. Pfoehler, H. Seyrffer, J. Vietmeyer, J. Tuer, W. Druschke and K. Schagerer (BASF AG), Eur. P Applic. 344 659 (1990); *Chem. Abstr*, **112**, 180 943 (1990)
175. A.C. Lavoie (Rohm & Haas Inc.), Eur. P Applic. 350 240 (1990); *Chem. Abstr*, **112**, 236 061 (1990)
176. M. Kashima and A. Nagasuga (Sekisui Chemical KK), JP Applic. 255 417 (1993); *Chem. Abstr*, **120**, 109 187 (1994)
177. Nitto Electric Industrial Co. KK, JP 102 795 (1982); *Chem. Abstr*, **98** 5 195 (1983)
178. K. Kinoshita, K. Fukukasa T. Fuzumi and N. Takahata (Nisshim KK), JP Applic. 49 108 (1994); *Chem. Abstr*, **121**, 109 913 (1994)
179. W.L. Morgan, R.J. Esser and D.P. Jensen (Johnson & Son Inc.), Eur. P Applic. 338 586 (1989); *Chem. Abstr*, **112**, 159 178 (1990)
180. L. Rios, N. Hidalgo, J.Y. Cavaille, A. Guyot, C. Pichot *et al.*, *Coll. Polym. Sci.*, **269**, 812, (1991); *Chem. Abstr*, **115**, 256 852 (1991)
181. H. Nishiike *et al.* (Showa HighPolymer Co. KK), JP Applic. 21 685 (1991); *Chem. Abstr*, **115**, 94 060 (1991)
182. R.A. Kiehlbauch *et al.*, PCT P 91/04 990; *Chem. Abstr*, **115**, 116 005 (1991)
183. S.J. Makower, *et al.* (Rohm & Haas Inc.), Eur. P Applic. 392 767 (1990); *Chem. Abstr*, **114**, 64 487 (1991)
184. C.-S. Chou, M.A. Kesselmayer and A.C. Lavoie (Rohm & Haas Inc.), Eur. P Applic. 576 128 (1993); *Chem. Abstr*, **121**, 10 930 (1994); see also reference [174]
185. T. Ugai, and H. Suzuki, (Toyo Ink KK), JP 117 033 (1978); *Chem. Abstr*, **90**, 73 018 (1979); see also R.F. Petella and E.J. Kuzma (Celanese Corp.), USP 4 113 540 (1978); *Chem. Abstr*, **90**, 55 762 (1979); Y. Ito, JP 134 833 (1978); *Chem. Abstr*, **90**, 105 233 (1979)
186. M. Okuso, Y. Nakamura and M. Asakura, *Nippon Setchaku*, **19**(9), 368(1983)
187. J.W. Hagan, C.B. Mallon and M.R. Rifi, *Adhes. Age*, **22**(3), 29 (1979)
188. Y. Kawachi and E. Hayashi (Sanyo Electrical KK), JP 41 335 (1978); *Chem. Abstr*, **89**, 111 299(1978)
189. N. Numata and K. Kawabeta (Sekisui Chemical Co. KK), JP Applic. 128 543 (1994); *Chem. Abstr*, **121**, 257 470 (1994)

190. M. Miura and N. Hanatani (Sekisui Chemical Co. KK), JP Applic. 145 624 (1992); *Chem. Abstr*, **121**, 282 093 (1994)
191. M. Nakamae and T. Sato (Kuraray Co. KK), JP Applic. 179 705 (1994); *Chem. Abstr*, **121**, 182 415 (1994); J P Applic. 220 414 (1994); *Chem. Abstr*, **121**, 302 715 (1994)
192. Rohm & Haas Inc., BP 1 317 931 (1973) (GP 2 020 496 (1970)); *Chem. Abstr*, **74**, 54 550 (1971)
193. H. Fukazaki et al. (Kao Soap Co. KK), JP 130 829 (1975); *Chem. Abstr*, **84**, 60 752 (1976)
194. N. Murakawa and S. Sano (Mitsui Toatsu KK), JP 112 431 (1975); *Chem. Abstr*, **84**, 18 461 (1976)
195. P. Mallya et al. (Avery Dennison Corp.), Eur. P Applic. 536 146 (1992); PCT Applic. WO 92/19691 (1992)
196. M.M. Bernard (Avery Dennison Corp.), PCT Applic. WO 92/18 072 (1992); *Chem. Abstr*, **120**, 165 829 (1994)
197. K. Kawabata (Sekisui Chemical KK), JP Applic. 311 425 (1996); *Chem. Abstr*, **126**, 118 878 (1997)
198. D.S. Gibbs and R.A. Wessling (Dow Chemical Co.), BP 1 229 510 (1971)
199. R.S. Steiner (Mobil Oil Co. Inc.), USP 4 058 649 (1977); *Chem. Abstr*, **88**, 24 373 (1978)
200. Anon, *Res. Discl.*, **157**, 76 (1977)
201. J.C. Fitch and A.M. Snow, *Adhes. Age*, **20**(10), 23 (1977)
202. A. Takahashi (Kobayashi KK), JP 37 099 (1978); *Chem. Abstr*, **90**, 7 297 (1979); see also H. Huhn et al. (Wolff Walstrode AG), GP 2 725 586 (1978)
203. M. Sato and K. Mochizuki (Denki KK) JP Applic. 256 738 (1994); *Chem. Abstr*, **122**, 189 226 (1995)
204. M. Sato et al. (Denki KK), Eur. P Applic. 626 398 (1995); *Chem. Abstr*, **122**, 267 568 (1995); JP Applic. 287 360 (1994); *Chem. Abstr*, **122**, 293 137 (1995)
205. A.M. Snow, *Adhes. Age*, **23**(7), 35–7 (1980)
206. C.M. Matulewicz and A.M. Snow, *Adhes. Age*, **24**(3), 40 (1981)
207. T. Tsuji and T. Sekya (Denki KK), JP Applic. 298 813 (1994); *Chem. Abstr*, **122**, 134 192 (1995)
208. R.S. Whitehouse, in *European Adhesives and Sealants*, FMJ Publications, Redhill, pp. 1–30
209. R. Satelmeyer and G. Wagner, *Kunstharz. Nachr.*, **35**(10), 29 (1976)
210. Aica KK, JP 69 273 (1983); *Chem. Abstr*, **99**, 177 325 (1983)
211. O. Ackermann, *Adhäsion*, **23**(6), 172 (1979)
212. T. Sando, and F. Tashiro (Hitachi Chemical KK), JP 24 949 (1979); *Chem. Abstr*, **91**, 108 799 (1979)
213. K. Franzen and D. Zimmerman (Beiersdorf AG), GP 3 028 693 (1982); *Chem. Abstr*, **96**, 105 629 (1982)
214. J. Ono and K. Ebihara (Aica Kogyo KK), JP Applic. 330 600 (1993); *Chem. Abstr*, **120**, 219 966 (1994)
215. A.M. Snow, *Adhes. Age*, **20**(6), 35 (1977)
216. Y. Tsutsumi et al. (Tosoh Corporation), JP Applic. 3 423 (1997); *Chem. Abstr*, **126**, 187 169 (1997)
217. M. Katz (Du Pont), Eur. P Applic. 72 223 (1983)
218. T. Watanabe (Hitachi Kasei Polymer KK), JP Applic. 302 315 (1996); *Chem. Abstr*, **126**, 118 862 (1997)
219. R.G. Jahn, *Adhes. Age*, **20**(6), 35 (1977)
220. A. Onoshi et al. (Tao Gosei KK), JP 27 633 (1978); *Chem. Abstr*, **89**, 111 712 (1978)
221. H. Tada et al. (Toyo Ink KK), JP 32 543 (1979); *Chem. Abstr*, **91**, 124 398 (1979)
222. S.G. Takemoto and O.J. Morrison (Avery International Inc.), USP 4 189 419 (1980)
223. A.C. Makati, D.I. Lee, B.W. Greene and R.T. Iwasama (Dow Chemical Co. Inc.), Eur. P Applic. 476 168 (1990); *Chem. Abstr*, **116**, 257 210 (1992)
224. Hirose, J. Hara, S. Suzuki et al. (Mitsui Toatsu KK), JP 137 031 (1979); *Chem. Abstr*, **92**, 130 196 (1980)

225. K. Tsuruoka and S. Nishida (Japan Synthetic Rubber Co.), JP Applic. 229 703 (1991); *Chem. Abstr*, **116**, 107 912 1992)
226. G. Suzuki, T. Morita, K. Sakata and M. Sogie (Japan Synthetic Rubber Co.), JP Applic. 23 730 (1992); *Chem. Abstr*, **117**, 9 483 (1992)
227. T. Hombach, R. Witkowski, O. Ganster and Gl Arend (Bayer AG), Eur. P Applic. 625 558 (1995); *Chem. Abstr*, **122**, 216 073 (1995)
228. W. Fujiwara, J. Hyoda, Y. Toyoda and S. Mishiba (Sumitomo Naugatuck KK), Eur. P Applic. 496 925 (1991); *Chem. Abstr*, **118**, 8 182 (1993)
229. Y. Watabe, A. Hiraharu, Y. Ikeda and T. Maeda (Japan Synthetic Rubber Co.), JP Applic. 331 440 (1993); *Chem. Abstr*, **121**, 85 489 (1994)
230. S. Hayano, Y. Maeda and A. Tanaka (Mitsui Toatsu Chemical Ltd), JP Applic. 192 997 (1994); *Chem. Abstr*, **122**, 12 399 (1995)
231. A. Nakagami, K. Tanako, E. Shiiyama and Y. Mizuta (Mitsui Toatsu Chemicals KK), JP Applic. 245 705 (1996); *Chem. Abstr*, **126**, 8 815 (1997)
232. A.C. Makati, C.S. Kan, R.J. Iwamasi and D.I. Lee (Dow Chemical Co. Inc.), Eur. P Applic. 345 385 (1989); *Chem. Abstr*, **112**, 160 069 (1990)
233. L.A. Spirin, R.I. Gorbunova and N.B. Geodakyan, USSRP 1 775 451 (1992); *Chem. Abstr*, **121**, 204 444 (1994)
234. K. Yoshii, H. Inagaki, M. Iwakiri and Y. Sato (Mitsubishi Petrochemical KK/Toyota Motor KK), JP Applic. 214 308 (1993); *Chem. Abstr*, **120**, 56 317 (1994)
235. K. Kitaoka and M. Tsuji (Takeda Chemical Industries KK), JP Applic. 26 775 (1991); *Chem. Abstr*, **115**, 73 309 (1991)
236. J.A. Fries, *Int. J. Adhesion Adhes.*, **1992**, 187 (1992)
237. See, especially, D. Satas (ed.), *Handbook of Pressure Sensitive Adhesive Technology*, Van Nostrand Reinhold, New York, 1982
238. T.F. Sanderson, *Adhes. Age*, **26**(1), 26 (1983)
239. J. Iwasaki, *Setchaku*, **27**(3), 126 (1983)
240. L.C. Graziano and S.E. Sjoostrand, *J. Plast. Film Sheeting*, **2**(2), 95 (1986)
241. C.B. Mallon (Union Carbide Inc.), BP 2 070 037 (1981)
242. T. Morikawa *et al.* (Toyo Ink Manufacturing Co. KK), JP 234 781 (1991); *Chem. Abstr*, **116**, 61 268 (1992)
243. P. Foreman and P. Mudge, *Proc of Adv. Pressure Sensitive Tape Technol.*, Illinois, 1989, pp. 203–15
244. H.Y. Chao (Moore Business Forms Inc.), Eur. P Applic. 590 839 (1994); *Chem. Abstr*, **121**, 257 423 (1994)
245. K.S.C. Lin (Avery International Inc.), PCT Int. Applic. 5 503 (1986)
246. National Starch and Chemical Co. Inc., JP Applic. 34 973 (1987); *Chem. Abstr*, **107**, 40 990 (1987)
247. T. Yoshikawa, T. Sugii, Y. Moroishi and K. Noda (Nitto Electrical Industrial Co. KK), JP Applic. 74 977 (1987); *Chem. Abstr*, **107**, 116 642 (1987)
248. T. Yoshikawa, I. Sakai, T. Sugii, and S. Wada (Nitto Electrical Industrial Co. KK), JP Applic. 254 678 (1986); *Chem. Abstr*, **106**, 157 621 (1987)
249. B. Bohnel (3M Co. Inc.), USP 4 786 696 (1989); *Chem. Abstr*, **111**, 8 735 (1989)
250. T. Yoshikawa, T. Sugii and I. Sakai (Nitto Electrical Industrial Co. KK), JP Applic. 264 076/7 (1986): *Chem. Abstr*, **106**, 260 076/7; *Chem. Abstr*, **106**, 215 134/5 (1987)
251. H. Nagatomo, I. Sakai, T. Sugii and I. So (Nitto Electrical Industrial Co. KK), JP Applic. 10 181 (1987); *Chem. Abstr*, **106**, 197 614 (1987)
252. A.J. Kielbania and S.S. Kim (Rohm & Haas Inc.), USP 4 619 964 (1986)
253. Y. Ishikawa and K. Yasuda (Kao Corp.), JP Applic. 192 341 (1994); *Chem. Abstr*, **122**, 57 831 (1995)
254. K. Suziki, A. Kunihiro and H. Ohhashi (New Oji Paper KK), PCT Applic. 96/18 703 (1996); *Chem. Abstr*, **125**, 250 078 (1996)

255. R. Tomioka, C. Kakinuma, Y. Shigematsu and S. Takano (DiaNippon Ink & Chemicals KK), JP Applic. 137 978 (1985); *Chem. Abstr*, **104**, 6 838 (1986)
256. H. Jaeger, *Adhäsion*, **29**(2), 35 (1985)
257. S. Benedeck, *Adhäsion*, **31**(3), 22–3, 26–8 (1987)
258. W. Druschke, *Adhäsion*, **31**(5), 29–32 (1987)
259. J.N. Longson (Lo Sound), BP 1 322 721 (1973)
260. R.G. Dolch and R.N. Kerr (Firestone Inc.), USP 4 145 327 (1979)
261. D.B. Farmer and B.H. Quadri (Unilever NV), Eur. P Applic. 332 175 (1989); *Chem. Abstr*, **111**, 155 501 (1989)
262. H. Frank and J.W. Louven (Jowat Lubers and Frank GmbH), Eur. P Applic. 429 998 (1991); *Chem. Abstr*, **114**, 248 947 (1991)
263. N. Hisada and N. Okinaga (Sanyo Chemical Industry KK), JP Applic. 238 078 (1990); *Chem. Abstr*, **114**, 63 832 (1991)
264. M.M. Bernard (Avery International Corp.), PCT P WO 2 759 (1990); *Chem. Abstr*, **115**, 996 (1991)
265. Sekisui Chemical Co. KK, JP Applic. 65 278 (1980); *Chem. Abstr*, **93**, 133 563 (1980)
266. K.S. Kumar, S. Rajadurai, K.G.A. Dev and G. Thyagarajan (CSIR India), Ind. P 170 449 (1992); *Chem. Abstr*, **121**, 207 204 (1994)
267. P.R. Mudge (National Starch & Chemical Inc.), Eur. P Applic. 222 541 (1987); USP 4 753 846 (1988); *Chem. Abstr*, **107**, 212 103 (1988)
268. T. Mihara, K. Yokobori and M. Taniguchi (Denki KK), JP Applic. 67 384 (1990); *Chem. Abstr*, **113**, 61 016 (1990)
269. R.A. Walch, *Proc. TAPPI Polymers Lamination & Coatings Conf.*, **2** 865 (1989)
270. Y. Kanafujii (Sekisui Chemical Co. KK), JP 80 341 (1976)
271. E. Witt (Kendall Co. Inc.), FP 2 467 872 (1981)
272. Soken Chemical Engineering KK, JP 25 378 (1983); *Chem. Abstr*, **99**, 89 236 (1983)
273. M. Atsuji and T. Hatsutori (Toa Gosei Chemical Ind. KK), JP Applic. 161 078 (1989); *Chem. Abstr*, **112**, 37 710 (1990)
274. Nitto Electrical Industrial KK, JP Applic. 49 263 (1981); *Chem. Abstr*, **95**, 188 288 (1981)
275. I. Morino and H. Hashimoto (Asahi Chemical Industries KK), JP Applic. 34 084 (1986); *Chem. Abstr*, **105**, 80 225 (1986)
276. S. Hori and Y. Nakamura (Sekisui Chemical Co. KK), JP Applic. 103 865 (1992); *Chem. Abstr*, **117**, 132 635 (1992)
277. T. Amano and H. Inoue (Sumitomo Chemical KK), JP Applic. 116 311 (1994); *Chem. Abstr*, **121**, 181 554 (1994)
278. M. Sato and A. Nishii (Soken Chemical Engineering Co. KK), JP Applic. 2 023 (1991); *Chem. Abstr*, **115**, 184 971 (1991)
279. S. Usii (Toyo Ink Manufacturing KK), JP Applic. 65 551 (1994); *Chem. Abstr*, **121**, 85 376 (1994)
280. S. Ori and Y. Nakamura (Sekisui Chemical Industry Co. KK), JP Applic. 146 578 (1991); *Chem. Abstr*, **116**, 22 428 (1992)
281. H. Nishiike, K. Abe, F. Kato and F. Tsukiyama (Showa HighPolymer Co. KK), JP Applic. 21 685 (1991); *Chem. Abstr*, **115**, 94 060 (1991)
282. H. Yang (Exxon Chemical Patents Ltd), Eur. P Applic. 530 300 (1993); PCT P 90/15 111 (1990); *Chem. Abstr*, **116**, 85 414 (1992); see also PCT P 91/18 070 (1991)
283. Nitto Electric KK, JP Applic. 22 371 (1981); *Chem. Abstr*, **95**, 26 117 (1981)
284. Y. Fuseya and H. Koyobashi (Mitsui Toatsu Chemicals KK), JP Applic. 91 277 (1986); *Chem. Abstr*, **105**, 192 463 (1986)
285. S. Takano, Y. Shigematsu and R. Tomioka (DaiNippon Ink & Chemicals KK), JP Applic. 73 781 (1986); *Chem. Abstr*, **105**, 173 952 (1986)
286. M.S. Mahil and J.M. Cruden (Harlow Chemical Co. Ltd), Eur. P Applic. 351 193 (1990); *Chem. Abstr*, **113**, 80 036 (1990)
287. P. Ball and R. Weissberger, (Wacker-Chemie GmbH), GP 3 446 565 (1986)

References

288. R. Weissberger, P. Ball, K. Marquardt, M. Selig et al. (Wacker-Chemie GmbH), GP 3 531 601 (1987); *Chem. Abstr*, **107**, 41 280 (1987)
289. M. Tomikanechara, M. Araki, H. Suzuki, S. Takeda et al. (Kanzaki Paper Manufacturing Co. KK), JP Applic. 64 385 (1991); *Chem. Abstr*, **115**, 234 115 (1991)
290. H. Yumama, F. Maekawa and K. Umekage (Kanebo NSC Co. KK), JP Applic. 124 786 (1991); *Chem. Abstr*, **115**, 209 809 (1991)
291. DaiNippon Ink KK, JP Applic. 12 084 (1982); *Chem. Abstr*, **97**, 24 853 (1982)
292. A. Harada, M. Nakano, K. Hirata and F. Tsukiyama (Showa HighPolymer Co. KK), JP Applic. 119 079 (1991); *Chem. Abstr*, **115**, 209 827 (1991)
293. Oji Paper KK, JP Applic. 79 068 (1983); *Chem. Abstr*, **99**, 213 729 (1983)
294. S. Hori and Y. Nakamura (Sekisui Chemical Industry Co. KK), JP Applic. 103 685 (1992); *Chem. Abstr*, **115**, 132 635 (1992)
295. See reference [276]
296. R. Tomioka and Y. Shigimatsu (DaiNippon Ink & Chemicals KK), JP Applic. 79 281 (1989); *Chem. Abstr*, **111**, 215 670 (1989)
297. I.S.P. Lee (Avery International Corp.), PCT Int. Applic. WO 90/10 028 (1990); *Chem. Abstr*, **114**, 44 488 (1991)
298. Hoechst Gosei KK, JP Applic. 7 486 (1983); *Chem. Abstr*, **99**, 106 445 (1983)
299. E. Witt (Kendall Co. Inc.), USP 4 204 023 (1980)
300. A.R.B. Furendal and H.U. Hellstrom (Bofors AB), GP 2 352 353 (1975)
301. Hoechst Gosei KK, JP Applic. 69 274 (1983); *Chem. Abstr*, **99**, 213 729 (1983)
302. Y. Kuroyama, E. Sumimura, T. Yasuda and S. Nakamura, JP Applic. 263 184 (1989); *Chem. Abstr*, **112**, 180 632 (1989)
303. T. Inada, T. Kato and T. Noguchi (Showa HighPolymer Co. KK), JP Applic. 322 345 (1994); *Chem. Abstr*, **122**, 189 963 (1995)
304. P. Anton and A. Mayer (Novacel SA), FP 2 664 280 (1992); *Chem. Abstr*, **117**, 132 177 (1992)
305. N. Bessho, H. Kurinomaru and J. Azuma (Ittsuposha Yushi KK), JP 75 625 (1975); *Chem. Abstr*, **84**, 122 816 (1976); S. Horiki and K. Ito (Nagoya Yukagaku KK), JP 6 235 (1976); *Chem. Abstr*, **84**, 165 839 (1976)
306. Y. Takahashi (Mitsubishi Yuka Badische KK), JP Applic. 3 404 (1990); *Chem. Abstr*, **113**, 7 020 (1990)
307. P. Gleichenhagen and W. Karmann (Beiersdorf AG), GP 2 455 133 (1976); *Chem. Abstr*, **85**, 79 055 (1976)
308. J. Durinda, P. Hudak, E. Moldo and J. Uhlar, Cz. P 211 948 (1982); *Chem. Abstr*, **98**, 162 081 (1983)
309. Y. Moroishi, Y. Tozaki, M. Chaarii, M. Natsume et al. (Nitto Electric Industry KK); JP Applic. 325 909 (1996); *Chem. Abstr,* **126**, 158 531 (1997)
310. N. Haraguchi and Y. Oyama (Sekisui Chemical Ind. Co. KK), JP Applic. 206 672 (1990); *Chem. Abstr*, **114**, 8 067 (1991)
311. J.M. Rosenki and A. Arora (National Starch & Chemical Inc.), USP 5 319 020 (1994); *Chem. Abstr*, **121**, 302 591 (1994)
312. P.K. Agarwal and T. Ougel (Exxon Research & Engineering Co. Inc.), USP 5 066 694 (1992); *Chem. Abstr*, **116**, 61 276 (1992)
313. T. Shinohara, T. Shimada and A. Sugiyama (Asahi-Dow KK), Eur. P Applic. 48 950 (1982); *Chem. Abstr*, **97**, 7 273 (1982)
314. A.C. Makati, B.W. Greene and R.T. Iwamasi (Dow Chemical Co. Inc.), USP 4 968 740 (1991); *Chem. Abstr*, **114**, 208 646 (1991); Eur. P Applic. 476 168 (1992); *Chem. Abstr*, **116**, 257 210 (1992)
315. E.S. Barabas, P. Mallya, S.J. Gromelski and E.J.C. Wotier (Polysar International SA), Eur. P Applic. 52 449 (1982); *Chem. Abstr*, **97**, 93 447 (1982)
316. T. Shimizu, N. Ito, K. Kasai and H. Hirai (Japan Synthetic Rubber Co. KK), JP Applic. 89 871 (1987); *Chem. Abstr*, **107**, 135 506 (1987)

317. Y. Toyada, S. Sekiguchi and H. Utsunomiya (Sumitomo Naugatuck KK), JP Applic. 141775 (1986); *Chem. Abstr*, **107**, 41 209 (1987)
318. T. Ozawa, M. Itagama and M. Sekiya (Nippon Zeon KK), JP Applic. 176 644 (1986); *Chem. Abstr*, **106**, 6 142 (1987)
319. T. Inoue, T. Kimi and T. Sogimura (Nippon Zeon KK), JP Applic. 124 988 (1990); *Chem. Abstr*, **113**, 124 988 (1990); JP Applic. 124 960 (1990); *Chem. Abstr*, **113**, 133 979 (1990)
320. H. Knoepfel (3M Corp.), GP 2 160 185 (1972) (BP 1 376 906 (1972)); *Chem. Abstr*, **77**, 115 595 (1972)
321. Kendall Co. Inc., BP 1 240 136 (1971)
322. 3M Co., BP 1 430 136 (1976); GP 2 327 452 (1973); *Chem. Abstr*, **80**, 134 452 (1973)
323. I.S.P. Lee, P. Keller, R.J. Norman, R.S. Dordick *et al.* (Avery Dennison Corporation), PCT P 91/12 290 (1991); *Chem. Abstr*, **116**, 84 922 (1992)
324. G.J. Antifinger and D.M. Yingling, *Proc. Adhesion Coat Technol., Pressure Sensitive Tape Ind. 80's Sem.*, 1980, pp. 34–54
325. B. Stanger (BASF AG), GP 2 459 160 (1976); *Chem. Abstr*, **85**, 178 389 (1976)
326. T. Sugii, I. Sakai and T. Yoshikawa (Nitto Electric Co. KK), JP Applic. 101 584 (1986); *Chem. Abstr*, **105**, 210 106 (1986); JP Applic. 101 585 (1986); *Chem. Abstr*, **106**, 6 142 (1987)
327. L.O. Ryrfors and S.O. Hassander (Hoechst-Perstorp AG), GP 2 906 968 (1979); *Chem. Abstr*, **91**, 176 156 (1979)
328. T. Yasuda and H. Tamara (National House Ind. KK), JP Applic. 123 881 (1989); *Chem. Abstr*, **111**, 215 749 (1989)
329. R. Hauber (Beiersdorf AG), GP 2 214 293 (1973) (Austral. P 328 972); *Chem. Abstr*, **80**, 84171 (1973)
330. M. Mori *et al.* (Toyo Ink KK), JP 93 458–60 (1974)
331. N. Numata and J. Shimaoaka (Sekisui Chemical Co. KK), JP Applic. 271 634 (1993); *Chem. Abstr*, **120**; 219 866 (1994)
332. N. Shiratori and S. Seki (Asahi Chemical Industries KK), JP Applic. 162 571 (1986); *Chem. Abstr*, **106**, 219 886 (1994)
333. F.W. Brown and L.E. Winslow (3M Co Inc), Eur. P Applic. 180 434 (1986)
334. O. Narimatsu, K. Komatsu, Y. Takemura and Y. Takeuchi (Mitsui Toatsu Chemicals KK), Eur. P Applic. 530 729 (1993); *Chem. Abstr*, **119**, 97 513 (1993)
335. A. Taga, Y. Ono and K. Ishibashi (Aica KK), JP Applic. 268 778 (1986); *Chem. Abstr*, **106**, 215 200 (1987)
336. G. Ley, O. Aydin and P. Ficketsen *et al.* (BASF AG), GP 3 542 367 (1987); *Chem. Abstr*, **107**, 218 818 (1987) (Eur. P Applic. 224 835 (1987))
337. T. Asano, Y. Moroishi and T. Sugii (Nitto Denko Corp. KK), JP Applic. 238 180 (1988); *Chem. Abstr*, **110**, 59 156 (1989)
338. S. Fujita, Y. Moroishi and T. Sugii (Nitto Denko Corp KK), JP Applic. 238 179 (1988); *Chem. Abstr*, **110**, 59 157 (1989)
339. Y. Ozawa and M. Sekya (Nippon Zeon KK), JP Applic. 176 645 (1986); *Chem. Abstr*, **106**, 6 141 (1987)
340. H. Miyasaki, K. Kitazaki, S. Masuda and S. Nagao (Nichiban KK), JP Applic. 261 381–3 (1987); *Chem. Abstr*, **106**, 197 531–3 (1987)
341. W.H. Kesky, J.E. Schuetz and A.D. Hitchman (Dow Chemical Co. Inc.), USP 4 644 032 (1987); *Chem. Abstr*, **106**, 157 704 (1987)
342. From Hercules Inc.
343. L.P. Ellinger (BP Chemicals Ltd), BP 1 314 517 (1973)
344. J.L. Walker and P.B. Foreman (National Starch & Chemical Inc.), Eur. P Applic. 224 795 (1987)
345. J.H.S. Chang (Merck & Co. Inc.), Eur. P Applic. 179 528 (1986)
346. R. Fukata, M. Mori and S. Tahara (Toyo Ink KK), JP Applic. 96 063 (1974); *Chem. Abstr*, **82**, 157 501 (1975)

References

347. T. Sasada, M. Nagano and N. Okinaga (Sekisui Chemical Co. KK), JP 239 425 (1993); *Chem. Abstr*, **120**, 136 093 (1993)
348. B.S. Snyder and D.A. Bors (Rohm & Haas Inc.), Eur. P Applic. 573 142 (1993)
349. J. Yoshii, Y. Kanejima, K. Nitsuta and T. Amano (Sumitomo Chemical Industries KK), JP Applic. 287 248 (1993); *Chem. Abstr*, **120**, 272 713 (1994)
350. B.J. Hutchinson, *J Coated Fabr.*, **12**(1), 46 (1983)
351. O. Shiono (Teraoka Seisakujo KK), JP Applic. 44 926 (1973); *Chem. Abstr*, **81**, 153 505 (1974)
352. M. Akahori (Kanebo KK), JP Applic. 295 335 (1993); *Chem. Abstr*, **120**, 193 734 (1994)
353. F.R. Baker, *TAPPI*, **58**(4), 123 (1975)
354. N.K. Henderson and E. Thomson (Smith & Maclaurin Ltd), GP 2 614 408 (1976); *Chem. Abstr*, **85**, 193 849 (1976)
355. H. Huhn, E. Reinhardt, W. Karsten and H. Rosenthal (Wolff Walstrode AG), GP 2 725 586 (1978); *Chem. Abstr*, **90**, 105 816 (1979)
356. Price KK, JP Applic. 13 867 (1985); *Chem. Abstr*, **102**, 205 376 (1985)
357. Tokyo Printing Ink Mfg Co. KK, JP Applic. 210 982 (1984); *Chem. Abstr*, **102**, 186 240 (1984)
358. Toyo Ink KK, JP 84 776 (1981); *Chem. Abstr*, **95**, 188 308 (1991)
359. Nitto Electric Industrial Co. KK, JP Applic. 68 320 (1984); *Chem. Abstr*, **101**, 73 995 (1984)
360. Daicel KK, JP 53 166 (1981); *Chem. Abstr*, **95**, 116 633 (1981)
361. H. Kigo and M. Koyama (Asahi-Dow KK), USP 4 211 684 (1980); *Chem. Abstr*, **93**, 169 762 (1980)
362. G.L. Duncam (Mobil Oil Co), Swiss P 634 591 (1983)
363. F. Kenji *et al.*, USP 4 291 090 (1981)
364. Aica KK, JP 59 883 (1981); *Chem. Abstr*, **95**, 116 886 (1981)
365. K. Fujisawa *et al.* (Daicel KK), GP 2 727 914 (1977)
366. S.L. Graham and D.O. Plunkett (Dow Chemical Inc.), USP 2 714 728 (1987)
367. L.I.J. Lui and R.H. Steiner (Mobil Oil Inc.), Eur. P Applic. 254 417 (1988)
368. K. Yoshii and S. Narisawa (Sumitomo Chemical KK), JP 61 284 (1979); *Chem. Abstr*, **91**, 176 132 (1979)
369. C. Tomizawa, H. Tanaka and M. Nitta (Sumitomo Chemical KK), JP 78 728 (1979); *Chem. Abstr*, **91**, 176 344 (1979)
370. H. Vollman and M. Siebel (Hoechst AG), GP 3 134 998 (1983); *Chem. Abstr*, **98**, 145 326 (1983) (Eur. P Applic. 74 528 (1981))
371. H. Frank and J.W. Louven (Jowat Lübers and Frank GmbH), Eur. P Applic. 420 998 (1991); *Chem. Abstr*, **114**, 248 947 (1991)
372. D.P. Knechtges and B.K. Kikofalvy (B.F. Goodrich Co. Inc.), USP 4 002 801 (1977)
373. K. Kimimura and M. Sugaya, JP 74 536 (1978); *Chem. Abstr*, **89**, 147 848 (1978)
374. Mitsubishi Petroleum Corp., JP 166 297 (1981); *Chem. Abstr*, **96**, 144 161 (1982)
375. Cemedine KK, JP Applic. 61 035 (1982); *Chem. Abstr*, **97**, 110 849 (1982)
376. Daicel Chemical KK, JP 23 465 (1985); *Chem. Abstr*, **103**, 23 486 (1985)
377. Mitsubishi Petrochemical KK, JP Applic. 8 777 (1984); *Chem. Abstr*, **101**, 8 352 (1984)
378. Asahi-Dow KK, JP Applic. 24 691 (1984); *Chem. Abstr*, **101**, 201 835-6 (1984)
379. Mitsubishi Petrochemical KK, JP Applic. 78 278 (1984); *Chem. Abstr*, **101**, 112 135 (1984)
380. J. Toda, H. Fukushima, S. Kiyama and T. Mizutanai (Kohjin KK), JP Applic. 215 039 (1985); *Chem. Abstr*, **104**, 131 663 (1986)
381. Aica KK, JP Applic. 204 078 (1983); *Chem. Abstr*, **100**, 193 178 (1984)
382. M. Tomikanchara, M. Akara, H. Suzuki and S. Takeda (Kanzaki Paper Manufacturing Co. KK), JP Applic. 64 385 (1991); *Chem. Abstr*, **115**, 234 135 (1991)
383. See reference [373]
384. See reference [375]
385. See reference [373]

386. Y. Tozaki et al. (Nitto Electric Industry KK), JP Applic. 319 469 (1996); *Chem. Abstr*, **126**, 132 486 (1997)
387. B.P. Barth (Union Carbide Inc.), GP 2 420 684 (1974)
388. G. Ackermann, *Adhäsion*, **23**(6), 172 (1979)
389. A.J. Azrak and B.P. Barth, *Adhes. Age*, **18**(6), 23–6, 27–33 (1975)
390. A.M. Snow, *Adhes. Age*, **23**(7), 35 (1980)
391. W. Huster, M. Krell and J. Fischer (Wacker-Chemie GmbH), GP 4 141 168 (1993); *Chem. Abstr*, **120**, 108 716 (1994)
392. E.G. Brugel and J.C. Chen (Du Pont), PCTInt WO 93/20161; *Chem. Abstr*, **120**, 247 465 (1994)
393. Showa HighPolymer Co. KK, JP Applic. 33 310 (1984); *Chem. Abstr*, **101**, 24 476 (1984)
394. Etablissment Soprim SA, FP 2 287 493 (1976)
395. K. Oshima, Y. Watabe, T. Nakajima and N. Sakurai (Japan Synthetic Rubber Co KK), JP Applic. 266 480 (1986); *Chem. Abstr*, **107**, 41 253 (1987)
396. Aica KK, JP Applic. 59 663 (1981); *Chem. Abstr*, **95**, 116 886 (1981)
397. Sekisui Chemical Co. KK, JP Applic. 113 224 (1983); *Chem. Abstr*, **100**, 69 492 (1984); JP Applic. 222 169 (1983); *Chem. Abstr*, **101**, 24 873 (1984)
398. K. Ryosuke and S. Ibata (DaiNippon Ink KK), JP 81 830 (1976); *Chem. Abstr*, **86**, 17 559 (1977)
399. M.L. Brown (Du Pont), GP 2 210 373 (1972)
400. B.P. Barth (Union Carbide Inc.), GP 2 420 684 (1974); *Chem. Abstr*, **82**, 74 124 (1975); See also *Adhes. Age*, **15**(12), 19 (1972); *Chem. Abstr*, **78**, 44 357 (1973)
401. B. Stanger, A. Müller, W. Druschke and H. Jaeger (BASF AG), GP 2 190 459 (1976); *Chem. Abstr*, **85**, 178 389 (1976)
402. R.W. Green (Formica Inc.), USP 4 260 942 (1981)
403. J.C. Padget, D.H. McIlrath and H. Donald (ICI Ltd), Eur. P Applic. 119 698–9 (1984); *Chem. Abstr*, **102**, 7 918–9 (1984) (USP 4 543 386–7)
404. M. Sato, M. Koga, S. Nagasawa, K. Asano et al. (Denki KK), Eur. P Applic. 648 788 (1995); *Chem. Abstr*, **123**, 172 019 (1995)
405. A.R. Furendal, H.U. Rand Hellstrom (Bofors AB), GP 2 355 364 (1974); *Chem. Abstr*, **83**, 60 941 (1975)
406. J.C. Padget and D.H. McIlrath (ICI Ltd), Eur. P Applic. 184 913 (1986); *Chem. Abstr*, **105**, 80 316 (1986)
407. L. Schwartz (Du Pont), Eur. P Applic. 62 343 (1982); *Chem. Abstr*, **98**, 17 815 (1983)
408. M.A. Mirzayan et al., USSRP 483 481 (1975)
409. L. Abramjian (Du Pont), GP 2 500 545 (1975)
410. Aica KK, JP Applic. 217 570 (1983); *Chem. Abstr*, **101**, 39 493 (1984)
411. H. Lactje et al. (BASF AG), Eur. P Applic. 50 226 (1982)
412. W. Perlinkski et al. (National Starch & Chemical Inc.), USP 4 342 843 (1982); S.B. Kiamil (Scholl PLC), Eur. P Applic. 498 524 (1992); *Chem. Abstr*, **117**, 192 566 (1992)
413. H. Dexheimer (Hoechst AG), GP 3 224 755 (1984)
414. R. Muller, *Polym. Paint Coll. J.*, **179**(4243), 571 (1989); *European Adhes. Sealants*, **7**(1), 13 (1990)
415. P. Foreman and P. Mudge, *Advances in Pressure Sensitive Tape Technology*, Illinois, 1989, pp. 203–215
416. Mitsubishi Petroleum Corp., JP Applic. 187 335 (1983); *Chem. Abstr*, **98**, 199 215 (1983)
417. Y. Yoshii and S. Narisawa (Sumitomo Chemical Co. KK), JP 61 284 (1979); *Chem. Abstr*, **91**, 176 132 (1979)
418. K. Sato, M. Yamamoto, A. Kikichi, Y. Yuwabara and H. Iida (Toyo Ink KK), JP Applic. 100 848 (1994); *Chem. Abstr*, **121**, 159 112 (1994)
419. K. Sato, M. Yamamoto, A. Kikichi, Y. Yuwabara and H. Iida (Toyo Ink KK), JP Applic. 100 847 (1994); *Chem. Abstr*, **121**, 135 900 (1994)
420. Cemidine KK, JP Applic. 61 035 (1982); *Chem. Abstr*, **97**, 110 849 (1982)

References

421. J. Ariyoshi and N. Yorimitsu (Arakawa Chemical Industries KK), JP 72 049 (1978); *Chem. Abstr*, **89**, 147 784 (1978)
422. Y. Murayama and K. Nishinaka (Sekisui Chemical Industry KK), JP Applic. 172 376 (1991); *Chem. Abstr*, **116**, 22 576 (1992)
423. Denki KK, JP 152 770 (1980); *Chem. Abstr*, **94**, 140 825 (1980)
424. S. Muroi, I. Morino and H. Hashimoto, *Nippon Setchaku Kyokaishi*, **19**(1), 5 (1982); *Chem. Abstr*, **99**, 39 420 (1983)
425. Kuraray Co. KK, JP Applic. 202 662 (1982); *Chem. Abstr*, **98**, 216 771 (1983)
426. Showa Union Gosei Co. KK, JP Applic. 217 779 (1984); *Chem. Abstr*, **102**, 186 263 (1985)
427. P.K. Agarwal *et al.* (Exxon Research & Engineering Co. Inc.), USP 4 517 250 (1985)
428. Toa Gosei Chemical Industry KK, JP Applic. 84 371 (1985); *Chem. Abstr*, **103**, 88 322 (1985)
429. H.Y. Chao (Moore Business Forms Inc.), Eur. P Applic. 590 839 (1994)
430. J.M. Rosenski and A. Arora (National Starch & Chemical Co. Inc.), USP 5 319 020 (1994); *Chem. Abstr*, **121**, 302 591 (1994)
431. W.F. Scholz and R.H. van Ham (Avery Dennison Corp.), Eur. P Applic. 608 371 (1994); USP 5 196 504 (1993); PCT Applic. 93/08 239 (1993); *Chem. Abstr.*, **119**, 119 100 (1993)
432. J.M. Rosenski and A. Arora (National Starch & Chemical Co. Inc.), Eur. P Applic. 608 371 (1994); PCT Applic. 93/08 239 (1993)
433. C.S. Chou *et al.* (Rohm & Haas Inc.), Eur. P Applic. 576 128 (1993); *Chem. Abstr*, **121**, 10 930 (1994)
434. A. Pizzi (ed.), *Wood Adhesives: Chemistry and Technology*, 2 Vols., Marcel Dekker Inc., New York and London, 1983
435. A. Hommaner, *Adhäsion*, **26**(11), 22–5 (1982); *Kunstharz Nachtrichten*, **10**, 26–34 (1982)
436. R. Kakusche, W. Fischer, I. Opitz, A. Meinecke *et al.* (VEB Chemische Werke Buna), DDR P 222 881 (1985); *Chem. Abstr*, **104**, 89 782 (1985)
437. V.A. Semenov and V.A. Voitovich, *Derevobrab. Prom. St.*, **1980**(9), 15 (1980); *Chem. Abstr*, **94**, 84 955 (1981)
438. For a discussion of theories of the adhesion of dried polymer films to wood and ceramics, see A.S. Freidin and M.G. Malyarik, *Koll. Zh. USSR*, **41**(5), 941–6 (1979) (English Trans: *Coll. J. USSR*, **41**(5), 794–6 (1979)
439. A. Takemura, B.L. Tomitz and H. Mizumachi, *Mokuzai Geekkaisha*, **32**(11), 883 (1986)
440. R. Lanthier (Shawinigan Chemicals Ltd), BP 1 092 690 (1972)
441. Borden, Inc., FP 2 154 837 (1973)
442. J.G. Iacovielli (Air Products & Chemicals Inc.), GP 2 211 690 (1972); *Chem. Abstr*, **78**, 44 506 (1973)
443. M. Ooshi (Sekisui Chemical Co. KK), JP Applic. 78 451 (1993); *Chem. Abstr*, **119**, 74 315 (1993)
444. P.S. Columbus, J. Anderson and Y.B. Patel (Borden Inc.), USP 5 306 749 (1994); *Chem. Abstr*, **121**, 159 141 (1994); and earlier Patents USP 3 442 645 (1969) and 4 251 400 (1981)
445. W.B. Armour and R.A. Fass (National Starch & Chemical Co.), BP 1 034 497 (1966); USP 3 300 430 (1967)
446. C.A. Finch and G. Barker (Croda Ltd), GP 2 206 571 (1972) (BP 1 383 494 (1975)
447. J.C. Iacoviello and D.W. Horwat (Air Products & Chemicals Inc.), USP 5 182 328 (1993); *Chem. Abstr*, **118**, 125 787 (1993)
448. T. Kondo, K. Saji, Y. Fujiwara and M. Hasegawa (Aica KK), JP 18 478 (1980); *Chem. Abstr*, **93**, 72 863 (1980)
449. Y. Taniguchi and T. Ike (Denki KK), JP Applic. 21 422 (1980); *Chem. Abstr*, **93**, 9 222 (1980)
450. Y. Kuzumaki, E. Nakamura and R. Sakaki (Mitsui Toatsu KK), JP Applic. 11 221 (1995); *Chem. Abstr*, **123**, 85 420 (1995)
451. T. Kondo (Aica KK), JP Applic. 34 246 (1980); *Chem. Abstr*, **93**, 72 863 (1980)
452. Naniwa Plywood KK, JP Applic. 67 456 (1980); *Chem. Abstr*, **93**, 134 127 (1980)

453. DaiNippon Ink Co. KK, JP Applic. 65 237 (1980); *Chem. Abstr*, **93**, 151 337 (1980)
454. A. Ichiwara *et al.* (Japan Synthetic Rubber KK), JP Applic. 16 045 (1980); *Chem. Abstr*, **93**, 9 218 (1980)
455. Y. Miyazaki, S. Sakurada and T. Tanaka (Koyo Sangyo KK), JP Applic. 5 937 (1980); *Chem. Abstr*, **92**, 199 494 (1980)
456. M.K. Lindemann (Charles S. Tanner Co. Inc.), USP 4 219 455 (1980); USP 4 001 160 (1977); *Chem. Abstr*, **86**, 107 401 (1977)
457. R.E. Porzel and C.D. Bator (DeSoto Inc.), USP 4 273 400 (1981)
458. Mitsui Toatsu KK, JP Applic. 156 055 (1980); *Chem. Abstr*, **94**, 158 602 (1980)
459. Mitsui Petrochemical KK, JP Applic. 147 523 (1980); *Chem. Abstr*, **94**, 122 746 (1981)
460. Kanebo KK, JP Applic. 155 269 (1981); *Chem. Abstr*, **96**, 124 861 (1982)
461. Nippon Print Color Gohan KK, JP Applic. 25 446 (1981); *Chem. Abstr*, **95**, 82 726 (1981)
462. Sumitomo Bakelite KK, JP Applic. 28 260 (1981); *Chem. Abstr*, **95**, 44 408 (1981)
463. Toyoda Gohsei KK, JP Applic. 136 867 (1981); *Chem. Abstr*, **96**, 163 807 (1982)
464. Showa HighPolymer KK, JP Applic. 129 242 (1981); *Chem. Abstr*, **96**, 36 292 (1982)
465. T. Yuki *et al.* (Kuraray Co. KK), JP Applic. 242 651 (1989); *Chem. Abstr*, **112**, 200 176 (1990)
466. N. Okinaga *et al.* (Sekisui Chemical Co. KK), JP Applic. 198 385–6 (1992); *Chem. Abstr*, **118**, 104 254–5 (1993); see also K. Hanashita and M. Ooishi (Sekisui Chemical Co. KK), JP Applic. 322 337 (1994); *Chem. Abstr*, **122**, 189 961 (1995)
467. Consortium für Elecktrochemische Industrie GmbH, GP 1 253 849 (1967)
468. S. Isobe and D. Asai (Aica KK), JP Applic. 179 074 (1991); *Chem. Abstr*, **115**, 281 778 (1991)
469. Hoechst AG, FP 2 210 651 (1974) (Austral. P 332 456); *Chem. Abstr*, **82**, 18 150 (1975)
470. E.K. Eisenhart (Rohm & Haas Inc.), Eur. P Applic. 590 886 (1994)
471. M. Jakob *et al.* (Hoechst AG), GO 19517777 (1996); *Chem. Abstr*, **126**, 60 502 (1997)
472. M. Nakamae and T. Juki (Kuraray Co. KK), JP Applic. 279 648 (1993); *Chem. Abstr*, **120**, 193 655 (1994)
473. M. Danno (Ushio KK), JP Applic. 238 377 (1985); *Chem. Abstr*, **104**, 169 733 (1986)
474. M. Iwasaki and I. Fujii (Oshika Shinko KK), JP Applic. 51 484 (1989); *Chem. Abstr*, **111**, 96 303 (1989)
475. C.L. Weidner and L.W. Eger (Industrial Tape Corporation), USP 2 474 292 (1949)
476. S. Asano (Toyo Bussan KK), BP 1 438 476 (1976)
477. Y. Kitakadu (Kowa KK and Imari Gohen KK), JP 42 166 (1976); *Chem. Abstr*, **86**, 141 254 (1977)
478. T. Maeda, H. Miki and A. Hiraharu (Japan Synthetic Rubber Co. KK), JP Applic. 182 558 (1991); *Chem. Abstr*, **115**, 281 645 (1991)
479. Borden Inc., USP 3 677 883 (1972); GP 2 011 725 (1971); FP 2 053 768–9 (1971)
480. K. Endo, T. Takahashi and T. Tsuukamoto (Mitsubishi Yuki Badische KK), JP Applic. 18 466 (1990); *Chem. Abstr*, **113**, 7 794 (1990)
481. M. Nakamae and T. Yuki (Kuraray Co. KK), JP Applic. 100 102 (1991); *Chem. Abstr*, **115**, 184 927 (1991)
482. W. Shimokawa (Hoechst Gosei KK), JP Applic. 202 176 (1985); *Chem. Abstr*, **104**, 110 995 (1986)
483. K. Ito *et al.*, JP 99 237 (1973); H. Isaki and M. Suzuki, JP 93 449 (1974); *Chem. Abstr*, **83**, 116 266 (1975)
484. I. Sugiyama and Y. Takaoka (Matsumoto Chemical KK), JP Applic. 94 768 (1074); *Chem. Abstr*, **82**, 99 174 (1975)
485. Y. Takemura, O. Narimatsu, K. Komatsu and Y. Takeuchi (Mitsui Toatsu Chemicals KK), JP Applic. 199 516 (1992); *Chem. Abstr*, **118**, 104 250 (1993)
486. Y. Yotsyanagi, T. Suga, and M. Shigemoto (Konishe Gisuke KK), JP 72 235 (1973); *Chem. Abstr*, **80**, 49 539 (1974)

487. H. Sakurai, E. Denjo, O. Fujita and T. Sakata, *Shizuoka-ken Kogyo Senta Kenkyu Hokoku*, **31**, 27 (1986); *Chem. Abstr*, **107**, 8193 (1987)
488. J.A. Sims (E I Du Pont de Nemours Inc.), USP 3261796 (1966); GP 1520590 (1971); N. Sakato and H. Nakamura (Shin-Etsu Chemical Co. KK), JP 14636 (1976) (Austral. P 408860); *Chem. Abstr*, **71**, 31317 (1969) (BP 1178933)
489. Y. Fujihara and J. Azoma (Itsuhosa Yushi KK), JP Applic. 25630 (1994); *Chem. Abstr*, **121**, 11308 (1994)
490. M. Yako, T. Kondo and H. Harino (Denki KK), JP Applic. 93243 (1994); *Chem. Abstr*, **121**, 11391 (1994)
491. Y. Takemura *et al.* (Mitsui Toatsu Chemicals KK), JP Applic. 199516 (1992); *Chem. Abstr*, **118**, 104250 (1993)
492. T. Hattori, S. Sakurada, T. Masada, M. Shiraishi, T. Inoue *et al.* (Kuraray Co. KK), JP 22435 (1974); *Chem. Abstr*, **81**, 92404/5; 122084 (1974)
493. M. Shiraishi *et al.* (Kuraray Co. KK), JP 69134/5, 69139, 69143, 69152 (1975); *Chem. Abstr*, **83**, 165175 (1975); JP 64342 (1975)
494. S. Sakurada *et al.* (Kuraray Co. KK), JP 69138 (1975), 69150; 140-69142 (1975) 69, 151 (1975); *Chem. Abstr*, **83**, 148456/7, 148162 (1975); JP Applic. 64341, 64342 (1975) and related patents. See also Refs. 39 and 40.
495. I. Honma, K. Kojima and A. Miyagawa (Okura Industrial KK), JP 109508 (1974); *Chem. Abstr*, **82**, 172138 (1975)
496. T. Hosogane, S. Shimazaki and E. Takiyama (Showa HighPolymer Co. KK), JP Applic. 276812 (1986); *Chem. Abstr*, **107**, 135440 (1987)
497. T. Hosogane, S. Shimazaki and E. Takiyama (Showa HighPolymer Co. KK), JP Applic. 18416 (1987); *Chem. Abstr*, **107**, 238159 (1987)
498. K. Tone and M. Takahashi (Shoei Chemical Industry KK), JP Applic. 12999 (1997); *Chem. Abstr*, **126**, 213111 (1997)
499. M. Nakamae *et al.* (Kuraray Co. KK), JP Applic. 269159 (1996); *Chem. Abstr*, **126**, 32723 (1997)
500. Y. Shibata and T. Juki (Sumitomo Chemical KK), Eur. P Applic. 783030
501. T. Juki *et al.* (Kuraray Co. KK), JP Applic. 11997594 (1993)
502. Y. Yamaguchi (Chuo Rika KK), JP Applic. 117611 (1993); *Chem. Abstr*, **120**, 32527 (1994)
503. K. Ryosuke and S. Ibata (DaiNippon Ink KK), JP 81830 (1976); *Chem. Abstr*, **86**, 17559 (1977)
504. M.L. Brown (E I Du Pont de Nemours), GP 2210373 (1972)
505. H. Nakasawa, T. Shintani and S. Okamoto, (Toyo Soda KK), JP 74648 (1977); *Chem. Abstr*, **87**, 137111 (1977)
506. H. Uehara and Y. Nagahara (Kuraray Co. KK), JP 93449 (1974); *Chem. Abstr*, **83**, 116366 (1975)
507. Asahi-Dow KK, JP Applic. 53575 (1975)
508. Toiritsu KK, JP 111367 (1982); *Chem. Abstr*, **97**, 217400 (1982)
509. O. Ohhara *et al.* (Kuraray Co. KK), JP Applic. 26341-4 (1974); *Chem. Abstr*, **81**, 78781-4 (1974); JP Applic. 26436 (1974); *Chem. Abstr*, **81**, 54323 (1974); S. Sakurada and Y. Miyaki (Kogyo Sanyo KK), JP 33431-2 (1974); *Chem. Abstr*, **81**, 78702-3 (1974)
510. T. Nakatsu (Konishi KK), JP 3969 (1977); *Chem. Abstr*, **86**, 191022 (1977)
511. N. Fujiwara *et al.* (Aica Chemical Co. KK), JP 129633 (1976)
512. Y. Yagi *et al.* (Sumitomo Rubber KK), JP 10844 (1976); *Chem. Abstr*, **84**, 151473 (1976); JP 10845 (1976); *Chem. Abstr*, **84**, 165563 (1976)
513. Showa High Polymer Co. KK, JP Applic. 12078 (1982); *Chem. Abstr*, **96**, 163707 (1982)
514. G.M. Knutson (Union Oil Co. Inc.), USP 3763073 (1973)
515. E. Inabe (Sumitomo Chemical KK), JP 18135 (1974); *Chem. Abstr*, **81**, 64728 (1974)
516. W.T. Tsen *et al.* (Ta-Lieng Chem Ind. Corp.), JP Applic. 297407 (1988); *Chem. Abstr*, **111**, 8560 (1989)

517. Y. Murayama and M. Oishi (Sekisui Chemical Co. KK), JP Applic. 172 375 (1991); *Chem. Abstr*, **116**, 153 415 (1992)
518. S.B. Luce (Swift Adhesives Inc.), USP 2 677 672 (1954)
519. Konishi KK, JP Applic. 55 575 (1985); *Chem. Abstr*, **103**, 23 359 (1985)
520. K. Kobayashi (Naniwa Plywood KK), JP 87 778 (1974); *Chem. Abstr*, **82**, 21 146 (1975)
521. E. Matsumoto and M. Yamamoto, JP 111 956 (1974)
522. M. Sakato and M. Yamamoto, JP Applic. 134 428–9 (1974); *Chem. Abstr*, **83**, 29 110 (1975)
523. E.F. Witucki and E.P. Plueddeman (Dow Corning Inc.), Eur. P Applic. 318 192 (1980); USP 4 818 779 (1989)
524. Y. Sakamoto, M. Yoshimoto and T. Manabe, JP 52 209–12 (1975); *Chem. Abstr*, **82**, 133 736–8 (1975)
525. T. Sakagami, JP 25 717 (1974)
526. K. Koyobashi *et al.*, J. P. Appl. 102 646 (1975); *Chem. Abstr*, **84**, 6 145 (1976)
527. M. Shorora and M. Suzuki (Senri KK), JP 44 253 (1975); *Chem. Abstr*, **83**, 98 750 (1975)
528. R. Hinterwaldner, *Adhäsion*, **11**, 252 (1967)
529. A. Chifor, J. Brandsch, S. Lungoci and A. Coneschi (Intreprindera Chimica Risnov), Rom. P 84 877 (1984); *Chem. Abstr*, **103**, 23 389 (1985)
530. Rohm & Haas Ltd, BP 1 045 837 (1966)
531. M. Nakamae and T. Sato, JP Applic. 179 851 (1994); *Chem. Abstr*, **121**, 232 773 (1994)
532. N. Kitamura and T. Usami (Sumitomo-Dow KK), JP Applic. 116 543 (1994); *Chem. Abstr*, **121**, 181 555 (1994)
533. F. Streit (Peelbond Products Inc.), USP 3 275 469 (1966)
534. T. Hayazaki and M. Ooishi (Sekisui Chemical Co. KK), JP Applic. 17 021 (1994); *Chem. Abstr*, **121**, 11 402 (1994)
535. BASF AG, GP 1 231 373 (1960)
536. J. Vietmeier *et al.* (BASF AG) GP 3 700 248 (1988); *Chem. Abstr*, **109**, 212 107 (1988)
537. S.A. Kizber, N.F. Jrasil'nikova and Z.P. Kalmyova, *Polygrafia*, **1971**(12), 30–1 (1971); *Chem. Abstr*, **77**, 20 585 (1972)
538. J.O. Geddes, *Adhes. Resins*, **7**, 101 (1959)
539. H. Yokoyama (Toyo Ink Seizo KK), JP Applic. 53 891 (1992); *Chem. Abstr*, **117**, 50 033 (1992)
540. T. Hayazaki *et al.* (Sekisui Chemical Co. KK), JP Applic. 172 728 (1994); *Chem. Abstr*, **121**, 159 193 (1994)
541. K. Azukma, JP 73 443 (1973); *Chem. Abstr*, **80**, 48 890 (1974)
542. M. Iseringhausen (Wolff-Walstrode AG), GP 2 111 841 (1972)
543. Henkel AG, GP 3 242 486 (1984)
544. Mitsui Petrochemical Industries KK, JP Applic. 109 516 (1984); *Chem. Abstr*, **102**, 7 873 (1985)
545. D. Yeung and K. Wai (Rhône-Poulenc Surfactants & Specialities Group), Eur. P Applic. 396 304 (1990); *Chem. Abstr*, **114**, 145 058 (1991)
546. N. Hamada and Y. Shimizu (Denki KK), JP 22 832 (1975); *Chem. Abstr*, **83**, 116 286 (1975); T. Toshigake and H. Narasawa (Denki KK), JP Applic. 22 833 (1975); *Chem. Abstr*, **83**, 80 688 (1975)
547. S.A. Hofman (International Paper Inc.), USP 3 984 275 (1976); *Chem. Abstr*, **85**, 193 815 (1976)
548. H.C. Lee and W.E. Lenney (Air Products & Chemicals Inc.), Eur. P Applic. 484 822 (1992); *Chem. Abstr*, **117**, 91 907 (1992)
549. G.E. Power and D. Wulfekotter (Formica Inc.), USP 3 983 307 (1976); *Chem. Abstr*, **85**, 193 822 (1976)
550. Y. Yokoyama (Nippon High Park KK), JP Applic. 135 931 (1976); *Chem. Abstr*, **86**, 91 961 (1975)
551. Y. Idota *et al.* (Fuji Film KK), GP 2 451 922 (1975) (JP Applic. 73 927 (1975)) *Chem. Abstr*, **83**, 132 541 (1975)

552. Koninlijke Papierfabrieken van Gelder Zonen NV, BP 1 329 468 (1973)
553. Components from Monsanto Inc.
554. A.E. Young, E.K. Stilbert and R.H. Lalk (Dow Chemical Co. Inc.), USP 2 539 236 (1952)
555. M.G. Wagner (Monsanto Inc.), GP 2 452 211 (1975)
556. M. Kishi, S. Sano and H. Koike et al. (Mitsui Toatsu Chemicals KK), JP Applic. 255 782 (1990); *Chem. Abstr*, **114**, 248 846 (1991)
557. W. Maier, *Adhäsion*, **16**(10), 13–15, 18–21 (1982)
558. E. Reckziegel, *Papier (Darmstadt)*, **36**(10A), 83 (1982)
559. N. Riley, *Soap Chem. Spec.*, **12**, 181 (1964)
560. J.S. Morphy, T.M. Santusosso D.J. Zimmer et al., *Adv. Polym. Blends Alloys Technol.*, **2**, 73 (1989); *Chem. Abstr*, **113**, 79 981 (1990)
561. A. Lannus and R. Zdanowski, *Resin Rev.*, **16**(4), 15 (1966)
562. R.R. Alexander (W.R. Grace Inc.), USP 3 275 589 (1965)
563. R.B. Tyler (W.R. Grace Inc.), FP 1 364 166 (1964)
564. J. Uchida, A. Sano, H. Nishiike and F. Tsukyama (Showa High Polymer Co.), JP Applic. 136 335 (1994); *Chem. Abstr*, **121**, 302 577 (1994)
565. A.E. Pierce, USP 3 311 561 (1967)
566. R.B. Tyler, USP 3 218 189 (1965)
567. A.N. Leadbetter and J.G. Galloway (Dow Chemical Co. Inc.), Eur. P Applic. 276 894 (1988); *Chem. Abstr*, **110**, 40 686 (1989)
568. R.M. Lambert and H.D. Sherman (Morton International), Eur. P Applic. 755 990 (1997); *Chem. Abstr*, **126**, 1887 122 (1997)
569. E.K. Eisenhardt, P.R. van Rheenan, L.C. Graziano, M.C. Bricker and L.J. Lin (Rohm & Haas Inc.), Eur. P Applic. 590 843 (1994)
570. T. Abe, T. Tsando, F. Tashiro and T. Tanno (Badische Petrochemical GmbH), JP Applic. 97 054 (1974)
571. T. Tanaka and K. Tokusumi (Mitsui Chemical KK), JP 5 089 (1955)
572. M. Yamamoto et al. (Arakawa KK), JP Applic. 277 680 (1991); *Chem. Abstr*, **116**, 195 823 (1992)
573. T. Adakia and M. Kishi (Shoei Chemical Industry KK), JP Applic. 116 539 (1994); *Chem. Abstr*, **121**, 181 556 (1994)
574. A.R.B. Furendal and H.U. Heilstrom (Bofors AB) GP 2 442 695 (1975)
575. N. Araki and K. Okazaki (Konishi Gisuke KK), JP 13 026 (1974); *Chem. Abstr*, **82**, 18 175(1975); T. Koshida and Y. Nakagawa, JP Applic. 17 965 (1976) (GP 2 532 728); *Chem. Abstr*, **84**, 151 607 (1974)
576. A.C. Makati (Dow Chemical Co. Inc.), Eur. P Applic. 345 385 (1989)
577. S. Ando, H. Akino and S. Miura (Toa Gosei KK), JP Applic. 44 581 (1974); *Chem. Abstr*, **79**, 137 692 (1973); S. Ando and M. Kanagawa (Toa Gosei KK), JP Applic. 8 313 (1973); *Chem. Abstr*, **78**, 149 078 (1973); S. Ando and S. Miura, JP Applic. 88 672 (1973); *Chem. Abstr*, **80**, 4 470 (1973)
578. I.J. Davis (National Starch & Chemical Inc.), Eur. P Applic. 56 452 (1982)
579. A.C. Makati and R.T. Iwamasa (Dow Chemical Co. Inc.), PCT Applic. 17 201 (1991); *Chem. Abstr*, **114**, 63 550 (1992)
580. T.C. Neubert (GenCorp Inc.), Eur. P Applic. 294 027 (1988); *Chem. Abstr*, **111**, 24 671 (1989)
581. W.E. Lenney (Air Products and Chemicals Inc.), Eur. P Applic. 389 893 (1990); *Chem. Abstr*, **114**, 63 550 (1991)
582. J. Saito, T. Nakai, S. Chiku and T. Kawamura et al. (Chisso Corporation and Toyota Motor Corporation), JP Applic. 252 784 (1990); *Chem. Abstr*, **115**, 73 241 (1991)
583. DuPont de Nemours Inc., Belg. P 697 800–1 (1967)
584. G.M. Knutsen and J. Biale (Union Oil Co. Inc.), USP 3 769 151 (1973)
585. R.B. Tyler (W R Grace Inc.), FP 1 364 166 (1964)

586. E. Higuchi and H. Mori (Nippon Zeon Co.), JP Applic. 145 413 (1994); *Chem. Abstr*, **121**, 281 923 (1994)
587. J. Yamamoto (Asahi Chemical Ind. KK), JP Applic. 269 292 (1996); *Chem. Abstr*, **126**, 48 163 (1997)
588. See International Standard ISO 4588, Guide to the Surface Preparation of Metals
589. C.R. Erikson (Arabol Manufacturing Inc.), USP 2 544 146 (1951)
590. United States Steel Co. Inc., Belg. P 697 899 (1967)
591. E.R. Parkinson (U S Steel Engineers & Consultants), GP 2 218 572 (1972)
592. T. Tanaka and T. Tolcuzumi (Mitsui Chemical KK), JP Applic. 6 374 (1957)
593. K.H. Kassner (Henkel GmbH), GP 2 529 523 (1977)
594. N. Kawahara *et al.* (Sumitomo Bakelite KK), JP Applic. 132 227 (1976)
595. V.D. Snezhko, S.P. Khainer, I.A. Mallenikov, V.F. Siderov *et al.* (Central Scientific Research Institute of Building Structures), USSRP 781 188 (1980); *Chem. Abstr*, **95**, 11 588 (1981)
596. K. Noguchi (Shin-Etsu Chemical KK), JP Applic. 278 169 (1993); *Chem. Abstr*, **120**, 136 673 (1994)
597. H. Ishizawa, K. Matsunaga and S. Sakakibara (Toppan Printing KK), JP Applic. 129 073 (1979); *Chem. Abstr*, **92**, 77 616 (1980)
598. R. Feldmann and K. Gorke (Hüls AG), GP 1 825 707 (1979)
599. A. Hayashi (Adachi Plywood KK), JP Applic. 19 925 (1974)
600. S. Onta and K. Kagawa (Chuo Rika KK), JP Applic. 92 156 (1974)
601. T. Iikuni (Toyo Soda KK), JP Applic. 66 949, 68 250, 72 791 (1977)
602. Mitsui Toatsu KK, JP Applic. 121 758 (1981)
603. P. Piatek and J. Kuber, *J. Polym.*, **56**(12), 362 (1976)
604. D. Kimura (Shin-Etsu Chemical KK), JP Applic. 47 829 (1977)
605. T. Isobe and A. Terada (Showa High Polymer KK), JP Applic. 163 481 (1993); *Chem. Abstr*, **120**, 56 247 (1994)
606. K. Machida, T. Tsuchida, K. Watanabe and M. Kawamoto (Mitsui Toatsu Chemical KK), JP Applic. 192 641 (1994); *Chem. Abstr*, **82**, 18 175 (1975); *Chem. Abstr*, **122**, 83 350 (1995)
607. J. Ono and H. Kido (Aica KK), JP Applic. 103 635 (1992); *Chem. Abstr*, **117**, 113 213 (1995)
608. K. Hayakawa and K. Kobayashi (Shin-Etsu Chemical Industry KK), JP Applic. 311 422 (1996); *Chem. Abstr*, **126**, 132 472 (1997)
609. J.R. Anghietti (American Cyanamid Inc.), USP 3 382 058 (1964)
610. L. Chroszcz, H.D. Erbs, W. Koenig and H. Schumann, DDR P 57 913 (1957); *Chem. Abstr*, **69**, 20 130 (1968)
611. A. Eckerman and A. Rothel, DDR P 15 485 (1958); *Chem. Abstr*, **54**, 6 500 (1960)
612. W.H. Church and D.B. Maney (Du Pont de Nemours Inc.), USP 2 128 229 (1938)
613. E.V. Reeves, *Adhes. Age*, **6**(3), 28 (1983)
614. W.B. Reynolds (Phillips Petroleum Inc.), USP 2 773 795 (1956)
615. W.D. Wolfe (Goodyear Inc.), USP 2 817 616 (1957)
616. D.B. Rahrig, *28th IUPAC Macromol Symp.*, Oxford, 1982, p. 623
617. D. Nowak, L.C. Kosno and M. Sadlowska (Inst. Ciezkiej Syntezy Oragnicznej 'Blachownia'), Pol. P 161 289 (1993); *Chem. Abstr*, **123**, 58 077 (1995)
618. G.M. Doyle, *Trans. Inst. Rubber Ind.*, **36**, 177 (1960); T. Takeyama and J. Matsui, *Rubber Chem. Technol.*, **42**, 159 (1969)
619. See, for example, I. Skeist (ed.), *Handbook of Adhesives*, 3rd edn, Van Nostrand Reinhold, New York, 1990, especially T.S. Solomon, Ch. 35, pp. 582–97 for a discussion of the mechanisms of fabric-to-rubber adhesion
620. T.A. Seberier, V.A. Bertsnev and I.L. Shmurak, *Kauch Rezina*, **32**(7), 30 (1973)
621. R.W. Kavchok (Goodyear Rubber Inc.), FP 2 478 114 (1981)
622. N.K. Porter, *J Coated Fabric*, **23**(1), 34 (1993)

623. I. Mitsuyoshi (Bridgestone Corporation), JP Applic. 210761 (1994); *Chem. Abstr*, **122**, 11745 (1995)
624. Y. Toyoda, T. Nishioka and A. Kamimura (Sumitomo-Dow KK), JP Applic. 184505 (1994); *Chem. Abstr*, **122**, 57854 (1995)
625. A.P. Kerwan, *J. Appl. Polym. Sci.*, **17**, 2027 (1973)
626. M.M. Girgis (PPG Industries Inc.), Belg. P 887834 (1981)
627. I C I Ltd, Belg. P 688424 (1967)
628. Y. Iyengar, *J. Appl. Polym. Sci*, **22**, 801 (1978)
629. O.C. Elmer (General Tire & Rubber Inc.), USP 4404055 (1983)
630. Y. Toyoda, T. Hishioka and A. Kamimua (Sumitomo-Dow KK), JP Applic. 184247 (1984); *Chem. Abstr*, **121**, 302686 (1994)
631. Sumitomo Naugatuck KK, JP Applic. 92371 (1985); *Chem. Abstr*, **103**, 88841 (1985)
632. Sumitomo Naugatuck KK, JP Applic. 96674 (1985); *Chem. Abstr*, **103**, 124806 (1985)
633. Sumitomo Naugatuck KK, JP Applic. 96675 (1985); *Chem. Abstr*, **103**, 124807 (1985)
634. Y. Toyoda *et al.* (Sumitomo Naugatuck KK), JP Applic. 81184 (1988); *Chem. Abstr*, **109**, 111784 (1988)
635. American Cyanamid Inc., JP Applic. 185643 (1983)
636. T.C.H. Tsai (Stamicarbon BV), Eur. P Applic. 402991 (1991); *Chem. Abstr*, **114**, 209042 (1991)
637. Teijin KK, JP Applic. 2156 (1981); *Chem. Abstr*, **94**, 210092 (1981)
638. Toyobo Petcord KK, JP Applic. 152815/6 (1981); *Chem. Abstr*, **94**, 122923/4 (1981)
639. R.S. Shakuni and G.W. Rye (Goodyear Inc.), USP 4031288 (1977)
640. A.M. Medvedeva B.V. Deryagin, K. Vand Zherebkov, *Koll. Zh.*, **19**, 552 (1957); *Koll. Zh.*, **22**, 217 (1960); *Chem. Abstr*, **55**, 6007f (1961)
641. Soc. Rhodiaceta SA, BP 1046166 (1966)
642. K. Kigane, H. Togawa and S. Yamada (Teikku Junzo KK), USP 3178547 (1965)
643. M. Tanaka (Mitsubishi Belting KK), JP 59184 (1974); *Chem. Abstr*, **82**, 32198 (1975)
644. T. Ohzeki, K. Chinura, T. Kimura, K. Ito *et al.* (Mitsubishi Rayon KK), JP 22570, 39230/1 (1972); *Chem. Abstr*, **78**, 137678/9 (1973) and later references
645. W.D. Timmons, *Adhes. Age*, **10**(10), 27 (1967)
646. R.J. Kelly, J.L. Witt and M.L. Tidmore (Uniroyal Inc.), USP 3307178 (1973); *Chem. Abstr*, **78**, 73395 (1973); R.J. Kelly, R. Miller and D. Adams, GP 2407408 (1985); *Chem. Abstr*, **82**, 18384 (1985)
647. R.J. Shimmel (Uniroyal Inc.), USP 3837993 (1974); *Chem. Abstr*, **82**, 18385 (1985)
648. M. Doki, H. Shimizu and J. Matsui, (Toray KK), JP 35142 (1971); *Chem. Abstr*, **77**, 76425 (1972)
649. L.W. Georges (Firestone Rubber Inc.), USP 3642553 (1972)
650. Y. Asada and T. Murakami (Teijin KK), JP Applic. 222118 (1971)
651. M. Hasegawa (Unitika KK), JP 30174 (1972) (GP 2115448; BP 1301603); *Chem. Abstr*, **76**, 112974 (1972)
652. R.S. Bhakuni and J.L. Cormany (Goodyear Rubber Inc), FP 2159297 (1973)
653. W.J. Van Ooij, *Rubber Chem. & Technol.*, **58**(3), (1985)
654. T. Fukuoka, M. Watanabe, I. Ogasawara and S. Kawashima (Unitika KK), JP 43196 (1972); *Chem. Abstr*, **80**, 38107 (1974)
655. Y. Aoki, M. Kohichi, T. Tateno and Y. Ueda (Toyobo Rubber KK), JP 49768 (1972); *Chem. Abstr*, **80**, 122282 (1974)
656. T. Oshima, Y. Oozumi, S. Fujiwara and T. Hashikura (Asahi Chemical KK), JP 107077 (1974); *Chem. Abstr*, **84**, 45479 (1976)
657. H. Ishida, H. Kuwayama and K. Katsushima (Yokohama Rubber KK), JP 8085 (1977) (JP 127990 (1977)); *Chem. Abstr*, **86**, 122724 (1977); **88**, 121290 (1978)
658. W.C. Pruitt and W.J. Schroder (Deering Milliken Research Inc), USP 3231412 (1966)
659. J.A. Cardine (Goodyear Tire Inc), USP 3247043 (1966)
660. Bridgestone Tire KK, FP 1513296 (1968)

661. Shell International Research NV, Neth. P Applic. 11 538 (1966)
662. Y. Kono and D.S. Takua (Toray KK), JP Applic. 70 934 (1976)
663. T. Fukuoka, M. Watanabe and T. Ogasawara (Unitika KK), JP Applic. 36 278 (1975); *Chem. Abstr*, **80**, 38 106 (1974); JP 17 028–32; *Chem. Abstr*, **82**, 18 388–92 (1975)
664. K. Chimura, T. Ohzeki, K. Ito, Y. Maeda *et al.* (Mitsubishi Rayon KK), JP 36 277 (1975); *Chem. Abstr*, **84**, 166 014 (1976)
665. M. Ikemoto, H. Enami, K. Kosaka and N. Nakajima (Mitsubishi Belting KK), JP Applic. 20 983 (1976); *Chem. Abstr*, **84**, 166 020 (1976)
666. R. Chapman (I C I Ltd), BP 1 091 206 (1967)
667. H.T. Adams (Uniroyal Inc.), USP 3 876 457 (1975)
668. O.C. Elmer (General Tire & Rubber Inc.), USP 4 285 756 (1984)
669. Bridgestone Tire KK, BP 1 254 719 (1971)
670. Inventa AG, BP 2 116 855 (1971); Swiss P 488 048 (1970); *Chem. Abstr*, **74**, 145 (1971); GP 2 520 733 (1961); *Chem. Abstr*, **84**, 60 913 (1976)
671. M. Hasegawa *et al.* (Unitika KK), JP 39 948 (1972); *Chem. Abstr*, **80**, 28 206 (1974)
672. E.B. McClary (Celanese Corporation), FP 2 085 780 (1972)
673. S. Ohbuchi *et al.* (Unitika KK), JP 12 187 (1973); *Chem. Abstr*, **80**, 71983 (1974)
674. I. Yoshida and M. Kato (Toray Industries KK), JP Applic. 181 369 (1985); *Chem. Abstr*, **104**, 90 274 (1986)
675. H. Ogawa and K. Ichimura (Toho Rayon KK), JP Applic. 215 772 (1986) *Chem. Abstr*, **106**, 19 778 (1987)
676. E. Sato (Asahi Chemical Ind KK), JP Applic. 215 042 (1987); *Chem. Abstr*, **107**, 238 426 (1987)
677. S. Toki (Bridgestone Corporation), JP Applic. 66 381 (1988); *Chem. Abstr*, **109**, 75 006 (1988)
678. T. Ohzeki, *et al.* (Mitsubishi Rayon KK), JP 22 555 (1974); *Chem. Abstr*, **82**, 87 441 (1975); *JP* 22 589 (1974)
679. Bridgestone Tire KK, JP Applic. 129 468 (1980); *Chem. Abstr*, **94**, 48 637 (1981); see also M. Onizawa, *USP* 4 254 016 (1980)
680. R.S. Bhakuni and J.L. Cormany (Goodyear Rubber Inc), USP 3 714 772 (1973)
681. G. Henneman (American Enka Inc., USP 305 430 (1967); K. Kigane, *et al.* (Teijin KK), GP 1 470 968 (1973); N. Shibata, *et al.* (Toyo Rubber Inc. KK), JP 41 450 (1971)
682. Hercules Inc. BP 1 275 120 (1976)
683. S. Nishido, *et al.* (Toyo Rubber KK), JP 38 328 (1971); T. Akamura and T. Eguchi (Kuraray Co. KK), JP 41 451 (1971)
684. Hercules Inc., BP 1 116 728 (1968)
685. Toyo Petcord KK, JP Applic. 118 855 (1980); *Chem. Abstr*, **94**, 31 915 (1981)
686. T. Takemura, N. Sakai, and H. Tagaki, M. Kikuchi (Bridgestone Tire KK), JP Applic. 14 555 (1976); *Chem. Abstr.*, **85**, 125 534 (1976)
687. Asahi Chemical KK, JP Applic. 32 933 (1982); *Chem. Abstr.*, **96**, 219 118 (1982)
688. P. Bourrain and R. Mingat (Rhone-Poulenc SA), FP Applic. 2 271 036 (1973); *Chem. Abstr.*, **84**, 166 021 (1976)
689. H. Enami, N. Yokota and M. Sakata (Mitsubishi Belting KK), JP Applic. 987 (1977); *Chem. Abstr.*, **86**, 122 723 (1977); H. Ishida, C. Kuwayama and A. Katsushima, JP Applic. 8 085 (1977); *Chem. Abstr.*, **86**, 122 724 (1977)
690. KK Toyo Petcord JP Applic. 118 855 (1980); *Chem. Abstr.*, **94**, 31 915 (1981)
691. See, especially, W.J. van Ooij, *Rubber Chem. Technol.*, **52**, 605 (1979); **57**, 421 (1984); *Chem. Abstr.*, **101**, 421 (1984).
692. J.J. Langer and W.J. McKillip (Ashland Oil Inc.), USP 3 933 833 (1976)
693. N.K. Kerakas (Monsanto Inc.), GP 2 636 611 (1976)
694. K.H. Moult (Koppers Inc.), USP 3 437 610 (1969)
695. M. Sakada (Mitsubishi Belting KK), JP Applic. 39 184 (1974); *Chem. Abstr*, **82**, 157 605 (1975)

References

696. I. Ogasawar and K. Kawashima (Unitika KK), JP Applic. 61 269 (1974); *Chem. Abstr*, **84**, 123 174 (1976)
697. T. Kurihara *et al.* (Teijin KK), USP 3 697 310 (1972)
698. T. Ohzeki, K. Ito, Y. Maeda and T. Kudo (Mitsubishi Rayon KK), JP 39 230/1 (1972); *Chem. Abstr*, **78**, 137 679/80 (1973); JP 56 730 *Chem. Abstr*, **79**, 137 839 (1973)
699. K. Chimura, T. Ohseki, K. Ito, Y. Maeda and T. Kudo (Mitsubishi Rayon KK), JP 154 385 (1975); R.M. Wise (General Tire and Rubber Inc.), USP 3 960 796 (1976); E.F. Kalafus (General Tire and Rubber Inc.), USP 3 857 730 (1975); USP 3 962 516 (1976)
700. M. Tanaka (Mitsubishi Belting KK), JP 80 134 (1973)
701. Firestone Tire Inc., BP 1 240 699 (1971)
702. K.L. Lewis (Dunlop Rubber Co. Ltd), S. Afr. P 75555
703. K. Shimizu (Toray KK) JP 42 975 (1971); *Chem. Abstr*, **77**, 103 064 (1970)
704. T. Fukuoka, T. Ogasawara and S. Kawashima (Unitika KK), JP 18 792 (1974); *Chem. Abstr*, **82**, 59 566 (1975)
705. E.F. Kalafus and S.C. Sharma (General Tire and Rubber Inc.), USP 4 040 999 (1977); *Chem. Abstr*, **87**, 119 106 (1977); GCP 2 719 220 (1976); *Chem. Abstr*, **88**, 106 291 (1978)
706. C.A. Aufdermarsch (Du Pont de Nemours Inc.), GP 2 316 034 (1973); *Chem. Abstr*, **80**, 48 977 (1975); GP 2 522 963 (1975); *Chem. Abstr*, **87**, 45 006 (1977); USP 3 977 592 (1977); *Chem. Abstr*, **87**, 50 045 (1977)
707. Toray Industries KK, J Pat Applic. 19 375 (1983); *Chem. Abstr*, **99**, 54 881 (1983)
708. R. Brosman (American Cyanamid Inc.), USP 3 951 723 (1976)
709. R.L. Wright (Monsanto Inc.), USP 3 951,723 (1976); GP 2 137 589 (1974)
710. K. Katsushima (Yokohama Rubber KK), GP 2 327 215 (1974)
711. S. Obuchi, K. Kageyama, Y. Suzuki, K. Katsushima *et al.* (Tunichika KK), JP 41 086 (1971)
712. H. Ishida, K. Kageyama, Y. Suzuki, K. Katsushima *et al.* (Yokohama Rubber KK), JP Applic. 28 579 (1977); *Chem. Abstr*, **87**, 7 263 (1977)
713. T. Yotsumoto and T. Akiyama (Bridgestone Corporation KK), GP 3 435 765 (1985)
714. Gen-Corp Inc., Technical Literature, 1992
715. R.J. van Valkenberg (Goodyear Inc.), USP 3 59 379 (1992)
716. T. Honda (Bridgestone Tire KK), USP 3 814 713 (1974)
717. T. Ohshima, Y. Uuzumi, S. Fujiwara and T. Hashikuta (Asahi Chemical KK), JP Applic. 107 077 (1974); *Chem. Abstr*, **84**, 45 479 (1976)
718. G.E. van Gils (General Tire and Rubber Inc.), USP 3 589 239 (1975)
719. K. Iwasaki (Daicel Chemical Ind. KK), JP Applic. 72 074 (1986); *Chem. Abstr*, **105**, 98 879 (1986)
720. W. Shimokawa and K. Fukumori (Hoechst Gosei KK), JP Applic. 78 883 (1986); *Chem. Abstr*, **105**, 192 456 (1986)
721. H. Hisaki and S. Suzuki, *Rubber World*, **201**(2), 19 (1989); *Chem. Abstr*, **112**, 180 950 (1990)
722. S.A. Voronov, *Kauch Rezina*, 1980(**10**), 34; *Chem. Abstr*, **94**, 31 871 (1981)
723. H. Kuki and M. Sekiya (Nippon Zeon KK), JP Applic. 26 629 (1986); *Chem. Abstr*, **105**, 44 325 (1986); H. Kuki (Nippon Zeon KK), JP Applic. 26 630 (1986); *Chem. Abstr*, **105**, 44 331 (1986)
724. H. Kuki (Nippon Zeon KK), JP Applic. 41 581 (1988); *Chem. Abstr*, **109**, 7 860 (1988)
725. G. Quack (Metallgeschaft AG), GP 3 035 181 (1982)
726. L.S. Geidysh *et al.*, USSRP 388 410 (1988); *Chem. Abstr*, **109**, 111 896 (1988)
727. R.F. Kelly (Goodyear Inc.), GP 2 254 483 (1973)
728. KK Teijin, JP Applic. 51 876/7 (1982); *Chem. Abstr*, **97**, 128 931/2 (1982)
729. KK Teijin, JP Applic. 228 078 (1984); *Chem. Abstr*, **102**, 169 026 (1985)
730. Toray Industries KK, JP Applic. 2 781 (1985); *Chem. Abstr*, **102**, 168 207 (1985)
731. Sumitomo Chemical KK, JP Applic. 83 077 (1983); *Chem. Abstr*, **100**, 8 021 (1984)
732. D.M. Zavisva (American Cyanamid Inc.), USP 4 263 190 (1981)

733. T.J. Meyrick and D.B. Wooton (ICI Ltd), BP 1 092 908 (1967)
734. Sumitomo Naugatuck KK, JP Applic. 2 730 (1983); *Chem. Abstr*, **98**, 216 820 (1983)
735. K. Ueno and Y. Sato (Takeda Chemical KK), GP 1 299 416 (1969)
736. T. Tashiro and T. Tanaka (Kogyo Samagyo KK), JP Applic. 160 598 (1975); *Chem. Abstr*, **84**, 160 013 (1976)
737. General Tire and Rubber Inc., JP Applic. 141 580 (1975); *Chem. Abstr*, **84**, 111 084 (1976)
738. T. Takemura *et al.* (Bridgestone Tire KK), also reference see [669]
739. M. Toki, T. Tanaka and T. Kikuchi (Toray KK), JP Applic. 94 289 (1975); *Chem. Abstr*, **84**, 19 099 (1976)
740. KK Toray, JP Applic. 45 043 (1982); *Chem. Abstr*, **97**, 25 019 (1982)
741. E.F. Kalafus and G.E. van Gils (General Tire and Rubber Inc.), USP 3 888 805 (1975); GP 2 463 320 (1976)
742. H. Nakamura, K. Kamono and H. Imagawa (Unitika KK), JP Applic. 82 185 (1975); *Chem. Abstr*, **83**, 180 823 (1975)
743. I. Saito, Y. Nakamura, Y. Komatsu and S. Ishizuka (Toray KK), JP Applic. 65 690 (1975); *Chem. Abstr*, **83**, 181 016 (1975); JP Applic. 65 691 (1975); *Chem. Abstr*, **83**, 195 115 (1975)
744. H.J. Boles (General Tire and Rubber Inc.), USP 3 964 950 (1976)
745. Honny Chemicals Co. Ltd, Belg. P 839 548 (1976); *Chem. Abstr*, **86**, 141 111 (1977)
746. G.M. Parker (Honny Chemicals Co. Ltd), JP Applic. 133 483 (1974); *Chem. Abstr*, **84**, 91 403 (1986)
747. W.E. Uffner (Owens-Corning Fiberglas Inc.), USP 3 707 399 (1972); *Chem. Abstr*, **78**, 73 393 (1973); P. Barrain and R. Mingat (Rhone-Poulenc SA), FP Applic. 2 271 036 (1973); *Chem. Abstr*, **84**, 166 021 (1976)
748. H.W. Kucera (Lord Corporation), FP 2 425 550 (1981)
749. Nippon Oils & Fats KK, JP Applic. 86 853 (1980); *Chem. Abstr*, **93**, 240 998 (1980)
750. H.P. Kohlstadt *et al.* (Henkel AG), GP 2 023 651 (1980)
751. D.M. Fahey (PPG Industrials Inc.), EP Applic. 53 797 (1982); *Chem. Abstr.* **98**, 77091 (1983)
752. M.M. Girgis (PPG Industries Inc.), USP 4 246 144 (1981)
753. PPG Industries Inc, JP Applic. 159 822 (1982); *Chem. Abstr*, **98**, 180 906 (1983)
754. M.M. Girgis (PPG Industries Inc.), USP 4 436 866 (1984)
755. T.C. Neubert (General Tire and Rubber Inc.), GP 3 001 838 (1980)
756. A. Okamura, *et al.* (Nippon Glass Fiber Co. KK), JP Applic. 212 572 (1994); *Chem. Abstr*, **122**, 83 457 (1995)
757. W.E. Uffner (Owens-Corning Fiberglas Inc.) USP 3 707 399 (1972); *Chem. Abstr*, **78**, 73 393 (1973)

Index

acetoacetoxyethyl methacrylate   337
2-acetoacetoxypropyl methacrylate   336
acetophenone derivatives   328
acetoxylated polyethlene latices   506
acid groups   367
aconitic acid   39
Acronal 5000   132
Acronal latices   369, 370
Acronal nomenclature   369
acrylamide   113
   derivatives   304–8
2-acrylamido-2-methylpropanesulfonic acid (AMPS)   40
acrylamidoglycollates   314–15
acrylate–acrylonitrile copolymer latices   558
acrylate-based homopolymer latices   657
acrylate copolymer latices   640
   crosslinking   558–61
acrylate latex copolymers   573
acrylic acid   22, 32, 113, 133–5, 284
acrylic copolymer latices
   adhesives   518–25
   with added polar groups   546–55
   with amphoteric/basic comonomers   557
acrylic copolymers   110, 135, 175
   formed in self-emulsified alkyd   408
acrylic ester latex, two-stage redox process   135, 137
acrylic ester–styrene copolymer   135, 137
acrylic esters   33–8, 167, 206, 277–8, 305–6
acrylic latices   133–40, 458, 645
   crosslinking   330, 525
   current production practice   522
   film formation   224–7
   pressure-sensitive applications   542
acrylic polymers   136
   properties   365–7
acrylic resin films, properties   367–71
acrylics   32–8
   based on amide group   38
   polymerisation   138–40
acrylonitrile   12, 38, 95
acrylonitrile–butadiene–styrene (ABS)   115
addition polymerisation   4–16, 404
addition polymers   3
addition reactions   268
additives   495–500
adherent   471
adhesion   471
   basic theory   473–7
   difficulties   480–1
   diffusion theories   475
   promoters   495–500
   role of chemisorption   476
   theories   470, 473–8
   types   476
adhesive tapes   569
adhesives   458–60
   acrylic latices   518–25
   applications   471, 606
   applying   479–80, 483
   barrier-type   602
   composition   480
   compounding   486–8
   corrosion   600
   corrugated board   504
   crosslinking   512–17
   ethylene-vinyl acetate copolymer   506–12

Index

adhesives (cont.)
  foaming 600
  foil laminating 529, 531
  formulations 470–667
  from natural products 470
  functions 471, 477–81
  glossary of terms 624–67
  inorganic substrates 603–6
  laminating 516–17, 530
  latex-based 478–82
  low setting speeds 600
  manufacturing cost 484
  non-pressure sensitive 651–2
  non-vinyl latices 537–8
  organic substrates 600–3
  packaging 511
  peelable 560
  performance behaviour requirements 483
  polyvinyl acetate 484–517
  polyvinyl chloride 502, 503
  polyvinyl propionate 517–18
  practical requirements 481–2
  pressure-sensitive see pressure-sensitive adhesives
  quick-break 528
  remoistenable 389
  specific applications 577–606
  specific functions 538–77
  specific types 482–538
  terminology 471–2
  test methods 624
  veneering 504
  water-based 499
  wet stick 481
  wood bonding 504
  wood veneer 512
  see also specific materials and applications
adsorption theory 474
alcohol ethoxylates 128
aldehydes, crosslinking 289
aliphatic chain sulfonates 63–4
aliphatic diglycidyl ethers 426
aliphatic hydrocarbons 385–8
alkali-soluble copolymers 363–4
alkali-soluble polymers 129–31, 136
alkali-soluble vinyl acetate copolymer latex 130

alkyds
  and allied products 410–17
  emulsion blends 413–15
  pigmented emulsions 413
  styrenated emulsions 415–16
alkyl (meth)acrylate esters 140
alkyl-substituted azosulfonates 93
alkylphenyl polyglycol emulsifiers 424
allyl acetoacetate 336
allyl acrylate–butyl acrylate copolymer 280
allyl alcohol 7, 43, 280
allyl derivatives 43
allyl dimethyl glycidate 43
allyl ethyl malonate 279–80
allyl glycidyl ethers 43, 295–6
allyl methacrylate 43
allyl polymers 278–80
amide derivatives with unsaturation 315–16
amide group 301
  acrylics based on 38
amine salt emulsification 423
aminimides 310–14
aminoplasts 316–21, 417–19, 583–4
aminosulfonates 455
ammonium oleate 374
ammonium perfluoro-octanoate 381
ammonium salts 72
amorphous materials 26–7
amorphous polymer 27
amphiphilic block copolymers 92
amphoteric emulsifiers 73
amylopectin 76–7, 122
Anacardium occidentale 283
anatase 247
anionic emulsification 454–6
anionic emulsifiers 440–1
anionic initiators 4
anionic polymerisation 4
anionic stabiliser 352
anionic surfactants 59–60, 165, 202
anti-foam agents 482
antimony oxide 248
antistatic properties 277
aqueous acrylic polymer adhesives 525
associative thickeners 88
atomic force microscopy (AFM) 213–15, 217
Avogadro number 168

Index

aziosobutyronitrile 94
aziridine derivatives 498
aziridines 586–7
azodiisobutyronitrile 5, 9, 123

barytes 249
bead polymerisation 29
Benzoflex 9-88 491
benzyl butyl phthalate 490
biocides, biodegradability 499
biodegradability of biocides 499
biodegradable emulsifiers 71
bis(1,3,2-dioxothioan-2-oxido-4-methyl)malonate 342
bisphenol A 299
blocked copolymers 18, 451–2, 457, 459
blocked isocyanates 614–15
blocked polymer 461
blocking 472, 517
blocking temperature 494
 versus fusion temperature 373
bond strength 472, 481
branched chains 24
bromine, monomers containing 148
buildings 447–8
bulk polymerisation 28
Bunatex F 2420, 385
butadiene 13, 42
butadiene copolymer latices 530, 533–7
 formulations 536–7
butadiene copolymers 42, 144, 176–7
butyl acrylate 111, 195
 copolymers 169–70
 emulsion polymerisation 169
butyl acrylate–methacrylic acid copolymer 522–3
butyl benzoate 240
butyl methacrylate latex (PBMA) 218

capillary forces 216
carbodi-imides 340–1
carboxyl function 284–8
carboxylated
 butadiene–styrene–vinylpyridine copolymer latex 619
carboxylated latex 281
 formulations 144–5

carboxylated styrene–butadiene latices 533–4
carboxylic acids 284
carboxylic groups 197
carboxymethyl cellulose 83
carnuaba–paraffin wax 567
cascade process 104–5
cationic acrylic monomers 38
cationic emulsification 452–4
cationic emulsifiers 143, 163, 439–40
cationic latex 513
cationic polymerisation 4
cationic surfactants 59, 71–3, 92
cationically charged latex 125
cellulose derivatives 78, 83–4, 138
cellulose ethers 77–84
cement 432, 433
chain branching 16–17, 20
chain reaction 4–6
chain transfer 6, 8–9
charge determination 401
chemical bonding 476
china clay 249, 251–2
chlorinated hydrocarbons 6
chlorinated monomers 31–2
chloroprene 32, 281
chlorotrifluoroethylene 150
coalescence aids 526
coalescing solvents 229–42
coatings 274, 428–9, 432, 433, 447–8, 461–2, 531
 kraft paper tapes 551
cohesive forces 471
colloid chemistry 49–58
colloid stabiliser 119–21, 352
colloids 1, 52, 198
 high molecular weight stabilisers 76–88
 see also specific types
colophony 496
condensation polymer 2
condensation polymerisation 2
condensation processes 404
construction adhesive applications 652–3
contact adhesives 568–72
 formulations 569–72
contact-grab adhesives 575–6
copolymer xxii, 15
copolymer emulsions 133

copolymer latices 115–33
copolymerisation xxii, 13–16, 269
  theoretical considerations 161
  vinyl acetate 173–5
cord adhesives 619
cord pre-pregnates 616–17
core–double shell latex 259
core latex 258
core–shell copolymerisation 111–12
core–shell copolymers 112, 131, 151, 258, 352, 562, 560
  surface and buried groups 196–7
core–shell latices 111–12, 197, 520, 564
core–shell morphology 278
core–shell particles 112, 197
core–shell polymer particles 212
core–shell polymerisation 138, 139
core–shell process 129
core–shell technique 277
corona-discharge treated polyolefins 559
corrosion-resistant primers 429
creaming 58, 192, 193
creep 472
critical humidity 472
critical micelle concentration (CMC) 52
critical pigment volume concentration (CPVC) 252–4
critical temperature 472
crosslinking 17, 95, 118, 134, 152, 256, 268–350, 453, 580, 586, 589–90
  acid and alkali induced 334
  acrylate copolymer latices 558–61
  acrylic latices 330, 525
  adhesives 514–17
  aldehydes 289
  cure studies 273–5
  cure tests 275
  degree 270
  density development 273
  diene polymers 280–1
  diketones 334
  estimation 273–5
  industrial practice 269
  β-ketonic acid esters 334–8
  kinetics 272, 277
  latices 270–2
  nitrogen compounds 338–41

reactions 272
  seed latex formulation for 318
  styrene–butyl acrylate copolymer latices 291
  survey 272–3
  two-pack 335
  two-pack latex system in single vessel 295–6
  vinyl polymers 270
  with residual unsaturation 282–3
crosslinking agents 512–13
crotonic acid 39, 129, 284, 285
crystallinity 22
curing 268–350, 316–21, 325–32, 472
curing agents 527
cycloaliphatic epoxides 426
cyclopentadienyl acrylate (CPDA) 282

Darzen's reaction 43
defoamers 499
degradative chain transfer 9
dehydro-caster oil (DCO) 416
delayed addition technique 135, 351
delayed tack adhesive 573–4
DER330 432
dextrin 76
diacetoneacrylamide (DAAM) 308–10
dialkyl peroxides 10
diallyl esters 279
diallyldimethylammonium chloride 498
diazonium salts 94
dibenzoyl peroxide 9, 93
dibenzoyl peroxide-cetyl pyridinium bromide redox initiation 94
dibutyl phthalate 234, 488
dichloroethylene 32
dicyclopentadiene–vinyl acetate copolymer 430
dicyclopentadienyl acrylate (DCPDA) 282–3
dicyclopentadienyl methacrylate (DCPDM) 282
dicyclopentadienyloxyethyl methacrylate (DPOMA) 283
dielectric cure analysis (DEA) 275
diene copolymers 143–5, 589–90
diene monomers 143, 561–2
diene polymers 42, 143–5
  crosslinking 280–1
dienes 41, 42
  non-rubber applications 144

Index

diffusion   475
diglycidyl ether   273, 299
di-isobutyl phthalate   488
di-isocyanates   300–1
diketones, crosslinking   334
dilatancy   200–2
dilatant flow   57
dilution test   398
dimethylolurea   513
dimethylpoly-siloxane   445
dinitrobenzenes   9
dinyl acetate   295
  copolymers   358–64
dioctyl sulfosuccinate esters   65
Diofan latices   379
dioxalane derivatives   290
dip additives   610–14, 616–17, 621–4
dipropyleneglycol dibenzoate   491
disc centrifuge   393–5
dispersing agents   29, 74–5
  see also specific types
dispersion polymerisation   xix
dispersions   xix, 364
disulfonated dodecyldiphenyl oxide
    sodium salt   140
divinylbenzene   30, 276–7
p-divinylbenzene   268
divinylbenzene, and allied monomers
    275–7
divinylethylene urea   45
divinylpropylene urea   45
t-dodecyl mercaptan   6
double bonds   41–2, 268
double shell layer latex   260
Dresinol   155, 571
dry strength   472
drying   479
  mechanism   217
  non-woven fibrous networks   213
Du-Nuoy ring tensionmeter   397

ecological studies   417
edge effects   210
electrical double layer   164
electron beam (EB) cure   331–2
electron microscopy   202
Elvace ethylene–vinyl acetate copolymer
    latices   644
emulsification
  principles   404
  silicones   439–44

emulsifier-free latices   112–13
emulsifiers   162, 482
  efficiency   54–5
  miscellaneous   425–6
  non-ionic   424–5
  see also specific types
emulsion copolymerisation   95
emulsion copolymers   134
emulsion particles   54–5
  and resultant properties   194
emulsion polishes   382, 386
emulsion polymerisation   15, 29,
    90–191, 279, 280, 451
  butyl acrylate   169
  cascade process   104–5
  chemistry   90–6
  complications   152, 153
  continuous processes   102–4
  conventional methods   382
  Fitch theory   158–61
  fundamentals   90–6
  Harkins theory   152, 153–5
  Harkins-Smith-Ewart theory   167
  kinetics   172
  laboratory-scale   96–8
  large-scale reactors   98–102
  mechanism   151–3
  Medvedev theory   157–8
  modifications by Gardon   157
  monomers   96–8
  on-line control   102
  pilot plants   98
  plant   101
  procedural variations   108
  reaction medium   195
  Smith–Ewart theory   155–7, 166,
    172
  special techniques   102–4
  stages   151
  supercritical carbon dioxide   178–9
  technique variations   108
  technology   96–115
  theoretical considerations   150–78
  three-stage   197
  Ugelstadt theory   158–61
  vinyl acetate   170–5
emulsion polymers   27, 116
emulsions   49–58
  oil-water interface   56
  water-in-oil   57, 408–9, 427, 516

environmental studies 417
epichlorhydrin 423
epithiopropyl methacrylate 343
epoxide compounds, external 298–300
epoxide emulsions, miscellaneous blends and reactions 429–32
epoxide esters, phosphated 431
epoxide function 290–300
epoxide groups 286–8
  internal 292–8
epoxide latex 424
epoxide–polyamide blends in coatings 428–9
epoxide–polyamide emulsions 427–9
epoxide–polyester emulsion 431
epoxide resins 404
  and blends 421–33
  applications 432–3
  non-ionic emulsifiers 424–5
  pigmented laquer emulsions 434
epoxides 615–16
  containing halogens 427
3,4-epoxycyclohexylmethyl-3,4-epoxycyclohexane carboxylate 426
esters for copolymerisation 40
ethyl acrylate 139, 277
ethylene 17, 30, 126, 358
ethylene–carboxylic acid copolymer 566
ethylene glycol dimethacrylate 278
ethylene oxide–propylene oxide block copolymer emulsifier 123
ethylene–vinyl acetate copolymers 562, 566, 572–3, 575
  adhesives 506–12
ethylene–vinyl acetate (EVA) polymers 333, 506–8
  preparation and use 509–10
ethylene–vinyl acetate resins 386
ethylenic compounds 404
2-ethylhexly acrylate 550
extenders 248–9
  organic 255–61
external plasticisation 131–3

fatty acid soaps 59–60
fatty acids 72
Ferranti–Shirley viscometer 274
fibreglass 532

fillers 500, 527, 578
  physical properties 501
film formation 209–15
  acrylic latices 224–7
  methods of study 213–15
  specific studies 221–7
  styrene–butadiene latices 227
  temperature and humidity effects 211
  see also minimum film-forming temperature (MFT)
film measurements 400–1
films 209–27, 508
  drying 210
  gas transmission rate from 378
  general properties 400–1
  theoretical developments 215–20
  see also specific materials
Flory's theory of network formation 272
fluorinated hydrocarbon 381
fluorine, vinyl compounds containing 148–50
fluoroalkyl aqueous solutions 68
fluorocarbon latices 380–1
fluorocarbon surfactants 68
fluorochemical emulsifiers 67–8
fluorochemical surfactants 68
foaming 482
foil laminating adhesives 529, 531
formaldehyde 273, 502, 512–13
Fourier transform infrared spectroscopy (FTIS) 213
free radical initiation 9–11, 93–5
free radical polymerisation 4–13, 41
free radicals 5–6, 11, 151–2, 155
freeze–thaw performance 504
freeze–thaw stability 141, 392
  general principles 203–5
  technical methods of control 205–7
  tests 208
Frenkel's equation 215
Freundlich adsorption isotherm 51
Friedel–Crafts catalysts 4
fumaric acid 39, 40

gas–liquid chromatography (GLC) 392
gas phase polymerisations 10
gas transmission rate from films 378
gel permeation chromatography 26, 275

gelation   499
glass fibre   446, 462, 623
glass transition temperature   374
glycidyl acrylate   292, 294
glycidyl esters   297–8
glycidyl groups   615–16
glycidyl methacrylate   140, 292–4
graft copolymerisation   114–15
graft copolymers   18–19, 147, 151
gutta percha   xx

halogens, polymers containing   321–5
hardeners   134, 433
head-to-tail addition   8
head-to-tail polymerisation   8
heat-resistant adhesives   592
heat-seal adhesives   562–7, 572–3
   for high-temperature-resistant bonds   567
   formulations   564–5
Helmholtz electrical double layer   52
Hevea brasiliensis   xx, 470
hexafluoroethylene   150
hexahydroxymethylmelamine   418
historical and general survey   xx–xxii
HLB   54, 162, 196
Hycar latex H5   202–3
hydrocarbon resins   496
hydrocarbons   30
hydrogen bonding   21
hydrogen peroxide   10–12, 113
hydrogen release   442–3
γ-hydroperfluoroalkyl phosphonate   68
hydrophilic monomers   95, 196
hydrophilic silicones   441–2
hydrophilic system   112
hydrophobic system   112
hydrophobically modified thickeners (HEUR and HMHEC)   218
hydrotropes   66
2-hydroxyethyl acrylate   277
hydroxyethyl acrylate   272
hydroxyethyl cellulose   79–82
N-(hydroxyethyl)acrylamide   113
hydroxyl function   288–90
hydroxyl groups   286–8
hydroxypropyl cellulose   82

imidazoline   423
impregnants   567

impregnations   432
inhibition of polymerisation   8
insolubility   17
inter-molecular reactions   341–2
internal plasticisation   131–3
interpenetrating networks (IPN)   180–1
intra-molecular reactions   341–2
inverse emulsions   178–80
inverted core–shell latices   520–2
ionic polymerisation   4
ionomeric core–shell latices   111
isobutoxyacrylamide   308
isobutoxymethacrylamide   306
isobutylacrylamide   308
isobutylmethacrylate   112
isocyanate adducts   449–50
isocyanates in wood adhesives   587–9
isoprene   470
isoprogylacrylamide   308
itaconic acid   39, 95, 284, 285, 367

Joyce–Loebl disc centrifuge   394

β-ketonic acid esters, crosslinking   334–8
kinetic equations   7
kraft paper tapes, coatings   551

laminate   471
lap joint   472
Laponite   76
latex particles
   adsorption of surfactants   197
   stabilization   164–5
latex properties relative to applications   192–267
latices   xx, 91
   historical development   115–16
   see also specific latices
leather treatment   462
Lifshitz–van der Waals interactions   474
light-decomposable emulsifiers   93
light scattering   26
liquid monomers   96–7
lithopone   248
London forces   21
loop reactor   104, 106–8

macromolecules   xix, 2
maleic acid   39, 40

maleic anhydride 14, 87
mechanical stability 398
metal complexes in wood adhesives
    584–6
metal oxides 288
metal-to-metal bonds 603
(meth)acrylamide, polymers
    containing 301–21
methacrylic acid 32, 111, 133, 135,
    177–8, 272, 277, 284, 364
methacrylic esters 33–8, 167, 206
methyl acrylate 111–12, 134, 366
methyl cellulose 78–9
N-methyldiethanolamine 452
methyl methacrylate 167–9, 177–8
methylene bisacrylamide (MBAM)
    310
methylphenylsiloxane emulsions 446
methyl-H-polysiloxanes 445
methylsilsesquioxane 446
2-methyl-5-vinylpyridine 44
mica 249
micelles 51–3
micro-emulsions 55, 140, 150
micro-foam 213
microgels 275, 278
microvoid latex 257
mini-emulsions 113–14
minimum film-forming temperature
    (MFT) 220–1, 271, 376, 399,
    472, 478
minimum heat-seal temperatures 493
molecular coil 20
molecular weight 19, 24–6
    measuring 25–6
monodisperse latices 108–11
monomers xxii, 3, 4, 24, 30–45
    containing nitrogen 44–5
    content estimation 390–2
    delayed addition 118–21
    physical properties 34–7, 45–6
    toxicity and handling 45
    with several double bonds 41–2
    see also specific monomers
mortar 432
Mowilith 357
Mowilith D 200
Mowilith dispersions 199
Mowilith DN 200

multiphase thermoplastic acrylic
    elastomers 522
multiple double bonds 275–83
multipolymers 360
multishell emulsion particle–core
    emulsion 260
multistage copolymerisation 276

naphthenic acids 60
natural rubber xx
neopentyl diacrylate 277
Neoprene 528–9, 620, 634–5
Nernst law of partition 239
Newtonian flow 57
nitrobenzene 9
nitrogen, monomers containing 44–5
nitrogen compounds, crosslinking
    338–41
NMAM 301, 303–7
non-ionic surfactants 58, 68–71, 92,
    164, 165, 441
non-pressure sensitive adhesives 651–2
non-vinyl emulsions 404–69
novolaks 434
nuclear magnetic resonance 8, 111,
    197, 205

osmometry 25
oxiranes 586–7

packaging adhesives 595–9
paint 250
paint films 470
paper applications 444–7, 460–1, 535,
    536
paper bonding 603–4
paper-to-paper bonding 593–5
    formulations 594–5
particle characterization 194–6
particle–particle contact 237
particle size distribution 194, 393–6
PBMA latex film 219
peelable adhesive tapes 567–8
pH effects 58
pH stability 392
phenol-formaldehyde novolak resin
    435
phenolic resins 435
phenoplasts 434–7, 583–4
1-phenoxy-2-propanol 240

Index

phosphate esters 493
phosphate surfactants 441
phosphated emulsifiers 163
phosphates 66–7
photoinitiators for radical cure 327, 328
photopolymerisation 525
pigment encapsulation 254–5
pigment volume concentration (PVC) 252–4
pigmentation 242–55, 373
  dried composition 244
  general properties of pigmented systems 252
  latices 249–52
  pigmentary power 245–6
Pilamec Vibracon 251
plastic flow 57
plasticisation
  external 361–3
  internal 361–3
  mechanism 230–1
  polystyrene 231
  polyvinyl acetate 118, 231–5, 354
  polyvinyl chloride 235–8
plasticisers 229–42, 372, 382, 578
  polyvinyl acetate 488–95
Polidene latices 379
polyamino hardener 433
polybutyl acrylate–polymethyl acrylate core–shell latices 197
polybutyl methacrylate latices 140
polychloroprene
  latex–resorcinol–formaldehyde dip formulation 620
polychloroprene latices 527, 530–3, 570
  contact adhesives 624
  formulations 526
polyelectrolytes 86–7, 164
polyester emulsions with vinyl monomers 407–8
polyesters 405–9
polyethylene glycol ethers 491
polymer chemistry
  definitions 2
  fundamentals 1–48
polymer colloids 90
  major features 193–4
polymer emulsion xix

polymer latices xix, 90
  applications, general survey xxii-xxiv
  commercial products 364–5
  principal applications xxiv-xxv
  technical 351–403
polymerisable acids and anhydrides 39
polymerisable surfactants 74, 163
polymerisation
  inhibition 8
  mechanism 4
  reaction 5
  retardation 8
  styrene 382
  technology 27–30
  vinylidene chloride copolymer 374
polymers xix, xx, xxii, 2
  concept 1–14
  historical introduction 1–2
  structure 19–24
polymethyl methacrylate 22
polymides, emulsions 419–21
polyorganopolysiloxane 440
polyorganosiloxanes 440
polyoxyethylene nonylphenol diglycidyl ether 425
polypropylene 387
polysiloxane oils 439
polystyrene 143, 381–3
  films 382
  plasticisation 231
polytetrafluoroethylene see PTFE
polyurethane latices 450–8
  applications 458–62
polyurethanes 300–1, 404, 449–62
  direct (chain extension) emulsification 456–7
  emulsification by salt formation 452–4, 454–6
  emulsions via blocked copolymers 451–2
polyvinyl acetate 22, 119, 173, 196, 207, 221, 353–5, 591
  adhesives 484–517
  and related copolymers 116–24
  commercially available 487
  copolymer latices 643, 656
  heel latex 120
  homopolymer latices 636, 643, 655, 658

polyvinyl acetate (*cont.*)
  latex-based wood adhesives 578–9
  plasticisation 231–5, 354
  plasticisers 488–95
polyvinyl alcohol 22, 52, 54, 84–6, 117, 118, 138
  as emulsifier 85–6
  chemically modified 86
polyvinyl chloride 371–4
  adhesives 502, 503
  copolymer latices 372–4
  plasticisation 235–8
polyvinyl propionate, adhesives 517–18
polyvinyl pyrrolidone 88
polyvinylidene chloride films 378
polyvinylpyrrolidone 340, 453
  films 388
polyvinyls 229–30
postemulsification 567
potassium laurate 53
potassium persulfate 11
pre-emulsion feed 125–7
pressure-sensitive adhesives 388, 481, 483–4, 521, 525, 528, 538–67, 642, 647–50, 653, 654
  bonding to polyolefin films 561
  copolymer latices with multiple unsaturation 555–7
  formulations 553
  metal surfaces 554–5
  non-tacky 545
  redispersible components 576–7
  release properties 548
  removable 559, 561, 567–8
  seed stage process 550
  tack-free 576
  terminology 538–40
  types of latices formulated into 540–67
  water resistance 552
Primal latices 649–52
priming 472
propane sultone 455
Propiofan polyvinyl propionate latices 632
pseudo-crosslinking 271–2
pseudoplastic flow 57
PTFE 96, 165, 380
  technical latices 380–1

Q,e scheme 15–16
quaternary ammonium salts 439
quick-tack adhesives 575–6

radiation curing 325–32
radiation-induced polymerisation 94–5
redox catalyst 502
redox initiators 97
redox polymerisation 11–13, 135
redox processes 352
redox systems 12, 94
relative humidity 212, 252
relative molecular weight 399
release coatings 448, 517, 561
remoistenable adhesives 574–5
Resad polyvinyl acetate homo and copolymer latices 646
residual monomers 100
resin xix, xx
resorcinol–formaldehyde condensate 608
resorcinol–formaldehyde resins 618–19
retardation of polymerisation 8
rheometers 200
rheopexy 200–2
Rhodopas 355, 356
Rhodotack base copolymer latices 647–8
rock wool 532
root mean square 20
rosin xix
Rovene carboxylated styrene–butadiene copolymer latices 641
rubber xx
rubber-to-fabric adhesives 606–24
Russell cascade stand model 105
Russell vibrating sieve 103
rutile 247

sailcloth 605
saponification value (SV) 84–5
scanning electron microscopy 218–19, 223
sedimentation 398
seed latex formulation for crosslinking 318
seed latices 108, 136, 138, 171, 257
seed particles 255
seed polymerisation 510

seeded emulsion polymerisation 145
self-crosslinking 340
self-emulsification 408, 440, 455
self-emulsifying monomers 39–40, 113
semi-continuous (gradual addition) methods 196
semi-synthetic colloids 84
setting rate 472
setting time 472
settlement 193
settling 398
shear thickening 202
shells, formation 261
shrinkage 478
side chain crystallinity 22
silanes 332
silica 220
silicates 249
silicones 447
　applications 444–9
　elementary chemistry 437–9
　emulsification 439–44
　self-emulsifying 440
　surface coating and building 447–8
siloxanes 134, 139, 332–3, 404
sizing 472
skinning 210
soap micelles 53
soap titration 396–8
soap-type surfactants 141–2
soaps 60, 382
　fatty acid 59–60
sodium methacrylate 177–8
sodium oleate 382
sodium polyphosphate 498
solid dispersants 75–6
solution polymerisation 28–9
solvents xxii, 7, 229–42, 437, 480, 527
specific gravity 392
stabilisers 58–9, 91–3, 138, 211, 355, 371, 399, 527
　mixed surfactant and colloid 119–21
　specificity 162–4
stability 192, 270
starch 76
　modified emulsifiers 121
steric hindrance 20
Stokes equation 393
stoving paint 268

streaming 395
stress–strain curves 201
styrene 13, 30, 110, 113, 134, 166–7, 177–8, 269, 277, 305–6, 366, 409
　polymerisation 382
styrene–acrylic ester copolymer latices 639
styrene-based latices 645
styrene–butadiene carboxylated copolymer 385
styrene–butadiene copolymers 94–5, 383–5
　with added solvent 535
styrene–butadiene latices 595
　film formation 227
styrene–butadiene rubber latex 145
styrene–butadiene–vinylidene chloride latex 602
styrene–butadiene–vinylpyridine latices 619
styrene–butyl acrylate–acrylonitrile 277
styrene–butyl acrylate copolymer latices, crosslinking 291
styrene copolymers 175–6, 381–5
styrene latex–vinyl acetate–maleic acid copolymer emulsifier 142
styrene latices 140–3
styrene polymers 381–5
substrate 471
sulfamic acid 455
sulfated fatty alcohols 61
sulfated natural oils and esters 61–2
sulfates 60–6, 440–1
　of ethoxylated alcohols 61
sulfonate emulsifiers 383
sulfonated aromatic and condensed ring compounds 62–3
sulfonates 60–6, 440–1
　general 62
sulfonic acid 67
sulfosuccinate emulsifier 125–6
sulfosuccinates 64–6
supercritical carbon dioxide 178–9
surface adsorption 51
surface coagulation 165
surface tension 51, 65, 68, 79, 208, 216
　measurements 396–8
surface wetting 480
surfaces, properties 50–1

698 Index

surfactant–polymer compatibility 218
surfactants 51–3, 58–9, 119–21, 208, 397
  adsorption 197
  amphoteric 58
  anionic 59–60
  ionised 58
  non-ionic 58, 68–71, 92, 164, 165, 441
  specificity 162–4
  survey 59–76
  see also specific types
suspension polymerisation xix, 29–30
swelling 270
  degree 274
synthetic resin (polymer) industry xix

tack 472, 495–500
tackifier emulsion 497
tackifiers 495–500
tackifying effect 491–3
tape coating, pressure-sensitive 524
Teksyn styrene–butadiene latices 642
terpolymers xxii, 15, 126, 131, 300, 360, 364, 510–11
test methods, adhesives 624
testing of latices 389–401
tetra-n-butyl titanate 286
tetraisopropyl titanate 286
tetramethoxymethylglycouril 339
Texicote latices 652–3
Texicryl latices 652–3
textiles 444–7, 459–60, 601, 606
thermosetting 268, 404
  coatings 274
thickeners 88, 372, 502, 526
thickness measurements 212
thixotropic recovery 220–2
thixotropy 57, 472
tile adhesives 606
time–temperature–transformation cure diagrams 273, 275
titanium chelates 203
titanium dioxide 242, 246–8, 254
Tolanate HCT 301
toluene di-isocyanate (TDI) 587
p-toluenesulfonamides 494
tolyl di-isocyanate (TDI) 614

torsion braid analysis (TBA) 275
total solids content 389
transient plasticisers 238–42
transition points 26–7
trichloroethylene 32
tricresyl phosphate 234
trimethylolpropane triacrylate 242
triphenylmethyl 5
Triton X-405 164, 197
Tromsdorf effect 167
Tromsdorf–Norrish gel effect 104
tropical stability 208–9
tyre bonding 612
tyre cord dip 622

Ucar latices 653
Ucecryly latices 654
ultrasound 95
ultraviolet curing 326–31
unsaturated fatty acid 285
unsaturated polyester 404–69
urea 273
UV cured systems 328–31

Van der Waals forces 76
Veova 10, 11, 132
Veova copolymers 126–9, 132, 223
Vinamul 9900 201
Vinavil group polyvinyl homopolymer latices 657–8
Vinnapas co- and terpolymer latices 662, 663
Vinnapas ethylene–vinyl acetate copolymer latices 660, 661
Vinnapas polyvinyl homopolymer latice 659
vinyl acetate 14, 15, 87, 105, 115–33, 195, 306–7, 358, 516–17
  and copolymers 119–21
  and ethylene copolymers 126
  content estimation 390–1
  copolymer latices 221–4
  copolymerisation 173–5
  copolymers 109, 124–8, 175, 637–8
  emulsion polymerisation 170–5
  polyvinyl alcohol stabilised 118
  redox method with heel 118–19
vinyl acetate–acrylic ester latex 202, 591

Index

vinyl acetate–crotonic acid copolymer latex 592
vinyl acetate–di-isobutyl maleate copolymer 592
vinyl acetate–ethyl acrylate copolymer 125–7
vinyl acetate-ethylene copolymer latex 128
vinyl acetate–ethylene copolymers 387
vinyl acetate–ethylene terpolymers 387
vinyl acetate–maleic acid stabiliser 142–3
vinyl acetate–Veova copolymer 129, 132, 223
vinyl benzoate 31
vinyl bromide 32
vinyl butyrate 31
vinyl caprolactam 45
vinyl chloride 31, 307–8, 351, 371
vinyl chloride latex formulation 147
vinyl chloride polymers and copolymers 146–7
vinyl chloride–vinylidene chloride copolymer 238
vinyl chloroacetate 31
vinyl compounds containing fluorine 148–50
vinyl ester polymer latices, pressure-sensitive adhesives 544–6
vinyl esters 31, 359
vinyl ethers 43–4
vinyl fluoride 32
vinyl halide polymers 371–81
vinyl halides 146–50
vinyl methyl ether 87
1-vinyl-2-methylimidazole 44
vinyl monomers 407–8
vinyl polymerisation in polyurethane latex 457–8
vinyl polymers
 crosslinking 270
 with residual unsaturation 282–3
vinyl propionate 31, 360
vinyl stearate 15
vinyl-type cures 300–1
vinyl-type monomers 429–30

vinyl-type polymers with residual unsaturation, crosslinking 282–3
vinyl–urethane block copolymers 457
vinylidene chloride 22, 32
 adhesives 526–33
 copolymers 148, 149, 374, 571
 films 377–8
 polymerisation 375
 polymers 148, 375
vinylidene fluoride, polymerisation 150
1-vinylimidazole 44
2-vinylpyridine 44
4-vinylpyridine 44
vinylpyridine copolymer adhesives 609–10
N-vinylpyrrolidone 44
vinylpyrrolidone, copolymers 388–9
viscoelastic behaviour 218
viscoelastic properties 227
viscometry 25
viscosity 55–8, 132, 198–202, 393, 399, 472
 characteristics 274
 definitions and theory 198–9
 measurement 199–200, 400
 phenomena with latices 202–3
viscous flow 216
volatile monomers 98
volatile organic compounds (VOC) 226
Vulcabond E 620

Wacker-Chemie products 506
wallpaper adhesives 594
water-based adhesives 624
water-based polymers 286
water-in-oil emulsions 408–9, 427, 516
water-insoluble monomers 114
water-resistant adhesives 592
water-soluble colloids 163
water-soluble monomers 95, 177–8
water-soluble plasticiser 239
water-soluble polymers 87–8, 197
water-vapour transmission (WVT) 214, 377
wetting agents 478, 527
wetting back 374

wicking 211–12
wood adhesives 504, 577–92
　additives 583–9
　bond formed in two phases 585
　formulations 580–2, 590–2
　glycidyl and epoxy additives 584
　isocyanates in 587–9
　metal complexes in 584–6
　modifications 583–9
　polymer latices 582–3
　water-resistant 578–9

Young's modulus 218

Ziegler–Natta catalysts 5, 21
zinc oxide 248, 526
zirconium ammonium carbonate 288
zirconium compounds 498